PROGRESS

IN

HETEROCYCLIC CHEMISTRY

Volume 9

PROGRESS
IN

HETEROCYCLIC CHEMISTRY

Volume 9

A critical review of the 1996 literature preceded by two chapters on current heterocyclic topics

Editors

G. W. GRIBBLE
Department of Chemistry, Dartmouth College, Hanover, New Hampshire, USA

and

T. L. GILCHRIST
Department of Chemistry, University of Liverpool, Liverpool, UK

PERGAMON

U.K. Elsevier Science Ltd. The Boulevard, Langford Lane,
Kidlington, Oxford, OX5 1GB, U.K.
U.S.A. Elsevier Science Inc., 660 White Plains Road, Tarrytown,
New York 10591-5153, U.S.A.

JAPAN Elsevier Science Japan, Higashi Azabu 1-chome Building 4F,
1-9-15 Higashi Azabu, Minato-ku, Tokyo 106, Japan

First Edition 1997

Library of Congress Cataloging in Publication Data

A catalog record for this book is available from the Library of Congress

British Library Cataloguing in Publication Data

A catalogue record for this book is available from the British Library

ISBN 0 08 0428010

Transferred to digital printing 2008

Contents

Foreword

Progress in Heterocyclic Chemistry (PHC) Volume 9 reviews critically the heterocyclic literature published mainly in 1996. The first two chapters are review articles. Chapter 1 by C.J. Moody and K.J. Doyle deals with "The Synthesis of Oxazoles from Diazocarbonyl Compounds," and Chapter 2 by J.A. Sikorski provides a detailed account of the heterocyclic chemistry surrounding the remarkable herbicide glyphosate ("Roundup"®). This latter chapter illustrates the role that heterocyclic chemistry plays in other areas of modern chemistry, since glyphosate is a far cry from being heterocyclic!

The remaining chapters deal with recent advances in the field of heterocyclic chemistry arranged by increasing ring size. Once again, the reference system follows the system employed in *Comprehensive Heterocyclic Chemistry* (Pergamon, 1984).

We thank all authors for providing camera-ready scripts and disks, and most especially for adopting our new uniform format. In this regard, we welcome comments from readers about the style, presentation, and coverage.

We are much indebted to David Claridge of Elsevier Science for his invaluable help with the presentation of Chapters and with his input on the new format.

Finally, we wish to acknowledge retiring editor Hans Suschitzky not only for his outstanding contributions in all previous volumes of this series as co-editor, but, jointly with Eric Scriven, for launching the series. Heterocyclic chemists owe Hans and Eric a debt of gratitude.

Once again, we hope that our readers will find PHC-9 to be a useful and efficient guide to the field of modern heterocyclic chemistry.

G. W. Gribble

T. L. Gilchrist

Editorial Advisory Board Members
Progress in Heterocyclic Chemistry

1997–1998

The International Society of Heterocyclic Chemistry is pleased to announce the establishment of its home page on the World Wide Web. Access can be gained from the following locations:

For USA, Americas, Japan:

http://euch6f.chem.emory.edu/ishc.html

for Europe:

http://www.ch.ic.ac.uk/ishc/

Chapter 1

The Synthesis of Oxazoles from Diazocarbonyl Compounds

Christopher J. Moody
University of Exeter, Devon, UK

Kevin J. Doyle
Loughborough University, Leicestershire, UK

1.1 INTRODUCTION

Oxazoles, which have been known for well over a hundred years, have been of considerable interest to organic chemists ever since the 1940's, when the intense research effort on penicillin led Cornforth and others to develop new routes to the oxazole ring. This work, summarised in the classic treatise in 1949,<B-49MI1> is the foundation of modern oxazole chemistry. The subsequent discovery during the 1950's by Kondrat'eva that oxazoles can function as azadienes in the Diels-Alder reaction, and by Huisgen that mesoionic oxazoles participate in 1,3-dipolar cycloaddition reactions prompted further research into the ring system.<86MI1> More recently the oxazole ring system has been found in an ever increasing range of natural products,<92JHC607, 93AG(E)1, 94NPR395, 95CRV2115, 95NPR135, 96NPR435> many of them "peptide alkaloids" in which the heterocyclic ring is most likely formed by a modification of a serine or threonine containing peptide.<96JOC778> The interesting biological activity associated with these natural products has not surprisingly prompted renewed interest in the synthesis of oxazoles. Although there are several methods available for the synthesis of oxazoles, this article focuses on just one route which has been used extensively in our own laboratory, namely that involving the reaction of diazocarbonyl compounds with nitriles (Scheme 1). Other aspects of diazocarbonyl chemistry have been widely reviewed.<86ACR348, 86CRV919, 87TCC(137)75, 91CRV263, 91T1765, 92T5385, 94AG(E)1797, 94CRV1091, 95T10811, 96AHC(65)93, 96CRV223>

Scheme 1

1.2 THERMAL AND PHOTOCHEMICAL REACTIONS

The formation of oxazoles from nitriles and diazocarbonyl compounds was investigated by Huisgen in the early 1960's during his classic studies on 1,3-dipolar cycloaddition reactions.<63AG(E)565, 64CB2628> He and co-workers found that the ketocarbene derived from diazoacetophenone **1a** by thermolysis at 150°C underwent formal cycloaddition with benzonitrile giving a 0.4% yield of 2,5-diphenyloxazole **2a** together with >50% of secondary products derived from a Wolff rearrangement (Scheme 2). The presence of electron withdrawing groups at the 2-position on the aromatic ring resulted in the formation of the oxazoles **2b** and **2c** in higher yield. The yield of oxazole **2a** was higher when the reaction was carried out in the presence of $Cu(acac)_2$.

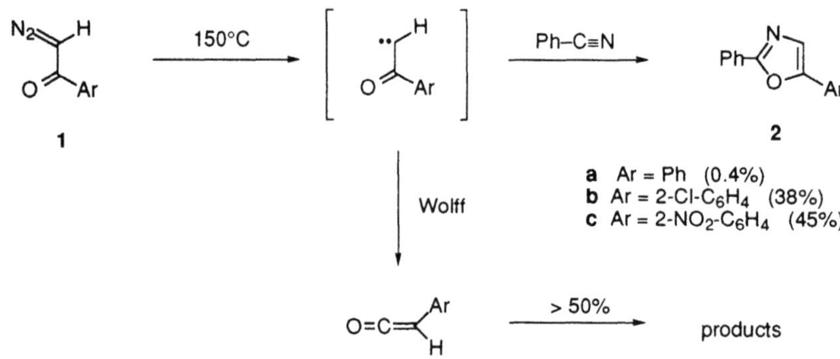

Scheme 2

Huisgen *et al.* also studied the thermal decomposition of ethyl diazoacetate in the presence of benzonitrile and phenylacetonitrile to give the corresponding 2-substituted-5-ethoxy oxazoles **3** in variable yields (Scheme 3).<64CB2864> The authors found that the solvent had an effect on the rate of decomposition of ethyl diazoacetate; in the polar solvent, nitrobenzene, the rate was found to be twice that in the hydrocarbon solvent, decalin.

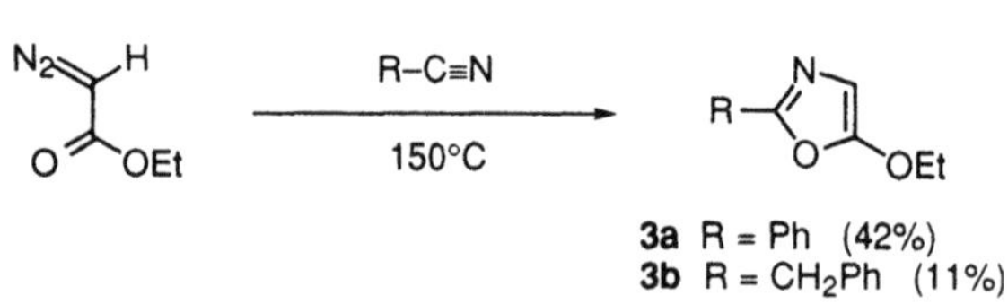

Scheme 3

Komendantov *et al.* found that thermal decomposition of methyl diazoacetate in the presence of benzonitrile yielded two products.<73JOU431> One is the expected 2-phenyl-5-methoxyoxazole **4** in about 35% yield and the other product was methyl 3-phenyl-2*H*-azirine-2-carboxylate **5** in around 1% yield (Scheme 4).

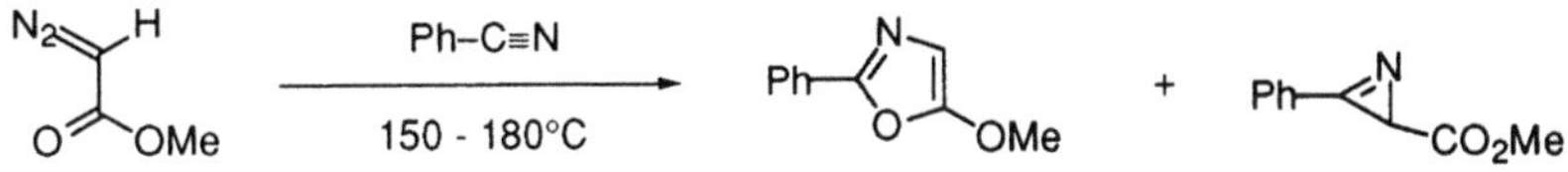

4 5

Scheme 4

In studies on 1-diazo-2-ketosulfones, Shioiri *et al.* found that the thermal decomposition of benzoyl(sulfonyl)diazomethanes **6** with benzyl alcohol in acetonitrile also gave two products.<82CPB526> One is the 4-sulfonyloxazole **7** whereas the other product **8** results from rearrangement and reaction with the alcohol. The ratio of products varies with the nature of the sulfone substituent with the benzyl group giving highest yields of oxazole (Scheme 5).

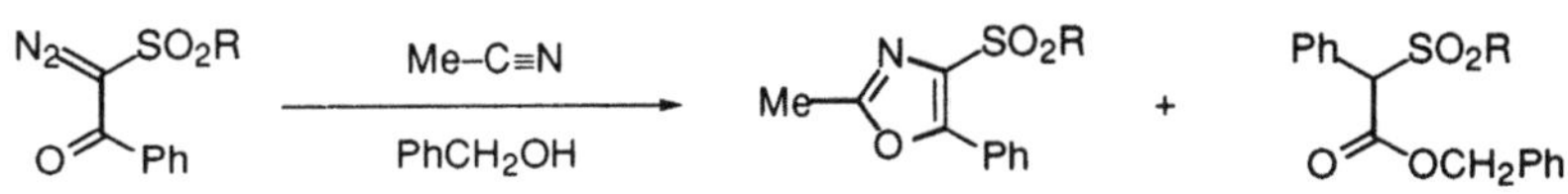

6 7 8

R	*Yield / %* **7** : **8**
$PhCH_2$	59 : 41
t-Bu	24 : 62
4-Me-C_6H_4	24 : 33
4-MeO-C_6H_4	00 : 37

Scheme 5

More recently, Williams has described the one pot synthesis of 2-substituted oxazoles **11** by the thermolysis of triazole amides **9**; the reaction does not proceed photochemically.<92TL1033> Although the reaction does not involve addition to a nitrile, it is an interesting application of a diazo compound since the proposed zwitterionic intermediate **10** is a resonance form of a diazo imine, so formally the reaction may be thought of as a thermal decomposition of a diazo imine (Scheme 6).

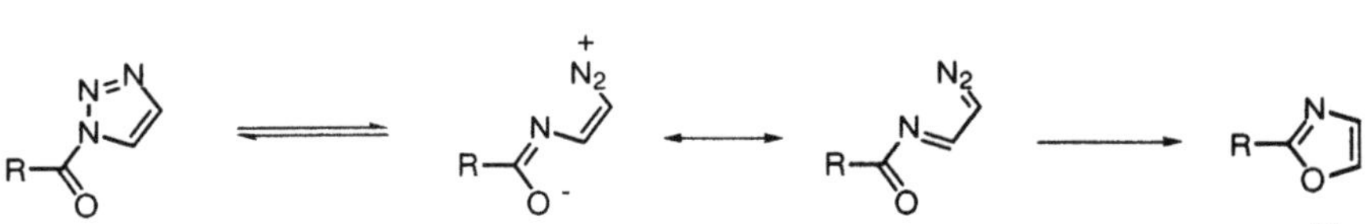

9 10 11

Scheme 6

The photochemical decomposition of ethyl diazoacetate, methyl diazoacetate and diazoacetophenones **1** in benzonitrile has been studied by Huisgen and Komendantov. <64CB2864, 73JOU431> Ethyl diazoacetate failed to give any oxazole, whilst methyl diazoacetate gave a 20% yield of the oxazole **4**. As in the thermal reaction, the 2*H*-azirine **5** was isolated in ~2% yield. The photochemical decomposition of diazoacetophenone **1a** gave the oxazole **2a** in extremely low yield. Huisgen also found that the cyclic diazo ketone, 4,7-

dimethyl-2-diazoindan-1-one **12** underwent photolysis in benzonitrile to give the oxazole **13** in 34% yield (Scheme 7).<63AG(E)565, 64CB2628>

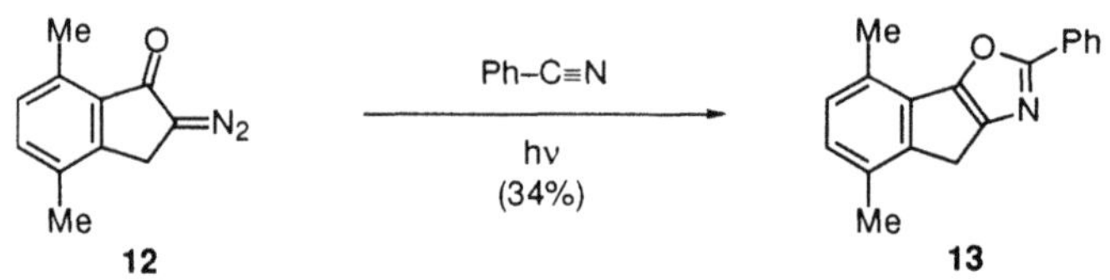

Scheme 7

The reaction of trifluoroacetyl diazoacetic ester **14** in acetonitrile has been studied by Weygand *et al.* who found that ethyl 2-methyl-5-trifluoromethyloxazole-4-carboxylate **15** could be formed photochemically in 60% yield. Further photolysis of the oxazole led to the formation of the dimeric species derived from a [2 + 2]-cycloaddition reaction in around 10% yield.<68CB302> The reaction has been exploited as a general approach for the preparation of 2-perfluoroalkylalanines **16**.<67AG(E)807> The oxazole ring is formed from the photolysis of the appropriate perfluoroacyl diazo esters in acetonitrile, and is then degraded under acid hydrogenolysis conditions to give the *N*-acetyl esters, which are then hydrolysed to the racemic 2-perfluoroalkylalanines **16** (Scheme 8).

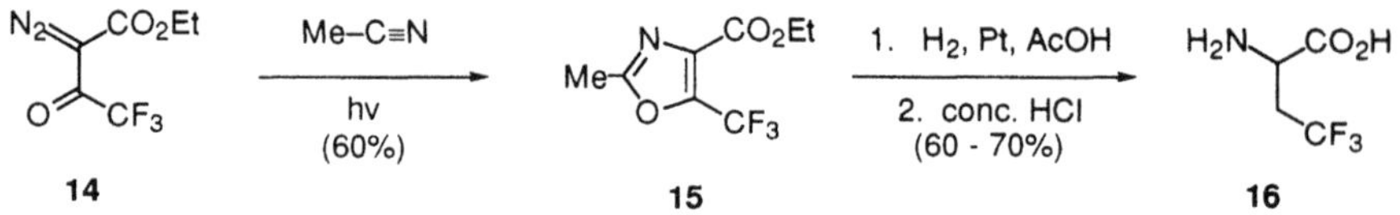

Scheme 8

1.3 MECHANISM

Oxazole formation can be envisaged as proceeding by three possible pathways: 1,3-dipolar cycloaddition of a free ketocarbene to the nitrile (Path A), the formation and subsequent 1,5-cyclisation of a nitrile ylide (Path B) or the formation and subsequent rearrangement of a 2-acyl-2*H*-azirine (Path C) (Scheme 9).

The mechanism of the thermal and photochemical formation of oxazoles from diazocarbonyls is often thought to involve the intermediacy of a free ketocarbene (Path A). In the thermal and photochemical decomposition of methyl diazoacetate in benzonitrile, the 2*H*-azirine **5** was formed along with the oxazole **4**.<73JOU431> However, when the photolysis was conducted in a 10 : 1 mixture of hexafluorobenzene and benzonitrile, the sole product was the oxazole in 20% yield. It was assumed that the formation of the 2*H*-azirine **5** and oxazole **4** was due to the reaction of methoxycarbonylcarbene in either its singlet or triplet state. The workers assumed that decomposition of the excited σ^2-singlet state led to the formation of the 2*H*-azirine, whilst the ground triplet state gave the oxazole. They rationalised the observed product ratio as being due to the presence of the inert solvent, hexafluorobenzene, and assumed it caused enhancement of the singlet-triplet transition, leading to more oxazole formation.

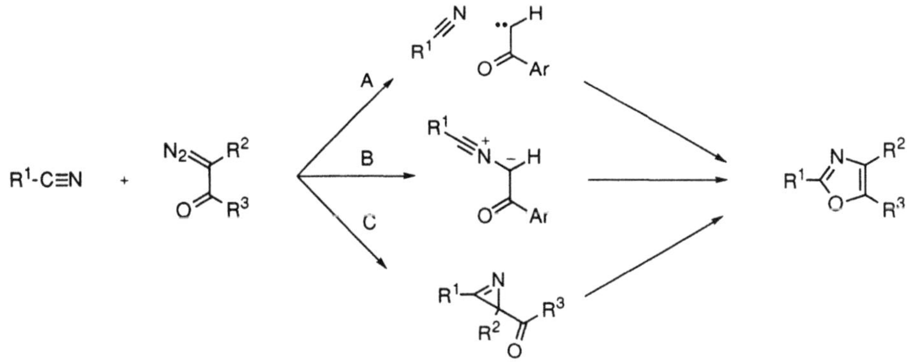

Scheme 9

However, an investigation into the photodecomposition of diazoesters in acetonitrile, conducted by Buu and Edward,<72CJC3730> led to a different conclusion for the reaction of carbenes in their singlet and triplet states. These investigators found that only singlet ethoxycarbonylcarbene reacts with nitriles to yield oxazoles. Upon benzophenone sensitisation of the reaction mixture, no oxazole formation takes place; instead the triplet carbene reacts with benzophenone to give the diradical, which adds to acetonitrile yielding ethyl 5,5-diphenyl-2-methyl-4,5-dihydro-oxazole-4-carboxylate.

Despite the above, there is also considerable evidence to suggest that oxazole formation proceeds *via* an intermediate nitrile ylide, particularly in the catalysed reactions (see below). Nitrile ylides have been detected in laser flash photolysis studies of diazo compounds in the presence of nitriles, and stable nitrile ylides can be isolated in some cases.<94CRV1091>

Although 2-acyl-2*H*-azirines are known to give oxazoles upon irradiation, the reaction is wavelength dependent, and isoxazoles are formed at some wavelengths, as they are in the thermal rearrangement of 2-acyl-2*H*-azirines.<74TL29, 75JA4682> Since the thermal reaction of diazocarbonyl compounds with nitriles leads to oxazole formation, it would seem that mechanistic path C is unlikely in these reactions.

1.4 LEWIS ACID CATALYSED REACTIONS

The role of Lewis acids in the formation of oxazoles from diazocarbonyl compounds and nitriles has primarily been studied independently by two groups. Doyle *et al.* first reported the use of aluminium(III) chloride as a catalyst for the decomposition of diazoketones.<78TL2247> In a more detailed study, a range of Lewis acids was screened for catalytic activity, using diazoacetophenone **1a** and acetonitrile as the test reaction.<80JOC3657> Of the catalysts employed, boron trifluoride etherate was found to be the catalyst of choice, due to the low yield of the 1-halogenated side-product **17** (X = Cl or F) compared to 2-methyl-5-phenyloxazole **18**. Unfortunately, it was found that in the case of boron trifluoride etherate, the nitrile had to be used in a ten-fold excess, however the use of antimony(V) fluoride allowed the use of the nitrile in only a three fold excess (Table 1).

1a 17 18

Lewis Acid	*Ratio* **17: 18**	*Isolated Yield / %*
$AlCl_3$	36 : 64	91
$SnCl_4$	24 : 76	41
TiF_4	5 : 95	99
$FeCl_3$	0 : 100	76
$BF_3.Et_2O$	0 : 100	99
SbF_5	0 : 100	99

Table 1

The group of Ibata has also reported the effectiveness of boron trifluoride etherate in the formation of oxazoles.<79BCJ3597> They found that not only diazoketones, as reported by Doyle, but also diazoketoesters could be decomposed in the presence of nitriles to give oxazoles (Table 2). They also studied the range of nitriles that could be employed, finding that substituted thiocyanates and cyanamides,<84BCJ2450> along with chloroacetonitrile <89BCJ618> also participate in the reaction (Table 2).

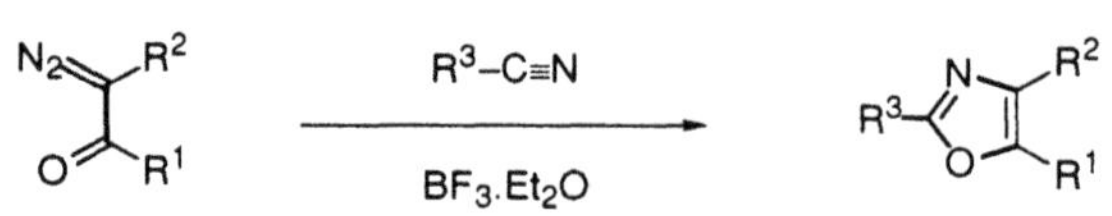

R[1]	*R*[2]	*R*[3]	*Yield / %*
Ph	H	Me	94
Ph	H	MeO_2CCH_2	46
Ph	H	MeS	78
Ph	H	EtS	66
Ph	H	Me_2N	29
Ph	H	$ClCH_2$	84
$4\text{-}NO_2\text{-}C_6H_4$	H	Me	84
Me	MeO_2C	Me	80

Table 2

The use of protic acids in oxazole formation from diazoketones and nitriles has also been reported. Holt and co-workers found that diazoacetophenone **1a** in the presence of trifluoromethanesulfonic acid and acetonitrile gave 2-methyl-5-phenyl oxazole **18**.<79JCS(P1)1485> It was assumed that protonation of the diazo compound occurred to give a diazonium ion which underwent nucleophilic attack by acetonitrile to give a nitrilium ion which subsequently cyclised. On the other hand, two mechanisms for the Lewis acid mediated process have been advanced. Ibata favours initial attack by the Lewis acid on the diazocarbonyl oxygen to give a diazonium betaine which suffers nucleophilic attack by the nitrile to give, with loss of nitrogen, a nitrilium betaine which subsequently cyclises (Scheme

10).<79BCJ3597> Doyle however favours a mechanism involving the initial formation of a Lewis acid-nitrile adduct which suffers nucleophilic attack by the diazocarbonyl oxygen to give a 2-imidatoalkenediazonium salt, which cyclises, with extrusion of nitrogen gas, to the oxazole (Scheme 10).<80JOC3657>

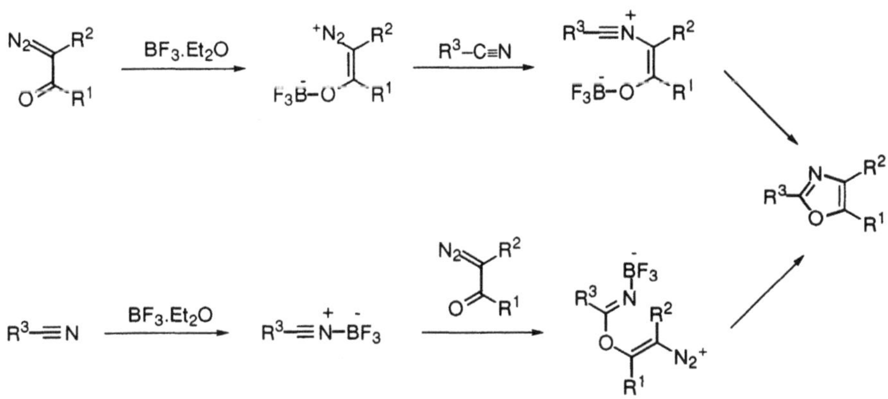

Scheme 10

The boron trifluoride etherate catalysed formation of oxazoles has been used in synthesis. Doyle has successfully employed the reaction in the synthesis of annuloline **20**, a disubstituted oxazole isolated from the roots of the annual rye grass. Thus, 1-diazo-4'-methoxy-acetophenone **19** was reacted with 3,4-dimethoxycinnamonitrile in the presence of boron trifluoride etherate to yield the natural product **20** in 48% yield (Scheme 11).<80JOC3657>

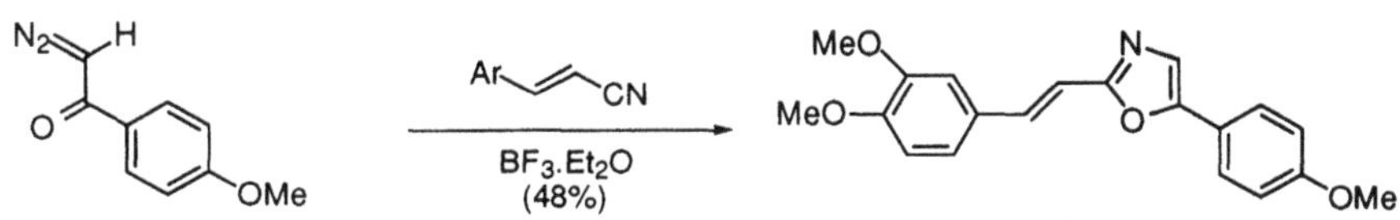

19 **20**

Scheme 11

Keehn and Mashraqui, in their studies on cyclophanes, used this ring-formation reaction to prepare the oxazole **21**, which was then elaborated to give the [2,2]-(2,5)oxazolophanes **22**, *via* a Hofmann elimination (Scheme 12).<82JA4461>

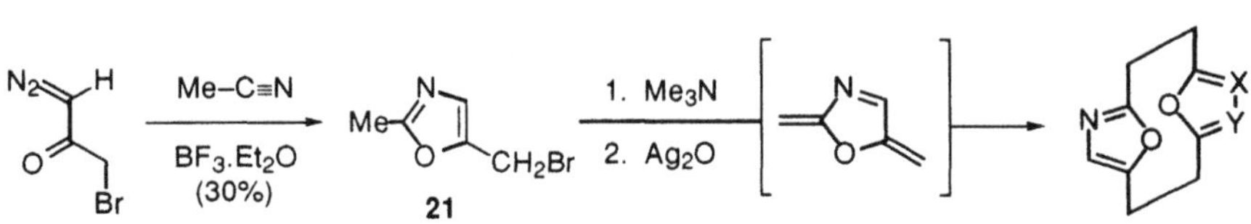

21

22a X = N, Y = CH
22b X = CH, Y = N

Scheme 12

The oxazole **23**, a key intermediate in an imaginative synthesis of indolequinones, was prepared similarly (Scheme 13),<93JOC1341> and even highly hindered diazo carbonyl compounds have been shown to give oxazoles using boron trifluoride etherate catalysis when other methods have failed. For example, the adamantyl diazoketoester **24** was shown to undergo oxazole formation in 76% yield using boron trifluoride etherate, whereas photochemical methods and rhodium(II) acetate catalysis (see Section 1.5.4) failed to give any desired product (Scheme 13).<93S793>

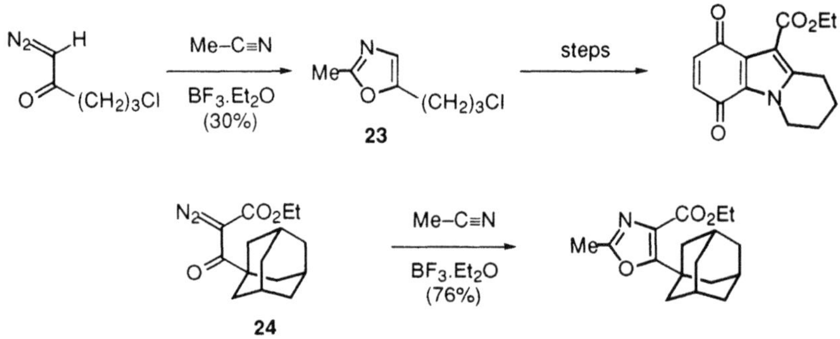

Scheme 13

Recently the use of the boron trifluoride catalysed reaction in the synthesis of the oxazolylindole fragment **25** of the natural product diazonamide A has been reported.<96SL609> Thus the BF_3-mediated reaction of the indolyl diazoketoester with acetonitrile gave oxazole **25** with simultaneous removal of the Boc-protecting group (Scheme 14).

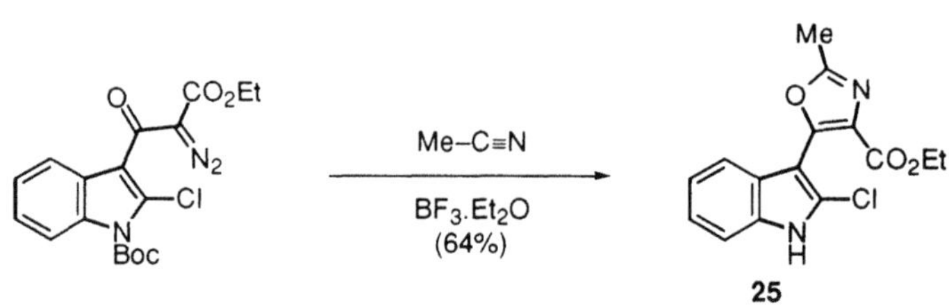

Scheme 14

1.5 TRANSITION-METAL CATALYSED REACTIONS

1.5.1 Copper

Many of the early workers who studied the thermal decomposition reactions of diazocarbonyl compounds found that the addition of copper metal or copper salts allowed the reaction to be achieved at a lower temperature,<63AG(E)565, 64CB2628, 73JOU431> although no detailed study of this catalytic effect was undertaken. Alonso and Jano studied the copper-salt reaction of ethyl diazopyruvate **26** with acetonitrile and benzonitrile. The corresponding oxazoles were found to be formed in 35% and 52% yield, respectively (Scheme 15).<80JHC721>

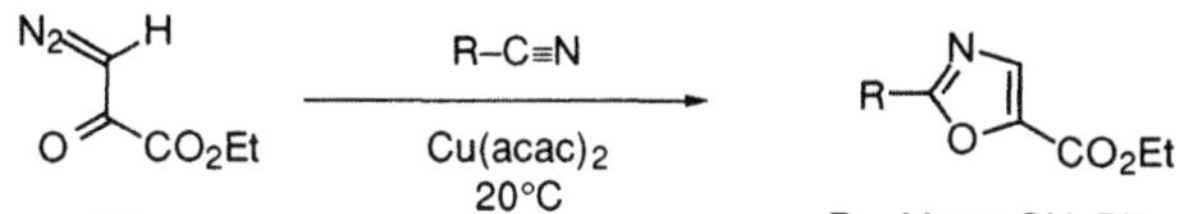

26

R = Me or CH_2Ph

Scheme 15

The use of copper complexes in the selective formation of oxazoles from unsaturated nitriles and *n*-butyl diazoacetate has been the subject of an investigation by Teyssié *et al.* (Scheme 16).<75JOM(88)115> Their studies pointed towards the copper undergoing two key changes in the catalytic process, the first being a reduction under the reaction conditions of copper(II) to copper(I), and the second being a change of ligands around the copper(I). The reactants, the nitrile and diazo compound, were found to play an important role in the formation of the most effective catalytic species.

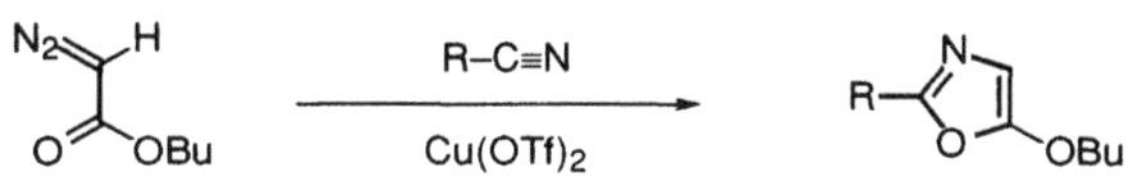

R = $CMe=CH_2$ (80%)
R = CH=CHPh (50%)

Scheme 16

1.5.2 Tungsten

Kitatani *et al.* found that tungsten(VI) chloride would catalyse the formation of a range of oxazoles from benzoyl(phenyl)diazomethane and nitriles (Scheme 17).<74TL1531, 77BCJ1647> The reaction with acetonitrile was studied with a range of other metal chlorides, but all proved less satisfactory than WCl_6. They attributed the catalytic nature of tungsten(VI) chloride to both its Lewis acidity and the affinity of tungsten for carbenes.

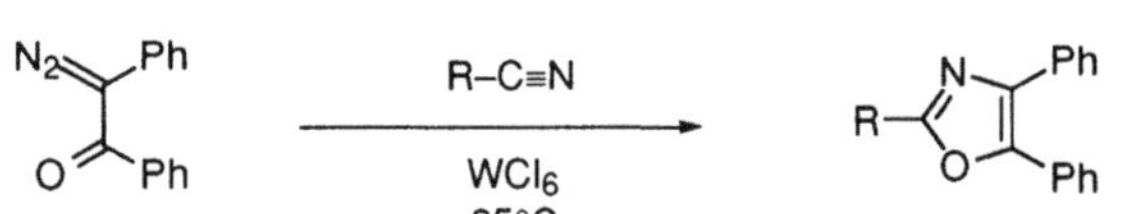

R = Me (65%)
R = Et (45%)
R = $CH=CH_2$ (50%)

Scheme 17

1.5.3 Palladium

Early studies into the decomposition of ethyl diazoacetate by a π-allyl palladium chloride complex in the presence of acetonitrile led to the isolation of 2-methyl-5-ethoxyoxazole in

16% yield.<66JOC618> Subsequently Teyssié *et al.* studied the reaction of ethyl diazoacetate with acrylonitrile, finding that in the presence of palladium(II) acetate, the oxazole **27** was formed in 30% yield (Scheme 18).<74TL3311> In the absence of any catalyst at room temperature a 2*H*-pyrazoline, formed by 1,3-dipolar cycloaddition of the diazo group to the alkene, was obtained, whilst at 100°C cyclopropanation of the double bond occurred. The authors claim that these observations fit with an oxazole formation mechanism involving the decomposition of the diazoester *via* co-ordination with the palladium. They assumed the nitrile formed an active complex with the catalyst, which decomposed the diazo compound.

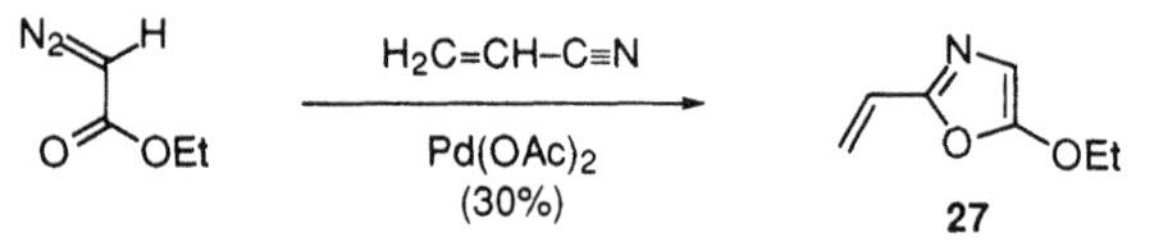

Scheme 18

1.5.4 Rhodium

In view of the highly successful use of rhodium catalysed reactions of diazocarbonyl compounds in synthesis,<86ACR348, 86CRV919, 87TCC(137)75, 91CRV263, 91T1765, 92T5385, 94AG(E)1797, 94CRV1091, 95T10811, 96CRV223> it is not surprising that the reaction with nitriles has also been widely studied.

Much of the early work into the rhodium(II)-catalysed formation of oxazoles from diazocarbonyl compounds was pioneered by the group of Helquist. They first reported, in 1986, the rhodium(II) acetate catalysed reaction of dimethyl diazomalonate with nitriles.<86TL5559, 93T5445, 96OS(74)229> A range of nitriles was screened, including aromatic, alkyl and vinyl derivatives; with unsaturated nitriles, cyclopropanation was found to be a competing reaction (Table 3).

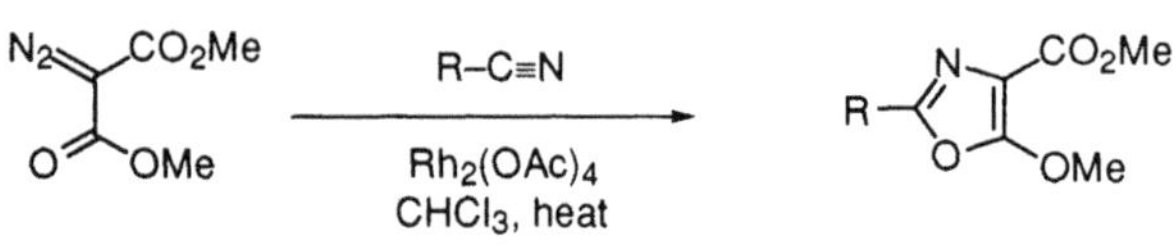

R	*Yield / %*
Ph	85
3-Cl-C_6H_4	96
Me	58
MeCH=CH	64*E*, 10*Z*
H_2C=$CHCH_2$	45 + 21[a]
EtOCH=CH	97

[a] = cyclopropanation product

Table 3

A series of catalysts was also screened in the reaction with benzonitrile to give methyl 2-phenyl-5-methoxyoxazole-4-carboxylate, with rhodium(II) acetate being the most effective (Table 4).<93T5445>

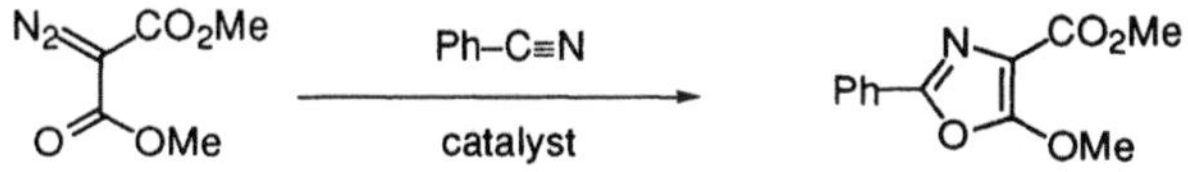

Catalyst	*Yield / %*
$Rh_2(OAc)_4$	99
$Rh_2(NHCOCH_3)_4$	83
$Cu(OTf)_2$	65
$Cu(Et\text{-}acac)_2$	44
$Rh_2(O_2CC_3F_7)_4$	35
$Rh_3(CO)_{16}$	23

Table 4

Helquist's work was primarily aimed towards the synthesis of the streptogramin antibiotics, such as virginiamycin M_1 and madumycin I, and this has involved the study of the rhodium(II) catalysed reactions of ethyl 1-formyldiazoacetate with nitriles.<91TL17, 93T5445> The oxazole **28** obtained from bromoacetonitrile was found to give a heteroaromatic benzylic organozinc derivative which underwent reaction with aldehydes and ketones leading to a range of alcohols **29** in good yield (Scheme 19).<92JOC4797> Utilising this reaction has led Helquist *et al.* to a non-racemic route to a protected form of the right hand portion of the type A streptogramin antibiotics.<93TL7371>

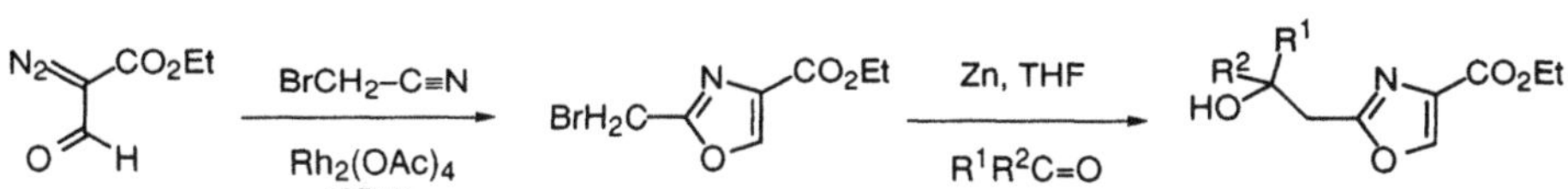

Scheme 19

Helquist's work on the use of diazomalonate in the synthesis of oxazoles has been extended to other diazocarbonyl compounds in our own laboratory.<92TL7769, 94T3761> Thus it was found that sulfonyl-, phosphonyl- and cyano-substituted diazoesters gave the corresponding 4-functionalised oxazoles **30** in acceptable yield (Scheme 20). In many cases the yield of oxazole was significantly improved by the use of rhodium(II) trifluoroacetamide as catalyst. The 4-cyano-oxazole **30** (R = Me, Z = CN) proved interesting in that it allowed the formation of a bis-oxazole **31** by a second rhodium catalysed reaction (Scheme 20).

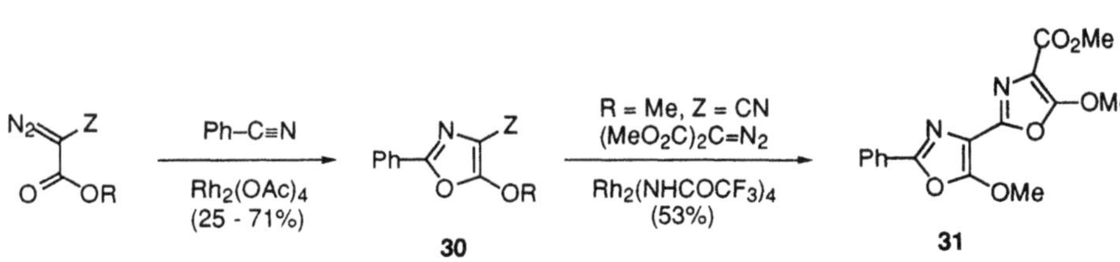

Scheme 20

Simultaneously with our own work, Yoo also utilised the iterative additions of diazocarbonyl compounds as a route to bis- and tris-oxazoles, although the fact that diazomalonate was used as the diazo component necessitated the conversion of the ester group to a nitrile in each case.<92TL2159> Thus the silyl-protected cyanohydrin **32** was reacted with dimethyl diazomalonate under rhodium(II) acetate catalysis to give the oxazole **33** in 57% yield. The ester was then reduced to the alcohol, with simultaneous removal of the 5-methoxy group, which was then transformed to the nitrile **34** *via* the aldehyde and oxime. The oxazole-4-carbonitrile **34** then underwent a second oxazole ring formation, again using dimethyl diazomalonate and rhodium(II) acetate, in 62% yield. The resulting bis-oxazole-4-carboxylate **35** was transformed to the corresponding bis-oxazole-4-carbonitrile which under the same conditions was taken to the substituted tris-oxazole **36** in 49% yield (Scheme 21).

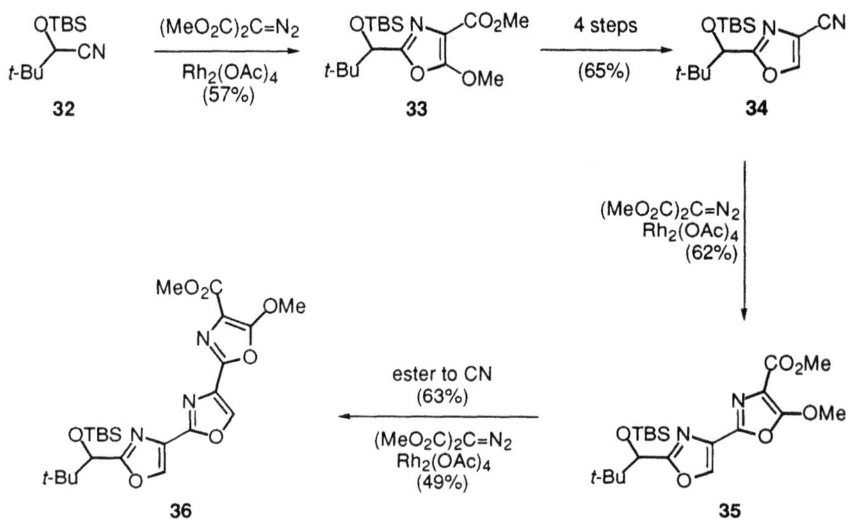

Scheme 21

Various other diazocarbonyl compounds have been shown to undergo oxazole formation under rhodium(II) catalysis. Shi and Xu have shown that ethyl 3,3,3-trifluoro-2-diazopropionate **37** will undergo oxazole formation in the presence of rhodium(II) acetate and a range of nitriles (Scheme 22).<89CC607> Likewise the rhodium(II) perfluorobutyrate catalysed reaction of the silyl diazo compound **38** with methyl cyanoformate in the presence of benzaldehyde gave the corresponding oxazole; no products resulting from carbonyl ylide formation with the aldehyde were isolated.<94T7435> In the rhodium(II) acetate catalysed decomposition of 2-diazo-1,3-cyclohexanedione in the presence of dihydrofuran and acetonitrile, two products were formed; the bis-tetrahydrofuran **39** and the oxazole **40** (Scheme 22).<91JOC6269>

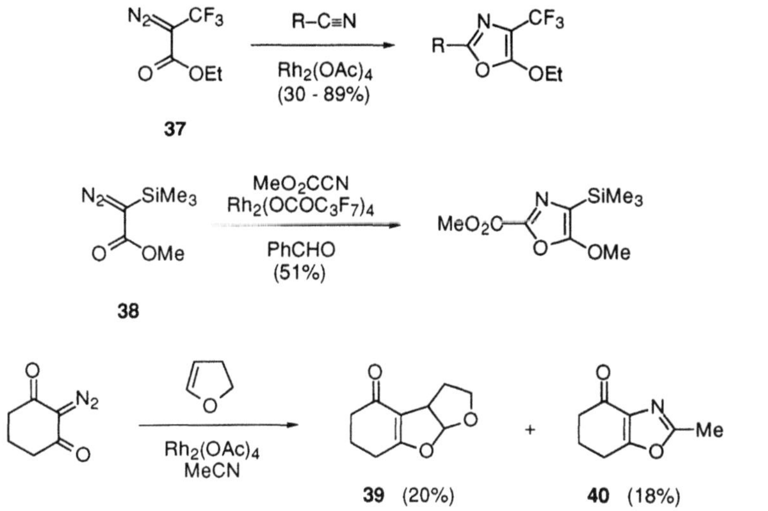

Scheme 22

The catalytic nature of these metal catalysts can be attributed to both their Lewis acidity and their ability to form electrophilic metal carbenoid intermediates. Apart from the Teyssié study on the role of copper in the catalytic process, the only other mechanistic study has been undertaken by Ibata and Fukushima.<92CL2197> They studied the rhodium(II) acetate catalysed decomposition of various substituted diazoacetophenones **1** in benzonitrile in the presence of the dipolarophile dimethyl acetylenedicarboxylate (DMAD). Two products were formed, the first being the expected 2,5-diaryloxazole **2**, whilst the second was the dimethyl 2,5-substituted pyrrole-3,4-dicarboxylate **41**. The results were explained by the formation of an intermediate nitrile ylide which can then either undergo 1,5-cyclisation to give the oxazole **2**, or be trapped by DMAD leading to the pyrrole **41**. The oxazole alone, in the presence of rhodium(II) acetate and DMAD, under thermal conditions, was found not to yield the pyrrole (Scheme 23). Similar studies using cyanamides as the nitrile component have been carried out.<95H(40)149>

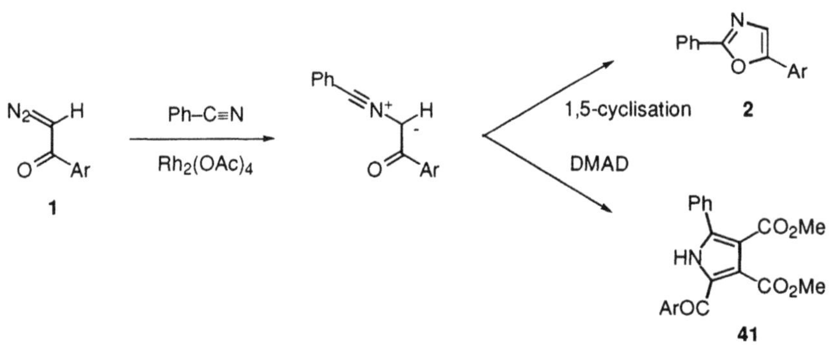

Scheme 23

The role of the rhodium is probably two-fold. Initially due to its Lewis acidity it reversibly forms a complex with the nitrile; nitriles are known to complex to the free axial coordination sites in rhodium(II) carboxylates as evidenced by the change of colour upon addition of a nitrile to a solution of rhodium(II) acetate, and by *X*-ray crystallography.<B-93MI1> Secondly the metal catalyses the decomposition of the diazocarbonyl compound to give a transient metallocarbene which reacts with the nitrile to give a nitrile ylide intermediate. Whether the nitrile ylide is metal bound or not is unclear.

Whatever the exact mechanism, the rhodium(II) catalysed reaction of diazocarbonyl compounds with nitriles is a useful route to oxazoles. A further example from our own laboratory illustrates the use of the reaction in the synthesis of the oxazolylindole alkaloids pimprinine **43a**, pimprinethine **43b**, and WS-30581A **43c**. Diazoacetylindole **42** reacted with simple nitriles in the presence of rhodium(II) trifluoroacetamide to give the corresponding oxazoles, deprotection of which gave the natural products **43** (Scheme 24).<94S1021>

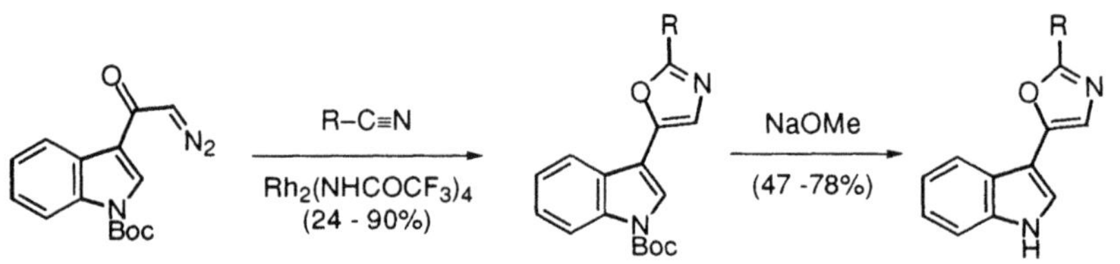

42

43a R = Me
43b R = Et
43c R = *n*-Pr

Scheme 24

To date most of the nitriles studied have been simple alkyl or aromatic derivatives with little other functionality. We recently attempted to extend the reaction to *N*-protected α-aminonitriles, derived by dehydration of α-aminoacid amides (Path A, Scheme 25), but this proved unsatisfactory, and therefore we investigated an alternative diazocarbonyl based route in which the order of steps was reversed, *i.e.* a rhodium catalysed N–H insertion reaction on the amide followed by cyclodehydration to the oxazole (Path B, Scheme 25).

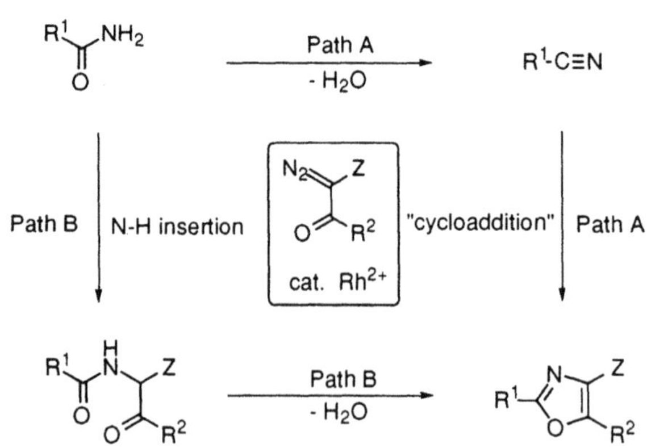

Scheme 25

This proved extremely effective, and a range of diazocarbonyl compounds was converted into the corresponding oxazoles **44** using this protocol (Scheme 26).<96SL825> The overall method is sufficiently mild that no racemisation occurs and the oxazoles **44** are obtained in excellent enantiomeric purity. We have recently completed the first synthesis of (+)-nostocyclamide **45** using a rhodium catalysed reaction to construct the chiral oxazole fragment of the natural product, further establishing the versatility of diazocarbonyl compounds in the synthesis of oxazoles.<96SL1171>

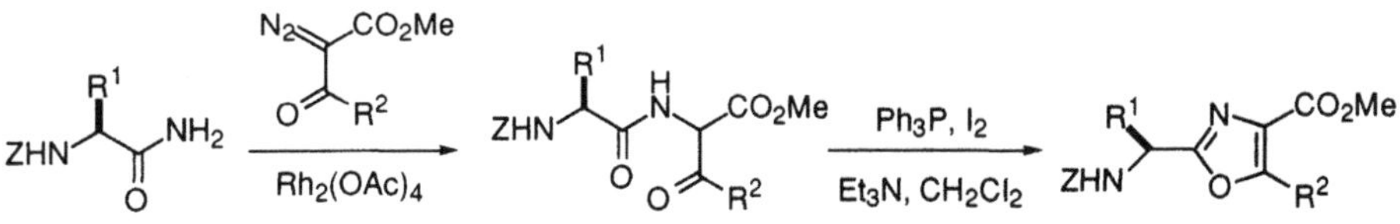

44

Scheme 26

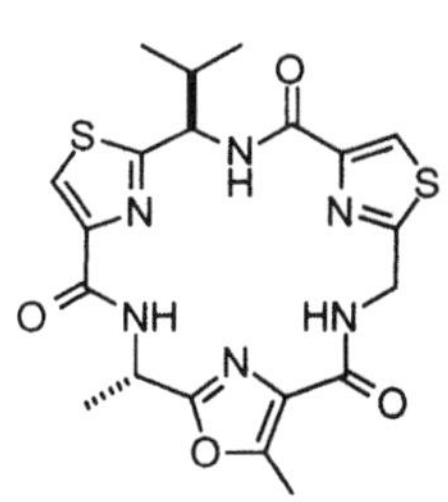

45

1.6 REFERENCES

<B-49MI1> J.W. Cornforth, in *The Chemistry of Penicillin*, Princeton University Press, Princeton, **1949**, 688.

<63AG(E)565> R. Huisgen, *Angew. Chem. Int. Ed. Engl.* **1963**, *2*, 565.

<64CB2628> R. Huisgen, G. Binsch, L. Ghosez, *Chem. Ber.* **1964**, *97*, 2628.

<64CB2864> R. Huisgen, H.J. Sturm, G. Binsch, *Chem. Ber.* **1964**, *97*, 2864.

<66JOC618> R.K. Armstrong, *J. Org. Chem.* **1966**, *31*, 618.

<67AG(E)807> W. Steglich, H.-U. Heininger, H. Dworschak, F. Weygand, *Angew. Chem. Int. Ed. Engl.* **1967**, *6*, 807.

<68CB302> H. Dworschak, F. Weygand, *Chem. Ber.* **1968**, *101*, 302.

<72CJC3730> N.T. Buu, J.T. Edward, *Can. J. Chem.* **1972**, *50*, 3730.

<73JOU431> M.I. Komendantov, V.N. Novinskii, R.R. Bekmukhametov, *J. Org. Chem. USSR* **1973**, *9*, 431.

<74TL29> A. Padwa, J. Smolanoff, A. Tremper, *Tetrahedron Lett.* **1974**, 29.

<74TL1531> K. Kitatani, T. Hiyama, H. Nozake, *Tetrahedron Lett.* **1974**, 1531.

<74TL3311> R. Paulissen, P. Moniotte, A.J. Hubert, P. Teyssié, *Tetrahedron Lett.* **1974**, 3311.

<75JA4682> A. Padwa, J. Smolanoff, A. Tremper, *J. Am. Chem. Soc.* **1975**, *97*, 4682.

<75JOM(88)115> P.G. Moniotte, A.J. Hubert, P. Teyssié, *J. Organomet. Chem.* **1975**, *88*, 115.

<77BCJ1647> K. Kitatani, T. Hiyama, H. Nozake, *Bull. Chem. Soc. Jpn.* **1977**, *50*, 1647.

<78TL2247> M.P. Doyle, M. Oppenhuizen, R.C. Elliott, M.R. Boelkins, *Tetrahedron Lett.* **1978**, 2247.

<79BCJ3597> T. Ibata, R. Sato, *Bull. Chem. Soc. Jpn.* **1979**, *52*, 3597.

<79JCS(P1)1485> W.T. Flowers, G. Holt, P.P. McCleery, *J. Chem. Soc., Perkin Trans. 1* **1979**, 1485.

<80JHC721> M.E. Alonso, P. Jano, *J. Heterocycl. Chem.* **1980**, *17*, 721.
<80JOC3657> M.P. Doyle, W.E. Buhro, J.G. Davidson, R.C. Elliott, J.W. Hoekstra, M. Oppenhuizen, *J. Org. Chem.* **1980**, *45*, 3657.
<82CPB526> Y.-C. Kuo, T. Aoyama, T. Shioiri, *Chem. Pharm. Bull.* **1982**, *30*, 526.
<82JA4461> S.H. Mashraqui, P.M. Keehn, *J. Am. Chem. Soc.* **1982**, *104*, 4461.
<84BCJ2450> T. Ibata, T. Yamashita, M. Kashiuchi, S. Nakano, H. Nakawa, *Bull. Chem. Soc. Jpn.* **1984**, *57*, 2450.
<86ACR348> M.P. Doyle, *Acc. Chem. Res.* **1986**, *19*, 348.
<86CRV919> M.P. Doyle, *Chem. Rev.* **1986**, *86*, 919.
<86MI1> I.J. Turchi, in *Oxazoles*, Wiley Interscience, New York, **1986**, Ch. 1.
<86TL5559> R. Connell, F. Scavo, P. Helquist, B. Åkermark, *Tetrahedron Lett.* **1986**, *27*, 5559.
<87TCC(137)75> G. Maas, *Top. Curr. Chem.* **1987**, *137*, 75.
<89BCJ618> T. Ibata, Y. Isogami, *Bull. Chem. Soc. Jpn.* **1989**, *62*, 618.
<89CC607> G. Shi, Y. Xu, *J. Chem. Soc., Chem. Commun.* **1989**, 607.
<91CRV263> A. Padwa, S.F. Hornbuckle, *Chem. Rev.* **1991**, *91*, 263.
<91JOC6269> M.C. Pirrung, J. Zhang, A.T. McPhail, *J. Org. Chem.* **1991**, *56*, 6269.
<91T1765> J. Adams, D.M. Spero, *Tetrahedron* **1991**, *47*, 1765.
<91TL17> R.D. Connell, M. Tebbe, P. Helquist, B. Åkermark, *Tetrahedron Lett.* **1991**, *32*, 17.
<92CL2197> T. Ibata, K. Fukushima, *Chem. Lett.* **1992**, 2197.
<92JHC607> G. Pattenden, *J. Heterocycl. Chem.* **1992**, *29*, 607.
<92J OC4797> A.R. Gangloff, B. Åkermark, P. Helquist, *J. Org. Chem.* **1992**, *57*, 4797.
<92T5385> A. Padwa, K.E. Krumpe, *Tetrahedron* **1992**, *48*, 5385.
<92TL1033> E.L. Williams, *Tetrahedron Lett.* **1992**, *33*, 1033.
<92TL2159> S. Yoo, *Tetrahedron Lett.* **1992**, *33*, 2159.
<92TL7769> K.J. Doyle, C.J. Moody, *Tetrahedron Lett.* **1992**, *33*, 7769.
<B-93MI1> F.A. Cotton, R.A. Walton, *Multiple Bonds between Metal Atoms*, Oxford University Press, Oxford, 1993.
<93AG(E)1> J.P. Michael, G. Pattenden, *Angew. Chem. Int. Ed. Engl.* **1993**, *32*, 1.
<93JOC1341> E. Vedejs, D.W. Piotrowski, *J. Org. Chem.* **1993**, *58*, 1341.
<93S793> M. Ohno, M. Itoh, T. Ohashi, S. Eguchi, *Synthesis* **1993**, 793.
<93T5445> R.D. Connell, M. Tebbe, A.R. Gangloff, P. Helquist, B. Åkermark, *Tetrahedron* **1993**, *49*, 5445.
<93TL7371> M. Bergdahl, R. Hett, T.L. Friebe, A.R. Gangloff, J. Iqbal, Y. Wu, P. Helquist, *Tetrahedron Lett.* **1993**, *34*, 7371.
<94AG(E)1797> A. Padwa, D.J. Austin, *Angew. Chem. Int. Ed. Engl.* **1994**, *33*, 1797.
<94CRV1091> T. Ye, M.A. McKervey, *Chem. Rev.* **1994**, *94*, 1091.
<94NPR395> J.R. Lewis, *Nat. Prod. Rep.* **1994**, *11*, 395.
<94S1021> K.J. Doyle, C.J. Moody, *Synthesis* **1994**, 1021.
<94T3761> K.J. Doyle, C.J. Moody, *Tetrahedron* **1994**, *50*, 3761.
<94T7435> M. Alt, G. Maas, *Tetrahedron* **1994**, *50*, 7435.
<95CRV2115> P. Wipf, *Chem. Rev.* **1995**, *95*, 2115.
<95H(40)149> K. Fukushima, T. Ibata, *Heterocycles* **1995**, *40*, 149.
<95NPR135> J.R. Lewis, *Nat. Prod. Rep.* **1995**, *12*, 135.
<95T10811> D.J. Miller, C.J. Moody, *Tetrahedron* **1995**, *51*, 10811.
<96AHC(65)93> A.F. Khlebnikov, M.S. Novikov, R.R. Kostikov, *Adv. Heterocycl. Chem.* **1996**, *65*, 93.
<96CRV223> A. Padwa, M.D. Weingarten, *Chem. Rev.* **1996**, *96*, 223.
<96JOC778> G. Li, P.M. Warner, D.J. Jebaratnam, *J. Org. Chem.* **1996**, *61*, 778.
<96NPR435> J.R. Lewis, *Nat. Prod. Rep.* **1996**, *13*, 435.
<96OS(74)229> J.S. Tullis, P. Helquist, *Org. Synth.* **1996**, *74*, 229.
<96SL609> J.P. Konopelski, J.M. Hottenroth, H.M. Oltra, E.A. Véliz, Z.-C. Yang, *Synlett* **1996**, 609.
<96SL825> M.C. Bagley, R.T. Buck, S.L. Hind, C.J. Moody, A.M.Z. Slawin, *Synlett* **1996**, 825.
<96SL1171> C.J. Moody, M.C. Bagley, *Synlett* **1996**, 1171.

Chapter 2

The Heterocyclic Chemistry Associated with the Herbicide Glyphosate

James A. Sikorski
Monsanto Company, St. Louis, MO 63198 USA

2.1 INTRODUCTION

As the active ingredient in the popular herbicide, ROUNDUP,® glyphosate [*N*-(phosphonomethyl)glycine, GLY, **1**] has achieved tremendous commercial success as a broad spectrum, non-selective herbicide that controls many of the world's worst weeds while exhibiting very low mammalian toxicity (1,2). Glyphosate is also essentially nontoxic to birds, fish, insects and most bacteria and is readily broken down in soil by microbes which eventually produce ammonia, inorganic phosphate, and carbon dioxide (3). In 1994, the editors of *Farm Chemicals* magazine highlighted the discovery of glyphosate's herbicidal properties as one of the top ten major inventions to impact US agriculture in the last century.

Glyphosate kills plants by specifically inhibiting one critical plant enzyme used in the biosynthesis of aromatic amino acids. As such, glyphosate was one of the first commercially successful herbicides to have a primary identified enzyme site of action in plants (4,5).

$$(HO)_2\overset{\overset{\displaystyle O}{\|}}{P}\text{-}C_\beta H_2NHC_\alpha H_2CO_2H$$

1

The biological activity of glyphosate has stimulated a worldwide search for simple and efficient methods for its preparation, particularly using methods that would be suitable on commercial scale. Similarly, other closely related derivatives (esters, amides, nitriles, etc.) and structurally modified analogs that would have improved or complementary activity have been avidly sought by virtually every major agrochemical company. The synthesis of glyphosate as well as its analogs and derivatives has therefore provided an important global stimulus to develop new synthetic methods leading to aliphatic phosphonic and phosphinic acids. The purpose of this review is to highlight the unique heterocyclic chemistry associated with this important product. The relevant published chemical and patent literature is reviewed through 1996.

At the time of its discovery in 1970, few chemical methods were known for the laboratory syntheses of such molecules. Fewer still were appropriate for their industrial scale production. Unlike many other amino acid derivatives, glyphosate is stable in strong acid or base, even at elevated temperatures. It can also tolerate strong reductants and some oxidants (1,2). This stability accounts for the diversity of synthetic methods that have been explored and developed to prepare glyphosate and its heterocyclic derivatives over the last 25+ years.

2.2 HETEROCYCLIC PRECURSORS TO GLYPHOSATE

2.2.1 Mannich Reactions

2.2.1.a Acidic Reactions with Phosphorous Acid. The most important methods for the commercial production of glyphosate involve construction of the P-C_β-N bond, usually via an acid-catalyzed modification of the well known Mannich reaction (6).

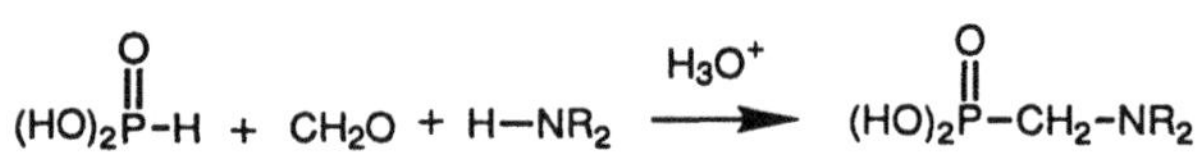

The first reported method for the direct phosphonomethylation of amino acids used phosphorous acid and formaldehyde (7). Typically, aqueous solutions of the amino acid, phosphorous acid, and concentrated (concd) hydrochloric acid were heated to reflux with excess aqueous formaldehyde or paraformaldehyde. The reaction proceeded equally well with either primary or secondary amines. However, with primary amines such as glycine, the yield of glyphosate was usually quite low, even at reduced temperature, and 1:1:1 stoichiometry. The resulting glyphosate acid ($GLYH_3$) reacted faster than glycine, so the bis-phosphonomethyl adduct **2** always predominated. With excess phosphorous acid and formaldehyde, good isolated yields of this 2:1 adduct **2** have been obtained (8).

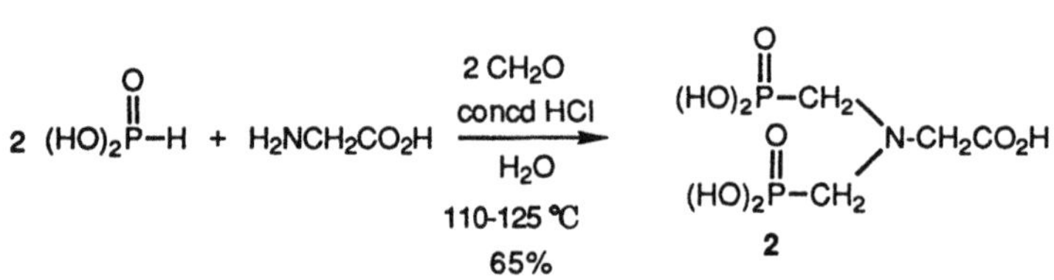

The propensity for glycine to form these 2:1 adducts under acidic Mannich conditions stimulated a search for alternative higher yielding routes to glyphosate. The rather vigorous conditions employed in this reaction limits the glycine-based starting materials to relatively simple, but stable analogs. Consequently, various *N*-substituted glycines have been used to prepare the corresponding *N*-substituted glyphosate intermediates that have been converted to glyphosate in dramatically improved yields. One of the most important methods for the commercial production of glyphosate utilizes iminodiacetic acid (IDA) **3** to make *N*-phosphonomethyliminodiacetic acid **4** (8). In essence, the second carboxymethyl group in **4** functions as an easily removable and inexpensive protecting group. This carboxymethyl fragment has been conveniently removed as carbon dioxide, formaldehyde, and formate with a variety of oxidizing agents to produce glyphosate in high yield (1,2).

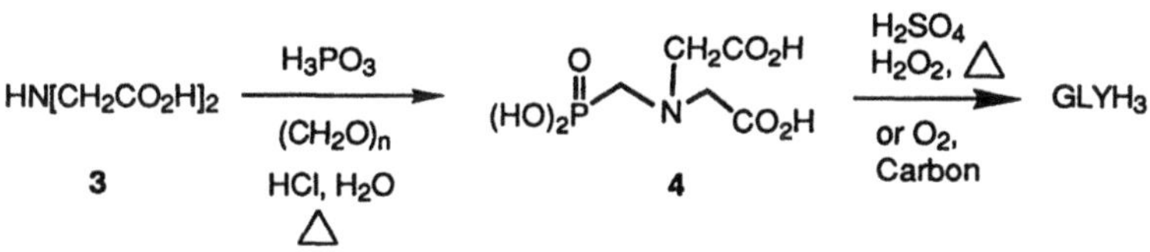

Perhaps the mildest and highest yielding deprotection sequence utilizes hydrogenolysis for the nearly quantitative removal of an *N*-benzyl protecting group as toluene. Dramatically improved yields of glyphosate were obtained using *N*-benzylglycine **5** via intermediate **6** (9,10). This procedure has recently been reported as a simple laboratory method for the production of $GLYH_3$ (11).

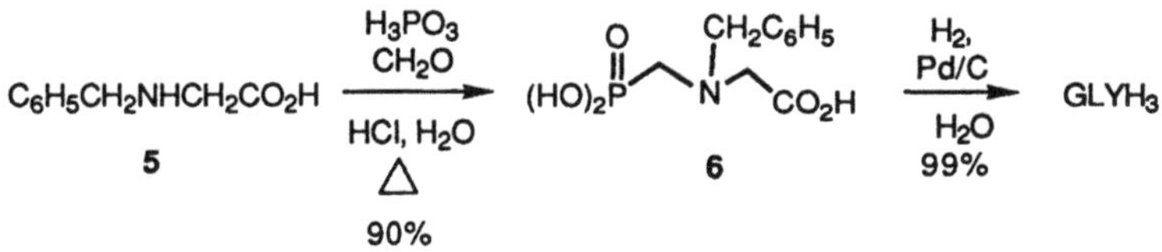

An alternative approach involved replacing glycine with a heterocyclic surrogate to eliminate bis adduct formation. Various cyclic protected forms of glycine have been used in these procedures. For example, diketopiperazine **7** represents a cyclic *N*-acylglycine derivative that is readily available by cyclodehydration of glycine or its esters (12). This derivative reacted well under the modified, acidic Mannich conditions to provide the corresponding *N*-phosphonomethyl derivative **8**, which was readily hydrolyzed under basic conditions to $GLYNa_3$ (13). Heating glyphosate to its decomposition point (200-230 °C) also produced **8** (1,11), which can be reconverted to glyphosate by heating in strongly acid media (1,2).

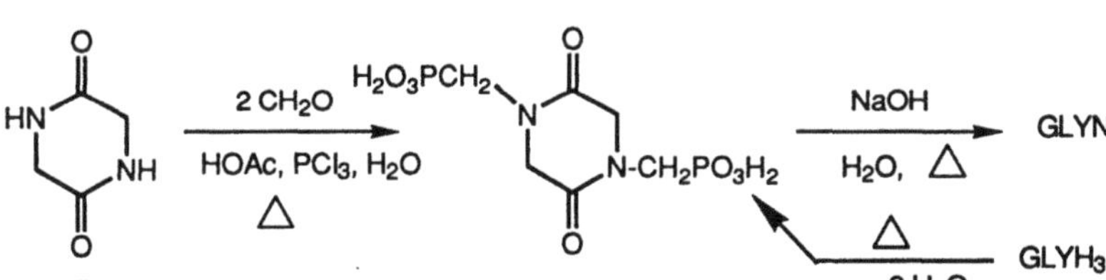

Hydantoin **9** has also been used as a heterocyclic starting material for glyphosate via intermediate **10** (14). Similarly, the phosphonomethylation of 3-*N*-alkyl hydantoins produced the corresponding unsymmetrically substituted cyclic analogs (14).

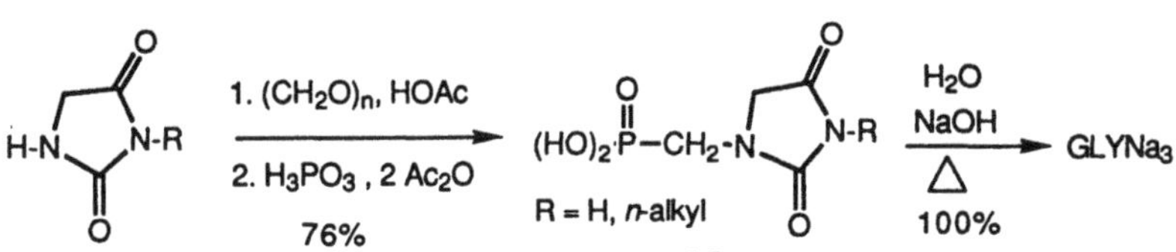

A different but complementary approach incorporated ethyl 2-aza-bicyclo[2.2.1]hept-5-ene acetate **11** to deliver the glycine aldimine component for the Mannich reaction through a thermal, retro Diels-Alder reaction. Thermal degradation of **11** in the presence of a suitable phosphorus acid precursor provided the glyphosate skeleton product in reasonable yield (15).

An alternative sequence utilized 2-oxazolidone, which was readily synthesized from urea and ethanolamine, as the glycine equivalent. Subsequent treatment with phosphorous acid and formaldehyde produced *N*-phosphonomethyl-2-oxazolidone **12** (16). Upon hydrolysis, and loss of CO_2, **12** provided the related derivative, *N*-phosphonomethylethanolamine **13**, which was oxidized at high temperature with a variety of metal catalysts including cadmium oxide (16) or Raney copper (17) to give $GLYH_3$, after acidification. A similar oxidation route has also been reported starting from *N*-phosphonomethyl-morpholine (18).

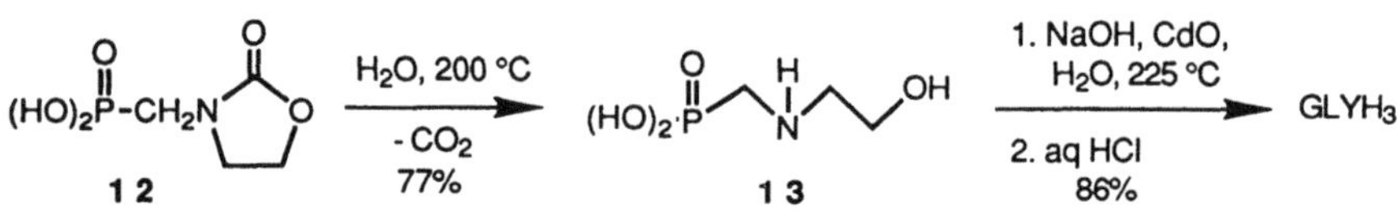

2.2.1.b Mannich Reactions with Phosphite Esters under Neutral Conditions. The Mannich reaction with glycine could be controlled to give glyphosate as the predominant product when the condensation reaction was conducted at a more neutral pH. With sodium glycinate under more conventional Mannich conditions, one must substitute an appropriate dialkyl (19,20) or trialkyl (21) phosphite for the phosphorous acid component described above. Consequently, this modification produced various glyphosate dialkyl phosphonate esters **14** and necessitated an additional acidic hydrolysis as the usual deprotection step to cleave the phosphonate esters in **14**. Reasonable isolated yields of $GLYH_3$ were reported using this two step, one-pot procedure. While these phosphites add significantly to the cost of any commercial process, they are quite convenient for common laboratory-scale operations.

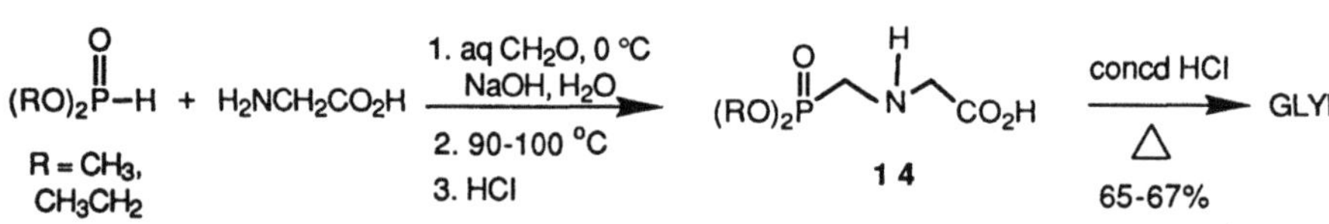

Alternatively, the reaction could be run at a lower temperature with better pH control under nonaqueous conditions using organic amines in alcohol solutions (22,23). The milder conditions produced $GLYH_3$ in high purity (97-99%) and improved yield (65-78%). Presumably, glycine combined with paraformaldehyde under these conditions to form the bis-hydroxymethylglycine intermediate **15**, which was not isolated but was immediately converted with dimethyl phosphite at reflux to give the reported *N*-hydroxymethyl-*N*-phosphonomethylglycine dimethyl phosphonate ester **16**. Sequential acidic hydrolysis produced first the phosphonate dimethyl ester **14a** then $GLYH_3$.

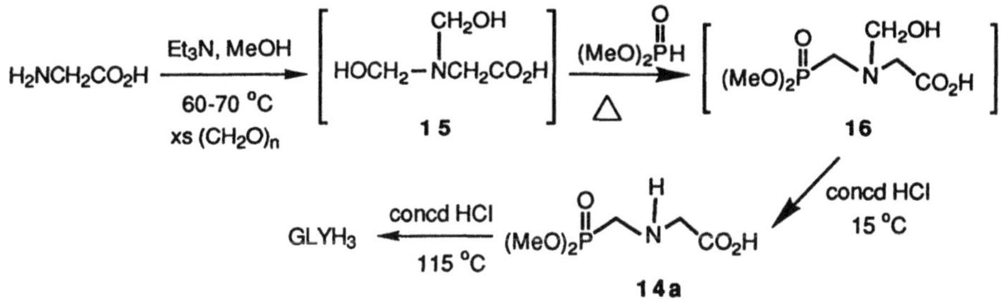

Suitable precautions should be taken in utilizing this procedure, since substantial quantities of the volatile, known carcinogen, chloromethyl methyl ether, as well as the volatile and flammable methyl chloride, form under the reaction conditions. While intermediates like **15** and **16** have been postulated in a number of publications and patents in this area, no experimental evidence has been reported that describes the presence or formation of detectable quantities of these species.

Similar reactions between diketopiperazine and either trialkyl phosphites or alkyl phosphinates produced the related cyclic analogs **17** and **18** (24).

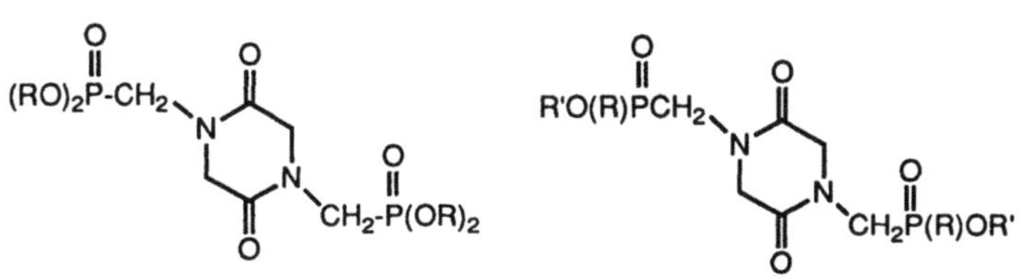

17 **18**

The related hydantoin monophosphinate **19**, bis-phosphonate **20**, or bis-phosphinate **21** systems have also been reported using these milder Mannich methods (24).

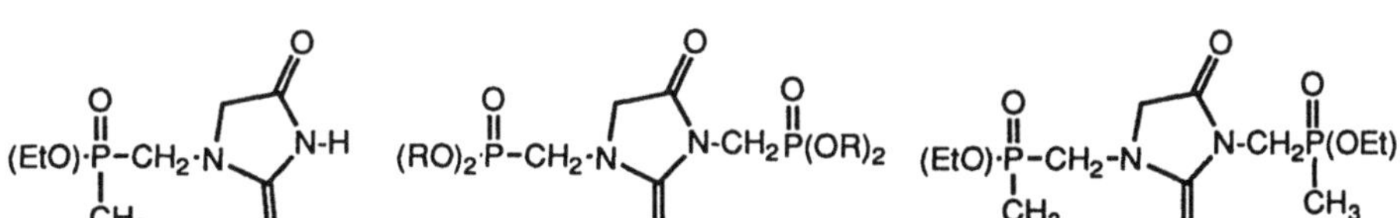

19 **20** **21**

2.2.2 Hexahydro-1,3,5-triazine Reactions

Another versatile approach, which nicely complements these Mannich-based procedures, incorporates a preformed symmetrical hexahydro-1,3,5-triazine (HHT) intermediate. In this case the phosphorus reagent reacts with HHT as a trimeric form of the normal aldimine species generated in situ between the amine or amino acid and formaldehyde. These HHT reagents can often be purified and isolated prior to the reaction with phosphites. They are reasonably stable under neutral or slightly basic conditions, but they can readily revert back to the original amine and formaldehyde after heating with aqueous acid (25). Several can be purchased commercially.

Stable HHTs such as the *N*-benzyl derivative **22**, when reacted with neat diethyl phosphite under fairly forcing conditions, gave the desired α-aminomethylphosphonate hydrochloride **23** (26). No significant product formation was observed at similar temperatures in organic solutions using toluene or acetonitrile (27).

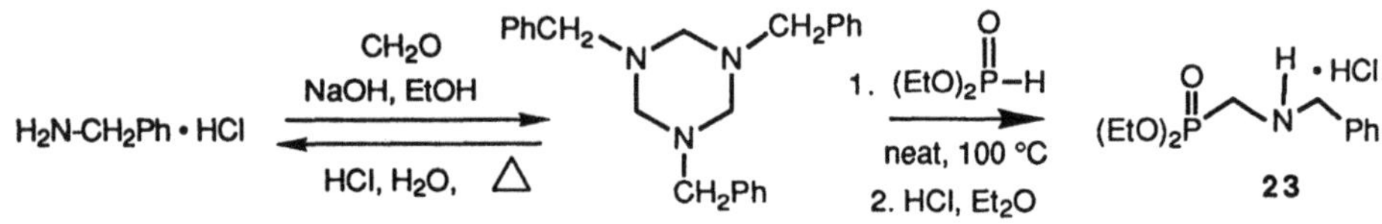

22 **23**

These reactions between HHTs and phosphites are believed to occur in an analogous fashion to those previously reported for reactive thiol moieties (28). It has been postulated that protonation by the weakly acidic, aliphatic phosphite activates the HHT to undergo a stepwise ring-opening (29). Alternatively, a series of activated iminium species, analogous to those proposed for the Mannich reaction, may be involved for the HHT ring-opening mechanism (2). While a thermal reversal to a transient aldimine may also be possible, all attempts to detect such a species under the reaction conditions or from thermal cracking of HHTs so far have been unsuccessful (27).

Glycine and its esters also readily form isolable trimeric HHT derivatives. These intermediates have also been successfully utilized in constructing the glyphosate backbone when they have sufficient thermal stability to tolerate the reaction conditions. For example, good overall yields of $GLYH_3$ were obtained under similar conditions with the stable HHT of sodium glycinate **24** via the phosphonate diester **14b** (30).

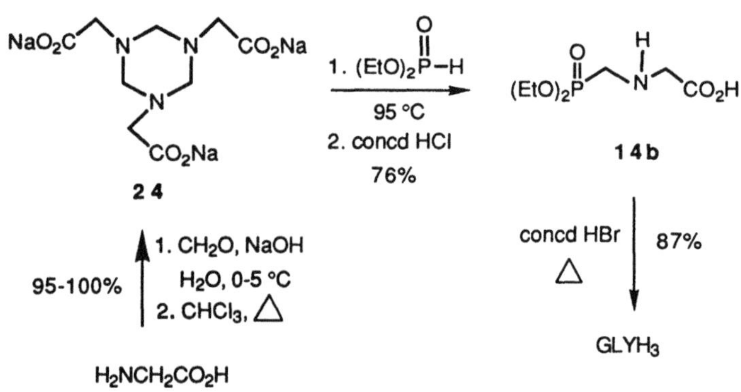

24 **14b**

The HHT of ethyl glycinate **25** also reacted quantitatively with aliphatic phosphites under comparable conditions to give the corresponding aliphatic glyphosate triester **26**, which was identical in all respects to the corresponding Mannich product (31). The product mixture from diethyl thiophosphite was much more complex and led to dramatically lower yields (27).

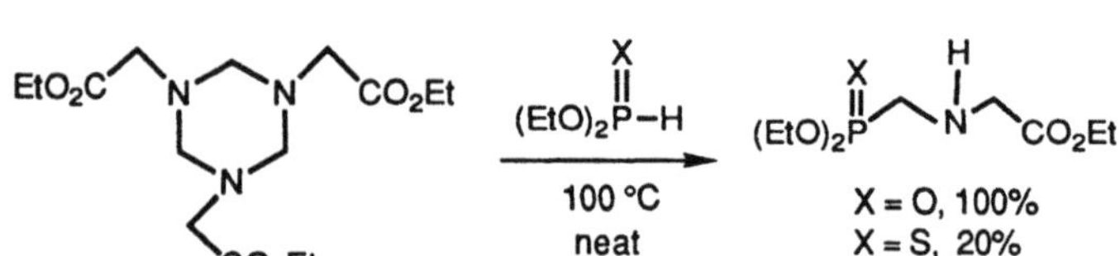

25 **26**

Subsequent hydrolysis of these glyphosate triesters **26** using hydrobromic acid again produced glyphosate (27,32).

Aliphatic trialkyl phosphites also reacted with HHT **25** under neat conditions and elevated temperatures (>100 °C) to produce glyphosate triesters such as **26** (27). However, the reaction proceeded at much lower temperatures (10 °C) when titanium tetrachloride was present in equimolar amounts (33).

Certain heteroatom-substituted aliphatic phosphites have also been employed in these sequences, depending on their stability. Generally, phosphites bearing α-substituents led to higher yields than were obtained with phosphites having β-substituents (27). For example, bis-β-cyanoethyl phosphite gave a relatively modest yield (~25%) of coupled triester product (34,35), whereas a very good yield (83%) of desired product was isolated from reactions with bis-α-cyanopropyl phosphite (36).

An interesting variation of this reaction that made use of a three-component, one-pot solventless procedure with the corresponding trialkyl phosphites gave dramatically improved yields of many heterosubstituted glyphosate phosphonate diesters (37). When exactly one equivalent of water, **25**, and tris-β-chloroethyl phosphite were mixed and heated under neat conditions for a few hours, nearly quantitative yields of displaced β-chloroethanol and the desired triester product **27** were obtained. If desired, the displaced alcohol was first removed by vacuum distillation, or the mixture could be hydrolyzed directly to $GLYH_3$. Various oxygen, sulfur, nitrogen, cyano, and carboxylate functionalities were similarly accommodated in the trialkyl phosphite.

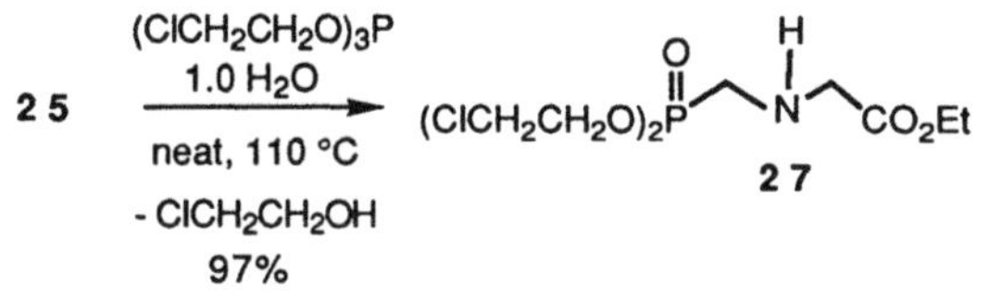

These results contrasted sharply with those obtained with a HHT that was relatively unstable to the reaction conditions. For example, the commercially available HHT of glycinonitrile **28** gave a very poor yield (6%) of coupled glyphosate product with diethyl phosphite because the reaction must be run with acid-catalysis at much lower temperatures (27,38). Somewhat higher yields were observed when **28** was used directly under the modified, acidic Mannich conditions to provide *N*-phosphonomethylglycinonitrile **29**, which was hydrolyzed directly to $GLYNa_3$ (39).

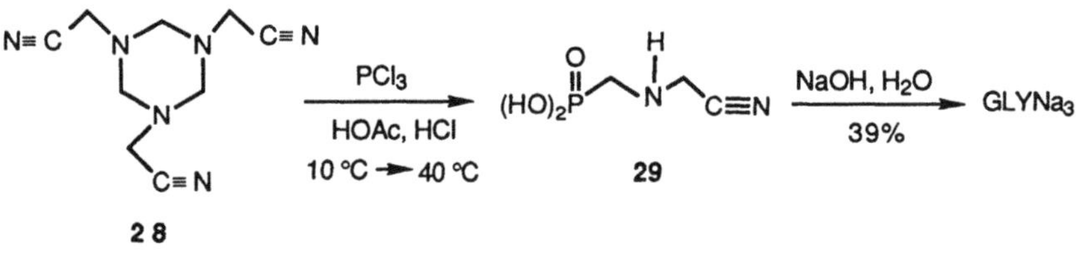

Considerably higher product yields resulted under much milder conditions when diaryl phosphites were condensed with HHTs. Whereas vigorous neat conditions and temperatures exceeding 100 °C were usually necessary with aliphatic phosphites, often quantitative conversions to the desired glyphosate derivatives were obtained in common organic solvents using aromatic or benzylic phosphites (40).

Presumably, the electron-withdrawing aryl groups made these phosphites sufficiently acidic so that they acted as internal catalysts for the required HHT ring-opening reactions. Essentially any nonreactive aprotic solvent could be used. Alcohol solvents should be avoided, however, because the aromatic phosphite ester groups can be easily exchanged. This reaction provides an extremely versatile, convenient, and general method for the laboratory synthesis of a number of useful glyphosate triesters **30** or diester nitrile intermediates **31**. While some variability in yields was observed with substituted aromatic phosphites, diphenyl phosphite reacted cleanly to give the corresponding glyphosate diphenyl phosphonate esters in very good isolated yields. The mildness of this procedure was easily demonstrated from the observation that the more unstable HHT **28** gave product conversions comparable to those obtained with **25** (41,42).

The value of using the preformed HHT with diphenyl phosphite in this procedure was readily apparent from the nearly quantitative conversion to glyphosate observed from **25**. A much lower yield (38%) of glyphosate was obtained after hydrolysis when the same components (ethyl glycinate hydrochloride and formaldehyde) were mixed and heated with neat triphenyl phosphite to give triester **30** (43).

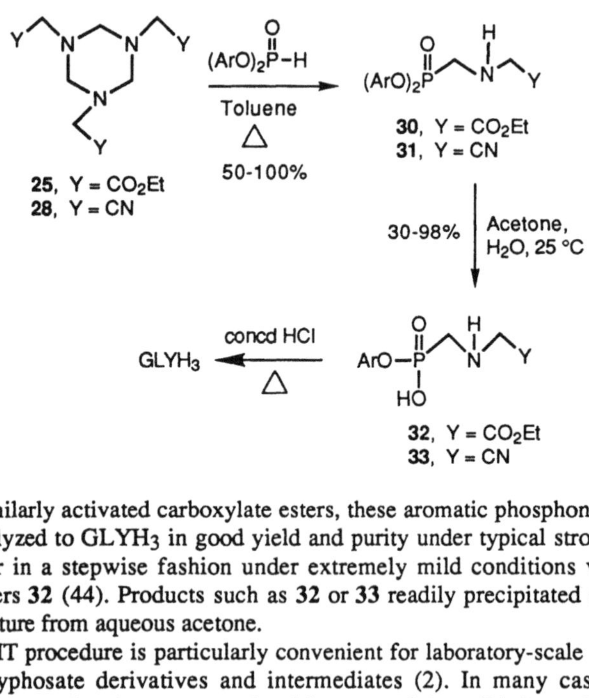

Like similarly activated carboxylate esters, these aromatic phosphonate diesters **30** were readily hydrolyzed to $GLYH_3$ in good yield and purity under typical strongly acidic or basic conditions, or in a stepwise fashion under extremely mild conditions via the zwitterionic monoaryl esters **32** (44). Products such as **32** or **33** readily precipitated after a few hours at room temperature from aqueous acetone.

This HHT procedure is particularly convenient for laboratory-scale syntheses of a wide variety of glyphosate derivatives and intermediates (2). In many cases, fairly sensitive functionalities can often be accommodated because of the mild thermal conditions and the complete absence of water. This method is therefore quite complementary to the other aqueous Mannich procedures, since groups that would not normally tolerate aqueous strongly acidic conditions would frequently decompose or would be difficult to recover in high yield.

Several variations and extensions of this HHT method have recently been reported. The mildness of this reaction was exemplified through the synthesis of glyphosate thiol ester derivatives **35**. The requisite thioglycinate HHT **34** was prepared in high yield by a novel, methylene-transfer reaction between *t*-butyl azomethine and the ethyl thioglycinate hydrobromides (45). The by-product *t*-butylamine hydrobromide was easily removed by filtration. Subsequent reaction with tris-trimethylsilyl phosphite in acetonitrile at room temperature followed by hydrolysis with cold, aqueous isopropanol gave the desired thiol carboxylate derivatives **35** in good yield. These were hydrolyzed under more vigorous conditions to $GLYH_3$ (46).

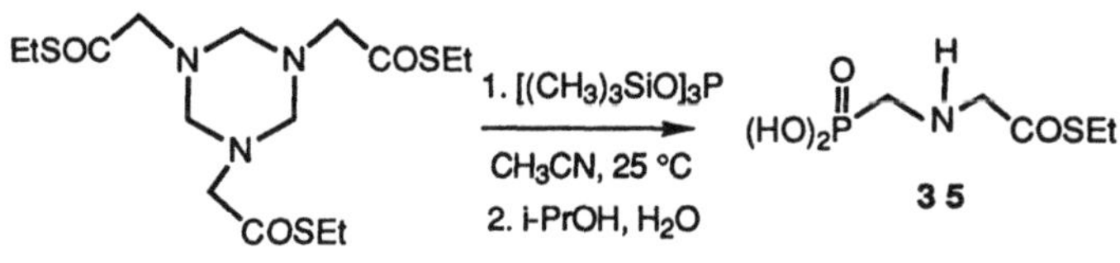

34 35

Examples of cyclic aliphatic or aryl phosphonate triesters, such as **36** or **37** have also been prepared by this HHT method from their cyclic phosphite precursors (2).

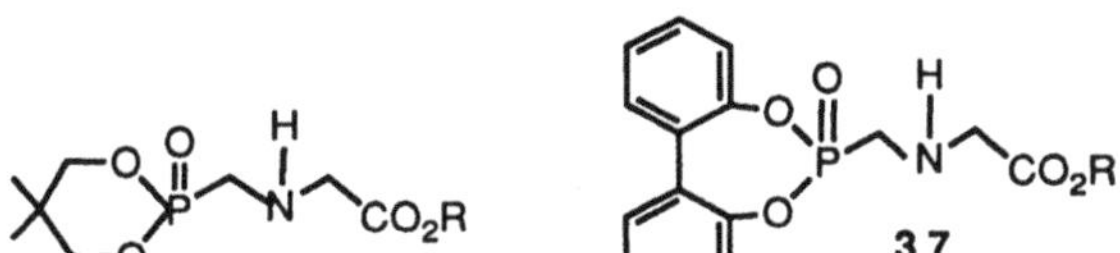

36 37

Another interesting application utilized an analogous ring-opening reaction between phosphites and a benzoxazine **38** to produce a protected *N*-benzyl-glyphosate derivative **39** directly. Hydrogenolysis and saponification then gave $GLYNa_3$. These benzoxazines were readily prepared from a base-catalyzed reaction between *p*-cresol, formaldehyde, and either ethyl or sodium glycinate (47).

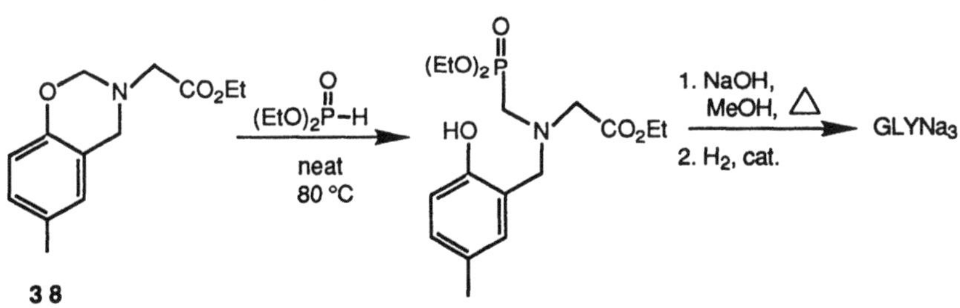

38 39

An interesting reaction between *N*-methyleneglycinonitrile **28** and phosphorous trichloride proceeded at low temperature under anhydrous conditions to form the glyphosate nitrile intermediate **40** as its hydrochloride salt (48). However, when the reaction was conducted in the presence of water, the glyphosate amide **41** was generated instead. Either intermediate **40** or **41** could be hydrolyzed directly to glyphosate under acidic conditions (49).

When excess amounts of the HHT of phenyl glycinate **42** were used with diphenyl phosphite, the preferred product was the novel cyclic derivative **45** (2). Presumably, ring-opening of the HHT produced intermediate **43** first, which lost an equivalent of glycinate formaldimine to give **44**. The proximity of the activated phenyl carboxylate ester to the N-H in **44** presumably promoted intramolecular cyclization to **45** with loss of phenol (2).

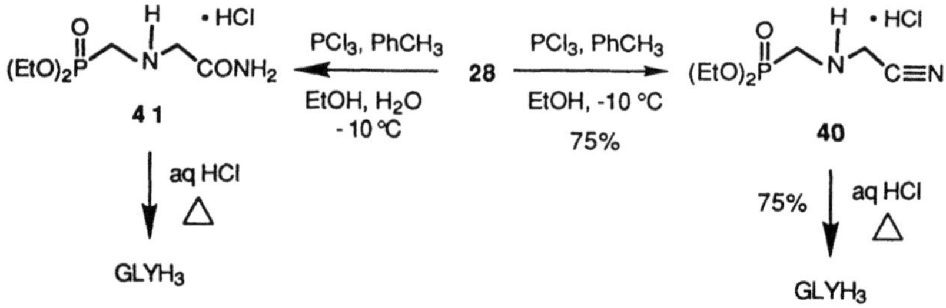

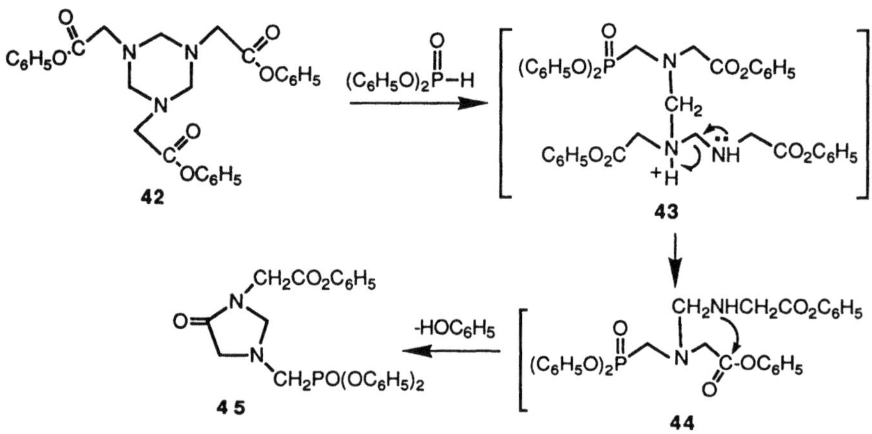

In contrast, when excess amounts of the HHTs derived from simple aliphatic glycinate esters, such as **25**, were used in excess in reactions with diaryl phosphites (50), the related glyphosate aminals containing aryl phosphonate esters **46** were isolated in low yield (5-15%). Like many aminals, these triesters **46** were acid-sensitive and were quantitatively converted to the corresponding triester strong acid salts **47** upon treatment with either HCl or methanesulfonic acid (27).

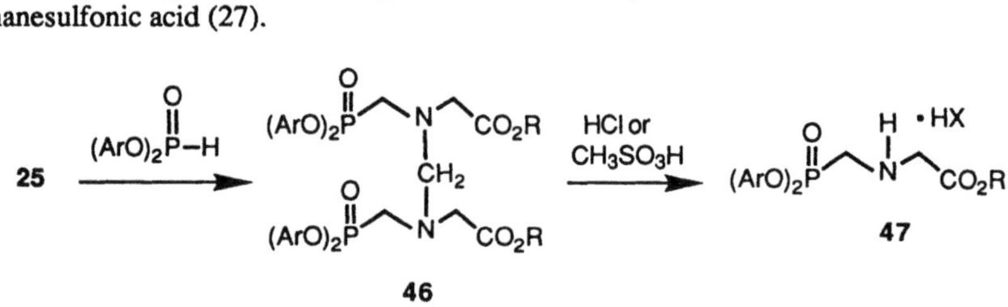

2.2.3 Michaelis-Arbuzov Reactions

This classical C-P bond-forming reaction (51) has seen limited application in the glyphosate arena, presumably for lack of suitable substrates that can tolerate the vigorous reaction conditions. Typically, C-P bond formation occurs when an alkyl halide reacts with excess neat trialkyl phosphite at temperatures exceeding 100 °C, near the boiling point of the phosphite. An Arbuzov-based strategy for glyphosate requires the synthesis of the corresponding protected *N*-halomethylglycine **48**. Several sequences have been reported that generate these species on the way to glyphosate, but for the most part, overall yields are lower than those obtained with other methods. These intermediates were usually prepared from ring-opening reactions of HHT **25** with acetyl chloride. Subsequent reaction with trimethyl phosphite gave the corresponding *N*-acetyl glyphosate triester **49**, which has been hydrolyzed to $GLYH_3$ (52).

Recently, good yields of glyphosate have been reported after hydrolysis using tris-trimethylsilyl phosphite in a similar sequence with **48** to generate the disilyl phosphonate triester intermediate **50** (53).

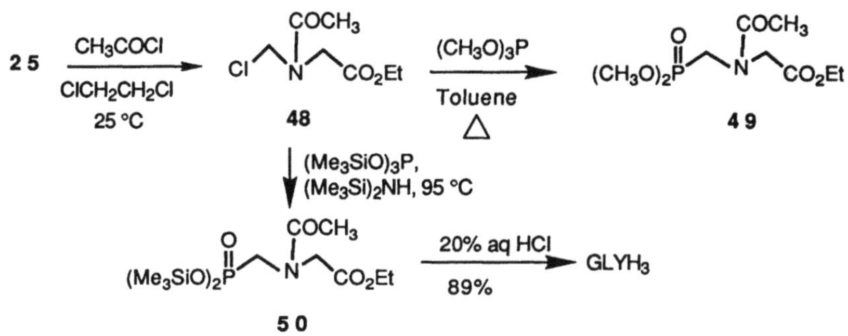

Surprisingly, the corresponding glycinonitrile HHT **28** also tolerated these reaction conditions and has been used in a similar process to generate $GLYH_3$ via the analogous *N*-acetyl-*N*-halomethylglycinonitrile intermediates (54).

Several earlier methods (55,56) utilized a piperazinedione derivative in an Arbuzov-based sequence as a more stable source of the requisite *N*-chloromethyl intermediate **51**. Treatment of piperazine-1,4-dione with formaldehyde and phosphorus trichloride provided convenient access to this starting material. Subsequent reaction with either trimethyl or triethyl phosphite produced the *N*-phosphonomethyl tetraester derivative **52**, which has been hydrolyzed to $GLYH_3$.

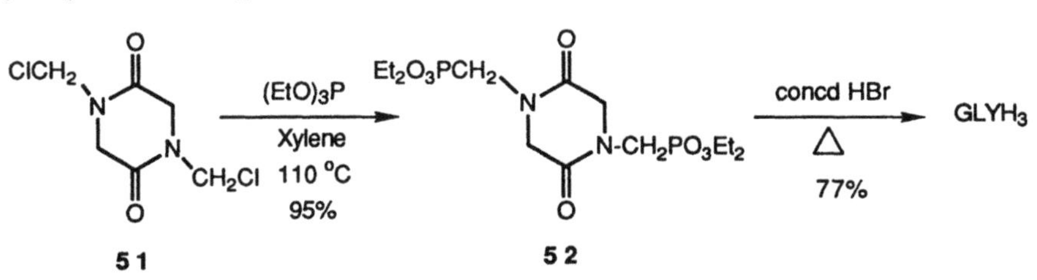

The electrophilic ring opening of *N*-allyl HHT **53** with chloroacetyl chloride gave *N*-allyl-*N*-chloromethyl-α-chloroacetamide **54**, which was then alkylated with the diethyl ester of α-aminomethylphosphonic acid (AMPA) to generate the imidazolone **55**. Subsequent hydrolysis of **55** gave $GLYH_3$ (57).

HHTs derived from AMPA diethyl ester **56** also reacted with acetyl chloride to generate glyphosate nitriles **58** following cyanide displacement with the resulting *N*-acetyl-*N*-chloromethyl-AMPA diethyl ester **57**. Subsequent acidic hydrolysis of **58** gave $GLYH_3$ (58).

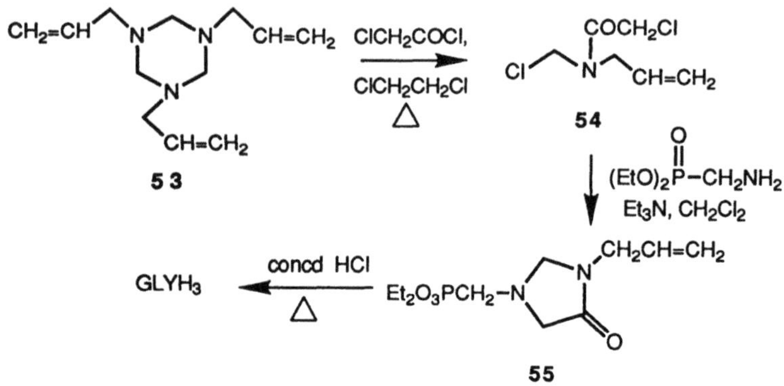

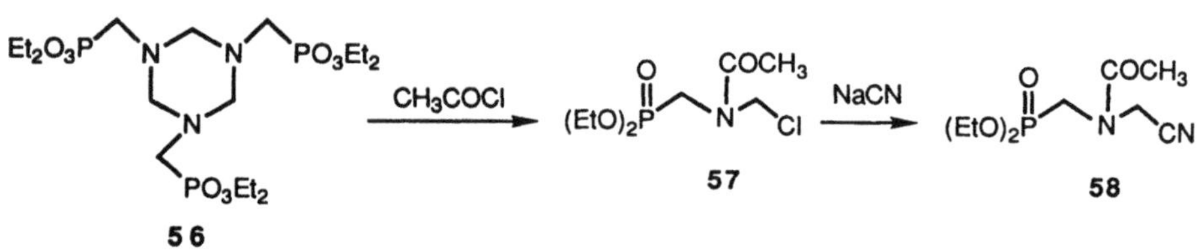

Another AMPA-derived procedure took advantage of the neat reaction between the *N*-carbamoyl-HHT **59** and diethyl phosphite catalyzed by boron trifluoride etherate to generate the AMPA carbamate **60**. Subsequent alkylation with ethyl bromoacetate and base produced the glyphosate triester carbamate **61**, which was hydrolyzed to $GLYH_3$ (59).

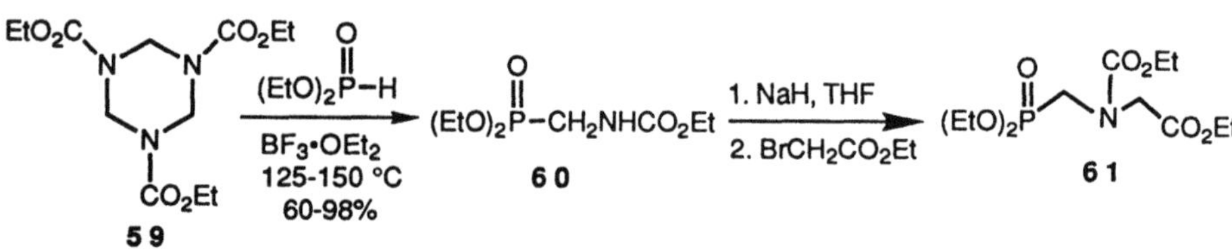

2.3 HETEROCYCLIC GLYPHOSATE DERIVATIVES

2.3.1 Heterocyclic "Masked" Carboxylates

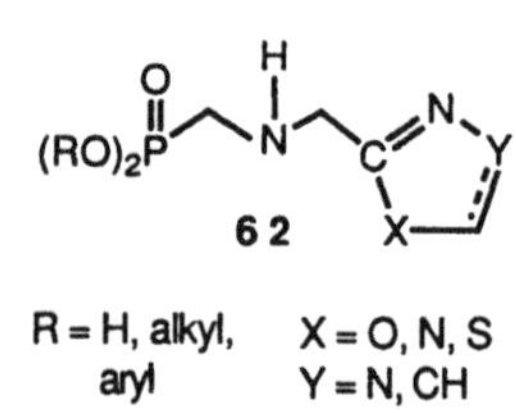

Various heterocyclic moieties have been incorporated at the glyphosate carboxylate center as potential "masked" carboxyl derivatives **62**. All of those reported to date contain a fully unsaturated heteroaromatic ring.

The herbicide activity ascribed to the 1,2,4-oxadiazole derivative **67** (60) has prompted the search for other biologically active derivatives. This example was prepared from the corresponding 5-chloromethyl-1,2,4-oxadiazole **63** via conversion to the HHT **65** through the aminomethyl analog **64**. Reaction of the HHT **65** with diethyl phosphite under neat conditions produced the desired diethyl ester intermediate **66**, which was hydrolyzed to **67**.

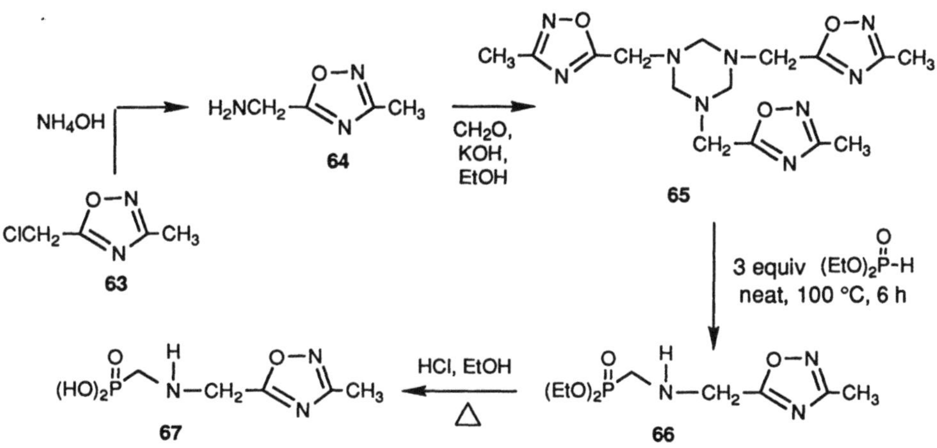

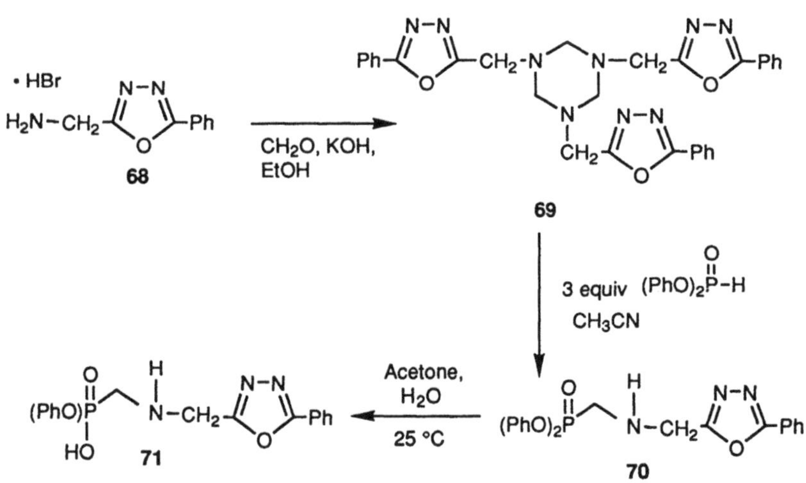

Many other examples have been prepared (2) from the corresponding aminomethyl heterocycles using the very versatile reaction between HHTs and diaryl phosphites, as demonstrated specifically above for the 5-phenyl-1,3,4-oxadiazole system. Conversion of the 2-aminomethyl-1,3,4-oxadiazole **68** to the required HHT intermediate **69** was accomplished under the usual conditions. The HHT phosphite addition products were isolated either as the diaryl phosphonate ester **70**, or as the zwitterionic monoaryl ester **71**, after mild hydrolysis in aqueous acetone (2).

Certain 1,3,4-oxadiazole and 1,2,4-triazole glyphosate derivatives have been conveniently prepared in a faster, more efficient manner by heating the thionoester intermediates **73** with the appropriate acid hydrazide (61). These versatile thionoesters **73** have been synthesized in nearly quantitative yield from the readily available nitrile **31a**, described previously, through the intermediate imidate ester **72**. The oxadiazole products such as **70** obtained using this procedure were identical to those obtained from the HHT approach.

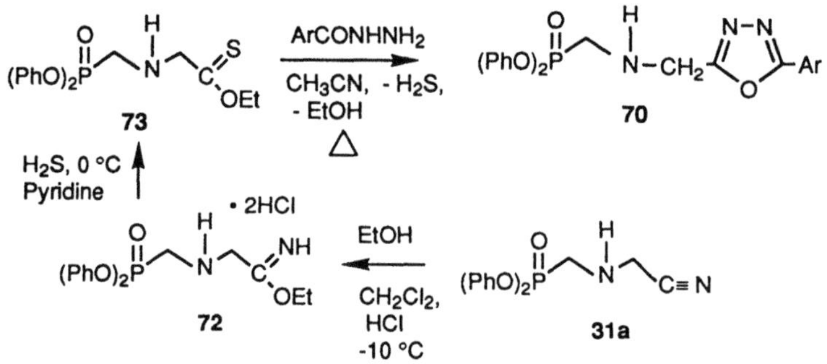

The analogous 1,2,4-triazoles **74** and **75** were prepared in a similar manner by heating **71** with either 2-hydrazinopyridine (28) or 2-hydrazinoquinuclidine (61).

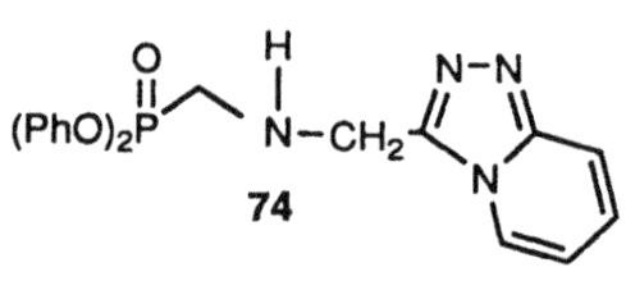

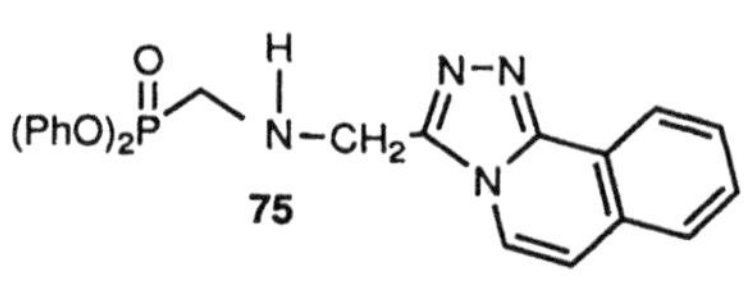

The parent tetrazole derivative of glyphosate **78** has been reported as a product of the 1,3-dipolar cycloaddition of n-Bu_3SnN_3 across the nitrile linkage in **76** and subsequent hydrolysis of the resulting diester **77** (62).

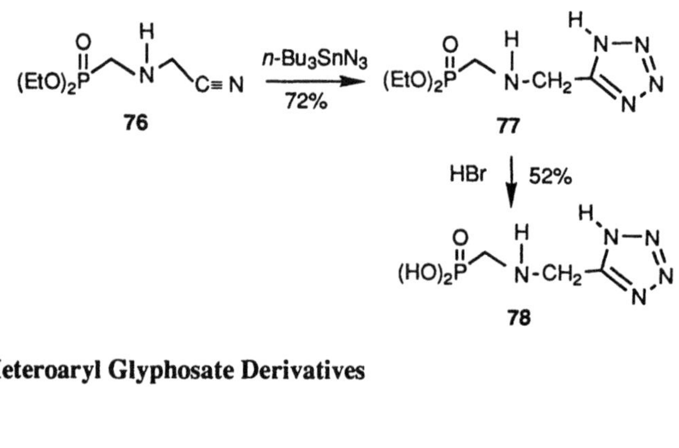

2.3.2 *N*-Heteroaryl Glyphosate Derivatives

Disodium glyphosate has been reacted with activated heteroaryl halides, such as 2-chlorobenzothiazole or 2-chlorobenzoxazole **79**, in aqueous alcohol at reflux to produce the *N*-heteroaryl glyphosate derivatives **80** (2). Improved yields have often obtained in these reactions using the more soluble bis-quaternary ammonium salt of glyphosate derived from tetrabutylammonium hydroxide (2).

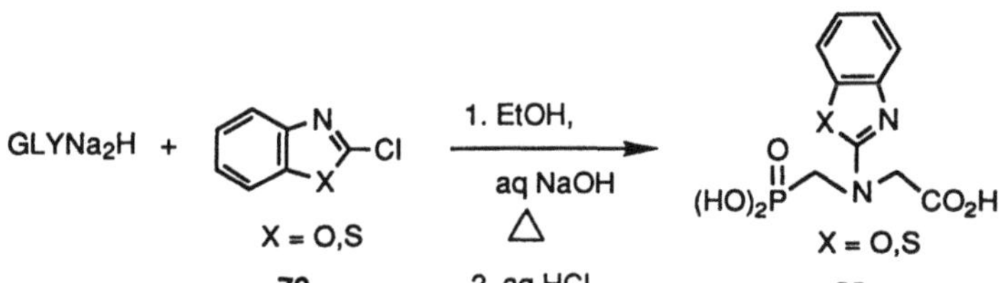

2.3.3 Quaternary *N*-Heteroalkyl Glyphosate Derivatives

The higher solubility of several quaternary ammonium salts of glyphosate in polar aprotic organic solvents such as acetonitrile was discovered (2), which permitted their reaction in solution with various alkyl halides. For example, GLY(n-Bu_4N)$_2$H reacted with either *o*-xylylene dichloride or 1,5-dibromopentane to produce the interesting quaternary glyphosate derivatives **81** and **82**, whose structures have been confirmed by x-ray analysis (2).

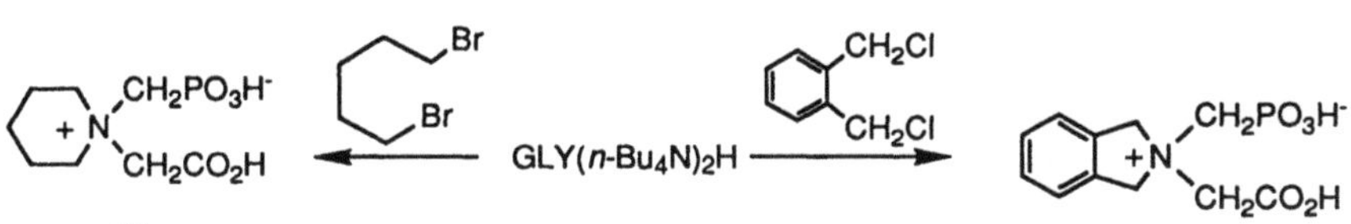

2.3.4 Cyclic Phosphonates

Various cyclic phosphonate esters **36** and **37** have been described previously as products from the HHT reaction of **25** with the appropriate cyclic phosphite. A complementary method has also been developed from the *N*-protected phosphonyl chloride **84**, which was readily prepared from the corresponding phosphonic acid **83**. Subsequent reaction of **84** with the appropriate diol produced the cyclic phosphonate esters **85** (63). Higher homologs of **85** have also been prepared from the analogous propane or butane diols.

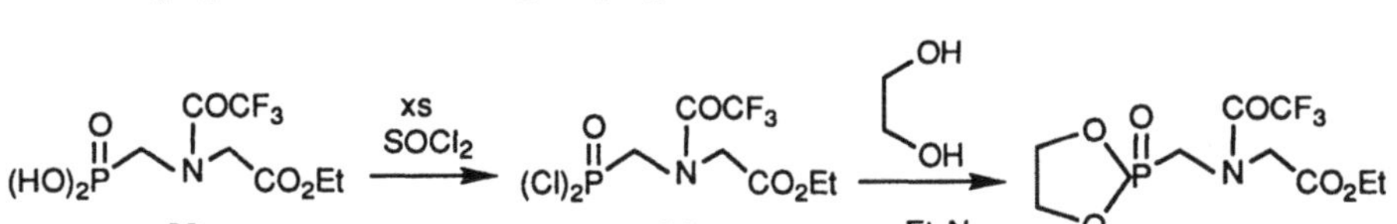

2.4 HETEROCYCLIC ANALOGS OF GLYPHOSATE

2.4.1 Connection Between the C_α and Nitrogen Centers

Linkages between the glyphosate α-carbon and nitrogen centers has been accomplished by using the readily available cyclic amino acid, proline, in the typical Mannich reaction. Treating proline with phosphorous acid and formaldehyde gave the cyclic acid analog **86** (2), while reaction of proline methyl ester, formaldehyde, and dialkyl phosphites generated the corresponding aliphatic triesters **87**.

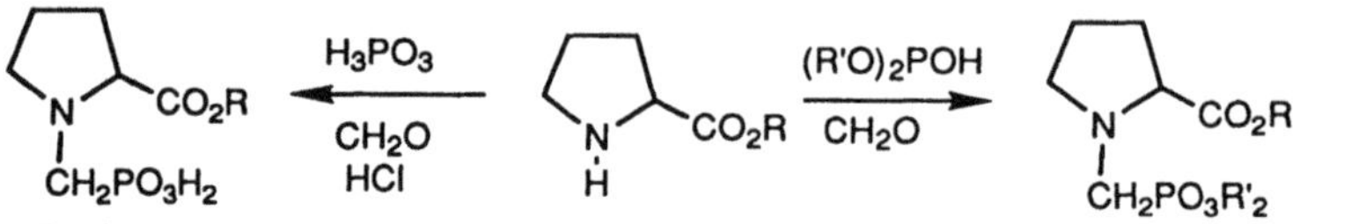

R = H

86

R = CH_3, R' = CH_3, Et

87

A similar procedure starting with commercially available piperidine-2-carboxylic acid produced the corresponding six-membered ring analog **88** (2).

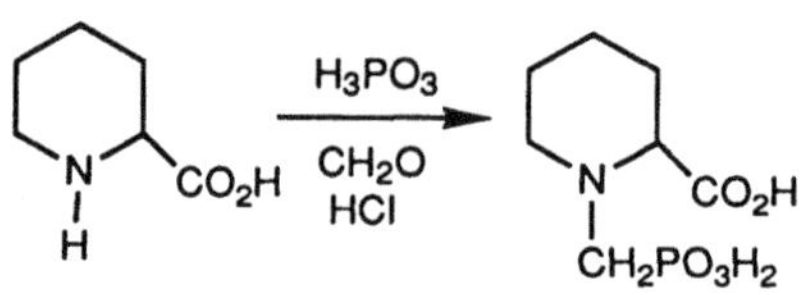

88

The unusual oxaziridine analog **91** has also been synthesized by oxidizing the glycolate imine **90** of diphenyl aminomethylphosphonate **89** with *m*-chloroperbenzoic acid (MCPBA). The resulting adduct was thermally quite labile but could be isolated after removing all of the volatile side components in vacuo (2).

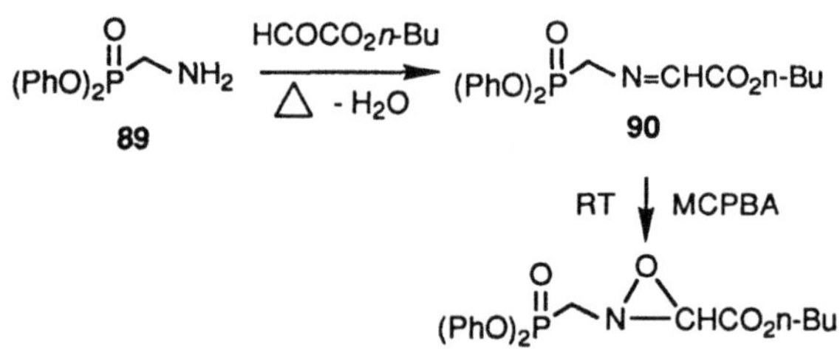

91

2.4.2 Connection Between the Carboxylate and Nitrogen Centers

By far the most widely represented class of cyclic glyphosate analogs in the chemical literature are those where cyclization has taken place between the nitrogen and carboxylate centers. For example, when alkyl glycinate esters were used instead of glycine in the basic Mannich reaction, an unusual phosphonomethyloxazolidinone **92** intermediate was isolated in moderate yield under these conditions (64). Acidic hydrolysis of **92** proceeded as expected in good yield to $GLYH_3$. Presumably, **92** formed via the previously described transient *N*-methylol species **15** and **16**. In this case, the carboxymethyl group may activate the carboxyl center to lactone formation.

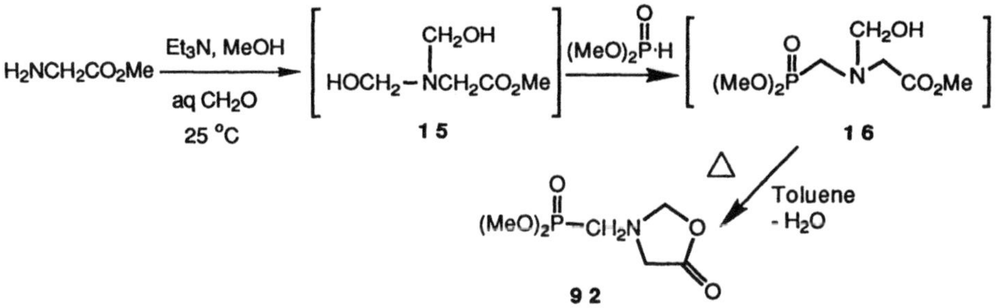

Glyphosate monocarboxylate esters **93** have been treated with various aryl isocyanates or isothiocyanates to produce the intermediate *N*-substituted derivatives **94** and **95**, respectively, which undergo spontaneous ring closure forming the cyclic analogs **96** and **97** (2). The thiono analog **97** was converted back to the oxo analog **96** by first alkylating with chloroacetic acid to give the intermediate thiaimidazolium species **98**, which was hydrolyzed with hot water to **96**.

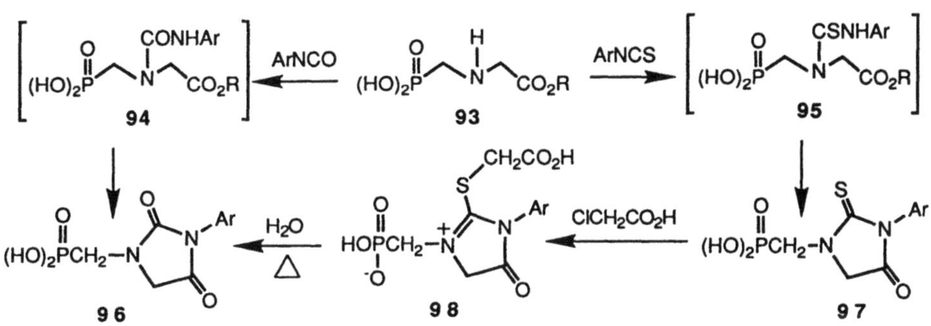

Alternatively, the reaction of **93c** (R = *n*-Bu) with ammonium thiocyanate in refluxing acetic acid gave the ammonium salt of *N*-phosphonomethylthiohydantoin **99** in moderate isolated yield (2).

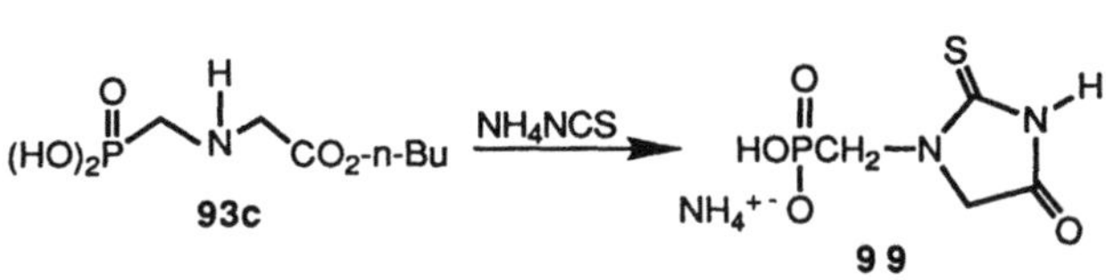

The benzyl phosphonate triesters **100** reacted with isocyanates under similar conditions to give the corresponding cyclic urea phosphonate diesters **101** (2).

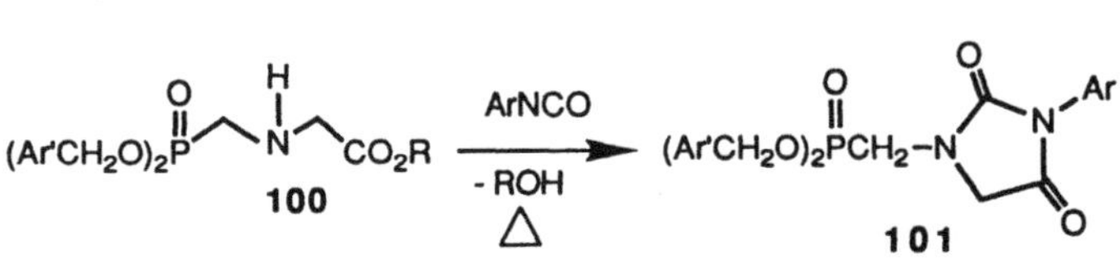

Glyphosate diarylphosphonate nitriles **31** reacted with aryl isocyanates to form the corresponding alicyclic ureas **102**, which were thermally quite labile and easily reverted back to starting material upon standing at room temperature. In contrast, the reaction of **31** with excess aryl isothiocyanates produced the unusual 2:1 diiminothiazoline adducts **105** as yellow to orange oils, presumably via intermediates **103** and **104** (2).

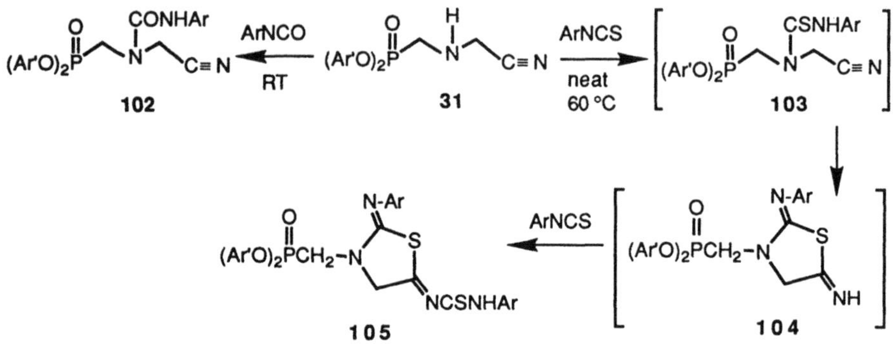

Several cyclic glyphosate analogs related to this series were described previously as intermediates to prepare glyphosate. These include various *N*-phosphonomethylhydantoins and diketopiperazines. A more extended glyphosate piperazine analog **106** has also been prepared from the Mannich reaction of ethylenediamine-*N,N'*-diacetic acid (65).

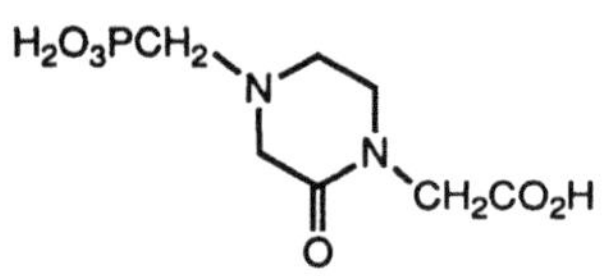

106

The morpholinone analogs **108** have been prepared from the reaction of sodium glyphosate with epoxides. Upon acidification and heating, the resulting *N*-hydroxyalkyl intermediates **107** produced the morpholinone monosodium salts **108** (66).

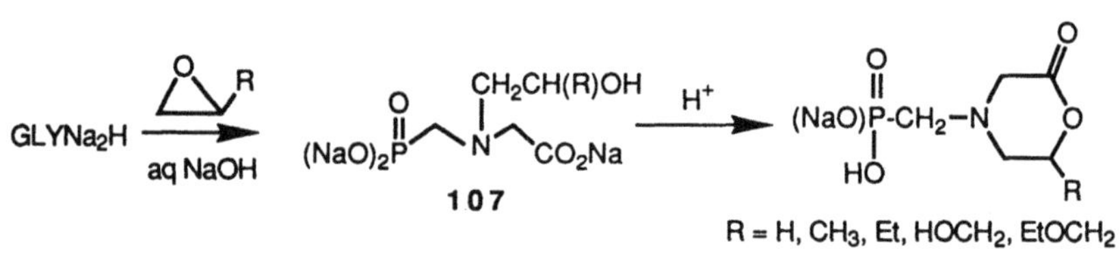

108

2.4.3 Connection Between the C_α and C_β Centers

The rigid, planar pyridine analog **111** was isolated in low yield by first hydrolyzing the known (67) pyridine diethyl phosphonate **109** to the corresponding free acid **110** followed by permanganate oxidation (2). An alternative synthesis of **111** has recently been reported (68). Alkylation of pyridine-2-carboxylate *N*-oxide with dimethylsulfate and subsequent reaction with the sodium salt of diethyl phosphite gave the triester **112**, which was readily converted to **111**.

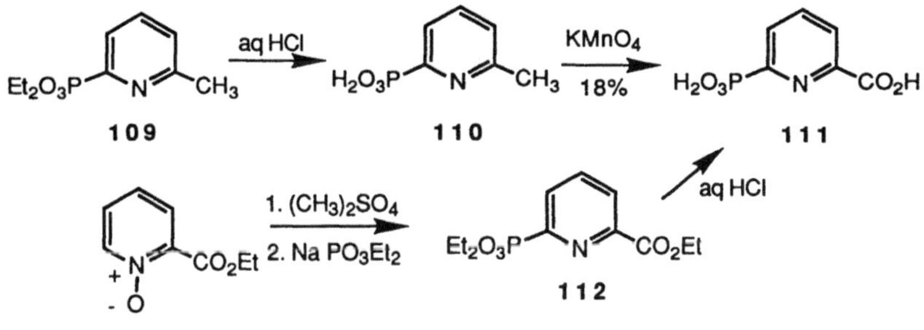

The related phosphonoproline **115** has also been reported as a mixture of diastereomers from the addition of diethyl phosphite to the imine **113** followed by acidic hydrolysis of the intermediate triester **114** (69).

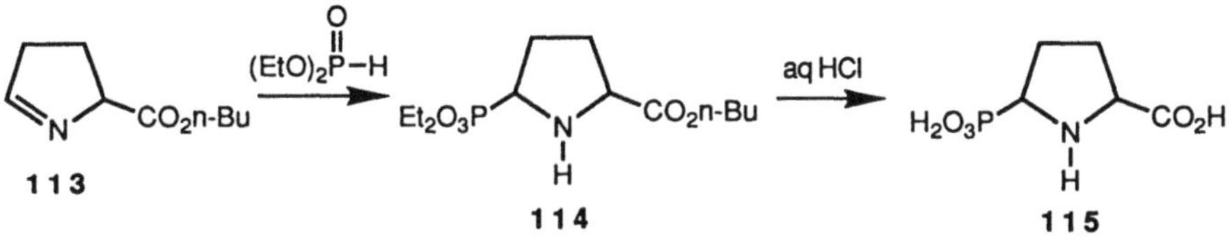

The related planar pyrrole analog **118** has also been prepared (2) from either ethyl or benzyl pyrrole-2-carboxylate **116**. Direct alkylation with diethyl phosphonomethyl triflate (70) and base produced the *N*-phosphonomethylpyrrole 2-carboxylate **117**, which was deprotected with trimethylsilyl bromide and saponified to the corresponding phosphonic acid **118**.

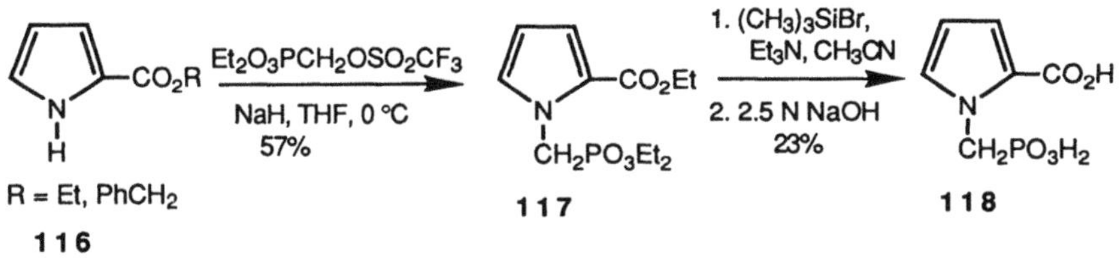

Alternatively, the *N*-BOC-protected 4-formyl pyrrole intermediate **119**, which was readily prepared from **116** by formylation with dichloromethyl methyl ether followed by acylation with BOC-anhydride and base, underwent smooth addition of diethyl phosphite to produce the α-hydroxymethylphosphonate **120**, which was subsequently halogenated in high yield with thionyl chloride to give the α-chloromethylphosphonate **121**. A radical reduction of **121** was accomplished in high yield using tri-*n*-butyltin hydride and 2,2'-azobis(isobutyronitrile) (AIBN) to give the related 4-phosphonomethyl pyrrole triester **122**. Deprotection of **122** occurred under the standard conditions with trimethylsilyl bromide in base, followed by hydrogenolysis of the benzyl ester, to give the fully deprotected free acid **123** in moderate yield (71).

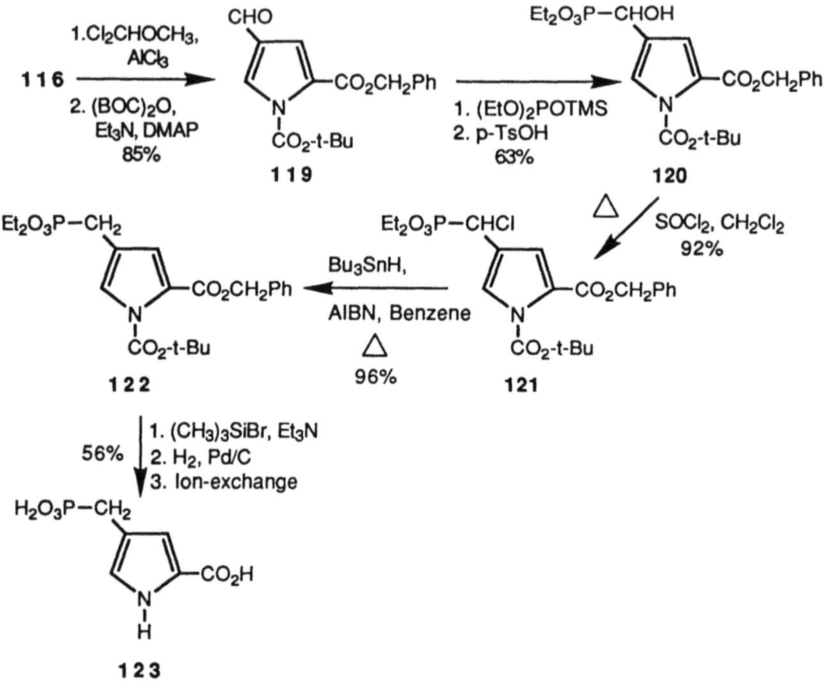

2.4.4 Cyclizations at Phosphorus

Glyphosate analogs formed through cyclizations at the phosphonate center are relatively rare in the chemical literature. The unusual oxazaphospholidine **124** has been isolated in high yield by carefully controlling the temperature below 60 °C during the basic Mannich reaction with glycine described previously via intermediate **15** (72). As expected, acidic hydrolysis of **124** proceeded cleanly to produce $GLYH_3$. In this case, cyclization at phosphorus to give the oxazaphospholidine **124** from **15**, may be favored because cyclization at the carboxylate center should be retarded by the ionized carboxylate.

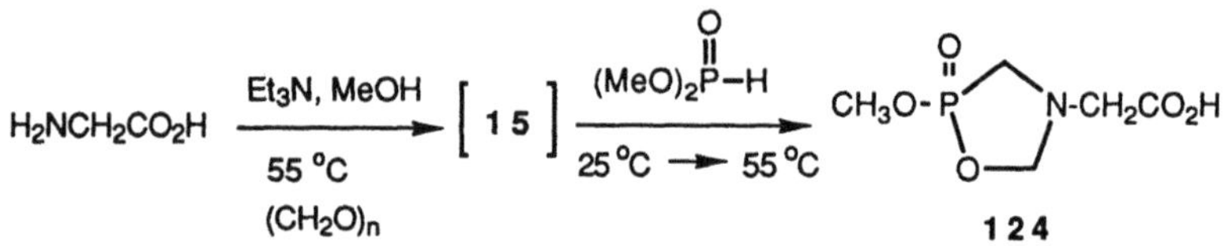

Other cyclizations at phosphorus have been observed when certain phosphinates were used in the acid-catalyzed Mannich reaction. As observed previously with various phosphonous acid derivatives, reaction of aliphatic phosphinic acids with primary amines favored the formation of 2:1 adducts (73). Thus, glycine and other α-amino acids reacted under the typical conditions with excess formaldehyde and alkyl phosphonous acids to give the bis-phosphinylmethyl adducts **125**.

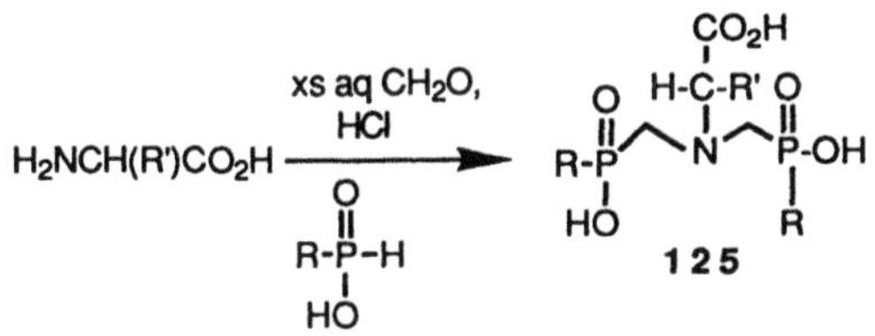

Many alkyl phosphonous acids have been successfully utilized in this procedure. The only reported limitation with these reagents occurred in systems prone to internal cyclization. For example, β-chloroethylphosphonous acid reacted with *N*-benzylglycine and formaldehyde to produce the unusual eight-membered ring adduct **126** via cyclization at the carboxylate oxygen (74). Hydrogenolysis of **126** gave the corresponding deprotected cyclic analog **127**.

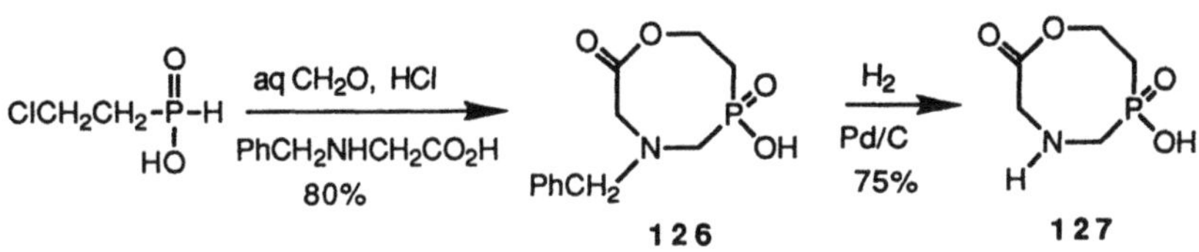

Interestingly, a different ring system was produced when the HHT of ethyl glycinate **25** was employed with ethyl β-chloroethylphosphinate to block cyclization at the carboxyl group. In this case, intramolecular cyclization occurred at nitrogen to give the unusual azaphospholidine oxide **128** in modest isolated yield (74).

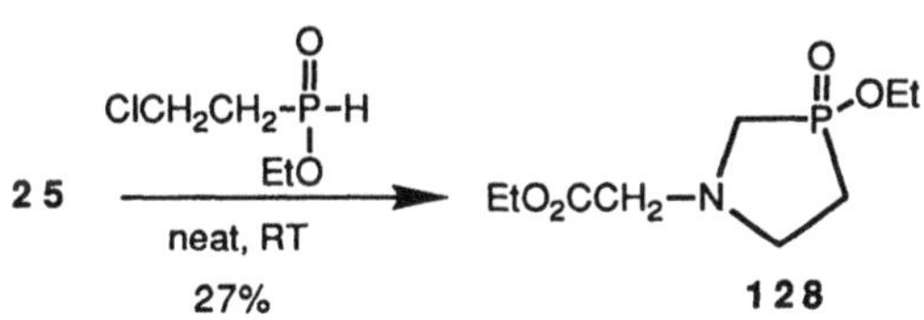

2.4.5 Cyclic Spatial Mimics

Two heterocyclic phosphonates have been designed and synthesized in an attempt to identify more spatially confined, planar analogs of glyphosate than obtained previously with the pyridine analog **111** (75). Molecular modeling experiments suggest that 5-phosphonothiazolin-2-one **133** and 5-phosphono-1,2,4-triazolin-3-one **137** each may overlap either with glyphosate or its known competitive substrate, phosphoenolpyruvate (PEP), very well (5).

The synthesis of 5-phosphonothiazolin-2-one **133** started with 2-bromothiazole **129**. Nucleophilic displacement of the 2-bromide proceeded cleanly with hot anhydrous sodium methoxide to give 2-methoxythiazole **130**. Low-temperature metalation of **130** with *n*-butyl lithium occurred selectively at the 5-position (76), and subsequent electrophilic trapping with diethyl chlorophosphate produced the 5-phosphonate **131**. Deprotection of **131** was accomplished either stepwise with mild acid to produce the thiazolin-2-one intermediate **132**, or directly with trimethylsilyl bromide to give the free phosphonic acid **133**, which was isolated as its cyclohexylammonium salt.

The synthesis of 5-phosphono-1,2,4-triazolin-3-one **137** began with the low-temperature metalation and phosphorylation (77) of the *t*-butyldimethylsilyl (TBDMS)-protected *N*-benzyltriazolinone **134**. The phosphonate diester **135** was obtained after the silyl protecting group

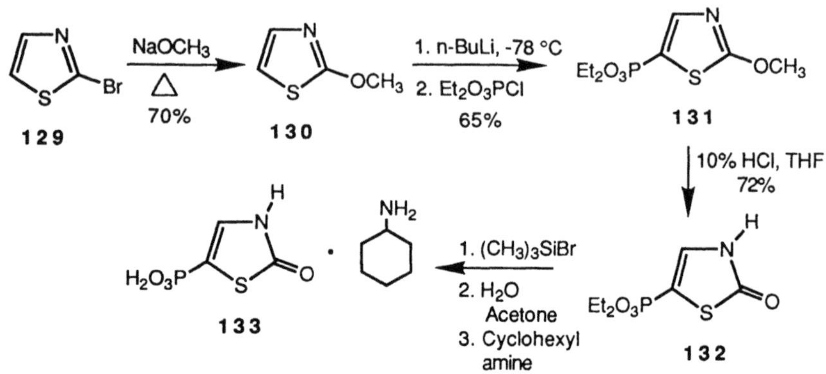

was removed with tetrabutylammonium fluoride. Subsequent treatment with trimethylsilyl bromide and hydrolysis cleaved the phosphonate esters and produced the free phosphonic acid **136** in high yield. Removal of the *N*-benzyl protecting group was accomplished in good isolated yield by hydrogenolysis to give the desired 5-phosphono-1,2,4-triazol-3-one target **137**. Presumably, in solution **137** can exist in tautomeric equilibrium with several corresponding 3-hydroxy-1,2,4-triazole isomers represented by **138**.

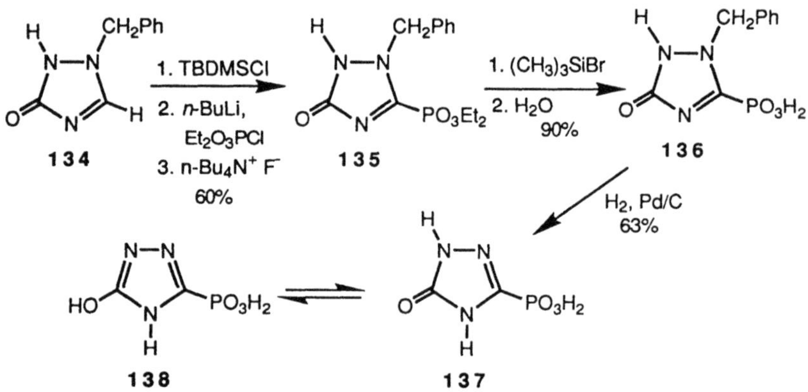

2.4.6 Miscellaneous Heterocyclic Glyphosate Analogs

The pyrazole analog **140** has been prepared from the reaction of the alkynyl phosphonate with ethyl diazoacetate, whereas the pyrazoline **139** was obtained under similar conditions with diethyl vinylphosphonate (78).

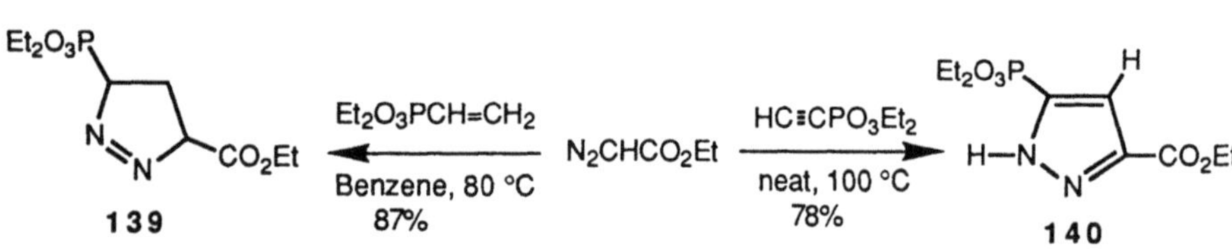

Similar pyrazoline adducts **142** and **143** have also been isolated from the reaction of β-phosphonoacrylate esters **141** with diazomethane (79).

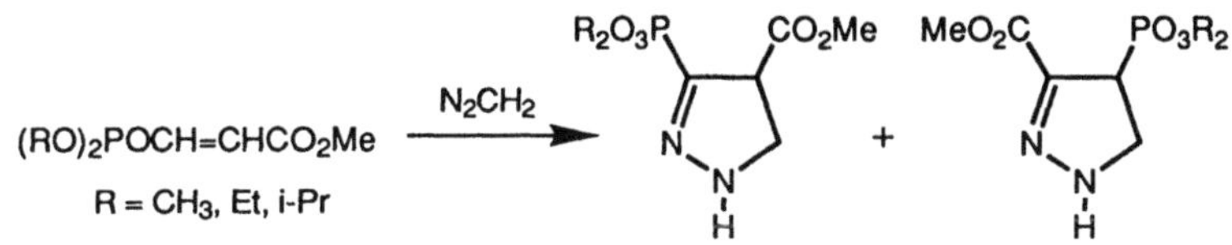

141 **142** **143**

The sydnone of glyphosate **145** has been prepared by heating *N*-nitroso-glyphosate **144** in acetic anhydride (2).

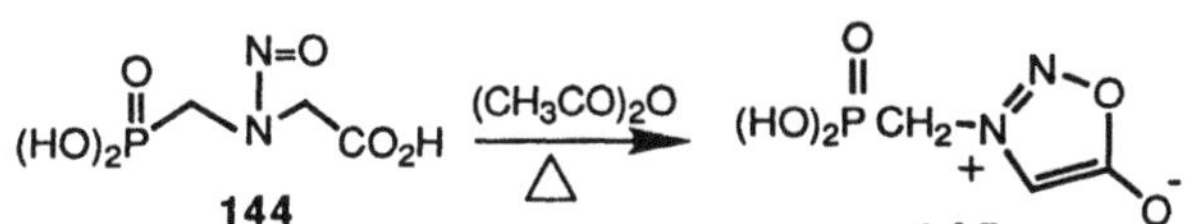

144 **145**

The related sydnone imine hydrochloride **147** has also been isolated in low yield by treating *N*-nitroso-*N*-diphenylphosphinylmethylglycinonitrile **146** with anhydrous HCl and ether at low temperature (2), in a manner similar to that described previously for other nitrosamines of aminoacetonitrile (80).

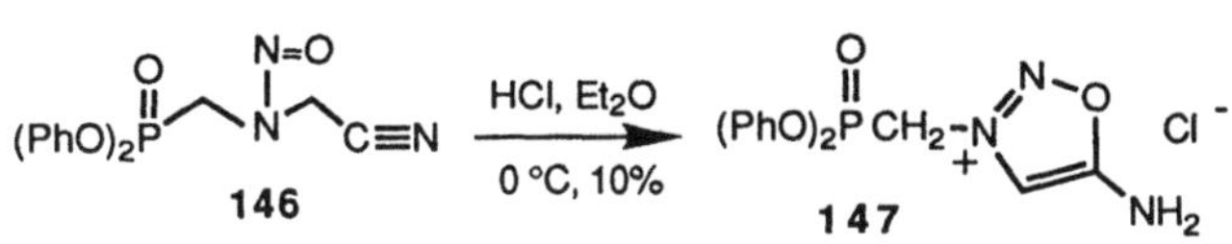

146 **147**

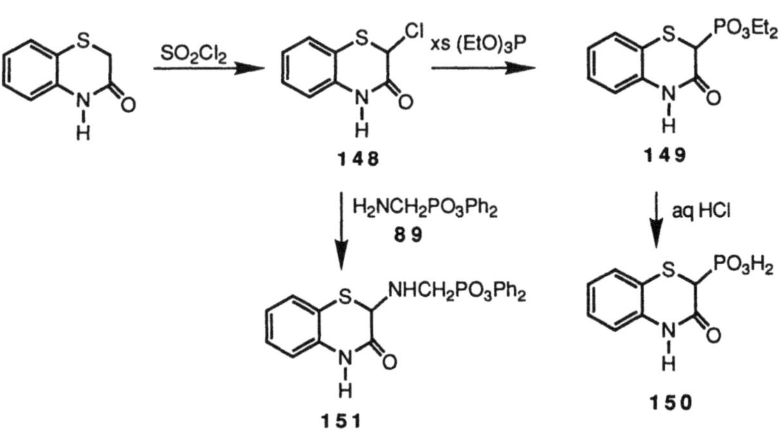

148 **149** **89** **151** **150**

Several benzothiazinone analogs have been synthesized in an attempt to introduce hetero substituents at the α-carbon center in these heterocyclic compounds. The required α-halo-benzothiazinone intermediate **148** was prepared by chlorination with sulfuryl chloride. This material was used successfully in an Arbuzov reaction to prepare the phosphonate diester **149**, which was hydrolyzed to the corresponding free phosphonic acid **150** (2). Alternatively, the chloro intermediate **148** also reacted cleanly with diphenyl aminomethylphosphonate **89** which gave the extended cyclic ester adduct **151** (2).

2.5 SUMMARY

The search for alternative new aminoalkyl phosphonates or phosphinates with similar or dramatically improved biological activity versus glyphosate has stimulated an enormous research effort in exploratory phosphorus and heterocyclic chemistry. Many new synthetic methods have been developed over the last two decades to prepare a variety of structural analogs of glyphosate containing a modified skeleton. Despite the large numbers of compounds evaluated, none has surpassed the overall biological properties of glyphosate itself, either on the enzymatic level or under field conditions. Consequently, the search for a second-generation, glyphosate-like herbicide continues (2).

2.6 REFERENCES

1. Franz, J. E. in *The Herbicide Glyphosate*; Grossbard, E.; Atkinson, D. Eds.; Butterworths: London, 1985; p 3-17.
2. Franz, J.E.; Mao, M. K.; Sikorski, J. A. in *Glyphosate: A Unique Global Herbicide*; American Chemical Society Monograph; ACS: Washington, DC, 1997.
3. Rueppel, M. L.; Brightwell, B. B.; Schaefer, J.; Marvel, J. T. *J. Agric. Food Chem.* **1977**, *25*, 517.
4. Steinrücken, H. C.; Amrhein, N. *Biochem. Biophys. Res. Commun.* **1980**, *94*, 1207-1212.
5. Sikorski, J. A.; Gruys, K. J. *Acc. Chem. Res.* **1997**, *30*, 2-8.
6. Kosolapoff, G. M. *Org. Reactions*, **1951**, *6*, 273.
7. Moedritzer, K.; Irani, R. R. *J. Org. Chem.* **1966**, *31*, 1603-1607.
8. Irani, R. R.; Moedritzer, K. U.S. Patent 3,288,846, 1966; to Monsanto.
9. Maier, L. European Patent 55,695, 1982; to Ciba-Geigy A.-G.
10. Suh, M. E.; Choi, D. M.; Kim, I. O. *Yakhak Hoechi*, **1988**, *32*, 1-5. CA 110:135668j.
11. Maier, L. *Phosphorus, Sulfur Silicon Relat. Elem.* **1991**, *61*, 65-67.
12. Svetkin, Y. V.; Abdrakhmanov, I. B.; Augustin, M. *Naturwiss. Reihe* **1973**, 22, 7-30. CA 80:107430u.
13. Wong, R. Y.; Bunker N. S. U.S. Patent 4,400,330, 1983; to Stauffer Chemical.
14. Bayer, A. C.; Robbins, J. D.; Bowler, D. J. U.S. Patent 4,578,224, 1986; to Stauffer Chemical.
15. Cortes, D. A. U.S. Patent 4,946,993, 1990; to American Cyanamid.
16. (a) Wong, R. Y.; Bunker, N. S. U.S. Patent 4,547,324, 1985; to Stauffer Chemical.
    (b) Fields, D. L.; Lee, L. F.; Richard, T. J. U.S. Patent 4,810,426, 1989; to Monsanto.
17. (a) Franczyk, T. S. U.S. Patent 5,292,936, 1994; to Monsanto.
    (b) Franczyk, T.S. U.S. Patent 5,367,112, 1994; to Monsanto.
18. Miller, W. H.; Neumann, T. E. U.S. Patent 4,587,061, 1986; to Monsanto.
19. Pfliegel, T.; Seres, J.; Gajory, A.; Daroczy, K.; Nagy, L. T. U.S. Patent 4,065,491, 1977; to Chinoin.
20. Spanish Patent ES 553,523, 1987; to Lerida Union Quimica S.A., CA 109:73659a.
21. Graziello, D. European Patent 402,887, 1990; to Finchimica S.p.a.
22. Brendel, M. H.; Gulyas, I.; Gyöker, I.; Zsupan, K.; Csorvassy, I.; Salamon, Z.; Somogyi, G.; Szentiralyi, I. Timar, T.; Biro, E. C.; Fodor, I.; Repasi, J. U.S. Patent 4,486,359, 1984; to Alkaloida.
23. Ehrat, R. U.S. Patent 4,237,065, 1980; to Biological Chemical Activities.
24. Natchev, I. A. *Synthesis* **1987**, 1079-1084.
25. Smolin, E.; Rapport, L. in *s-Triazines and Derivatives*, Interscience Publishers Inc.: New York, NY; 1959, p 473-544.

26. Ratcliffe, R. W.; Christensen, B. G. *Tetrahedron Lett.* **1973**, (46), 4645-4648.
27. Sikorski, J. A.; Logusch, E. W. in *Handbook of Organophosphorus Chemistry*; Engel, R. Ed.; Marcel Dekker: New York, NY; 1992, p 739-805.
28. Reynolds, D. D.; Cossar, B. C. *J. Heterocycl. Chem.* **1971**, *8*, 597-604 & 605-615.
29. Soroka, M. *Pr. Nauk. Inst. Chem. Org. Fiz. Politech. Wroclaw.* **1987**, 3-92, CA 110:172353y.
30. Issleib, K.; Balszuweit, A.; Wetzke, G.; Moegelin, W.; Kochmann, W.; Guenther, E. East German Patent 141,929, 1980; to Martin Luther University, CA 95: P42377v.
31. Dutra, G. A. U.S. Patent 4,053,505, 1977; to Monsanto.
32. Issleib, K.; Balszuweit, A.; Wetzke, G.; Moegelin, W.; Kochmann, W.; Guenther, E. East German Patent 141,930, 1980; to Martin Luther University, CA 95:81520f.
33. Ha, H.-J.; Park, K. P. U.S. Patent 5,053,529, 1991; to Korean Institute of Science & Technology.
34. Leber, J.-P. U.S. Patent 4,025,331, 1977; to Sandoz Ltd.
35. Purdum, W. R. U.S. Patent 4,395,275, 1983; to Monsanto.
36. Purdum, W. R. U.S. Patent 4,391,625, 1983; to Monsanto.
37. Purdum, W. R. U.S. Patent 4,442,044, 1984; to Monsanto.
38. Barton, J. E. D. U.S. Patent 3,923,877, 1975; to Imperial Chemicals Industries.
39. Robbins, J. D. U.S. Patent 4,415,503, 1983; to Stauffer Chemical.
40. Dutra, G. A. U.S. Patent 4,120,689, 1978; to Monsanto.
41. Dutra, G. A. U.S. Patent 4,067,719, 1978; to Monsanto.
42. Dutra, G. A. U.S. Patent 4,083,898, 1978; to Monsanto.
43. Kuwahara, M; Mutsukada, M.; Kawamura, Y.; Ohya, T.; Hirai, Y. Japanese Patent 52108955, 1977; to Nissan Chemical Industries, CA 88:90039.
44. Dutra, G. A. U.S. Patent 4,053,505, 1977; to Monsanto.
45. Mao, M. K.; Franz, J. E. *Synthesis* **1991**, 920-922.
46. Franz, J. E. U.S. Patent 4,659,860, 1987; to Monsanto.
47. Corbet, J.-P. U.S. Patent 4,755,614, 1988; to Rhone-Poulenc Agrochimie.
48. Soroka, M. Polish Patent 107,789, 1980; to Politechnika Wroclawska; CA 95:43347x.
49. Soroka, M. Polish Patent 107,688, 1980; to Politechnika Wroclawska; CA 95:43348y and 100:192277y.
50. Sikorski, J. A.; Mischke, D.; Dutra, G. A. U.S. Patent 4,601,744, 1986; to Monsanto.
51. Engel, R. *Synthesis of Carbon-Phosphorus Bonds*; CRC Press: Boca Raton, FL; 1988, and references cited therein.
52. Felix, R. A. U.S. Patent 4,425,284, 1984; to Stauffer Chemical.
53. Moegelin, W.; Stiebitz, B.; Balszuweit, A.; Issleib, K. East German Patent 286,590, 1991; to Martin Luther University.
54. Felix, R. A. U.S. Patent 4,427,599, 1984; to Stauffer Chemical.
55. Pfliegel, T.; Seres, J.; Gajary, A.; T. Nagy, L.; Daroczy-Csuka, K. Hungarian Patent 13027, 1977; to Chinoin Gyogyszer, CA 87:136832.
56. Coloma, F. M.; Ibanez, P. P. Brazilian Patent 8103684, 1981; to Saxttin Consulting Associates, CA 96:143079.
57. Felix, R. A. U.S. Patent 4,552,968, 1985; to Stauffer Chemical.
58. Felix, R. A. U.S. Patent 4,454,063, 1984; to Stauffer Chemical.
59. Maier, L. *Phosphorus, Sulfur Silicon Relat. Elem.* **1990**, *47*, 361-365.
60. Kondo, M.; Okabe, T.; Takase, M.; Yoshida, A.; Hino, S. Japanese Patent 5936688, 1984; to Sumitomo, CA 101:38655j.
61. Dutra, G. A.; Finkes, M. J.; Sikorski, J. A. *Heteroat. Chem.* **1992**, *3*, 279-291.
62. Kraus, J. L. *Synth. Commun.* **1986**, *16*, 827-832.
63. Franz, J. E. U.S. Patent 4,199,345, 1980; to Monsanto.
64. Gaertner, V. R. U.K. Patent 1,482,377, 1975; to Monsanto.
65. Shkol'nikova, L. M.; Sotman, S. S.; Tsirul'nikova, N. V.; Egorushkina, N. A.; Karandeeva, I. V.; Dyatlova, N. M. *Kristallografiya* **1990**, *35*, 1154-1159, CA 114:62646.

66. Gaertner, V. R. U.S. Patent 4,105,432, 1978; to Monsanto.
67. Redmore, D. *J. Org. Chem.* **1970**, *35*, 4114-4117.
68. Boduszek, B. *J. Prakt. Chem./Chem.-Ztg.* **1992**, *334*, 444-446.
69. Diehl, P. J.; Maier, L. *Phosphorus Sulfur*, **1983**, *18*, 482.
70. Phillion, D. P.; Andrew, S. S. *Tetrahedron Lett.* **1986**, *27*, 1477-1480.
71. Peterson, M. L.; Corey, S. D.; Font, J. L.; Walker, M. C.; Sikorski, J. A. *Bioorg. Med. Chem. Lett.* **1996**, *6*, 2853-2858.
72. Fields, D. L.; Grabiak, R. C. U.S. Patent 4,889,906, 1989; to Monsanto.
73. Issleib, K.; Mögelin, W. *Synth. React. Inorg. Met.-Org. Chem.* **1986**, *16*, 645-662.
74. Maier, L. *Phosphorus Sulfur*, **1981**, *11*, 149-56.
75. Anderson, D. K.; Duewer, D. L.; Sikorski, J. A. *J. Heterocycl. Chem.* **1995**, *32*, 893-898.
76. Knaus, G.; Meyers, A. I. *J. Org. Chem.* **1974**, *39*, 1192-1195.
77. Anderson, D. K.; Sikorski, J. A.; Reitz, D. B.; Pilla, L. T. *J. Heterocycl. Chem.* **1986**, *23*, 1257-1262.
78. Matoba, K.; Yonemoto, H.; Fukui, M.; Yamazaki, T. *Chem. Pharm. Bull.* **1984**, *32*, 3918-3925.
79. Gareev, R. D.; Pudovik, A. N. *Zh. Obshch. Khim.* **1982**, *52*, 2637-2638.
80. Vohra, S. K.; Harrington, G. W.; Swern, D. *J. Org. Chem.* **1978**, *43*, 1671-73.

Chapter 3

Three-Membered Ring Systems

S. Shaun Murphree
Bayer Inc., Charleston, SC, USA

Albert Padwa
Emory University, Atlanta, GA, USA

3.1 INTRODUCTION

Three-membered ring systems contain a wealth of worthy synthetic targets, as well as compact and efficient intermediates and reagents. Accordingly, this field offers fertile ground for innovation and exploration, which is manifested by the body of work continually generated by physical organic chemists and synthetic methodologists. So replete is the literature with examples of three-membered ring chemistry that a comprehensive survey would be far beyond the scope of the following pages. Rather, as in past editions, this review is intended to be a sampling of the previous year's advances from a synthetic chemist's point of view, with an emphasis on methodology. The organization is much the same as that of previous years.

3.2 EPOXIDES

3.2.1 Preparation of Epoxides

Synthetic chemists have always been in search of new and improved preparative routes to epoxides, since they provide versatile intermediates for natural product synthesis. As many synthetic targets are of biological interest, it follows that particular emphasis has been placed on practical pathways to enantiopure epoxides, whether derived from the chiral pool or by asymmetric methods. Of course, one of the most seminal protocols in this regard is the Sharpless asymmetric epoxidation (AE), a type of chemistry both fundamental and contemporary. In a recent application, Honda and coworkers <96JCS(P1)1729> have reported an enantioselective synthesis of (+)-asperlin (**4**) *via* asymmetric epoxidation of furylbutenol **1** under Sharpless conditions. Curiously, the sense of the tartrate affects the regioselectivity of the key step. Thus, L-(+)-DIPT promotes formation of pyranone **2**, a product of electrophilic attack at the more reactive furan double bond, whereas D-(-)-DIPT induces the oxidation of the less electron-rich olefin in an entirely regio- and diastereoselective manner to give the epoxide **3**.

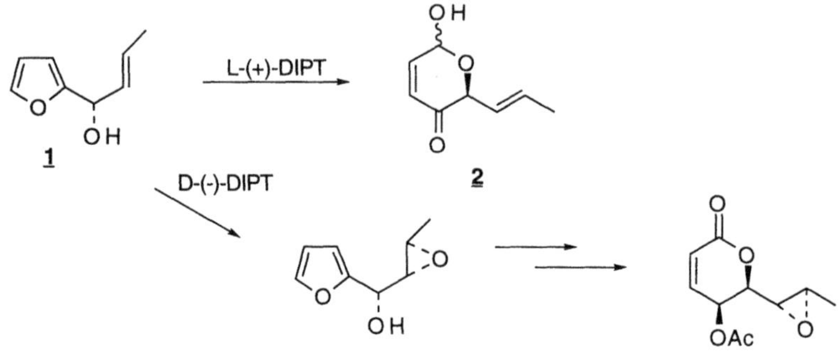

1 **2** **3** **4**

Such user-definable, if not always predictable, control over the regio- or stereochemical outcome of epoxidation methodologies greatly increases their flexibility. Toward this end, Katsuki and coworkers <96TL4533> have investigated the reagent-dependent stereoselectivity of another well-established approach. In this study of the Mn-salen catalyzed asymmetric epoxidation of (+)-3-alkylindenes **7**, it was found that the location and nature of the C3- and C3'-substituents of the Mn-catalysts (**5** and **6**) strongly influenced the stereochemistry of the epoxidation. Continuing the dialectic in this somewhat controversial area, Katsuki proposes a non-planar conformation of the oxo Mn-salen complex and rationalizes the enantioface differentiation by invoking a combination of steric and π - π interactions.

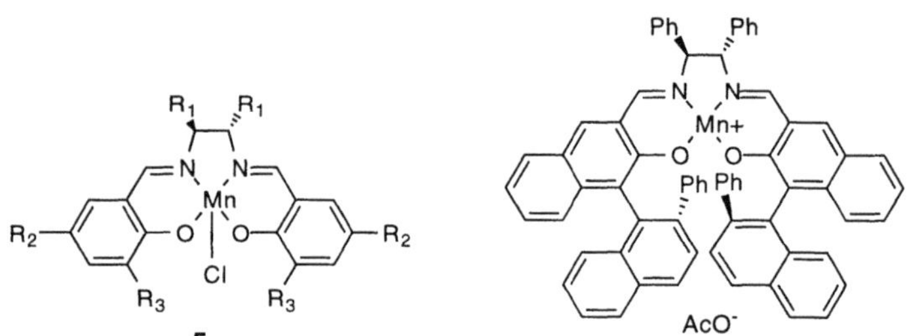

5 **6**

Figure 1

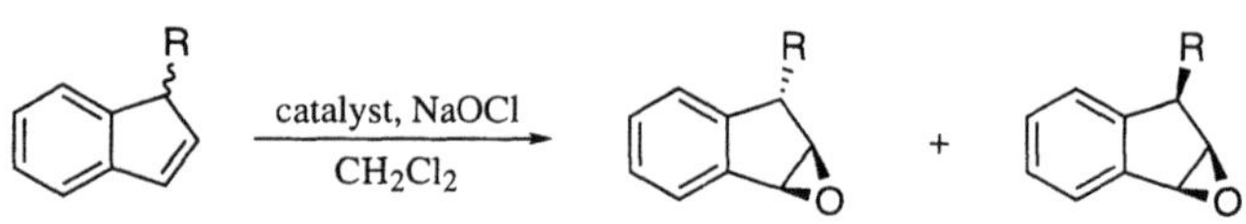

7 **8** **9**

Illustrating the practical utility of these reagents, Geen and coworkers <96TL3895> have optimized the chiral epoxidation of chromene **10** using the readily available Mn(III) salen catalyst **11** in their synthesis of the novel potassium channel activator BRL55834. Some work has also been carried out towards adapting these catalysts to solid-support systems <96TL3375>; however, while such polymer-bound catalysts offer high stability and ease of handling, reaction rates and enantioselectivities remain low compared to the corresponding homogeneous systems.

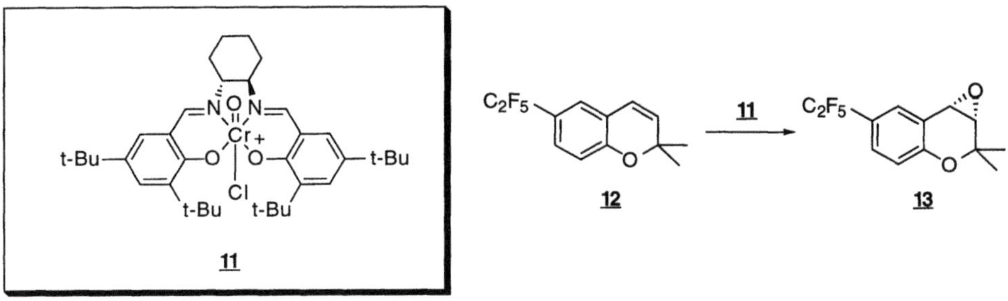

Fueled by the success of the Mn (salen) catalysts, new forays have been launched into the realm of hybrid catalyst systems. For example, the Mn-picolinamide-salicylidene complexes (*i.e.*, **13**) represent novel oxidation-resistant catalysts which exhibit higher turnover rates than the corresponding Jacobsen-type catalysts. These hybrids are particularly well-suited to the low-cost--but relatively aggressive--oxidant systems, such as bleach. In fact, the epoxidation of trans-β-methylstyrene (**14**) in the presence of 5 mol% of catalyst **13** and an excess of sodium hypochlorite proceeds with an *ee* of 53%. Understanding of the mechanistic aspects of these catalysts is complicated by their lack of C_2 symmetry. For example, it is not yet clear whether the 5-membered or 6-membered metallocycle plays the decisive role in enantioselectivity; however, in any event, the active form is believed to be a manganese oxo complex <96TL2725>.

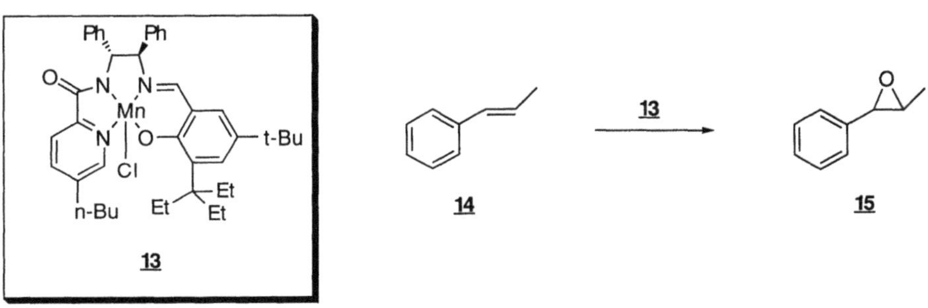

Other metals can also be used as a catalytic species. For example, Feringa and coworkers <96TET3521> have reported on the epoxidation of unfunctionalized alkenes using dinuclear nickel(II) catalysts (*i.e.*, **16**). These slightly distorted square planar complexes show activity in biphasic systems with either sodium hypochlorite or t-butyl hydroperoxide as a terminal oxidant. No enantioselectivity is observed under these conditions, supporting the idea that radical processes are operative. In the case of hypochlorite, Feringa proposed the intermediacy of hypochlorite radical as the active species, which is generated in a catalytic cycle (Scheme 1).

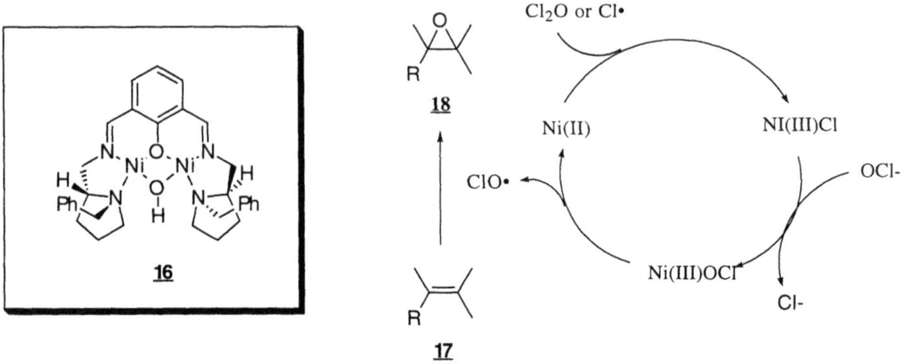

Scheme 1

Also drawing inspiration from the Jacobsen-type catalysts, but in a completely different manner, is the catalytic asymmetric epoxidation of olefins mediated by the chiral iminium salt **19,** which can be viewed somewhat as a disembodied ligand from a Mn(salen) complex. A stoichiometric amount of Oxone and 5 mol% of **19** is sufficient for epoxidation of unfunctionalized alkenes in a bicarbonate/acetonitrile/water system. The oxidations were stereospecific, suggesting a concerted formation of both C-O bonds, in contrast to the analogous Mn(salen) reactions. The active oxidizing species in this case is believed to be oxaziridinium salt **20**. Enantioselectivities were generally low to moderate for mono- and disubstituted alkenes, but markedly higher for the trisubstituted variety . For example, 1-phenylcyclohexene (**21**) provided the corresponding epoxide (**22**) in 80% yield and 71% *ee* <96CC191>.

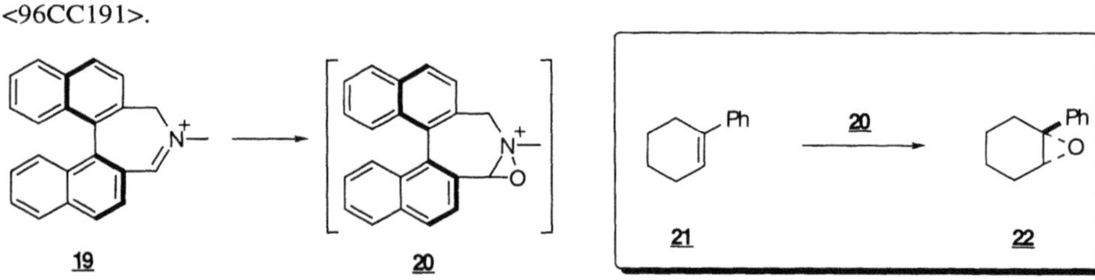

While the Sharpless and Jacobsen approaches are certainly predominant, especially in the area of asymmetric synthesis, myriad other methodologies are being continually developed. For example, it has been known for some time that dioxiranes are efficient stoichiometric reagents for the epoxidation of alkenes. More recently, catalytic protocols have been developed in which the active dioxirane is produced and regenerated *in situ* from a suitable ketone and an auxiliary terminal oxidant. The latest twist to this development is the use of chiral ketones in catalytic amounts for the induction of asymmetry during the epoxidation. However, the resultant dioxiranes have two reacting sites, a fact which makes the prediction and execution of asymmetric induction problematic. Yang and coworkers <96JA491> addressed this problem by designing the C_2 symmetric, 11-membered ring ketone **23** as a chiral dioxirane precursor. In this case, both active sites are characterized by the same chiral environment. In fact, a catalytic amount of ketone **23**, in the presence of Oxone as a terminal oxidant, was capable of epoxidizing *trans*-disubstituted and trisubstituted alkenes in modest to good enantiomeric excess (*e.g.*, **24** → **25**, 98% yield, 50% e.e.).

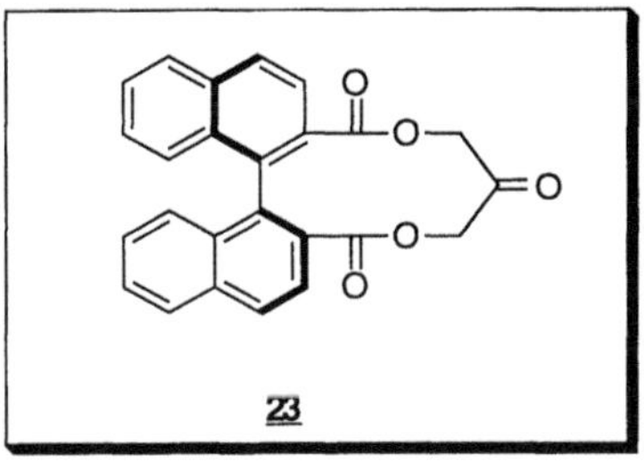

Shi and coworkers <96JA9806> took a slightly different tack. Their method relies on the ketalized D-fructose derivative **26**, the corresponding dioxirane of which functions as an excellent reagent for the asymmetric epoxidation of unfunctionalized alkenes (*e.g.*, **27** → **29**, 81% yield, 90% *ee*). There are several important features of this reagent. First, the stereogenic centers are in close proximity to the ultimate reactive site of the dioxirane; second, the carbonyl group is flanked by a fused ring on one side and a quaternary center on the other, preventing epimerization (*i.e.*, **28**); and, finally, only one face of approach is available, since the other is sterically blocked. As to the actual transitions state, the results are consistent with a spiro configuration (**28**) which is directed by steric interactions. Ketone **26** is, however, unstable under the reaction conditions, so that an excess must be used to achieve high selectivity.

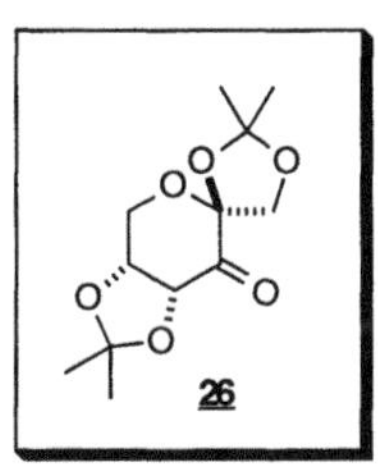

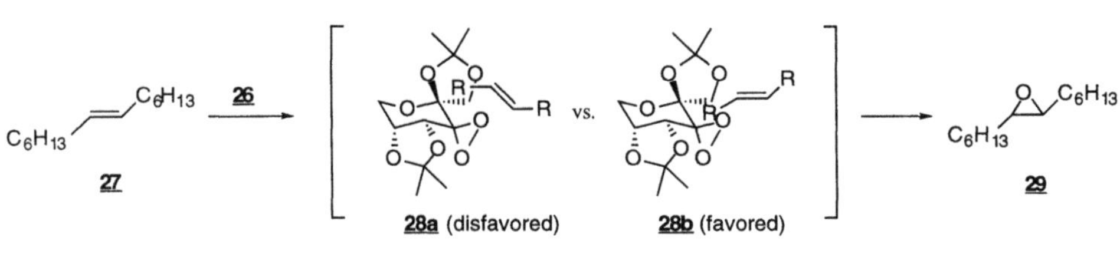

Enders and coworkers <96AG(E)1725> have developed an interesting general one-pot method for the asymmetric epoxidation of enones with oxygen in the presence of diethylzinc and (*R,R*)-N-methylpseudoephedrine (**30**), which provides α, β - epoxyketones in very high yield and high enantiomeric excess (*e.g.*, **33** → **34**). The actual reactive species is believed to be the chirally modified alkoxy(ethylperoxy)zinc **31**, which attacks the *si* face of the s-*cis* conformation of the (E) enones (*cf.* **32**).

30 31

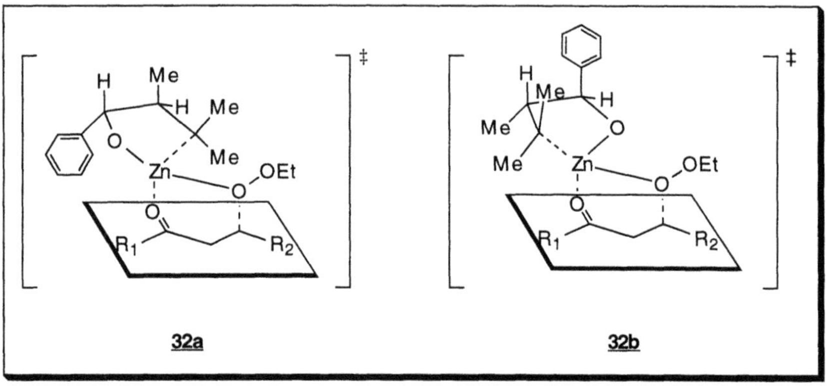

32a 32b

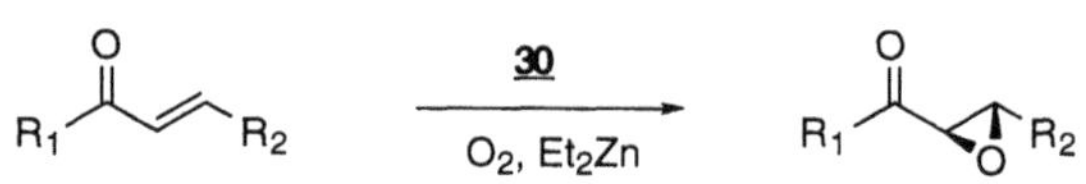

33 34

Asymmetric epoxidation can also be achieved by taking advantage of chirality on the substrate itself. For example, under carefully selected conditions, vinyl sulfoxides **35** undergo clean nucleophilic epoxidation with moderate to excellent diastereofacial selectivity to give the synthetically useful sulfinyl oxiranes **36** <96JOC3586>. Similarly, 2-silyl-3-alkenols undergo epoxidation with high 1,2-asymmetric induction in a type of auto-Sharpless reaction, where the quaternary center provides the asymmetric bias under Ti- or V-catalysis. Thus, *Z*-olefins provide *syn* epoxides and *E*-olefins yield the *anti* counterparts (*cf.* **37** → **38**). This outcome is in sharp contrast to previous reports on the analogous allylsilanes (**39**), where both *E*- and *Z*-isomers lead to *anti* epoxides. The stereoselectivity in the former series is rationalized on the basis of a chair-like transition state (*i.e.*, **40**, not shown) <96TL1205>.

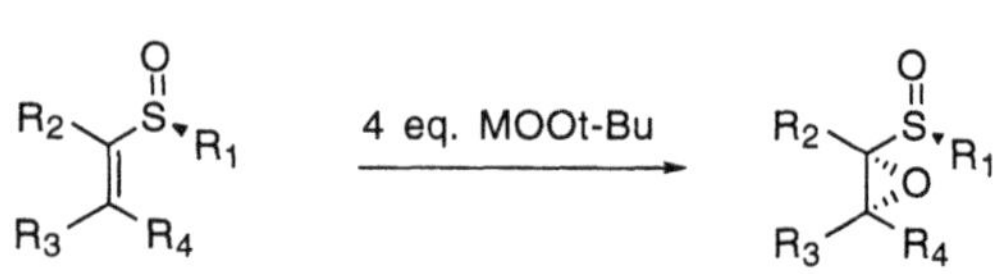

35 36

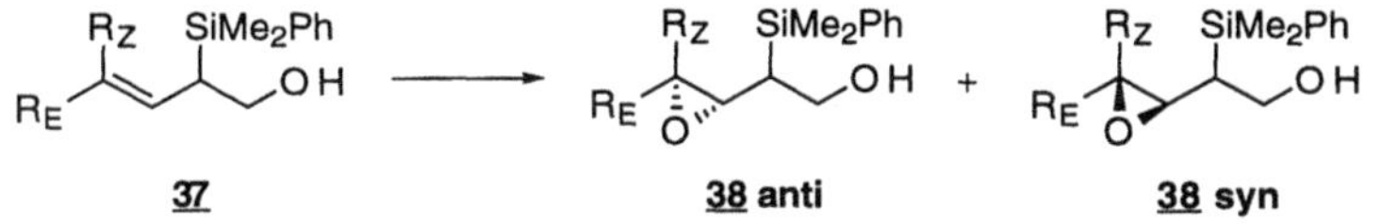

	38 anti		38 syn
R_Z = n-C_5H_{11}, R_E = H	98	:	2
R_Z = H, R_E = n-C_5H_{11}	18	:	82

Other structural stereochemical factors may also guide the course of selectivity in epoxidations. Vedejs and coworkers <96JA3556> have carried out detailed studies into the torsional, rotor, and electronic effects in the dimethyldioxirane-mediated epoxidation of 4-*t*-butylmethylenecyclohexanes **41**, which show a tendency for equatorial attack. This behavior appears to be largely steric-driven, torsional and σ - σ* effects being negligible. In a similar vein, remarkable stereoselectivity has been observed in the epoxidation of certain flexible cyclic olefins (*e.g.*, **42** and **45**). Dynamic conformational analysis according to the method of Toromanoff indicates that this selectivity is largely attributable to a 1,2-diplanar conformation of the cyclohexene ring in the transition state (*cf.* **46**, not shown) <96TET6699>.

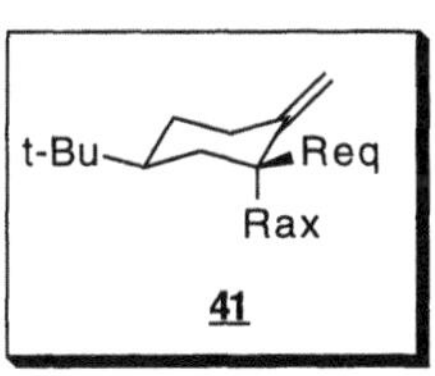

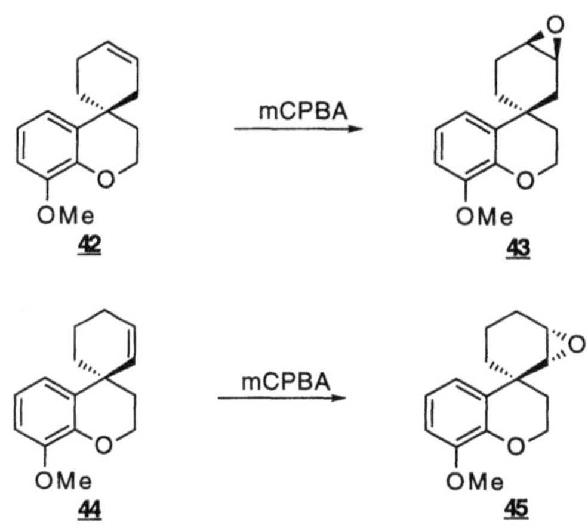

Another viable approach for the preparation of enantiomerically pure epoxides is the kinetic resolution of the corresponding racemic mixtures. This can be a microbiological transformation, as in the fungal epoxide hydrolase mediated hydrolysis of indene oxide (**47**), whereby one enantiomer (*1R, 2S*) can be prepared in modest yield and excellent optical purity. Obviously, kinetic resolution of the racemate is also the consequence of any enantioselective ring-opening of epoxides, interesting examples of which have been recently reported. For example, Jacobsen has shown that chiral Cr(III) salen complexes (**50**) effectively catalyze the asymmetric ring opening of epoxides by TMS azide (*i.e.*, **51** → **52**). Detailed kinetic studies suggest that the catalytic cycle includes a bimetallic intermediate in the rate-determining step <96JA10924>. The analogous enantioselective addition of TMSCN (*i.e.*, **53** → **54**) can be carried out using a Ti catalyst equipped with dipeptide chiral dipeptide ligands which were developed using combinatorial techniques <96AG(E)1668>.

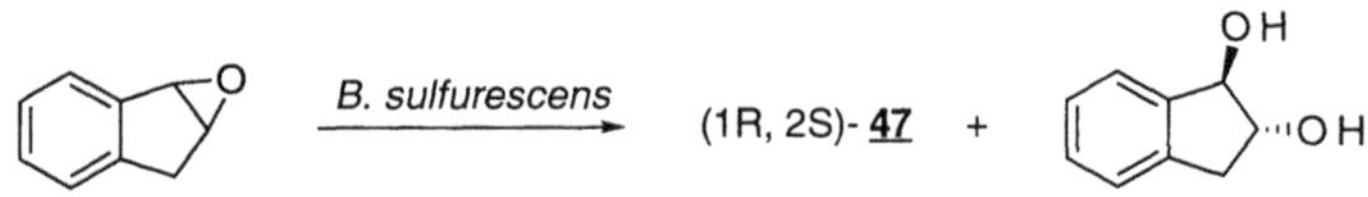

47 (1R, 2R)- 48

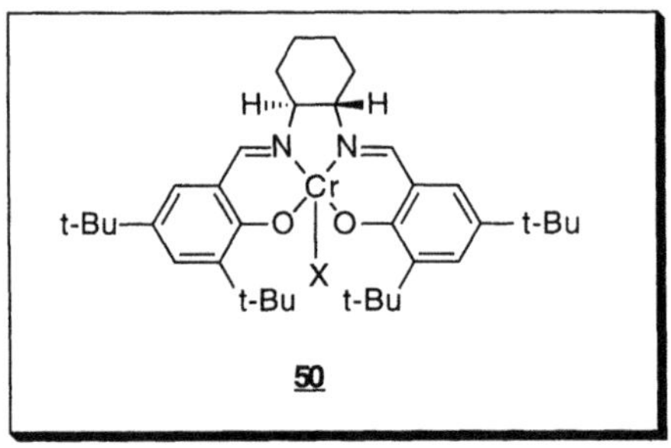

50

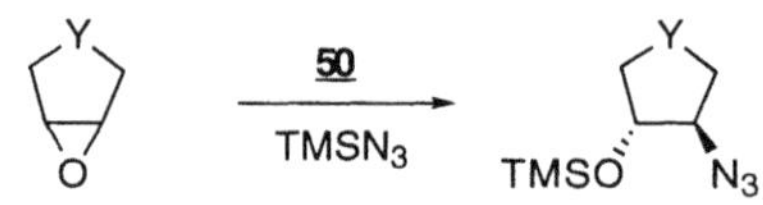

51 52

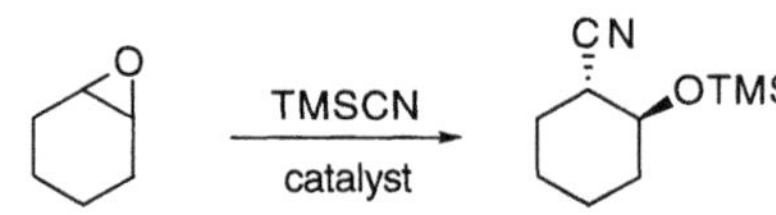

53 54

While asymmetric approaches are certainly important, other synthetically significant epoxidation protocols have also been reported. For example, buffered two-phase MCPBA systems are useful for epoxidations in which the alkenes and/or resultant epoxides are acid-sensitive. Bicarbonate works quite well for cinnamate derivatives (*e.g.*, **55**) <96SC2235>; however, 2,6-di-*t*-butyl-pyridine was shown to give superior results in the case of certain allyl acetals (*e.g.*, **57**) <96SC2875>.

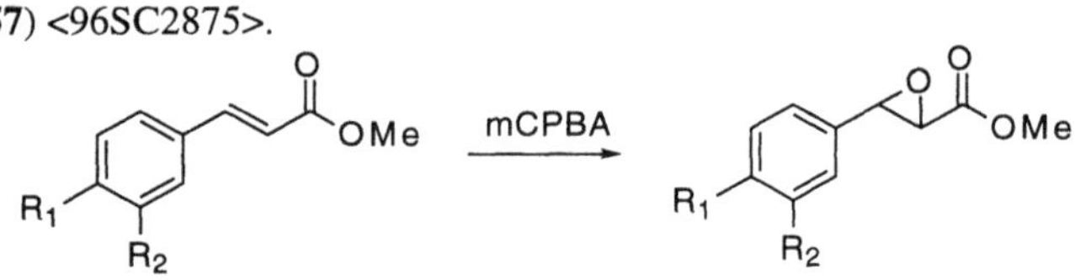

55 56

57 58

buffer = $NaHCO_3$: 77%
buffer = amine : 99%

Olefins containing free hydroxyl groups or carboxylic acid moieties can be oxidized rapidly and efficiently at room temperature using an easily prepared acetonitrile complex of hypofluorous acid (HOF·CH_3CN). The reagent does not induce formation of peroxides with free hydroxy groups, nor do aromatic rings interfere with the reaction. Thus, oleic acid (**59**) was epoxidized in 10 min and in 90% yield <96TL531>.

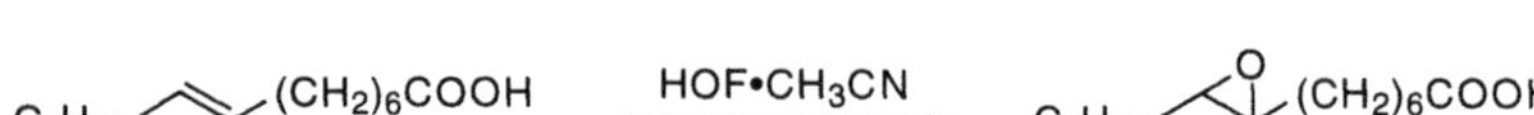

59 60

Majetich and Hicks <96SL649> have reported on the epoxidation of isolated olefins (*e.g.*, **61**) using a combination of 30% aqueous hydrogen peroxide, a carbodiimide (*e.g.*, DCC), and a mildly acidic or basic catalyst. This method works best in hydroxylic solvents and not at all in polar aprotic media. Type and ratios of reagents are substrate dependent, and steric demand about the alkene generally results in decreased yields.

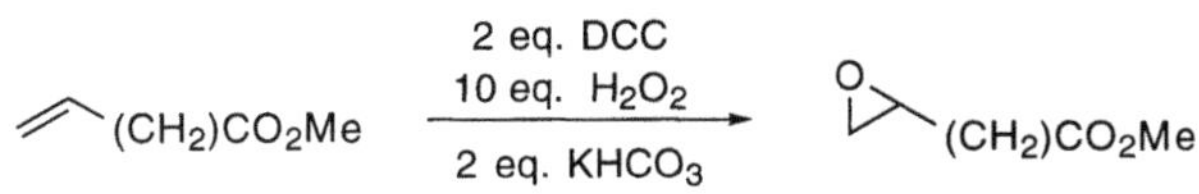

61 62

Spilling and Boehlow <96TL2717> have developed an anhydrous system consisting of catalytic methyl trioxorhenium (MTO) and using urea-hydrogen peroxide adduct (UHP) as a terminal oxidant. The selectivity of this system is roughly equivalent to that of the common peracids, with a predilection for the formation of *syn* epoxides, but the anhydrous conditions are advantageous for particularly labile products. The method is particularly successful in the oxidation of guaiol (**63**).

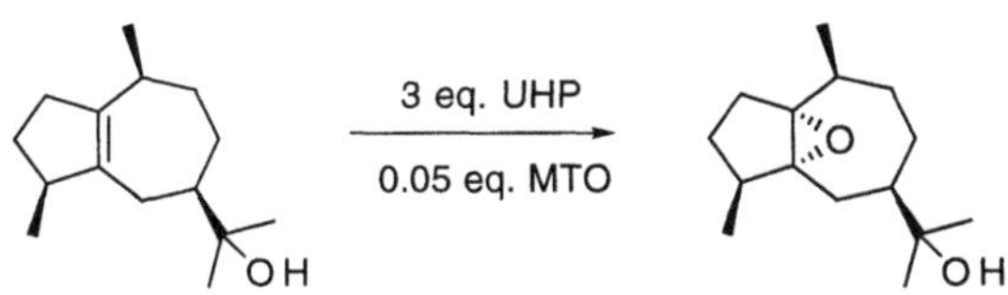

63 64

Sulfonic peracids (**66**) have also been applied recently to the preparation of acid sensitive oxiranes and for the epoxidation of allylic and homoallylic alcohols, as well as relatively unreactive α , β - unsaturated ketones. These reagents, prepared *in situ* from the corresponding sulfonyl imidazolides **65**, promote the same sense of diastereoselectivity as the conventional peracids, but often to a higher degree. In particular, the epoxidation of certain Δ^4-3-ketosteroids (*e.g.*, **67**) with sulfonic peracids **66** resulted in the formation of oxirane products (*e.g.*, **68**) in remarkably high diastereomeric excess. This increased selectivity is most likely the result of the considerable steric requirements about the sulfur atom, which enhances non-bonded interactions believed to be operative in the diastereoselection mechanism <96TET2957>.

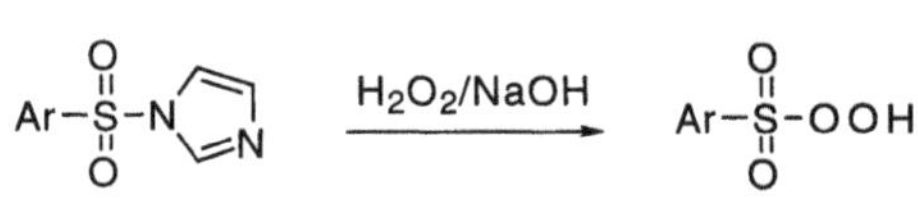

65 66

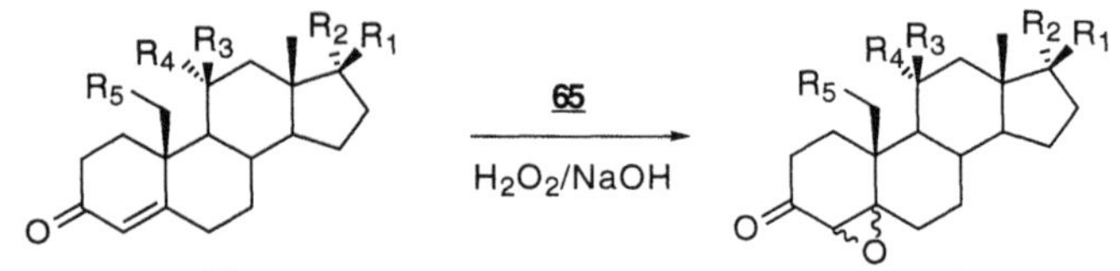

67 68

3.2.2 Reactions of Epoxides

Epoxides are highly reactive species which provide products containing versatile oxygen functionality in the products, a combination highly coveted for synthetic applications. However, the challenge presented in epoxide methodology is often one of selectivity, whether it be inducing, enhancing, or overcoming a bias for a certain regio-, diastereo-, and/or enantiochemical outcome. Jacobsen and Leighton <96JOC389> provide an illustrative example of such efforts in their reported synthesis of (R)-4-((trimethylsilyl)oxy)-2-cyclopentenone (**73**)--an important prostaglandin intermediate--by the enantioselective ring opening of epoxide **71**. This reaction, which provides a convenient alternative to the Nogori method, is promoted by the Mn(salen) catalyst **69,** actually believed to be a pre-catalyst which is converted under the reaction conditions to the active azido species (**70**).

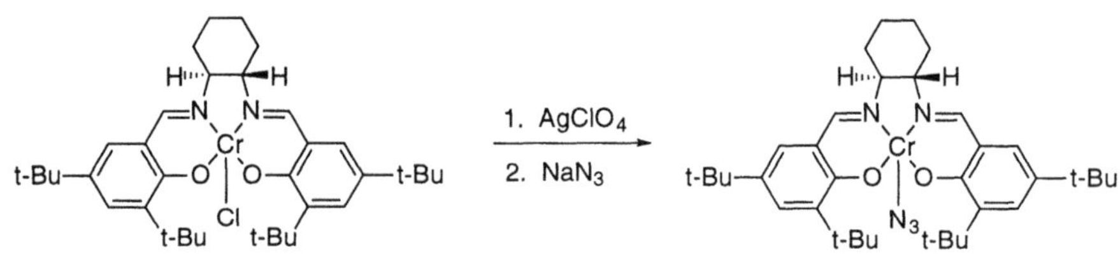

69 70

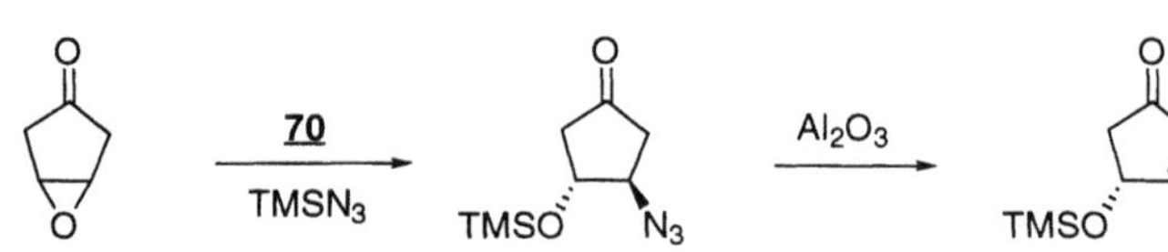

71 72 73

Selectivity is often conferred by on-board functionality, as in the α-sulfenyl-directed ring-opening of 1-phenylthio-2,3-epoxyalkanes (*e.g.*, **74**) with trimethylaluminum, a reaction which gives exclusively C-2 alkylated products with complete retention of configuration about C-2.

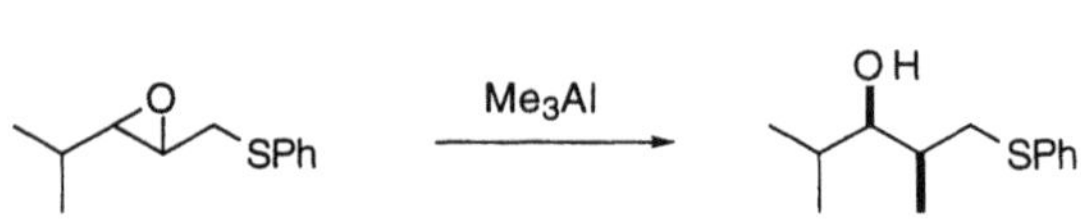

74 75

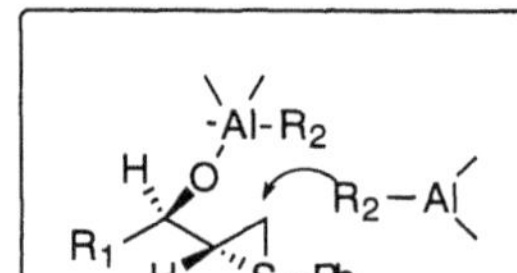

76

The process is thought to involve active neighboring group participation and the intermediacy of episulfonium **76** <96BCSJ2095>. Similar selectivity is observed in the ring-opening of 2,3-epoxy amines **77** with organo-aluminum reagents, a process in which the same type of intermediate (*e.g.*, **79**) is believed to be operative <96TL6177>.

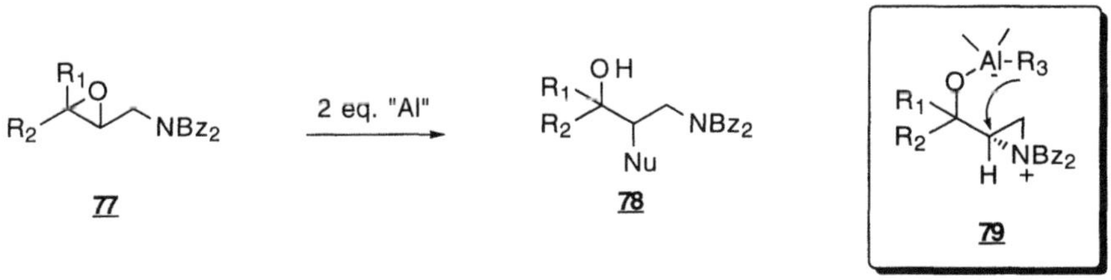

Slightly more exotic carbon-centered nucleophiles can also participate in the ring-opening of epoxides. For example, the vinyl metallate **81**, prepared by the treatment of the alkenyl acetal **86** with Schlosser's reagent, attacks mono-substituted epoxides **82** at the C-2 position to give the labile homoallyl alcohols **83** in fair to very good yields <96TET1433>.

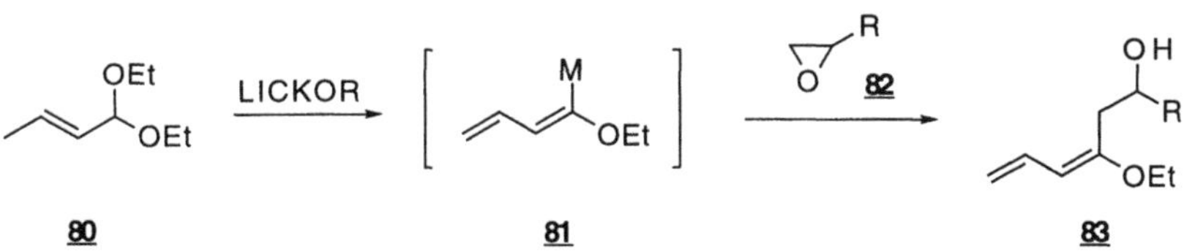

An interesting alcoholysis of epoxides has been reported by Masaki and coworkers <96BCSJ195>, who examined the behavior of epoxides in the presence of a catalytic amount of the π-acid tetracyanoethylene (TCNE, **85**) in alcoholic media. Ring-opening is very facile under these conditions, typically proceeding *via* normal C-2 attack, as exemplified by styrene oxide (**86**). Certain epoxy ethers (*e.g.*, **89**) undergo C-1 attack due to anchimeric assistance. Analysis of the reaction mixtures revealed the presence of captodative ethylenes (*e.g.*, **85**) formed *in situ*, which were shown to be active in catalyzing the reaction. The proposed mode of catalysis is represented by the intermediate **87**. The affinity of these captodative olefins for

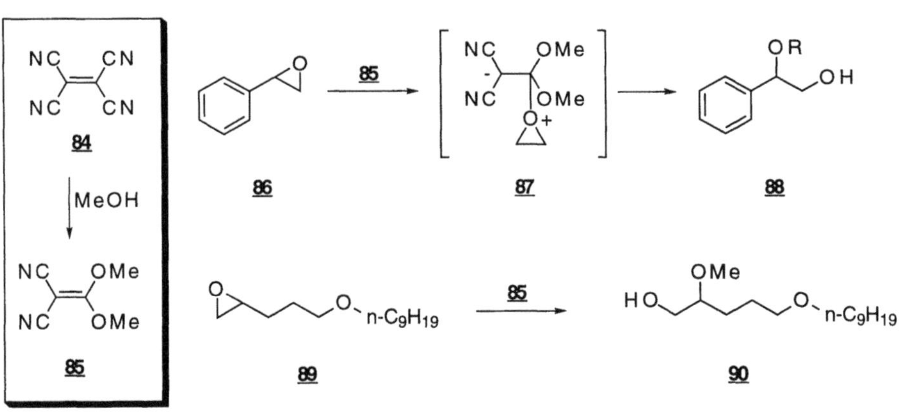

the oxirane oxygen is specific enough to allow for selective epoxide alcoholysis in the presence of THP ethers and ethylene acetals.

Other examples of nucleophilic attack on the oxirane ring include the formation of β-halohydrins with silica-gel supported lithium halides <96TL1845>, the addition of amines catalyzed by lithium triflate, an ersatz for lithium perchlorate <96TL7715>, and the addition of pyrroles, indoles and imidazoles under high pressure (*i.e.*, **91** → **93**) <96JOC984>.

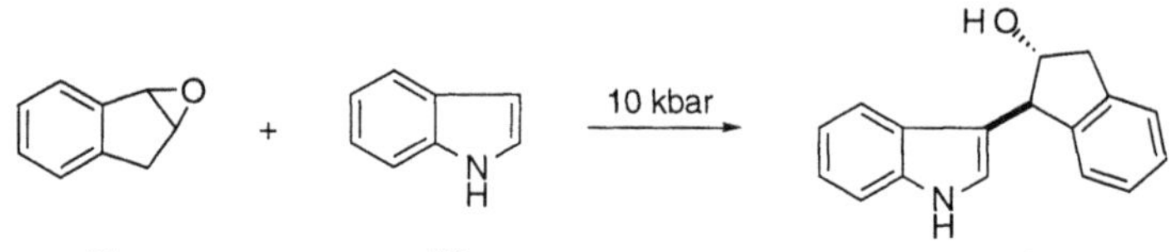

91 92 93

The oxirane ring also undergoes cleavage by α-eliminative mechanisms. For example, the amino alkoxy epoxides **94**, derived from amino acids, undergo stereoselective isomerization to amino alcohols **95** upon treatment with the superbasic mixture butyllithium/diisopropylamine/potassium *tert*-butoxide (LIDAKOR) <96TL5209>. Singh and coworkers have applied novel chiral lithium amide bases for the enantioselective deprotonation of epoxides. In a recent report <96JOC6108>, this method was improved and extended into the preparation of a cyclopentanoid core unit for prostaglandin synthesis. Thus, treatment of epoxide **96** with the lithiate of chiral ligand **97** at 0°C in a non-coordinating solvent gives the chiral allylic alcohol **98** in 97% *ee*. The high degree of asymmetric induction points to a highly ordered transition state, which has been represented by the chelated structure **99**.

94 95

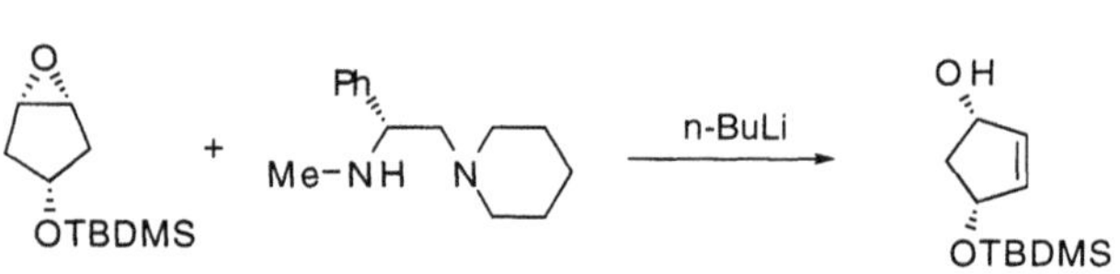

96 97 98

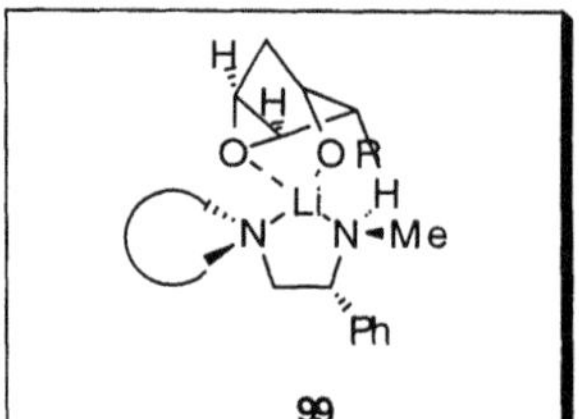

99

Deprotonation can occur not only at the α–position, but also under certain conditions at the C-1 or C-2 position to give an oxiranyl anion (**100**). This fascinating species can be thought of as having significant carbenoid nature, as indicated by resonance structure **101**. Indeed, α-alkyloxyepoxide **102** can be regioselectively deprotonated (presumably chelation control) to form an oxiranyl anion which undergoes α-eliminative ring opening and alkyl insertion to give cyclic allylic alcohols **103** in good to excellent yield. The carbenoid nature of the intermediates was supported by the isolation of the tricyclic alcohol **107**, the product of intramolecular trapping by an olefin <96CC549>. Hodgson and Lee <96CC1015> have devised a clever method for accessing enantiopure bicyclic alcohols from meso-epoxides by such a reaction. For example, treatment of cyclononene oxide (**108**) with isopropyllithium in the presence of an excess of (-)-sparteine leads to an enantioselective α-deprotonation, followed by intramolecular C-H insertion, to give bicyclononanol **109** in 77% yield and 83% enantiomeric excess.

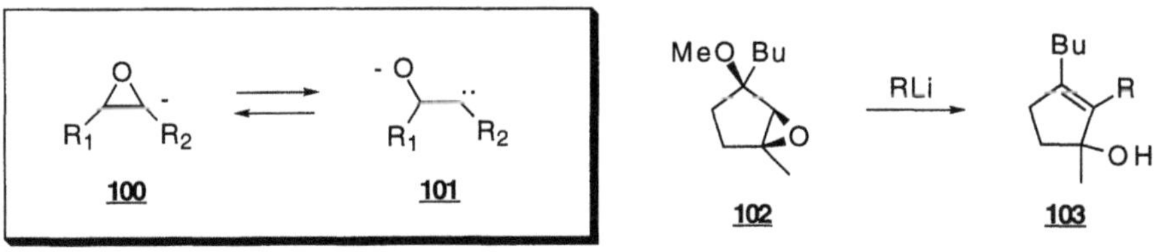

100 101 102 103

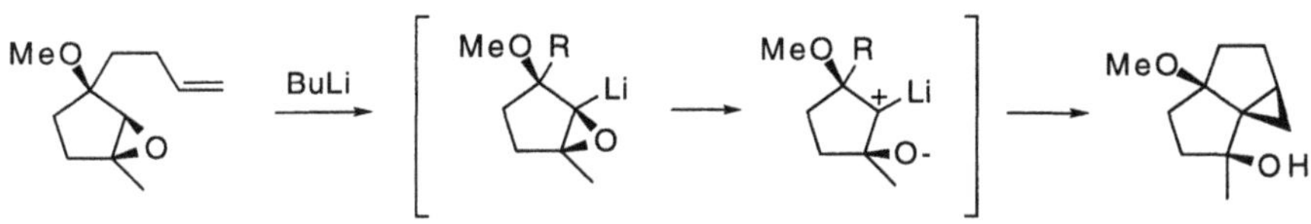

104 105 106 107

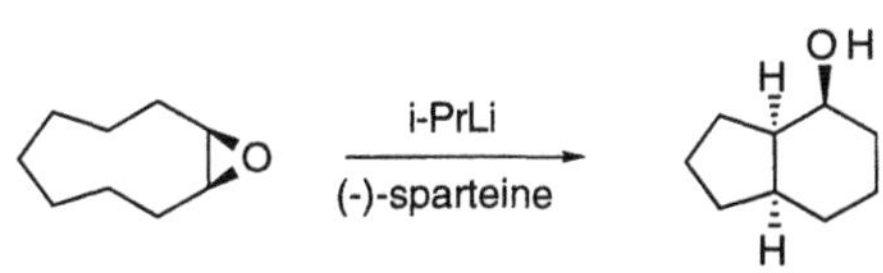

108 109

When the oxiranyl anion is stabilized by an electron withdrawing group in the α-position, it can be alkylated by electrophiles without ring-opening. Thus, treatment of a mixture of sulfonyl epoxide **110** and tetrahydropyranyltriflate **111** with n-BuLi at low temperature leads to the smooth formation of adduct **112** in excellent yield <96TL2605>. Mori and coworkers <96JA8158> have taken this concept even one step further with the development of the protected sulfonyl epoxy alcohol **114,** which functions as an umpoled carbonyl attached to a double electrophile. This multifunctional reagent was applied to the reiterative synthesis of *trans*-fused tetrahydropyrans (*e.g.*, **117**) of interest as a major structural feature of certain marine toxins. Reaction of **114** with triflate **113** under analogous conditions to the example above (**110** + **111**) yields the alkylated tetrahydropyran **115**; treatment with PTSA results in deprotection of the TBS ether and electrophilic epoxide-ring opening, followed by ejection of phenylsulfinate to give the bicyclic ketone **116** in remarkably good yield (80%). Conversion of the remaining silyl ether to a triflate sets the stage for the first iteration. A total of three such cycles yields the fused tetrahydropyran **117**.

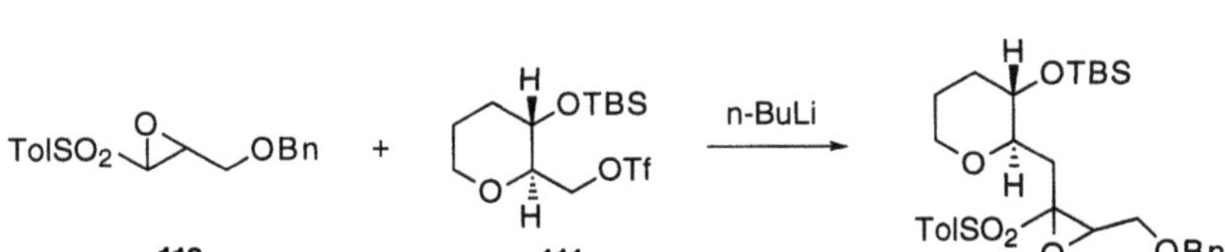

110 111 112

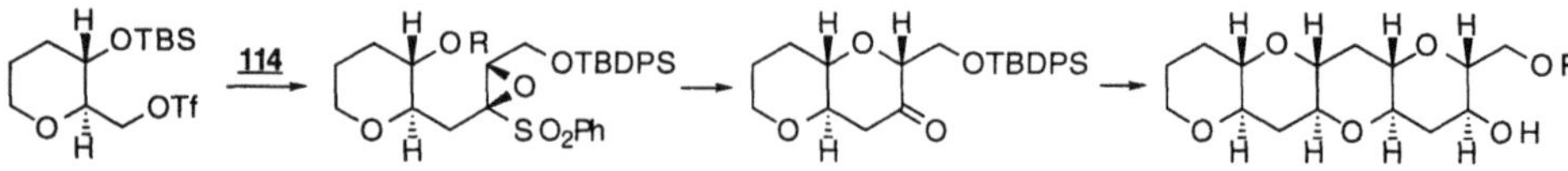

113 115 116 117

The reactivity of epoxides can be modified by various proximal functionality. For example, 2,3-epoxy sulfides **118** are converted to the corresponding TMS-thiiranium species **119** upon treatment with TMS triflate. This intermediate reacts with O-silyl amides regiospecifically to form 1-substituted-3-hydroxy-2-thioethers (*e.g.*, **120**). Simple primary amines undergo polyalkylation, but imines can be used as an indirect amine equivalent <96TET3609>.

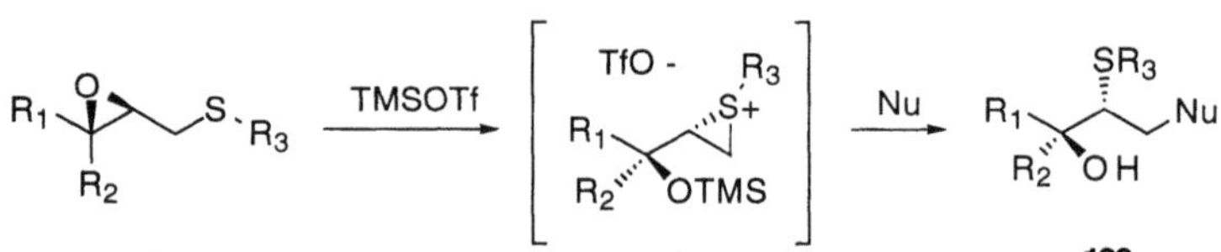

118 119 120

In the case of vinyl epoxides, the double bond is activated toward attack by the vicarious reactivity of the oxirane ring. Taking advantage of this special reactivity, Molander and Shakya <96JOC5885> have developed a reductive annulation method for vinyl epoxides bearing a distal ketone functionality. For example, treatment of the substituted cyclopentanone **121** with SmI_2 induces an intramolecular ketyl olefin coupling reaction, which leads to the bicyclooctanol **122** with very high diastereoselectivity. The cyclization is believed to proceed *via* intermediates of type **123** and is generally limited to the formation of a 5-membered ring.

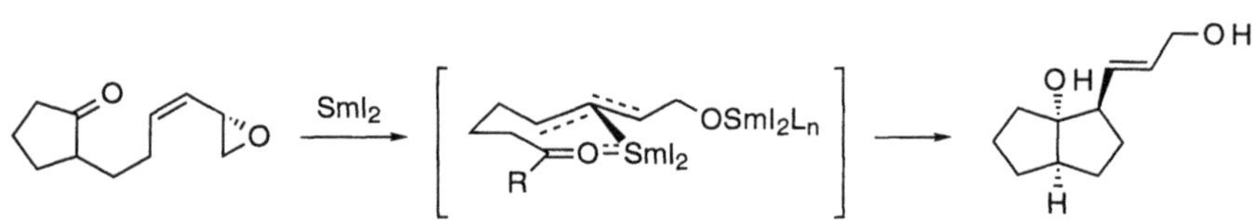

121 123 122

Finally, epoxides can be converted into other functional groups under certain well-defined conditions. For example, ceric ammonium nitrate (CAN) catalyzes the efficient conversion of epoxides to thiiranes (*i.e.*, **124** → **125**) at room temperature in *tert*-butanol <96SYN821>. Lithium perchlorate-diethyl ether promotes the chemo- and regioselective conversion of epoxides to carbonyl compounds (*e.g.*, **126** → **127**), a reaction which is thought to proceed *via* a sequence of lithium coordination to the epoxide oxygen, C-O bond cleavage to give the most stable carbenium ion, and hydride migration <96JOC1877>. Epoxides also react with triphenyl-phosphonium anhydride (POP) and triethylamine to give dienes by a double elimination (*e.g.*, **128** → **129**). Phosphorus NMR studies suggest a mechanism involving a rapidly equilibrating initial complex which can form a diol *bis*-phosphonium ether (**132**) prior to elimination <96SL661>.

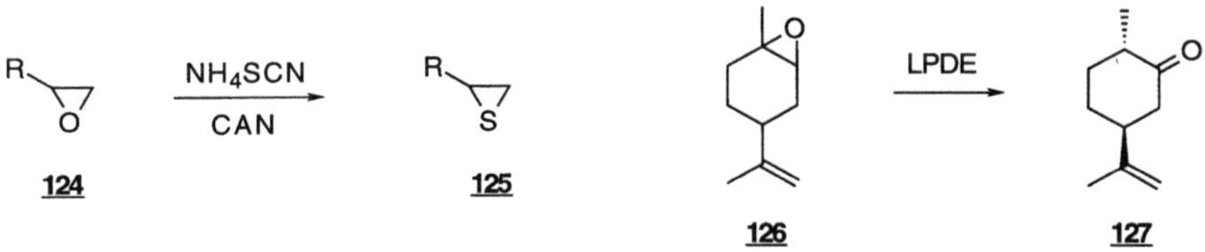

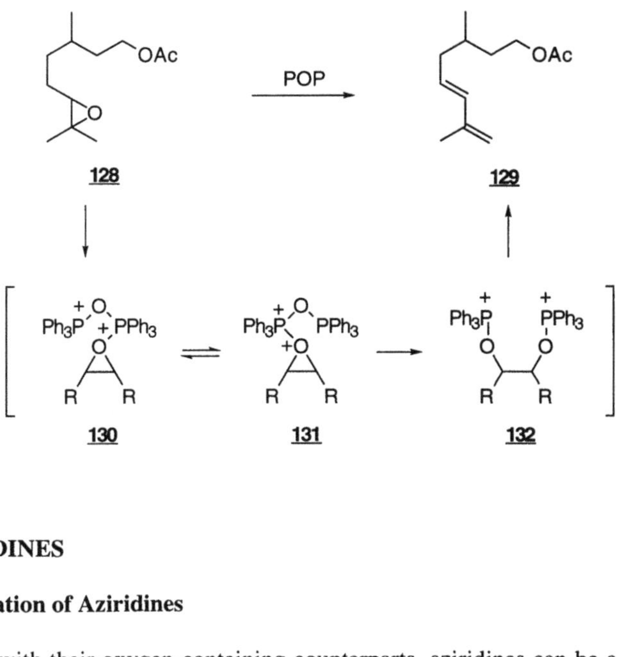

3.3 AZIRIDINES

3.3.1 Preparation of Aziridines

In contrast with their oxygen-containing counterparts, aziridines can be accessed *via* two different basic approaches. Like the epoxides, the aziridines can be prepared by the addition of the heteroatom to the corresponding olefin (N + C=C); however, the cyclization of a carbon center onto an imine (C + C=N) is also an option for this class of heterocycles. Dai and coworkers <96JOC4641, 96CC491> have utilized this less common strategy in developing a direct route to C-vinylaziridines **135**. Thus, allylic sulfonium salts (*i.e.*, **134**) react with aromatic, heteroaromatic, and α,β-unsaturated N-sulfonylimines (*i.e.*, **133**) under solid-liquid phase-transfer conditions in the presence of KOH at room temperature to produce vinyl aziridines **135**. Yields are excellent, but *cis/trans* selectivity is modest. Interestingly, however, it has been demonstrated that the isomerization of such mixtures is feasible under palladium(0) catalysis. Under these conditions N-arylsulfonyl-*trans*-3-alkyl-2-vinylaziridines are converted almost quantitatively to the corresponding *cis*-isomers through the intermediacy of a palladium-allyl complex (**Scheme 2**). The observed apparent thermodynamic preference of the *cis*-isomer is in agreement with *ab initio* calculations <96CC351>.

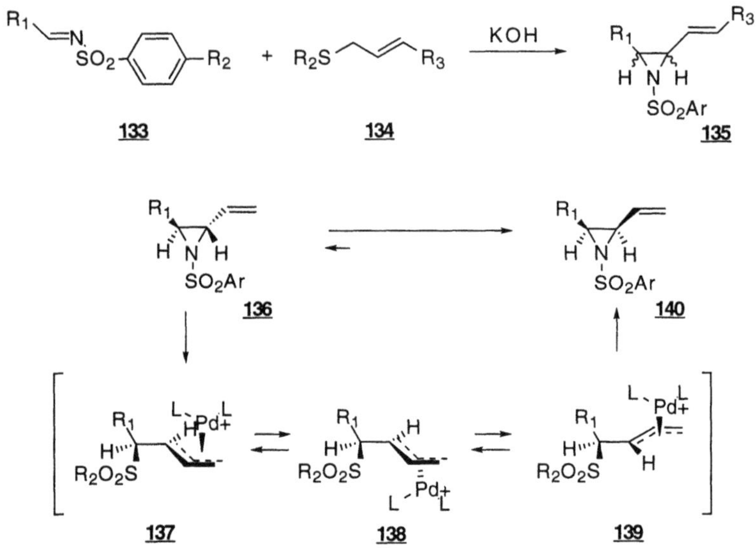

Scheme 2

Of course, new variants of the (N + C=C) approach continue to be reported. Müller and coworkers, who recently reviewed the field of rhodium(II)-catalyzed aziridinations with [N-(p-nitrobenzenesulfonyl)imino]phenyliodinane <96JPO341>, have explored the application of this technology to asymmetric synthesis. Thus, treatment of *cis*-β-methylstyrene (**141**) with PhI=NNs and Pirrung's catalyst [Rh_2\{(-)(R)-bnp\}$_4$] in methylene chloride medium afforded the corresponding aziridine (**142**) in 75% yield and 73% *ee* <96TET1543>.

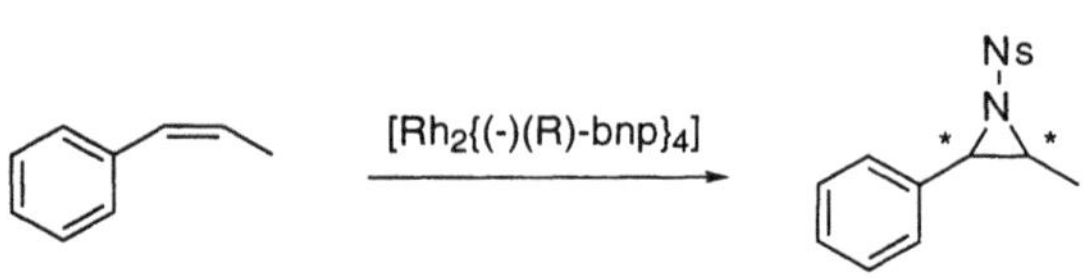

141 142

Alkenes undergo diastereoselective aziridination in the presence of chiral 3-acetoxyamino-quinazolinones (*e.g.*, **143**), prepared *in situ* by acetoxylation of the corresponding 3-amino-quinazolinones. Thus, trimethylsilylstyrene **144** is converted to the aziridine **145** with a diastereomeric ratio of 11:1. The diastereoselectivity is rationalized by a transition state which maximizes *endo* overlap of the substrate phenyl ring the with reagent π-system, while minimizing non-bonded interactions, a factor largely dominated by the relative steric volumes of the two substituents on the chiral carbon center. This is illustrated by a sharp drop in the diastereoselectivity of this reaction (4:1) when the bulkier *t*-butyl group is substituted for the methyl group (*i.e.*, **146**) <96TL5179>. After aziridination, the chiral auxiliary can be removed by desilylative elimination to give an intermediate azirine (**147**), which can be trapped *in situ* by the addition of cyanide, providing the NH-aziridine **148** in 83% *ee* <96CC789>.

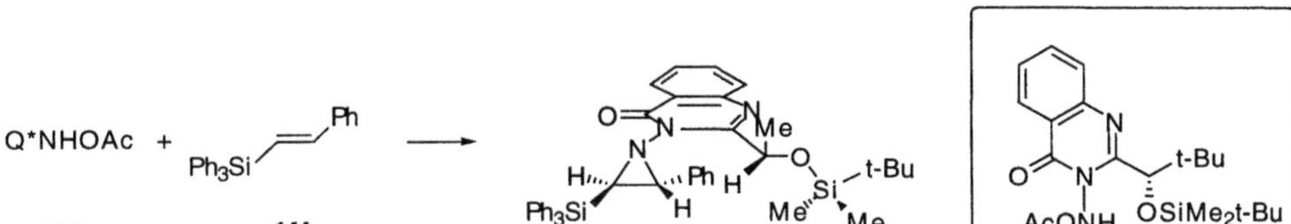

143 144 145 146

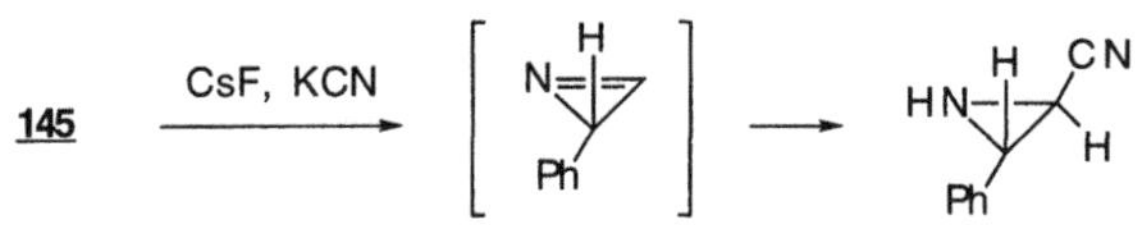

147 148

3.3.2 Reactions of Aziridines

As with epoxides, aziridines undergo a variety of useful ring-opening reactions, some interesting examples of which have been reported in the last year. For example, chiral dialkyl tartrate-diethylzinc complexes catalyze the asymmetric ring opening of symmetrical N-acylaziridines (*e.g.*, **149**) with thiols to give thioamides (*e.g.*, **151**) in up to 93% *ee*. The enantioselectivity is dependent upon the stoichiometry of the reactants and the nature of the tartrate <96TET7817>.

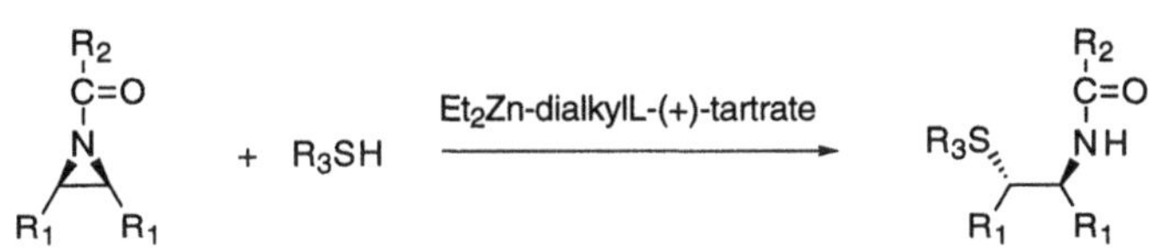

149 150 151

If the starting aziridine is already optically pure, then it is desirable to preserve the enantiomeric excess during any subsequent ring openings. In this vein, 2-substituted aziridine **152** can be cleaved in a regio- and stereocontrolled manner upon heating to 73°C in 50% trifluoroacetic acid to give (2*R*, 3*R*)-(+)-α-methyl-β-phenylserine (**153**) in 75% yield and 96% *de* <96TL5473>. In an interesting intramolecular example of this process, aziridinylmethanol **154** reacts with formaldehyde in the presence of cesium carbonate to form a hemiacetal intermediate, which cyclizes with concomitant aziridine ring opening to furnish acetal **155**, which was then used to synthesize a key synthetic intermediate for bestatin <96H(42)701>.

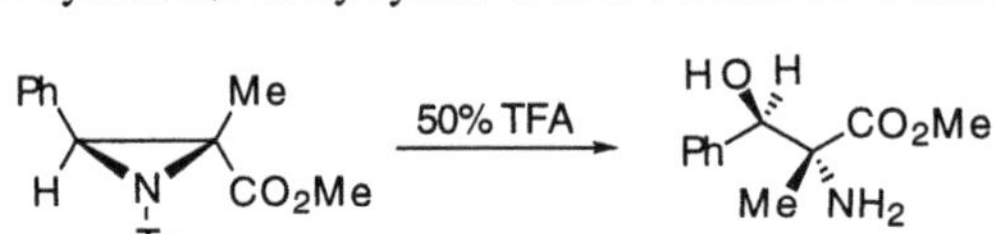

152 153

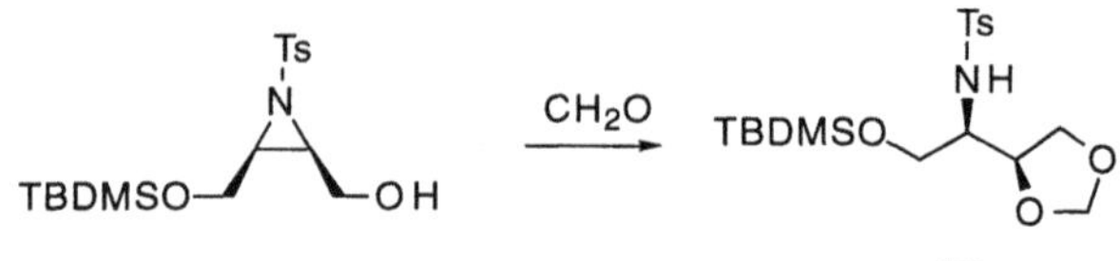

154 155

The regioselectivity of such nucleophilic ring-opening reactions can sometimes be controlled by the reaction conditions. A striking example of such reaction steering is given by the cleavage of 3-substituted N-ethoxycarbonyl aziridine-2-carboxylates (*e.g.*, **156**) with metal halides. Thus, treatment of **156** with sodium bromide leads to exclusive C-2 attack, providing amino acid derivative **157** as the sole product. On the other hand, use of magnesium bromide results in a complete crossover of reactivity to give isomer **158** *via* C-3 attack, presumably due to chelation effects <96TL6893>.

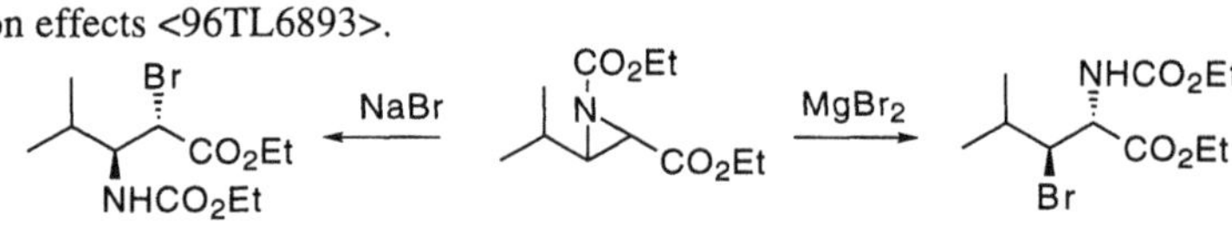

157 156 158

Certain reagents promote ring opening and subsequent cyclization to give other heterocycles. For example, di-*tert*-butyl dicarbonate induces the stereoselective ring transformation of N-alkyl aziridines **159** into oxazolidin-2-ones **160** <96TET2097>.

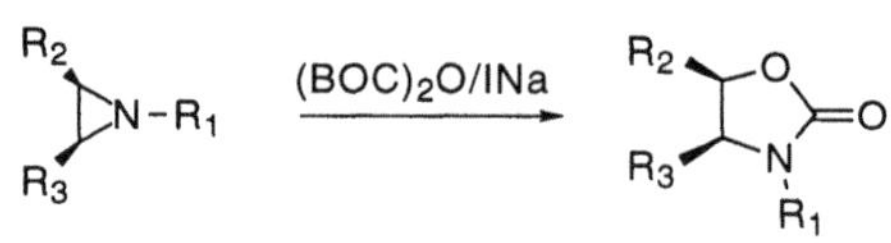

159 160

Rearrangements of complex aziridines can also result in interesting cyclic structures. For example, Zwanenburg and coworkers <96TET12253> have applied the Michael-reaction induced ring closure (MIRC) reaction to aziridinyl-methylenemalonates **161**. Interestingly, this produces *cis*-cyclopropane derivatives **162** preferentially, in contrast to the analogous epoxide reactions. The results suggest that steric interactions between the nucleophilic reagent and the substituents of the aziridine ring direct the stereochemistry of the process.

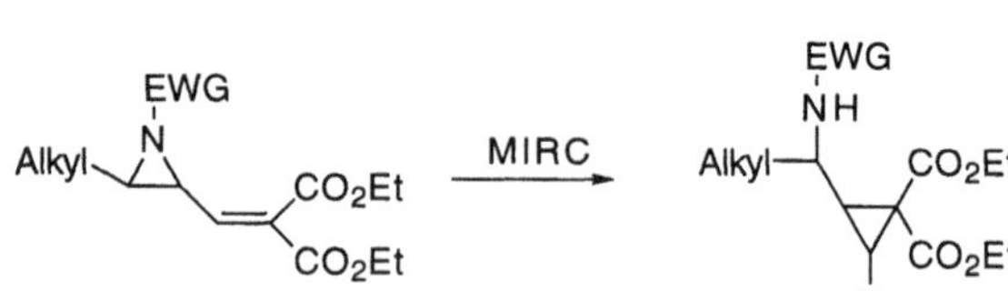

161 162

Aziridinocyclopropanes **163** derived from 2-phenylsulfonyl-1,3-dienes undergo BF_3-induced rearrangement to bicyclic amines **165**, which feature the skeleton of the tropane alkaloids. The reaction proceeds *via* cyclopropyl carbinyl cation **164**, an intermediate also invoked in the analogous epoxide rearrangements. Trapping by fluoride ion is a competing pathway <96TL3371>.

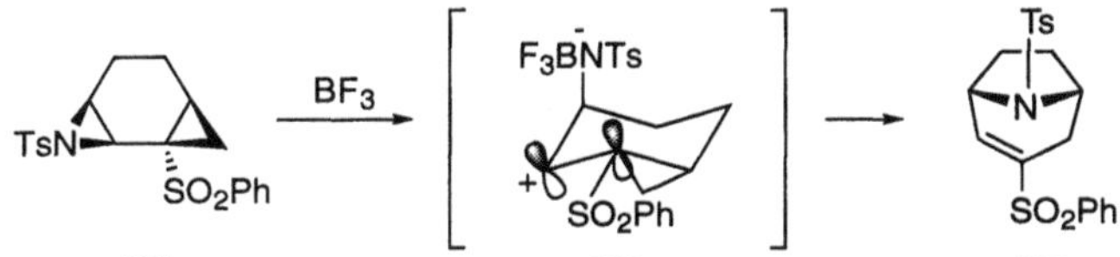

163 164 165

A novel rearrangement of N-propargyl vinylaziridines **166** under Wittig rearrangement conditions has been reported. Thus, treatment of **166** with *s*-BuLi led to the formation of the expected tetrahydropyridines **167** and **168**, products of an aza-[2,3]-Wittig rearrangement, along with significant amounts of pyrroline **169**. The formation of this latter product was surprising, and studies were carried out to elucidate the reaction pathway. Deuterium labeling experiments indicate that the mechanism involves opening of the aziridine ring by an initially formed propargylic anion (*cf.* **170**) to give the corresponding allylic anion (*cf.* **171**), which then undergoes a 5-*exo*-dig cyclization to form a vinylic anion (*cf.* **172**) <96TL2495>.

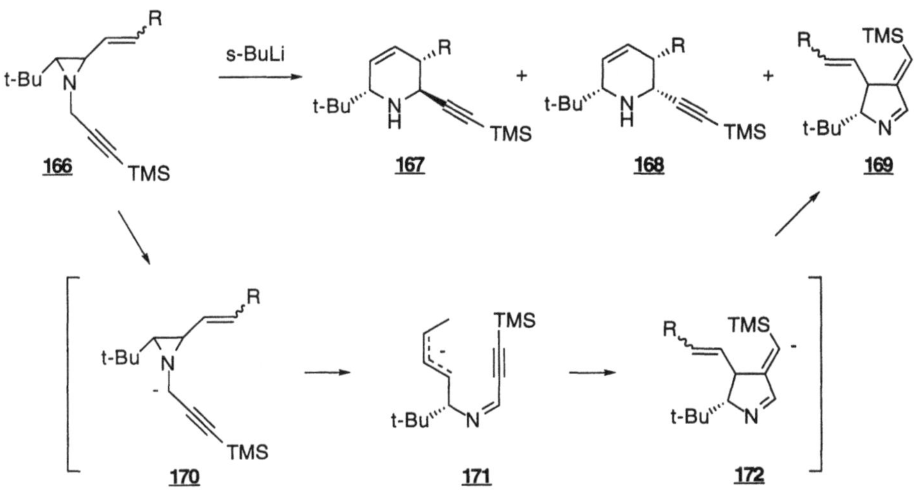

3.4 AZIRINES

3.4.1 Preparation of Azirines

Azirines can be prepared in optically enriched form by the asymmetric Neber reaction mediated by *Cinchona* alkaloids. Thus, ketoxime tosylates **173**, derived from 3-oxocarboxylic esters, are converted to the azirine carboxylic esters **174** in the presence of a large excess of potassium carbonate and a catalytic amount of quinidine. The asymmetric bias is believed to be conferred on the substrate by strong hydrogen bonding *via* the catalyst hydroxyl group <96JA8491>.

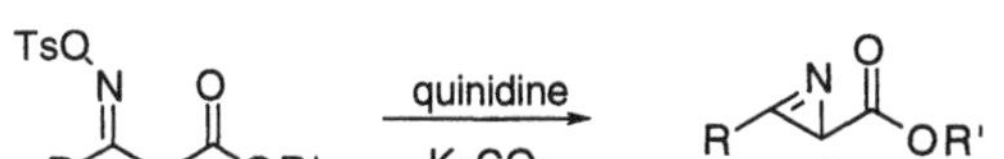

173 174

3.4.2 Reactions of Azirines

Azirines which have pendant electron-withdrawing functionality undergo an interesting reaction with aldehydes and acetone *via* a so-called *"3-X mode"*, a reactivity arising from the pushing effect on the azirine ring by the active methylene center. Thus, azirine ester **175** reacts with acetone in the presence of DABCO to give the 3-oxazoline **176** <96JOC3749>.

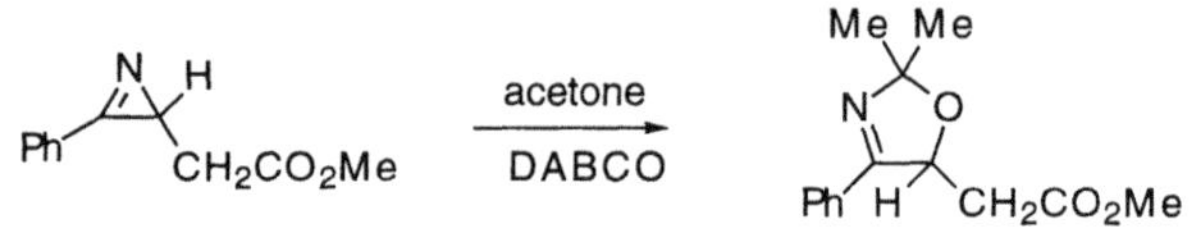

175 176

3.5 REFERENCES

96AG(E)1668 B.M. Cole, K.D. Shimizu, C.A. Krueger, J.P.A. Harrity, M.L. Snapper, A.H. Hoveyda, *Angew. Chem. Int. Ed. Engl.* **1996**, *35*, 1668.

96AG(E)1725 D. Enders, J. Zhu, G. Raabe, *Angew. Chem. Int. Ed. Engl.* **1996**, *35*, 1725.

96BCSJ195 Y. Masaki, T. Miura, M. Ochiai, *Bull. Chem. Soc. Jpn.* **1996**, *69*, 195.

96BCSJ2095 C. Liu, Y. Hashimoto, K. Kudo, K. Saigo, *Bull. Chem. Soc. Jpn.* **1996**, *69*, 2095.

96CC191 V.K. Aggarwal, M.F. Wang, *J. Chem. Soc., Chem. Commun.* **1996**, 191.

96CC351 N. Mimura, T. Ibuka, M. Akaji, Y. Miwa, T. Taga, K. Nakai, H. Tamamura, N. Fujii, Y. Yamamoto, *J. Chem. Soc., Chem. Commun.* **1996**, 351.

96CC491 A.-H. Li, L.-X. Dai, X.-L. Hou, *J. Chem. Soc., Chem. Commun.* **1996**, 491.

96CC549 L. Dechoux, E. Doris, C. Mioskowski, *J. Chem. Soc., Chem. Commun.* **1996**, 549.

96CC789 R.S. Atkinson, M.P. Coogan, I.S.T. Lochrie, *J. Chem. Soc., Chem. Commun.* **1996**, 789.

96CC1015 D.M. Hodgson, G.P. Lee, *J. Chem. Soc., Chem. Commun.* **1996**, 1015.

96H(42)701 K. Fuji, T. Kawabata, Y. Kiryu, Y. Sugiura, *Heterocycles* **1996**, *42*, 701.

96JA491 D. Yang, Y.-C. Yip, M.-W. Tang, M.-K. Wong, J.-H. Zheng, K.-K. Cheung, *J. Am. Chem. Soc.* **1996**, *118*, 491.

96JA3556 E. Vedejs, W.H. Dent, III, J.T. Kendall, P.A. Oliver, *J. Am. Chem. Soc.* **1996**, *118*, 3556.

96JA8158 Y. Mori, K. Yaegashi, H. Furukawa, *J. Am. Chem. Soc.* **1996**, *118*, 8158.

96JA8491 M.M.H. Verstappen, G.J.A. Ariaans, B. Zwanenburg, *J. Am. Chem. Soc.* **1996**, *118*, 8491.

96JA9806 Y. Tu, Z.-X. Wang, Y. Shi, *J. Am. Chem. Soc.* **1996**, *118*, 9806.

96JA10924 K.B. Hansen, J.L. Leighton, E.N. Jacobsen, *J. Am. Chem. Soc.* **1996**, *118*, 10924.

96JCS(P1)157 R.S. Atkinson, M.P. Coogan, C.L. Cornell, *J. Chem. Soc., Perkin Trans. I* **1996**, 157.

96JCS(P1)1729 T. Honda, H. Mizutani, K. Kanai, *J. Chem. Soc., Perkin Trans. I* **1996**, 1729.

96JOC389 J.L. Leighton, E.N. Jacobsen, *J. Org. Chem.* **1996**, *61*, 389.

96JOC984 H. Kotsuki, K. Hayashida, T. Shimanouchi, H. Nishizawa, *J. Org. Chem..* **1996**, *61*, 984.

96JOC1877 R. Sudha, K. Malola Narasimhan, V. Geetha Saraswathy, S. Sankararaman, *J. Org. Chem.* **1996**, *61*, 1877.

96JOC3586 R. Fernández de la Pradilla, S. Castro, P. Manzano, J. Priego, A. Viso, *J. Org. Chem.* **1996**, *61*, 3586.

96JOC3749 M.C.M. Sá., A. Kascheres, *J. Org. Chem.* **1996**, *61*, 3749.

96JOC4641 A.-H. Li, L.-X. Dai, X.-L. Hou, M.-B. Chen, *J. Org. Chem.* **1996**, *61*, 4641.

96JOC5885 G.A. Molander, S.R. Shakya, *J. Org. Chem.* **1996**, *61*, 5885.

96JOC6108 D. Bhuniya, A. DattaGupta, V. K. Singh, *J. Org. Chem.* **1996**, *61*, 6108.

96JPO341 P. Müller, C. Baud, Y. Jacquier, M. Moran, I. Nägeli, *J. Phys. Org. Chem.* **1996**, *9*, 341.

96SC2235	G. Moyna, H.J. Williams, A.I. Scott, *Synth. Commun.* **1996**, *26*, 2235.
96SC2875	A. Svensson, U.M. Lindström, P. Somfai, *Synth. Commun.* **1996**, *26*, 2875.
96SL649	G. Majetich, R. Hicks, ***Synlett*** **1996**, 649.
96SL661	J.B. Hendrickson, M.A. Walker, A. Varvak, Md. Sajjat Hussoin, *Synlett* **1996**, 661.
96SYN821	N. Iranpoor, E. Kazemi, *Synthesis* **1996**, 821.
96TET1433	A. Deagostino, C. Prandi, P. Venturello, *Tetrahedron.* **1996**, *52*, 1433.
96TET1543	P. Müller, C. Baud, Y. Jacquier, *Tetrahedron* **1996**, *52*, 1543.
96TET2097	J. Sepúlveda-Arques, T. Armero-Alarte, A. Acero-Alarcón, E. Zaballos Garcia, B. Yruretagoyena Solesio, J. Ezquerra Carrera, *Tetrahedron* **1996**, *52*, 2097.
96TET2957	R. Kluge, M. Schulz, S. Liebsch, *Tetrahedron* **1996**, *52*, 2957.
96TET3521	M.T. Rispens, O.J. Gelling, A.H.M. de Vries, A. Meetsma, F.V. Bolhuis, B.L. Feringa, *Tetrahedron* **1996**, *52*, 3521.
96TET3609	D.M. Gill, N.A. Pegg, C.M. Rayner, *Tetrahedron.* **1996**, *52*, 3609.
96TET6699	P. Ducrot, R. Bucourt, C. Thal, B. Marçot, J. Mayrargue, H. Moskowitz, *Tetrahedron.* **1996**, *52*, 6699.
96TET7817	M. Hayashi, K. Ono, H. Hoshimi, N. Oguni, *Tetrahedron* **1996**, *52*, 7817.
96TET12253	I. Funaki, R.P.L. Bell, L. Thijs, B. Zwanenburg, *Tetrahedron* **1996**, *52*, 12253.
96TL531	S. Rozen, Y. Bareket, S. Dayan, *Tetrahedron Lett.* **1996**, *37*, 531.
96TL1205	Y. Landais, L. Parra-Rapado, *Tetrahedron Lett.* **1996**, *37*, 1205.
96TL1845	H. Kotsuki, T. Shimanouchi, *Tetrahedron Lett.* **1996**, *37*, 1845.
96TL2495	J. Åhman, P. Somfai, *Tetrahedron Lett.* **1996**, *37*, 2495.
96TL2605	Y. Mori, K. Yaegashi, K. Iwase, Y. Yamamori, H. Furukawa, *Tetrahedron Lett.* **1996**, *37*, 2605.
96TL2717	T.R. Boehlow, C. D. Spilling, *Tetrahedron Lett.* **1996**, *37*, 2717.
96TL2725	S.-H. Zhao, P.R. Ortiz, B.A. Keys, K.G. Davenport, *Tetrahedron Lett.* **1996**, *37*, 2725.
96TL3371	C.M.G. Löfström, J.-E. Bäckvall, *Tetrahedron Lett.* **1996**, *37*, 3371.
96TL3375	F. Minutolo, D. Pini, P. Salvadori, *Tetrahedron Lett.* **1996**, *37*, 3375.
96TL3895	D. Bell, M.R. Davies. F.J.L. Finney, G.R. Geen, P.M. Kincey, I.S. Mann, *Tetrahedron Lett.* **1996**, *37*, 3895.
96TL4533	Y. Noguchi, R. Irie, T. Fukuda, T. Katsuki, *Tetrahedron Lett.* **1996**, *37*, 4533.
96TL5179	R.S. Atkinson, M.P. Coogan, I.S.T. Lochrie, *Tetrahedron Lett.* **1996**, *37*, 5179.
96TL5209	A. Mordini, M. Valacchi, S. Pecchi, A. Degl'Innocenti, G. Reginato, *Tetrahedron Lett.* **1996**, *37*, 5209.
96TL5473	F.A. Davis, H. Liu, G.V. Reddy, *Tetrahedron Lett.* **1996**, *37*, 5473.
96TL6177	C. Liu, Y. Hashimoto, K. Saigo, *Tetrahedron Lett.* **1996**, *37*, 6177.
96TL6893	G. Righi, R. D'Achille, C. Bonini, *Tetrahedron Lett.* **1996**, *37*, 6893.
96TL7715	J. Augé, F. Leroy, *Tetrahedron Lett.* **1996**, *37*, 7715.

Chapter 4

Four-Membered Ring Systems

J. Parrick and L. K. Mehta
Brunel University, Uxbridge, UK

4.1 INTRODUCTION

Several recently published reviews include the following aspects of the chemistry of four-membered heterocycles: preparative routes to fluorine containing heterocycles <94ZOR1704>, saturated nitrogen heterocycles <95MI209, 96MI259> and combinatorial syntheses <96ACR144>. Reviews having four-membered heterocycles as their main theme describe the uses of benzothiete in synthesis <96JPR383>, silacyclobutenes <96MI41>, β-lactam chemistry (530 references) <95MI330>, strategies for the stereoselective synthesis of β-lactams and functionalization at the α-carbon atom of the ring system <96MI101>, synthesis and application of fluoro-β-lactams <95MI727>, the use of β-lactams as synthetic intermediates <95MI95> and syntheses leading to two pivotal monocyclic β-lactam intermediates in routes to 1β-methylcarbapenems <96T331>. Of major interest to all heterocyclic chemists is the appearance of the second edition of Comprehensive Heterocyclic Chemistry and volume 1B is of particular significance in the context of this chapter <96MI102>.

As usual, this review is very selective due to limitations of space, and a large number of publications are not mentioned.

4.2 AZETINES AND AZETIDINES

Mixtures of the *cis* and *trans* isomers of 2,3,4-trisubstituted azetine **2**, together with the isoquinoline **3**, are obtained on photolysis of α-dehydrophenylalanine amides **1** <96TL5917>. The mechanism suggested involves photoacetyl migration.

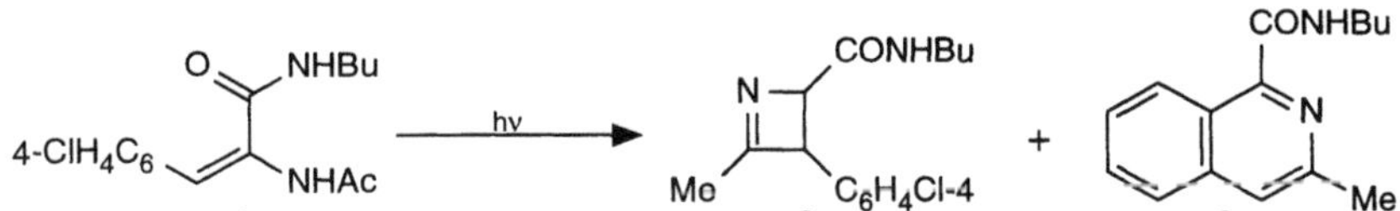

1 **2** **3**

Irradiation of methacrylonitrile and 2-alkoxy-3-cyano-4,6-dimethylpyridine mixture causes [2+2]photocycloaddition across the C(2)-C(3) bond of the latter to give the bicyclic azetine **4** (45-55%) via an azacyclooctatetraene intermediate <96CC1349>. The [2+2]cycloaddition of *N*-acylaldimines and allyltriisopropylsilane to give 2,4-disubstituted azetidines is catalysed by Lewis acids <95CL789>.

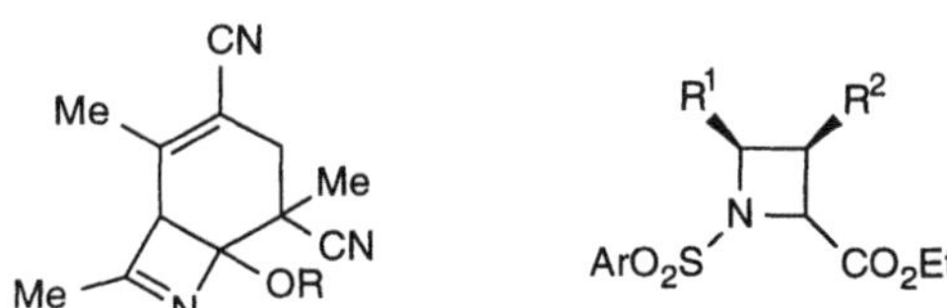

4 **5**

Other reactions leading to azetidines include the dialkylation of chromium or tungsten complexes of aminocarbenes with 1,3-diiodopropane under phase-transfer conditions <96CL827> and the regio- and stereo-specific reaction of dimethylsulfoniumethoxy-carbonylmethylide with 2-substituted or 2,3-disubstituted *N*-arylsulfonylaziridines to afford **5** ($R^1 \neq$ H, R^2 = H) or **5** (R^1 and $R^2 \neq$ H) respectively, generally in useful yields <95JCS(P1)2605>.

Interesting developments in simple azetidine chemistry continue to be reported. The apparently general acetylative dealkylation of *N-tert*-butyl-3-substituted azetidines **6** (R^1 = Bu^t) in the presence of boron trifluoride provides a two-step route to azabicyclobutane **7** from **6** (R^1 = Bu^t, R^2 = Cl). An aqueous solution of **7** reacts with ethyl chloroformate to give **8**. Relatively unexplored 3-azetidinones **9** (R = Ac or NO_2) are available from 3-acetoxyazetidine **6** (R^1 = Ac, R^2 = OAc) which is obtained by acetylative dealkyation of **6** (R^1 = Bu^t, R^2 = OAc) <96JOC5453>. 3-Substituted azetidines can be utilized in the synthesis of polyfunctional γ- and δ-aminophosphonic acid derivatives <95TL9201>.

6 **7** **8** **9**

4.3 OXETANES, 2-OXETANONES, THIETANES AND BENZOTHIETES

Head-to-head [2+2]photocycloaddition of 1,2-diarylethanediones and 2-aminopropene nitriles (CH_2:C(CN)NR_2) occurs to yield oxetanes **10** in moderate to good yields. The formation of only one diastereoisomer in each of the cases investigated is rationalized in terms of the most easily accessible and stabilized 1,4-diradical intermediate <95RTC498>. 2,3,4-Trisubstituted oxetanes **11** are obtained in high yield by intramolecular nucleophilic attack of the anion from certain 2-(1-alkoxyethyl)-3-substituted oxiranes <96JOC4466>.

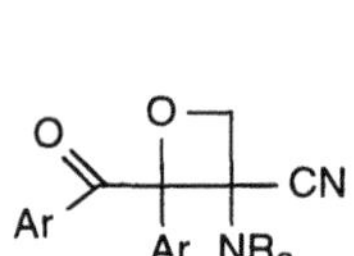

10

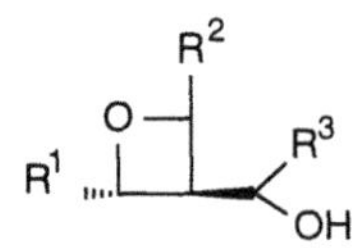

11

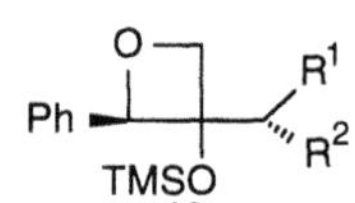

12

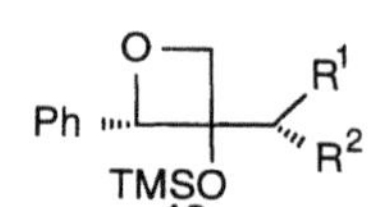

13

The facial diastereoselectivity shown in the formation of **12** and **13** in the Paterno-Buechi [2+2]photocycloaddition of benzaldehyde with silyl enol ethers (TMSOC(:CH_2)CHR^1R^2) can be controlled and reversed by suitable choice of R^1 and R^2. For instance when R^1 = Pr^i and R^2 = OCH_2Ph the ratio of isomers **12** to **13** is 67:33 but when R^1 = Pr^i and R^2 = Cl the same isomer ratio is 15:85 <95AG(E)2271, 96T10861>. The Paterno-Buechi reaction of benzaldehyde and 1-methoxycarbonyl-2-pyrrolenine gives the bicyclic oxetane **14** <96AG(E)884>. A thermal [2+2]cycloaddition of dichloroketene with α-oxyaldehydes and α-aminoaldehydes is used to obtain the 2-oxetanones (β-lactones) **15** (X = O) and **15** (X = NR^1), respectively <95CC1735>. Lithium enolates from thiol esters undergo addition to ketones <96OS61> or homochiral aldehydes to yield β-lactones, in the latter case with stereochemical control at C(4) <96TL245>.

An unexpected, one-step synthesis of α-chloro-β-lactones in 40-83% yield resulted when phenyl esters of α-chlorocarboxylic acids were treated under the conditions of the Darzens reaction with strong base in the presence of ketones or aldehydes <95AG(E)2028>. Previously, phenyl esters do not appear to have been used in this reaction.

α-Methylene β-lactones **16** (R^1 = H) are produced in poor to moderate yield by cyclization of α-methylene 3-hydroxypropanoic acid with mesyl chloride <96MI51>. In a similar way, α-iodomethylene β-lactones **16** (R^1 = I) are obtained in moderate to good yield from the corresponding iodomethylene acids <96S586>.

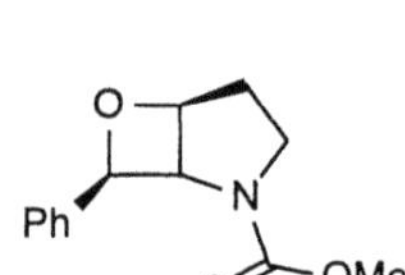

14

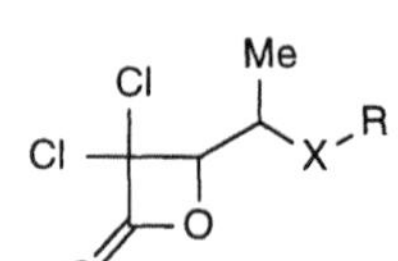

15

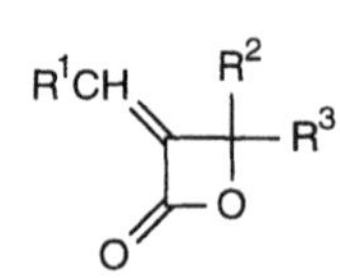

16

β-Lactones have also been obtained from reactions involving loss of dinitrogen from other heterocycles. For example, the benzotriazolide **17** affords a β-lactone <96LA881> and the rhodium catalysed decomposition of the diazo compound **18** (X = O) gives a carbene which undergoes a ring contraction process to yield **19** <96ZN(B)1325>. The corresponding four-membered ring products are obtained from **18** when X = S or X = NH.

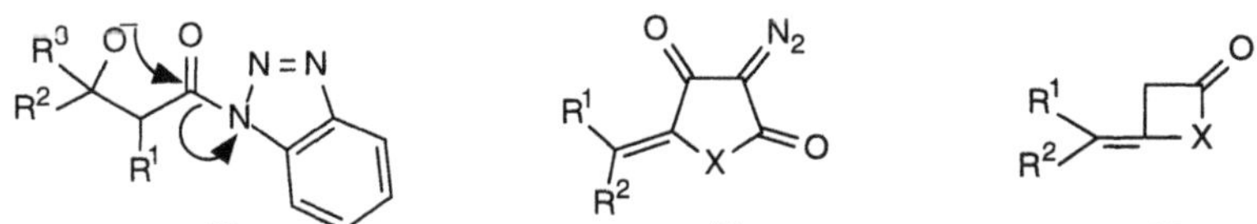

17 **18** **19**

Other thietane derivatives have been obtained by isomerization of nitrogen-containing heterocycles. The reaction of an acyl isothiocyanate (RCONCS) with diphenyldiazomethane gave **20** and this isomerized in solution to the tetraphenyl-3-thietanone **21** <96BSB253>. Additionally, the isoxazolidine **22** was converted into **23** by the action of trimethylsilyl iodide and zinc iodide <96H1211>.

The naphthothietes **24** (R^1R^2 = SCH_2, R^3 = H) and **24** (R^1 = H, R^2R^3 = SCH_2) are available from F.V.P. of the corresponding mercapto-naphthalenemethanols. The isomers show interesting differences in their ring opening behaviour under thermal and photochemical activation <95LA2221>.

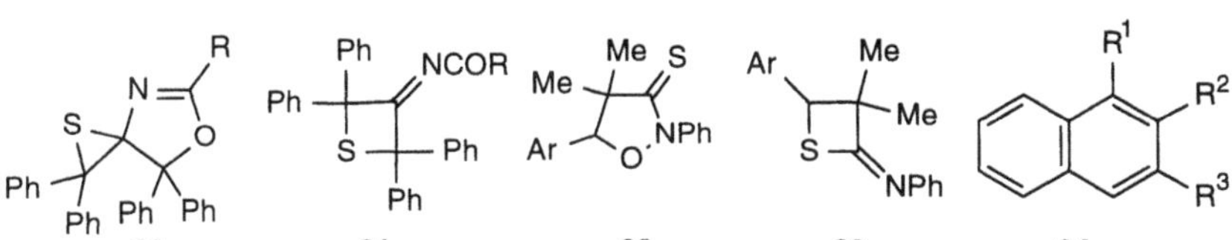

20 **21** **22** **23** **24**

4.4 DIOXETANES, DIAZETANES AND DITHIETANES

The blue fluorescence of dioxetane **25** (X = O and CH_2, R = alkyl) in the presence of fluoride ion has been studied <96TL5939>.

Thermolysis of the dicarbamic acid silyl ester ($CH_2\{N[SiMe_3]CO_2SiMe_3\}_2$) gives the 1,3-diazetane derivative **26** <96JOM93>. Alkylation of the 1,3-dithietane tetraoxide **27** with α,ω-dihaloalkanes yields the dispiro compounds **28** (n = 1-4) <95ZOR589>. The first 1,2-dithiete *S*-oxide **29** is reported <95TL8583>.

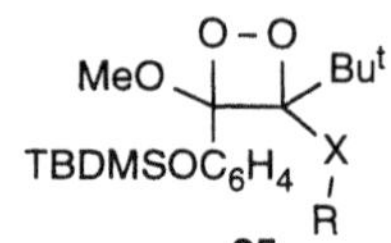

25

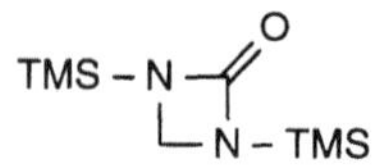

26

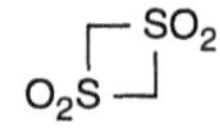

27

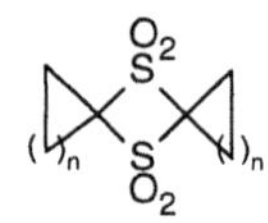

28

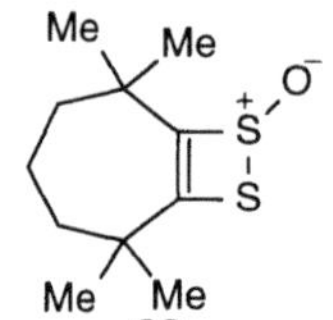

29

4.5 THIAZETIDINE AND THIAZETIDINONE

Thermolysis of the thiadiazabicyclo[3.1.0]hexene **30**, obtained from reaction of an azide with an isothiazole dioxide affords the 1,2-thiazetidine-1,1-dioxide **31**, but in poor yield and as part of a three-component mixture <96T7183>. Isocyanates **32** are available in moderate yield <94ZOR1700>.

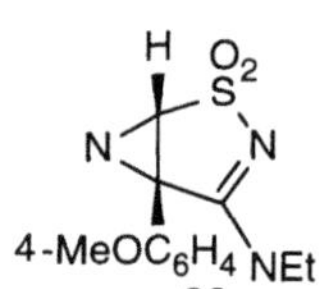

30

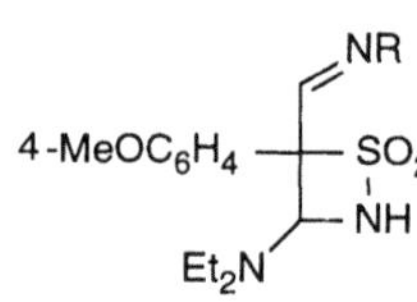

31

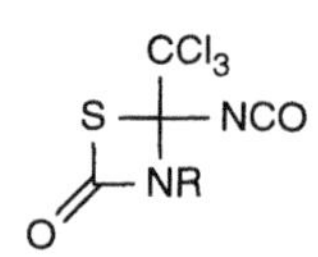

32

4.6 PHOSPHORUS AND SILICON HETEROCYCLES

The diphosphetene **33** and diphosphetane **34** have been obtained. X-Ray crystal structures are available for **33** <96S265> and the novel diazaphosphete **35** <96IC2458, 96JA1060>.

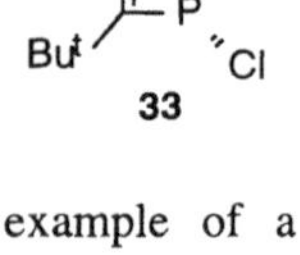

33

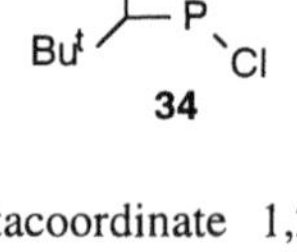

34

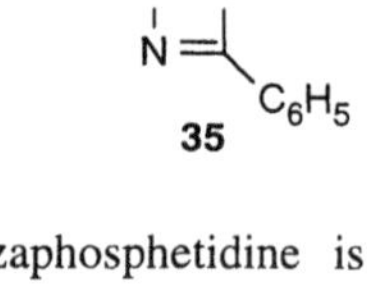

35

The first example of a pentacoordinate 1,2λ^5-azaphosphetidine is reported. The compound **36** may be regarded as potential intermediate in the reaction of a phosphorus ylide with a Schiff base, and thermolysis of **36** gave 1,1-diphenylethene and the iminophosphorane **37**. This result supports the argument that olefin formation in this type of Wittig reaction occurs via a *C*-apical pseudorotamer <96AG(E)1096>.

The migratory aptitude of substituents on the disilylcarbene generated by photolysis of **38** has been studied <96JPO619>. 1-Silabutadienes ((TMS)$_2$Si:CHCH:CHPh) undergo head-to-head [2+2]dimerization to give 1,2-disilacyclobutanes **39** <96MI801>. The tetrahydro-2,3-disilanaphthalenes **40** decompose to yield 3,4-diaryl-1,2-disilacyclobutanes <95CB1083, 96CB15, 96ZN(B)370>. Alkynes react with **41** in the presence of nickel catalyst to give both 5,6-benzo-1,4-disilacyclohexa-2,5-dienes and 5,6-benzo-1,2-disilacyclohexa-3,5-dienes <95JOM35>.

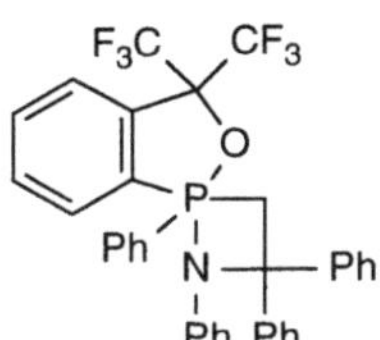

36

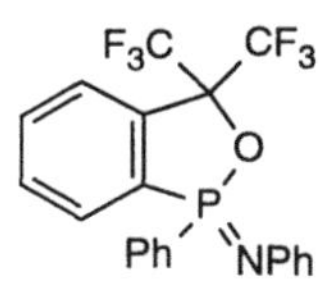

37

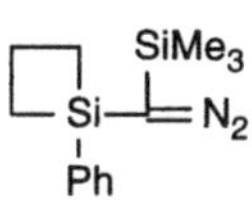

38

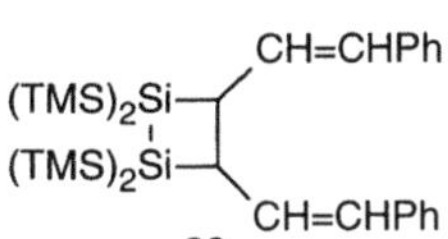

39

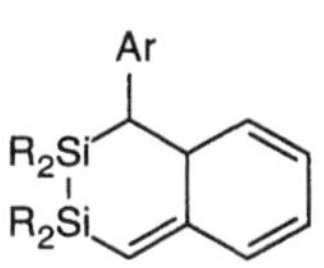

40

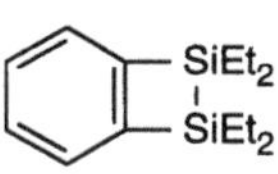

41

The dispiro-1,3-disilacyclobutanes **42** (n = 2 or 3) are available from **43** (n = 2 or 3) <96PO1545>.

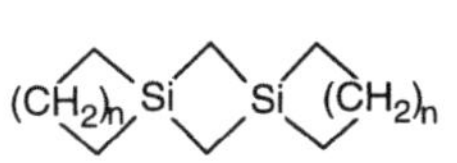

42

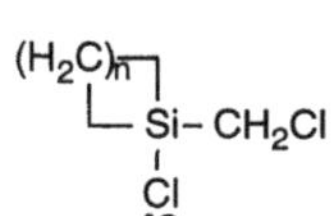

43

4.7 AZETIDINONES (β-LACTAMS)

The intramolecular cyclization route to β-lactams still provides interest. β-Amino esters (obtained by a Reformatsky-type reaction of an imine and bromoacetates derived from chiral alcohols) are cyclized by the action Grignard reagents to 4-substituted β-lactams with impressive e.e. <96TL4095>. A similar approach through a Reformatsky-type reaction uses tricarbonyl(η^6-benzaldimine)chromium complexes and ultrasound <96T4849>. 3-Methylazetidin-2-ones (obtained from 3-amino-2-methylpropionates) have been resolved and their absolute configuration established <96TA699>. Mitsunobu cyclization provides (*S,S*)-3-(1-prolyl)azetidin-2-one from 2-bromo-3-hydroxypropionamide <96TL965>. β-Aminoacids are cyclized in good to excellent yield in the presence of *N,N*-(diethoxyphosphinyl)benzo-1,2,5-thiadiazolidine 1,1-dioxide <96MI290> and 3,3-di-substituted β-lactams are available from 3-chloro-*N*-arylpropionamides by cyclization in a one-phase system of sodium carbonate in DMF <96JHC427>. Other 3-substituted β-lactams have been obtained from 3-amino-2-substituted propionic acid <96TL5565> and from chiral, non-racemic, sulfinylpropionamides **44** (R^1 = (*S*)-1-phenylethyl) by a highly asymmetric Pummerer-type cyclization in the presence of *O*-methyl-*O*-*tert*-butyldimethyl-silylketene acetal <95JCS(P1)2405> to give **45**. In the latter case, the stereoinduction is governed by the absolute configuration of the sulfoxide. Exceptional enantiocontrol (up to 97% e.e.) is obtained when chiral dirhodium(II) carboxamidates are present during the intramolecular C-H insertion reaction of *N*-(diazoacetyl)azacycloalkanes to give a bicyclic β-lactam <95SL1075>. *Trans*-4-acetoxy-3-substituted β-lactams are obtained in high yield from the 3-substituted compounds by the Murahashi procedure <96TL5565>. Details are provided for the solid-state, asymmetric photocyclization of α-oxoamides <96JCS(P2)61>.

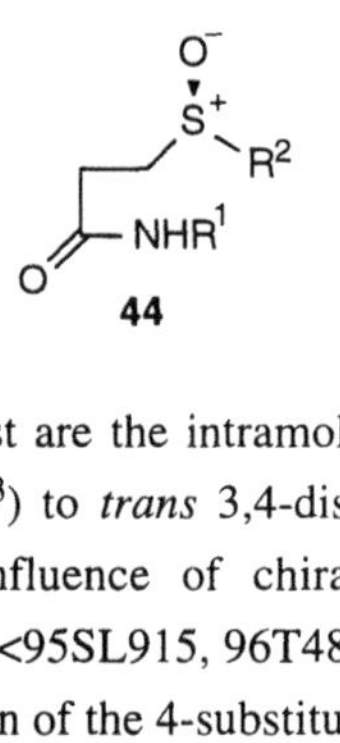

44

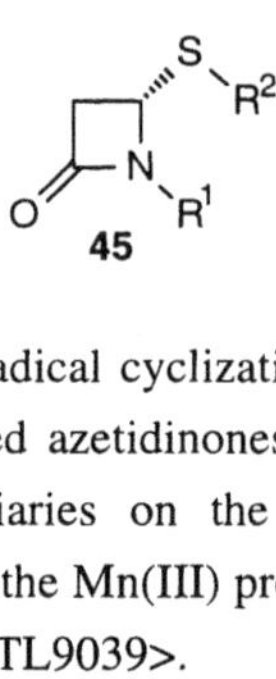

45

Of increasing interest are the intramolecular radical cyclizations of *N*-vinylacetamides ($R^1CH_2CONR^2CH:CHR^3$) to *trans* 3,4-disubstituted azetidinones <95SL912, 96TL1397>. Investigations of the influence of chiral auxilliaries on the stereocontrol of radical cyclizations are reported <95SL915, 96T489> as is the Mn(III) promoted radical cyclization and concomitant oxidation of the 4-substitutent <95TL9039>.

Lithium enolates from 2-fluoropropanethioates add to imines to give 4-substituted 3-fluoroazetidinones in good yield with at least 97% of the product being the *trans* isomer in the six cases reported <96T255>. New conditions for a Staudinger-type reaction include isolation of a 2-azabuta-1,3-diene and its thermal cyclization to give *trans* 1-unsubstituted 3-phthalimido-β-lactams with complete stereoselectivity but in yields which vary from poor to good <96TL4409>. A stereoselective one-pot synthesis of *trans*-3,4-disubstituted β-lactams from 2-pyridyl thioesters and imines in the presence of aluminium tribromide or ethyl aluminium dichloride is available <96T2583>. Lewis acids are used to promote a similarly stereoselective addition of silylketene thioacetals to imines <96T2573>.

The use of chiral auxilliaries in the Staudinger reaction has been explored extensively. Chiral imines derived from (1*S*)-(+)-camphor 10-sulfonic acid <96TA2733>,

(+)-(1*S*,2*S*)-2-amino-1-phenylpropane-1,3-diol <96T8989>, bicyclic terpenes <96T3741>, L-malic acid <95SL1067> and phenylethylamines <95TL8821> have been exploited, as have chiral oxazolidineacetyl chlorides <96AG(E)1239> and Oppolzer sultam substituted acetyl chlorides <96T5579> as the source of the ketene. Resin bound imines have been used in solid-supported combinatorial synthesis of β-lactams <96JA253>.

Degradation of bicyclic β-lactams has been used as a route to the monocyclic system. Highly functionalised β-lactams have been obtained in good yield by ozonolysis of Δ^2-cephems <96T10205> and enantiomerically pure 4-hydroxymethylazetidin-2-ones have been produced by degradation of a bicyclic oxazolidine, though in poor to moderate yield <96JCS(P1)227>.

The anomalous behaviour observed in attempted 1-deprotection of certain 4-heteroarylmethyl-1-(4-methoxyphenyl)-2-azetidinones by the action of CAN has been investigated further <96T771> and evidence obtained to support the mechanism proposed, which involves the intermediate **47** when the 4-substituent is the tetrazolylmethyl group as in **46** <96T10169>. Use of the novel enzyme, *o*-phthalyl amidase, as a deprotection agent for β-lactams has been developed <96MI875>.

The influence of the solvent on the basic hydrolysis of β-lactams has been studied by theoretical calculation and a close fit found with experimental results when the β-lactam is considered to be enclosed in a shell of 20 molecules of water <96HCA353>. The electronic and vibrational circular dichroism of 3-methyl- and 4-methyl-azetidin-2-one have been studied <96MI630>. There is remarkable difference in the ease of decarboxylation of the stereoisomers of the methyl malonate **48**. When R^1 = H and R^2 = Me, decarboxylation occurs at 80°C to give **49** but no reaction occurs at 80°C for **48** (R^1 = Me and R^2 = H) <95JOC8367>, and on further heating to 120°C the latter undergoes ring-opening and decomposition.

The β-thiolactams **50** are obtained by the action of Lawesson's or Davy's reagent on 3-substituted 1-methoxy- or 1-benzyloxy-azetidinones followed by reductive *N*-deprotection. The simple β-thiolactam can be modified to give the thiolactam analogues of monobactams <96LA141>.

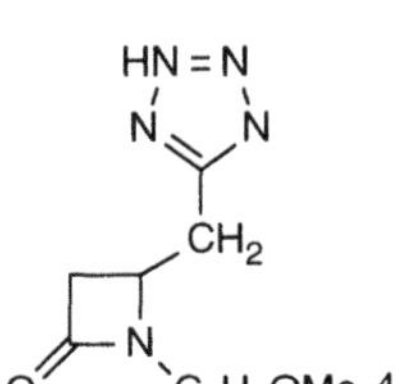

46

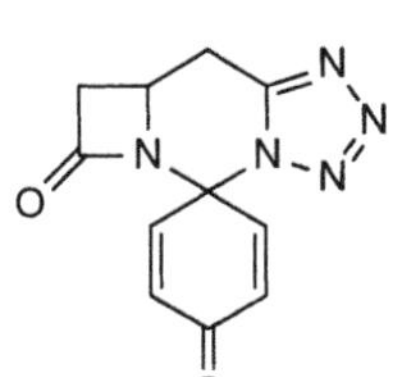

47

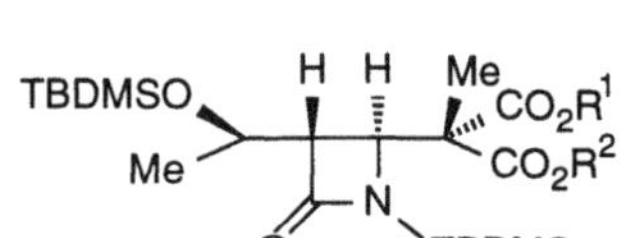

48

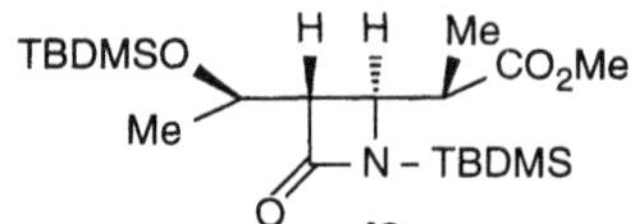

49

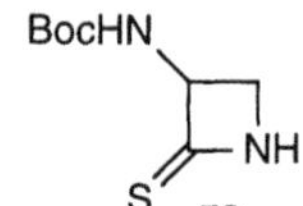

50

Fluoride ion induced desilylative α-hydroxylation of 3-alkylidene-4-trimethylsilylazetidinone with aldehydes (RCHO) to give **51** (R^3 = CH(OH)R) occurs in good yields <96CPB466>. A 4-acetoxy substituent **52** (R = OAc) may be replaced with high β-selectivity and high yield by reaction with the anion **53** and subsequent hydrolysis of the resultant imide with aqueous lithium hydroxide and hydrogen peroxide to give **52** (R = CH(Me)CO_2H) <96TL4967>.

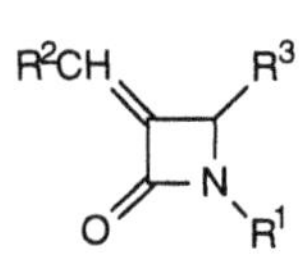

51

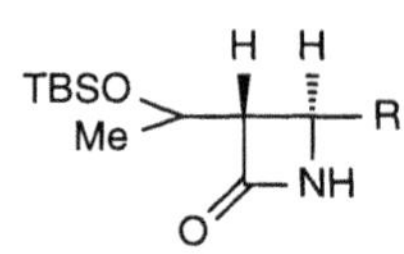

52

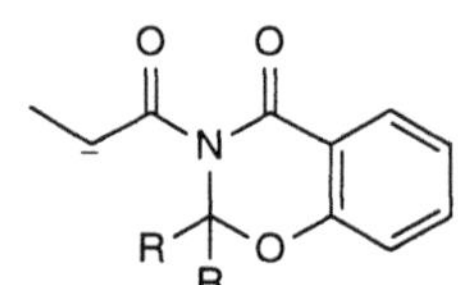

53

The anticancer activity of complex natural products having a cyclodecenediyne system [for a review see <96MI93>] has prompted the synthesis of **54** (X = CH_2 and OCH_2) <96CC749> and **55** (R = α-OH and β-OH) <95AG(E)2393> on the basis that such compounds are expected to develop anticancer activity as the β-lactam ring opens. This is because cycloaromatization can only occur in the monocyclic enediyne and the diradical intermediate in the cyclization is thought to be the cytotoxic species.

The synthesis of **54** (X = CH_2 or OCH_2) involved Pd(0) mediated coupling of 5-chloro-1-hydroxypent-4-ene-2-yne with **56** (R = H) to give **56** (R = CH:CH.C:CH_2OH), conversion of the alcohol to a chloride, and cyclization on to the β-lactam <96TL2475>.

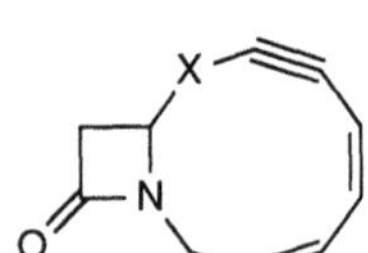

54

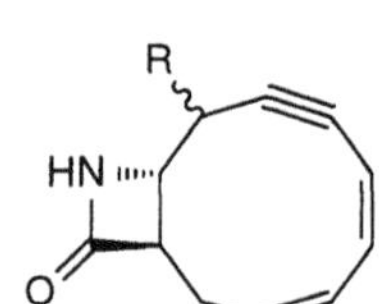

55

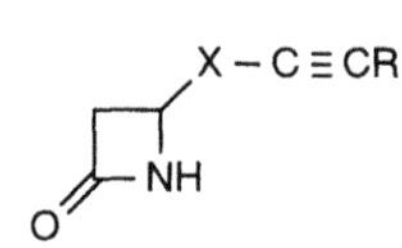

56

The *cis* β-lactams **57** are shown to act as cholesterol absorption inhibitors <96BMCL1947> and **58**, an analogue of the dipeptide Phe-Gly methyl ester, is a protease inhibitor <96BMCL983>. A straightforward synthesis of proclavaminic acid **59**, a biosynthetic precursor of clavulanic acid, is reported <96TA2277>.

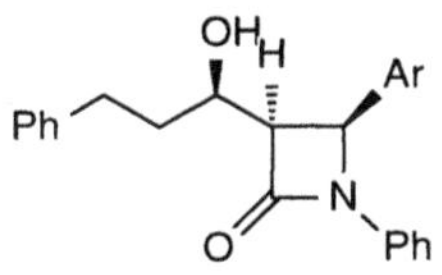

57

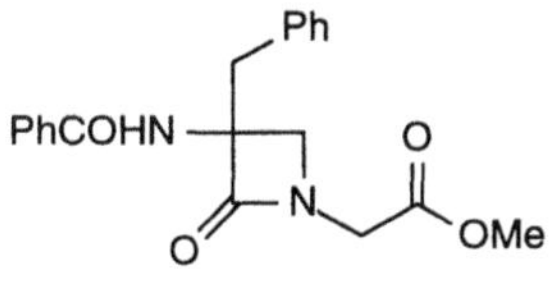

58

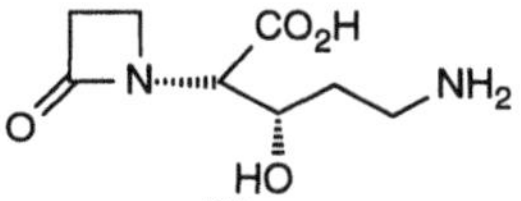

59

REFERENCES

94ZOR1700 M.V. Vovk, *Zh. Org. Khim.* **1994**, *30*, 1700.

94ZOR1704 G.G. Furin, *Zh. Org. Khim.* **1994**, *30*, 1704.

95AG(E)2028 C. Wedler, A. Kunath, H. Schick, *Angew. Chem., Int. Ed. Engl.* **1995**, *34*, 2028.

95AG(E)2271 T. Bach, K. Joedicke, K. Kather, J. Hecht, *Angew. Chem., Int. Ed. Engl.* **1995**, *34*, 2271.

95AG(E)2393 L. Banfi, G. Guanti, *Angew. Chem., Int. Ed. Engl.* **1995**, *34*, 2393.

95CB1083 C. Krempner, H. Reinke, H. Oehme, *Chem. Ber.* **1995**, *128*, 1083.

95CL789 T. Uyehara, M. Yuuki, H. Masaki, M. Matsumoto, M. Ueno, T. Sato, *Chem. Lett.* **1995**, 789.

95CC1735 C. Palomo, J.I. Miranda, C. Cuevas, J.M. Odriozola, *J. Chem. Soc., Chem. Commun.* **1995**, 1735.

95JCS(P1)2405 Y. Kita, N. Shibata, N. Kawano, T. Tohjo, C. Fujimori, K. Matsumoto, S. Fujita, *J. Chem. Soc., Perkin Trans. 1* **1995**, 2405.

95JCS(P1)2605 U.K. Nadir, A. Arora, *J. Chem. Soc., Perkin Trans. 1* **1995**, 2605.

95JOC8367 W.-B. Choi, H.R.O. Churchill, J.E. Lynch, R.P. Volante, P.J. Reider, I. Shinkai, D.K. Jones, D.C. Liotta, *J. Org. Chem.* **1995**, *60*, 8367.

95JOM35 A. Naka, S. Okazaki, M. Hayashi, M. Ishikawa, *J. Organomet. Chem.* **1995**, *499*, 35.

95LA2221 A. Mayer, N. Rumpf, H. Meier, *Liebigs Ann.* **1995**, 2221.

95MI95 I. Ojima, *Adv. Asymmetric Synth.* **1995**, *1*, 95.

95MI209 T. Harrison, *Contemp. Org. Synth.* **1995**, 2, 209.

95MI330 C.J. Schofield, N.J. Westwood, *Amino-Acids, Pept., Proteins* **1995**, *26*, 330.

95MI727 J.T. Welch, R. Kawecki, *Stud. Nat. Prod. Chem.* **1995**, *16*, 727.

95RTC498 D. Doepp, M.-A. Fischer, *Recl. Trav. Chim. Pays-Bas* **1995**, *114*, 498.

95SL912 H. Ishibashi, K. Kodama, C. Kameoka, H. Kawanami, M. Ikeda, *Synlett* **1995**, 912.

95SL915 H. Ishibashi, C. Kameoka, K. Kodama, M. Ikeda, *Synlett* **1995**, 915.

95SL1067 T. Fujisawa, A. Shibuya, D. Sato, M. Shimizu, *Synlett* **1995**, 1067.

95SL1075 M.P. Doyle, A.V. Kalinin, *Synlett* **1995**, 1075.

95TL8583 J. Nakayama, A. Mizumura, Y. Yokomori, A. Krebs, K. Schuetz, *Tetrahedron Lett.* **1995**, *36*, 8583.

95TL8821 Y. Hashimoto, A. Kai, K. Saigo, *Tetrahedron Lett.* **1995**, *36*, 8821.

95TL9039 A. D'Annibale, S. Resta, C. Trogolo, *Tetrahedron Lett.* **1995**, *36*, 9039.

95TL9201 J. Helinski, Z. Skrzypczynzki, J. Michalski, *Tetrahedron Lett.* **1995**, *36*, 9201.

95ZOR589 V.M. Neplyuev, I.M. Bazavova, A.N. Esipenko, M.O. Lozinskii, *Zh. Org. Khim.* **1995**, *31*, 589.

96ACR144 E.M. Gordon, M.A. Gallop, D.V. Patel, *Acc. Chem. Res.* **1996**, 144.

96AG(E)884 T. Bach, *Angew. Chem., Int. Ed. Engl.* **1996**, *35*, 884.

96AG(E)1096 T. Kawashima, T. Soda, R. Okazaki, *Angew. Chem., Int. Ed. Engl.* **1996**, *35*, 1096.

96AG(E)1239 C. Palomo, J.M. Aizpurua, M.Legido, R. Galarza, P.M. Deya, J.Dunogues, J.P. Picard, A. Ricci, G. Seconi, *Angew. Chem., Int. Ed. Engl.* **1996**, *35*, 1239.

96BMCL983 Z. Wu, G.I. Georg, B.E. Cathers, J.V. Schloss, *Bioorg. Med. Chem. Lett.* **1996**, *6*, 983.

96BMCL1947 B.A. McKittrick, Ke Ma, S. Dugar, J.W. Clader, H. Davis Jr., M. Czarniecki, A.T. McPhail, *Bioorg. Med. Chem. Lett.* **1996**, *6*, 1947.

96BSB253 G. L'abbe, A. Francis, W. Dehaem, J. Bosman, *Bull. Soc. Chim. Belg.* **1996**, *105*, 253.

96CB15 F. Luderer, H. Reinke, H. Oehme, *Chem. Ber.* **1996**, *129*, 15.

96CC749 A. Basak, U.K. Khamrai, U. Mallik, *J. Chem. Soc., Chem. Commun.* **1996**, 749.

96CC1349 M. Sakamoto, T. Sano, M. Takahashi, K. Yamaguchi, T. Fujita, S. Watanabe, *J. Chem. Soc., Chem. Commun.* **1996**, 1349.

96CL827 L. Zhao, H. Matsuyama, M. Iyoda, *Chem. Lett.* **1996**, 827.

96CPB466 K. Hotoda, M. Aoyagi, T. Takanami, K. Suda, *Chem. Pharm. Bull.* **1996**, *44*, 466.

96IC2458 G. Alcaraz, A. Baceiredo, M. Nieger, W.W. Schoeller, G. Bertrand, *Inorg. Chem.* **1996**, *35*, 2458.

96H1211 K.S. Yoon, S.J. Lee, K. Kim, *Heterocycles* **1996**, *43*, 1211.

96HCA353 J. Frau, J. Donoso, F. Munoz, F. Garcia-Blanco, *Helv. Chim. Acta* **1996**, *79*, 353.

96JA253 B. Ruhland, A. Bhandari, E.M. Gordon, M.A. Gallop, *J. Am. Chem. Soc.* **1996**, *118*, 253.

96JA1060 G. Alcaraz, V. Piquet, A. Baceiredo, F. Dahan, W.W. Schoeller, G. Bertrand, *J. Am. Chem. Soc.* **1996**, *118*, 1060.

96JCS(P1)227 M.D. Andrews, A.G. Brewster, M.G. Moloney, K.L. Owen, *J. Chem. Soc., Perkin Trans. 1* **1996**, 227.

96JCS(P2)61 D. Hashizume, H. Kogo, A. Sekine, Y. Ohashi, H. Miyamoto, F. Toda, *J. Chem. Soc., Perkin Trans. 2* **1996**, 61.

96JHC427 J. C. Lancelot, C. Saturnino, H. El-Kashef, D. Perrine, C. Mahatsekake, H. Prunier, M. Robba, *J. Heterocycl. Chem.* **1996**, *33*, 427.

96JOC4466 A. Mordini, S. Bindi, S. Pecchi, A. Capperucci, A. Degl'Innocenti, G. Reginato, *J. Org. Chem.* **1996**, *61*, 4466.

96JOC5453 P.R. Dave, *J. Org. Chem.* **1996**, *61*, 5453.

96JOM93 R. Szalay, Zs. Boecskei, D. Knausz, Cs. Lovasz, K. Vjszaszy, L. Szakacs, P. Sohar, *J. Organomet. Chem.* **1996**, *510*, 93.

96JPO619 A. Oku, T. Miki, Y. Ose, *J. Phys. Org. Chem.* **1996**, *9*, 619.

96JPR383 H. Meier, *J. Prakt. Chem./Chem.-Ztg.* **1996**, *338*, 383.

96LA141 J. Nieschalk, E. Schaumann, *Liebigs Ann.* **1996**, 141.

96LA881 C. Wedler, K. Kleiner, A. Kunath, H. Schick, *Leibigs Ann.* **1996**, 881.

96MI41 M. Backer, M. Grasmann, W. Ziche, N. Auner, C. Wagner, E. Herdtweck, W. Hiller, M. Heckel in *Organosilicon Chem. II*, [Muench. Silicontage], N. Auner, J. Weis eds.; VCH, Weinheim, Germany, **1994** (Pub. **1996**), 41.

96MI51 G. Roso-Levi, I. Amer, *J. Mol. Catal. A: Chem.* **1996**, *106*, 51.

96MI93 H. Lhermitte, D.S. Grierson, *Contemp. Org. Synth.* **1996**, *3*, 93.

96MI101 P.R. Guzzo, M.J. Miller in *Adv. Nitrogen Heterocycles,* C.J. Moody ed.; JAI Press Inc., Greenwich, Connecticut, USA and London, **1996**, *2*, 1.

96MI102 Several Authors, *Comprehensive Heterocyclic Chemistry* A.R. Katritzky, C.W. Rees, E.F.V. Scriven eds.; Elsevier, Amsterdam, **1996**, *1B*.

96MI259 T. Harrison, *Contemp. Org. Synth.* **1996**, *3*, 259.

96MI290 Y.H. Lee, C.-H. Lee, W.S. Choi, *Bull. Korean Chem. Soc.* **1996**, *17*, 290.

96MI630 J. McCann, A. Rauk, G.V. Shustov, H. Wieser, D. Yang, *Appl. Spectrosc.* **1996**, *50*, 630.

96MI801 C. Wendler, H. Oehme, *Z. Anorg. Allg. Chem.* **1996**, *622*, 801.

96MI875 T.D. Black, B.S. Briggs, R. Evans, W.L. Muth, S. Vangala, M.J. Zmijewski, *Biotechnol. Lett.* **1996**, *18*, 875.

96OS61 R.L. Danheiser, J.S. Nowick, J.H. Lee, R.F. Miller, A.H. Huboux, *Org. Synth.* **1996**, *73*, 61.

96PO1545 A.J. Kotlar, W.A. Kriner, T.E. Mueller, L.A. Falvello, *Polyhedron* **1996**, *15*, 1545.

96S265 F. Knoch, S. Kummer, U. Zenneck, *Synthesis* **1996**, 265.

96S586 C. Zhang, X. Lu, *Synthesis* **1996**, 586.

96T255 T. Ishihara, K. Ichihara, H. Yamanaka, *Tetrahedron* **1996**, *52*, 255.

96T331 A.H. Berks, *Tetrahedron* **1996**, *52*, 331.

96T489 H. Ishibashi, C. Kameoka, K. Kodama, M. Ikeda, *Tetrahedron* **1996**, *52*, 489.

96T771 A. Sapi, F. Bertha, J. Fetter, M. Kajtar-Peredy, G.M. Keseru, K. Lempert, *Tetrahedron* **1996**, *52*, 771.

96T2573 R. Annunziata, M. Cinquini, F. Cozzi, V. Molteni, O. Schupp, *Tetrahedron* **1996**, *52*, 2573.

96T2583 R. Annunziata, M. Benaglia, M. Cinquini, F. Cozzi, O. Martini, V. Molteni, *Tetrahedron* **1996**, *52*, 2583.

96T3741 M. Jayaraman, V. Srirajan, A.R.A.S. Deshmukh, B.M. Bhawal, *Tetrahedron* **1996**, *52*, 3741.

96T4849 C. Baldoli, P. Del Buttero, E. Licandro, A. Papagni, T. Pilati, *Tetrahedron* **1996**, *52*, 4849.

96T5579 V. Srirajan, V.G. Puranik, A.R.A.S. Deshmukh, B.M. Bhawal, *Tetrahedron* **1996**, *52*, 5579.

96T7183 F. Clerici, F. Galletti, D. Pocar, P. Roversi, *Tetrahedron* **1996**, *52*, 7183.

96T8989 M. Jayaraman, A.R. Deshmukh, B.M. Bhawal, *Tetrahedron* **1996**, *52*, 8989.

96T10169 LeT. Giang, J. Fetter, K. Lempert, M. Kajtar-Peredy, A. Gomory, *Tetrahedron* **1996**, *52*, 10169.

96T10205 M. Botta, M. Crucianelli, R. Saladino, C. Mozzetti, R. Nicoletti, *Tetrahedron* **1996**, *52*, 10205.

96T10861 T. Bach, K. Joedicke, B. Wibbeling, *Tetrahedron* **1996**, *52*, 10861.

96TA699 G.V. Shustov, A. Rauk, *Tetrahedron: Asymmetry* **1996**, *7*, 699.

96TA2277 M.C. Di Giovanni, D. Misiti, C. Villani, G. Zappia, *Tetrahedron: Asymmetry* **1996**, *7*, 2277.

96TA2733 V. Srirajan, A.R.A.S. Deshmukh, V.G. Puranik, B.M. Bhawal, *Tetrahedron: Asymmetry*, **1996**, *7*, 2733.

96TL245 I. Arrastia, B. Lecea, F.P. Cossio, *Tetrahedron Lett.* **1996**, *37*, 245.

96TL965 M.N. Qabar, M. Kahn, *Tetrahedron Lett.* **1996**, *37*, 965.

96TL1397 B. Quiclet-Sire, J.-B. Saunier, S.Z. Zard, *Tetrahedron Lett.* **1996**, *37*, 1397.

96TL2475 A. Basak, U.K. Khamrai, *Tetrahedron Lett.* **1996**, *37*, 2475.

96TL4095 B.B. Shankar, M.P. Kirkup, S.W. McCombie, J.W. Calder, A.K. Ganguly, *Tetrahedron Lett.* **1996**, *37*, 4095.

96TL4409 E. Bandini, G. Martelli, G. Spunta, A. Bongini, M. Panunzio, *Tetrahedron Lett.* **1996**, *37*, 4409.

96TL4967 T. Yamanaka, M. Seki, T. Kuroda, H. Ohmizu, T. Iwaski, *Tetrahedron Lett.* **1996**, *37*, 4967.

96TL5565 M. Seki, T. Yamanaka, T. Miyake, H. Ohmizu, *Tetrahedron Lett.* **1996**, *37*, 5565.

96TL5917 K. Kubo, S. Yaegashi, K. Sasaki, T. Sakurai, H. Inoue, *Tetrahedron Lett.* **1996**, *37*, 5917.

96TL5939 M. Matsumoto, N. Watanabe, H. Kobayashi, H. Suganuma, J. Matsubara, Y. Kitani, H. Ikawa, *Tetrahedron Lett.* **1996**, *37*, 5939.

96ZN(B)370 D. Hoffmann, H. Reinke, H. Oehme, *Z. Naturforsch., B: Chem. Sci.* **1996**, *51*, 370.

96ZN(B)1325 H.D. Stachel, H. Poschenrieder, J. Redlin, *Z. Naturforsch., B: Chem. Sci.* **1996**, *51*, 1325.

Chapter 5.1

Five-Membered Ring Systems: Thiophenes & Se, Te Analogs

Jeffery B. Press
Galenica Pharmaceuticals, Inc., Frederick, MD, USA

Erin T. Pelkey
Dartmouth College, Hanover, NH, USA

5.1.1 INTRODUCTION

Thiophenes continue to play a major role in commercial applications as well as basic research. In addition to its aromatic properties that make it a useful replacement for benzene in small molecule syntheses, thiophene is a key element in superconductors, photochemical switches and polymers. The presence of sulfur-containing components (especially thiophene and benzothiophene) in crude petroleum requires development of new catalysts to promote their removal (hydrodesulfurization, HDS) at refineries. Interspersed with these commercial applications, basic research on thiophene has continued to study its role in electrocyclic reactions, newer routes for its formation and substitution and new derivatives of therapeutic potential. New reports of selenophenes and tellurophenes continue to be modest in number.

This chapter follows the organization used in the past. A summary of the electronic properties leads into reports of electrocyclic chemistry. Recent reports of studies of HDS processes and catalysts are then summarized. Thiophene ring substitution reactions, ring-forming reactions, the formation of ring-annelated derivatives, and the use of thiophene molecules as intermediates are then reported. Applications of thiophene and its derivatives in polymers and in other small molecules of interest are highlighted. Finally, the few examples of selenophenes and tellurophenes reported in the past year are noted.

The literature search for this chapter has uncovered a typically large number of references for papers published in 1996. These references were culled to remove repetition and to provide the most interesting examples of thiophene chemistry and its applications. As in the past, the authors apologize for omissions and oversights.

One review covering the synthesis, chemistry and biological properties of thienopyrmidines has appeared <96AHC193>.

5.1.2 ELECTRONICS, AROMATICITY, CYCLOADDITIONS AND DESULFURIZATION (HDS)

The unique electronic and aromatic properties of the thiophene ring produce properties that make thiophene derivatives important for commercial applications. A new method to estimate the aromaticity of five-membered heterocycles using hybrid B3LYP density functional theory gives predictions in good accord with experimental results <96JHC1079>. CASSCF structures and vibrational frequencies of 2,3- and 3,4-didehydrothiophene indicate that the π-bonds are biradical in character <96JCR(S)458. Kerr dynamic studies of thiophene show an unusual increase in relaxation suggestive of complex formation between thiophene and carbon tetrachloride <96CPL329>. Semiempirical calculations of oligothiophenes for ground, singlet and triplet excited states show that Franck–Condon factors are not important in intersystem crossing <96CPL31>. Conformers of 2,2':5',2"-terthiophene were determined by ab initio MO full geometry calculations and interconversion rates were estimated <96CPL73>.

Rotational equilibria of 2-carbonyl substituted thiophene and furan derivatives were calculated and show that the 2-substituent favors the anti-isomer in thiophene <96MI199>. ^{13}C NMR shifts of 35 alkyl 3-hydroxythiophene-2-carboxylates and 3-alkylamino-1-(3-thienyloxy)-2-propanols have been compiled and analyzed <96HC17>.

Oxidation of thiophene with Fenton-like reagents produces 2-hydroxythiophene of which the 2(5*H*)-one isomer is the most stable (Eq. 1) <96JCR(S)242>. In contrast, methyltrioxorhenium (VII) catalyzed hydrogen peroxide oxidation of thiophene and its derivatives forms first the sulfoxide and ultimately the sulfone derivatives <96IC7211>. Anodic oxidation of aminated dibenzothiophene produces stable radical cation salts <96BSF597>. Reduction of dihalothiophene at carbon cathodes produces the first example of an electrochemical halogen dance reaction (Eq. 2) <96JOC8074>.

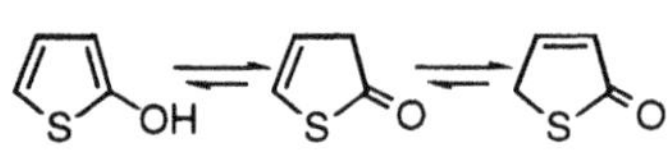

(Eq. 1)

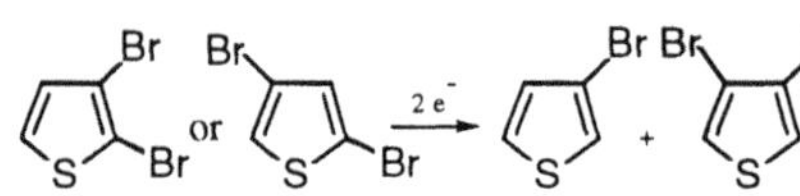

(Eq. 2)

Thiophene can be induced to undergo [4+2] or [2+2] cycloadditions as well as electrocyclizations by proper functionalization. While thiophene itself is relatively inert to Diels-Alder reactions, activation by formation of a sulfoxide with Lewis Acids (Eq. 3) or *S*-methyl thiophenium ion (Eq. 4) allows cycloaddition reactions to occur <96SL461, 96JCS(P2)455>. Substitution of thiophene by 3-acetyl (Eq. 5) or 2-benzoyl moieties (Eq. 6) activates the ring to photochemical cycloaddition reactions <96LA697, 96M529>. Unactivated thiophene undergoes cycloaddition reaction with 1,2-dihydro-*o*-carborane with concomitant loss of sulfur under relatively severe conditions to form benzo-*o*-carboranes <96IC7311>. Thiophenes activated by ring fusion also undergo Diels-Alder reaction; thieno[3,4-*c*]quinoline and its pyranone isostere react with a variety of electron-poor olefins <96T11915>. Electroactive diarylethenes act as photoswitches by undergoing photo-reversible ring closure (Eq. 7) <96CL817, 96MI1399>.

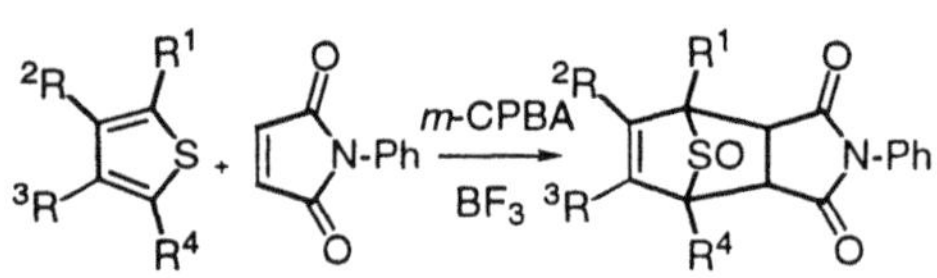

(Eq. 3)

(Eq. 4)

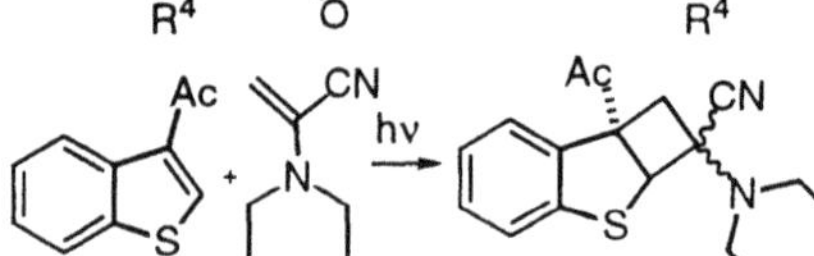

(Eq. 5)

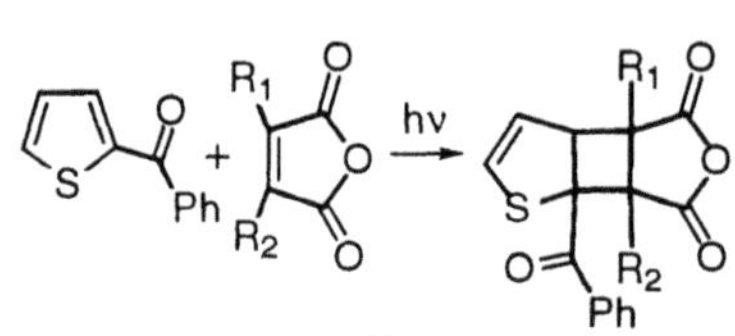

(Eq. 6)

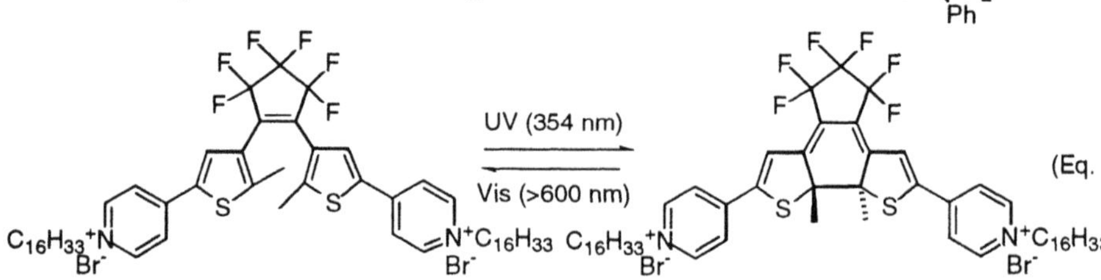

(Eq. 7)

The study of hydrodesulfurization (HDS) continues to find new transition metal catalysts and new intermediates. The variety of metal complexes includes rhodium, iridium, iron, manganese and ruthenium. Irradiation of *trans*-RH(PMe)$_2$(CO)Cl with 2,5-dimethylthiophene produces C–H insertion products. On the other hand, rhodium phosphine complexes insert in the C–S bond of dibenzothiophenes (Eq. 8) <96O2905>. Reaction of Tp*Rh(C_2H_4)(PMe_3) with thiophene leads to mixture of C–H and C–S insertion <96O2679>. Homogeneous HDS of benzo[*b*]thiophene uses Rh as a metal promoter <96AC(E)1706>. Soluble rhodium and iridium complexes react with thiophene and benzothiophene derivatives to give C–S insertion (Eq. 9) <96O4604, 96O1223>. Reaction of Cp*Ir(η^4-2,5-Me_2T)$^{+2}$ gives coordination through sulfur <96JOM21, 96O2727>.

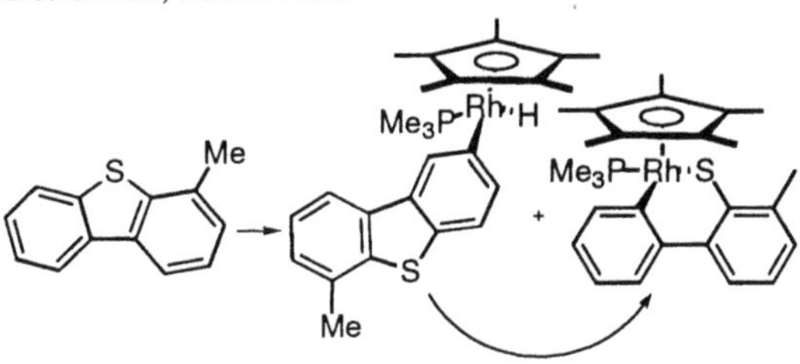

(Eq. 8)

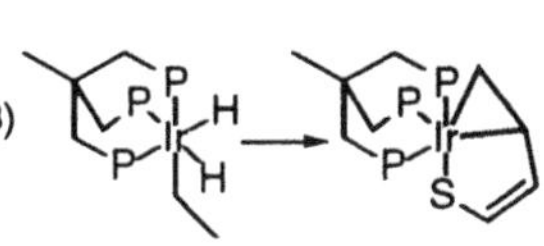

(Eq. 9)

Diphenyl-2-thiophene reacts with tetraruthenium clusters via cyclometalation to produce **1** <96O786>. Manganese inserts into the S–aryl bond as a model for HDS (Eq. 10) <96AC(E)212, 96O325, 96JOM149>. Iron complexes of thiophene have also been reported <96O4352, 96MI2825>. Tetrakis(2-thienyl)borate (**2**) does not readily coordinate with Mo(CO)$_3$ in contrast to polythioether analogues <96IC7095>. Cracking of thiophene derivatives by palladium <96CL829>, ruthenium <96CL743>, molybenum <96MI135>, cobalt <96MI212>, and cobalt molybdenum <96MI1078, 96MI236> on zeolite catalysts has been reported.

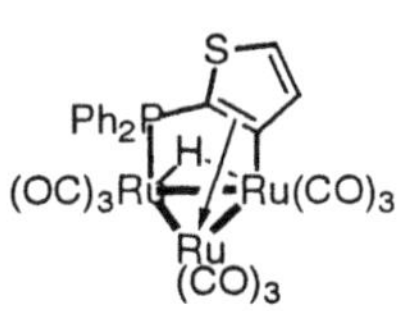

1

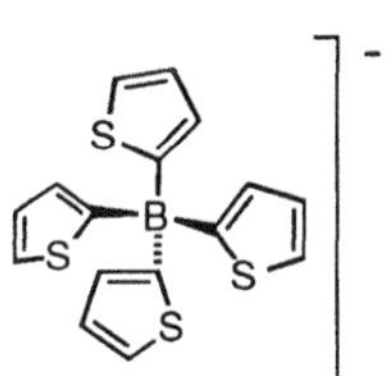

2

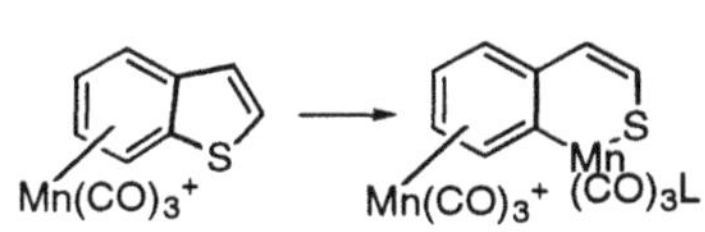

(Eq. 10)

5.1.3 RING SUBSTITUTION

The thiophene ring can be elaborated using standard electrophilic, nucleophilic, and organometallic chemistry. A variety of methods have been developed to exploit the tendency for the thiophene ring (analogous to that of furan and pyrrole) to favor electrophilic substitution and metallation at its α-carbons. Substitution at the β-carbons is more challenging, but this problem can also be solved by utilizing relative reactivity differences.

Although less electron rich than its heteroaromatic counterparts furan and pyrrole, thiophene easily undergoes a variety of electrophilic substitution reactions under mild conditions. Treatment of thiophene with $SnCl_4$/Pb(OAc)$_4$ gives **3** in 75% yield <96T8863>. Tetrathiomethylation of thiophene is achieved with excess dimethyl disulfide in the presence of $ZnCl_2$-modified montmorillonite clay catalyst to give **4** in 50% isolated yield. In a similar manner, benzo[*b*]thiophene is converted to 2,3-bis(methylthio)benzo[*b*]thiophene in 73% yield which may be followed by oxidation to the disulfone with potassium permanganate (Eq. 11) <96PS99>. Treatment of 3-bromothiophene with chlorosulfonic acid produces a 1:10 mixture of 2,4- and 2,3-substituted thiophenes with the latter being utilized as a building block for the synthesis of endothelin receptor antagonists (Eq. 12) <96BML2651>. Synthesis of β-substituted thiophenes is achieved utilizing the ability of silicon to direct ipso substitution. Selective mono-ipso-iodination of 3,4-bis(trimethylsilyl)thiophene forms 3-iodo-4-trimethylsilylthiophene in 96% yield which is utilized in an iterative fashion to synthesize a variety of 3,4-substituted thiophenes (Eq. 13) <96CC339>.

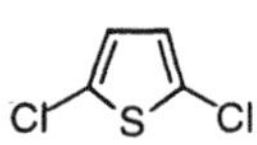

3

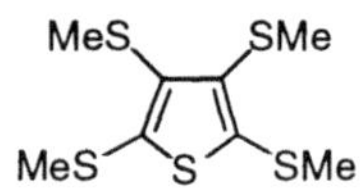

4

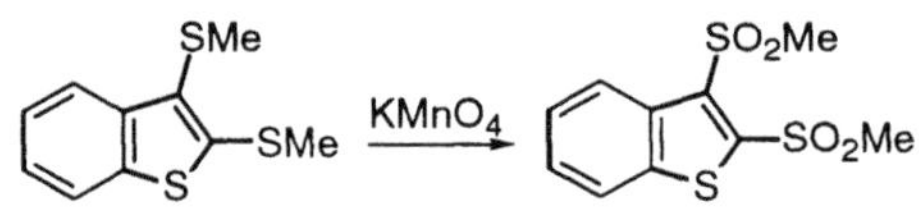

(Eq. 11)

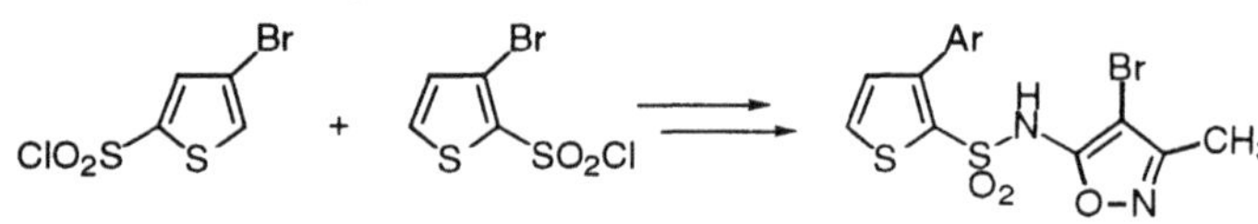

(Eq. 12)

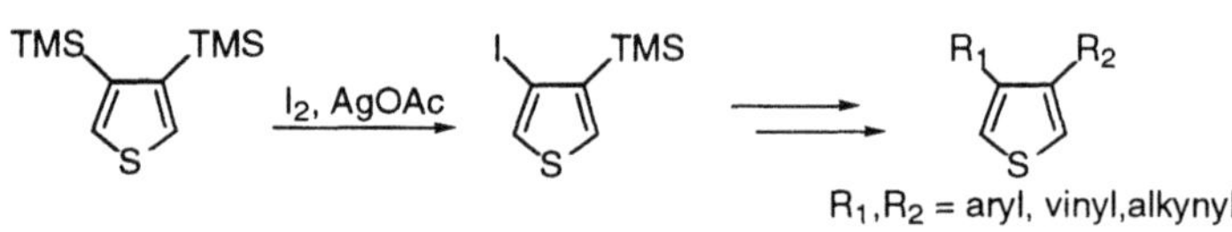

(Eq. 13)

An alternative method to achieving selective β-substitution of thiophenes involves excessive α-substitution followed by removal of the α-substituents. Tetrabromination of 2,2'-bithiophene with four equivalents of bromine in acetic acid followed by selective α-debromination with zinc dust gives 3,3'-2,2'-bithiophene (Eq. 14) <96JCR(S)232>. Tetraselenation of thiophene with 7.5 equivalents of benzeneselenyl sulfate followed by selective α-hydroselenation with two equivalents of *n*-BuLi forms 3,4-diphenylselenylthiophene in 68% overall yield (Eq. 15) <96H861>. Readily available via the Hinsberg thiophene synthesis, 3,4-dimethoxythiophenedicarboxylic acid can be decarboxylated by simple heating to give 3,4-dimethoxythiophene in 65% yield (Eq. 16) <96MI672>. In a similar fashion or by heating with copper/quinoline, **5** and **6** can be obtained from the corresponding carboxylic acids <96SC2205, 96H1767>. Finally, regioselective debromination of 2,3,5-tribromothiophene with sodium borohydride has been optimized to give either 2,3-dibromothiophene or 2,4-dibromothiophene depending on the presence or absence of a palladium catalyst (Eq. 17) <96JCR(S)150>.

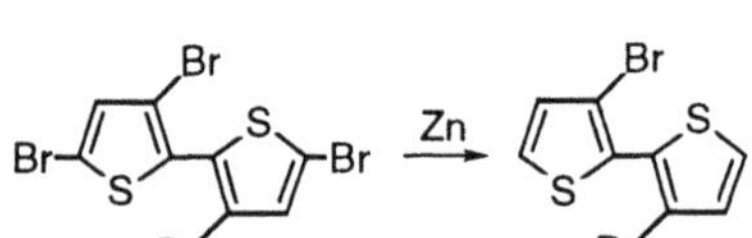

(Eq. 14)

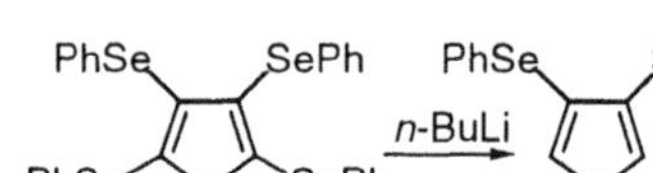

(Eq. 15)

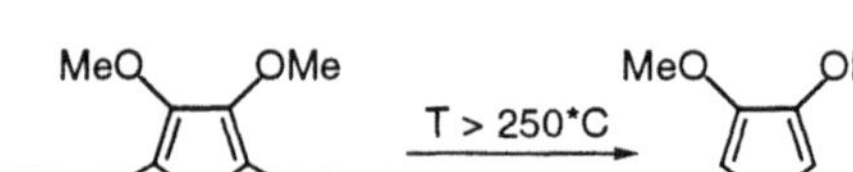

(Eq. 16)

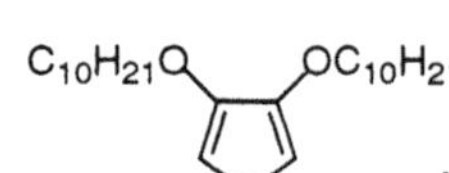

5

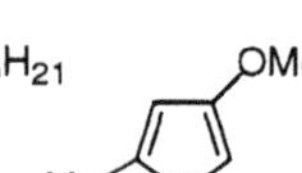

6

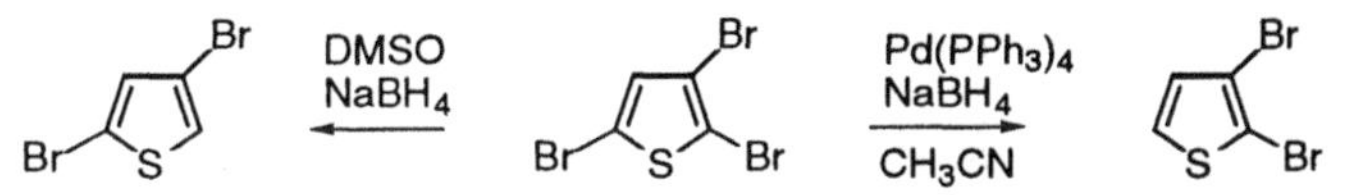

(Eq. 17)

Thiophenes containing electron withdrawing groups (i.e., nitro) are capable of reacting with nucleophiles. Vicarious nucleophilic substitution (VNS) of 2-nitrothiophene with the anion derived from chloroform followed by formic acid hydrolysis produces 2-nitrothiophene-3-carboxaldehyde, which is utilized as a key starting material in the synthesis of complex tetrazoles (Eq. 18) <96P396>. The anion derived from ethyl isocyanoacetate is capable of adding in a Michael fashion to 3-nitrobenzo[*b*]thiophene to give a 60% yield of ethyl 2*H*-benzothieno[2,3-*c*]pyrrole-2-carboxylate (Eq. 19) <96JCS(P1)417>. The resulting fused pyrrole was utilized in the synthesis of a novel benzo[*b*]thiophene containing porphyrin. Alternatively, the addition of ethyl isocyanoacetate to 2-nitrothiophene unexpectedly gives the fused pyrimidine **7** in 40% yield <96JCS(P1)1403>.

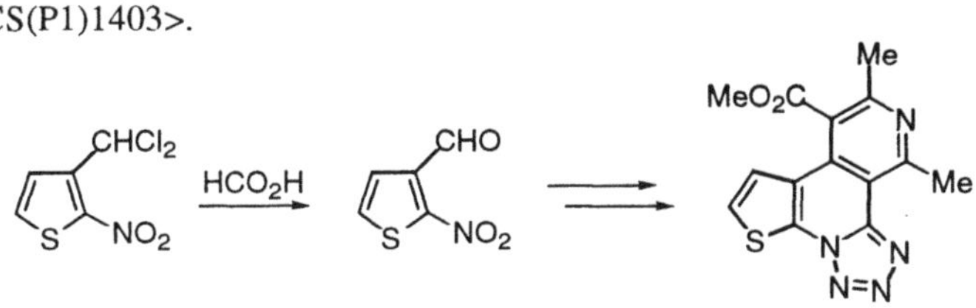

(Eq. 18)

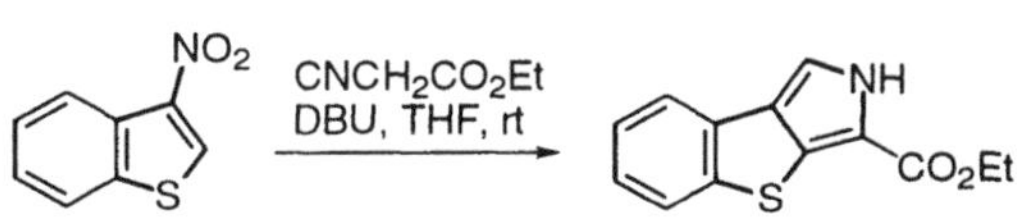

(Eq. 19)

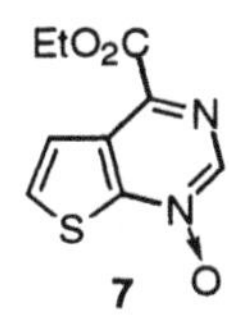

7

An effective method for preparing α-substituted thiophenes is direct deprotonation with alkyllithiums followed by quenching with various electrophiles. Generation of lithium 2-thiophenethiolate by adding sulfur to 2-lithiothiophene followed by trapping with (*R*)-β-butyrolactone gave the corresponding carboxylic acid which was cyclized in the presence of trifluoroacetic anhydride (Eq. 20) <96TA2721>. Quenching 2-lithiothiophene with 2,2-dichlorohexamethyltrisilane forms dithiophene **8** in 80% yield <96HA45>. Deprotonation of the thiophene ring followed by sequential quenching with sulfur or selenium and alkyl halides occurs preferentially in the presence of a fused pyridine to give **9** in 81% yield and in the presence of a pyridyl substituent to give **10** in 65-75% yield <96JCS(P2)1377>.

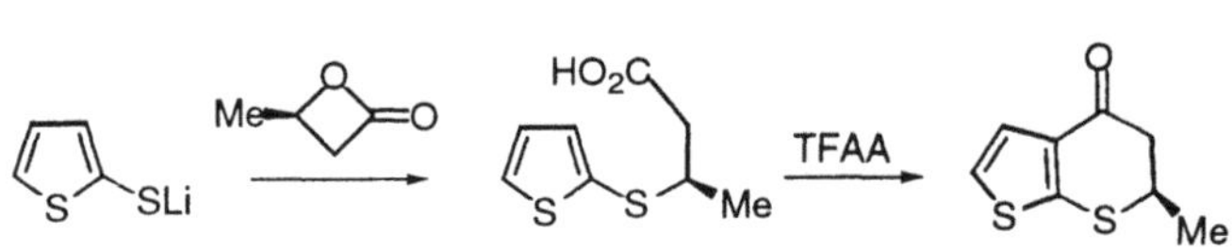

(Eq. 20)

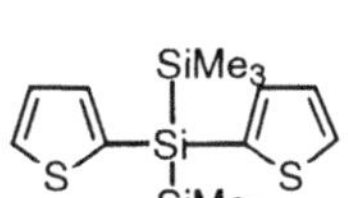

8

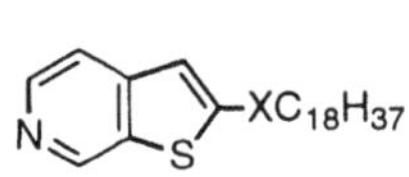

9 X=S,Se

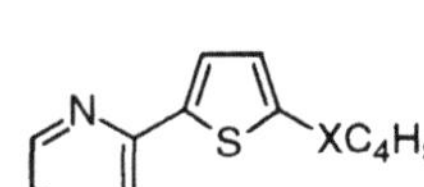

10 X=S,Se

Preparation of both α- and β-substituted thiophenes can be achieved by halogen-metal exchange. The synthesis of 3,4-bis(isopropylthio)thiophene is achieved in 60-70% yield from

3,4-dibromothiophene by two successive halogen-lithium exchanges each followed by quenching with isopropyl disulfide <96JOC4833>. Halogen-lithium exchange of 2,3-dibromothiophene occurs preferentially at the 2-position. This reactivity difference is exploited to regiospecifically prepare various 2,3-disubstituted thiophenes by successive halogen-lithium exchanges (Eq. 21) <96JCS(P1)963>. Direct deprotonation of a lactam-containing thiophene followed by quenching with carbon dioxide gives the undesirable thiophene-4-carboxylic acid as the major product (Eq. 22) <96SC1363>. The corresponding thiophene-5-carboxylic acid is obtained in 50% overall yield by electrophilic bromination, halogen-lithium exchange, and quenching with carbon dioxide.

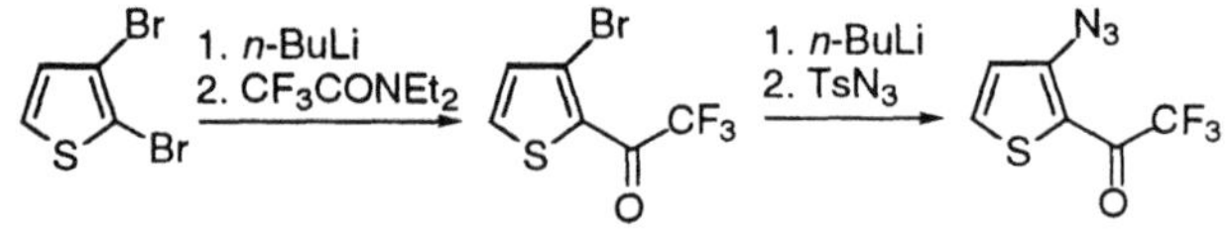

(Eq. 21)

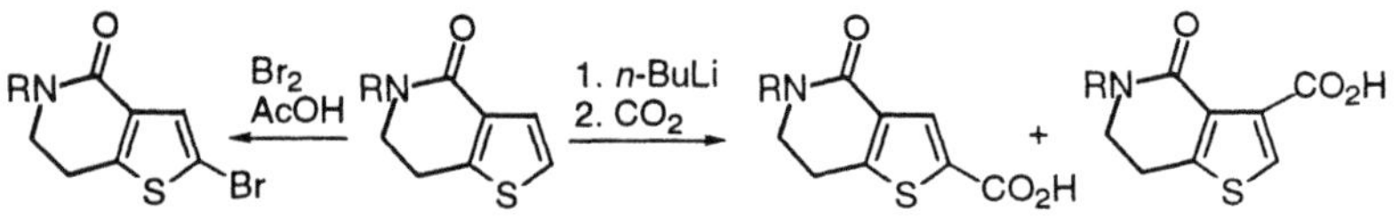

(Eq. 22)

Finally, organometallic cross-coupling continues to be one of the most powerful methods of synthesizing highly substituted thiophenes. Heteroatom-substituted thiophenes can be made by coupling alcohols, thiols, and amines with halothiophenes in the presence of copper. A variety of 3,3'-substituted-2,2-bithiophenes **11** have been prepared utilizing copper-catalyzed coupling <96JCR(S)232>. A stoichiometric amount of copper is required to synthesize 2-diphenylaminothiophene in 40% yield, which is transmetallated into a stannane (Eq. 23) <96JOC2242>. Stille cross-coupling reactions produce quinone **12** <96CL139> and naphthyridine **13** <96MI411>. A new method for coupling vinyl iodides and arylstannanes with copper(I) thiophene-2-carboxylate is utilized to synthesize bithiophene **14** in 89% yield (Eq. 24) <96JA2748>. Isoquinoline **15** is produced by the nickel-catalyzed cross coupling of 2-thienylmagnesium bromide with 3-bromoisoquinoline <96JHC1123>. Allylation followed by palladium-catalyzed Heck cyclization of *N*-boc-4-iodo-3-aminothiophene gives a 58% yield of a thieno[2,3-*c*]pyrrole (Eq. 24) <96T14975>. Using the same method, a thieno[3,2-*b*]pyrrole **16** and a thieno[2,3-*b*]pyrrole **17** are synthesized in 81% and 83% yields, respectively. Polyethynyl-substituted 2,2'-bithiophenes have been synthesized by coupling (trimethylsilyl)acetylene and polybrominated 2,2'-bithiophenes utilizing palladium and copper catalysts <96HCA755>. Regioselective palladium cross-coupling of 2,3-dibromothiophene with 2-propenylzinc chloride followed by halogen-lithium exchange and addition of dichlorosilane gives a dihydrosilane (Eq. 25) <96JA12469>.

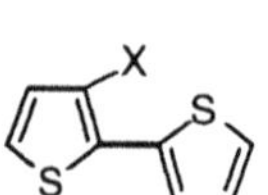

X = O-alkyl,S-alkyl

11

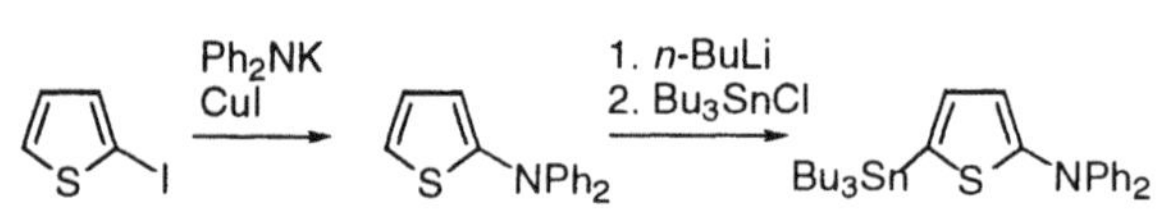

(Eq. 23)

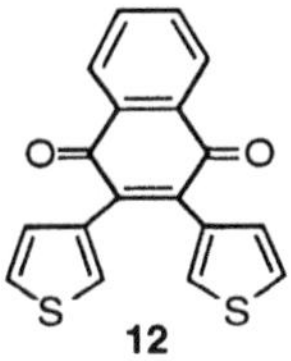

12

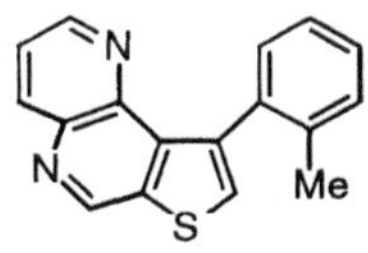

13

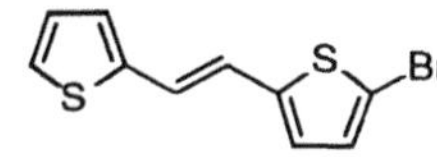

14

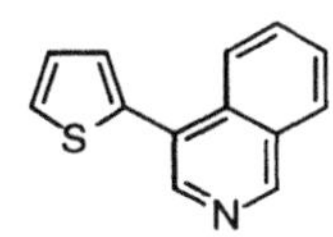

15

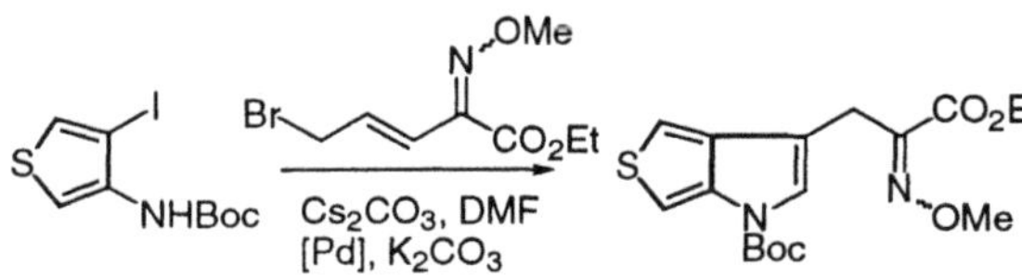

(Eq. 24)

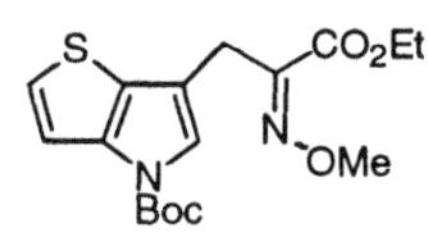

16

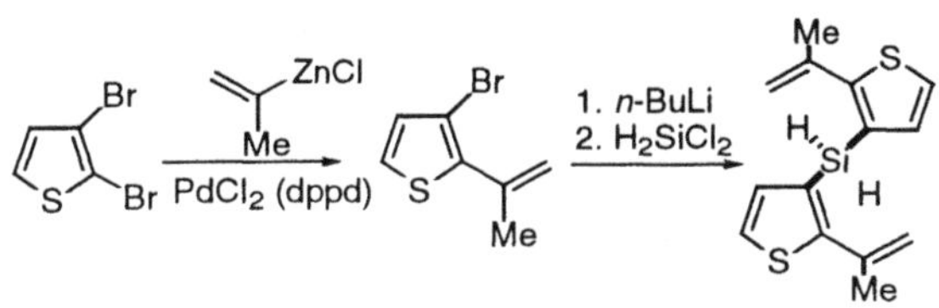

(Eq. 25)

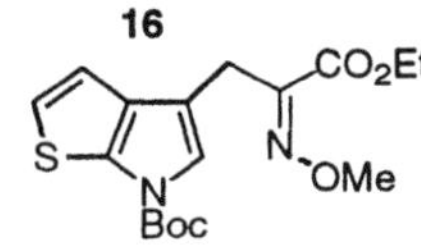

17

5.1.4 THIOPHENE RING FORMATION

The thiophene ring can be synthesized by a variety of methods including sulfur condensations, classical cyclizations, electrocyclizations, and unusual rearrangements. Sulfur atoms can be transferred to carbon fragments from a variety of sources including sodium sulfide, elemental sulfur, and carbon disulfide. The condensation of carbon disulfide onto 1,3-cyclohexanedione followed by reaction with ethyl bromoacetate and iodomethane gives a 6,7-dihydrobenzo[*c*]thiophene. Vilsmeier-Haack conversion of the resulting ketone to the α-chlorobenzaldehyde in 61% yield followed by condensation with sodium sulfide and ethyl chloroacetate gives a dithiophene in 63% yield after base-catalyzed cyclization (Eq. 26) <96LA239>. Similarly, sodium sulfide is utilized in an acid-catalyzed formation of ketone **18** <96CC177>. The condensation of elemental sulfur, 2-aminoprop-1-ene-1,1,3-tricarbonitrile, and acetylacetone in the presence of triethylamine gives a 79% yield of thieno[2,3-*b*]pyridine **19** <96JCR(S)356>.

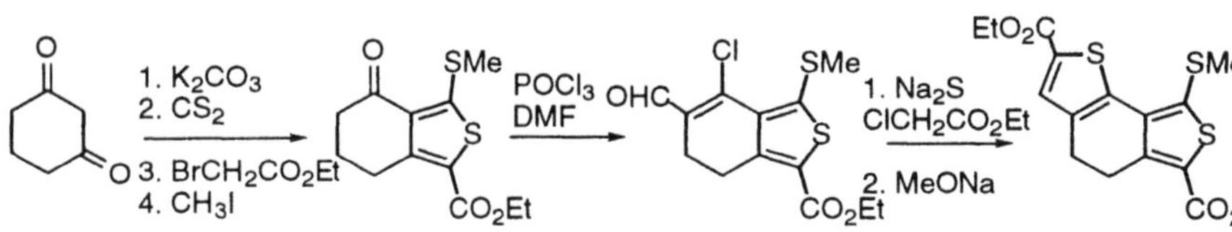

(Eq. 26)

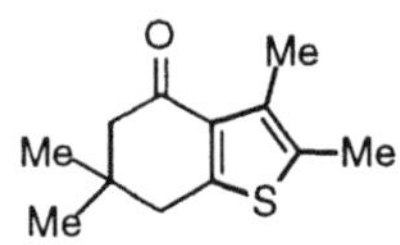

18

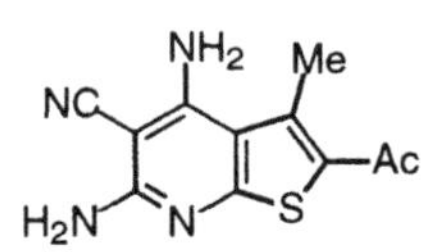

19

A number of interesting strategies for synthesizing highly functionalized thiophene rings have been developed. Michael addition by a 2-mercapto-2-arylethanol onto a vinyl phosphonate followed by oxidation and intramolecular Horner-Wadsworth-Emmons reaction gives a dihydrothiophene which then can be oxidized to a thiophene with DDQ (Eq. 27) <96JHC687>. Deprotonation of the *S*-methyl group of an acyclic ketene *N*,*S*-acetal leads to an anionic cyclization providing a 69% yield of 4-(*p*-methoxyphenyl)-2-pyrrolidinothiophene (Eq. 28) <96SC4157>. Treatment of 2-[(carboxymethyl)mercapto]benzoic acid with the Vilsmeier-Haack reagent gives 3-chlorobenzo[*b*]thiophene-2-carboxaldehyde in 51% yield (Eq. 29) <96JOC6523>. Base-catalyzed intramolecular condensations of mercaptoacetic acid derivatives are utilized to synthesize 3-hydroxythiophene **20** <96H775> and C_{14}-labelled (*) benzo[*b*]thiophene **21** <96H1189>. Condensation of a mercaptopyrimidine with an α-chloroketone followed by acid-catalyzed cyclization gives thienopyrimidine **22** <96H349>.

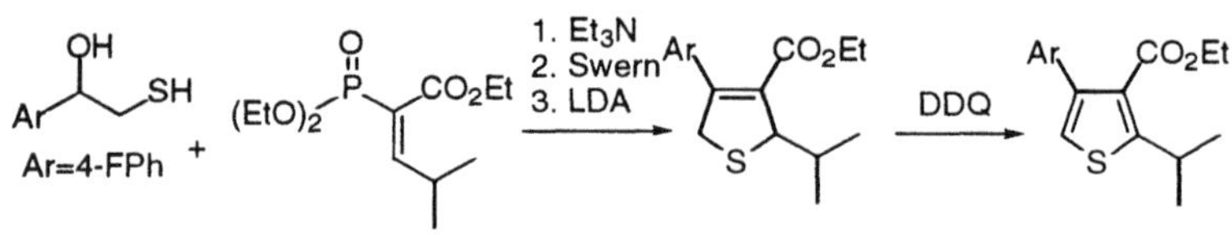

(Eq. 27)

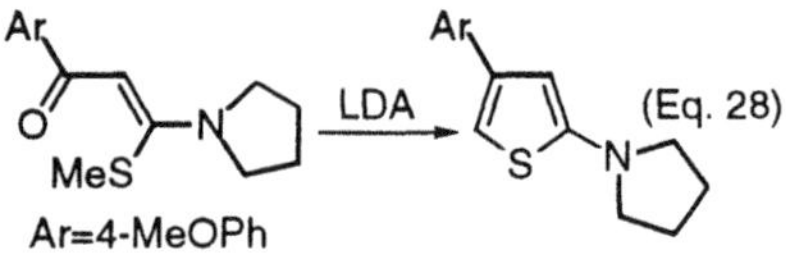

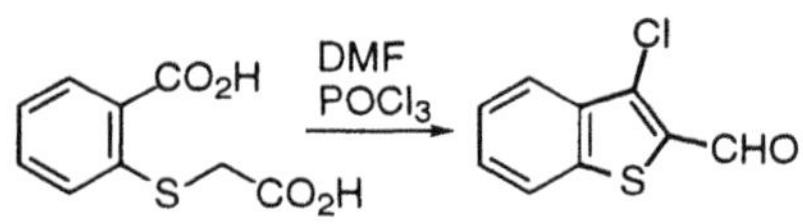

(Eq. 29)

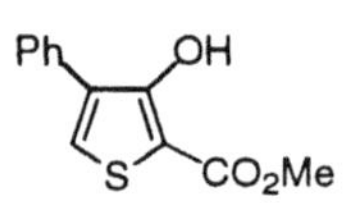

20

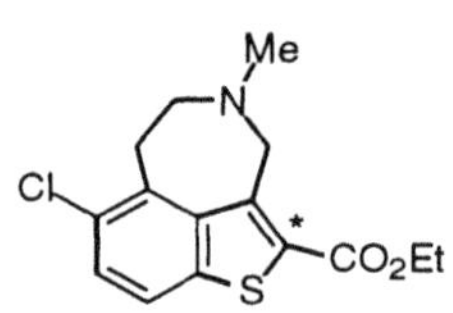

21

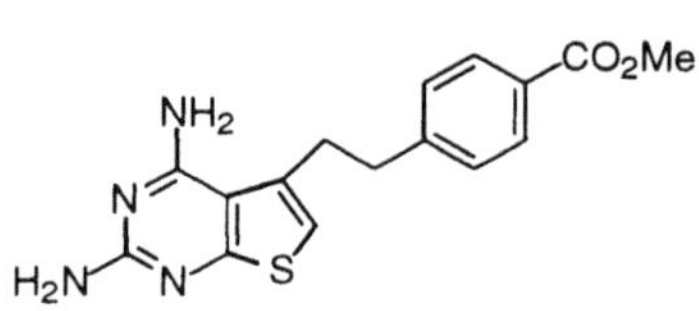

22

The classical Gewald thiophene synthesis involving condensation of sulfur with nitriles is still a widely utilized strategy for the synthesis of α-aminothiophenes. This method was applied to the syntheses of 2-amino-3-benzoylthiophene **23** <96MI683>, 4-isobutylthiophene **24** <96M297>, and saturated carbocyclic fused thiophenes **25** <96P4>. Alternatively, the cyclization onto a nitrile by the activated methylene moiety of a mercaptoacetic acid derivative gives 3-aminothieno[2,3-*b*]quinoline **26** <96JHC431>.

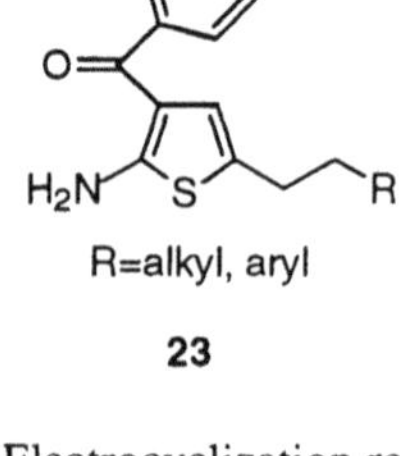

R=alkyl, aryl

23

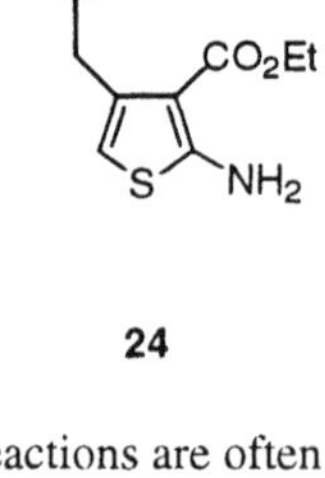

24

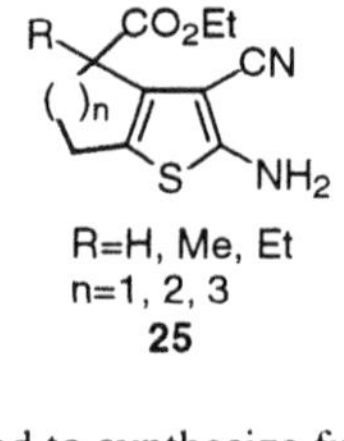

R=H, Me, Et
n=1, 2, 3

25

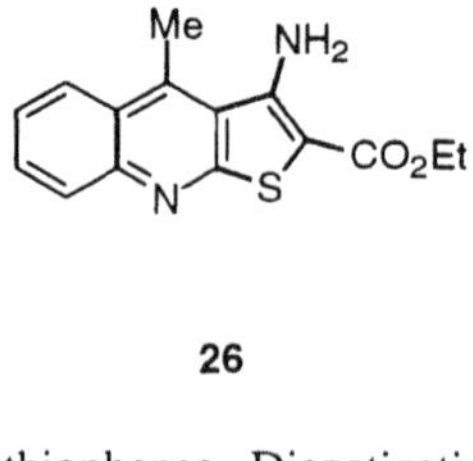

26

Electrocyclization reactions are often used to synthesize fused thiophenes. Diazotization of an aminoisoquinoline gave a mixture of products including a 6-thiaellipticine and a pyrazolo[3,4-*h*]isoquinoline in 34% and 41% yields, respectively (Eq. 30) <96CC2711>. A diazotization approach was used successfully to synthesize isomerically pure 4,6-dimethyldibenzothiophene (**27**) <96T3953>. A new enantioselective photocyclization method has been developed. Reaction of thiaketones with complex chiral diols gives 1:1 inclusion complexes which then undergo photocyclizations to give dihydrobenzothiophenes with good enantioselective efficiency (Eq. 31) <96JA11315>. Thermal isomerization of a bisallenylsulfone gives a thiophene 1,1-dioxide (Eq. 32) <96LA171>. A rare cycloaddition approach to 3,4-substituted thiophenes was also reported. A Diels-Alder reaction between 4-phenylthiazole and diphenylacetylene required 320-360°C and is followed by extrusion of benzonitrile to give 3,4-diphenylthiophene (Eq. 33) <96CC339>. This method was applied to the synthesis of 3,4-bis(trimethylsilyl)thiophene which is a precursor to a variety of 3,4-disubstituted thiophenes.

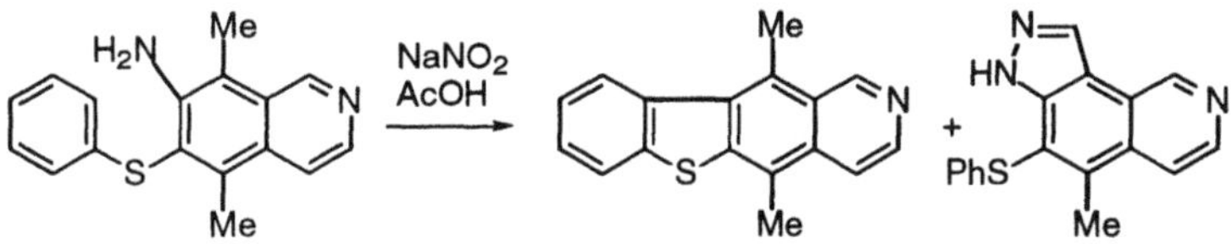

(Eq. 30)

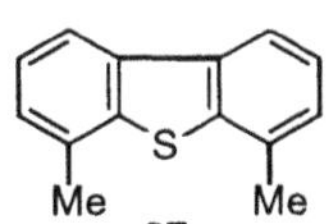

27

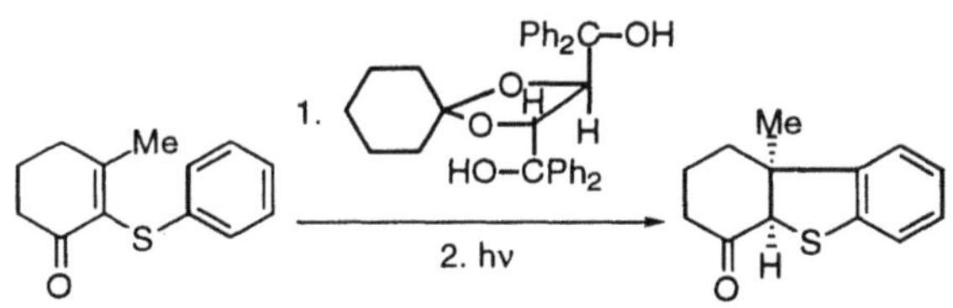

(Eq. 31)

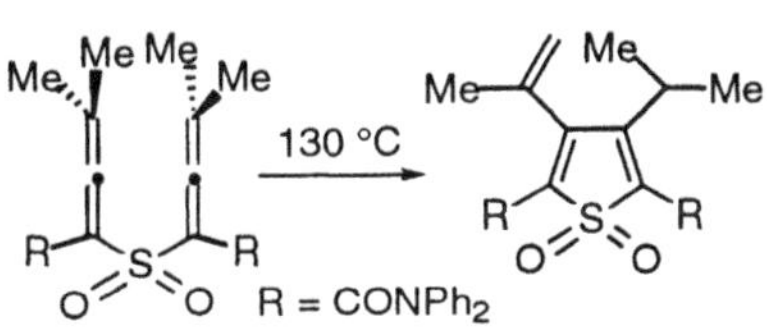

(Eq. 32)

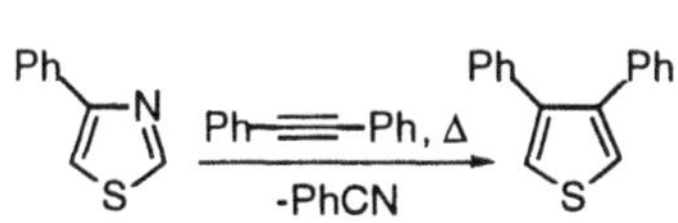

(Eq. 33)

Some interesting rearrangements lead to the thiophene ring system. Treatment of cyclopropanethiones with triphenylphosphine gives stable thieno[3,4-*c*]thiophenes in 21-30% yield (Eq. 34) <96CL421>. The reaction of tris(isopropylthio)cyclopropenyl perchlorate with various thiolate derivatives gives the corresponding tris(isopropylthio)thiophenes in 55-94% yield (Eq. 35) <96S1193>. The product of the reaction between a 1,8-diarylketone with Lawesson's reagent depends on the nature of aromatic substituents. A six-membered ring product is obtained when the aryl rings are substituted with electron-withdrawing groups, while electron-donating groups favor the formation of a fused thiophene (Eq. 36) <96TL2821>.

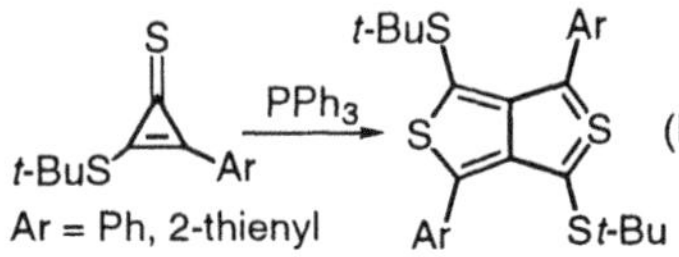

(Eq. 34)

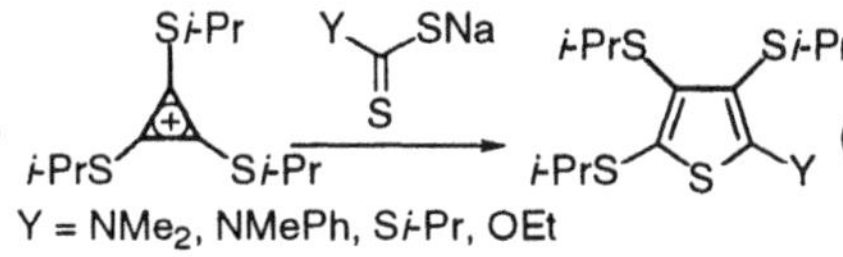

(Eq. 35)

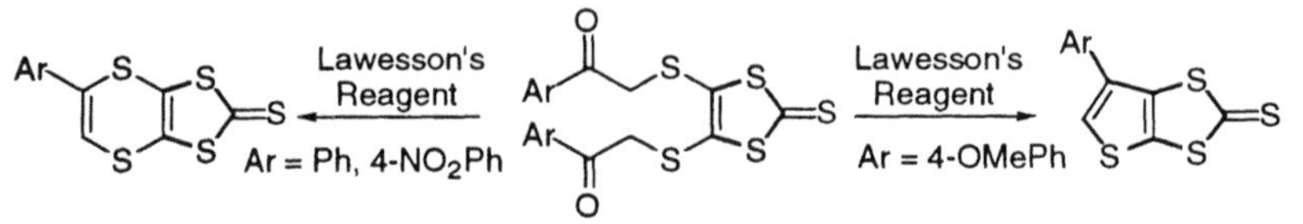

(Eq. 36)

5.1.5 RING ANNELATION ON THIOPHENE

The electron rich nature of the thiophene ring makes it ideal for electrophilic ring annelation reactions especially at the α-positions. A Friedel-Crafts acylation onto either the α- or β-position of the thiophene nucleus is utilized to synthesize a variety of fused thiophene lactams including an oxopyrrolodinothieno[2]azepinone (Eq. 37) <96JHC1251>, thienoazepino[2,1-*a*]isoindoledione **28** <96JHC129>, and [1]benzothienoindolizidinone **29** <96JHC873>. A short synthesis of 5,6-dihydrobenzo[*b*]thiophen-7(4*H*)-one involves an acid-catalyzed cyclization as the final step (Eq. 38) <96JCR(S)128>. A similar cyclization is mentioned in a previous section (Eq. 20) <96TA2721>. A Mannich-type cyclization is utilized in a synthesis of a 4,5,6,7-tetrahydrothieno[3,2-*c*]pyridine (Eq. 39) <96SC1363>. A free-radical cyclization approach to fused thiophenes is shown by treatment of a bromothiophene enamide with tributyltin hydride gave a complex mixture of products including the expected indolizidine product (Eq. 40) <96TL7707>.

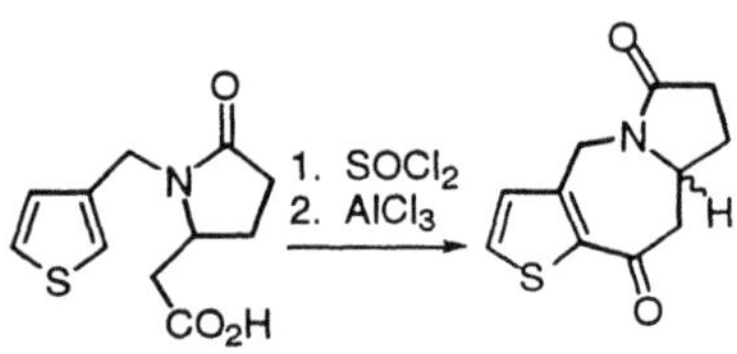

(Eq. 37)

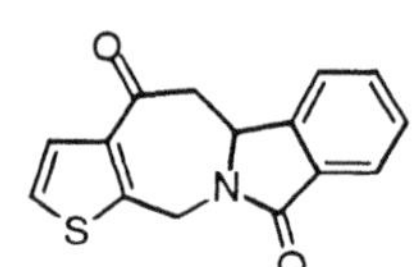

28

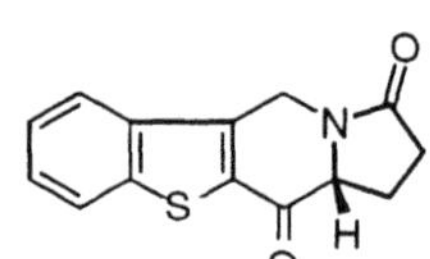

29

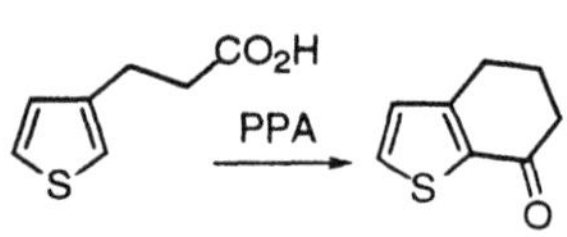

(Eq. 38)

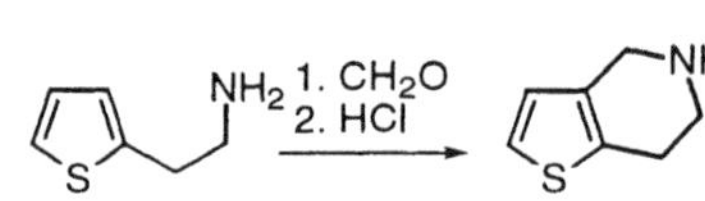

(Eq. 39)

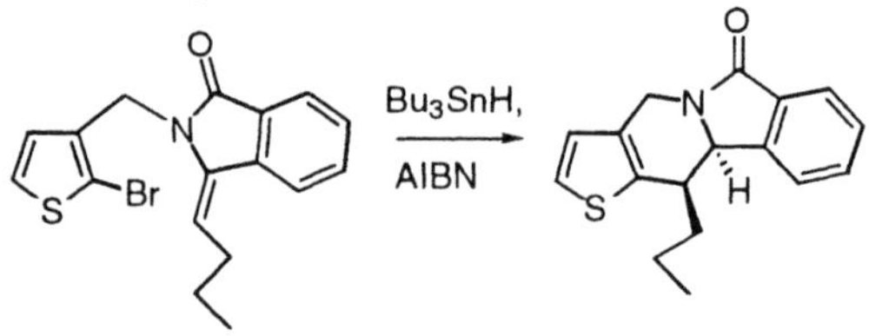

(Eq.40)

A variety of electrocyclization strategies have been employed to synthesize novel fused thiophenes. Treatment of the iminophosphorane derived from 3-azidobenzo[*b*]thiophene with acrylaldehyde in toluene at 70°C gives benzo[*b*]thieno[3,2-*b*]pyridine in 70% yield (Eq. 41) <96JCS(P1)2561>. Using the same method, benzo[*b*]thieno[2,3-*b*]pyridine is synthesized from 2-azidobenzo[*b*]thiophene. The electrocyclization of a ruthenium-substituted dienylvinylidene is utilized to make a variety of heterocyclic compounds including a fused thiophene (Eq. 42) <96JA11319>. Flash-vacuum pyrolysis of a thiophene epoxide proceeds to form a carbonyl ylide which then undergoes 1,7-electrocyclization followed by a [1,5] sigmatropic hydrogen shift to give a dihydrothieno[*d*][2]benzooxepine (Eq. 43) <96JCS(P1)515>.

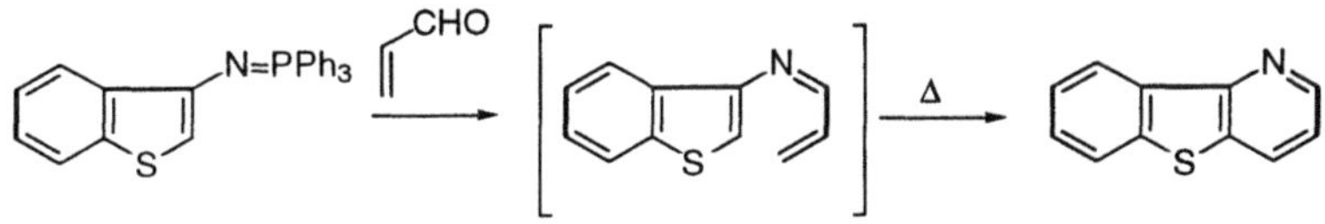

(Eq. 41)

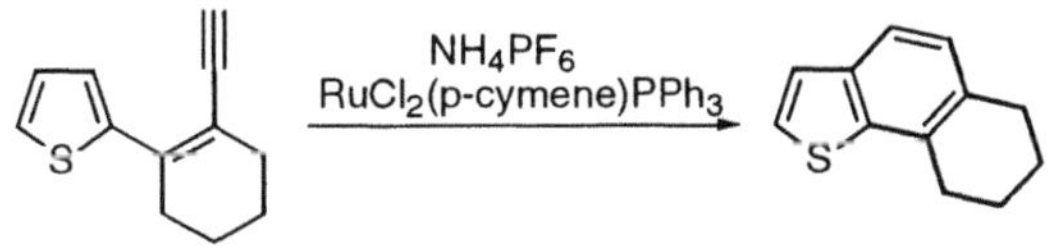

(Eq. 42)

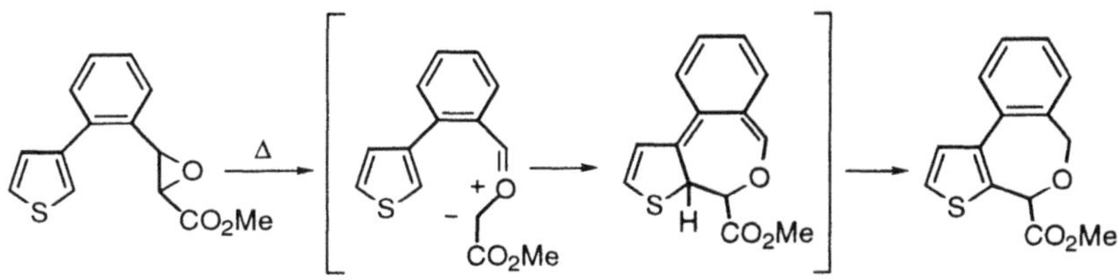

(Eq. 43)

Photocyclization remains a popular strategy for synthesizing a variety of complex fused thiophenes. The synthesis of a thieno[3,2-*a*]quinolizinium salt is accomplished by irradiating 1-(2-thienylvinyl)pyridinium bromide in the presence of iodine (Eq. 44) <96JHC57>. This method is also utilized to synthesize the regioisomer **30**. Photocyclization of a 3-chlorobenzo[*b*]thiophene produces a benzo[*b*]thieno[2,3-*c*]quinolinone (Eq. 45) <96HC213>. Similarly, photocyclization of the chlorothiophene moiety has been widely utilized to synthesize a variety of fused thiophenes including **31** <96JHC119>, **32** <96JHC923>, as well as many additional ring systems <96JHC179, 96JHC185, 96JHC727>. Photocyclization also proceeds in the absence of chlorine substituents to give phenanthro[2,1-*b*]thiophene **33** <96JHC1017, 96JHC1319>.

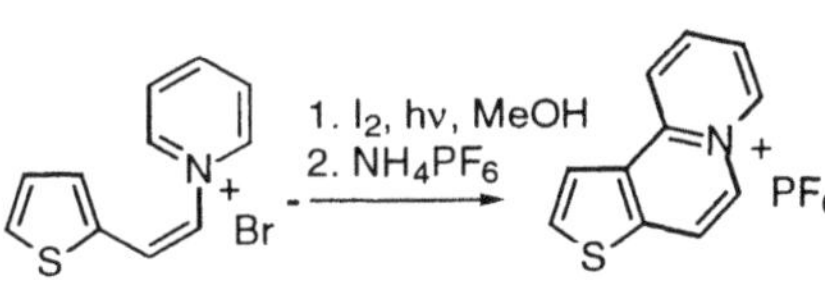

(Eq. 44)

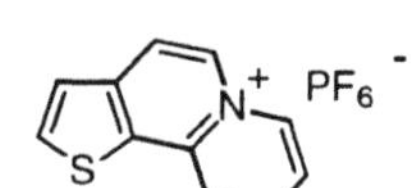

30

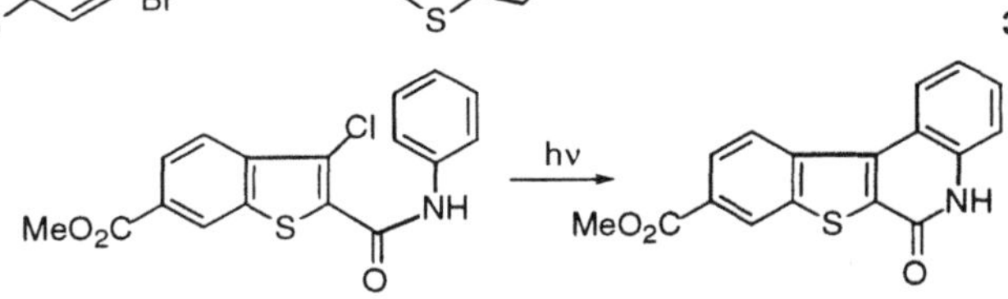

(Eq. 45)

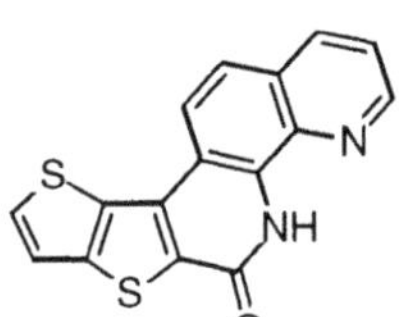

31

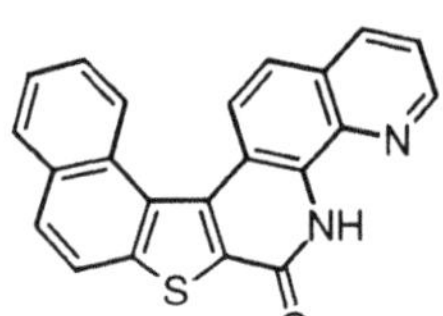

32

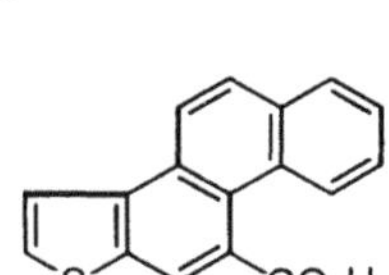

33

Finally, many [4+2] cycloaddition strategies utilizing various substituted thiophenes as dienes have been reported. The reaction between 2-(prop-1-enyl)-1-benzo[*b*]thiophene and 3-chlorocyclobut-3-ene-1,2-dione followed by bromination and loss of HBr gives a benzo[*b*]thieno-fused benzocyclobutenedione (Eq. 46) <96JCS(P1)497>. Two thieno-*o*-quinodimethane derivatives react with C_{60} to give annulated fullerenes <96JCS(P1)1077>. Likewise, electrophilic substitution of thieno-3-sulfolenes followed by pyrolysis and reaction of the resulting thieno-*o*-quinodimethanes with dimethyl fumarate produces the corresponding tetrahydrobenzo[*b*]thiophenes <96T12459>. The reaction between 4-(ethylthio)-6-phenylthieno[2,3-*c*]furan, an oxygen-stabilized *o*-quinodimethane derivative, and methyl acrylate in the presence of $Sc(OTf)_3$ forms a benzo[*b*]thiophene in 78% yield (Eq. 47) <96JOC6166>.

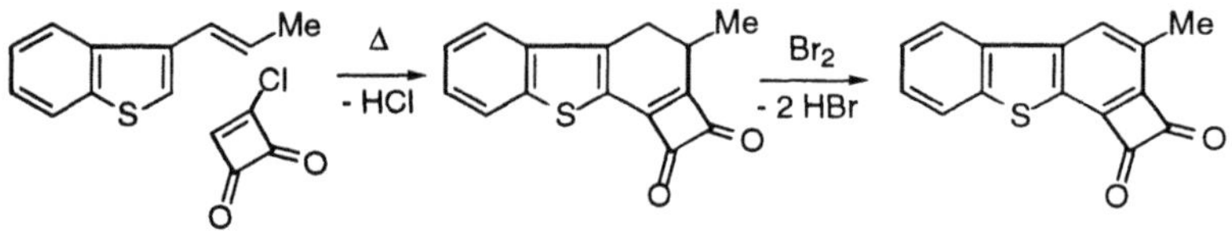

(Eq. 46)

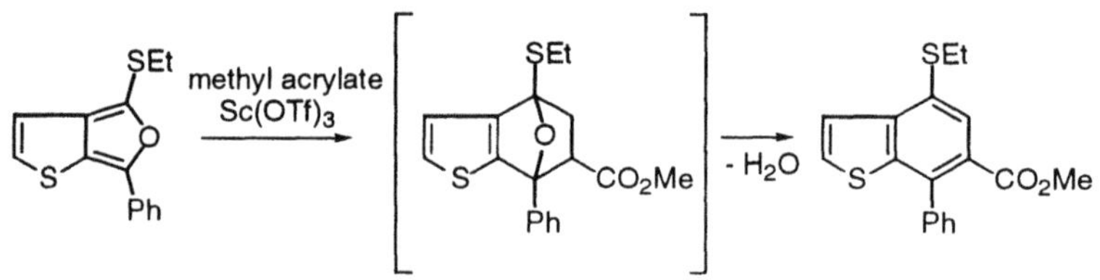

(Eq. 47)

5.1.6 THIOPHENES AS INTERMEDIATES

The formation of a thiophene derivative with subsequent extrusion of sulfur is a powerful synthetic tool. An intramolecular Pauson-Khand reaction gives a cyclic enone containing a saturated thiophene ring. Conjugate addition by *p*-tolyllithium in the presence of copper iodide followed by reductive desulfurization with Raney nickel gives a cyclopentanone which can be converted into (+)-β-cuparenone (Eq. 48) <96JOC9016>. Ring-opening reaction of 3,4-dinitrothiophene with diethylamine has been exploited to make a variety of 2,3-diaminobutanes (Eq. 49) <96T3313>.

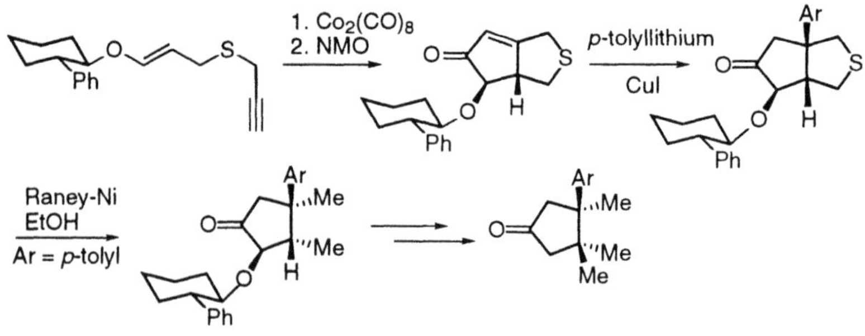

(Eq. 48)

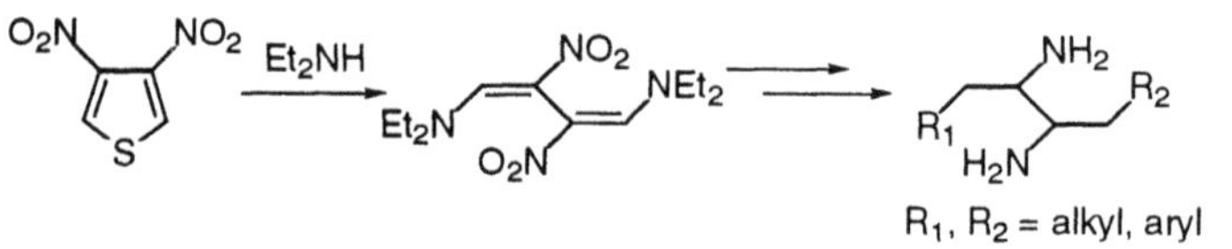

(Eq. 49)

Extrusion of sulfur dioxide to form *o*-quinodimethane and its analogues is a useful synthetic tool for cycloaddition chemistry. The solvent effect associated with the cheletropic extrusion of sulfur dioxide has been recently examined <96T6241>. Heating 1,3-dihydrobenzo[*c*]thiophene 2,2-dioxide (**34**) in the presence of benzo[*b*]thiophene gives **35** <96JHC109>. Various heterocyclic fused thiophene dioxides have been utilized as *o*-quinodimethane precursors includes 3-ethyl-1-tosyl-4,6-dihydrothieno[3,4-*d*]pyrazole (**36**) <96SC569> and 4-methoxy-2-methyl-5,7-dihydrothieno[3,4-*b*]pyrimidine (**37**) <96T1723, 96T1735>.

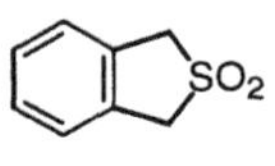

34

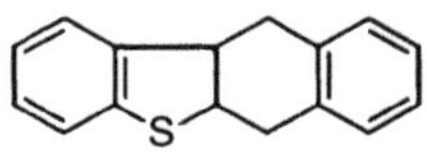

35

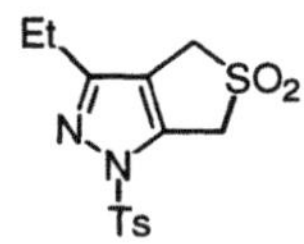

36

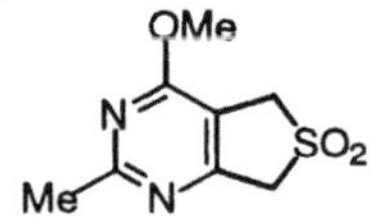

37

5.1.7 POLYMERS AND SUPERCONDUCTORS

As discussed earlier, thiophene's electronic and physicochemical properties make it highly useful as a component in a variety of polymers. Methods to prepare thiophene monomers, oligomers and thiophene-containing polymers provide some highly specialized examples of thiophene ring formation and substitution. Palladium acetate catalyzes the cycloaddition copolymerization of diynes with elemental sulfur to produce poly(thiophene)s (Eq. 50) <96CC1317>. Other methods to prepare poly(thiophene)s include plasma polymerization of 2-iodothiophene <96MI916> and palladium-catalyzed coupling of thienyl mercuric chlorides <96MI936>. Cross-conjugation of zirconocene alkynyl(benzyne) derivatives with sulfur dichloride forms poly(benzo[*b*]thiophene)s <96JOM425>.

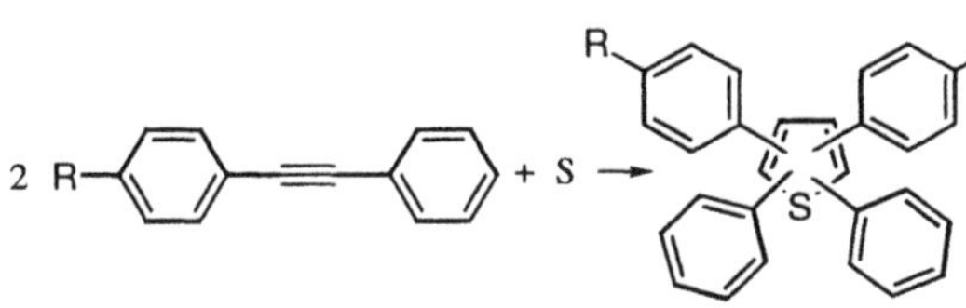

(Eq. 50)

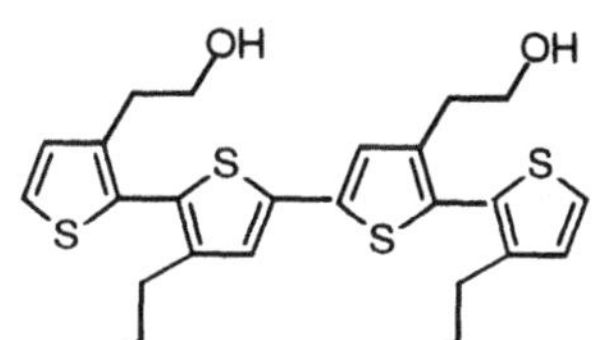

38

α-Oligothiophenes with two to seven rings have been evaluated for absorption, photophysical, energy transfer, structural and theoretical properties <96JPC18683, 96JPC11029, 96JA6453>. α-Hexathienyl has great promise as a thin-film-transistor material <96MI4952>. Methylated bi- and terthiophenes can be oxidized with $FeCl_3$ to form dimers <96MI3370>. Conformational properties of 2,2'-bithiophene and its dimethyl derivatives have been studied as model systems for conducting polymers <96JPC1524>. Poly(3-hexylthiophene-2,5-diyl) has unique temperature dependent NMR and UV properties <96CL703>.

Compound **38** is the building block for head-to-head/ tail-to-tail polyhydroxyoligothiophenes. In contrast to other bithiophenes, **38** has a noncoplanar *anti* conformation with an inter-ring twist angle of 67.5° <96JOC4708, 96MI161>. Poly{3,4-di[(*S*)-2-methylbutoxy]thiophene} has chiral properties that cause a split in excited state exciton levels <96JA4908>. Acceptor-substituted poly(3-butylthiophene)s have been prepared to study the role of acceptor molecules as intrinsic charge traps during photochemical excitation <96JA2980>. Heterocyclic heptamers containing thiophene and pyrrole units (**39**) are excellent model compounds for charge-transfer studies and electrical conductivity <96MI54>.

Poly[(tetraethyldisilanylene)oligo(2,5-thienylenes)] **40** were synthesized; irradiation of **40** causes Si–Si bond cleavage <96O2000>. Electrochemical methodology is one method to prepare polysilyl thiophene derivatives <96O2041, 96JOM213>. 5,5'-Bis[2,2,5,5-tetramethyl-1-aza-2,5-disila-1-cyclopentyl)-methyl]-2,2'-bithiophene has bulky terminal groups that causes peculiar crystallographic properties as compared to other bithiophenes <96CL501>. Several analogous boron bithiophene derivatives have non-linear optical properties <96MI555>. Oligo(vinylthiophenes) with "push-pull" substitution such as **41** have liquid crystal properties <96JCS(P2)713>. Push-pull dihexylbithiophenes have electronic absorption spectra affected by solvent polarity <96TL1617>. π-Conjugated copolymers containing a π-excessive thiophene, selenophene or furan with a π-deficient pyridine or quinoxaline form semiconductors by oxidation or reduction <96JA10389>. Methylthio oligothiophenes have good film-making properties with interesting electroconductive properties <96LA279, 96MI7671>. Polymers containing quaterthiophene with limited π-conjugation due to ethane bridging were conductive only in oxidized form <96CL495>.

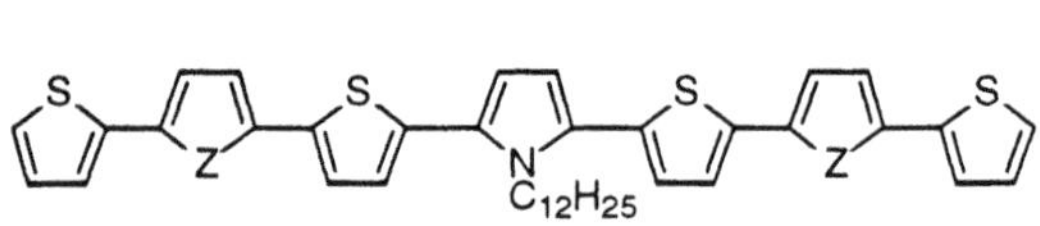

39

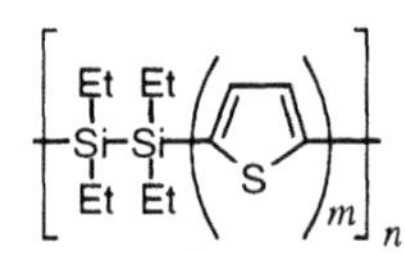

40

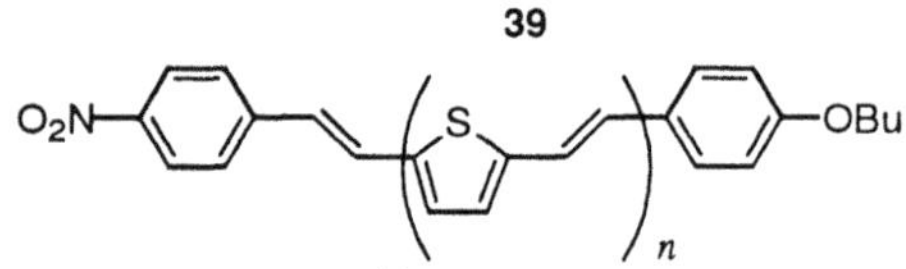

41

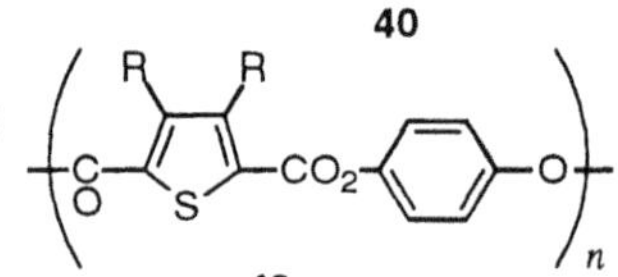

42

Polyesters containing thiophenes (**42**) were prepared in attempts to improve thermal properties of these materials; unfortunately, **42** displayed poor liquid crystallinity upon melting <96MI675>. Orthogonally fused conjugated oligomers (**43**) contain two conducting chains held perpendicular by a spirobithiophene moiety with a central silicon atom and are potential molecular switches <96JOC6906>. Dithienothiophenes polymerize photochemically to provide the basis for two-dimensional and in-depth directional polymer films <96CL285>. Thiophene imino dyes (**44**) have second-order nonlinear optical properties <96CC1045>. Thiophene-fused DCNQI compounds (**45**) provide zig-zag ladder CuI complexes with unique π-stacked acceptors <96CC2275>.

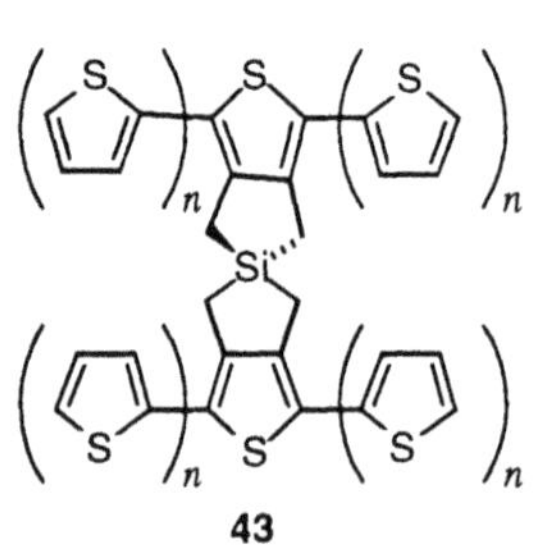

43

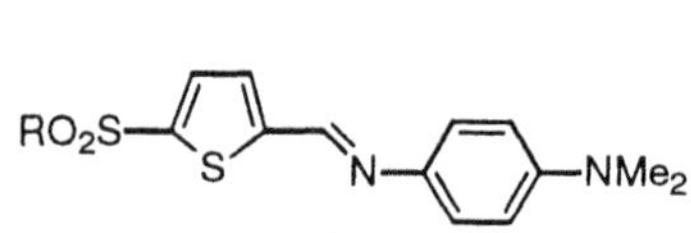

44

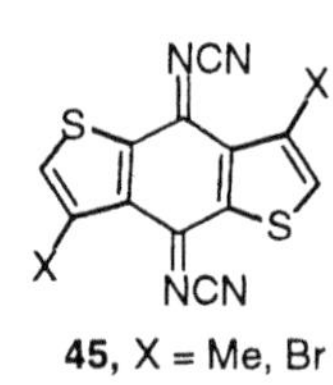

45, X = Me, Br

5.1.8 INTERESTING THIOPHENE DERIVATIVES

In addition to useful polymers, thiophene derivatives have a variety of other interesting properties. Novel benzodiazepine analogues such as **46** and related compound **47** continue to provide interesting leads for CNS research <96T13547>. Compound **48** is a partial 5-HT_3 receptor agonist and has potent *in vivo* anti-anxiety activity <96JMC2068>. Tetrahydrothieno[3,2-*c*]pyridine **49** is among the best antagonists of NMDA-induced seizures in mice and has good anticonvulsant activity <96CPB778>. Some α-terthienyl-5-derivatives may be conjugated to carrier proteins and antibodies to serve as directed, selective phototoxic agents <96T11281>. Piperidinomethylthiophene **50** is a potent H_2 antagonist and inhibits *H. pylori* <96BMCL1795>.

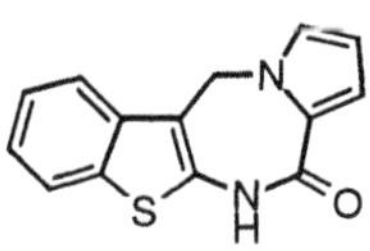

46

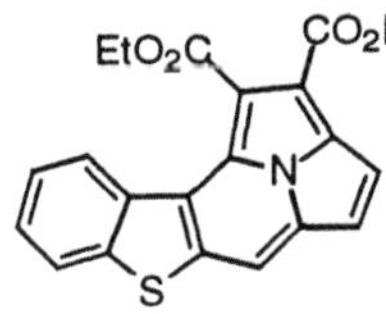

47

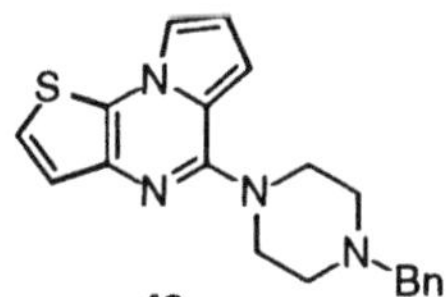

48

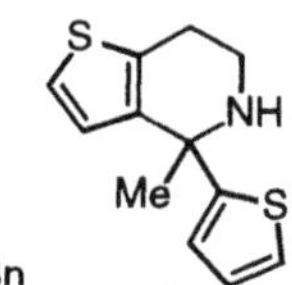

49

The thiophene analogue (**51**) of the natural antifeedant product tonghaosu has been prepared <96TL893>. Several acetylenic thiophene derivatives including **52** have been isolated from *Blumea obliqua* and characterized <96MI733>.

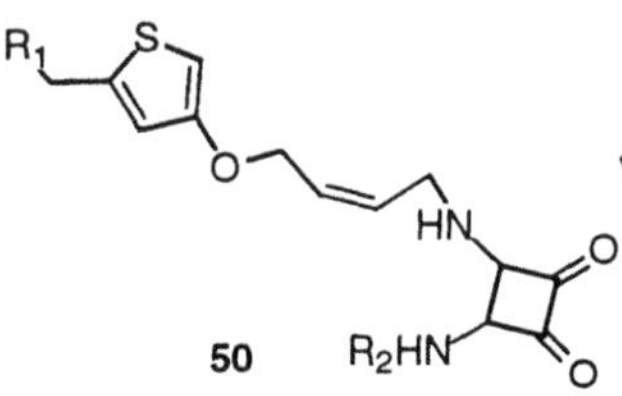

50

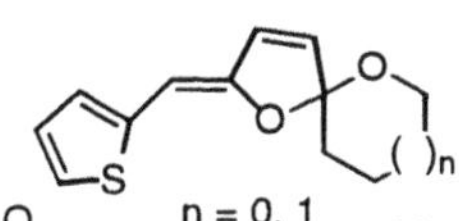

51

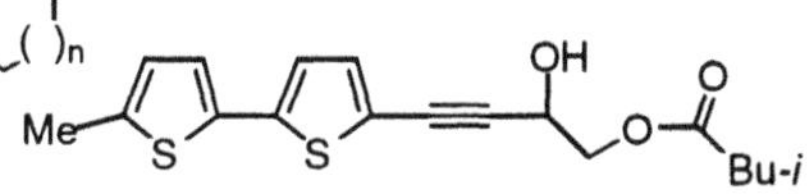

52

Moving away from biologically active compounds, compound **53** is a chiral bifunctional helicene terminated with π-deficient pyridine and π-excessive thiophene and is a promising candidate as a ligand for asymmetric syntheses <96TL5925>. π-Conjugated polyradicals such as **54** are under study to understand bulk ferromagnetism in molecular solids <96T6893>. A series of thiophene-pyrrole annulenes including **55** do not show overall aromaticity or antiaromaticity properties <96JOC3657>. Thiophene-linked azacryptands complex with copper and silver ions but show no evidence of thiophene–*S* coordination <96JCS(D)3021>.

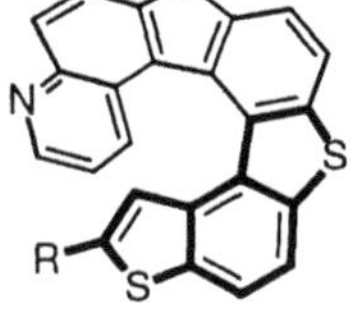

53

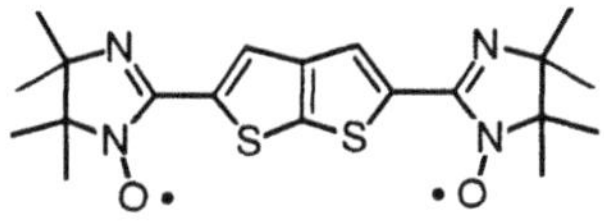

54

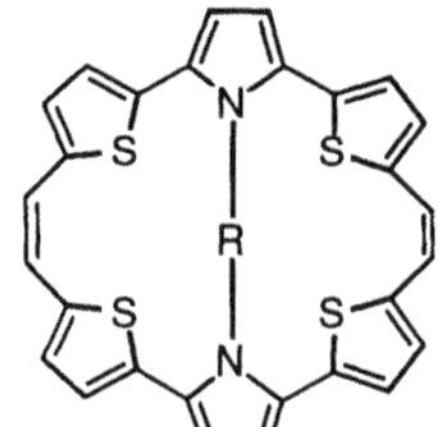

55

Porphyrinoid macrocycles containing thiophene, such as **56**, have been prepared as potential advanced materials; its dication deviates from planarity while the trication of the analogous pentathiophene analogue of **56** is planar <96AC(E)1520>. In a study of "conformational" polymorphs, **57** crystallizes to form varying colored crystals depending on solvent <96CL363>. Heteroaromatic triptycenes such as thiophene **58** have interesting properties; hydrogen bonding of the bridgehead hydroxyl of **58** is hampered by steric hindrance from the three thiophene ring sulfur atoms <96JCS(P1)1277, 96JA12836>. Thienocarbazoles may be formed using Fischer indole synthetic techniques <96H367>.

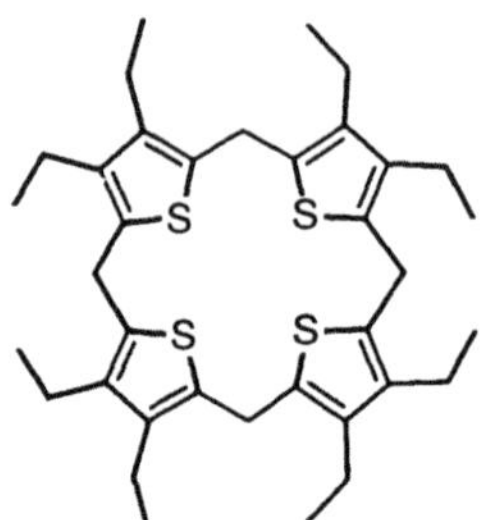

56

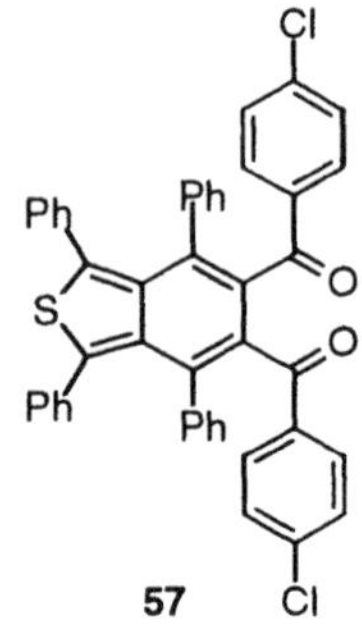

57

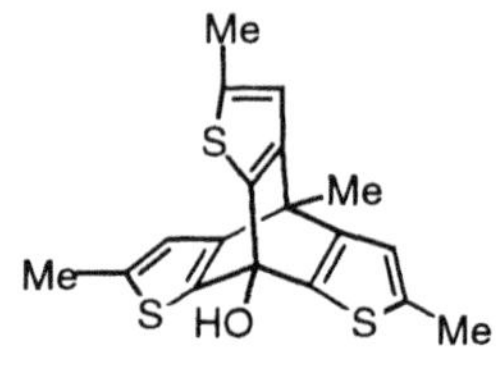

58

5.1.9 SELENOPHENES AND TELLUROPHENES

Several reports of chemistry of selenophenes and tellurophenes have appeared during 1996. In the first synthesis of selenophene 1,1-dioxides (**59**), dimethyldioxirane oxidises selenophenes in high yields <96CL269>. An efficient synthesis of 2-cyanomethyl thiophene, selenophene and tellurophene derivatives utilizes Pd-catalyzed carbon-carbon bond formation from the corresponding 2-iodo compounds <96H1927>. The dimer of 3,6-dimethylselenolo[3,2-*b*]selenophene has an *S-cis* structure with a large dihedral angle of 69.5° (Eq. 50) <96T471>. Reaction of γ,δ-unsaturated selenothioic acid *S*-esters with cadmium acetate form selenophenes (Eq. 51) <956CL877> while reductive cyclization with sodium borohydride produces tetrahydroselenophenes <96CC1461>. Iodocyclization produces 3-iodotellurophenes (Eq. 52) <96JOC9503>. Both tellurophene and dibenzotellurophene detellurate upon treatment with $Fe_3(CO)_{12}$; in the latter case, a dibenzoferrole is the product <96JCS(D)1545>.

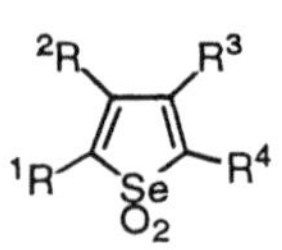

59

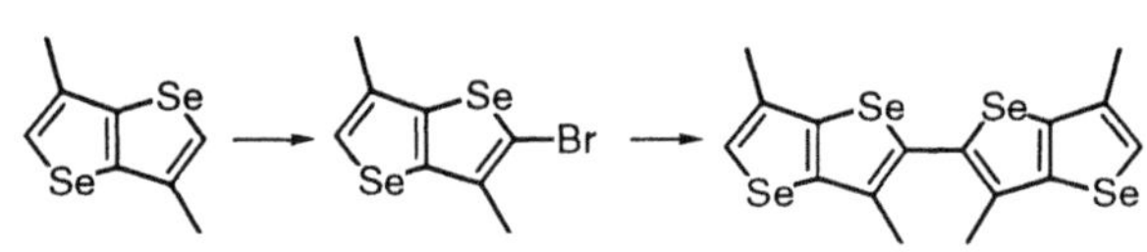

(Eq. 50)

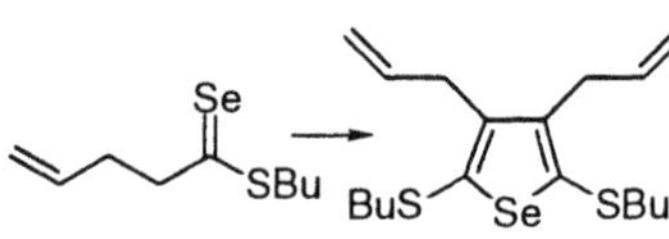

(Eq. 51)

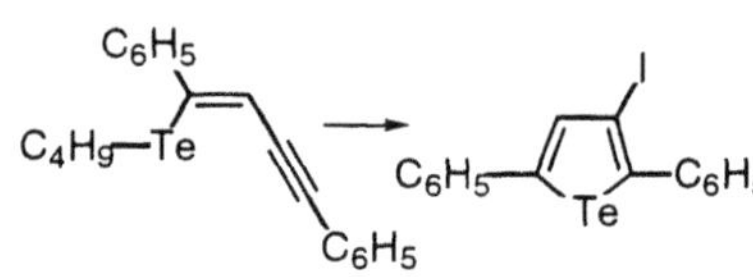

(Eq. 52)

5.1.10 REFERENCES

96AC(E)212 C.A. Dullaghan, S. Sun, G.B. Carpenter, B. Weldon and D.W. Sweigart, *Angew. Chem. Int. Ed. Engl.*, **1996**, *35*, 212.

96AC(E)1520 E. Vogel, M. Pohl, A. Herrmann, T. Wiss, C. König, J. Lex, M. Gross and J.P. Gisselbrecht, *Angew. Chem. Int. Ed. Engl.*, **1996**, *35*, 1520.

96AC(E)1706 C. Bianchini, M.V. Jiménez, C. Mealli, A. Meli, S. Moneti, V. Patinec and F. Vizza, *Angew. Chem. Int. Ed. Engl.*, **1996**, *35*, 1706.

96AHC193 G. Varounis and T. Giannopoulos, *Adv. Het. Chem.*, **1996**, *66*, 153.

96BML1795 K. Kojima, K. Nakajima, H. Kurata, K. Tabata and Y. Utsui, *Bioorg. Med. Chem. Lett.*, **1996**, *6*, 1795.

96BML2651 B. Raju, C. Wu, A. Kois, E. Verner, I. Okun, F. Stavros and M.F. Chan, *Bioorg. Med. Chem. Lett.*, **1996**, *6*, 2651.

96BSF597 M. Cariou, T. Douadi and J. Simonet, *Bull. Soc. Chim. Fr.*, **1996**, *133*, 597.

96CC177 W. Adam and S. Weinkotz, *J. Chem. Soc., Chem. Commun.*, **1996**, 177.

96CC339 X.-S. Ye and H.N.C. Wong, *J. Chem. Soc., Chem. Commun.*, **1996**, 339.

96CC1045 S.-S.P. Chou, D.-J. Sun, H.-C. Lin and P.-K. Yang, *J. Chem. Soc., Chem. Commun.*, **1996**, 1045.

96CC1167 B.S. Kang, M.-L. Seo, Y.S. Jun, C.K. Lee and S.C. Shin, *J. Chem. Soc., Chem. Commun.*, **1996**, 1167.

96CC1317 T. Tsuda and A. Takeda, *J. Chem. Soc., Chem. Commun.*, **1996**, 317.

96CC1461 T. Murai, M. Maeda, F. Matsuoka, T. Kanda and S. Kato, *J. Chem. Soc., Chem. Commun.*, **1996**, 1461.

96CC2275 K. Takahashi, Y. Mazaki and K. Kobayashi, *J. Chem. Soc., Chem. Commun.*, **1996**, 2275.

96CC2711 R.B. Miller, J.G. Stowell, C.W. Jenks, S.C. Farmer, C.E. Wujcik and M.M. Olmstead, *J. Chem. Soc., Chem. Commun.*, **1996**, 2711.

96CL139 S. Yoshid, H. Kubo, T. Saika and S. Katsumura, *Chem. Lett.*, **1996**, 139.

96CL269 J. Nakayama, T. Matsui, Y. Sugihara, A. Ishii and S. Kamakura, *Chem. Lett.*, **1996**, 269.

96CL285 M. Fujitsuka, T. Sato, A. Watanabe, O. Ito and T. Shimidzu, *Chem. Lett.*, **1996**, 285.

96CL363 S. Mataka, H. Moriyama, T. Sawada, K. Takahashi, H. Sakashita and M. Tashiro, *Chem. Lett.*, **1996**, 363.

96CL421 N. Matsumura, Y. Yagyu, H. Tanaka, H. Inoue, K. Takada, M. Yasui and F. Iwasaki, *Chem. Lett.*, **1996**, 421.

96CL495 M. Sato and M. Hiroi, *Chem. Lett.*, **1996**, 495.

96CL501 H. Muguruma, S. Hotta and I. Karube, *Chem. Lett.*, **1996**, 501.

96CL703 T. Yamamoto, *Chem. Lett.*, **1996**, 703.

96CL743 A. Ishihara, M. Yamaguchi, H. Godo, W. Qian, M. Godo and T. Kabe, *Chem. Lett.*, **1996**, 743.

96CL817 N. Nakashima, Y. Deguchi, T. Nakanishi, K. Uchida and M. Irie, *Chem. Lett.*, **1996**, 817.

96CL829 K. Kaneda, T. Wada, S. Murata and M. Nomura, *Chem. Lett.*, **1996**, 829.

96CL877 T. Murai, M. Fujii, T. Kanda and S. Kato, *Chem. Lett.*, **1996**, 877.

96CPB778 M. Ohkubo, A. Kuno, K. Katsuta, Y. Ueda, K. Shirakawa, H. Nakanishi, T. Kinoshita and H. Takasugi, *Chem. Pharm. Bull.*, **1996**, *44*, 778.

96CPL31 M. Belletête, N. D. Césare, M. Leclerc and G. Durocher, *Chem. Phys. Lett.*, **1996**, *250*, 31.

96CPL73 M. Ciofalo and G. La Manna, *Chem. Phys. Lett.*, **1996**, *263*, 73.

96CPL329 K. Kamada, M. Ueda, T. Sakaguchi, K. Ohta and T. Fukumi, *Chem. Phys. Lett.*, **1996**, *249*, 329.

96H349 E.C. Taylor, H.H. Patel, G. Sabitha and R. Chaudhari, *Heterocycles*, **1996**, *43*, 349.

96H367 L. Martarello, D. Joseph and G. Kirsch, *Heterocycles*, **1996**, *43*, 367.

96H775 J. Lissavetzky and I. Manzanares, *Heterocycles*, **1996**, *43*, 775.

96H861 L. Engman and P. Eriksson, *Heterocycles*, **1996**, *43*, 861.

96H1189 S.W. Landvatter, *Heterocycles*, **1996**, *43*, 1189.

96H1767 J. Lissavetzky and I. Manzanares, *Heterocycles*, **1996**, *43*, 1767.

96H1927 K. Takahashi and S. Tarutani, *Heterocycles*, **1996**, *43*, 1927.

96HA45 S. Wu, Y. Luo, F. Liu and S. Chen, *Heteroatom Chem.*, **1996**, *7*, 45.

96HC17 R. Donosa, J. Elguero, P. Goya, M.P. Jordán de Urries and J. Lissavetszky, *Heterocyclic Commun.*, **1996**, *2*, 17.

96HC213 J. Dogan, G. Karminski-Zamola and D.W. Boykin, *Heterocyclic Commun.*, **1996**, *2*, 213.

96HCA755 U. Dahlmann and R. Neidlein, *Helv. Chim. Acta*, **1996**, *79*, 755.

96IC7095 A.L. Sargent, E.P. Titus, C.G. Riordan, A.L. Rheingold and P. Ge, *Inorg. Chem.*, **1996**, *35*, 7095.

96IC7211 K.N. Brown and J.H. Espenson, *Inorg. Chem.*, **1996**, *35*, 7211.

96IC7311 L. Barnett-Thamattoor, G. Zheng, D.M. Ho, M. Jones, Jr., J.E. Jackson, *Inorg. Chem.*, **1996**, *35*, 7311.

96JA2748 G.D. Allred and L.S. Liebeskind, *J. Am. Chem. Soc.*, **1996**, *118*, 2748.

96JA2980 Y. Greenwald, G. Cohen, J. Poplawski, E. Ehrenfreund, S. Speiser and D. Davidov, *J. Am. Chem. Soc.*, **1996**, *118*, 2980.

96JA4908 B.M.W. Langeveld-Voss, R.A.J. Janssen, M.P.T. Christiaans, S.C.J. Meskers, H.P.J.M. Dekkers and E. W. Meijer, *J. Am. Chem. Soc.*, **1996**, *118*, 4908.

96JA6453 D. Beljonne, J. Cornil, R.H. Friend, R.A.J. Janssen and J.L. Brédas, *J. Am. Chem. Soc.*, **1996**, *118*, 6453.

96JA10389 T. Yamamoto, Z. Zhou, T. Kanbara, M. Shimura, K. Kizu, T. Maruyama, Y. Nakamura, T. Fukuda, B.-L. Lee, N. Ooba, S. Tomaru, T. Kurihara, T. Kaino, K. Kuboto and S. Sasaki, *J. Am Chem. Soc.*, **1996**, *118*, 10389.

96JA11315 F. Toda, H. Miyamoto, S. Kikuchi, R. Kuroda and F. Nagami, *J. Am. Chem. Soc.*, **1996**, *118*, 11315.

96JA11319 C.A. Merlic and M.E. Pauly, *J. Am. Chem. Soc.*, **1996**, *118*, 11319.

96JA12469 K. Tamao, K. Nakamura, H. Ishii, S. Yamaguchi and M. Shiro, *J. Am. Chem. Soc.*, **1996**, *118*, 12469.

96JA12836 A. Ishii, K. Komiya and J. Nakayama, *J. Am. Chem. Soc.*, **1996**, *118*, 12836.

96JCR(S)128 S.S. Samanta, S. Mukherjee and A. De, *J. Chem. Res. (S)*, **1996**, 128.

96JCR(S)150 Y. Xie, S.-C. Ng, T.S.A. Hor and H.S.O. Chan, *J. Chem. Res. (S)*, **1996**, 150.

96JCR(S)232 S.-C. Ng, H.S.-O. Chan, H.-H. Huang and R.S.-H. Seow, *J. Chem. Res. (S)*, **1996**, 232.

96JCR(S)242 D.W. Allen, M.R. Clench, A.T. Hewson and M. Sokmen, *J. Chem. Res. (S)*, **1996**, 242.

96JCR(S)356 S.M. Sherif, W.W. Wardakhan and R.M. Mohareb, *J. Chem. Research (S)*, **1996**, 356.

96JCR(S)458 H.N.C. Wong, X.-S. Ye, Y.-S. Cheung and W.-K. Li, *J. Chem. Res. (S)*, **1996**, 458.

96JCS(D)1545 K. Singh, W.R. McWhinnie, H.L. Chen., M. Sun and T.A. Hamor, *J. Chem. Soc., Dalton Trans.*, **1996**, 1545.

96JCS(D)3021 M.G.B. Drew, C.J. Harding, O.W. Howarth, Q. Lu, D.J. Marrs, G.G. Morgan, V. McKee and J. Nelson, *J. Chem. Soc., Dalton Trans.*, **1996**, 3021.

96JCS(P1)417 N. Ono, H. Hironaga, K. Ono, S. Kaneko, T. Murashima, T. Ueda, C. Tsukamura and T. Ogawa, *J. Chem. Soc., Perkin Trans. 1*, **1996**, 417.

96JCS(P1)497 A.H. Schmidt, K.O. Lechler, T. Pretz and I. Franz, *J. Chem. Soc., Perkin Trans. I*, **1996**, 497.

96JCS(P1)515 D.F. O'Shea and J.T. Sharp, *J. Chem. Soc., Perkin Trans. 1*, **1996**, 515.

96JCS(P1)963 P. Spagnolo and P. Zanirato, *J. Chem. Soc., Perkin Trans. 1*, **1996**, 963.

96JCS(P1)1077 U.M. Fernandez-Paniagua, B.M. Illescas, N. Martin and C. Seoane, *J. Chem. Soc., Perkin Trans. I*, **1996**, 1077.

96JCS(P1)1277 A. Ishii, K. Maeda, M. Kodachi, N. Aoyagi, K. Kato, T. Maruta, M. Hoshino and J. Nakayama, *J. Chem. Soc. Perkin Trans . 1*, **1996**, 1277.

96JCS(P1)1403 T. Murashima, K.-i. Fujita, K. Ono, T. Ogawa, H. Uno, N. Ono, *J. Chem. Soc., Perkin Trans. 1*, **1996**, 1403.

96JCS(P1)2561 A. Degl'Innocenti, M. Funicello, P. Scafato, P. Spagnolo and P. Zanirato, *J. Chem. Soc., Perkin Trans. I*, **1996**, 2561.

96JCS(P2)455 B.S. Jursic, Z. Zdravkovski and S.L. Whittenburg, *J. Chem. Soc., Perkin Trans. 2*, **1996**, 455.

96JCS(P2)713 C. Maertens, J.-X. Zhang, P. Dubois and R. Jérôme, *J. Chem. Soc., Perkin Trans. 2*, **1996**, 713.

96JCS(P2)1377 M. Blenkle, P. Boldt, C. Brauchle, W. Grahn, I. Ledoux, H. Nerenz, S. Stadler, J. Wichern and J. Zyss, *J. Chem. Soc., Perkin Trans. 2*, **1996**, 1377.

96JHC57 K. Sato, S. Arai and T. Yamagishi, *J. Heterocyclic Chem.*, **1996**, *33*, 57.

96JHC109 J.H. Markgraf, and D.E. Patterson, *J. Heterocyclic Chem.*, **1996**, *33*, 109.

96JHC119 A.P. Halverson, L.W. Castle and R.N. Castle, *J. Heterocyclic Chem.*, **1996**, *33*, 119.

96JHC129 P. Pigeon and B. Decroix, *J. Heterocyclic Chem.*, **1996**, *33*, 129.

96JHC179 A.P. Halverson, L.W. Castle and R.N. Castle, *J. Heterocyclic Chem.*, **1996**, *33*, 179.

96JHC185 J.-K. Luo, R.F. Federspiel, R.N. Castle and L.W. Castle, *J. Heterocyclic Chem.*, **1996**, *33*, 185.

96JHC431 A.A.A. Hafez, A.K. El-Dean, A.A. Hassan, H.S. El-Kashef, S. Rault and M. Robba, *J. Heterocyclic Chem.*, **1996**, *33*, 431.

96JHC687 G.M. Coppola, R.E. Damon and H. Yu, *J. Heteroyclic Chem.*, **1996**, *33*, 687.

96JHC727 A.P. Halverson and L.W. Castle, *J. Heterocyclic Chem.*, **1996**, *33*, 727.

96JHC873 A. Daich and B. Decroix, *J. Heterocyclic Chem.*, **1996**, *33*, 873.

96JHC923 J.-K. Luo, R.F. Federspiel and R.N. Castle, *J. Heterocyclic Chem.*, **1996**, *33*, 923.

96JHC1017 Y. Tominaga, L.W. Castle and R.N. Castle, *J. Heterocyclic Chem.*, **1996**, *33*, 1017.

96JHC1079 B.S. Jursic, *J. Heterocyclic Chem.*, **1996**, *33*, 1079.

96JHC1123 D.E. Tupper, T.M. Hotten and W.G. Prowse, *J. Heterocyclic Chem.*, **1996**, *33*, 1123.

96JHC1251 A. Mamouni, A. Daich, B. Decroix, *J. Heterocyclic Chem.*, **1996**, *33*, 1251.

96JHC1319 Y. Tominaga, L.W. Castle and R.N. Castle, *J. Heterocyclic Chem.*, **1996**, *33*, 1319.

96JMC2068 S. Rault, J.-C. Lancelot, H. Prunier, M. Robba, P. Renard, P. Delagrange, B. Pfeiffer, D.-H. Caignard, B. Guardiola-Lemaitre and M. Hamon, *J. Med. Chem.*, **1996**, *39*, 2068.

96JOC2242 P.V. Bedworth, Y. Cai, A. Jen and S.R. Marder, *J. Org. Chem.*, **1996**, *61*, 2242.

96JOC3657 M. Kozaki, J.P. Parakka and M.P. Cava, *J. Org. Chem.*, **1996**, *61*, 3657.

96JOC4708 G. Barbarella, M. Zambianchi, A. Bongini and L. Antolini, *J. Org. Chem.*, **1996**, *61*, 4708.

96JOC4833 J.L. Reddinger and J.R. Reynolds, *J. Org. Chem.*, **1996**, *61*, 4833.

96JOC6166 C.O. Kappe and A. Padwa, *J. Org. Chem.*, **1996**, *61*, 6166.

96JOC6523 V.J. Majo and P.T. Perumal, *J. Org. Chem.*, **1996**, *61*, 6523.
96JOC6907 R. Wu, J.S. Schumm, D.L. Pearson and J.M. Tour, *J. Org. Chem.*, **1996**, *61*, 6906.
96JOC8074 M.S. Mubarak and D.G. Peters, *J. Org. Chem.*, **1996**, *61*, 8074.
96JOC9016 J. Castro, A. Moyano, M.A. Pericas, A. Riera, A.E. Greene, A. Alvarez-Larena and J. F. Piniella, *J. Org. Chem.*, **1996**, *61*, 9016.
96JOC9503 M.J. Dabdoub, V.B. Dabdoub, M.A. Pereira and J. Zukerman-Schpector, *J. Org. Chem.*, **1996**, *61*, 9503.
96JOM21 J. Chen, C.L. Day, R.A. Jacobson and R.J. Angelici, *J. Organomet. Chem.*, **1996**, *522*, 21.
96JOM149 J. Chen, Y. Su, R.A. Jacobson and R.J. Angelici, *J. Organomet. Chem.*, **1996**, *512*, 149.
96JOM213 C. Moreau, F. Serein-Spirau, M. Bordeau, C. Biran and J. Dunoguès, *J. Orgaonmet. Chem.*, **1996**, *522*, 213.
96JOM425 S.S.H. Mao and T.D. Tilley, *J. Organomet. Chem.*, **1996**, *521*, 425.
96JPC1524 C. Alemán and L. Julia, *J. Phys. Chem.*, **1996**, *100*, 1524.
96JPC11029 L.E. Bolívar-Martinez, M.C. dos Santos and D.S. Galvão, *J. Phys. Chem.*, **1996**, *100*, 11029.
96JPC18683 R.S. Becker, J. S. de Melo, A.L. Macanita and F. Elisei, *J. Phys. Chem.*, **1996**, *100*, 18683.
96LA171 R.W. Saalfrank, A. Welch, M. Haubner and U. Bauer, *Liebigs Ann.*, **1996**, 171.
96LA239 D. Prim, D. Joseph and G. Kirsch, *Liebigs Ann.*, **1996**, 239.
96LA279 P. Bäuerle, G. Götz, A. Synowczyk and J. Heinze, *Liebigs Ann.*, **1996**, 279.
96LA697 D. Döpp, A.A. Hassan, and G. Henkel, *Liebigs Ann.*, **1996**, 697.
96M297 M. Gutschow, H. Schroter, G. Kuhnle and K. Eger, *Monatsh. Chem.*, **1996**, *127*, 297.
96M529 C. Rivas, R.A. Bolivar, R. González and F. Vargas, *Monatsh. Chem.*, **1996**, *127*, 529.
96MI54 J.P. Parakka, J.A. Jeevarajan, A.S. Jeevarajan, L.D. Kispert and M.P. Cava, *Adv. Mater.*, **1996**, *8*, 54.
96MI135 Y. Zhang, Z. Wei, W. Yan, P. Ying, C. Ji, X. Li, Z. Zhhou, X. Sun and Q. Xin, *Catalysis Today*, **1996**, *30*, 135.
96MI161 K.A. Murray, A.B. Holmes, S.C. Moratti and R.H. Friend, *Syn. Metals*, **1996**, *76*, 161.
96MI199 I.-S. Han, C.K. Kim, H.J. Jung and I. Lee, *Theo. Chim. Acta*, **1996**, *93*, 199.
96MI212 D.K. Lee, I.C. Lee, S.K. Park, S.Y. Bae and S.I. Woo, *J. Catalysis*, **1996**, *159*, 212.
96MI236 M.V. Landau, D. Berger and M. Herskowitz, *J. Catalysis*, **1996**, *158*, 236.
96MI411 P. Bjork, A.B. Hornfeldt, S. Gronowitz and U. Edvardsson, *Eur. J. Med. Chem.*, **1996**, *31*, 411.
96MI555 C. Branger, M. Lequan, R.M. Lequan, M. Barzoukas and A. Fort, *J. Mater. Chem.*, ***1996***, *6*, 555.
96MI672 A. Merz and C. Rehm, *J. Prakt. Chem. / Chem-Zeit.*, **1996**, *338*, 672.
96MI675 N. Desrosiers, J.-Y. Bergeron, M. Belletête, G. Durocher and M. Leclerc, *Polymer*, **1996**, *37*, 675.
96MI683 Y. Kawakami, H. Kitani, S. Yuasa, M. Abe, M. Moriwaki, M. Kagoshima, M. Terasawa and T. Tahara, *Eur. J. Med. Chem.*, **1996**, *31*, 683.
96MI733 V.U. Ahmad and N. Alam, *Phytochem.*, **1996**, *42*, 733.
96MI916 M.E. Ryan, A.M. Hynes, S.H. Wheale, J.P.S. Badyal, C. Hardacre and R.M. Ormerod, *Chem. Mater.*, **1996**, *8*, 916.
96MI1078 T. Isoda, S. Nagao, X. Ma, Y. Korai and I. Mochida, *Energy Fuels*, **1996**, *10*, 1078.
96MI1399 G.M. Tsivgoulis and J.-M. Lehn, *Chem. Eur. J.*, **1996**, *2*, 1399.96MI2825 C. Diaz and C. Leal, *Polyhedron* , **1996**, *15*, 2825.
96MI7671 X. Wu, T.-A. Chen and R.D. Riecke, *Macromolecules*, **1996**, *29*, 7671.
96MI3370 G. Engelmann, W. Jugelt, G. Kossmehl, H.-P. Welzel, P. Tschuncky and J. Heinze, *Macromolecules*, **1996**, *29*, 3370.
96MI4952 A.J. Lovinger, D.D. Davis, A. Dodabalapur, H.E. Katz and L. Torsi, *Macromolecules*, **1996**, *29*, 4952.
96MI7671 X. Wu, T.-A. Chen and R.D. Riecke, *Macromolecules*, **1996**, *29*, 7671.
96O325 J. Chen, V.G. Young, Jr. and R.J. Angelici, *Organometallics*, **1996**, *15*, 325.
96O786 A.J. Deeming, S.N. Jayasuriya, A.J. Arce and Y. De Sanctis, *Organometallics*, **1996**, *15*, 786.
96O872 M.G. Partridge, L.D. Field and B.A. Messerle, *Organometallics*, **1996**, *15*, 872.
96O1223 J. Chen, L.M. Daniels and R.J. Angelici, *Organometallics*, **1996**, *15*, 1223.
96O2000 A. Kunai, T. Ueda, K. Horata, E. Toyoda, I. Nagamoto, J. Ohshita, M. Ishikawa, K. Tanaka, *Organometallics*, **1996**, *15*, 2000.
96O2041 D. Deffieux, D. Bonafoux, M. Bordeau, C. Biran and J. Dunoguès, *Organometallics*, **1996**, *15*, 2041.
96O2679 M. Pasneque, S. Taboada and E. Carmona, *Organometallics*, **1996**, *15*, 2678.
96O2727 J. Chen, V.G. Young, Jr. and R.J. Angelici, *Organometallics*, ***1996***, *15*, 2727.
96O2905 A.W. Myers and W.D. Jones, *Organometallics*, **1996**, *15*, 2905.
96O4352 E. Viola, C. L. Sterzo and F. Trezzi, *Organometallics*, **1996**, *15*, 4352.

96O4605 C. Bianchini, D. Fabbri, S. Gladiali, A. Meli, W. Pohl and F. Vizza, *Organometallics*, **1996**, *15*, 4604.
96P4 B. Naumann, R. Bohn, F. Fulop, G. Bernath, *Pharmazie*, **1996**, *51*, 4.
96P396 K. Gorlitzer and P.-M. Dobberkau, *Pharmazie*, **1996**, *51*, 386.
96PS99 P.D. Clark, S.T.E. Mesher and A. Primak, *Phosphorus, Sulfur, and Silicon*, **1996**, *114*, 99.
96S1193 H. Kojima, K. Nakamura, K. Yamamoto and H. Inoue, *Synthesis*, **1996**, 1193.
96SC569 H.-H. Tso, Y.-M. Chang and H. Tsay, *Synth. Commun.*, **1996**, *26*, 569.
96SC1363 W. Halczenko and G.D. Hartman, *Synth. Commun.*, **1996**, *26*, 1363.
96SC2205 M. Coffey, B.R. McKellar, B.A. Reinhardt, T. Nijakowski and W.A. Feld, *Synth. Commun.*, **1996**, *26*, 2205.
96SC4157 K.R. Reddy, M.V.B. Rao, H. Illa and H. Junjappa, *Synth. Commun.*, **1996**, *26*, 4157.
96SL461 Y. Li, M. Matsuda, T. Thiemann, T. Sawada, S. Mataka and M. Tashiro, *Synlett*, **1996**, 461.
96T471 J. Nakayama, H. Dong, K. Sawada, A. Ishii and S. Kumakura, *Tetrahedron*, **1996**, *52*, 471.
96T1723 A.C. Tome, J.A.S. Cavaleiro and R.C. Storr, *Tetrahedron*, **1996**, *52*, 1723.
96T1735 A.C. Tome, J.A.S. Cavaleiro and R.C. Storr, *Tetrahedron*, **1996**, *52*, 1735.
96T3313 C. Dell'Erba, M. Novi, G. Petrillo, D. Spinelli and C. Tavani, *Tetrahedron*, **1996**, *52*, 3313.
96T3953 V. Meille, E. Schulz, M. Lemaire, R. Faure and M. Vrinat, *Tetrahedron*, **1996**, *52*, 3953.
96T6241 G. Desimoni, G. Faita, S. Garau and P. Righetti, *Tetrahedron*, **1996**, *52*, 6241.
96T6893 T. Akita and K. Kobayashi, *Tetrahedron*, **1996**, *52*, 6893.
96T8863 H.A. Muathen, *Tetrahedron*, **1996**, *52*, 8863.
96T11281 B. Cosimelli, D. Neri and G. Roncucci, *Tetrahedron*, **1996**, *52*, 11281.
96T11915 R. Al-Omran, M.M.A. Khalik, H. Al-Awadhi and M.H. Elnagdi, *Tetrahedron*, **1996**, *52*, 11915.
96T12459 T.-s. Chou and C.-T. Wang, *Tetrahedron*, **1996**, *52*, 12459.
96T13547 P. Ohier, A. Daïch and B. Decroix, *Tetrahedron*, **1996**, *52*, 13547.
96T14975 D. Wensbo and S. Gronowitz, *Tetrahedron* , **1996**, *52*, 14975.
96TA2721 O. Tempkin, T.J. Blacklock, J.A. Burke and M. Anastasia, *Tetrahedron: Asymmetry*, **1996**, *7*, 2721.
96TL893 Y. Gao, W.-L. Wu, B. Ye, R. Zhou and Y.-L. Wu, *Tetrahedron Lett.*, **1996**, *37*, 893.
96TL1617 H. Higuchi, H. Koyama, H. Yokota and J. Ojima, *Tetrahedron Lett.*, **1996**, *37*, 1617.
96TL2821 T. Ozturk, *Tetrahedron Lett.*, **1996**, *37*, 2821.
96TL5925 K. Tanaka, Y. Kitahara, H. Suzuki, H. Osuga and Y. Kawai, *Tetrahedron Lett.*, **1996**, *37*, 5925.
96TL7707 P. Pigeon and B. Decroix, *Tetrahedron Lett.*, **1996**, *37*, 7707.

Chapter 5.2

Five Membered Ring Systems: Pyrroles and Benzo Derivatives

Daniel M. Ketcha
Wright State University, Dayton, OH, USA

5.2.1 INTRODUCTION

A recent monograph, *Indoles,* by Richard J. Sundberg for the *Best Synthetic Methods* series presents a comprehensive but concise summary of the reactions and synthesis of this heterocycle <96MI1>. Theoretical studies of pyrroles and indoles include structural analyses of indole <96JCS(PI)2653> and indolyl radicals <96JCS(PI)2663> by density functional and hybrid Hartree-Fock/density functional methods, and the use of group additivity parameters for the prediction of the enthalpies of formation of these species <96T14335>. Reviews on the photochemical reactions of pyrroles <96H(43)1305, 96H(43)1529> and the chemistry of 7-azabicyclo[2.2.1]hepta-2,5-dienes and related pyrrole Diels-Alder chemistry <96CRV1179> have appeared during the past year as have reviews on β-carbolines <96OPP1>, indoloquinazolines <96H(42)453>, and the Pictet-Spengler reaction <96MI2>.

5.2.2 SYNTHESIS OF PYRROLES

Some advances have been made in the Paal-Knorr synthesis of pyrroles by the condensation of primary amines with 1,4-dicarbonyl species. For instance, a new synthetic route to monosubstituted succinaldehydes allows for the facile preparation of 3-substituted pyrroles <96TL4099>. Additionally, a general method for the synthesis of 1-aminopyrroles has been devised by the condensation of commercially available 2,2,2-trichloroethyl- or 2-(trimethylsilyl)ethylhydrazine with 1,4-dicarbonyl compounds <96JOC1180>. A related route to such compounds involves the reaction of α-halohydrazones with β-dicarbonyl compounds <96H(43)1447>. Finally, hexamethyldisilazane (HMDS) can be utilized as the amine component in the Paal-Knorr synthesis in the presence of alumina, and this modification has been employed in the synthesis of an azaprostacyclin analog <96S1336>.

Katritzky has developed a new synthesis of pyrroles by the reaction of alkynyloxiranes **1** with primary amines yielding highly versatile 2-[(benzotriazol-1-yl)methyl]pyrroles (**2**) <96JOC1624>. Interestingly, the Bt-methyl side chain can be further elaborated by nucleophilic substitution or *via* utilization of the anion stabilizing capability of this moiety to effect lithiation and substitution of the methyl group. These features have been utilized in a new route to indoles **3**, *via* a sequence involving lithiation, reaction with α,β-unsaturated aldehydes or ketones, and dehydrobenzotriazolylation-cyclodehydration <96TL5641>.

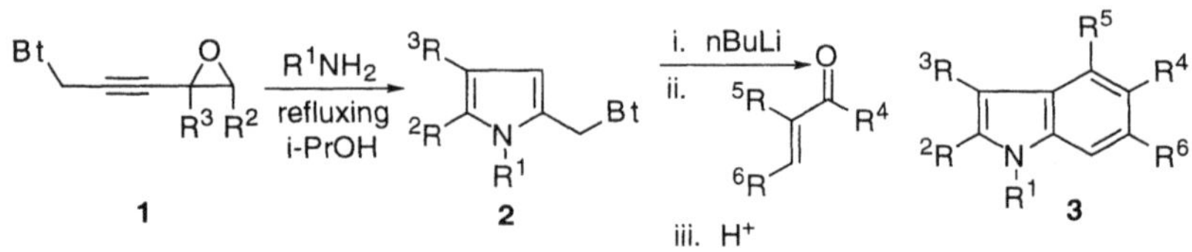

In a deceivingly simple process apparently involving a butatriene intermediate, a one-pot preparation of ethyl 5-methylpyrrole-2-carboxylate (**6**) from diethyl acetamidomalonate (**4**) and 1,4-dichloro-2-butyne (**5**) has been described <96JOC9068>.

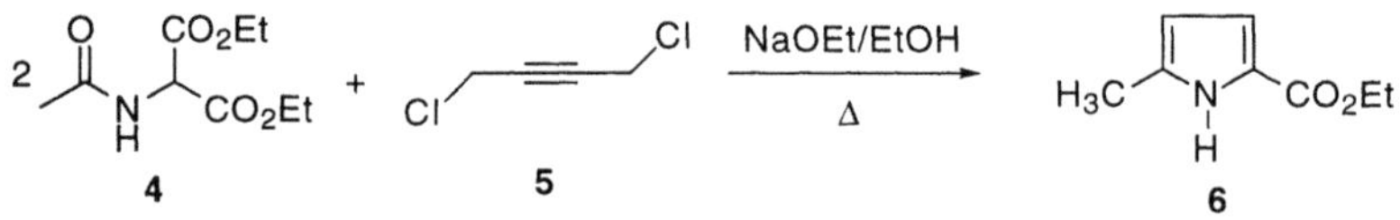

In another route employing alkynes, it was found that heating propargyl azadienes **7** in toluene at 25-60 °C produces pyrrolic imines **8** which hydrolyze upon work-up to afford 3-acylpyrroles **9** <96JOC2185>. This *exo-dig* cyclization occurs with complete chemoselectivity wherein the more substituted nitrogen is involved in the cyclization.

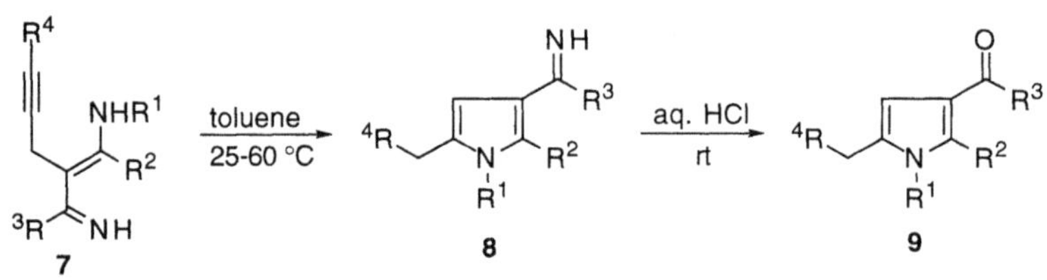

In a process related to the Knorr pyrrole synthesis, condensation of β-amino alcohols **10** with β–dicarbonyl compounds **11** affords β–hydroxy enamines **12** which are then oxidized to the pyrroles **13** <96TL9203>.

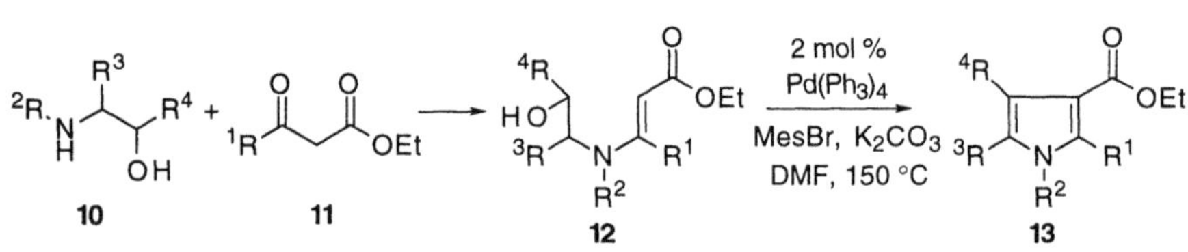

Alternatively, Cushman has devised a facile route to pyrroles by the reaction of Boc-α-amino aldehydes or ketones **14** with the lithium enolates of ketones **15** to afford aldol intermediates **16** which cyclize to pyrroles **17** under mild acidic conditions <96JOC4999>. This method offers several advantages over the Knorr since it employs readily available Boc-α-amino aldehydes or ketones and utilizes simple ketones instead of the β-diketo compounds or β-keto esters normally used in the Knorr.

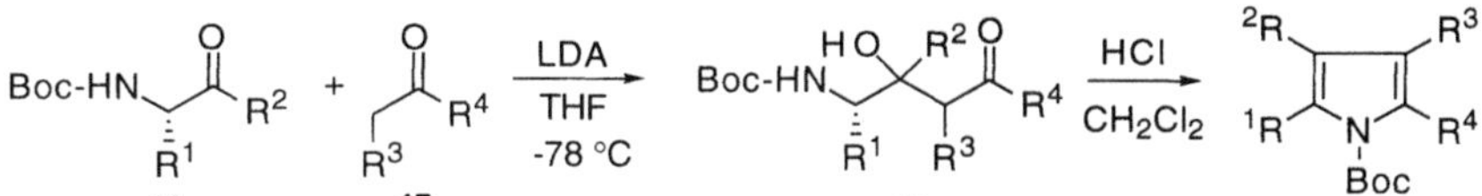

14 **15** **16** **17**

Vinamidinium salts have been used for the preparation of 2,3- or 2,5-disubstituted pyrroles. Thus, reaction of sarcosine ethyl ester with **18** results in an amine-exchange reaction at the least hindered position. Anion mediated cyclization and elimination of dimethylamine leads to **19** <96T6879>.

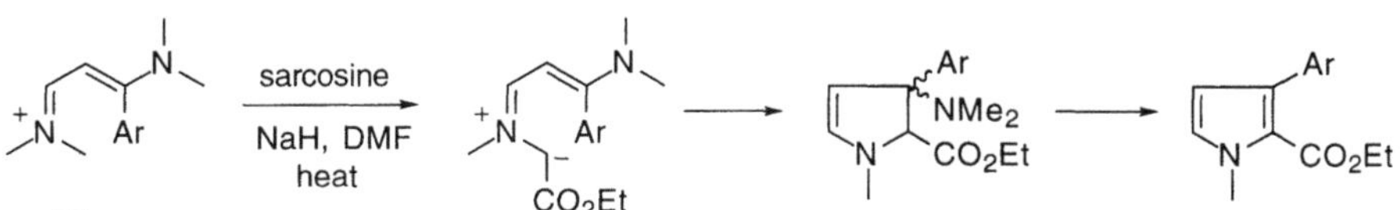

18 **19**

Base induced addition of tosylmethyl isocyanide (TosMIC) <96JOC8730, 96S851, 96SC1839> or ethyl isocyanoacetate <96JCS(P1)183, 96JCS(P1)417> to alkenes continues to represent one of the most popular approaches to pyrroles. In fact, a one-pot, three component condensation of aldehydes, nitromethane and TosMIC has been developed by van Leusen as a reliable route to β-nitropyrroles <96S871>. Interestingly, Gribble reported an abnormal Barton-Zard process in the reaction of 3-nitro-1-phenylsulfonylindole (**20**) with ethyl isocyanoacetate which affords the pyrrolo[2,3-*b*]indole **21** rather than the anticipated pyrrolo[3,4-*a*]indole **22** <96CC1909>.

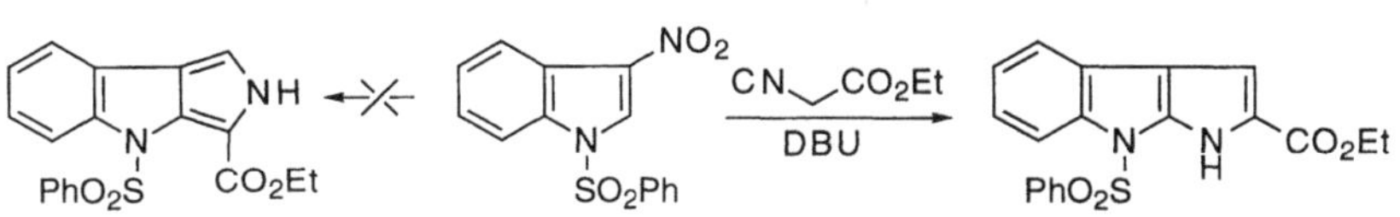

22 **20** **21**

The intramolecular cycloaddition of munchnone intermediates (derived from the cyclodehydration of *N*-acyl amino acids) with 1,3-dipolarophiles was employed to construct the mitomycin skeleton. Thus, heating alkynyl acids **23** with acetic anhydride forms the intermediates **24** which undergo cyclization with loss of carbon dioxide to afford the 4-oxo-tetrahydroindoles **25** <96TL2887>

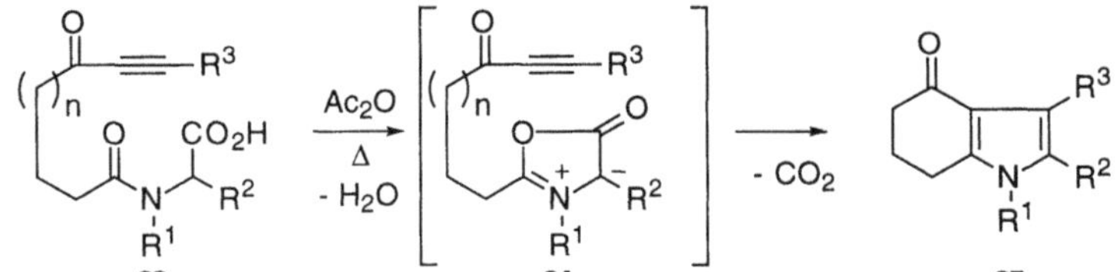

23 **24** **25**

Application of the Ugi reaction to the synthesis of diverse pyrrole libraries has been limited by a lack of commercially available isocyanides. To circumvent this problem, Armstrong has developed the concept of a "universal isocyanide": an input for the Ugi reaction that can be converted, post-condensation, into other functionalities. In this approach, 1-isocyanocyclohexene is used as the isocyanide input along with a carboxylic acid, a primary amine, and an aldehyde in this multiple component condensation (MCC). The resulting α-(acylamino)-amides **26** undergo acid-activated conversion of the cyclohexenamide moiety to munchnone intermediates **27** which react with dipolarophiles affording the pyrroles **28**. <96JA2574>. Solid-phase versions of this reaction have appeared <96TL1149, 96TL2943> as well as a review on MCC strategies in combinatorial synthesis <96ACR123>.

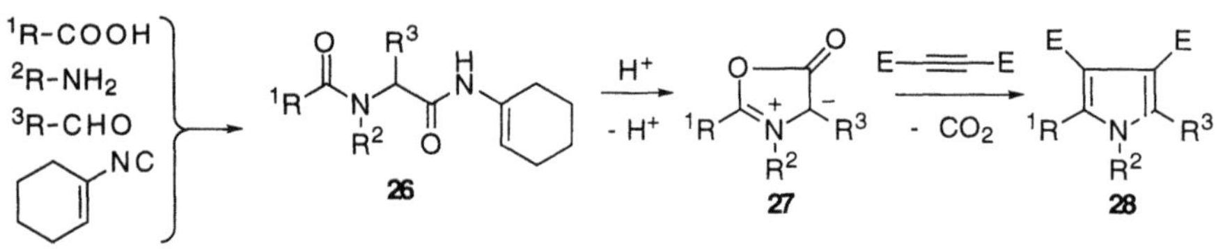

Titanium-acetylene complexes **29** generated *in situ* from acetylenes, Ti(O-*i*-Pr)4 and *i*-PrMgX react with imines to form azatitanacyclopentenes **30** which then react with carbon monoxide under atmospheric pressure to provide pyrroles **31** <96TL7787>. This reaction, which utilizes commercially available reagents is an improvement over a related procedure *via* the corresponding zirconium complexes under 1500 psi CO <89JA776>.

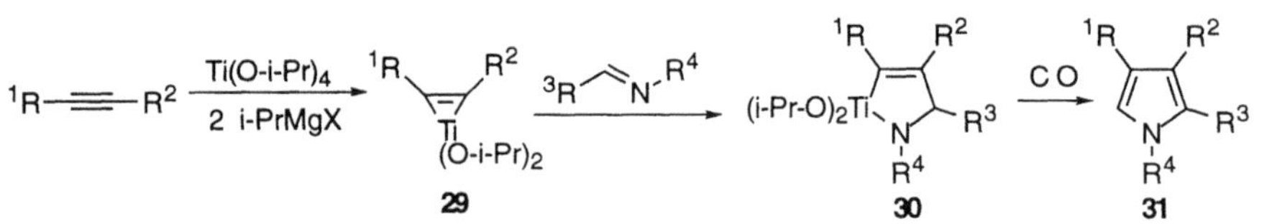

5.2.3 REACTIONS OF PYRROLES

Although the synthesis of multiple-pyrrolyl compounds can be achieved by S_NAr reactions of perfluoroaromatics with pyrrolylsodium at ambient temperature <96JOC9012>, deleterious side reactions are often observed during attempted *N*-alkylations of the alkali salts of pyrrole. A protocol has therefore been developed for the preparation of *N*-arylmethylenepyrroles by reduction of the corresponding *N*-acyl derivatives by treatment with sodium borohydride/boron trifluoride etherate in a sealed tube <96S457> .

While the propensity for reaction with electrophiles at the C-2 position is perhaps the most characteristic reaction of the pyrrole nucleus, much of the new C-2 chemistry of pyrrole published during the last year centers on the radical chemistry of this site. For instance, heterocyclic aromatic sulfones such as **32** undergo a novel radical *ipso*-substitution upon treatment with tri-*n*-butyltin hydride to afford heteroaromatic stannanes **33** <96T11329>. Interestingly, the synthesis of some 2-tosylpyrroles can be accomplished either by direct Friedel-Crafts reaction of 3,4-disubstituted 2-pyrrolecarboxylates, or by a sequence involving iodination using *N*-iodosuccinimide (NIS) followed by treatment of the α-iodospecies with sodium *p*-toluenesulfinate (TsNa) in the presence of a copper salt <96BCJ3339>.

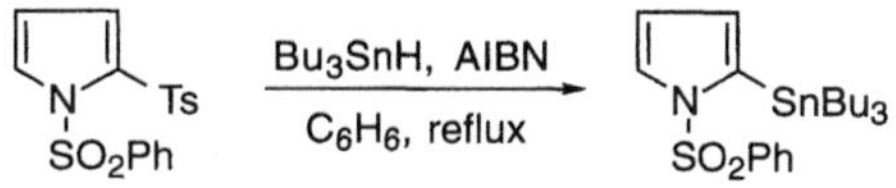

32 33

In another example of a radical process at the pyrrole C-2 position, it has been reported that reductive radical cycloaddition of 1-(2-iodoethyl)pyrrole and activated olefins, or 1-(ω-iodoalkyl)pyrroles **34** lead to cycloalkano[*a*]pyrroles **35** *via* electroreduction of the iodides using a nickel(II) complex as an electron transfer catalyst <96CPB2020>. Thus, it appears the radical chemistry of pyrroles portends to be a fertile area of research in the immediate or near future.

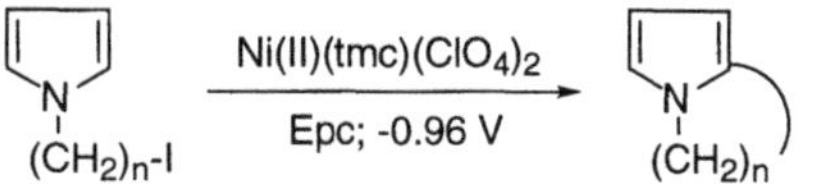

34 35

Whereas pyrroles normally undergo substitution at the C-2 position, 1-arylsulfonylpyrroles display a *tunable reactivity* in Friedel-Crafts acylations, wherein substitution occurs at C-3 in the presence of "hard acids" such as aluminum chloride but mainly at C-2 when catalyzed by weaker acids <81TL4899, 81TL4901>. An alternative route to 3-aroylpyrroles **38** has been introduced *via* irradiation of 1-acetyl- or 1-phenylsulfonylpyrrole (**36**) with arenethiocarboxamides <96H(43)463>. The proposed mechanism involves formation and scission of a thietane intermediate **37**, followed by hydrolysis of the resultant imine.

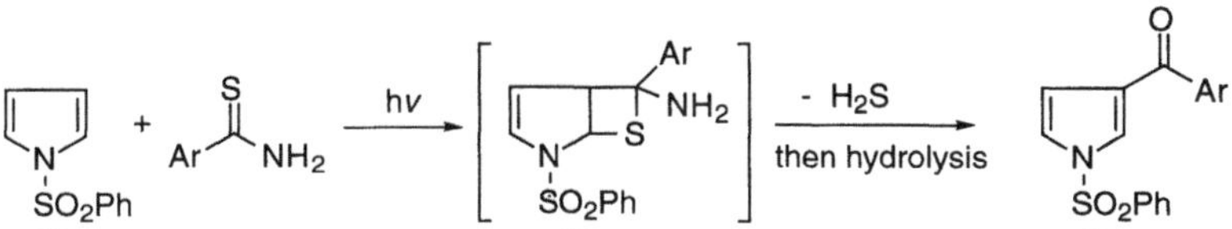

36 37 38

In a modified version of a previously published procedure <83JOC3214>, the thallium induced transposition of 3-acetyl-1-tosylpyrrole to the corresponding 2-(3-pyrrolyl)acetic acid has been reported employing thallium(III) nitrate supported on acidic montmorillonite K-10 <96SC1289>. In this sequence, *N*-tosylpyrrole (**39**) undergoes regioselective C-3 acylation to afford **40** which is rearranged with Tl(III)/K-10 in methanol to yield the ester **41**.

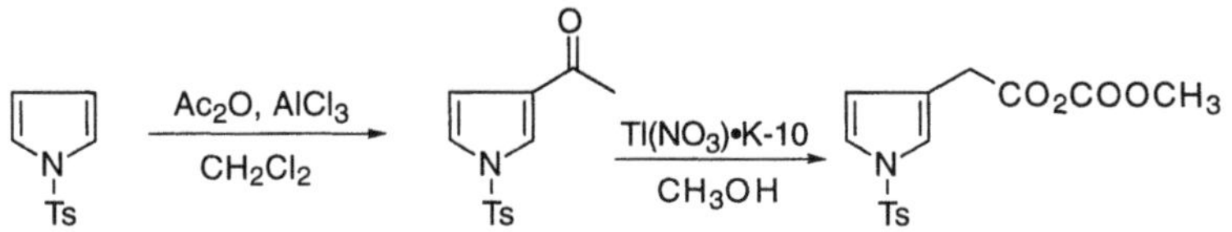

39 40 41

Although the regioselectivity of Friedel-Crafts acylations upon 1-phenylsulfonylpyrrole is ostensibly determined by the "hard-soft" nature of the catalyst <83JOC3214>, this paradigm may not be the controlling principle in determining the regioselectivity of acylations upon alkyl substituted derivatives. For instance, acylation of 3-alkyl-1-phenylsulfonylpyrroles **42** appears to be kinetically favored at the adjacent C-2 position affording **43** regardless of catalyst, whereas rearrangement to the presumably more stable C-5 isomer occurs after prolonged reaction times affording **44** <96TL1523>.

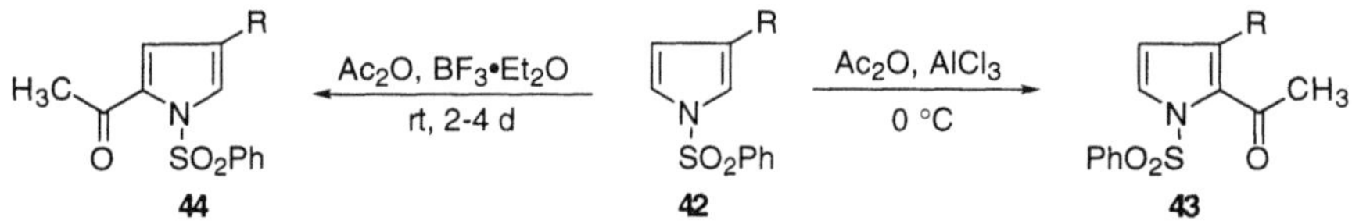

In what may be an alternate example of the same underlying principles, Murakami finds that Friedel-Crafts acylations of ethyl pyrrole-2-carboxylate with various acid chlorides occurs with high β-selectivity in the presence of aluminum chloride affording 4-acylpyrrole-2-carboxylates, whereas weaker Lewis acids produce mixtures of the 4- and 5-acyl derivatives <96CPB48>. This finding has been utilized in a new strategy for the synthesis of indoles, wherein the keto-acid **45**, formed by regioselective acylation of ethyl pyrrole-2-carboxylate with succinic anhydride, was reduced to **46** with triethylsilane in trifluoroacetic acid (TFA). Cyclization to the 7-oxo-tetrahydroindole **47** was then effected with trifluoroacetic anhydride (TFAA) <96CPB55>.

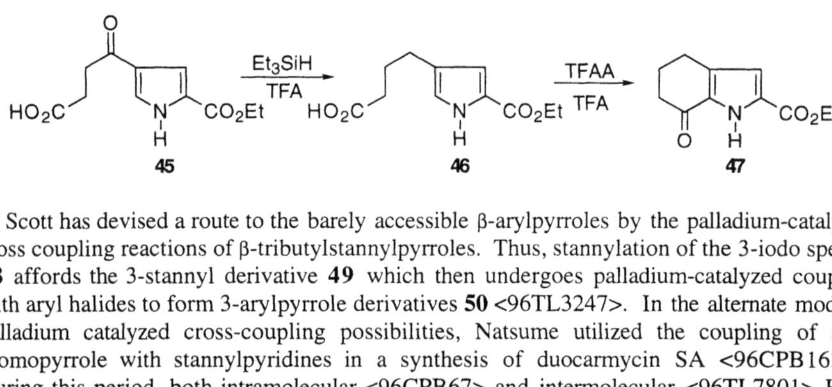

Scott has devised a route to the barely accessible β-arylpyrroles by the palladium-catalyzed cross coupling reactions of β-tributylstannylpyrroles. Thus, stannylation of the 3-iodo species **48** affords the 3-stannyl derivative **49** which then undergoes palladium-catalyzed coupling with aryl halides to form 3-arylpyrrole derivatives **50** <96TL3247>. In the alternate mode of palladium catalyzed cross-coupling possibilities, Natsume utilized the coupling of a 3-bromopyrrole with stannylpyridines in a synthesis of duocarmycin SA <96CPB1631>. During this period, both intramolecular <96CPB67> and intermolecular <96TL7801> Heck reactions of 3-bromo pyrroles were also reported.

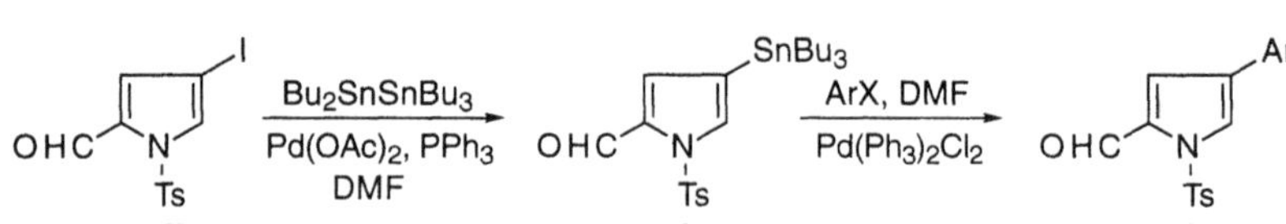

5.2.3.1 Cycloaddition Reactions of Pyrroles

N-Ethoxycarbonyl Dewar pyrrole (**51**), generated *via* a photofragmentation reaction at -60 deg. C either spontaneously rearranges to *N*-ethoxycarbonylpyrrole (**52**) or can be trapped as its cycloadduct with 1,3-diphenylisobenzofuran <96CC1519>.

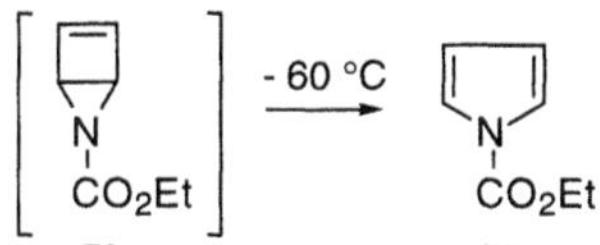

51 52

UV irradiation of heterocycles such as **53** in ethanol containing catalytic iodine leads to oxidative cyclization affording tosylphenanthrenoids **54** <96JOC1188>. The synthesis of a heterocycle related to **53** was accomplished by a titanium coupling reaction <96TL2721>.

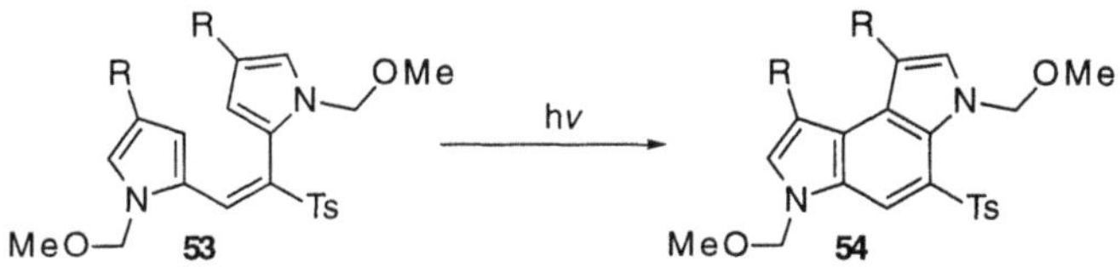

53 54

The analgesic alkaloid epibatidine (**57**) continues to receive much synthetic interest <96JOC4600, 96T11053, 96TL7845> and Trudell has devised a novel approach to the synthesis of this alkaloid, the key step of which utilizes a [4 + 2] cycloaddition of methyl 3-bromopropiolate with *N*-Boc-pyrrole (**55**) to afford the 7-azabicyclo[2.2.1]heptane skeleton **56** characteristic of this alkaloid <96JOC7189>.

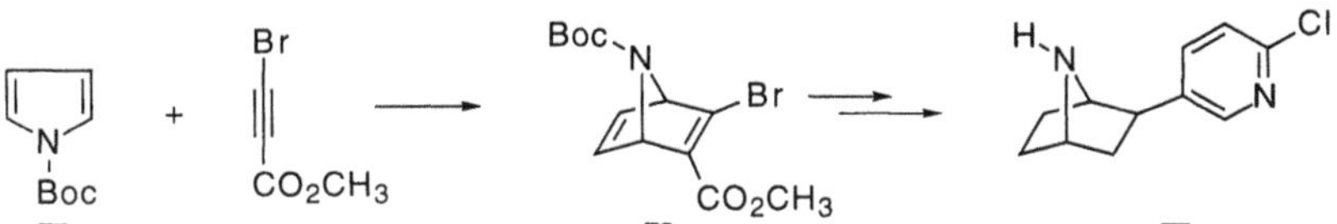

55 56 57

Utilizing an alternate mode of Diels-Alder reactivity, Harman has examined the cyclo-addition reactions of 4,5-η^2-Os(II)pentaammine-3-vinylpyrrole complexes with suitably activated dienophiles <96JA7117>. For instance, cycloaddition of the β-vinylpyrrole complex **58** with 4-cyclopentene-1,3-dione, followed by DDQ oxidation affords **59**, possessing the fused-ring indole skeleton of the marine cytotoxic agent, herbindole B.

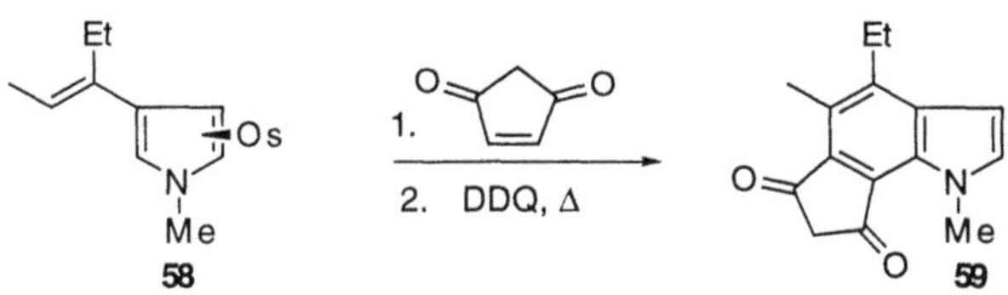

58 59

5.2.3.2 Reduction-Oxidation Chemistry of Pyrroles

Under the conditions of the Birch reduction, *N*-Boc amides such as **60** can be reductively alkylated in high yields, presumably *via* a dianion intermediate which is protonated by ammonia at C-5 leaving an enolate anion at C-2 <96JOC7664>. Quenching the reaction with alkyl halides or ammonium chloride then affords the 3-pyrrolines **61**.

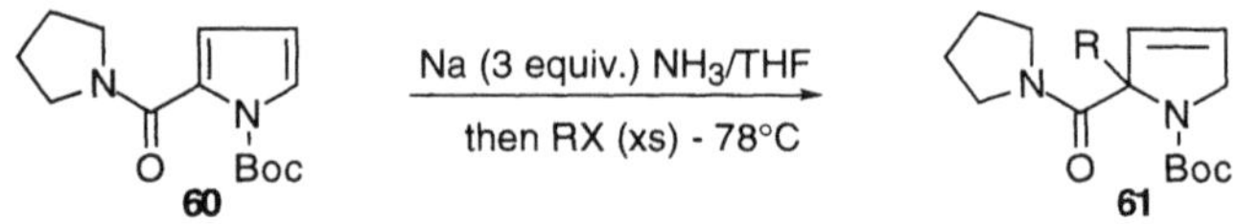

A rhodium catalyzed reduction of pyrrole 2,5-bis(propanoate) **62** afforded the *cis*-pyrrolidine **63** en route to the unusual heterocycle, azatriquinane (**64**) <96TL131>.

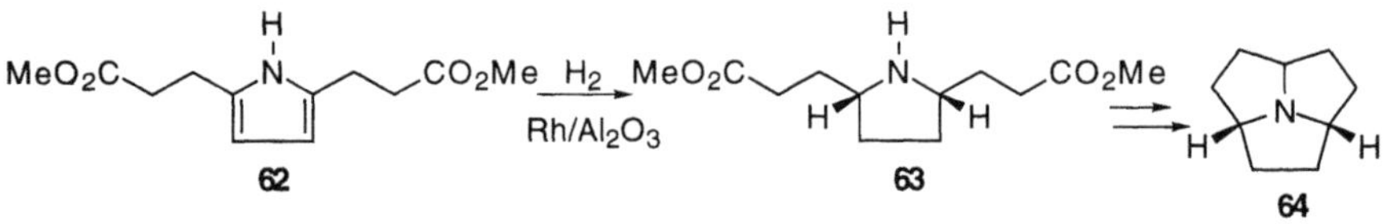

Selective oxidation of methyl pyrroles **65** possessing an α-carboxylic ester and sensitive β-substituents can be accomplished using cerium triflate in methanol <96TL315>. Moreover, the resultant α-methoxymethylpyrroles **66** may be converted to dipyrrylmethanes **67** in a "one-pot" sequence by treatment with 48% HBr. The dipyrrylmethanes, in turn, can be further oxidized to dipyrryl ketones by ceric ammonium nitrate <96JHC221>.

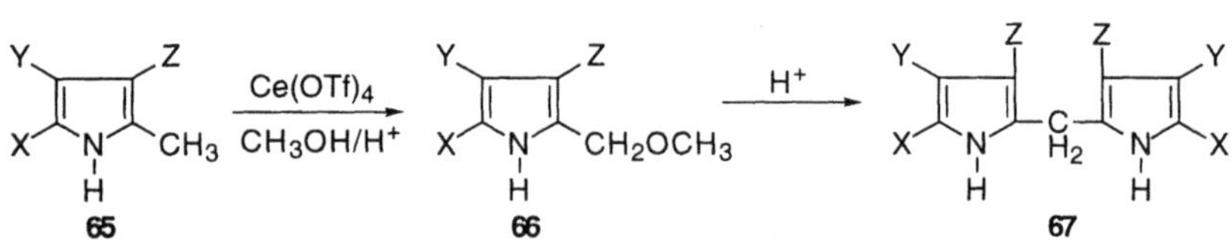

5.2.4 SYNTHESIS OF INDOLES

Modest improvements of most of the name indole syntheses have been made in the past year. Most notable amongst these involves the reductive amination of early intermediates in the Leimgruber-Batcho synthesis which allows for the preparation of indoles bearing branched *N*-alkyl substituents <96TL6045>. Other uses of this process were employed in the synthesis of conformationally restricted indolmycin <96TL3067> analogs and for the synthesis of linear pyrroloquinazolines <96JOC1155>. A novel use of the Reissert synthesis was utilized in a synthesis of the mitocene skeleton <96JOC816> and for a precursor to the antibiotic nosiheptide <96H(43)891>. Further work on the Engler variation of the Nenitzescu synthesis has been reported <96JOC9297>, while the original approach was utilized in a synthesis of the drug LY311727 <96JOC9055>. An example of the Fukuyama <94JA3127> indole synthesis employing intramolecular cyclization of aryl isocyanides was reported <96TL7099>. Finally, the Smith synthesis was utilized for the synthesis of 2-trifluoromethylindole <96H(43)1471>.

In the most economically efficient route to indoles reported this past year, *N*-substituted anilines **68** were reacted with triethanolamine in the presence of a homogeneous ruthenium catalyst to give the corresponding 1-substituted indoles **69** <96SC1349>.

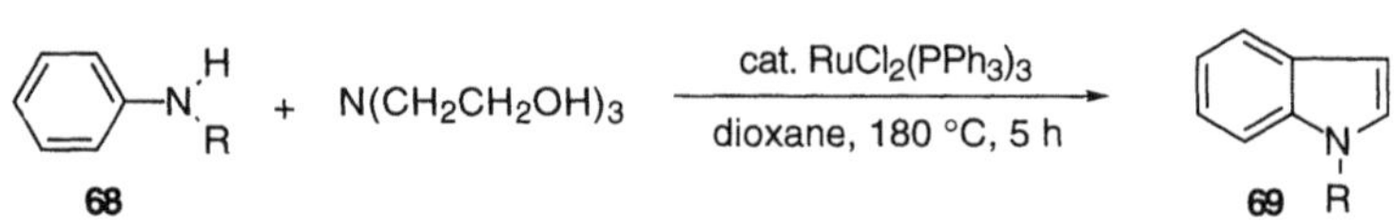

Buchwald has developed a route to indolines by the Pd-catalyzed intramolecular amination of aryl halides <96T7525> and applied this method to the synthesis of natural products. Thus, cyclization of tetrahydroquinoline **70** provided **71** which was elaborated to a key intermediate in syntheses of damirones A and B and makuluvamine <96JA1028>.

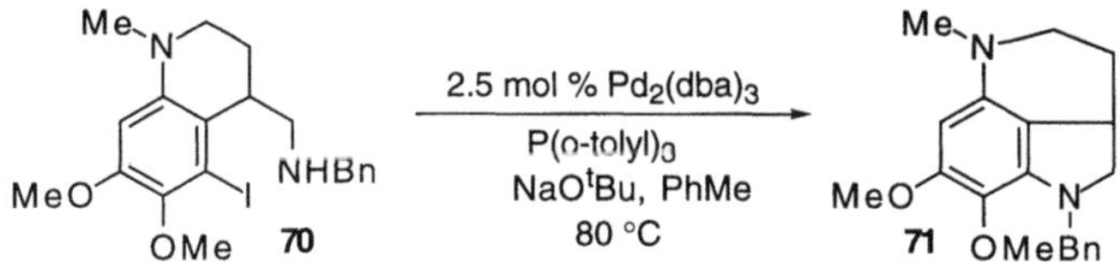

Benzene ring substituted indole derivatives are available by novel strategies involving CuI mediated nitrogen cyclization onto alkynes. One approach utilizes the regioselective *ortho* iodination of 4-substituted aromatic amines **72** with bis(pyridine)iodonium(I) tetrafluoroborate (IPy_2BF_4) <96JOC5804>. Palladium-catalyzed cross-coupling of the iodo species **73** with trimethylsilylacetylene affords the *o*-alkynylanilines **74**, which are cyclized to the indoles **75** using CuI. In a related process, *N*-Boc protected *o*-alkynylanilines were lithiated adjacent to the nitrogen prior to CuI mediated cyclization to afford 7-substituted indoles <96H(43)2741>.

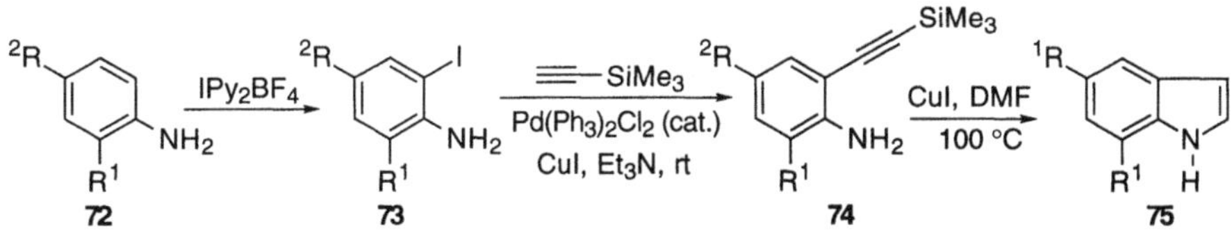

Larock has developed a new catalyst system for the Pd-catalyzed cyclization of olefinic tosylamides. Whereas typical conditions require either stoichiometric amounts of Pd(II) salts or catalytic amounts of Pd(II) in the presence of benzoquinone as a reoxidant, the new catalyst system utilizes catalytic $Pd(OAc)_2$ under an atmosphere of O_2 in DMSO with no additional reoxidant <96JOC3584>. Although *o*-vinylic tosylamides **76** can be cyclized to *N*-tosylindoles **77** using this catalyst system, $PdCl_2$/benzoquinone is more effective for such cyclizations. Interestingly, in the case of *o*-allylic tosylanilides, the cyclization can be modulated to afford either dihydroindole or dihydroquinoline products. In a related approach involving a common π-allyl Pd-intermediate, 2-iodoanilines were coupled with vinylic cyclopropanes or cyclobutanes in the presence of a Pd catalyst to afford dihydroindoles <96T2743>.

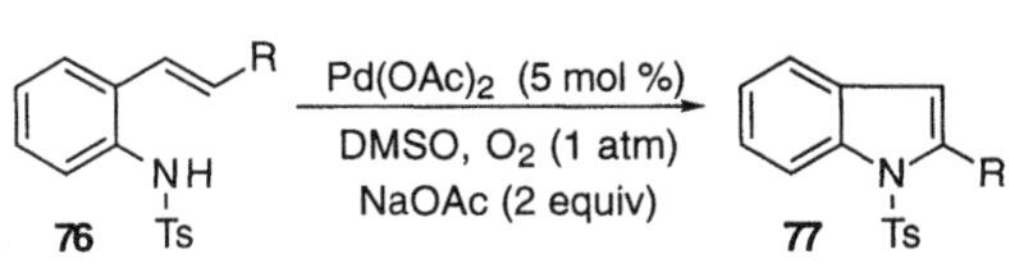

A novel approach to 3-substituted indolines and indoles *via* the anionic cyclization of 2-bromo-*N,N*-diallyanilines has been developed simultaneously by Bailey <96JOC2596> and Liebeskind <96JOC2594>. Thus, treatment of 2-bromo-*N,N*-diallylanilines **78** with 2 equivalents of ^{t}BuLi at -78 °C leads to the formation of the intermediate **79** which may be trapped with an electrophile to afford 3-substituted indolines **80**. Aside from ease of preparation, an additional benefit of the intramolecular carbolithiation of *o*-lithio-*N,N*-diallyl-anilines is the production of *N*-allyl-protected indolines, which are easily deprotected using

$Pd_2(dba)_3$/1,4-bis(diphenylphosphino)butane (DPPB) in the presence of 2-mercaptobenzoic acid <95TL1267>. The *N*-allylindolines can be easily oxidized to the corresponding indoles at room temperature with *o*-chloranil. Additionally, *N*-allylanilines were also found to undergo aromatic 3-aza-Cope rearrangements in the presence of Zeolite catalysts to give indoline derivatives as the major product <96TL5281>.

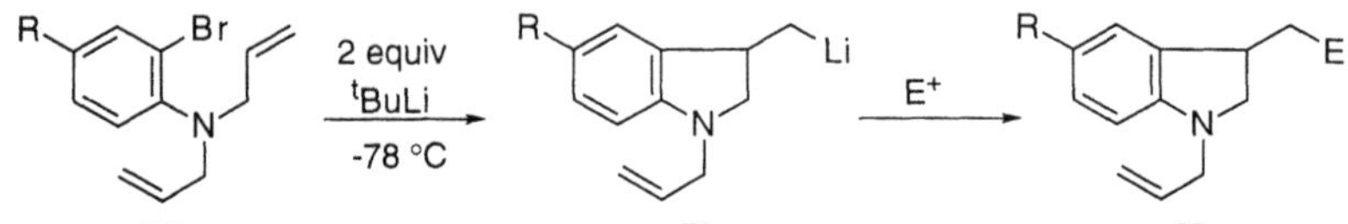

78 79 80

The mode of intramolecular ring opening of epoxides was found to be dependent on the nature of the ate complex formed upon halogen-metal exchange <96JA8733, 96AG(E)736>. Thus, while treatment of **81** with $Me_3Zn(SCN)Li_2$ afforded the *exo*-cyclized indoline derivative **82**, reaction with Me_3ZnLi gave the *endo*-cyclized 1,2,3,4-tetrahydroquinoline derivative **83**.

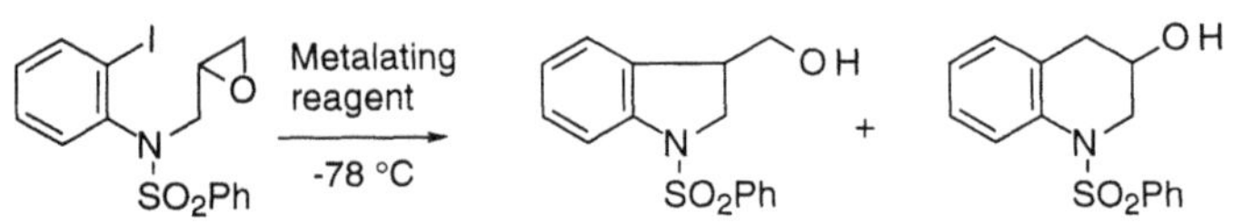

81 82 83

The intramolecular Heck reaction of polymer bound aryl halides such as **84** affords indole analogs **85** after cleavage of the final product from the resin with TFA <96TL4189>. Other notable uses of the Heck cyclization include a synthesis of an antimigraine agent <96TL4289>, and thia-tryptophans <96T14975>.

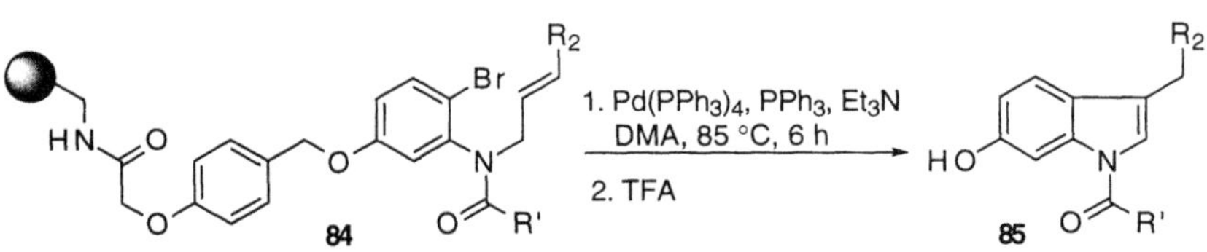

84 85

A novel route to indoles and quinolines has been developed by sequential Wittig and Heck reactions <96CC2253>. Thus, treatment of *o*-bromo- or iodo-*N*-trifluoroacetylanilines (**86**) with a stabilized phosphorane affords the corresponding enamines **87** as a mixture of isomers. Cyclization to **88** is effected by heating with palladium acetate, triphenylphosphine, and base.

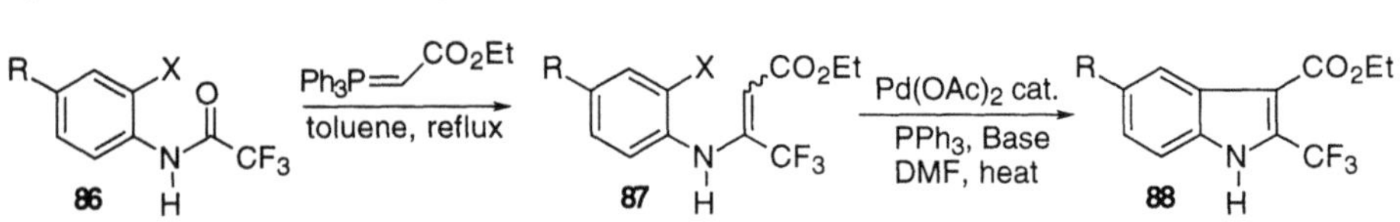

86 87 88

Grigg has developed radical and Pd-catalyzed cascade reactions to polycyclic indole derivatives <96T13441, 96TL4221, 96TL4413, 96TL6565>. In one example, substrates **89** (X = $C(CO_2Et)_2$, NSO_2Ph, Y = NR_2) undergo palladium catalyzed [2+2+2] cycloaddition to furnish systems such as **90** <96TL3399>.

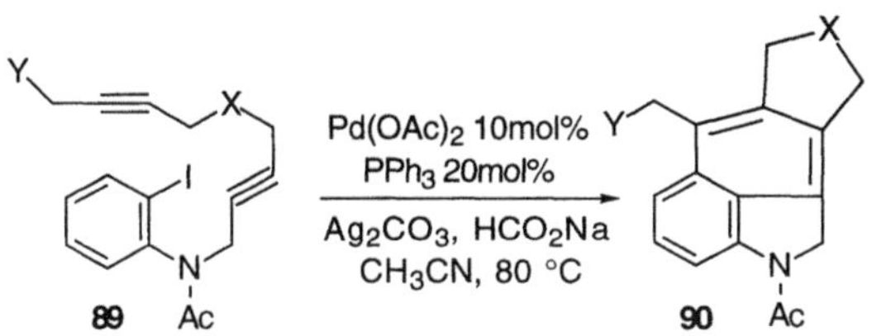

Feldman reported a route to dihydropyrroles, pyrroles, and indoles *via* the reaction of sulfonamide anions with alkynyliodonium triflates <96JOC5440>. Thus, upon nucleophilic addition of the anion of **91** to the β-carbon of the alkynyliodonium salt, the alkylidene carbene **92** is generated which can the undergo C-H insertion to the desired product **93**.

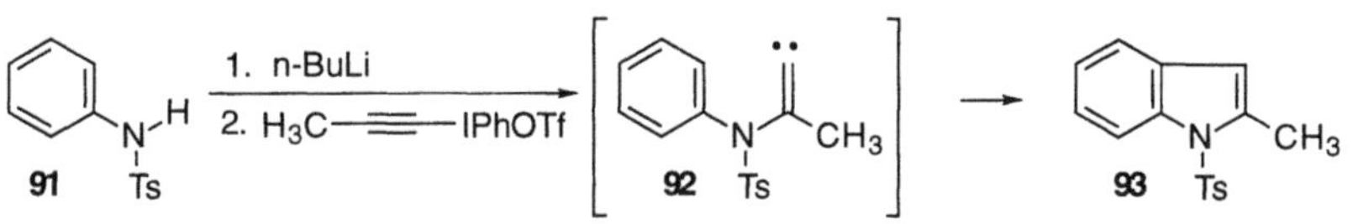

The Fischer indole synthesis of 2-arylindoles on a solid support has been reported using support bound 4-benzoylbutyric acid and variously substituted phenylhydrazine hydrochlorides <96TL4869>. While treatment of ketone **94** with $PhNHNH_2•HCl$ and $ZnCl_2$ in glacial acetic acid at 50 °C for 18 hours produced the desired indole **95** in excellent purity, reactions employing electron deficient hydrazines required slightly higher temperatures and yields declined due to decomposition of the PEG-PS support. Therein, switching to a polystyrene resin improved the yield and purity of the products.

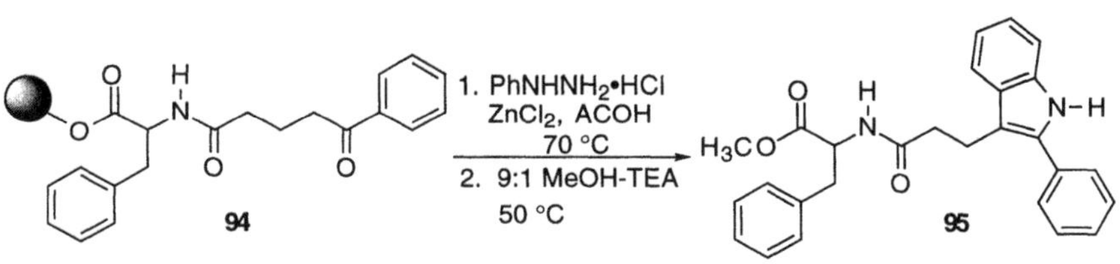

In a process related to the Fischer indole synthesis, arenesulfinamides **96** underwent thermal conversion into the corresponding indoles **97** *via* a [3.3] sigmatropic rearrangement followed by cyclization and loss of HSOH <96BSF329>.

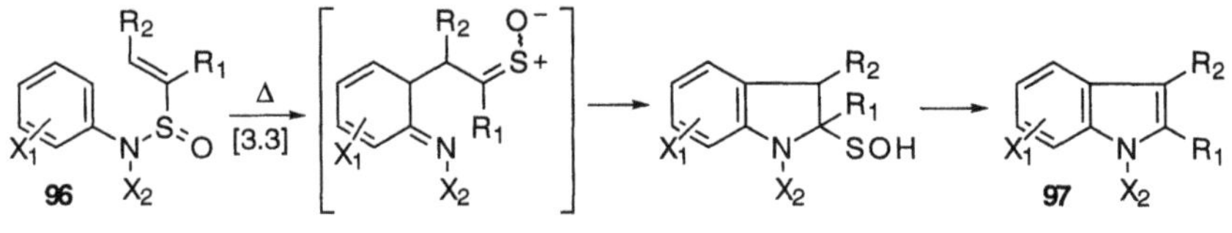

Imanishi and coworkers reported that thermolysis of the benzyl methyl ethers **98** (or benzyl-phosphonium salts) leads to high yields of indoles **99** in the absence of strong base. In the case of the methyl ethers, heating in the presence of an acid and catalyst and PPh_3 presumably involves *in situ* formation of a phosphonium salt intermediate <96JCS(P1)1261, 96H(42)513>.

However, thermolysis of the phosphonium salts (X=$^{+}PPh_3$) leads directly to the indolic products without need of acid catalyst or PPh_3, and thus may not proceed *via* a normal Wittig pathway. Alternatively, Hughes has effected a solid-phase version of this reaction employing a polymer-bound phosphonium salt and potassium tert-butoxide as base <96TL7595>. In this case, the phosphine oxide by-product remains bound to the polymer resin.

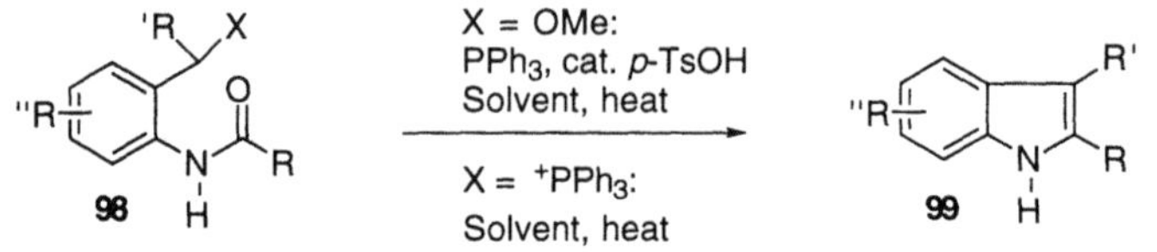

A few examples of the photocyclization of *N,N*-diarylamines or related species have been applied to the synthesis of spirocyclic indoline precursors <96TL37> and carbazole derivatives <96SC657, 96JCS(P1)669>.

A convenient modification of the Gassman oxindole synthesis was reported using ethyl (methylsulfinyl)acetate (**101**) activated by oxalyl chloride to generate the same chlorosulfonium salt **102** normally generated from ethyl (methylthio)acetate **100** and elemental chlorine <96TL4631>. Thus, treatment of the sulfoxide **101** with oxalyl chloride, followed by the addition of the desired aniline, triethylamine, and finally acid cyclization of **103** affords the oxindoles **104**. This procedure is particularly convenient for reactions carried out on smaller scales and for anilines that are susceptible to electrophilic halogenation.

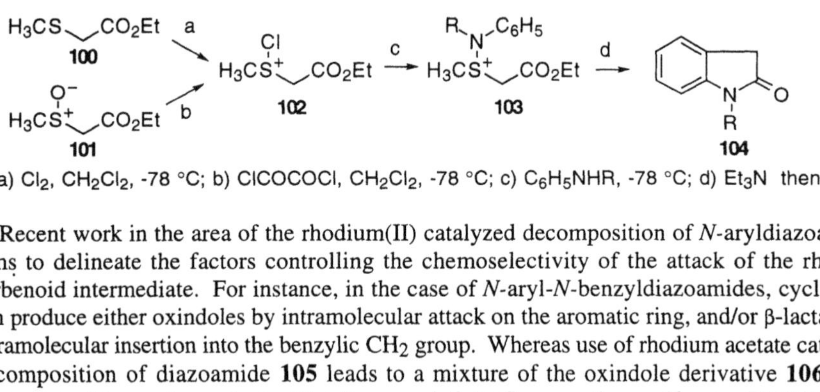

a) Cl_2, CH_2Cl_2, -78 °C; b) ClCOCOCl, CH_2Cl_2, -78 °C; c) C_6H_5NHR, -78 °C; d) Et_3N then HCl

Recent work in the area of the rhodium(II) catalyzed decomposition of *N*-aryldiazoamides aims to delineate the factors controlling the chemoselectivity of the attack of the rhodium carbenoid intermediate. For instance, in the case of *N*-aryl-*N*-benzyldiazoamides, cyclization can produce either oxindoles by intramolecular attack on the aromatic ring, and/or β-lactams by intramolecular insertion into the benzylic CH_2 group. Whereas use of rhodium acetate catalyzed decomposition of diazoamide **105** leads to a mixture of the oxindole derivative **106** (after treatment of the crude product with triisopropylsilyltrifluoromethanesulfonate, TIPSOTf) and the β-lactam, rhodium perfluorobutyramide leads to exclusive formation of the oxindole <96T2489>. Beyond ligand effects, addition of inorganic solids (e.g., zeolite Kβ) to the rhodium acetate decompositions also favors oxindole products <96JCS(P1)2793>.

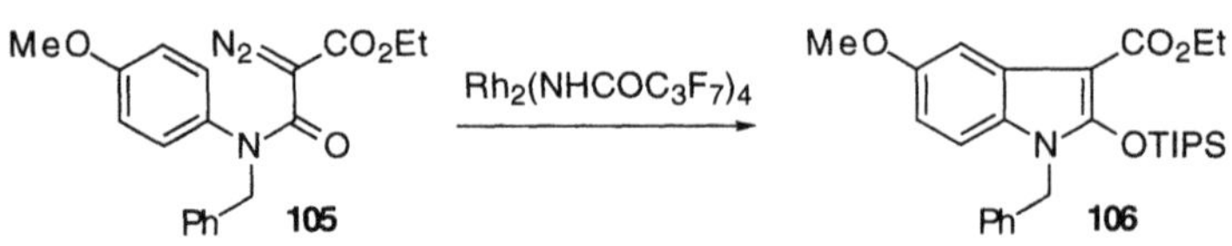

Meth-Cohn has developed an efficient one-pot synthesis of isatins **108** by the sequential treatment of *N*-substituted formanilides **107** with oxalyl chloride, Hunig's base, and bromine

<96TL9381, 96CC1395>. The reaction involves deprotonation of the Vilsmeier reagent, dimerization of the carbene formed and electrophilic cyclization of the dimer by bromonium ion action followed by aqueous hydrolysis.

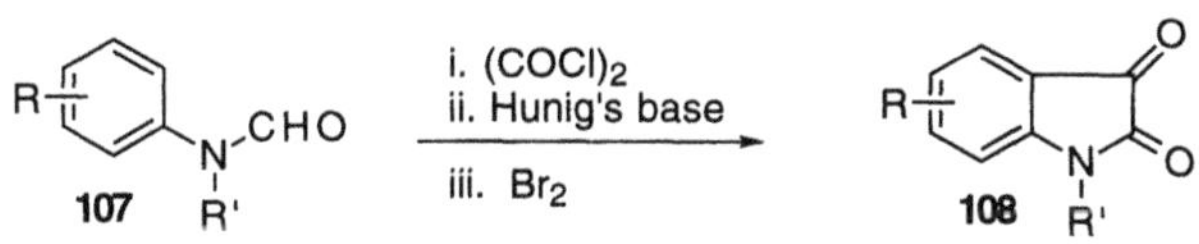

5.2.5 REACTIONS OF INDOLES

A novel method has been developed for the selective *N*-arylation of indoles with aryl electrophiles in the presence of potassium fluoride adsorbed onto basic alumina <96TL299>.

Although electrophilic substitution reactions of indoles normally take place at the C-3 position, 2-substituted indoles are commonly prepared by the lithiation of indoles possessing directed metallation groups (DMG) at the *N*-1 position. A new *N*-protecting group capable of serving this function is the diethoxymethyl group (DEM) which is easily introduced by heating the indole in neat triethyl orthoformate <96S1196> and removed by treatment with aqueous HCl and NaOH.

In addition to the preparation of 2-lithio species, recent advances in this area have concentrated on the generation and reactions of indolylzinc and indolylmagnesium derivatives. For instance, magnesiation at the 2-position of indole can be effected either by metal-halogen exchange of 2-iodo-1-phenylsulfonylindole with ethylmagnesium bromide <96H(42)105>, or by the direct deprotonation of 1-phenylsulfonylindole with Hauser bases (R_2NMgBr) or magnesium diamides [$(R_2N)_2Mg$] <96JCS(P1)2331>. Alternatively, *N*-protected indolylzinc halides can be prepared from either by transmetallation of indolyllithiums with zinc chloride or by direct oxidative addition of active zinc to iodoindoles <96JCS(P1)1927>. The resulting indolylzinc or -magnesium anions tolerate a broader range of functionalities than the corresponding lithio species, and moreover can undergo Pd-catalyzed cross-coupling reactions to afford 2-arylindoles. Thus, the Pd-catalyzed cross coupling of 1-phenylsulfonyl-2-indolylzinc chloride (**109**) with 2-halopyridines provided a key step in a synthesis of the indolo[2,3-a] quinolizidine system <96TL3071>, while coupling with acetate **110** led to intermediate **111** in a synthesis of inverto-yuehchukene <96JCS(P1)1213>. Alternatively, the Pd-catalyzed reaction 1-phenylsulfonylindol-2-yl triflate with a variety of aryl- and heteroaryl boronic acids or vinyl tributylstannane also provides a general route for the synthesis of 2-aryl indoles <96SC3289>.

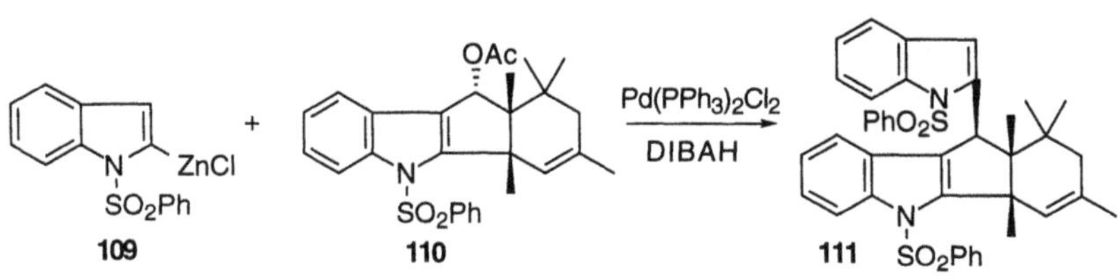

In addition to the radical *ipso*-substitution of indolyl sulfones producing stannanes described earlier <96T11329>, Caddick has also reported an approach to fused [1,2-*a*]indoles based on the intramolecular cyclization of alkyl radicals. Thus, treatment of **112** with Bu_3SnH leads to the fused ring derivatives **113** (n = 1-4) <96JCS(P1)675>.

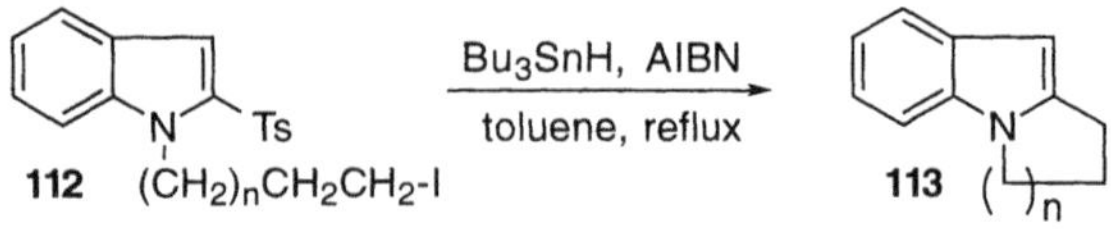

Gribble had previously reported that 3-lithio-1-phenylsulfonylindole (prepared by halogen-metal exchange of the corresponding 3-iodo derivative) is unstable and rearranges to the C-2 lithio species at temperatures above -100 °C <82JOC757>. Sakamoto and coworkers now report that halogen-metal exchange of 3-iodo-1-phenylsulfonylindole with ethylmagnesium bromide in THF at (0 °C to rt) affords a stable 3-magnesio derivative which can undergo reaction with typical electrophiles as well as Pd(0)-catalyzed arylations <96H(42)105>.

Alternatively, Bosch has employed the bulky *tert*-butyldimethylsilyl protecting group on the indole nitrogen to prepare stable (-78 °C) 3-lithio species *via* halogen-metal exchange <94JOC10, 96OS248>. This methodology was recently utilized by Otha to generate indole-3-boronic acids for aryl coupling reactions in syntheses of the marine alkaloids, nortopsentins A-D <96CPB1831>. Bosch and Amat now report that stable 3-lithio-1-triisopropylsilylindoles can also be generated from the corresponding 3-bromo species at -78 °C <96H(43)1713>.

The use of lanthanide triflates as stable, reusable Lewis acid catalysts in aqueous solution has been applied to the functionalization of indoles. Thus, treatment of indoles with electron-deficient olefins under the influence of $Yb(OTf)_3$•$3H_2O$ leads to alkylation of the indoles at the 3-position <96SL1047>. This catalyst has also been employed to catalyze regioselective opening of epoxides with indoles in an enantioselective synthesis of (+)-diolmycin A2 <96TL3727>. Finally, lanthanide triflates have been found to catalyze reaction of indoles **114** with aldehydes or ketones affording bisindolyl-methane adducts **115** <96TL4467, 96SL871>.

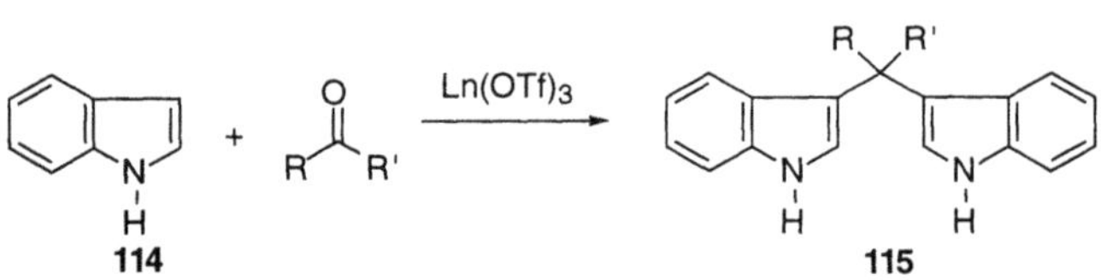

The Pictet-Spengler (P-S) reaction involving the condensation of tryptamines or tryptophan esters with aldehydes or reactive ketones affording tetrahydro-β-carbolines has been conducted on a variety of solid supports <96TL3963, 96TL5041, 96TL5633>. Additionally, P-S reaction of chiral *N*-(β-3-indolyl)ethyl-α-methylbenzylamine <96H(42)347> or the alkaloid *Abrine* [(S)-*N*-methyltryptophan] <96TL5971> with aldehydes has been found to proceed with high diastereoselectivity and the P-S reaction of an L-tryptophan derivative was employed as a key step in the synthesis of the alkaloid (-)-suaveoline <96CC1479>. Finally, evidence was offered to suggest that epimerization of kinetically formed *cis*-1,3-disubstituted tetrahydro-β-carbolines into the *trans*-isomers occurs *via* scission of the C(1)-N(2) bond <96CC2477>.

A clever artifice has been devised for circumventing the general lack of phenylacetaldehyde derivatives available for the synthesis of β-carbolines possessing CH_2-aryl groups at C-1 <96JOC7937>. Therein, azalactones **117** (prepared by condensation of *N*-acetylglycine with aryl aldehydes), serve as "phenylacetaldehyde equivalents" in the P-S reaction with tryptamines **116** or conformationally restrained tryptamines <96TL5813> affording β-carbolines **118**.

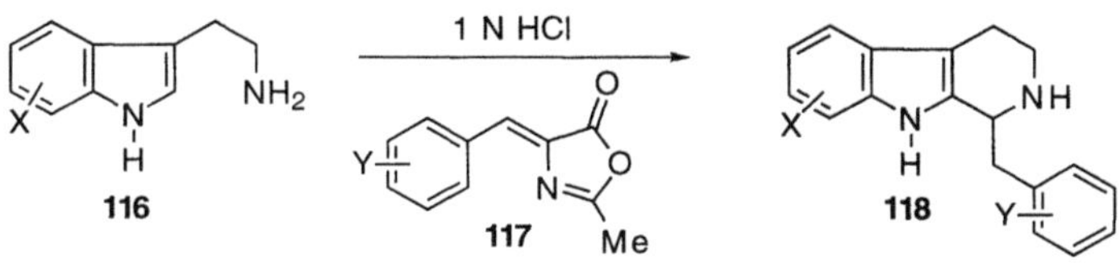

Whereas triethyl (1-methyl-2-yl)indolyl borate (**119**) undergoes Pd-catalyzed coupling with propargyl carbonates in an S_N2' manner yielding 2-allenylindoles <96H(43)1591>, conducting this reaction in the presence of CO (10 atm) affords cyclopenta[*b*]indoles **120** <96CC2409>.

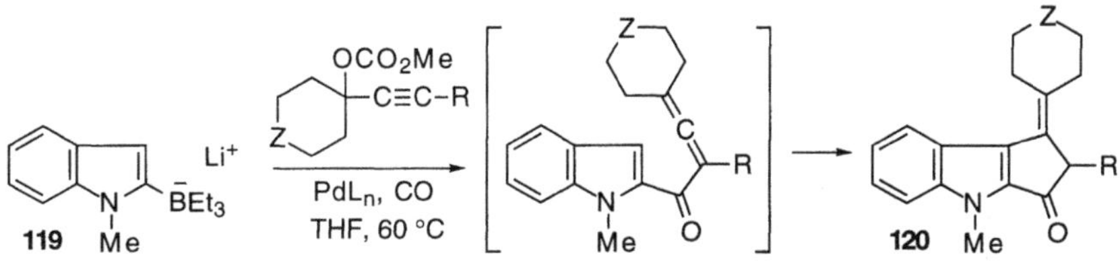

Tetrahydrocarbazoles have been prepared in one-flask syntheses from indoles, ketones and maleic anhydride, with acid catalysis. The reactions involve a condensation of the indole **121** with the ketone or aldehyde, followed by *in situ* trapping of the vinylindole **122** with maleic anhydride to afford tetrahydrocarbazoles **123** after double bond isomerization <96T4555>.

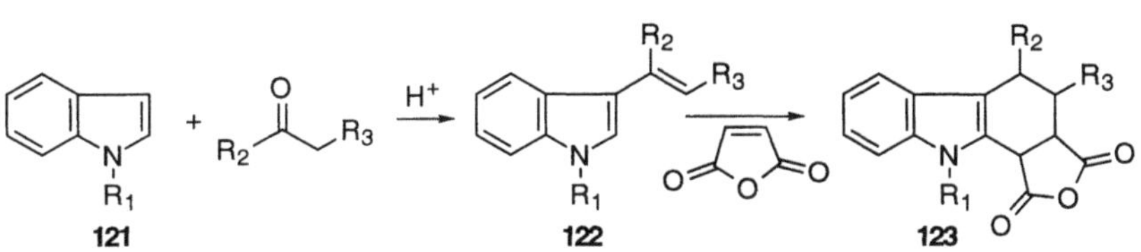

The inverse electron demand Diels-Alder reaction of 3-substituted indoles with 1,2,4-triazines and 1,2,4,5-tetrazines proceeds in excellent yields both inter- and intramolecularly, The cycloaddition of tryptophan **124** with a tethered 1,2,4-triazine produced a diastereomerically pure cycloadduct **125** <96TL5061>.

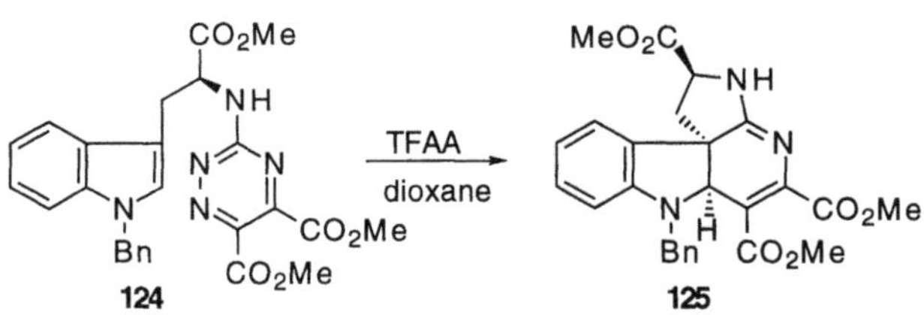

5.2.5 INDOLE ALKALOIDS

Due to the prevalence and diversity of indole alkaloids and the limitations of space, no attempt will be made to describe all of the synthetic efforts in this area. However, noteworthy synthetic achievements include a synthesis of gelsemine <96JA7426>, (+)-duocarmycin by

Boger <96JA2301>, and the synthesis of deethylibophyllidine *via* a novel Pummerer rearrangement/thionium ion mediated indole cyclization <96JOC7106>. Syntheses of representatives of the *Strychnos* family of alkaloids have been reported by Bosch <96JOC4194>, Kuehne <96JOC7873> and Martin <96JA9804>, while Aspidosperma alkaloids were prepared by Kuehne <96JOC6001> and Rubiralta <96JOC7882>.

By far, the most studied indole alkaloids this past year were the indolocarbazole family. Thus, total syntheses of the protein kinase C inhibitory indolocarbazole alkaloid (+)-staurosporine (**126**) have been reported by the groups of Danishefsky <96JA2828> and Wood <96JA10656>. In chemistry related to this alkaloid, Van Vranken has developed a two-step synthesis of 2,2'-bisindoles by TFA induced coupling followed by oxidation using either DDQ <96JA1225> or air/light <96TL5629>, and has reported a total synthesis of a related alkaloid (+)-tjipanazole F2 <96JA5500>. Although the coupling of 2,3-dibromo-*N*-methylmaleimide with indolylmagnesium bromides constitutes the most common route to these alkaloids and was employed in a synthesis of arcyroxocin A by Steglich <96TL4483>, Ohkubo has demonstrated that indolyl anions generated using lithium hexamethyldisilazide (LiHMDS) can serve the place of Grignard reagents for this purpose <96T8099>. Finally, Pindur has examined some new additions of acyl chlorides and arynes to 2,2'-bisindolyls <96JHC623>, while novel oxidation chemistry of some related species have been investigated <96JOC413>.

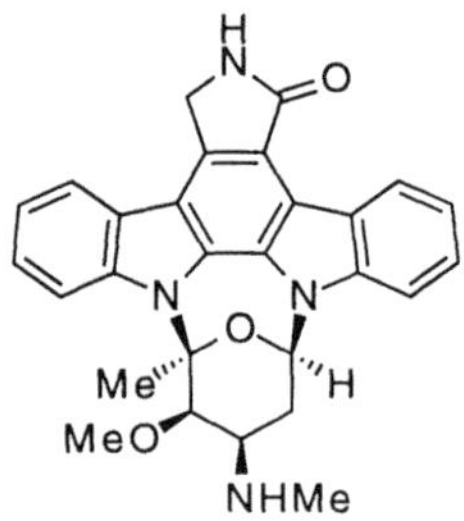

126

5.2.6 REFERENCES

<64AG(E)135> R. Huisgen, H. Gotthardt, H.O. Bayer, F.C. Schaefer, *Angew. Chem. Int, Ed. Engl.* **1964**, *3*, 135.

<81TL4899> R.X. Xu, H.J. Anderson, N.J. Gogan, C.E. Loader, R. McDonald, *Tetrahedron Lett.* **1981**, *22*, 4899.

<81TL4901> J. Rokach, P. Hamel, M. Kakushima, G.M. Smith, *Tetrahedron Lett.* **1981**, *22*, 4901.

<82JOC757> M.G. Saulnier, G.W. Gribble, *J. Org. Chem.* **1982**, *47*, 757.

<83JOC3214> M. Kakushima, P. Hamel, R. Frenette, J. Rokach, *J. Org. Chem.* **1983**, *48*, 3214.

<89JA776> S.L. Buchwald, M.W. Wanamaker, B.T. Watson, *J. Am. Chem. Soc.* **1989**, *111*, 776.

<94JA3127> T. Fukuyama, X. Chen, G. Peng, *J. Am. Chem. Soc.* **1994**, *116*, 3127.

<94JOC10> M. Amat, S. Hadida, S. Sathyanarayana, J. Bosch, *J. Org. Chem.* **1994**, *59*, 10.

<95TL1267> S. Lemaire-Audoire, M. Savignac, M.P. Genet, J.-M. Bernard, *Tetrahedron Lett.* **1995**, *36*, 1267.

<96ACR123> R.W. Armstrong, A.P. Combs, P.A. Tempest, S.D. Brown, T.A. Keating, *Acc. Chem. Res.* **1996**, *29*, 123.

<96AG(E)736> Y. Kondo, T. Matsudaira, J. Sato, N. Murata, T. Sakamoto, *Angew. Chem., Int. Ed. Engl.* **1996**, *35*, 736.

<96BCJ3339> Y. Murata, H. Kinoshita, K. Inomata, *Bull. Chem. Soc. Jpn.* **1996**, *69*, 3339.

<96BSF329> J.-B. Baudin, M.-G. Commenil, S.A. Julia, R. Lorne, L. Mauclaire, *Bull. Soc. Chim. Fr.* **1996**, *133*, 329.

<96CC1395> Y. Cheng, S. Goon, O. Meth-Cohn, *Chem. Commun.* **1996**, 1395.
<96CC1479> P.D. Bailey, K.M. Morgan, *Chem. Commun.* **1996**, 1479.
<96CC1519> R.N. Warrener, A.S. Amarasekara, R.A. Russell, *Chem. Commun.* **1996**, 1519.
<96CC1909> E.T. Pelkey, L. Chang, G.W. Gribble, *Chem. Commun.* **1996**, 1909.
<96CC2253> E.J. Latham, S.P. Stanforth, *Chem. Commun.* **1996**, 2253.
<96CC2409> M. Ishikura, Y. Matsuzaki, *Chem. Commun.* **1996**, 2409.
<96CC2477> E.D. Cox, J. Li, L.K. Hamaker, P. Yu, J.M. Cook, *Chem. Commun.* **1996**, 2477.
<96CPB48> M. Tani, T. Ariyasu, C. Nishiyama, H. Hagiwara, T. Watanabe, Y. Yokoyama, T. Murakami, *Chem. Pharm. Bull.* **1996**, *44*, 48.
<96CPB55> M. Tani, T. Ariyasu, M. Ohtsuka, T. Koga, Y. Ogawa, Y. Yokoyama, Y. Murakami, *Chem. Pharm. Bull.* **1996**, *44*, 55.
<96CPB67> H. Muratake, A. Abe, M. Natsume, *Chem. Pharm. Bull.* **1996**, *44*, 67.
<96CPB1631> H. Muratake, M. Tonegawa, M. Natsume, *Chem. Pharm. Bull.* **1996**, *44*, 1631.
<96CPB1831> I. Kawasaki, M. Yamashita, S. Ohta, *Chem. Pharm. Bull.* **1996**, *44*, 1831.
<96CPB2020> S. Ozaki, S. Mitoh, H, Ohmori, *Chem. Pharm. Bull.* **1996**, *44*, 2020.
<96CRV1179> Z. Chen, M.L. Trudell, *Chem Rev.* **1996**, *96*, 1179.
<96H(42)105> Y. Kondo, A. Yoshida, S. Sato, T. Sakamoto, *Heterocycles*, **1996**, *42*, 105.
<96H(42)347> T. Soe, T. Kawate, N. Fukui, T. Hino, *Heterocycles* **1996**, *42*, 347.
<96H(42)453> A.D. Billimoria, M.P. Cava, *Heterocycles* **1996**, *42*, 453.
<96H(42)513> K. Miyashita, K. Tsuchiya, K. Kondoh, H. Miyabe, T. Imanishi, *Heterocycles* **1996**, 42, 513.
<96H(43)463> K. Oda, R. Hiratsuka, M. Machida, *Heterocycles* **1996**, *43*, 463.
<96H(43)891> C. Shin, Y. Yamada, K. Hayashi, Y. Yonezawa, K. Umemura, T. Tanji, J. Yoshimura, *Heterocycles* **1996**, *43*, 891.
<96H(43)1305> M. D'Auria, *Heterocycles* **1996**, *43*, 1305.
<96H(43)1447> O. A. Attanasi, L. De Crescentini, R. Giorgi, A. Perone, S. Santeusanio, *Heterocycles* **1996**, *43*, 1447.
<96H(43)1471> K.E. Henegar, D.A. Hunt, *Heterocycles* **1996**, *43*, 1471.
<96H(43)1529> M. D'Auria, *Heterocycles* **1996**, *43*, 1529.
<96H(43)1591> M. Ishikura, I. Agata, *Heterocycles* **1996**, *43*, 1591.
<96H(43)1713> M. Amat, S. Sathyanarayana, S. Hadida, J. Bosch, *Heterocycles* **1996**, *43*, 1713.
<96H(43)2741> Y. Kondo, S. Kojima, T. Sakamoto, *Heterocycles* **1996**, *43*, 2741.
<96JA1028> A.J. Peat, S.L. Buchwald, *J. Am. Chem. Soc.* **1996**, *118*, 1028.
<96JA1225> S.J. Stachel, R.L. Habeeb, D.L. Van Vranken, *J. Am. Chem. Soc.* **1996**, *118*, 1225.
<96JA2301> D.L. Boger, J.A. McKie, T. Nishi, T. Ogiku, *J. Am. Chem. Soc.* **1996**, *118*, 2301.
<96JA2574> T.A. Keating, R.W. Armstrong, *J. Am. Chem. Soc.* **1996**, *118*, 2574.
<96JA2828> J.T. Link, S. Raghavan, M. Gallant, S.J. Danishefsky, T.C. Chou, L.M. Ballas, *J. Am. Chem. Soc.* **1996**, *118*, 2828.
<96JA5500> E.J. Gilbert, D.L. Van Vranken, *J. Am. Chem. Soc.* **1996**, *118*, 5500.
<96JA7117> L.M. Hodges, M.L. Spera, M.W. Moody, W.D. Harman, *J. Am. Chem. Soc.* **1996**, *118*, 7117.
<96JA7426> T. Fukuyama, G. Liu, *J. Am Chem. Soc.* **1996**, *118*, 7426.
<96JA8733> M. Uchiyama, M. Koike, M. Kameda, Y. Kondo, T. Sakamoto, *J. Am. Chem. Soc.* **1996**, *118*, 8733.
<96JA9804> S.F. Martin, C.W. Clark, M. Ito, M. Mortimore, *J. Am. Chem. Soc.* **1996**, *118*, 9804.
<96JA10656> J.L. Wood, B. M. Stoltz, S.N. Goodman, *J. Am. Chem. Soc.* **1996**, *118*, 10656.
<96JCS(P1)183> Y. Furusho, A. Tsunoda, T. Aida, *J. Chem. Soc., Perkin Trans. 1* **1996**, 183.
<96JCS(P1)417> N. Ono, H. Hironaga, K. Ono, S. Kaneko, T. Murashima, T. Ueda, C. Tsukamura, T. Ogawa, *J. Chem. Soc., Perkin Trans. 1* **1996**, 417.
<96JCS(P1)669> P.W. Groundwater, D. Hughes, M.B. Hursthouse, R. Lewis, *J. Chem. Soc., Perkin Trans. 1*, **1996**, 669.
<96JCS(P1)675> S. Caddick, K. Aboutayab, K. Jenkins, R.I. West, *J. Chem. Soc., Perkin Trans 1* **1996**, 675.
<96JCS(P1)1213> K.-F. Cheng, M.-K. Cheung, *J. Chem. Soc., Perkin Trans. 1* **1996**, 1213.
<96JCS(P1)1261> K. Miyashita, K. Kondoh, K. Tsuchiya, H. Miyabe, T. Imanishi, *J. Chem. Soc., Perkin Trans. 1* **1996**, 1261.
<96JCS(P1)1927> T. Sakamoto, Y. Kondo, N. Takazawa, N.; H. Yamanaka, *J. Chem. Soc. Perkin Trans. 1* **1996**, 1927.
<96JCS(P1)2331> Y. Kondo, A. Yoshida, T. Sakamoto, *J. Chem. Soc., Perkin Trans. 1*, **1996**, 2331.
<96JCS(P1)2653> S.E. Waldem, R.A. Wheeler, *J. Chem. Soc., Perkin Trans 1* **1996**, 2653.
<96JCS(P1)2663> S.E. Waldem, R.A. Wheeler, *J. Chem. Soc., Perkin Trans 1* **1996**, 2663.
<96JCS(P1)2793> K. Smith, D. Bahzad, *J. Chem. Soc., Perkin Trans. 1* **1996**, 2793.

<96T2489> S. Miah, A.M.Z. Slawin, C.J. Moody, S.M. Sheehan, J.P. Marino, Jr., M.A. Semones, A. Padwa, I.C. Richards, *Tetrahedron* **1996**, *52*, 2489.

<96T2743> R.C. Larock, E.K. Yum, *Tetrahedron* **1996**, *52*, 2743.

<96T4555> W.E. Noland, G.-M. Xia, K.R. Gee, M.J. Konkel, M.J. Wahlstrom, J.J. Condoluci, D.L. Rieger, *Tetrahedron* **1996**, *52*, 4555.

<96T6879> J.T. Gupton, S.A. Petrich, L.L. Smith. M.A. Bruce, P.Vu, K.X. Du, E.E. Dueno, C.R. Jones, J.A. Sikorski, *Tetrahedron* **1996**, 52, 6879.

<96T7525> J.P. Wolfe, R.A. Rennels, S.L. Buchwald, *Tetrahedron* **1996**, *52*, 7525.

<96T11053> C. Szantay, Z. Kardos-Balogh, I. Moldvai, C. Szantay, Jr., E. Temesvari-Major, G. Blasko, *Tetrahedron* **1996**, *52*, 11053.

<96T11329> K. Aboutayab, S. Caddick, K. Jenkins, S. Joshi, S. Khan, *Tetrahedron* **1996**, **52**, 11329

<96T13441> R. Grigg, J.M. Sansano, *Tetrahedron* **1996**, *52*, 13441.

<96T14335> C.W. Bird, *Tetrahedron* **1996**, *52*, 14335.

<96T14975> D. Wensbro, S. Gronowitz, *Tetrahedron*, **1996**, *52*, 14975.

<96TL37> M. Ibrahim-Ouali, M.-E. Sinibaldi, Y. Troin, J.-C. Gramain, *Tetrahedron Lett.* **1996**, *37*, 37.

<96TL131> M. Mascal, N.M. Hext, O.V. Shishkin, *Tetrahedron Lett.* **1996**, *37*, 131.

<96TL299> W.J. Smith, Jr., J.S. Sawyer, *Tetrahedron Lett.* **1996**, 37, 299.

<96TL315> T. Thyrann, D.A. Lightner, *Tetrahedron Lett.* **1996**, *37*, 315.

<96TL1149> A.M. Strocker, T.A. Keating, P.A. Tempest, R.W. Armstrong, *Tetrahedron Lett.* **1996**, *37*, 1149.

<96TL1523> D. Xiao, J.A. Schreier, J.H. Cook, P.G. Seybold, D.M. Ketcha, *Tetrahedron Lett.* **1996**, *37*, 1523.

<96TL2721> J. Cheng, J.E. Garo, A.R. Morgan, *Tetrahedron Lett.* **1996**, *37*, 2721.

<96TL2887> D.R. Hutchinson, N.K. Nayyar, M.J. Martinelli, *Tetrahedron Lett.* **1996**, *37*, 2887.

<96TL2943> A.M.M. Mjalli, S. Sarshar, T.J. Baiga, *Tetrahedron Lett.* **1996**, *37*, 2943.

<96TL3067> D.R. Witty, G. Walker, J.H. Beatson, P.J. O'Hanlon, D.S. Eggleston, R.C. Haltiwanger, *Tetrahedron Lett.* **1996**, *37*, 3067.

<96TL3071> M. Amat, S. Hadida, S. Sathyanarayana, J. Bosch, *Tetrahedron Lett.* **1996**, *37*, 3071.

<96TL3247> J. Wang, A.I. Scott, *Tetrahedron Lett.* **1996**, *37*, 3247.

<96TL3399> R. Grigg, V. Loganathan, V. Sridharan, *Tetrahedron Lett.* **1996**, *37*, 3399.

<96TL3727> H. Kotsuki, M. Teraguchi, N. Shinomoto, M. Ochi, *Tetrahedron Lett.* **1996**, *37*, 3727.

<96TL3963> R. Mohan, Y.-L. Chou, M.M. Morissey, *Tetrahedron Lett.* **1996**, *37*, 3963.

<96TL4099> J.M. Mendez, B. Flores, F. Leon, L.E. Martinez, A. Vazquez, G.A. Garcia, M. Salmon, *Tetrahedron Lett.* **1996**, *37*, 4099.

<96TL4189> W. Yun, R. Mohan, *Tetrahedron Lett.* **1996**, *37*, 4189.

<96TL4221> R. Grigg, V. Sridharan, C. Terrier, *Tetrahedron Lett.* **1996**, *37*, 4221.

<96TL4289> J.E. Macor, R.J. Ogilvie, M.J. Wythes, *Tetrahedron Lett.* **1996**, *37*, 4289.

<96TL4413> A. Casachi, R. Grigg, J.M. Sansano, D. Wilson, J. Redpath, *Tetrahedron Lett.* **1996**, *37*, 4413.

<96TL4467> D. Chen, L. Yu, P.G. Wang, *Tetrahedron Lett.* **1996**, *37*, 4467.

<96TL4483> G. Mayer, G. Wille, W. Steglich, *Tetrahedron Lett.* **1996**, *37*, 4483.

<96TL4631> S.W. Wright, L.D. McClure, D.L. Hageman, *Tetrahedron Lett.* **1996**, *37*, 4631.

<96TL4869> S.M. Hutchins, K.T. Chapman, *Tetrahedron Lett.* **1996**, *37*, 4869.

<96TL5041> L. Yang, L. Guo, *Tetrahedron Lett.* **1996**, *37*, 5041.

<96TL5061> S.C. Benson, L. Lee, J.K. Snyder, *Tetrahedron Lett.* **1996**, *37*, 5061.

<96TL5281> R. Sreekumar, R. Padmakumar, *Tetrahedron Lett.* **1996**, *37*, 5281.

<96TL5629> D.S. Carter, D.L. Van Vranken, *Tetrahedron Lett.* **1996**, *37*, 5629.

<96TL5633> J.P. Mayer, D. Bankaitis-Davis, J. Zhang, G. Beaton, K. Bjergarde, Andersen, C.M.; B.A. Goodman, C.J. Herrera, *Tetrahedron Lett.* **1996**, *37*, 5633.

<96TL5641> A.R. Katritzky, J.R. Levell, J. Li, *Tetrahedron Lett.* **1996**, *37*, 5641.

<96TL5813> J. Ezquerra, C. Lamas, A. Pastor, P. Alvarez, J.J. Vaquero, W.G. Prowse, *Tetrahedron Lett.* **1996**, *37*, 5813.

<96TL5971> W.-M. Dai, H.J. Zhu, X.-J. Hao, *Tetrahedron Lett.* **1996**, *37*, 5971.

<96TL6045> J.W. Coe, M.G. Vetelino, M.J. Bradley, *Tetrahedron Lett.* **1996**, *37*, 6045.

<96TL6565> R. Grigg, V. Savic, *Tetrahedron Lett.* **1996**, *37*, 6565.

<96TL7099> T. Shinada, M. Miyachi, Y. Itagaki, H. Naoki, K. Yoshihara, T. Nakajima, *Tetrahedron Lett.* **1996**, *37*, 7099.

<96JHC221> T. Thyrann, D.A. Lightner, *J. Heterocycl. Chem.* **1996**, *33*, 221.
<96JHC623> U. Pindur, Y.-S. Kim, *J. Heterocyclic Chem.* **1996**, *33*, 623.
<96JOC413> S.W. McCombie, S.F. Vice, *J. Org. Chem.* **1996**, *61*, 413.
<96JOC816> L. Wang, L.S. Jimenez, *J. Org. Chem.* **1996**, *61*, 816.
<96JOC1155> H.D.H. Showalter, L. Sun, A.D. Sercel, R.T. Winters, W.A. Denny, B.D. Palmer, *J. Org. Chem.* **1996**, *61*, 1155.
<96JOC1180> M. McLeod, N. Boudreault, Y. Leblanc, *J. Org. Chem.* **1996**, *61*, 1180.
<96JOC1188> B. Antelo, L. Castedo, J. Delamano, A. Gomez, C. Lopez, G. Tojo, *J. Org. Chem.* **1996**, *61*, 1188.
<96JOC1624> A.R. Katritzky, J. Li, *J. Org. Chem.* **1996**, *61*, 1624.
<96JOC2185> J. Barluenga, M. Tomas, V. Kouznetsov, A. Suarez-Sobrino, E. Rubio, *J. Org. Chem.* **1996**, *61*, 2185.
<96JOC2594> D. Zhang, L.S. Liebeskind, *J. Org. Chem.* **1996**, *61*, 2594.
<96JOC2596> W.F. Bailey, X.-L.Jiang, *J. Org. Chem.* **1996**, *61*, 2596.
<96JOC3584> R.C. Larock, T.R. Hightower, L.A. Hasvold, K.P. Peterson, *J. Org. Chem.* **1996**, *61*, 3584.
<96JOC4194> D. Sole, J. Bonjoch, J. Bosch, *J. Org. Chem.* **1996**, *61*, 4194.
<96JOC4600> D. Bai, R. Xu, G. Chu, X. Zhu, *J. Org. Chem.* **1996**, *61*, 4600.
<96JOC4999> P. Nagafuji, M. Cushman, *J. Org. Chem.* **1996**, *61*, 4999.
<96JOC5440> K.S. Feldman, M.M. Bruendl, K. Schildknegt, A.C. Bohnstedt, *J. Org. Chem.* **1996**, *61*, 5440.
<96JOC5804> J. Ezquerra, C. Pedregal, C. Lamas, J. Barluenga, M. Perez, M.A. Garcia-Martin, J.M. Gonzalez, *J. Org. Chem.* **1996**, *61*, 5804.
<96JOC6001> M.E. Kuehne, T. Wang, P.J. Saeton, *J. Org. Chem.* **1996**, *61*, 6001.
<96JOC7106> J. Bonjoch, J. Catena, N. Valls, *J. Org. Chem.* **1996**, *61*, 7106.
<96JOC7189> C. Zhang, M.L. Trudell, *J. Org. Chem.* **1996**, *61*, 7189.
<96JOC7664> T.J. Donohoe, P.M. Guyo, *J. Org. Chem.* **1996**, *61*, 7664.
<96JOC7873> M.E. Kuehne, T. Wang, D. Seraphin, *J. Org. Chem.* **1996**, *61*, 7873.
<96JOC7882> P. Forns, A. Diez, M. Rubiralta, *J. Org. Chem.* **1996**, *61*, 7882.
<96JOC7937> J.E. Audia, J.J. Droste, J.S. Nissen, G.L. Murdoch, D.A. Evrard, *J. Org. Chem.* **1996**, *61*, 7937.
<96JOC8730> C.Y. De Leon, B. Ganem, *J. Org. Chem.* **1996**, *61*, 8730.
<96JOC9012> H.A.M. Biemans, C. Zhang, P. Smith, H. Kooijman, W.J.J. Smeets, A.L. Spek, E.W. Meijer, *J. Org. Chem.* **1996**, *61*, 9012.
<96JOC9055> J.M. Prawlak, V.V. Khau, D.R. Hutchinson, M.J. Martinelli, *J. Org. Chem.* **1996**, *61*, 9055.
<96JOC9068> T.P. Curran, M.T. Keaney, *J. Org. Chem.* **1996**, *61*, 9068.
<96JOC9297> T.A. Engler, W. Chai, K.O. LaTessa, *J. Org. Chem.* **1996**, *61*, 9297.
<96MI1> R.J. Sundberg, *Indoles*, Best Synthetic Methods Series, Academic Press, San Diego, 1996.
<96MI2> K.M. Czerwinski, J.M. Cook, *Advances in Heterocyclic Natural Product Synthesis.* Vol. 3, Pearson, W.H., Ed., JAI Press, Greenwich, CT, 1996.
<96OPP1> B.E. Love, *Org. Prep. Proced. Int.* **1996**, *28*, 1.
<96OS248 M. Amat, S. Hadida, S. Sathyanarayana, J. Bosch, *Org. Synth.* **1996**, *74,* 248.
<96S457> C. D'Silva, R. Iqbal, *Synthesis* **1996**, 457.
<96S851> L.F. Tietze, G. Kettschau, K. Heitmann, *Synthesis* **1996**, 851.
<96S871> R. ten Have, F.R. Leusink, A.M. van Leusen, *Synthesis* **1996**, 871.
<96S1196> P. Gmeiner, J. Kraxner, B. Bollinger, *Synthesis* **1996**, 1196.
<96S1336> B. Rousseau, F. Nydegger, A. Gossauer, B. Bennua-Skalmowski, H. Vorbruggen, *Synthesis* **1996**, 1336.
<96SC657> M. Ibrahim-Ouali, A. Missoumi, M.-E. Sinibaldi, Y. Troin, J.-C. Gramain, *Synth. Commun.* **1996**, *26*, 657.
<96SC1289> A. Ho-Hoang, F. Fache, M. Lemaire, *Synth. Commun.* **1996**, *26*, 1289.
<96SC1349> S.C. Shim, Y.Z. Youn, D.Y. Lee, T.J. Kim, C.S. Cho, S. Uemura, Y. Watanabe, *Synth. Commun.* **1996**, *26*, 1349.
<96SC1839> R. DiSanto, R. Costi, S. Masa, M. Artico, *Synth. Commun.* **1996**, *26*, 1839.
<96SC3289> B. Joseph, B. Malapel, J.-Y. Merour, *Synth. Commun.* **1996**, *26*, 3289.
<96SL871> M.T. El Gihani, H. Heaney, K.F. Shuhaibar, *Synlett* **1996**, 871.
<96SL1047> P.E. Harrington, M.A. Kerr, *Synlett* **1996**, 1047.

<96TL7485> B.M. Trost, G.R. Cook, *Tetrahedron Lett.* **1996**, *37*, 7485.
<96TL7595> I. Hughes, *Tetrahedron Lett.* **1996**, *37*, 7595.
<96TL7787> Y. Gao, M. Shirai, F. Sato, *Tetrahedron Lett.* **1996**, *37*, 7787.
<96TL7801> Y. Yamamoto, T. Kimachi, Y. Kanaoka, S. Kato, K. Bessho, T. Matsumoto, T. Kusakabe, Y. Sugiura, *Tetrahedron Lett.* **1996**, *37*, 7801.
<96TL8099> M. Ohkubo, T. Nishimura, H. Jona, H. Morishima, *Tetrahedron Lett.* **1996**, *37*, 8099.
<96TL9203> Y. Aoyagi, T. Mizusaki, A. Ohta, *Tetrahedron Lett.* **1996**, *37*, 9203.
<96TL9381> O. Meth-Cohn, S. Goon, *Tetrahedron Lett.* **1996**, *37*, 9381.

Chapter 5.3

Five-Membered Ring Systems

Furans and Benzo Derivatives

Stephan Reck and Willy Friedrichsen

Institute of Organic Chemistry, University of Kiel, Germany

5.3.1 INTRODUCTION

As in previous years there was much activity in the field of furans and their benzo derivatives. Again, a great variety of natural products was isolated. Most of this work was collected - as in the past - in "*Heterocycles*" and will not be repeated here. Only a few examples will be given: saricandin <96JAN596>, glabrocoumarone A <96CPB1218>, vibsanol, 9'-*O*-methyl-vibsanol, benzo[b]furans <96CPB1418, 96P1167>, heyneanol A <96P1163>, scutalbin A, B <96P1059>, caesaldekarin A <96CPB1157>, melianolide <96H(43)1477>, mulberrofuran V <96H(43)425>, astrotrichilin <96P1239>, steroids <96P1065>, and others <96JAN505, 96P1235>. 9-Hydroxyfuranoeremophilanes (**1**) are the main compounds of freshly harvested rhizomes of *Petasites hybridus*.

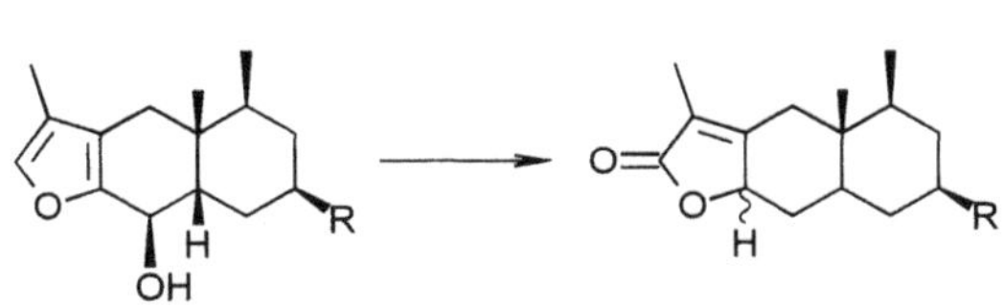

1 **2**

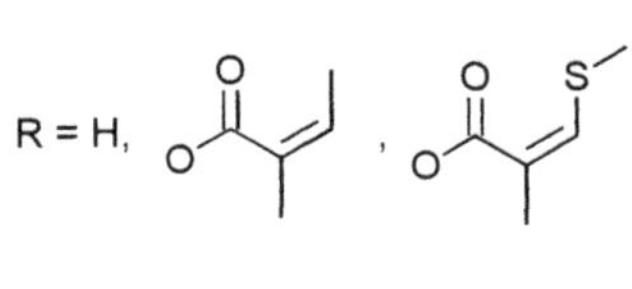

They easily and quantitatively rearrange in the presence of traces of acids to give an epimeric mixture of 8-*H*-eremophilanlactones (**2**) <96HCA1592>. Novel mono-THF acetogenins (gigantransenin A, B, C) were isolated from *Goniothalamus giganteus* (Annonaceae) <96TL5449>. A review on recent advances in Annonaceous Acetogenins covering the literature up to January 1996 was published <96MI1>.

5.3.2 REACTIONS

A review on furan and its derivatives in the synthesis of other heterocycles was published <95CHE1034>. Furan decomposes on Pd(111) at 300 K to form H, CO and C_3H_3, which can dimerize to benzene at 350 K <96JA907>. Again, a considerable number of Diels-Alder reactions with furan and furan derivatives was reported. The synthesis of 2-pyridinyl-7-oxabicyclo[2.2.1]heptanes (*e.g.*, **3**, **4**) was accomplished *via* zinc chloride-mediated Diels-Alder reaction of furan with 2-vinylpyridines <96SL703>.

3 **4**

py = 2-pyridyl

The reaction of triflates **6** (available *via* Diels-Alder reaction of **5** with furan or cyclopentadiene) with alkynylstannanes proceeds smoothly and with selectivity to afford good yields of bicyclic ene-diynes <96JOC6162>.

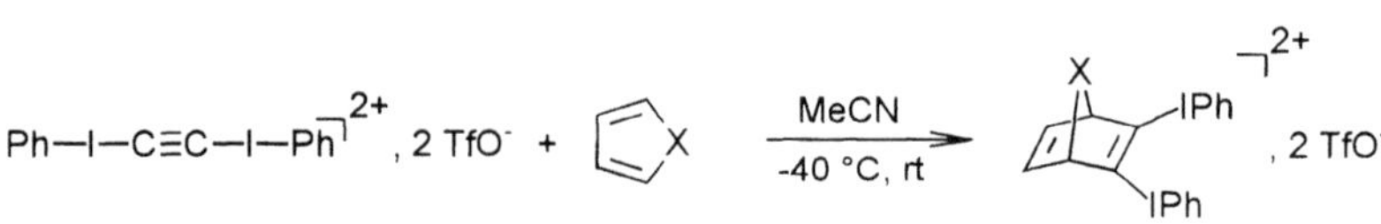

5 **6**

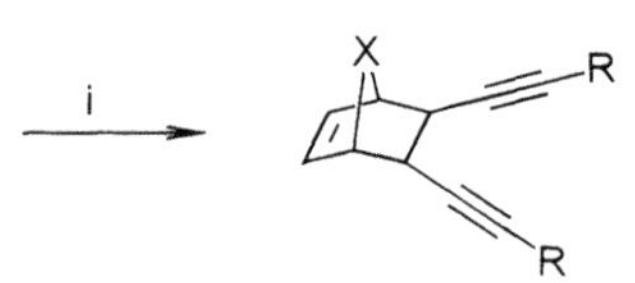

X = CH_2, O

i, 2.4 eq $Bu_3Sn{-}C{\equiv}C{-}R$, 8% CuI, DMF, rt, 5% $Bu(PPh_3)_2PdCl$

Furan was used as building block in the total synthesis of (±)-palasonin <96JOC4816, 96TL1129>. A synthetic protocol to assemble a benzene ring mimicking the [4+2] cycloaddition of benzene and furan (or cyclopentadiene) was reported (Scheme 1, <96T14247>).

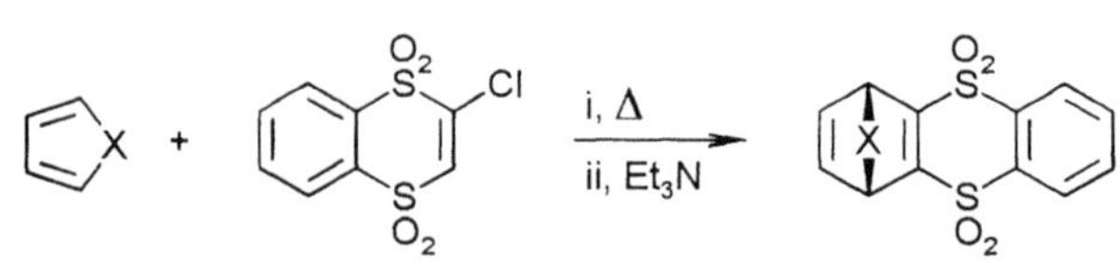

Scheme 1

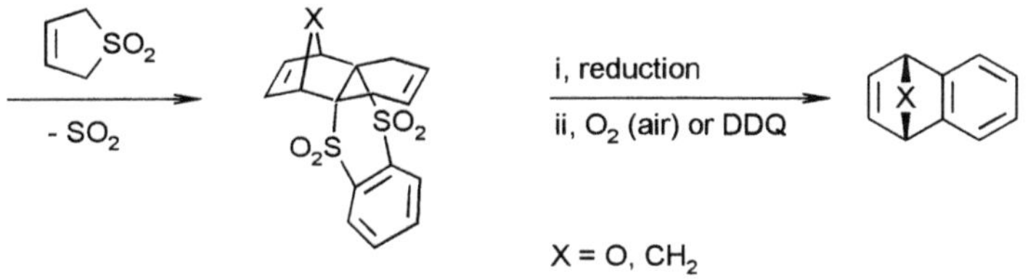

Scheme 1, contd.

The synthesis and Diels-Alder reactions of enantiopure (-)-*trans*-benzo[d]thiin-S,S'-dioxide **7** were described. Whereas with cyclopentadiene and 1,3-cyclohexadiene a high *endo*/*exo* ratio (> 99/1) was obtained, with furan these values are lower depending on the reaction conditions <96TA369>.

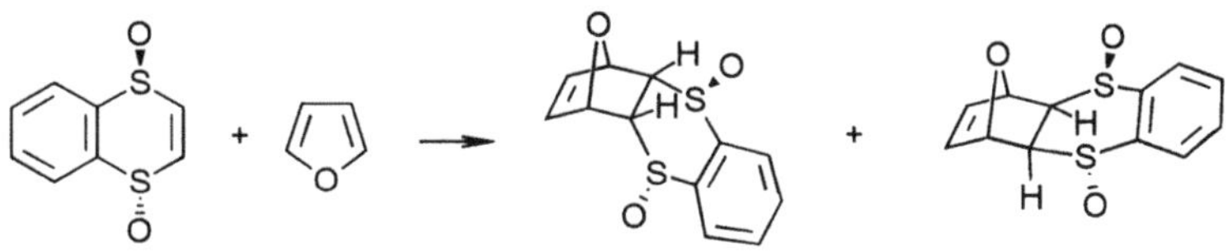

7

conditions : -60 °C, 15h, 60%, 28:72
$AlEt_2Cl$ (0.5eq.), 0 °C, 0.25h, 73%, 71:29

Asymmetric Diels-Alder reactions of chiral sulfinyl acrylate derivatives with furans proceed under high pressure (1.2 GPa) conditions to give *endo* cycloadducts <96H(42)129>. For further asymmetric Diels-Alder reactions of furan and chiral acrylates see <96JOC9479>. Furan reacts with hexachloronorbornadiene to give the relatively stable *endo-exo* adduct. Besides, the *endo-endo* adduct and two 1:2 adducts were isolated in minor amounts <96JCS(P2)1233>. The Diels-Alder reaction of ketovinylphosphonate with furan was performed with and without Lewis acid assistance. The acetyl group directed *endo* in all cases <96TL2149, 96TL2153>. A new strategy for the stereocontrolled construction of decalins and fused polycycles *via* a tandem Diels-Alder ring-opening sequence was published (Scheme 2, <96JOC7994>).

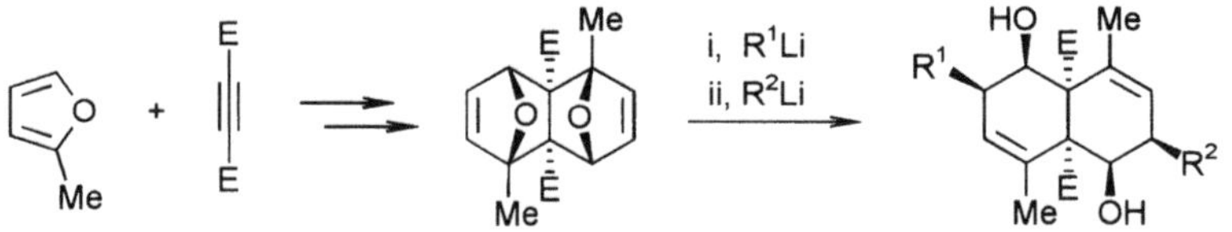

Scheme 2

The preparation of 1-substituted 4*H*,6*H*-dihydrothieno[3,4-c]furan 5,5-dioxide and some intermolecular Diels-Alder reactions with typical dienophiles (*e.g.*, DMAD) were reported (Scheme 3, <96JCS(P1)2699>).

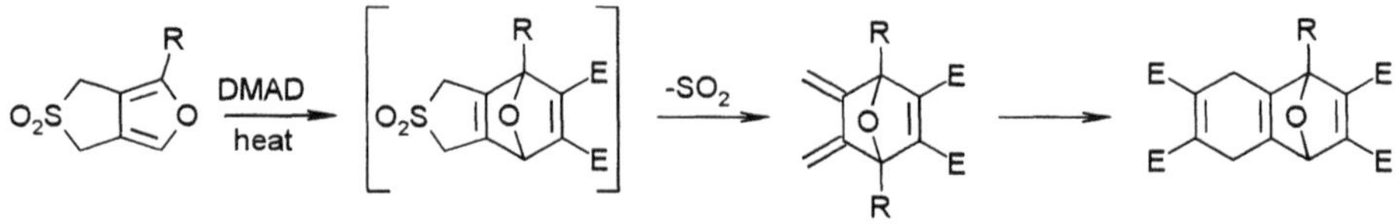

R = Br, (4-OMe)C_6H_4, Ac, NO_2

Scheme 3

The simultaneous double Diels-Alder addition of 1,1-bis(3,5-dimethylfur-2-yl)ethane (**8**) with a bis-dienophile such as diethyl (*E*,*E*)-4-oxohepta-2,5-diene-1,7-dioate was proposed as new, asymmetric synthesis of long-chain polypropionate fragments and analogues <96TL4149>.

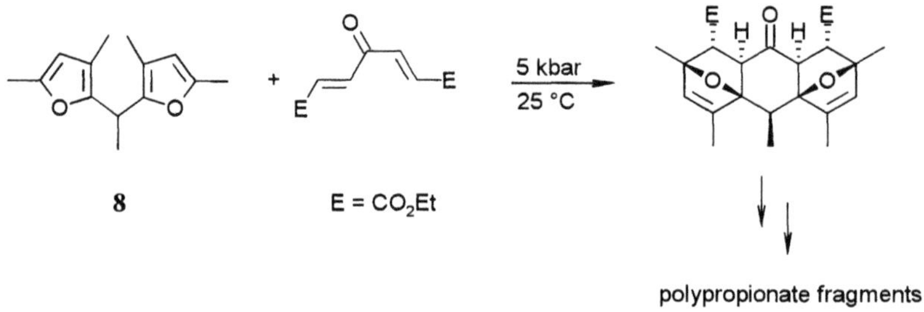

8 E = CO_2Et

Further reactions with **8** are described <96HCA1393, 96HCA1415>. $CpW(CO)_3$ (W) can be considered as an electron-donating group. Treatment of **9** with dimethyl fumarate, *N*-phenylmaleimide, ethyl propiolate, and DMAD affords **10-13** in excellent yields (> 90%) <96CC1041> (Scheme 2 of <96CC1041> contains a typographical error).

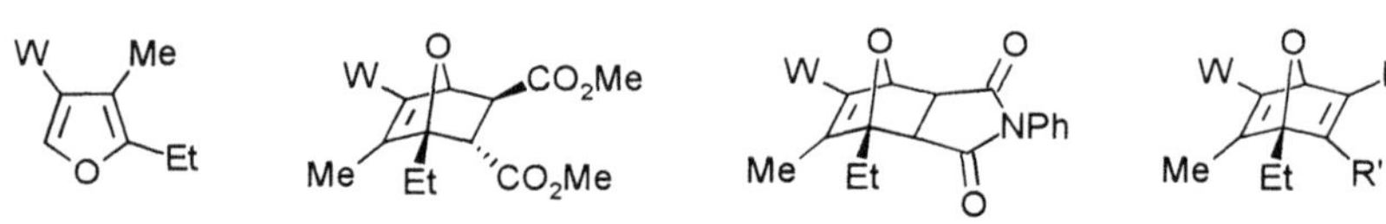

9 **10** **11** **12, 13**

R = H, R' = CO_2Et : 90%
R = R' = CO_2Me : 95%

Methyl 5-aminofuroate undergoes a facile [4+2] cycloaddition with a variety of dienophiles to ring-opened cycloadducts which are readily dehydrated using $BF_3{\cdot}OEt_2$ to give polysubstituted anilines <96TL2903>. Polyfluorinated furans have been obtained by Diels-Alder reaction of 2-methylfuran and hexafluorobut-2-yne with a subsequent reduction/retro Diels-Alder step <96JCS(P1)1095>. Acid catalyzed (camphorsulfonic acid) acetalisation of furfural with (2*S*,3*S*)-butanediol gives **14** which on heating with a 6-7 fold excess of maleic anhydride without solvent (50 °C; 1 d, then 55 °C, 7 d) and extraction with toluene yields an enantiomerically pure 1:1 complex of **15** and maleic anhydride in good yields (78%) <96TA3153>.

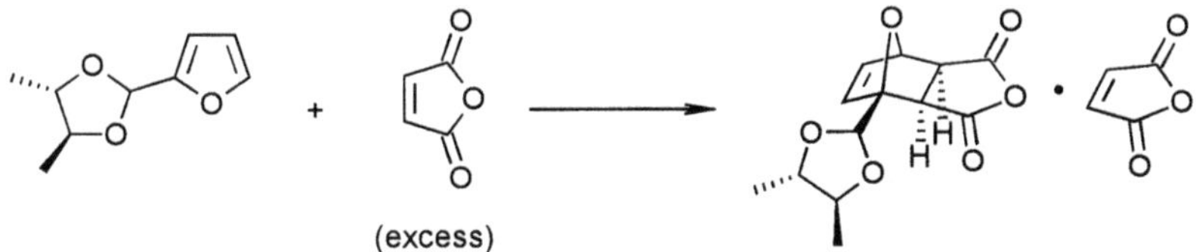

14 **15**

Vinyl sulfonamides of furan-containing *N*-benzylamines cyclize at room temperature to give δ-sultams with high diastereoselectivity (Scheme 4, <96SL741>).

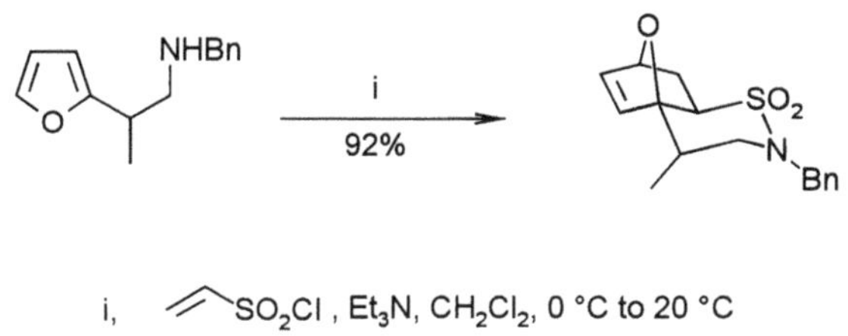

Scheme 4

Bicyclopropylidene derivatives with furan moieties on tethers of various length and nature all undergo clean intramolecular Diels-Alder reactions with complete *endo*-diastereoselectivities when heated to 70-130 °C under 10 kbar pressure (Scheme 5, <96T12185>).

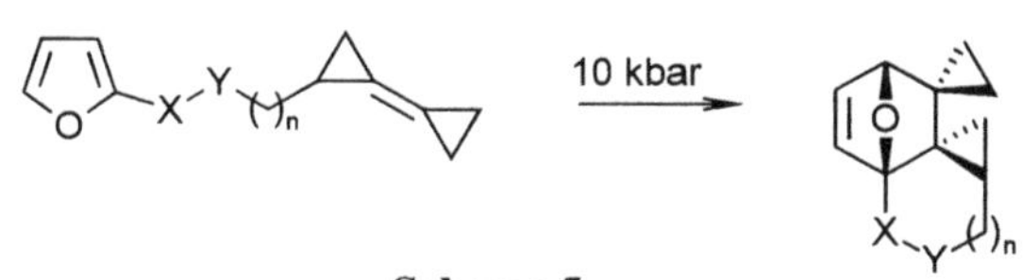

Scheme 5

The synthesis of compound **16** was reported. Valence isomerisation of **16** to the isoannulenofuran **17** could be achieved either photochemically or thermally with **16** as the thermodynamically more stable isomer <96JOC935>.

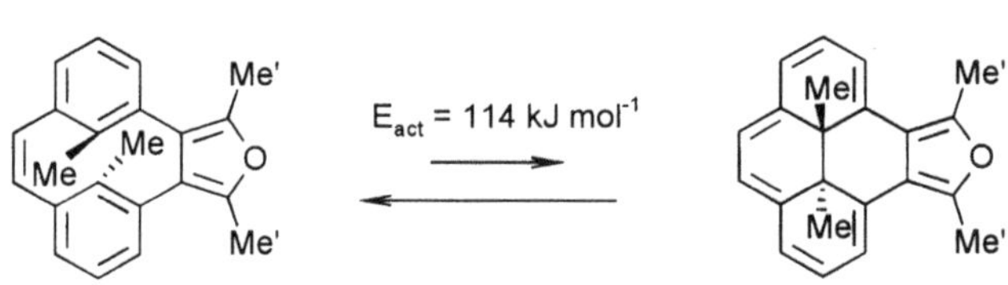

16 **17**

δ_{Me} = 1.38 δ_{Me} = 0.63

Oxyallyl carbocations can be generated efficiently using diethylzinc together with polybromoketones. These intermediates can be trapped with furan-alcohols to give good yields of polysubstituted 8-oxabicyclo[3.2.1]oct-6-en-3-ones <96S31, 96JCS(P1)1101>. A route to enantiomerically pure oxabicyclo[3.2.1]octenes using a "chelate-controlled" facially selective [4+3] cycloaddition reaction has been developed <96JA10930>. For further [4+3] cycloadditions with furans see <96TL783, 96JOC1478>. The Diels-Alder reaction of 3-vinylfurans with DMAD, *N*-phenylmaleimide, and dimethyl maleate afforded products derived from addition to the furan ring diene system (*intra*annular addition) and to the furan 2,3-double bond 3-vinyl group diene system (*extra*annular addition) <96JOC1487>. A highly stereoselective addition of 2-trimethylsilyloxyfuran to sultam **18** (synthesis <96TA1385>) in the presence of $Eu(fod)_3$ (2 to 3 mol %) was reported <96TA981>.

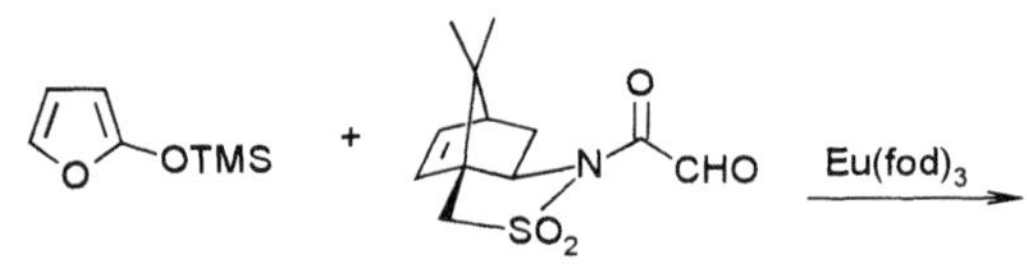

18

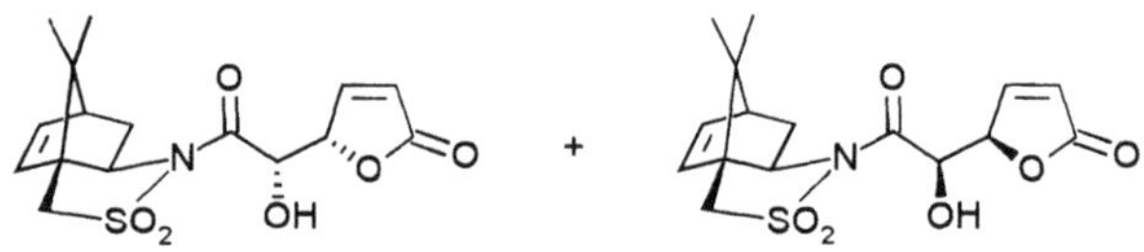

94 : 6

The synthesis of several cyclic oligomers of furan and acetone containing four or more furan units (Scheme 6) has been reexamined. The structure of [1.6](2,5)furanophane (n=2) has been investigated by X-ray crystallography. The macrocycle is self-filling, four of the furan rings being oriented approximately orthogonally with respect to the mean plain of the macrocycle whilst the other two lie almost within this plain. The first synthesis of [1.9](2,5)furanophan (n=5) was also reported <96TL4593>.

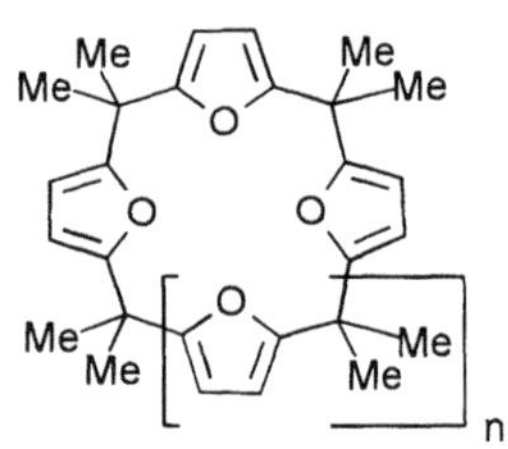

Scheme 6

The conversion of the cyclic hexamer of furan and acetone into naphthafurophane was reported <96TL6201>. Stepwise synthesis of core-modified, *meso*-substituted porphyrins (*e.g.*, **19**) was achieved by acid catalyzed condensations <96TL197>.

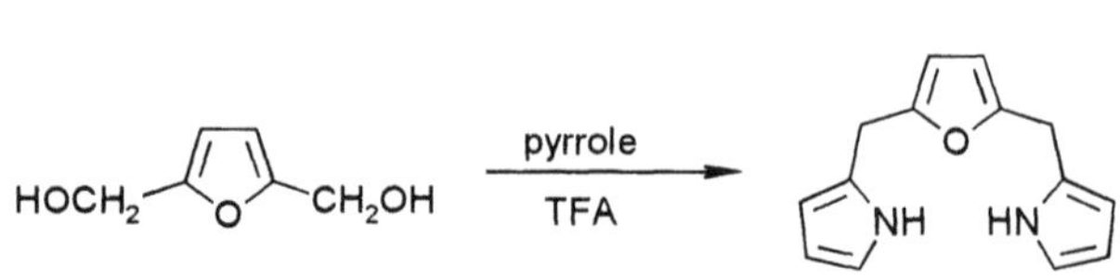

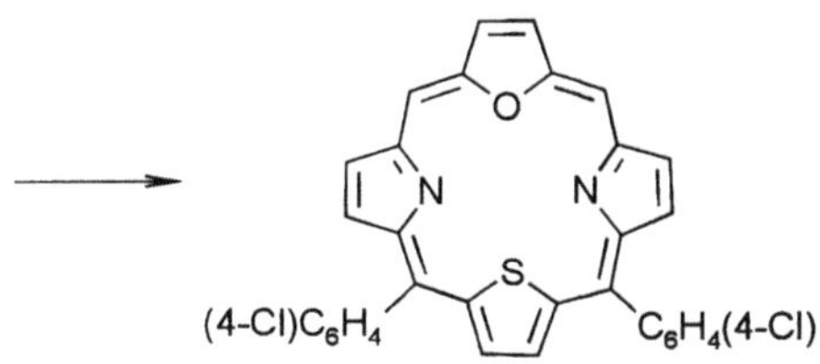

19

3-Alkoxy-2,5-diphenylfurans were selectively converted into *cis*-2-alkoxy-1,4-diphenyl-2-butene-1,4-diones with DDQ in CH_2Cl_2 <96H(43)1371>. A novel method based upon the kinetic resolution of a α-furfurylamine derivative has been reported for the asymmetric synthesis of deoxymannojirimycin <96TL1461>. A new approach to the synthesis of aromatic stannanes *via* a radical substitution reaction of aromatic sulfones was reported. Thus, α-heterocyclic aromatic sulfines derived from furan (indole, pyrrole, pyrazole, thiophene) undergo rapid and high yielding *ipso*-substitution to furnish organostannanes <96T11329>. Nickel-catalyzed coupling reactions of cyclopentenyl acetate and lithium 2-furylborate are possible <96TL6125>. Pd-catalyzed cross-coupling reactions of furylstannanes using both Koser's and Zefirov's reagent were reported <96TL3723>. In spite of the theoretical importance of benz[a]azulene, which is a well known polycyclic nonbenzenoid hydrocarbon, synthetic difficulties have precluded progress in this area. This situation may change in the near future as a novel synthesis of β-(10-benz[a]azulenyl)-α,β-unsaturated ketones by intramolecular cyclization of *o*-[2-furyl]cycloheptatrienylbenzene was reported (Scheme 7, <96TL4965>).

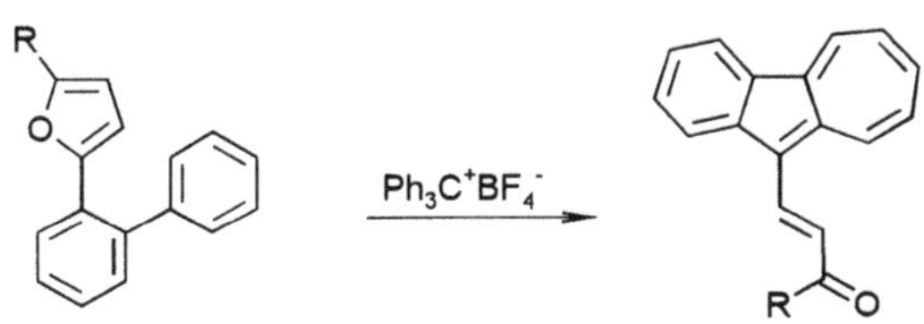

Scheme 7

Starting with aldehyde **20** and 2-alkoxyfurans an enantioselective synthesis of bioactive melodorinol and acetylmelodorinol was achieved <96TA3141>.

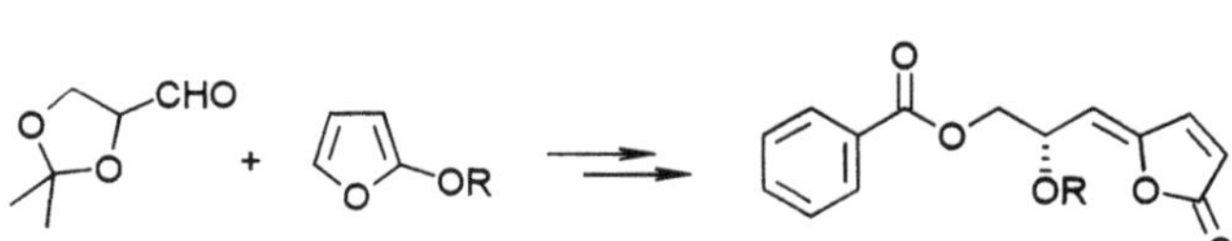

20

R = t-Bu, Me R = H, Ac

Homochiral (*S*)- and (*R*)-1-(2-furyl)ethanols were prepared from **21** by lipase-catalyzed transesterification with vinyl acetate. The pure enantiomers are precursors for the synthesis of L- and D-daunomycin <96TA907>.

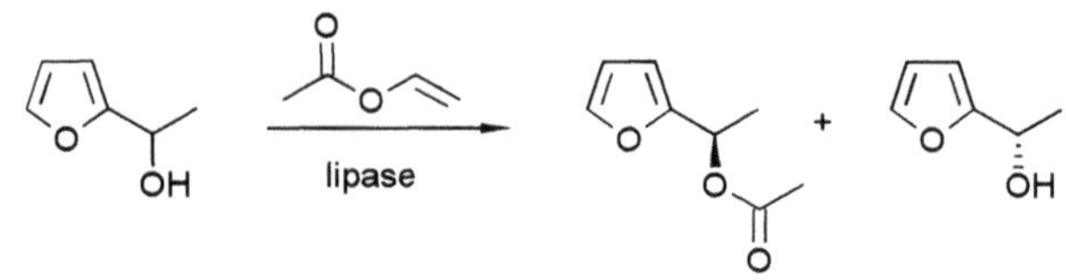

21

Highly stereoselective hetero Diels-Alder reactions of a chiral furylaldehyde (**22**) with Danishefsky's diene were reported. In the presence of $Ln(OTf)_3$ exclusively **23** was obtained, while the reaction in the presence of $Eu(thd)_3$ produced **24** as the major product <96TA1199>.

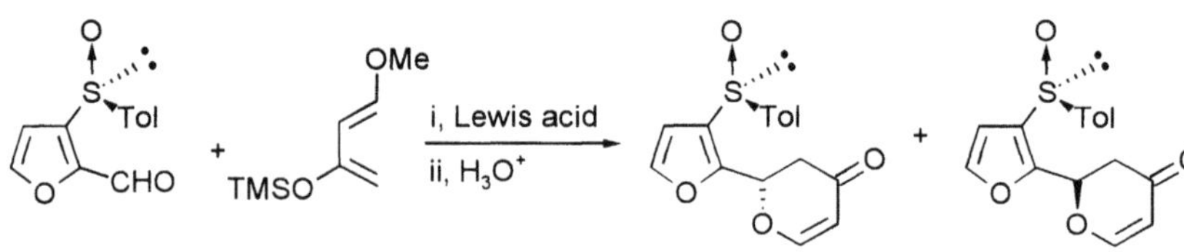

22 **23** **24**

The diastereoselective hydrogenation of 2,5-disubstituted furans on a Raney nickel contact provides an easy access to tetrahydrofurylcarbinols. Due to the alcohol used as the solvent it is possible to influence the direction of the stereoselection process (*erythro* vs. *threo*). The highest diastereoselectivities reached till now are in the range of 70% <96S349>. The (±)-C_{15}-C_{23} portion of the venturicidins was synthesized stereoselectively in 17 steps from 2-furaldehyde in an overall yield of 7% <96SL135>.

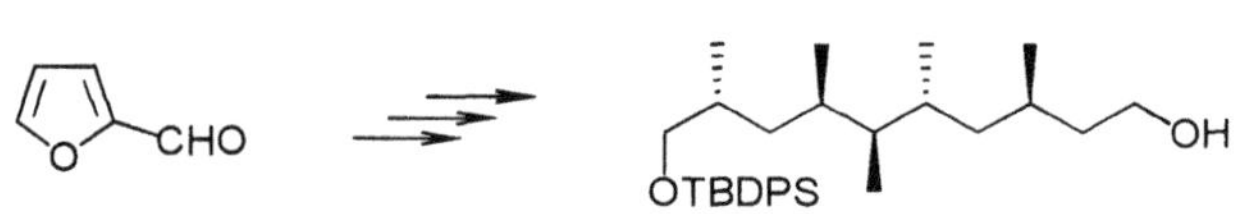

17 steps, 7% yield

Polysubstituted 3-thiofurans which are receiving a great interest as flavour and odour chemicals have been obtained by mono-*ipso*-substitution and *ortho*-metallation from 3,4-dibromofuran <96T4065>. Dihydrofuran is used in a new synthesis of ketones from acids *via* acyl hemiacetals (Scheme 8, <96JOC6071>).

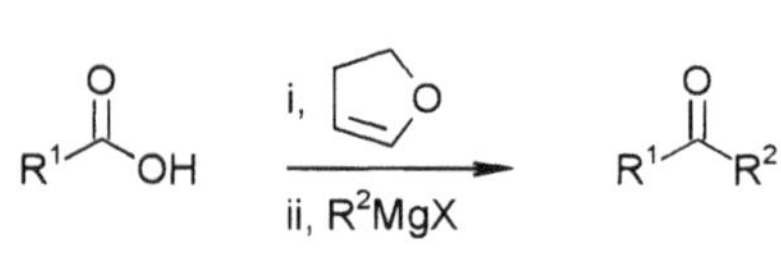

Scheme 8

The synthesis of novel heterocycle-fused furo[3,4-d]isoxazoles *via* ring transformation of 2-isoxazoline-2-oxides by Lewis acids was reported <96H(42)289>. A practical application of the photooxygenation of 3-substituted furans to construction of the zaragozic acid/squalestatin backbone was described <96JOC6685>. The photochemical dimerization of methyl 3-(2-furyl)acrylate in acetonitrile in the presence of benzophenone as triplet sensitizer was reinvestigated in order to understand regio- and stereochemical control <96H(43)959>. The photocycloaddition with 1-naphthalenecarbonitrile was reinvestigated <96TL9329>. Efficient construction of bi- and tricyclic cyclooctanoid systems *via* crossed [4+4]-photocycloadditions of pyran-2-ones was achieved (Scheme 9, <96SL1173>).

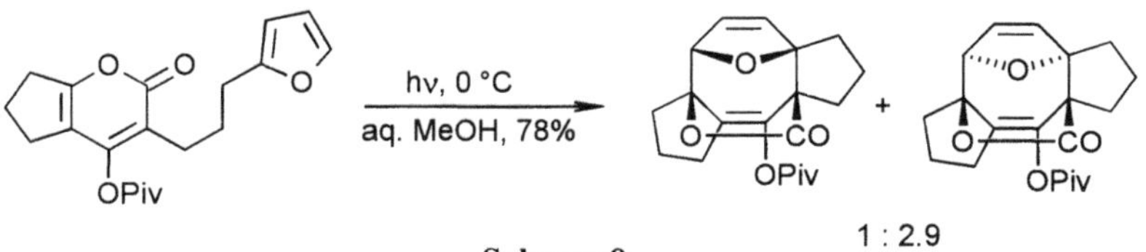

Scheme 9

Participation of aromatic groups in oxy-Cope reaction sequences enables the synthesis of highly substituted polyquinane ring systems (Scheme 10, <96JOC7976>).

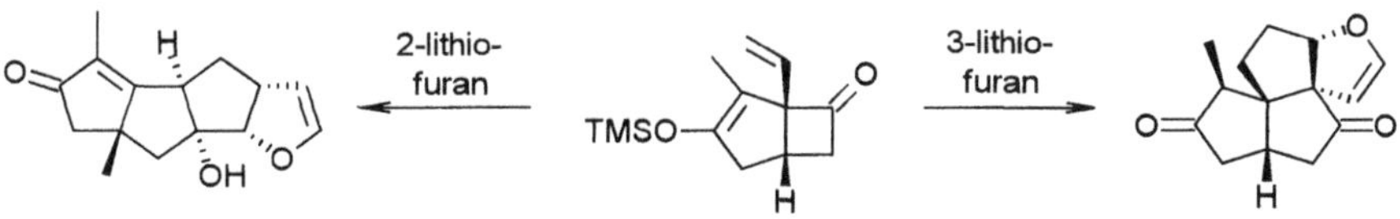

Scheme 10

Levoglucosenone and its enantiomer have been prepared in an enantiocontrolled manner from a non-carbohydrate prochiral precursor 2-vinylfuran *via* the corresponding isolevoglucosenone precursors employing the Sharpless asymmetric dihydroxylation (AD reaction) as key step <96SL971>. A selective electrochemical oxidation of 5-hydroxymethylfurfural to 2,5-furandicarbaldehyde was performed. Yields were various depending on the time of the reaction and on the kind of salt used as a supporting electrolyte <96SL1291>. A ring expansion strategy for the enantioselective synthesis of the medium ring ethers oxepene, oxocane, and oxonene starting with a 3,4-dimethylenedihydrofuran was reported <96S1165>. An asymmetrization of tetrahydrofuran-2,2-dimethanol using L-menthone as a chiral template was achieved <96JOC384>. Employing an intramolecular Michael addition as the key step, hispanolone (**26**) has been converted to prehispanolone (**25**) and 14,15-dihydroprehispanolone (**27**) <96T12137>.

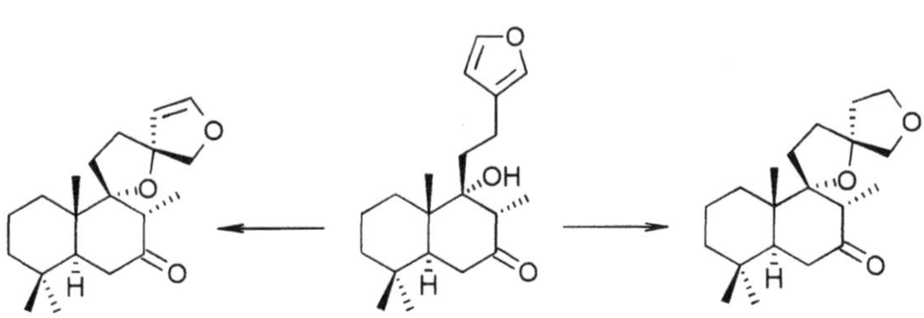

25 **26** **27**

A tandem ring opening-ring closure methodology that exploits the 2-tetrahydrofuryl synthon for elaborating all the carbon atoms and the oxygen atom of the α-hydroxy-δ-valerolactam system is described. This route provides access to lactam based low osmolar X-ray contrast agents (Scheme 11, <96T11177>).

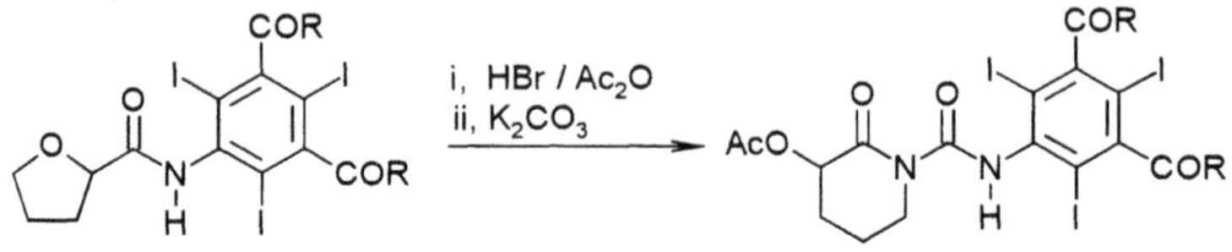

Scheme 11

The ene reaction of fullerene (C_{60}) with 3-methylene-2,3-dihydrofuran gives an easily isolated addition product in good yield <96JOC2559>. There is a continuous need for chiral acrylate esters for asymmetric Diels-Alder reactions with high diastereoselectivity. Lewis acid promoted Diels-Alder reactions of acrylate esters from monobenzylated isosorbide **28** (or isomannide) and cyclopentadiene provided exclusively *endo*-adducts with good yields and high diastereoselectivity <96TL7023>.

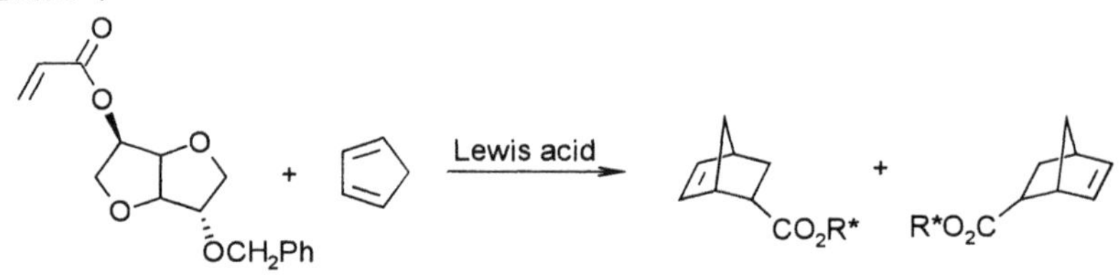

28

+ exo-adduct (endo/exo up to > 99:1)

R_{endo} / S_{endo} up to 4:96

A new synthesis of monomeric *o*-amino thioaldehydes of thiophene, benzo[b]thiophene, furan and benzo[b]furan by reacting the corresponding *o*-azido aldehydes with hexamethyldisilathiane in neat acetonitrile and/or in methanol in the presence of hydrochloric acid at room temperature is reported <96S1185>. (±)-Arthographol - a dihydrobenzofuran derivative - and related compounds were synthesized and their inhibitory activities against 5-lipoxygenase were investigated <96H(43)665>. The synthesis of furoacridines **29a** and furopyranoacridines **29b** was reported. By the saponification of the ester function under aqueous sodium hydroxide basic conditions, an abnormal ring opening reaction of the ethoxycarbonylated furan was observed <96H(43)641>.

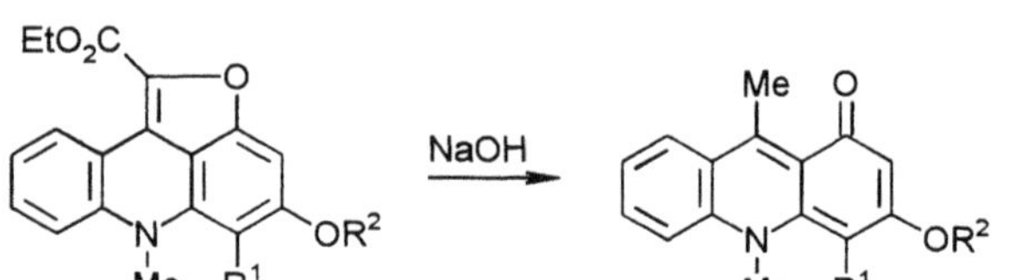

29a, b **a**; R^1 = H, R^2 = CH_3

b; R^1, R^2 = -CH=CH-C(CH_3)$_2$-

The synthesis of a new sulfur analogue of angelicin, a thiopyrano[2,3-e]benzofuran, was reported <96JOC4842>. The reaction of 3(2*H*)-benzofuranone with S_2Cl_2 yields bis-2-spirocoumaranoylidenetetrathiane instead of oxindigo (an oxidative coupling of a coumaranone) (Scheme 12, <96T1961>).

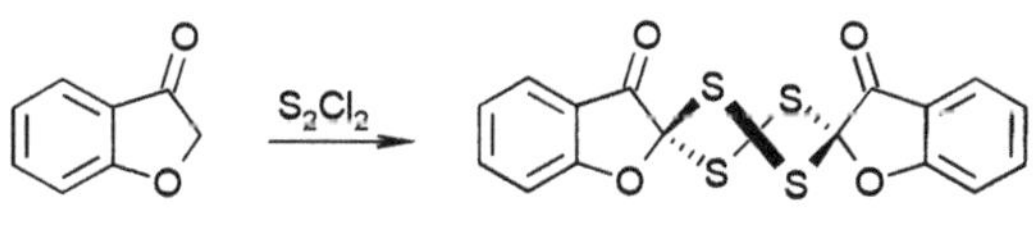

Scheme 12

The photochemical properties of 5-azido-8-methoxypsoralen in the presence of alcohols and water were investigated <96JOC398>. A dramatic reversal of π-facial selectivity in the osmium-catalyzed asymmetric dihydroxylation (AD) of a sterically hindered 3-methylidenebenzofuran was reported; switching from a phthalazine-linked ligand to a pyrimidine-linked ligand led to the opposite enantiomer using the same pseudo-enantiomer of the chinchona alkaloid <96TL1375>. *Ortho*-quinodimethanes have been widely utilized in the construction of polycyclic compounds. Isobenzofurans were generated in a tandem Pummerer-Diels-Alder reaction sequence and were utilized for the preparation of 1-arylnaphthalene lignans <96JOC3706>. The study of heterocyclic analogues has also attracted some attention. A facile synthesis of thieno[2,3-c]furans and furo[3,4-b]indoles *via* a Pummerer-induced cyclization reaction was reported (Scheme 13, <96JOC6166, 96MI2>).

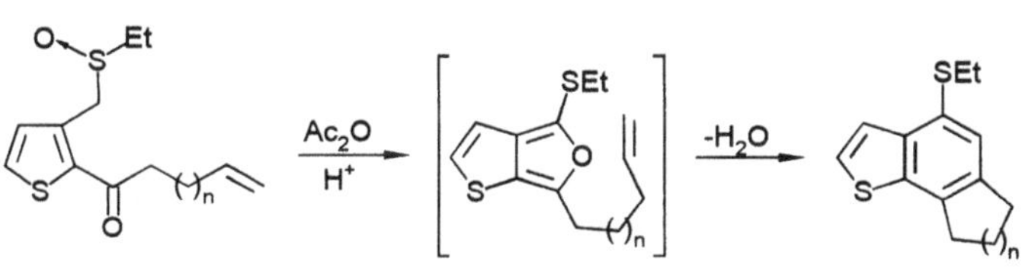

Scheme 13

This methodology has also been used as a very interesting entry into the field of erythrinanes <96JOC4888>. The Ito-Saegusa method has been used to generate benzofuran-2,3-quinodimethane from the corresponding silyl acetate in solution at -4 °C. The intermediate reacts efficiently with a range of dienophiles <96JCS(P1)2827>. 2,3-Dimethylene-2,3-dihydrofuran was also generated *in situ* by a boron trifluoride induced 1,4-elimination. Its Diels-Alder reaction with some dienophiles was reported <96CC2251>. Diels-Alder reactions of 1,3-diphenylisobenzofuran with ferrocenyl- and (η^6-phenyl)tricarbonylchromium analogues of chalcone have been investigated. Unexpected products were isolated in the $AlCl_3$-catalyzed reaction <96JOC3392>. Crown ether isobenzofurans have been prepared and used in supramolecular chemistry studies <96MI3, 97SL47>. Ladder polymer synthesis has received considerable attention in recent years. Acenequinone model compounds (*e.g.*, **31**) are accessible starting with *in situ* prepared derivative **30b** of Hart's benzo[1,2-c:4,5-c']furan (**30a**) <96JOC7304>.

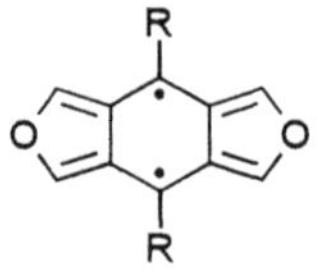

30a; R = H

30b; R = $C_{12}H_{24}$

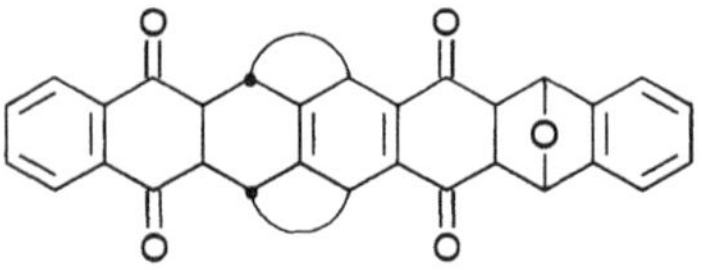

31

5.3.3 SYNTHESIS

5.3.3.1 Furans, Dihydro- and Tetrahydrofurans

The photochemical rearrangement of acylcyclopropenes to furans is well known. The prototype (Scheme 14) has been studied theoretically using CAS-SCF calculations with a 6-31G* basis set. The topology and reaction funnels of the singlet ($^1(n\pi^*)$ and S_0) and higher ($^3(n\pi^*)$ and $^3(\pi\pi^*)$) potential energy surfaces have been characterized along four possible reaction coordinates <96JA4469>.

Scheme 14

The synthesis of different substituted furans by cyclization of 4-pentynones using potassium *tert*-butoxide in DMF was reported <96TL3387>. Dihydrofuran **32** can be prepared by a destannylative acylation of 1-[(2-methoxyethoxy)methoxy]-2-(phenylsulfonyl)-2-(tributylstannyl)-cyclopropane. Treatment of **32** with $BF_3{\cdot}Et_2O$ yields 3-acylfurans *via* an intramolecular Prins-type reaction of the resulting oxonium ion intermediate <96TL4585>.

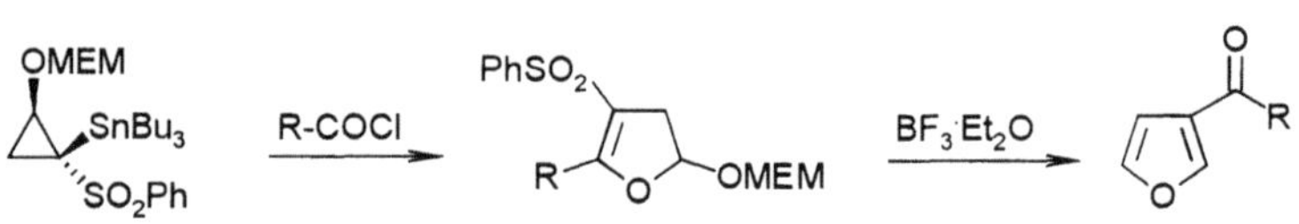

32

A novel route to 2-fluoromethyl- and 2-hydroxymethyl-4-alkylfurans was reported. Treating α-alkylacroleins with 1-bromo-1-trimethylsilylethylene in the presence of butyllithium yields **33**. Oxidation of both double bonds followed by reaction with MsCl provides the key intermediate **34** which on treatment with TBAF produces the desired compounds <96TL7437>.

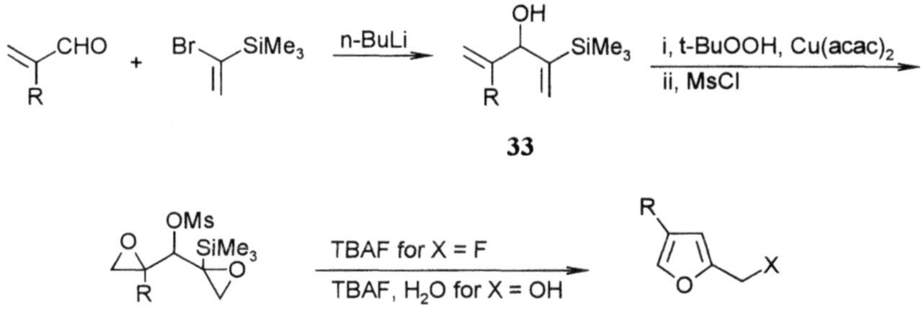

33

34

The nucleophilic vinylic substitution reaction of (*E*)-α-haloenyne sulfones with sodium alkoxides proceeds regioselectively to give (*E*)-α-alkoxyenyne sulfones. These compounds are versatile intermediates for the preparation of furans (Scheme 15, <96TL7381>).

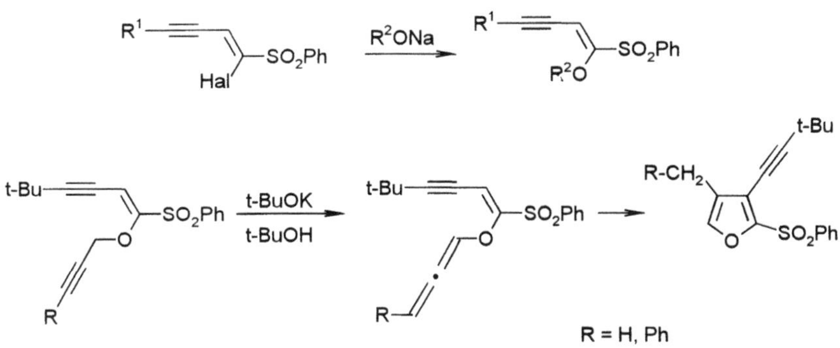

Scheme 15

The acid-catalyzed cyclodehydration of (*Z*)- and (*E*)-δ-hydroxy-α,β-unsaturared ketones offers a mild synthesis of substituted furans. In the case of (*E*)-olefins, photochemical isomerisation was found to accelerate the reaction <96TL6065>. Reaction of alkynyl(phenyl)iodonium tetrafluoroborates with tropolone in the presence of a base yields 2-substituted furotropones (Scheme 16, <96TL5539>).

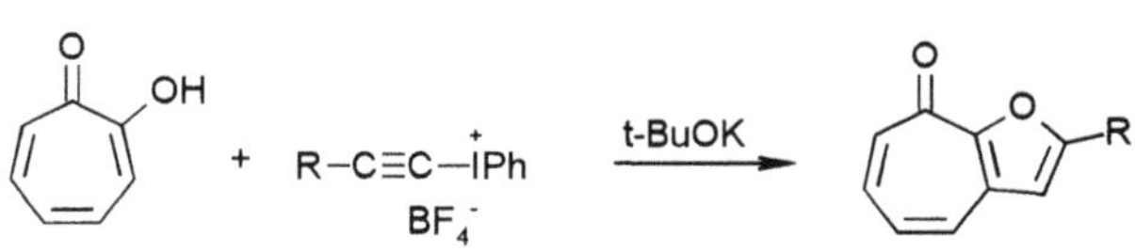

Scheme 16

The Lewis acid-catalyzed ene reaction of 3-methylene-2,3-dihydrofuran (available by Wolff-Kishner reduction of 3-furaldehyde <93TL5221>) with aldehydes gives the corresponding alcohols in good to excellent yields <96TL7893>. 5-*Endo*-dig iodocyclizations of alk-3-yn-1,2-diols, followed by *in situ* dehydration, lead to good yields of β-iodofurans. These compounds serve as good starting material for a wide range of 3-substituted furans (using transition metal-catalyzed coupling reactions or halogen-metal exchange) <96CC1007>. Dihydrofurans were obtained by selective 1,4-nucleophilic additions to 3,4-epoxy-2-methylene oxolanes. Aromatization gave the corresponding substituted furans <96TL2781>.

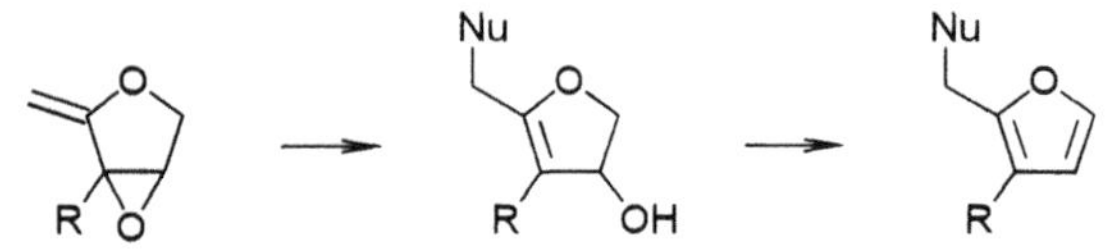

R = H, CH_3

2,5-Diaryl-3-halofurans can be prepared *via* regioselective ring cleavage of aryl 3-aryl-2,2-dihalocyclopropyl ketones <96S388>. An intramolecular radical approach to furanoditerpenes was reported (Scheme 17, <96JOC1806>).

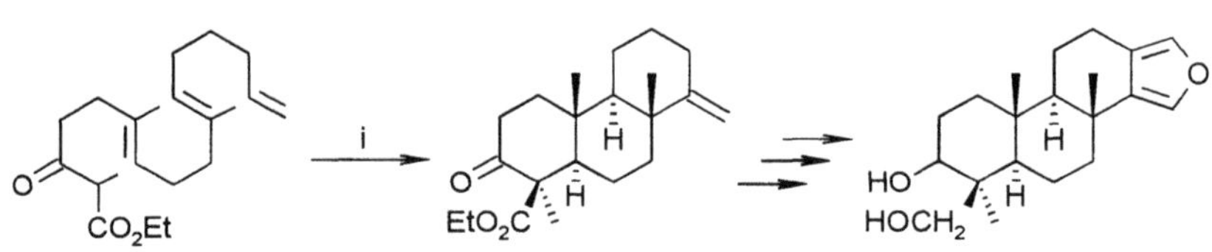

i, $Mn(OAc)_3$, $Cu(OAc)_2$

Scheme 17

2-Silylfurans are available from acylsilane dicarbonyl compounds <96JOC1140>. Photocycloaddition of **35** with alkenes leads cleanly to tetrasubstituted furans **38** in yields of 85%. A mechanism is proposed involving an alkyl propargyl biradical (as **36**) that closes first to a vinyl carbene (as **37**) and than to **38** <96JOC3388>.

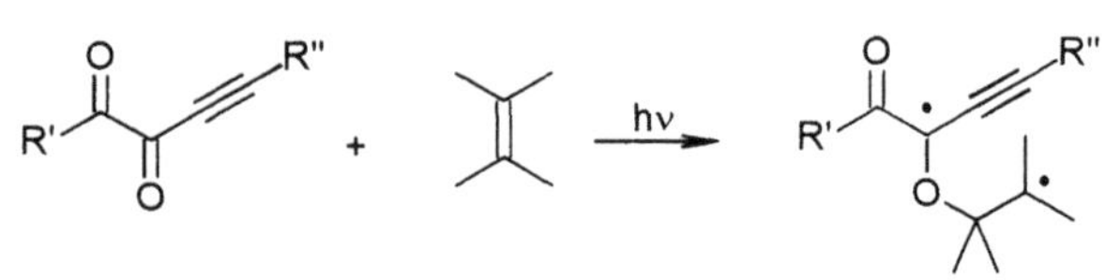

35 **36**

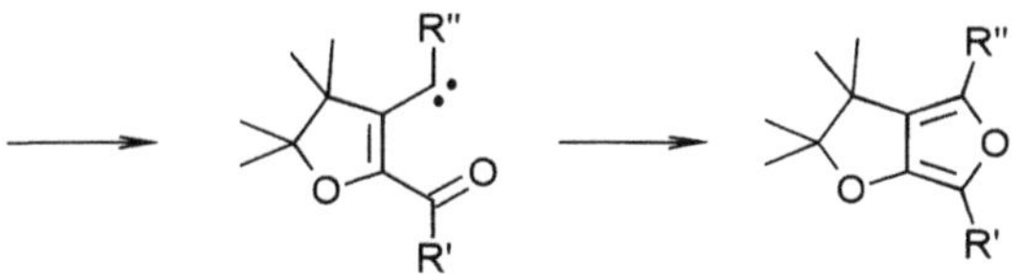

37 **38**

A simple and general synthesis of 2,2,4,5-tetrasubstituted furan-3(2*H*)-ones from 4-hydroxyalk-2-ynones and alkyl halides *via* tandem CO_2 addition-elimination protocol is described <96S1431>. Palladium-mediated intramolecular cyclization of substituted pentynoic acids offers a new route to γ-arylidenebutyrolactones <96TL1429>. The first total synthesis of (-)-goniofupyrone **39** was reported. Construction of the dioxabicyclo[4.3.0]nonenone skeleton was achieved by tosylation of an allylic hydroxy group, followed by exposure to TBAF-HF <96TL5389>.

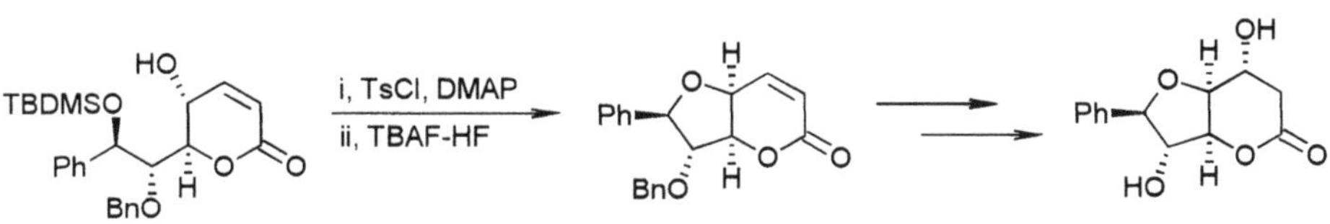

39

(±)-Ambrox is available from (*E*)-nerolidol and β-ionone *via* allylic alcohol [2,3] sigmatropic rearrangement <96JOC2215>. The synthesis of (+)-goniofurfurone, a bioactive furolactone isolated from the stem bark of *Goniothalamus giganteus* Hook. f. & Thomes (Annonaceae) was reported <96TL4373>. Starting with 2-silylfuroic acids α-alkoxy silanes can be prepared quite easily. These compounds can be used for the synthesis of a wide variety of tetrahydrofurans and related compounds. Although the Birch reduction of furan-3-carboxylate was first reported in 1975 by Kinoshita synthetic applications of this reduction sequence have received only scant attention. The Birch reduction of (2-trialkylsilyl)furan-3-carboxylic acids was used for the preparation of methyl (2-trialkylsilyl)tetrahydrofuran-3-carboxylates. These compounds may be considered as synthons of type **40** <96TL9119>.

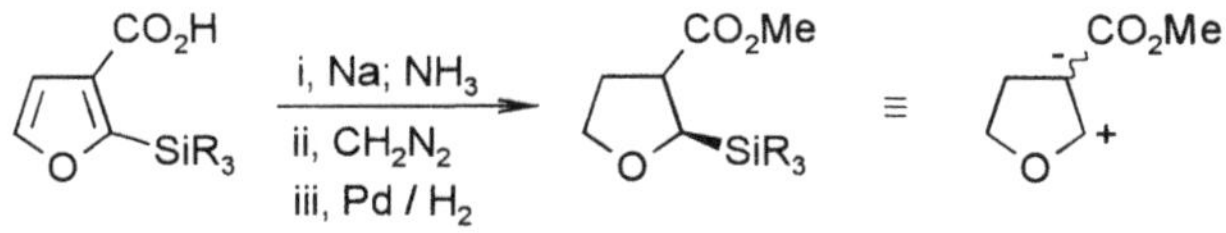

40

The intramolecular reductive coupling of diene, enynes, and diynes by low-valent group 4 metal reagents is a valuable method for the construction of carbo- and heterocycles. It was reported that low-valent titanium and zirconium reagents prepared *in situ* from metallocene dichlorides and magnesium powder react with allyl propargyl ethers to give 3-methylenetetrahydrofurans in good yields <96TL9059>.

reagents: i, Cp_2MCl_2-M_3 (M = Ti or Zn), THF, 0 °C; ii, H_2SO_4

The reaction of fullerene (C_{60}) with α-diazoketones leads to 6,6-bridged, closed 1,2-methano[60]fullerenes and 2'-substituted 6,6-bridged, closed 1,2-dihydro-(4',5'-dihydrofurano)-[60]fullerenes <96SL729>. The reaction of arenesulfonyl iodides with alkynols generally provides adducts in good yields. Cyclization of a functionalized pentenol with $KN(SiMe_3)_2$ results in the formation of *exo*-allylidene tetrahydrofurans <96T7779>. An efficient and highly stereoselective synthesis of either *cis* or *trans* 2,5-disubstituted dihydrofurans was reported. As starting material the same lactol was used <96TL63>.

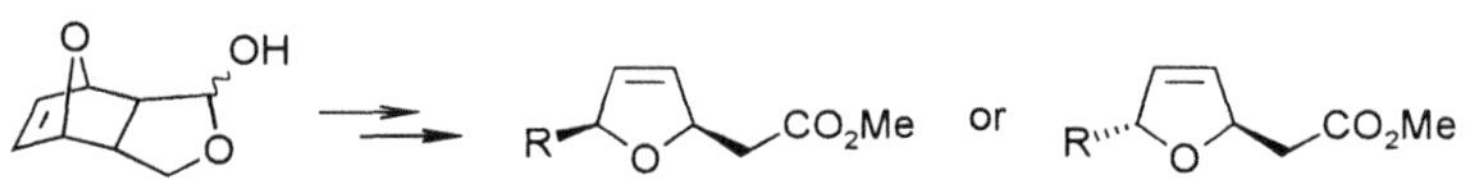

Nonstabilized carbonyl ylides (**41**) prepared by reaction of α-iodosilyl ethers with SmI_2, can be trapped with various alkenes, alkynes and allenes to form furans of type **42**, **43**, and **44** <96JA3533>.

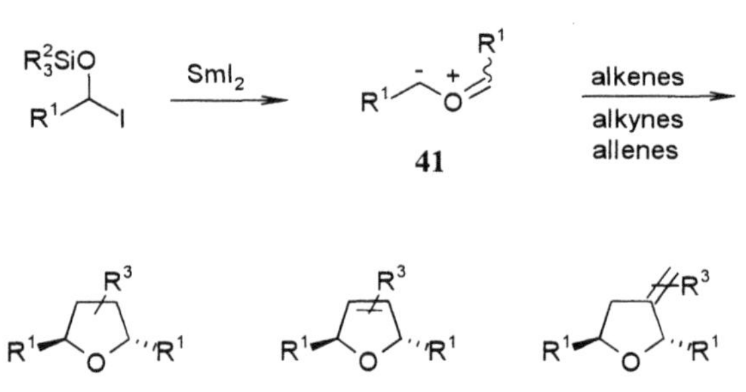

42 **43** **44**

Carbonyl ylides bearing only alkyl substituents or no substituents were efficiently generated from α-chloroalkyl α'-chloroalkyl ethers in the presence of samarium reagents. Subsequent [3+2] cycloaddition with alkenes (and analogues) offers a rapid access to tetrahydrofurans and dihydrofurans <96TL9241>. α,α'-Diactivated ketones undergo a facile stereoselective tandem C-O-cycloalkylation process with *trans*-1,4-dibromo-2-butene leading to functionalized enol ethers (Scheme 18, <96SL339>).

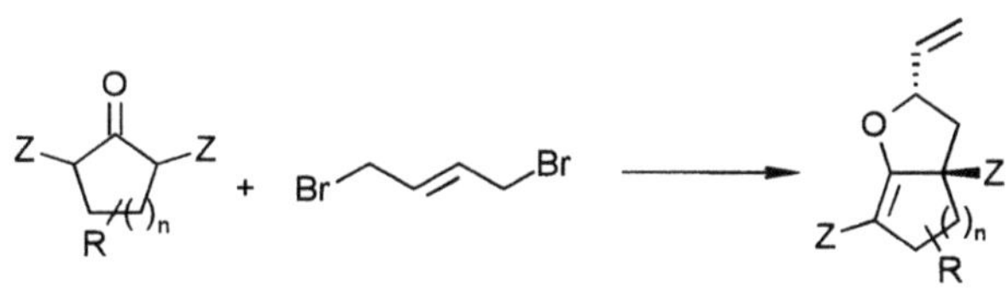

conditions: 3 eq. K_2CO_3, THF, reflux, 3-25 h, 87-100%
Z = CO_2Me, n = 1, 2, 3

Scheme 18

A reasonable asymmetric induction can be obtained in the vinyl epoxide-dihydrofuran rearrangement through the use of established chiral auxiliaries. The diastereomeric products are readily separated to provide access to enantiomerically pure tetrahydrofuran derivatives <96SL401>. The rearrangement of bis[(2,4,6-trimethylphenyl)methanoyl]bicyclo[2.2.1]hept-5-enes to 2-oxabicyclo[3.3.0]octa-3,7-dienes was reported <96T5699>. Benzo- and naphtho-dihydrofuran derivatives were obtained by the reaction of glycosyl fluorides with phenol and naphthol derivatives in the presence of Cp_2MCl_2/AgOTf (M=Hf, Zr). This process includes a glycosylation, 1,2-migration and an intramolecular cyclization (Scheme 19, <96T7797>).

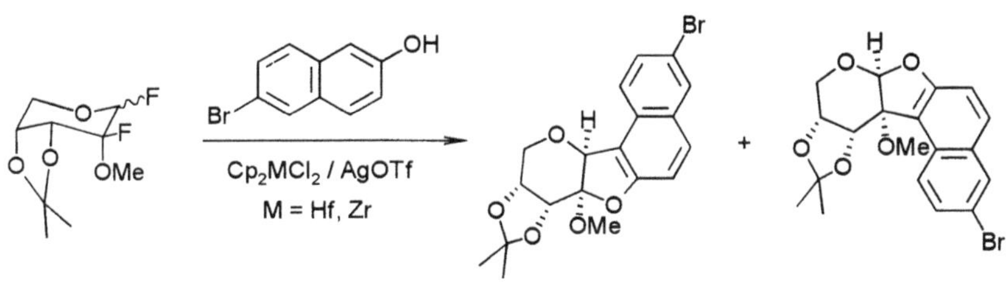

Scheme 19

A new strategy for the synthesis of oxide-containing fragments of morphine has been developed. Thus, the tricyclic (ANO) morphine fragment **45** was obtained as the sole product *via* an intramolecular radical cyclization. The tetracyclic (ACNO) fragment **46** was synthesized in a similar fashion starting from 5,6,7,8-tetrahydroisoquinoline <96T10935>.

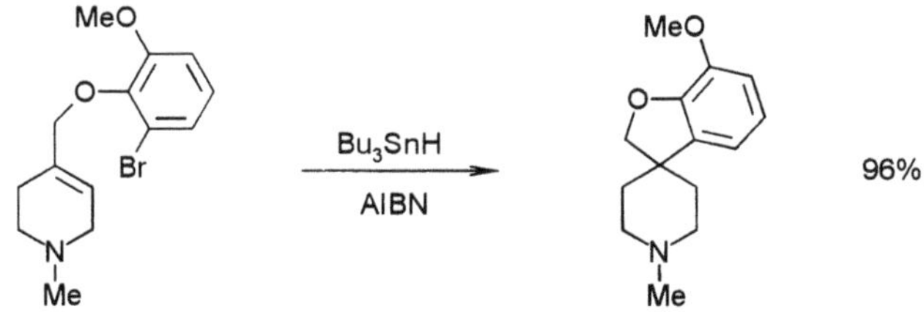

45

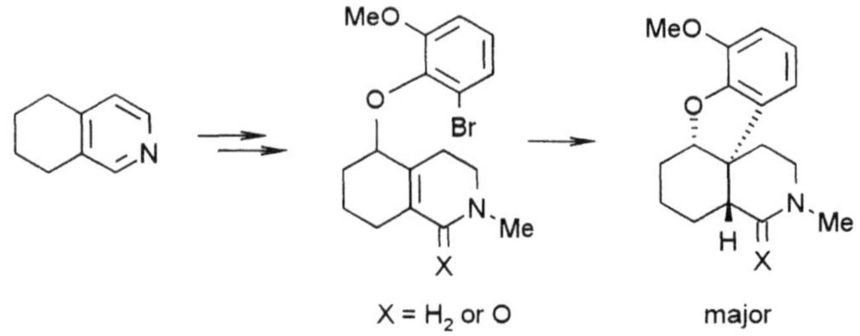

46

An easy and convenient preparation of substituted oxolenes and furans by metal-assisted cyclization of alkynols using labile pentacarbonylchromium and pentacarbonylmolybdenium complexes was described <96T1617>. Metal carbonyl salts ($CpW(CO)_3Na$, $Re(CO)_5Na$, $CpFe(CO)_2Na$) were used for the intramolecular cyclization of compound **47**. Among these salts, $CpW(CO)_3Na$ was found to be the most effective in yielding a metallated fused η^1-2,5-dihydro-3-furyl complex. Demetalations of these organometallic products by $(NH_4)_2Ce(NO_3)_6$ (CAN) in $MeOH/CH_2Cl_2$ under flowing CO provided fused 3-(methoxycarbonyl)-2,5-dihydrofurans; the yields were 50-60% for most cases <96JOC3245>.

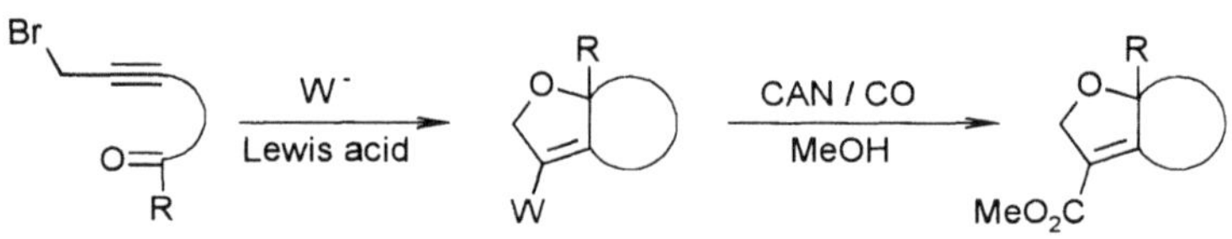

47 W = $CpW(CO)_3$; R = H, Me

Novel heteroquaterphenoquinones were synthesized by a stepwise cross-coupling reaction or by a more convenient one-pot oxidative homocoupling reaction of the heterocycle-substituted phenols (Scheme 20, <96JOC4784; see also 95TL8055>).

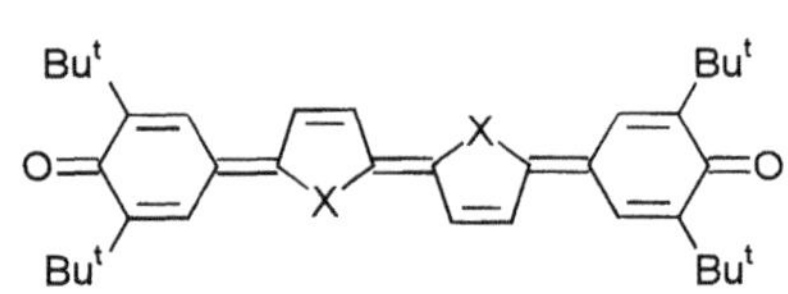

X = S, Se, O, NMe

Scheme 20

Treatment of phenylchalkogen substituted alkenyl alcohols with *t*-BuOK provided useful tetrahydrofurans stereoselectively <96JOC8200>. A concise synthesis of *cis*- and *trans*-theaspirones *via* oxonium ion-initiated pinacol ring expansion was developed <96JOC1119>.

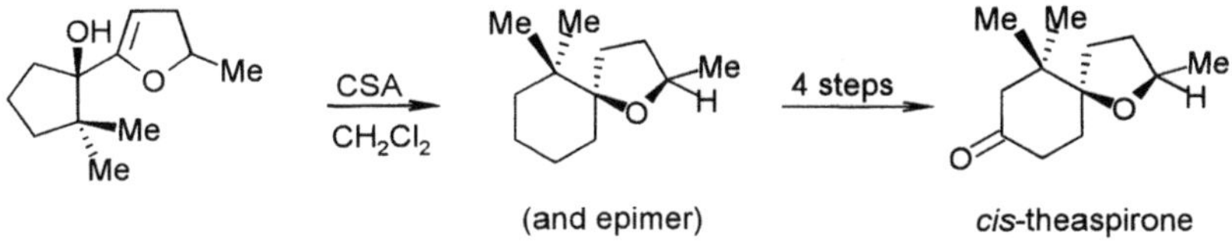

The synthesis of a C1'-C11' synthon **49** of pamamycin-607 starting from alcohol **48** in 10 steps was reported <96TL3751>.

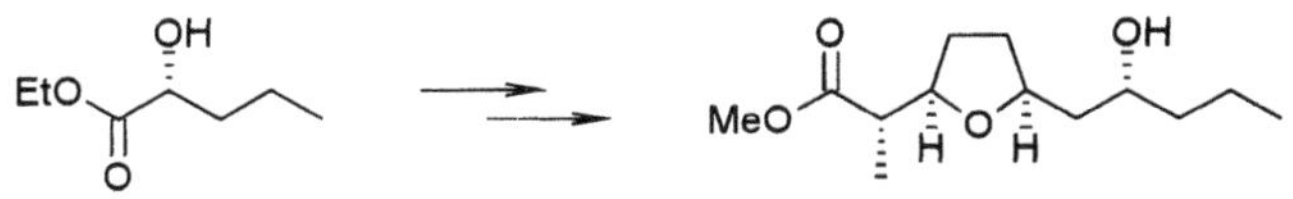

48 **49**

Bis-pyranoside alkenes are reported as novel templates for the stereoselective synthesis of highly substituted, adjacently linked tetrahydrofurans <96TL3619>.

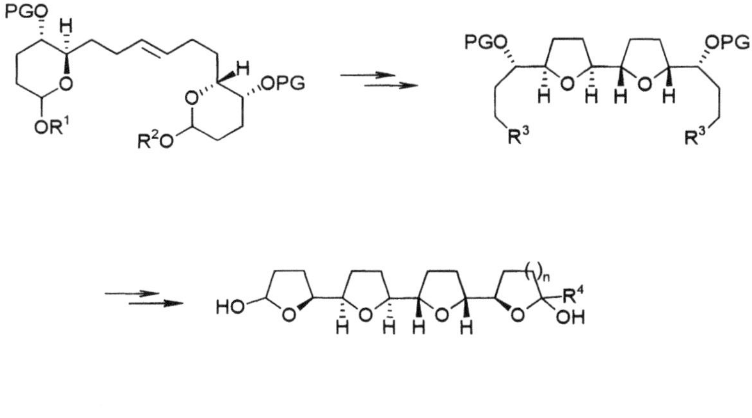

(+)-Ipomeamarone, a furanosesquiterpene isolated from mold-damaged sweet potato (*Ipomea batatas*) as one of the phytoalexins was synthesized starting from (*S*)-lactic acid as the chiral source using Seebach's chiral self-reproduction method <96H(43)1287>. The first total synthesis of the 15-epimer of the naturally occuring acetogenin annonin I was reported <96TL7001>. Unsaturated alkyl bromides of type **50** undergo a stereoselective ring closure when treated with a $MnBr_2$ (5 mol%)/CuCl (3 mol%) catalytic mixed metal system and diethylzinc at 60 °C in DMPU to give five-membered carbo- and heterocycles. Bicyclic derivatives of type **51** are also available on this route <96TL5865>.

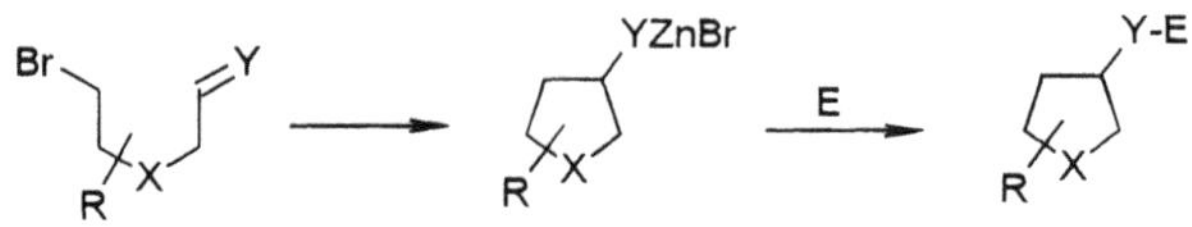

50

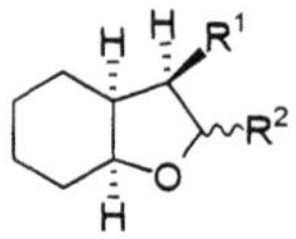

X, Y = CH_2, O

E = I_2, H_2O, allylic bromide, ethyl propiolate, aryl iodide

51

Unsaturated iodo or bromo acetals undergo a smooth cyclization mediated by diethylzinc and $Ni(acac)_2$ as catalyst. This cyclization proceeds *via* a radical mechanism affording a (tetrahydrofuranyl)zinc halide, which can be reacted with various electrophiles after transmetalation with CuCN · 2 LiCl <96JOC5743>. An asymmetric route to 2,2,4-trisubstituted tetrahydrofurans *via* chiral titanium imide enolates was reported. Thus treatment of **52** with $TiCl_4$ (CH_2Cl_2, 0 °C) and subsequently with $BnOCH_2Cl$ (Et_3N, 0 °C) yields **53**, which after reduction (LAH, THF) and iodocyclization (iodine, MeCN, pyridine) gives **54** <96TL5657, 96TL6821>.

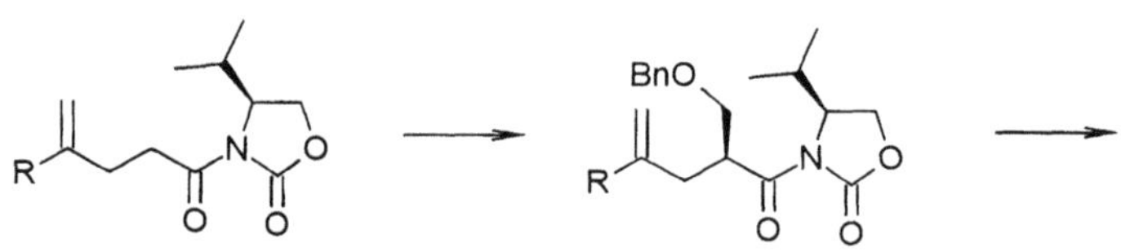

52 **53**

R = $(2,4\text{-}F_2)C_6H_3$

54

A new route to 2,5-dihydrofurans and tetrahydro[3,2-b]furans *via* ring contraction of pyranoside C-glycosides was reported <96CC1663>. (±)-Homononactic acid and its 8-epimer were synthesized by using a *cis*-selective iodoetherification as the key step (Scheme 21, <96SL777>).

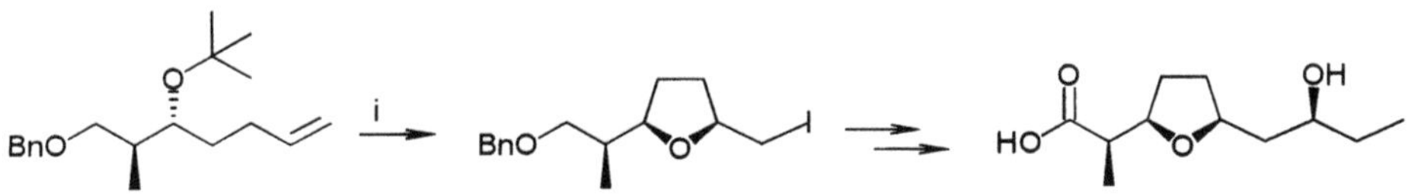

i, I_2, $NaHCO_3$, MeCN, 99.4% ee, 100% de

Scheme 21

A stereocontrolled synthesis of trans-2,5-diaryl dihydro- and tetrahydrofurans starting with the readily available lactol **55** was reported <96JCR(S)309, 96JCR(M)1746; see also 96TL63>.

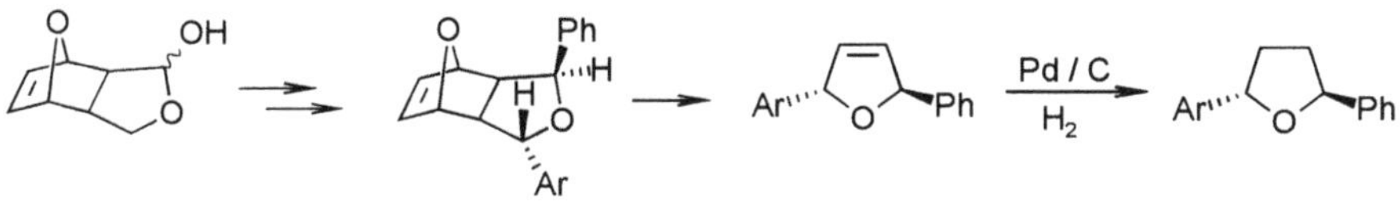

55

Ar = Ph, (4-MeO)C_6H_4, 2-naphthyl

The synthesis of diol **56** from L-lyxose is reported. Compound **56** constitutes a subunit of the toxin erythroskyrine <96JCS(P1)1323>.

L-lyxose **56**

The well-known vanadium(V)-catalyzed oxidation-cyclization of homoallyl alcohols to 2,5-disubstituted tetrahydrofurans has been applied to the synthesis of (+)-eurylene, a cytotoxic bicyclic squalenoid isolated from *Eurycoma longifola* <96TL2039>. Highly functionalized 2,4-dioxohydrindans can be obtained in a stereocontrolled synthesis starting from 2,3,6-tri-*o*-benzylglucopyranosides <96S131>. An enantioselective synthesis of bicyclic tetrahydrofuran carbaldehyde from chiral 3-stannylbut-1-enyl carbamates by tandem homoaldol/aldol reaction was reported <96S145>. The stereochemistry of the condensation of 2-cyclohexenones, α-arylidencyclohexanones carrying one or two (both *syn* and *anti*) spirotetrahydrofuran units adjacent to the carbonyl with allyl organometallics (especially indium) and with the Normant reagent ($ClMgO(CH_2)_3MgCl$) was described <96JOC7492>. A novel route to substituted tetrahydrofurans based on the Lewis acid-promoted Prins cyclization with side chain formation of carbon-carbon bond was reported. Bishomoallylic silyl ethers, rather than the (chloro) benzyl ethers and esters, provide selectively tetrahydrofurans, indicating the siloxy effect for facilitation of the cyclization <96T7287>. Treatment of β-diketones and β-ketoesters with ceric ammonium nitrate and sodium hydrogencarbonate in acetonitrile leads to the formation of intermediates which add efficiently to cyclic enol ethers to furnish fused acetals in good yields <96T12495>. The diastereoselectivity in intramolecular oxymercurations of γ-hydroxyalkenes bearing a remote allylic oxy substituent has been investigated. Cyclization, using mercuric acetate in dichloromethane or acetonitrile, of the (*Z*)-alkenols gave the *syn* diastereomer as the major product <96JOC2109>.

Dirhodium (II) carboxylate catalyzed cyclization of a series of γ-alkoxy-α-diazoesters has been shown to proceed with substantial diastereoselectivity, producing the 2,3,5-trisubstituted tetrahydrofurans. The diastereoselectivity of the cyclization improved as the electron-withdrawing ability of the substituent R increased (Scheme 22, <96JOC6706>).

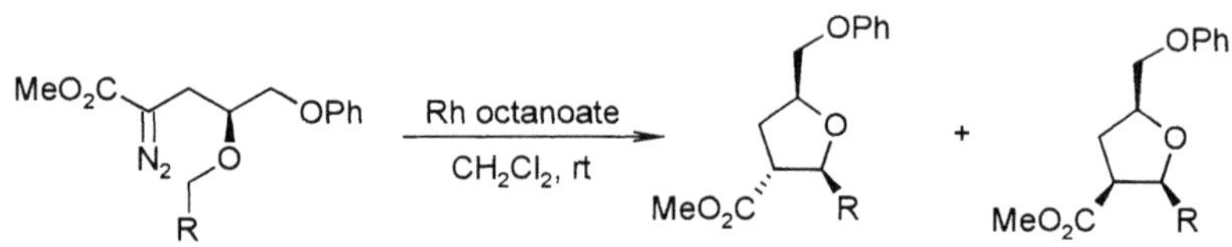

ratio 2/3

R = (4-MeO)C_6H_4; 1.7:1

Scheme 22 R = CH_2OPh; 11.4:1

The thermally induced Pauson-Khand intramolecular cyclization of **57** leads to tricyclic enones in moderate yields and with variable diastereoselectivities <96T14021>.

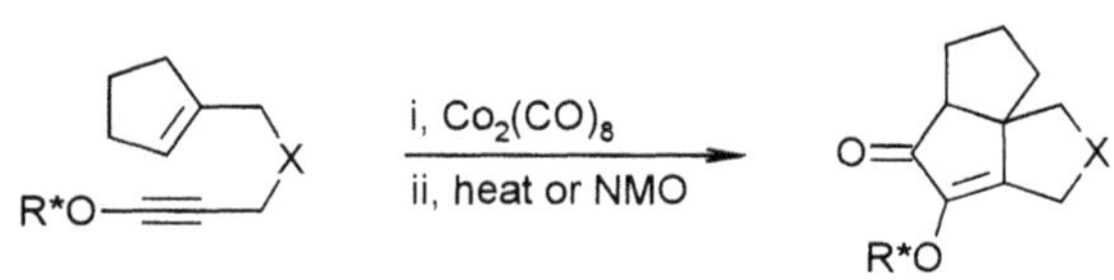

57 X = O, CH_2

Enantiospecific ring expansion of oxetanes to tetrahydrofurans with diazoacetic acid ester was found to be catalyzed by a chiral dipyridine copper complex <96H(42)305>. Bromo γ-oxygenated-α,β-unsaturated sulfones underwent efficient 5- and 6-*exo* radical cyclizations by reaction with Bu_3SnH/AIBN. The presence of a double bond joined to the oxygen allowed the preparation of [3.3.0] and [4.3.0] bicyclic compounds according to a tandem sequence based on two consecutive radical cyclizations <96SL640>. Compounds **59** were synthesized by PET activation of substrates **58** which is achieved through a photosystem comprised of light absorbing DMN as electron donor and ascorbic acid as cooxidant <96JOC6799>.

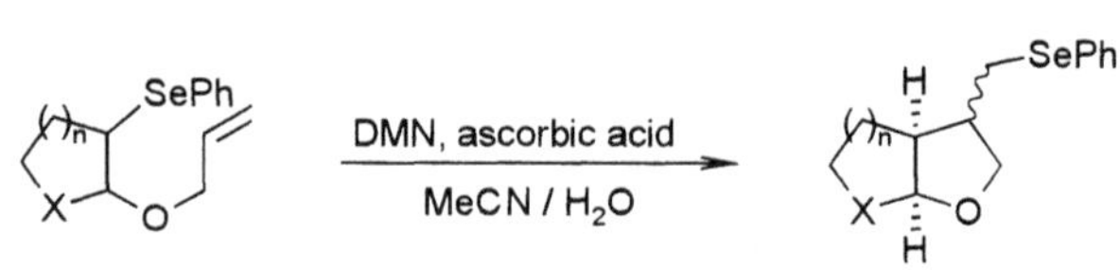

58 **59**

a, n = 1; X = CH_2

b, n = 2; X = O DMN = 1,5-dimethoxynaphthalene

The conversion of vicinal azido selenides into tetrahydrofurans by PhSeOTf in MeCN at room temperature is reported <96JOC7085>. 3-Butadienyl tetrahydrofurans and α-butadienyl γ-butyrolactones can be prepared by radical cyclization of β-bromopent-4-en-2-ynyl ethers and mixed acetals <96SL391>. The total synthesis of trilobacin was reported <96JOC7642>.

5.3.3.2 Benzo- and Dihydrobenzofurans

A great variety of benzo[b]furans of type **60** were synthesized by conversion of the corresponding *o*-(3-hydroxy-3-methylbutynyl)phenyl tosylates in the presence of base (KOH, K_2CO_3) <96H(43)101>.

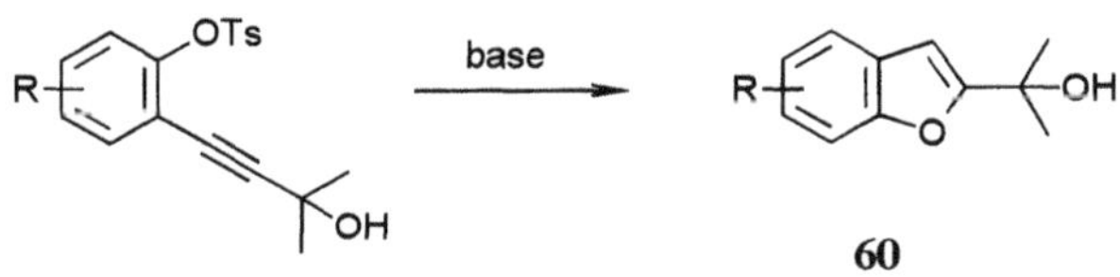

60

Highly substituted annulated benzo[b]furans are available from 2-dienylcyclobutenones <96JOC2584>. Propargyl naphthyl ethers have been efficiently rearranged to naphthofurans under microwave irradiation <96JCR(S)338>. The first total synthesis of furostifoline, an indole-benzo[b]furan alkaloid obtained from the root bark of *Murraya euchrestifolia*, was achieved using an iron-mediated construction of the carbazole nucleus <96TL9183>. The synthesis of alkylfuropyridines *via* Pd-catalyzed cyclization of iodopyridinyl allyl ethers was reported <96H(43)1641>. Substituted and fused furo[3,2-c]pyridines were synthesized starting from 5-aryl-2-furancarbaldehyde (Scheme 23, <96KGS1390>).

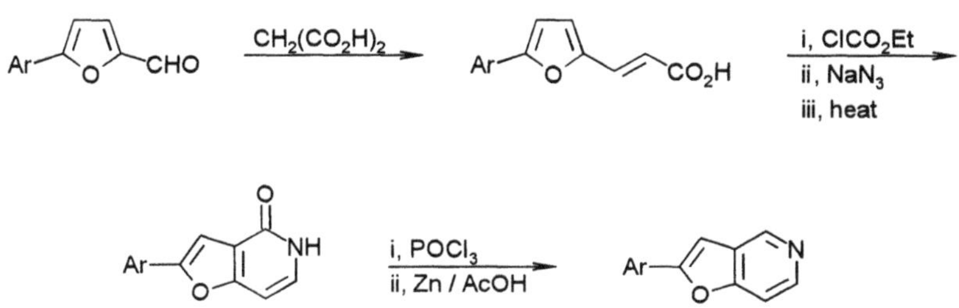

Scheme 23

Furoquinones, such as naphtho[2,3-b]furan-4,9-dione, naphtho[1,2-b]furan-4,5-dione, benzofuran-4,7-dione and benzofuran-4,5-dione derivatives are available by the ceric ammonium nitrate mediated [3+2] cycloaddition of 2-hydroxy-1,4-naphthoquinones and 2-hydroxy-1,4-benzoquinones with alkenes or phenylacetylene <96CL451>.

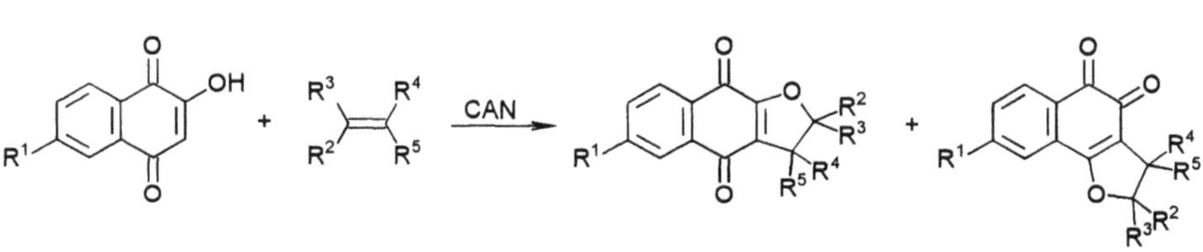

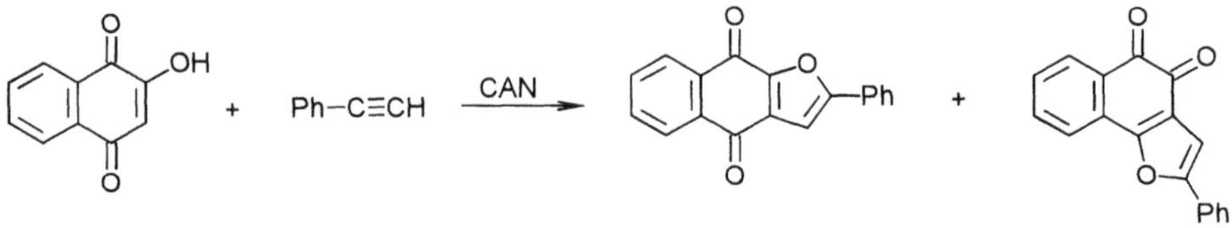

Treatment of methylated diphenyl ethers with BuLi/TMEDA in cyclohexane/diethyl ether followed by addition of $CuCl_2$ resulted in excellent yields of the corresponding dibenzofurans <96JCR(S)362, 96JCR(M)2016>. An efficient route to *meso* benzoxabicyclo[2.2.1]heptyl derivatives beginning from piperonal has been developed (Scheme 24).

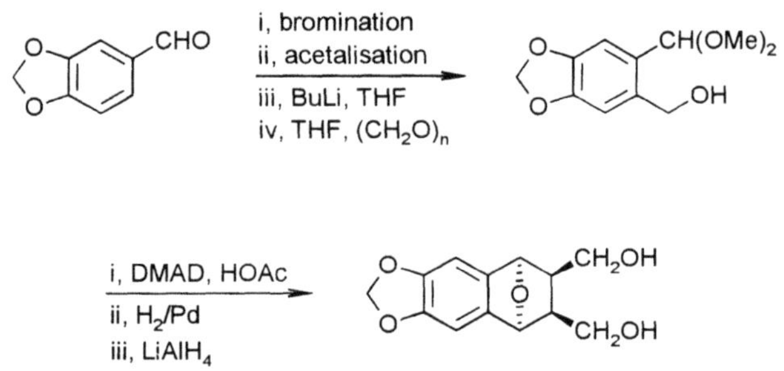

Scheme 24

An isobenzofuran is an intermediate in this reaction sequence <96TA1577>. The reactivity of chalcone and its metalloorganic and other analogues as dienophiles in Diels-Alder reactions with 1,3-diphenylisobenzofuran was examinated under various conditions. The best results were achieved using SiO_2 and very acidic montmorillonite clay, KSF, as the catalyst <96JOC3392>. 2-Hydroxybutenolide is used to homologate naphtho[2,3-c]furan to anthra[2,3-c]furan (trapped as Diels-Alder adducts) <96S77>.

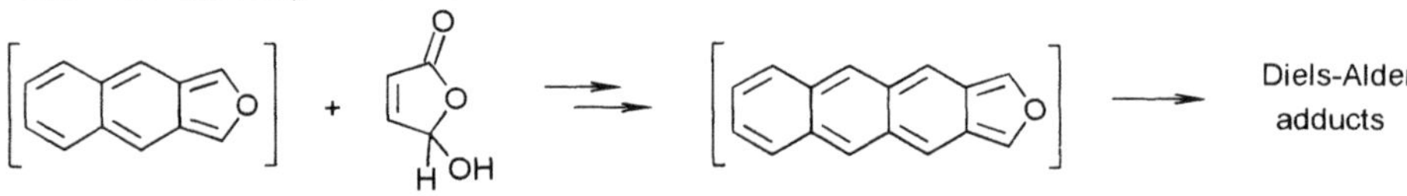

The synthesis of furo[3,4-d]isoxazoles (**61**) and inter- and intramolecular cycloaddition reactions (to **62**, **63**) were reported <96H(43)1165>.

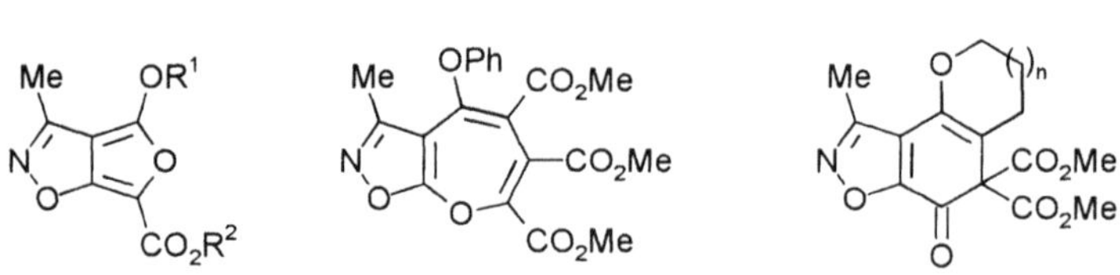

61 **62** **63**, n = 0, 1

Transition metal catalyzed decomposition of diazoester **64b**, which is available from **64a** with TsN_3/base, results in the formation of furo[3,4-b]indole **65**. This intermediate is trapped intramolecularly *in situ* to give **66**. Furo[3,4-b]indoles of type **67** can be prepared similarly <96MI4>.

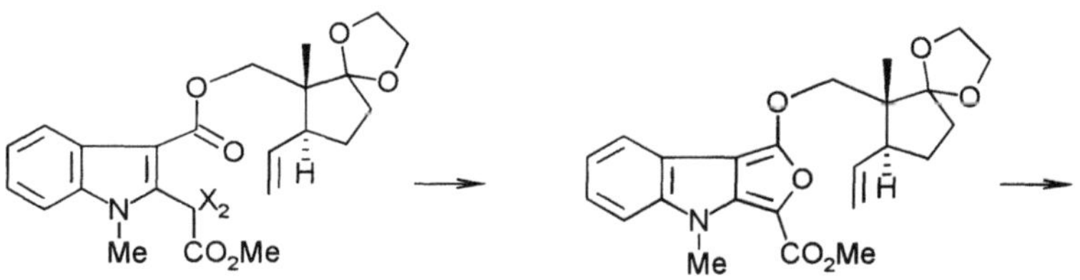

64a, b **65**

a, X = H; **b**, X = N

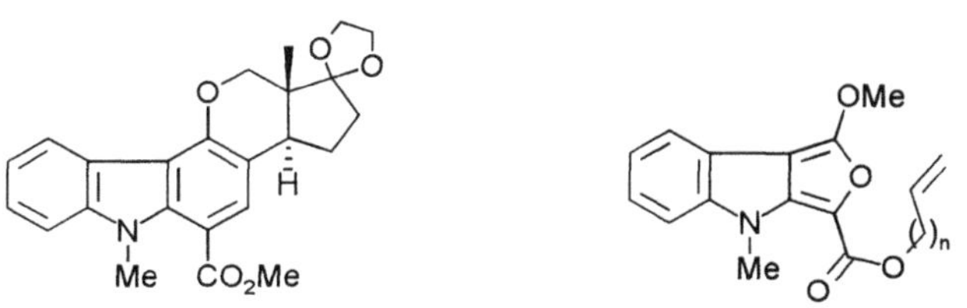

66 **67**, n = 1, 2

Dihydrofurans are accessible by a palladium catalyzed cyclization-allene insertion-anion capture cascade starting with *o*-iodoarylethene <96TL4221, 96T13441>. The base-catalyzed intramolecular Diels-Alder reaction of 5-trimethylsilyl-2-furyl propargyl ethers **68** gave dihydrobenzo[c]furans, presumably *via* the corresponding allenyl ether. Under the reaction conditions, the primary cycloadducts undergo ring opening with subsequent 1,2-shift of the trimethylsilyl group or Brook rearrangement <96TL7395>.

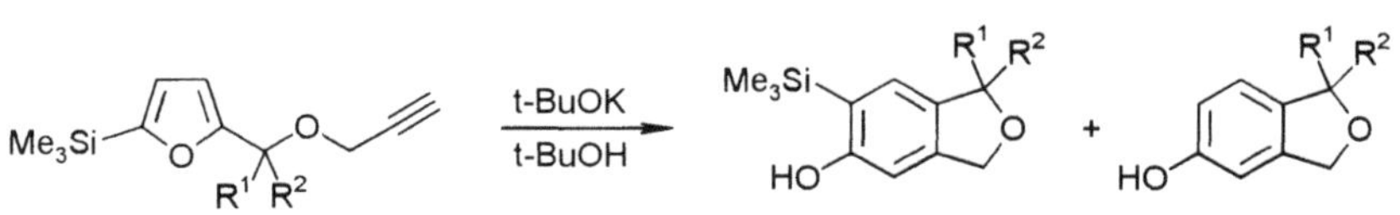

68 R^1, R^2 = H, alkyl

The rhodium-mediated reaction of **69** with 2,3-dihydrofuran (a formal dipolar cycloaddition of a cyclic diazo dicarbonyl compound with a vinyl ether) yields **70**. Compound **70** can be transformed in a number of steps to **71a,b** <96TL2391>.

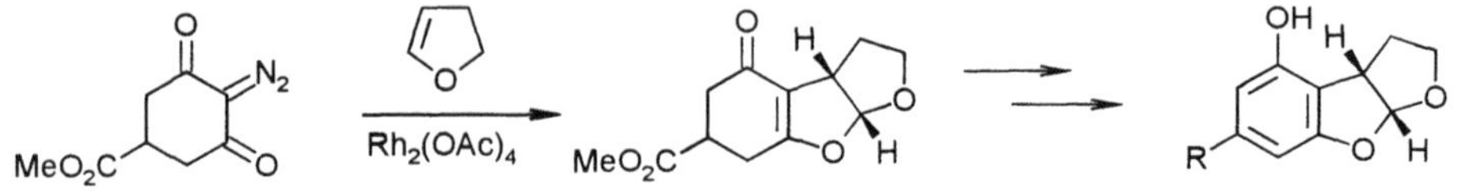

70 **70** **71a,b**

a; R = CHO
b; R = OMe

The Lewis acid promoted reactions of styrene with *N*-phenylsulfonyl-1,4-benzoquinone monoimine to the 2-arylbenzofuranoid ring system have been reported previously. This offers a new route to benzofuranoid neolignans ((±)-licarin B, eupomatenoid-1, eupomatenoid-12) (Scheme 25, <96TL6969>).

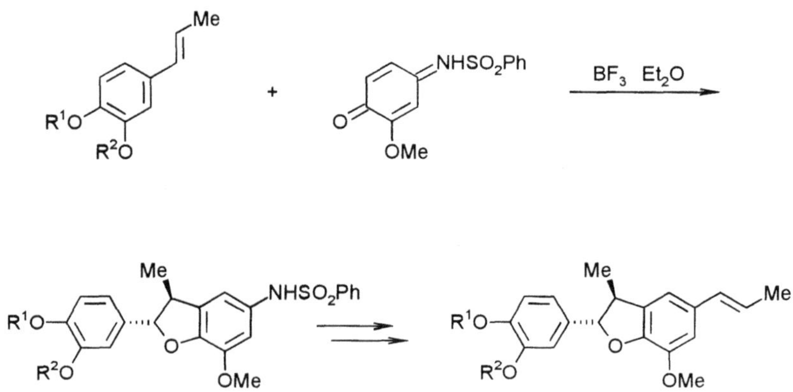

Scheme 25

Treatment of **69** with oxalic acid and subsequently with base yields benzofuran **70**. This constitutes a new formal synthesis of lycoramine **71** <96TL6283>.

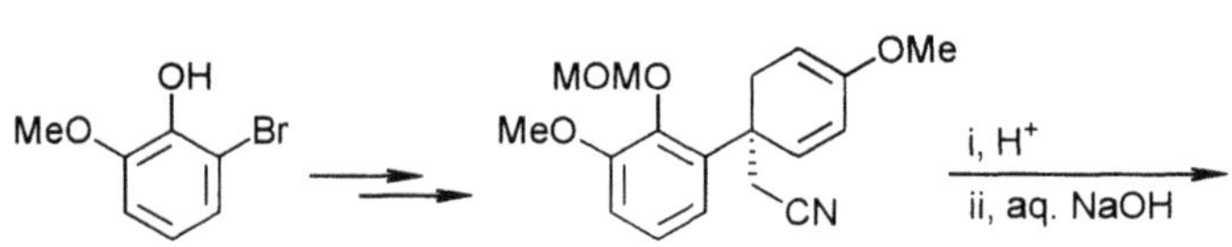

69

70 **71**

Intramolecular cycloadditions between cyclobutadiene and an oxygen-tethered unactivated alkene (alkyne) offers an attractive route to benzo[c]furans (Scheme 26, <96JA9196>).

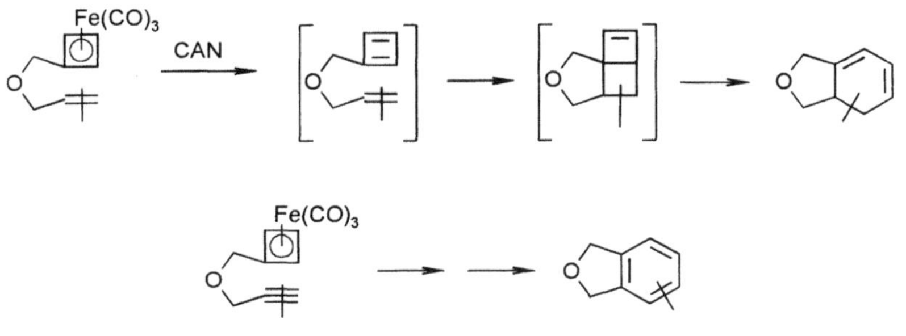

Scheme 26

Photoaddition of electron donor olefins such as vinyl ethers and stilbene to variously methyl and halogeno-substituted 1,4-benzoquinones resulted in the formation of dihydrobenzofurans *via* a dienone-phenol rearrangement of the primary product spirooxetanes <96H(43)619>. High-temperature water seems to be an alternative to use of acid catalysts or organic solvents by the cyclization of allyl phenyl ethers to dihydrobenzofurans <96JOC7355>.

5.3.4 MISCELLANEOUS

The index of aromaticity introduced by C. W. Bird has been calculated for a variety of furans and benzofurans using a density functional theoretical (DFT) methodology (Becke3LYP/6-311+G**) <96AG2824, 96AG(E)2638>. For energetic and geometric contributions to the aromaticity of furan and related compounds see <96T10255>. The relationship of classical and magnetic criteria has also been studied for the furan system <96T9945>. Absolute magnetic shieldings computed at *ring centers* have been proposed as a new aromaticity/antiaromaticity criterion and applied to a variety of five-membered heterocyclic systems (and other compounds), furan included <96JA6317>. 4-Penten-1-oxyl radicals cyclize regioselectively in an 5-*exo* manner. The preference of this reaction path has been studied theoretically. Whereas semiempirical methods are unable to give a correct picture the results of *ab initio* and DFT methods are in accord with experimental observations <96AG3056, 96AG(E)2820; see also <96JOC9264>. The smaller molecular dipole moment of furan relative to tetrahydrofuran has been investigated using the HF/6-311++G**//HF/6-311++G** level of theory. The decreased dipole moment results from a perturbation of the butadiene fragment rather then a delocalization of the heteroatom lone pairs into the π system <96CJ1215>. In a theoretical investigation the question was treated why catalytic quantities of Lewis acid generally yield more product than 1.1 eq. in the intramolecular Diels-Alder reaction of furans <96JOC751>. An NMR-spectroscopic investigation (^{1}H, ^{13}C, ^{29}Si) of 2,5-disubstituted silylfurans was reported <96KGS449>. DFT studies on furo[3,4-d]isoxazoles and furo[3,4-b]indoles were reported <96H(43)1165, 96MI4>. The rotational equilibria of 2-substituted furan (and thiophene) carbonyl derivatives have been investigated by *ab initio* methods (HF/6-31+G**; MP2/6-31+G**//HF/6-31+G**, SCRF calculations) <96TCA199>. The structure of 2-cyanofuran has been determined by microwave spectroscopy. Theoretical calculations (MP2/6-311G**) are in very good agreement with these data <96UP1>. The molecular dynamics of antiaromatic [28]tetraoxaporphyrines was studied in detail <96T11763>. A review of furan derivatives of group VI elements has been published <96KGS867>. The halochromism of polyfuryl (aryl) alkanes has been studied <96KGS73>.

5.3.5 REFERENCES

93TL5221	W.H. Miles, C.L. Berreth, P.M. Smiley, *Tetrahedron Lett.* **1993**, *34*, 5221.
95CHE1034	T.I. Gubina, V.G. Kharchenko, *Chemistry of Heterocylic Compounds* **1995**, 1034 (in Russian).
95TL8055	K. Takahashi, A. Gunji, K. Yanagi, M. Miki, *Tetrahedron Lett.* **1995**, *36*, 8055.
96AG(E)2638	G. Subramanian, P.v.R. Schleyer, H. Jiao, *Angew. Chem. Int. Ed. Engl.* **1996**, *35*, 2638.
96AG(E)2820	J. Hartung, R. Stowasser, D. Vitt, G. Bringmann, *Angew. Chem. Int. Ed. Engl.* **1996**, *35*, 2820.
96AG2824	G. Subramanian, P.v.R. Schleyer, H. Jiao, *Angew. Chem.* **1996**, *108*, 2824.
96AG3056	J. Hartung, R. Stowasser, D. Vitt, G. Bringmann, *Angew. Chem.* **1996**, *108*, 3056.
96CC1007	S.P. Bew, D.W. Knight, *J. Chem. Soc., Chem. Commun.* **1996**, 1007.
96CC1041	L.-H. Shiu, H.-K. Shu, D.-H. Cheng, S.-L. Wang, R.-S. Liu, *J. Chem. Soc., Chem. Commun.* **1996**, 1041.
96CC1663	A. Tenaglia, J.-Y. Le Brazidec, *J. Chem. Soc., Chem. Commun.* **1996**, 1663.
96CC2251	G.-B. Liu, H. Mori, S. Katsumura, *J. Chem. Soc., Chem. Commun.* **1996**, 2251.
96CJ1215	K.E. Laidig, P. Speers, A. Streitwieser, *Can. J. Chem.* **1996**, *74*, 1215.
96CL451	K. Kobayashi, M. Mori, T. Uneda, O. Morikawa, H. Konishi, *Chem. Lett.* **1996**, 451.
96CPB1157	I. Kitagawa, P. Simanjuntak, T. Mahmud, M. Kobayashi, S. Fuji, T. Uji, H. Shibuya, *Chem. Pharm. Bull.* **1996**, *44*, 1157.
96CPB1218	T. Kinoshita, K. Kajiyama, Y. Hiraga, K. Takahashi, Y. Tamura, K. Mizutani, *Chem. Pharm. Bull.* **1996**, *44*, 1218.
96CPB1418	Y. Fukuyama, M. Nakahara, H. Minami, M. Kodama, *Chem. Pharm. Bull.* **1996**, *44*, 1418.
96H(42)129	Y.N. Yamakoshi, W-S. Ge, J. Sugita, K. Okayama, T. Takahashi, T. Koizumi, *Heterocycles* **1996**, *42*, 129.
96H(42)289	K. Harada, E. Kaji, K. Sasaki, S. Zen, *Heterocycles* **1996**, *42*, 289.
96H(42)305	K. Ito, M. Yoshitake, T. Katsuki, *Heterocycles* **1996**, *42*, 305.
96H(43)101	A. Sogawa, M. Tsukayama, H. Nozaki, M. Nakayama, *Heterocycles* **1996**, *43*, 101.
96H(43)425	T. Fukai, Y.-H. Pei, T. Nomura, C.-O. Xu, L.-J. Wu, Y.-J. Chen, *Heterocycles* **1996**, *43*, 425.
96H(43)619	T. Oshima, Y.-i. Nakajima, T. Nagai, *Heterocycles* **1996**, *43*, 619.
96H(43)641	C. Jolivet, C. Rivalle, A. Croisby, E. Bisagni, *Heterocycles* **1996**, *43*, 641.
96H(43)665	M. Miyake, Y. Hanaoka, Y. Fujimoto, Y. Sato, N. Taketomo, I. Yokota, Y. Yoshiyama, *Heterocycles* **1996**, *43*, 665.
96H(43)959	M. D'Auria, *Heterocycles* **1996**, *43*, 959.
96H(43)1165	S. Reck, K. Bluhm, T. Debaerdemaeker, J.-P. Declercq, B. Klenke, W. Friedrichsen, *Heterocycles* **1996**, *43*, 1165.
96H(43)1287	K. Matsuo, T. Arase, S. Ishida, Y. Sakaguchi, *Heterocycles* **1996**, *43*, 1287.
96H(43)1371	S. Sayama, Y. Inamura, *Heterocycles* **1996**, *43*, 1371.
96H(43)1477	R.C. Huang, Y. Minami, F. Yagi, Y. Nakamura, N. Nakayama, K. Tadera, M. Nakatani, *Heterocycles* **1996**, *43*, 1477.
96H(43)1641	S.Y. Cho, S.S. Kim, K.-H. Park, S.K. Kang, J.-K. Choi, K.-J. Hwang, E.K. Yum, *Heterocycles* **1996**, *43*, 1641.
96HCA1393	J. Ancerewicz, P. Vogel, *Helv. Chim. Acta* **1996**, *79*, 1393.
96HCA1415	J. Ancerewicz, P. Vogel, K. Schenk, *Helv. Chim. Acta* **1996**, *79*, 1415.
96HCA1592	P. Siegenthaler, M. Neuenschwander, *Helv. Chim. Acta* **1996**, *79*, 1592.
96JA907	T.E. Caldwell, I.M. Abdelrehim, D.P. Land, *J. Am. Chem. Soc.* **1996**, *118*, 907.
96JA3533	M. Hojo, H. Aihara, A. Hosomi, *J. Am. Chem. Soc.* **1996**, *118*, 3533.
96JA4469	S. Wilsey, M.J. Bearpak, F. Bernardi, M. Olivucci, M.A. Robb, *J. Am. Chem. Soc.* **1996**, *118*, 4469.
96JA6317	P.v.R. Schleyer, C. Maerker, A. Dransfeld, H. Jiao, N.J.R. van Eikema Hommes, *J. Am. Chem. Soc.* **1996**, *118*, 6317.
96JA9196	J.A. Tallarico, M.L. Randall, M.L. Snapper, *J. Am. Chem. Soc.* **1996**, *118*, 9196.
96JA10930	M. Lautens, R. Aspiotis, J. Colucci, *J. Am. Chem. Soc.* **1996**, *118*, 10930.
96JAN505	M.A. Hayes *et al.*, *J. Antibiot.* **1996**, *49*, 505.
96JAN596	R.H. Chen *et al.*, *J. Antibiot.* **1996**, *49*, 596.
96JCR(M)1746	H. Shi, G. Mandville, M. Ahmar, C. Girard, R. Bloch, *J. Chem. Res. (M)* **1996**, 1746.
96JCR(M)2016	F. Radner, L. Eberson, *J. Chem. Res. (M)* **1996**, 2016.
96JCR(S)309	H. Shi, G. Mandville, M. Ahmar, C. Girard, R. Bloch, *J. Chem. Res. (S)* **1996**, 309.
96JCR(S)338	F.M. Moghaddam, A. Sharifi, M.R. Saidi, *J. Chem. Res. (S)* **1996**, 338.
96JCR(S)362	F. Radner, L. Eberson, *J. Chem. Res. (S)* **1996**, 362.
96JCS(P1)1095	R.D. Chambers, A.J. Roche, M.H. Rock, *J. Chem. Soc., Perkin Trans. 1* **1996**, 1095.
96JCS(P1)1101	L.-C. de Almeida Barbosa, D. Cutler, J. Mann, M.J. Crabbe, G.C. Kirby, D.C. Warhust, *J. Chem. Soc., Perkin Trans. 1* **1996**, 1101.

96JCS(P1)1323 R. Rossin, P.R. Jones, P.J. Murphy, W.R. Worsley, *J. Chem. Soc., Perkin Trans. 1* **1996**, 1323.
96JCS(P1)2699 T. Suzuki, H. Fuchii, H. Takayama, *J. Chem. Soc., Perkin Trans. 1* **1996**, 2699.
96JCS(P1)2827 S.B. Bedford, M.J. Begley, P. Cornwall, J.P. Foreman, D.W. Knight, *J. Chem. Soc., Perkin Trans. 1* **1996**, 2827.
96JCS(P2)1233 K. Mackenzie, E.C. Gravett, J.A.K. Howard, K.B. Austin, A.M. Tomlins, *J. Chem. Soc., Perkin Trans. 2* **1996**, 1233.
96JOC384 K. Prasad, R.L. Underwood, O. Repic, *J. Org. Chem.* **1996**, *61*, 384.
96JOC398 K. Feng, Y. Li, *J. Org. Chem.* **1996**, *61*, 398.
96JOC751 I.R. Hunt, A. Rauk, B.A. Keay, *J. Org. Chem.* **1996**, *61*, 751.
96JOC935 Y.-H. Lai, P. Chen, *J. Org. Chem.* **1996**, *61*, 935.
96JOC1119 L.A. Paquette, J.C. Lauter, H.-L. Wang, *J. Org. Chem.* **1996**, *61*, 1119.
96JOC1140 C.S. Siedem, G.A. Molander, *J. Org. Chem.* **1996**, *61*, 1140.
96JOC1478 M.A. Walters, H.R. Arcand, *J. Org. Chem.* **1996**, *61*, 1478.
96JOC1487 A. Benítez, F.R. Herrera, M. Romero, F.X. Talamás, *J. Org. Chem.* **1996**, *61*, 1487.
96JOC1806 P.A. Zoretic, M. Wang, Y. Zhang, Z. Shen, A.A. Ribeiro, *J. Org. Chem.* **1996**, *61*, 1806.
96JOC2109 K. Bratt, A. Garavelas, P. Perlmutter, G. Westman, *J. Org. Chem.* **1996**, *61*, 2109.
96JOC2215 A.F. Barrero, J. Altarejos, E.J. Alvarez-Manzaneda, J M. Ramos, S. Salido, *J. Org. Chem.* **1996**, *61*, 2215.
96JOC2559 W.H. Miles, P.M. Smiley, *J. Org. Chem.* **1996**, *61*, 2559.
96JOC2584 P. Turnbull, M.J. Heileman, H.W. Moore, *J. Org. Chem.* **1996**, *61*, 2584.
96JOC3245 S.-J. Shieh, T.-C. Tang, J.-S. Lee, G.-H. Lee, S.-M. Peng, R.-S. Liu, *J. Org. Chem.* **1996**, *61*, 3245.
96JOC3388 A.K. Mukherjee, P. Margaretha, W.C. Agosta, *J. Org. Chem.* **1996**, *61*, 3388.
96JOC3392 M. Prokešová, E. Solčániová, Š. Toma, K.W. Muir, A.A. Torabi, G.R. Knox, *J. Org. Chem.* **1996**, *61*, 3392.
96JOC3706 A. Padwa, J.E. Cochran, C.O. Kappe, *J. Org. Chem.* **1996**, *61*, 3706.
96JOC4784 K. Takahashi, A. Gunji, K. Yanagi, M. Miki, *J. Org. Chem.* **1996**, *61*, 4784.
96JOC4816 W.G. Dauben, J.Y.L. Lam, Z. R. Guo, *J. Org. Chem.* **1996**, *61*, 4816.
96JOC4842 A.E. Jacobs, L. Christiaens, *J. Org. Chem.* **1996**, *61*, 4842.
96JOC4888 A. Padwa, C.O. Kappe, T. S. Reger, *J. Org. Chem.* **1996**, *61*, 4888.
96JOC5743 A. Vaupel, P. Knochel, *J. Org. Chem.* **1996**, *61*, 5743.
96JOC6071 M.N. Mattson, H. Rapoport, *J. Org. Chem.* **1996**, *61*, 6071.
96JOC6162 J.H. Ryan, P.J. Stang, *J. Org. Chem.* **1996**, *61*, 6162.
96JOC6166 C.O. Kappe, A. Padwa, *J. Org. Chem.* **1996**, *61*, 6166.
96JOC6685 N. Maezaki, H.J.M. Gijsen, L.-Q. Sun, L.A. Paquette, *J. Org. Chem.* **1996**, *61*, 6685.
96JOC6706 D.F. Taber, Y. Song, *J. Org. Chem.* **1996**, *61*, 6706.
96JOC6799 G. Pandey, K.S. Sesha Poleswara Rao, K.V. Nageswar Rao, *J. Org. Chem.* **1996**, *61*, 6799.
96JOC7085 M. Tingoli, L. Testaferri, A. Temperini, M. Tiecco, *J. Org. Chem.* **1996**, *61*, 7085.
96JOC7304 O. Knitzel, W. Münch, A.-D. Schlüter, A. Godt, *J. Org. Chem.* **1996**, *61*, 7304.
96JOC7355 L. Bagnell, T. Cablewski, C. R. Strauss, R. W. Trainor, *J. Org. Chem.* **1996**, *61*, 7355.
96JOC7492 L.A. Paquette, M. Stepanian, U.V. Mallavadhani, T.D. Cutarelli, T. . Lowinger, H.J. Klemeyer, *J. Org. Chem.* **1996**, *61*, 7492.
96JOC7642 S.C. Sinha, A. Sinha, A. Yazbak, E. Keinan, *J. Org. Chem.* **1996**, *61*, 7642.
96JOC7976 V.J. Santora, H.W. Moore, *J. Org. Chem.* **1996**, *61*, 7976.
96JOC7994 M. Lautens, E. Fillion, *J. Org. Chem.* **1996**, *61*, 7994.
96JOC8200 M. Yoshimatsu, M. Naito, H. Shimizu, O. Muraoka, G. Tanabe, T. Kataoka, *J. Org. Chem.* **1996**, *61*, 8200.
96JOC9264 Y. Yamamoto, M. Ohno, S. Eguchi, *J. Org. Chem.* **1996**, *61*, 9264.
96JOC9479 J.M. Fraile, J.I. García, D. Gracia, J.A. Mayoral, E. Pires, *J. Org. Chem.* **1996**, *61*, 9479.
96KGS449 E. Lukevics, I.E. Demicheva, Yu.Yu. Popelis, *Khim. Geterotsikl. Soedin.* **1996**, 449.
96KGS738 A.V. Butin, T. Stroganova, V.G. Kulnevich, *Khim. Geterotsikl. Soedin.* **1996**, 738.
96KGS867 E. Lukevics, O. Pudova, *Khim. Geterotsikl. Soedin.* **1996**, 867.
96KGS1390 A. Krutošíková, *Khim. Geterotsikl. Soedin.* **1996**, 1390.
96MI1 L. Zeng, Q. Ye, N.H. Oberlies, G. Shi, Z.-M. Gu, K. He, J.L. McLaughlin, *Nat. Prod. Rep.* **1996**, *13*, 275.
96MI2 C.O. Kappe, A. Padwa, Article 16, "Electronic Conference in Heterocyclic Chemistry (Echet 96)", H.S. Rzepa, J. Snyder, C. Leach (Eds), Royal Society of Chemistry 1997, ISBN 0-85404-894-4, in press.
96MI3 R.-N. Warrener, R.A. Russell, S. Wang, Article 90, "Electronic Conference in Heterocyclic Chemistry (Echet 96)", H.S. Rzepa, J. Snyder, C. Leach (Eds), Royal Society of Chemistry 1997, ISBN 0-85404-894-4, in press.

96MI4 O. Peters, W. Friedrichsen, *Het. Commun.* **1996**, *2*, 202.
96P1059 M. Bruno, F. Piozzi, B. Rodriguez, M.C. de la Torre, N. Vassallo, O. Servettaz, *Phytochemistry* **1996**, *42*, 1059.
96P1065 Y. Mimaki, T. Kanmoto, M. Kuroda, Y. Sashida, Y. Satomi, A. Nishino, H. Nishino, *Phytochemistry* **1996**, *42*, 1065.
96P1163 W.-W. Li, L.-S. Ding, B.-G. Li, Y.-Z. Chen, *Phytochemistry* **1996**, *42*, 1163.
96P1167 N.H. Anh, H. Ripperger, T. . Sung, G. Adam, *Phytochemistry* **1996**, *42*, 1167.
96P1235 B. Torto, A. Hassanali, E. Nyandat, M.D. Bentley, *Phytochemistry* **1996**, *42*, 1235.
96P1239 D.A. Mulholland, J.J. Nair, D.A.H. Taylor, *Phytochemistry* **1996**, *42*, 1239.
96S31 L.-C. de Almeida Barbosa, J. Mann, *Synthesis* **1996**, 31.
96S77 N.P.W. Tu, J.C. Yip, P.W. Dibble, *Synthesis* **1996**, 77.
96S131 Abu T. Khan, H. Dietrich, R.R. Schmidt, *Synthesis* **1996**, 131.
96S145 H. Paulsen, C. Graeve, R. Fröhlich, D. Hoppe, *Synthesis* **1996**, 145.
96S349 A. Gypser, H.-D. Scharf, *Synthesis* **1996**, 349.
96S388 Y. Tanabe, K.-i. Wakimura, Y. Nishii, Y. Muroya, *Synthesis* **1996**, 388.
96S1185 A. Capperucchi, A. Degl'Innocenti, M. Funicello, P. Scafato, P. Spagnolo, *Synthesis* **1996**, 1185.
96S1291 R. Skowroński, L. Cottier, G. Descotes, J. Lewkowski, *Synthesis* **1996**, 1291.
96S1431 T. Kawaguchi, S. Yasuta, Y. Inoue, *Synthesis* **1996**, 1431.
96SL135 S. Woo, B.A. Keay, *Synlett* **1996**, 135.
96SL339 T. Lavoisier, J. Rodriguez, *Synlett* **1996**, 339.
96SL391 J.-P. Dulcère, E. Dumez, R. Faure, *Synlett* **1996**, 391.
96SL401 A.S. Batsanov, A.L. Byerley, J.A.K. Howard, P.G. Steel, *Synlett* **1996**, 401.
96SL640 J. Adrio, J.C. Carretero, R.G. Arrayás, *Synlett* **1996**, 640.
96SL703 B. Sundermann, H.-D. Scharf, *Synlett* **1996**, 703.
96SL729 H.J. Bestmann, C. Moll, C. Bingel, *Synlett* **1996**, 729.
96SL741 P. Metz, D. Seng, R. Fröhlich, B. Wibbeling, *Synlett* **1996**, 741.
96SL777 M. Abe, H. Kiyota, M. Adachi, T. Oritani, *Synlett* **1996**, 777.
96SL971 T. Taniguchi, K. Nakamura, K. Ogasawara, *Synlett* **1996**, 971.
96SL1165 T. Oishi, M. Shoji, K. Maeda, *Synlett* **1996**, 1165.
96SL1173 C.E. Chase, J.A. Bender, F.G. West, *Synlett* **1996**, 1173.
96T1617 B. Schmidt, P. Kocienski, G. Reid, *Tetrahedron* **1996**, *52*, 1617.
96T1961 H. Langhals, B. Wagner, K. Polborn, *Tetrahedron* **1996**, *52*, 1961.
96T4065 C. Alvarez-Ibarra, M. L. Quiroga, E. Toledano, *Tetrahedron* **1996**, *52*, 4065.
96T5699 F. Freeman, J.D. Kim, M.Y. Lee, X. Wang, *Tetrahedron* **1996**, *52*, 5699.
96T7287 K. Mikami, M. Shimizu, *Tetrahedron* **1996**, *52*, 7287.
96T7779 G.L. Edwards, CA. Muldoon, D.J. Sinclair, *Tetrahedron* **1996**, *52*, 7779.
96T7797 M.I. Matheu, R. Echarri, C. Domènech, S. Castillón, *Tetrahedron* **1996**, *52*, 7797.
96T9945 C.W. Bird, *Tetrahedron* **1996**, *52*, 9945.
96T10255 T.M. Krygowski, M. Cyrański, *Tetrahedron* **1996**, *52*, 10255.
96T10935 C.-Y. Cheng, L.-W. Hsin, J.-P. Liou, *Tetrahedron* **1996**, *52*, 10935.
96T11177 R. Marinelli, T. Arunachalam, G. Diamantidis, J. Emswiler, H. Fan, R. Neubeck, K.M R. Pillai, T. R. Wagler, C.-K. Chen, K. Natalie, N. Soundararajan, R.S. Ranganathan, *Tetrahedron* **1996**, *52*, 11177.
96T11329 K. Aboutayab, S. Chaddick, K. Jenkins, S. Koshi, S. Khan, *Tetrahedron* **1996**, *52*, 11329.
96T11763 G. Märkl, Th. Knott, P. Kreitmeier, Th. Burgemeister, F. Kastner, *Tetrahedron* **1996**, *52*, 11763.
96T12137 E.S. Wang, Y.M. Choy, H.N.C. Wong, *Tetrahedron* **1996**, *52*, 12137.
96T12185 T. Heiner, S.I. Kozhushkov, M. Noltemeyer, T. Haumann, R. Boese, A. de Meijere, *Tetrahedron* **1996**, *52*, 12185.
96T12495 S.C. Roy, P.K. Mandal, *Tetrahedron* **1996**, *52*, 12495.
96T13441 R. Grigg, J.M. Sansano, *Tetrahedron* **1996**, *52*, 13441.
96T14021 J. Tormo, X. Verdaguer, A. Moyano, M.A. Pericàs, A. Riera, *Tetrahedron* **1996**, *52*, 14021.
96T14247 S. Cossu, O. De Lucchi, *Tetrahedron* **1996**, *52*, 14247.
96TA369 E. Cecchet, F. Di Furia, G. Licini, G. Modena, *Tetrahedron Asym.* **1996**, *7*, 369.
96TA907 J. Kamińska, I. Górnicka, M. Sikora, J. Góra, *Tetrahedron Asym.* **1996**, *7*, 907.
96TA981 T. Bauer, *Tetrahedron Asym.* **1996**, *7*, 981.
96TA1199 Y. Arai, T. Masuda, Y. Masaki, M. Shiro, *Tetrahedron Asym.* **1996**, *7*, 1199.
96TA1385 T. Bauer, A. Jeżewski, C. Chapuis, J. Jurczak, *Tetrahedron Asym.* **1996**, *7*, 1385.
96TA1577 D.B. Berkowitz, J.-H. Maeng, *Tetrahedron Asym.* **1996**, *7*, 1577.
96TA3141 C.-C. Shen, S.-C. Chou, C.-J. Chou, L.-K. Ho, *Tetrahedron Asym.* **1996**, *7*, 3141.
96TA3153 A. Guidi, V. Theurillat-Moritz, P. Vogel, A. A. Pinkerton, *Tetrahedron Asym.* **1996**, *7*, 3153.
96TCA199 I. Han, C.K. Kim, H.J. Jung, I. Lee, *Theoretica Chimica Acta* **1996**, *93*, 199.

96TL63 C. Girard, G. Mandville, H. Shi, R. Bloch, *Tetrahedron Lett.* **1996**, *37*, 63.
96TL197 P.-Y. Heo, K. Shin, C.-H. Lee, *Tetrahedron Lett.* **1996**, *37*, 197.
96TL783 M. Harmata, D.E. Jones, *Tetrahedron Lett.* **1996**, *37*, 783.
96TL1129 D.B. Rydberg, J. Meinwald, *Tetrahedron Lett.* **1996**, *37*, 1129.
96TL1375 D. J. Krysan, *Tetrahedron Lett.* **1996**, *37*, 1375.
96TL1429 M. Cavicchioli, D. Bouyssi, J. Goré, G. Balme, *Tetrahedron Lett.* **1996**, *37*, 1429.
96TL1461 Y.-M. Xu, W.-S. Zhou, *Tetrahedron Lett.* **1996**, *37*, 1461.
96TL2039 K. Ujihara, H. Shirahama, *Tetrahedron Lett.* **1996**, *37*, 2039.
96TL2149 C.K. McCluve, K.B. Hansen, *Tetrahedron Lett.* **1996**, *37*, 2149.
96TL2153 C.K. McCluve, K.J. Herzog, K.B. Hansen, *Tetrahedron Lett.* **1996**, *37*, 2153.
96TL2391 M.C. Pirrung, Y.R. Lee, *Tetrahedron Lett.* **1996**, *37*, 2391.
96TL2781 V. Dalla, P. Pale, *Tetrahedron Lett.* **1996**, *37*, 2781.
96TL2903 J.E. Cochran, T. Wu, A. Padwa, *Tetrahedron Lett.* **1996**, *37*, 2903.
96TL3387 A. Arcadi, F. Marinelli, E. Pini, E. Rossi, *Tetrahedron Lett.* **1996**, *37*, 3387.
96TL3619 Z. Ruan, P. Wilson, D.R. Mootoo, *Tetrahedron Lett.* **1996**, *37*, 3619.
96TL3723 S.-K. Kang, H.-W. Lee, J.-S. Kim, S.-C. Choi, *Tetrahedron Lett.* **1996**, *37*, 3723.
96TL3751 I. Mavropoulos, P. Perlmutter, *Tetrahedron Lett.* **1996**, *37*, 3751.
96TL4149 C. Marchionni, P. Vogel, P. Roversi, *Tetrahedron Lett.* **1996**, *37*, 4149.
96TL4221 R. Grigg, V. Sridharan, C. Terrier, *Tetrahedron Lett.* **1996**, *37*, 4221.
96TL4373 J.-P. Surivet, J.-M. Vatèle, *Tetrahedron Lett.* **1996**, *37*, 4373.
96TL4585 M. Pohmakotr, A. Takampon, *Tetrahedron Lett.* **1996**, *37*, 4585.
96TL4593 F.H. Kohnke, G.L. La Torre, M.F. Parisi, S. Menzer, D.J. Williams, *Tetrahedron Lett.* **1996**, *37*, 4593,
96TL4965 K. Yamamura, T. Yamane, M. Hashimoto, H. Miyake, S.-i. Nakatsuji, *Tetrahedron Lett.* **1996**, *37*, 4965.
96TL5389 C. Mukai, S. Hirai, I.J. Kim, M. Hanaoka, *Tetrahedron Lett.* **1996**, *37*, 5389.
96TL5449 L. Zeng, Y. Zhang, J.L. McLaughlin, *Tetrahedron Lett.* **1996**, *37*, 5449.
96TL5539 T. Shu, D.-W. Chen, M. Ochiai, *Tetrahedron Lett.* **1996**, *37*, 5539.
96TL5657 A.K. Saksena, V.M. Girijavallabhan, H. Wang, Y.-T. Liu, R.E. Pike, A.K. Ganguly, *Tetrahedron Lett.* **1996**, *37*, 5657.
96TL5865 E. Riguet, I. Klement, Ch. Kishan Reddy, G. Cahiez, P. Knochel, *Tetrahedron Lett.* **1996**, *37*, 5865.
96TL6065 D.M. Sammond, T. Sammakia. *Tetrahedron Lett.* **1996**, *37*, 6065.
96TL6125 Y. Kobayashi, K. Watatani, Y. Kikori, R. Mizojiri, *Tetrahedron Lett.* **1996**, *37*, 6125.
96TL6201 P. Fonte, F.H. Kohnke, M.F. Parisi, D.J. Williams, *Tetrahedron Lett.* **1996**, *37*, 6201.
96TL6283 E.J. Sandoe, G.R. Stephenson, S. Swanson, *Tetrahedron Lett.* **1996**, *37*, 6283.
96TL6821 Correction of <96TL5657>.
96TL6969 T.A. Engler, W. Chai, *Tetrahedron Lett.* **1996**, *37*, 6969.
96TL7001 I. Wöhrle, A. Claßen, M. Peterek, H.-D. Scharf, *Tetrahedron Lett.* **1996**, *37*, 7001.
96TL7023 A. Loupy, D. Monteux, *Tetrahedron Lett.* **1996**, *37*, 7023.
96TL7381 M. Yoshimatsu, J. Hasegawa *Tetrahedron Lett.* **1996**, *37*, 7381.
96TL7395 H.-J. Wu, C.-H. Yen, C.-T. Chuang, *Tetrahedron Lett.* **1996**, *37*, 7395.
96TL7437 M.M. Kabat, *Tetrahedron Lett.* **1996**, *37*, 7437.
96TL7893 W.H. Miles, C. L. Berreth, C.A. Anderton, *Tetrahedron Lett.* **1996**, *37*, 7893.
96TL9059 K. Miura, M. Funatsu, H. Saito, H. Ito, A. Hosomi, *Tetrahedron Lett.* **1996**, *37*, 9059.
96TL9119 R.L. Beddoes, M.L. Lewis, P. Gilbert, P. Quayle, S.P. Thomson, S. Wang, K. Mills, *Tetrahedron Lett.* **1996**, *37*, 9119.
96TL9183 H.-J. Knölker, W. Fröhner, *Tetrahedron Lett.* **1996**, *37*, 9183.
96TL9241 M. Hojo, H. Aihara, H. Ito, A. Hosomi, *Tetrahedron Lett.* **1996**, *37*, 9241.
96TL9329 T. Noh, D. Kim, *Tetrahedron Lett.* **1996**, *37*, 9329.
96UP1 K. Lukoschus, Diploma Thesis, University of Kiel 1996; K. Lukoschus, D.H. Sutter, *Z. Naturforsch. A*, to be published.
97SL47 R.N. Warrener, S. Wang, R.A. Russell, M.J. Gunter, *Synlett* **1997**, 47.

Chapter 5.4

Five-Membered Ring Systems: With More than One N Atom

Michael A. Walters
Dartmouth College, Hanover, NH, USA

J. Ramón Vargas
Eastman Kodak Company, Rochester, NY, USA

5.4.1 INTRODUCTION

Azoles continue to be of interest for a variety of biological applications. Synthetic incorporation of azoles into potentially active agents has been reported for antibiotics <95CPB2123>, enzyme inhibitors <96JMC918>, and anticonvulsants <95JMC3884>. In general synthetic applications, alkylation of azoles with 4-bromophenacyl bromide in the absence of base and solvent using microwave irradiation gave excellent yields (>95%) of mono alkylated products. Reactions of pyrazole, 3,5-dimethyl pyrazole, 1,2,4-triazole, indazole and benzotriazole were investigated. The reactions were >96% selective for 1-alkylation except for benzotriazole (*ca.* 80%) <96H539>. Stable carbenes derived from 1*H*-imidazole and 4*H*-1,2,4-triazole were shown to be efficient catalysts for benzoin-type condensations of formaldehyde <96HCA61>. Reaction of a variety of azoles with 2,5-dihydro-1,2,5-azoniasilaboratoles gave the respective N-azoyl derivatives <96CB147>. 1-Acylazoles were converted to 1-vinylazoles by reaction with acceptor-stabilized phosphoranes <96TL7249>. The catalysis and regioselectivity of the Michael-addition of several azoles with α,β-unsaturated nitriles and esters was investigated <96TL4423>.

5.4.2 PYRAZOLES AND RING FUSED DERIVATIVES

Structurally novel pyrazole derivatives include the propellene 2,3,4,5,6-pentakis(pyrazol-1'-yl)pyridine **1** and the corresponding 3',5'-dimethylpyrazole derivative **2** <96T11075>. Poly(pyrazol-1-ylmethyl)benzenes, such as **3**, have been prepared as multidentate ligands <95AJC1587>. Solid phase synthesis of structurally diverse 1-phenylpyrazolones was reported, with application to combinatorial synthesis <96SL667>.

1: R = H
2: R = Me

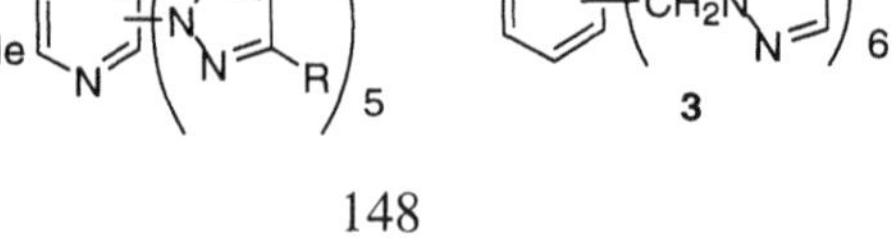

3

Among the more interesting reports of pyrazole ring formation is the treatment of *cis*-aziridinyl ketone tosylhydrazones **4A** with sodium ethoxide to afford the pyrazoles **5** <96H1759>, while treatment of the *trans*-aziridinyl ketones **4B** with tosyl hydrazine and polyphosphoric acid provides 2-pyrazolines **6** <95NKK713>. Improved pyrazole syntheses include the one-pot procedure for 3(5)-carboethoxypyrazoles **8** from **7** and hydrazine hydrochloride <95S1491>. Sulfonamide conjugate addition to the azoalkenes **9**, directed by the azo substituent, followed by cyclization, provides pyrazolones **10** <96S533>. Dihydrothieno[3,4-*c*]-pyrazole **12** was prepared from **11** by reaction with hydrazine <96T9035>. Tandem 1,4-conjugate addition-Michael annulation of **13** effectively provides a [3 + 2] annulation to the pyrazolo[1,5-*a*]quinoline ring system **14** <96T8471>.

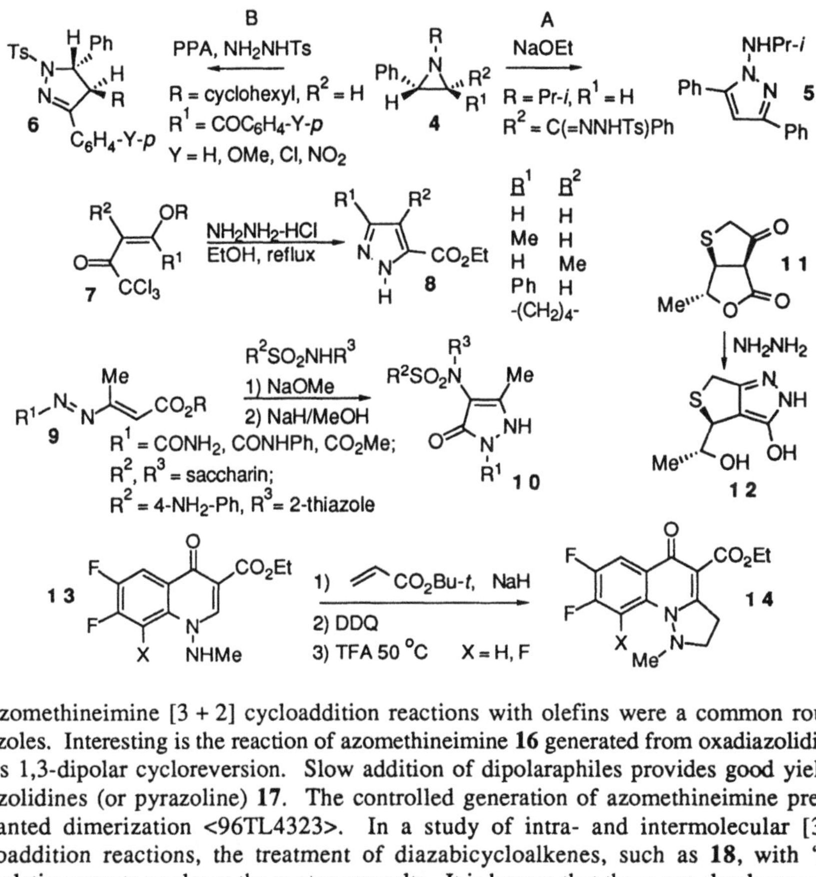

Azomethineimine [3 + 2] cycloaddition reactions with olefins were a common route to pyrazoles. Interesting is the reaction of azomethineimine **16** generated from oxadiazolidine **15** by its 1,3-dipolar cycloreversion. Slow addition of dipolaraphiles provides good yields of pyrazolidines (or pyrazoline) **17**. The controlled generation of azomethineimine prevents unwanted dimerization <96TL4323>. In a study of intra- and intermolecular [3 + 2] cycloaddition reactions, the treatment of diazabicycloalkenes, such as **18**, with 'hard' methylating agents produces the quaternary salts. It is known that these can slowly convert to azomethineimines. However, treatment of diazabicycloalkenes with soft methylating agents readily affords the azomethineimines. The reactions between olefins and these unstabilized azomethineimines (from **18**) afford pyrazolidines **19** <95T13197>. Generation of azomethineimines from hydrazines under neutral conditions is proposed to proceed via intramolecularly assisted proton transfer. Thermal cyclization provides pyrazoline **21** from **20** via the pyrazolidine <96T901>.

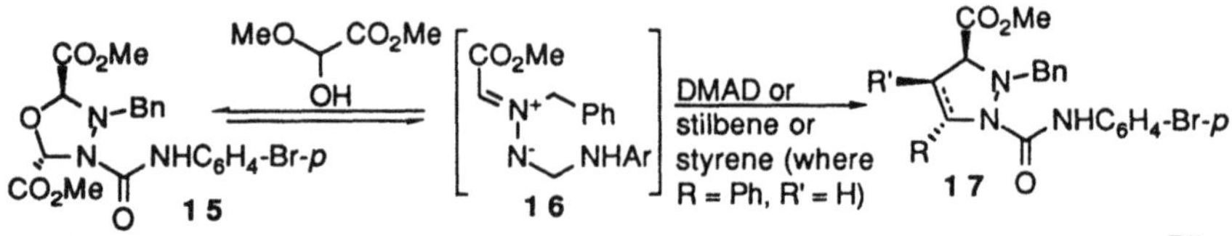

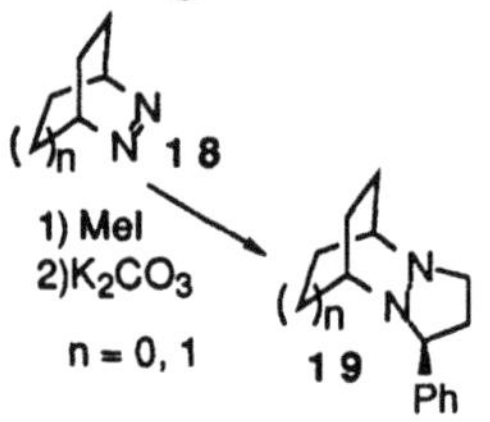

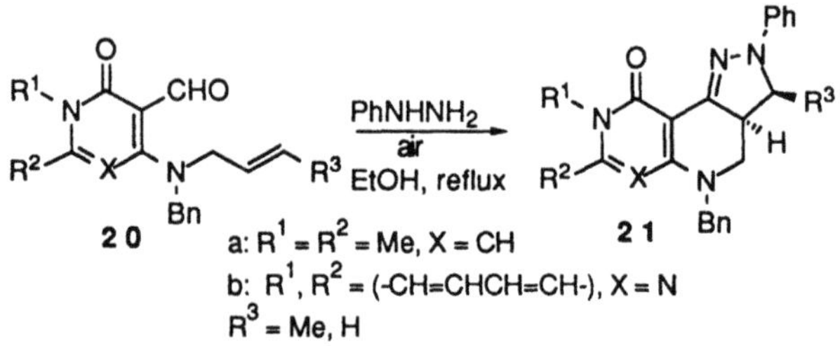

Favorski-like ring contraction of oxadiazinone **22** followed by dimerization provides bimane **23** <96TL5039>. The pyrazolo[1,5-*a*]quinoxalines **26** are prepared via intramolecular nitrileimine cycloaddition. Coupling of diazoarene **24** with methyl chloroacetoacetate provides the highly reactive hydrazonyl chloride **25**. Treatment with triethylamine leads to near quantitative conversion to **26** <96S1076>. Approaches to indazoles include additions of γ-enolate **27** to keteneacetals **28** followed by acid-catalyzed cyclizations to indazolones **29** <95T10941>. Reactions of *o*-acetylenylchloroarenes **30** with hydrazine also provide indazoles **31** via an initial S_NAr adduct <96MC98>.

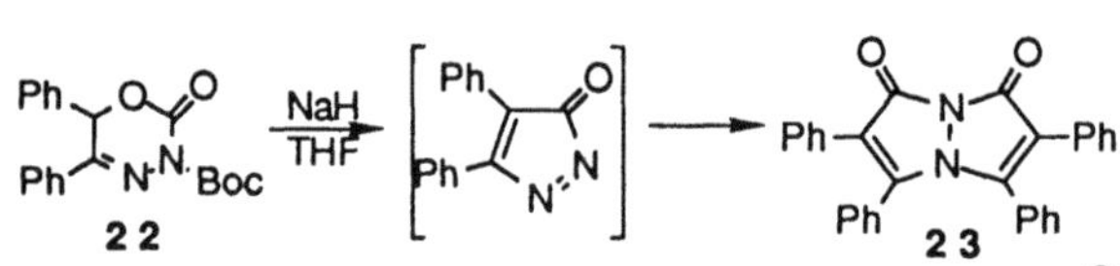

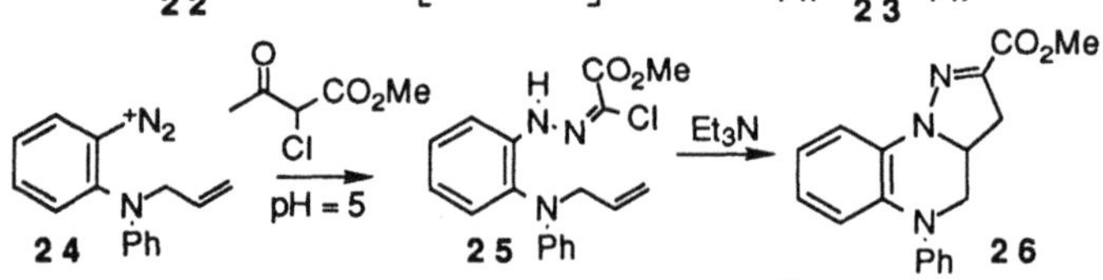

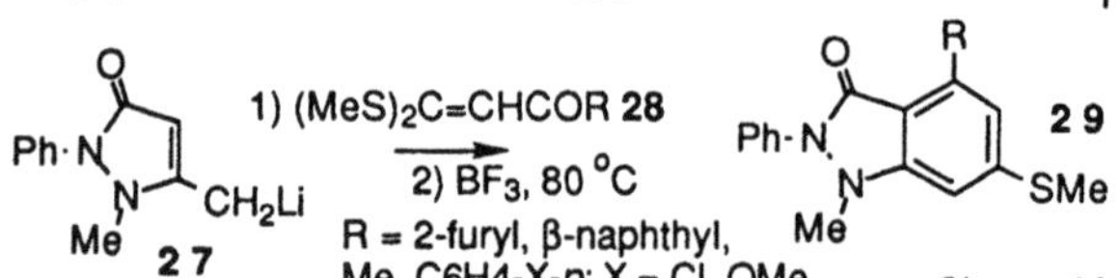

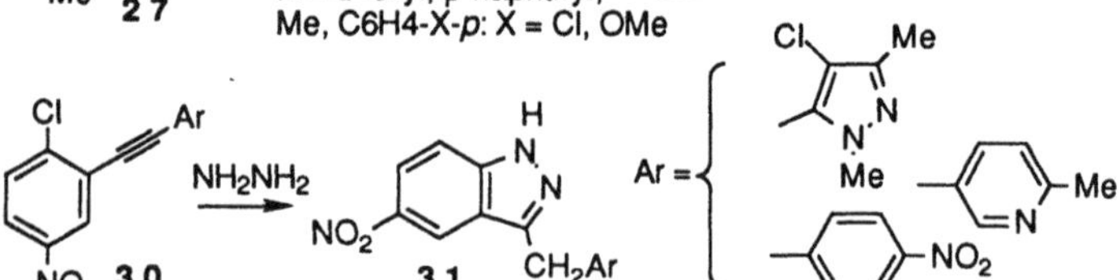

N-Hydroxypyrazoles are of interest as acylation and phosphorylation catalysts. Pyrazoles with strongly electron withdrawing groups (pKa < 9) **32** can be oxidized under buffered $KHSO_5$ conditions to provide hydroxypyrazoles **33** in fair to good yields <96MC139>.

Alternatively, the 3-carboethoxy derivatives **35** are synthesized by the reaction of nitroenamines **34** with ethyl isocyanoacetate <96JCR(S)76>.

N-Nitration of pyrazoles **36** to **37** is reported via *kyodai* nitration (NO_2/O_3) <96JCR(S)244>. *N*-Methyl-4-(nitro)iodopyrazoles are obtained by *ipso*-nitration of the 4-iodo derivatives. This nitrodehalogenation fails for chloro- and bromoderivatives <95MC233>. The previously 'not possible' N-2 alkylation of indazolinones **38** can be achieved by heating with the halide without base in an inert solvent <96LA683>. A highly regioselective one-pot synthesis of 1-alkyl-5-amino-3-aryl-4-cyanopyrazoles **39** is described from aryl acid chlorides, malononitrile, and alkylhydrazines. Yields are comparable to the normal three-step process <96JCS(P1)1545>. Highly *trans*-stereoselective addition of hydrazines to optically active butenolides and pentenolides **40**, followed by cyclization, leads to pyrazolidinones **41** with typical d.e. > 90% <96LA1581>. Optically pure *trans*-2-(1-pyrazolyl)cyclohexan-1-ol **42** is obtained by kinetic resolution with lipase B from *candida antarctica*. Compound **42** is potentially useful as a chiral ligand <96TA1717>.

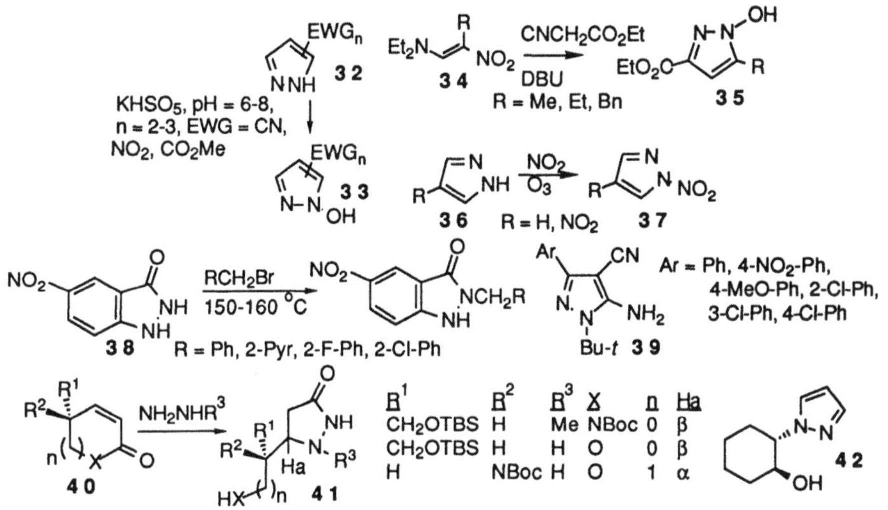

Syntheses of fluoro-substituted pyrazoles continue to be of interest. Both 3- and 5-fluoropyrazoles (**44** and **45**, respectively) can be prepared from **43** <96JOC2763>. Treatment of **43** with hydrazine followed by N-alkylation provides **44**, whereas reactions with monosubstituted hydrazines afford **45**. The 4-(trifluoromethyl)pyrazoles **47** are obtained from β-trifluoromethyl vinamidinium salt **46** <96TL1829>. The 5-trifluoromethyl-3-carboethoxypyrazoles **49** are obtained from the 1,3-dipolar cycloadditions of trifluoromethyl alkenes **48** with ethyl diazoacetate <96T4383>.

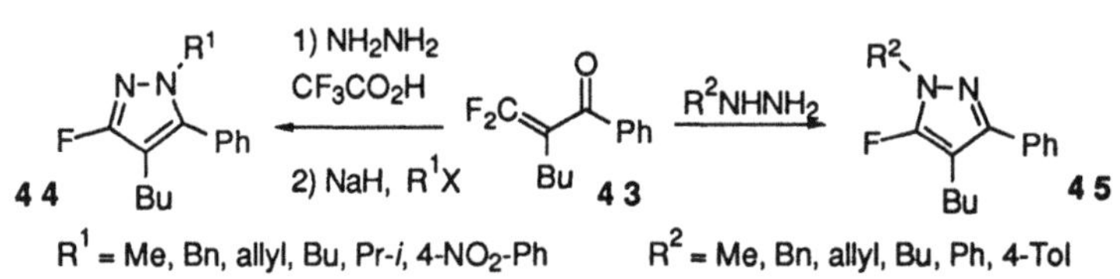

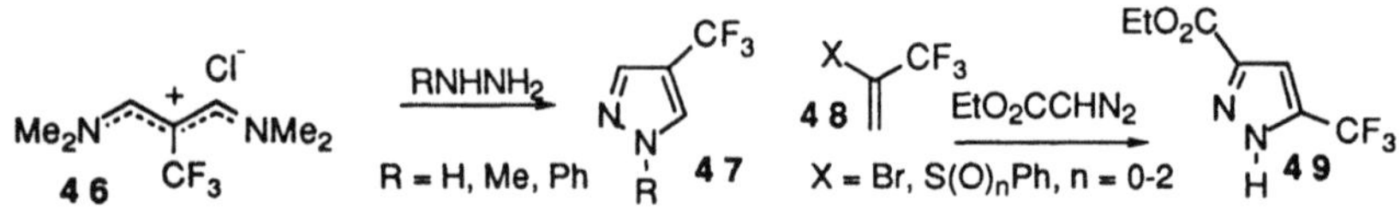

The use of aminopyrazoles **51** as dieneophiles in reactions with **50** offers a new route to pyrazolo[3,4-*d*]pyrimidines **52**. Initial inverse electron demand Diels-Alder reaction is followed by a retro Diels-Alder, then loss of ammonia to provide **52** in one pot <96JOC5204>. The pyrazole '*o*-quinodimethane' dienes **54** are available from *N*-tosyl sulfolenes **53**. Heating **53** with *N*-phenylmaleimide **55** at 180 °C gave the [4 + 2] adduct **56** <96SC569>. Although vinylpyrazole is a poor diene in Diels-Alder reactions, under solvent-free conditions with microwave irradiation cycloaddition to good dieneophiles occur. The yields, while generally fair, are superior to previous conventional attempts <96T9237>.

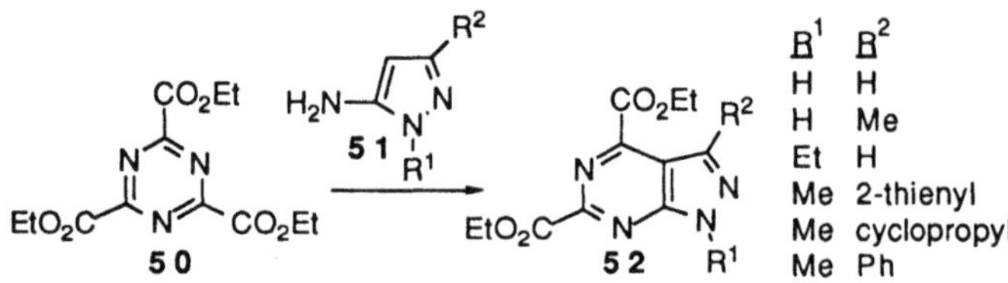

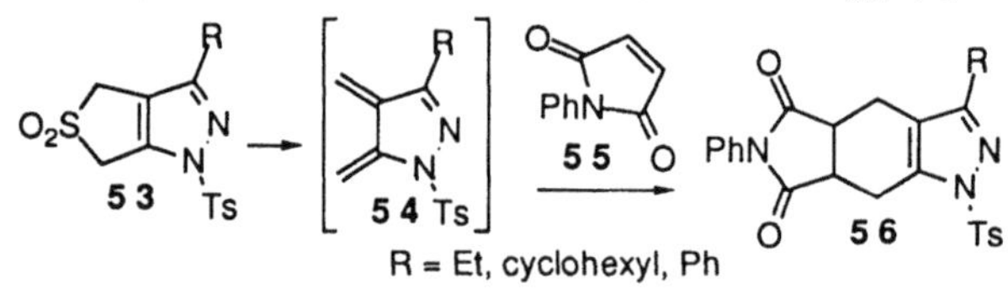

5.4.3 IMIDAZOLES AND RING FUSED DERIVATIVES

Imidazoles were incorporated as key elements in the structures of potential antihypertensive agents <96TA1641>, a derivative of the antitumor agent temozolomide <96BML185>, the α_2-adrenoceptor agonist dexmedetomidine <96SC1585>, and selective histamine H_3 receptor antagonists <96BML833>.

The total synthesis of a wide variety of imidazole-containing natural products was also reported in 1996. Notable among these syntheses was the preparation of the natural products topsentin (**1**) <96TL5503>, stevensine (**2**) <96TL8121>, cynometrine (**3**) <96TL3915>, xestomanzamine A (**4**) <96TL9353>, and (+)-hydantocidin (**5**) <96T1177>. The structure of a new, cytotoxic, dimeric, disulfide alkaloid, polycarpine dihydrochloride (**6**) was also reported <96TL2369>.

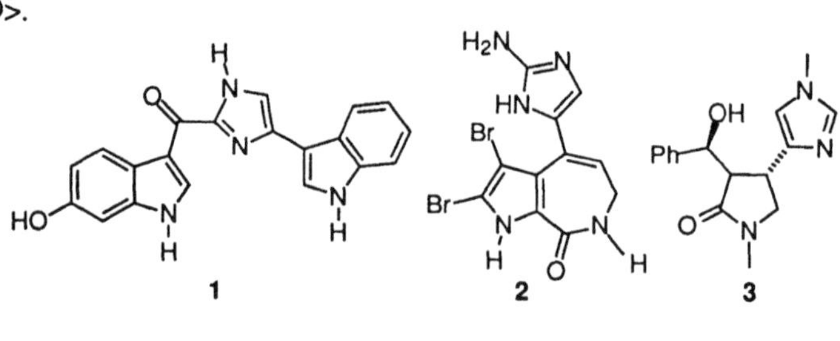

1 2 3

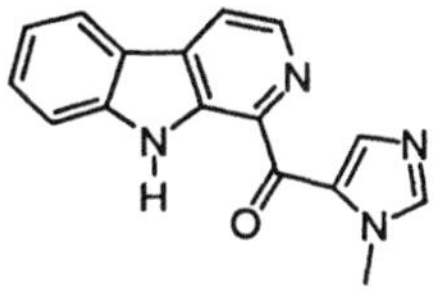

4

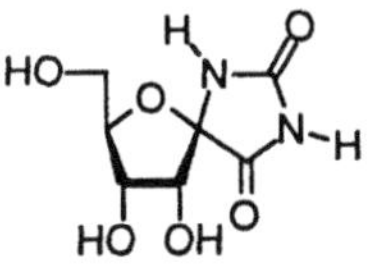

5

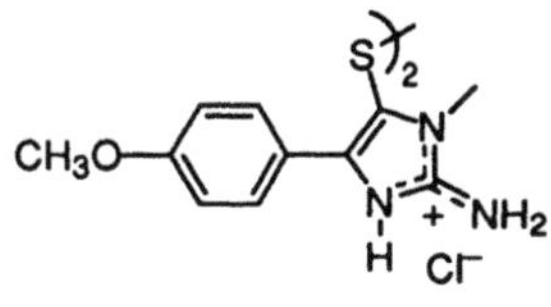

6

Numerous uses of imidazole derivatives as useful leaving groups were reported during the year. The inexpensive and efficient preparation of imidazole-1-sulfonates (imidazylates; Imd, e.g., **8**) from alcohols was reported <96T10557>. These imidazylate esters were shown to function as excellent leaving groups in a variety of displacement reactions. A general α-keto ester synthesis was developed that employed the deprotonation of imidazolium salts **9** (R = Et, *t*-Bu) with LDA followed by reaction with Grignard reagents <96JOC9009>. Similar N-imidazolium-N-methyl amides could be used in the synthesis of mono- and diketones and aldehydes <96T14297>. Diketones were also prepared from bis-benzimidazole methiodide salts **10** <96SC3175>.

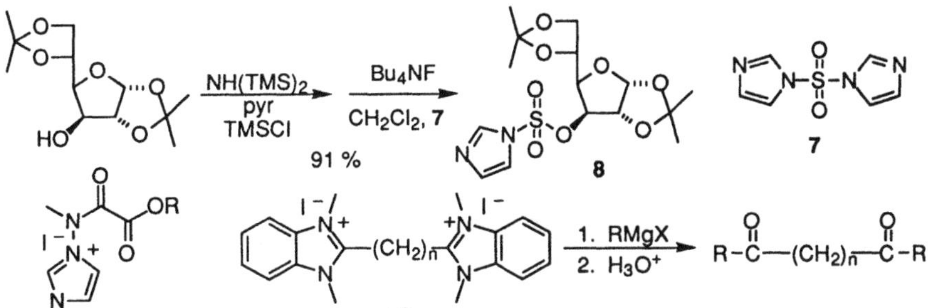

7 **8** **9** **10**

A few new or improved methods for the preparation of imidazole and its reduced derivatives were reported. Potentially useful chloroimidazoles **11** were prepared by the reaction of N-chloro-N'-arylbenzamidines with 1,1-diaminoethenes <96T7939>. Novel 1-aryl-2-(tosylamino)-1H-imidazole derivatives **13** were prepared by heating N-tosylguanidines **12** in refluxing acetic acid <96JOC2202>. Treatment of α-keto hemithioacetals with ureas gave substituted hydantoins in good yield <96H49> and the kinetics and mechanisms of the racemization of 5-substituted hydantoins was also investigated <96HCA767>. An efficient synthesis of cis-imidazoline carboxylates **14** via the reaction of N-tosylaldimines and methyl isocyanoacetate in the presence of a Au(I) catalyst was reported <96TL4969>. An unusual entry into bisimidazole derivatives like **15** was developed involving the reaction of 1,2-diamines with hexachloroacetone <96TL5265>. Photochemical ring-cleavage of the imidazole ring in bredinin **16**, an imidazole nucleoside antibiotic, was accomplished by irradiation in 0.1N HCl with a low-pressure (60 W) mercury lamp <96TL187>. Two methods were reported for the conversion of imines to imidazoline derivatives. A variety of imidazolines **17** were prepared by the low-valent, titanium-induced dimerization of aryl imines <96TL4767>. A stereoselective synthesis of trans-4,5-disubstituted-1,3-imidazolin-2-ones **18** was reported that featured the reaction of an α-nitrogen carbanion with a variety of aryl imines (An = *p*-$MeOC_6H_4$-) <96JOC428>.

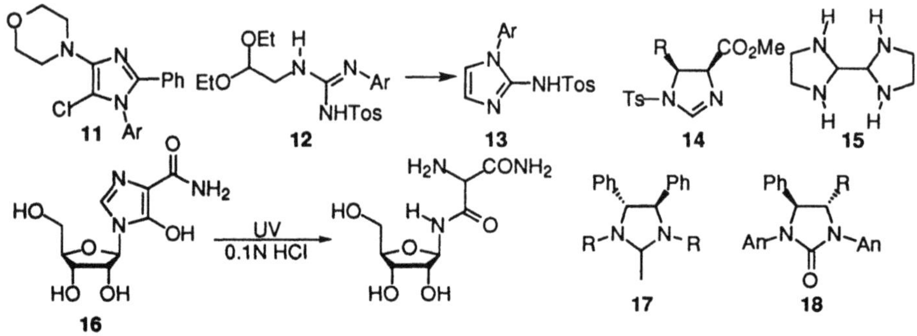

11 12 13 14 15 16 17 18

A new route to imidazo-2-ylacetic acids from hetrocyclic pyruvic acid derivatives was developed <96TL1901>. The synthesis of 2-(2-arylethyl)-1-methylimidazoles was found to be best accomplished by Wittig reaction of aryltriphenylphosphonium chlorides with 2-formyl-1-methylimidazole followed by Raney nickel reduction of the resulting alkenes <96JHC671>. Imidazole and benzimidazole were reported to undergo reactions with various silyl enol ethers and ketene silyl acetals in the presence of alkyl chloroformates to give 2-substituted imidazolines in good yields <96CC1217>. Certain 1,2,4-trisubstituted imidazoles were found to undergo electrophilic attack at C-5 when heated with azodicarbonyl compounds <96JHC41>. An efficient procedure for the introduction of the amino group into the 2-position of imidazole was developed using the palladium-mediated displacement of a 2-bromo substituent by various nitrogen nucleophiles <96H1375>. An approach to 2-, 4-, and 5-cyano imidazoles was reported which featured the reaction of 3-substituted-imidazole N-oxides with TMSCN <96JOC6971>. Judicious choice of the reaction conditions allowed the selective formation of each isomer.

Reduction of 1-substituted 4-nitroimidazoles by the system $NaBH_4/CH_3OH/CH_3ONa$ at 25 °C was found to give the oximes of 1-substituted 4-imidazolidinones **19** <96T9541>. Reactions of 1,4-dinitroimidazoles with hydrazines were also investigated <96T14905>. The preparation of 1-R-5-[(2-nitro-2-phenyl)ethenyl]imidazoles **20** (R = Bn, CH_3, H) <96JHC1345> was reported. The dinitroimidazoles **21** and **22** were each prepared in three steps from a common precursor <96SC837>.

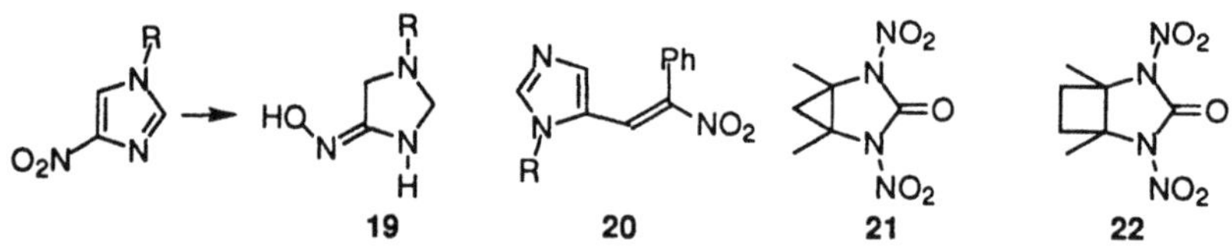

19 20 21 22

4-Alkenyl and 4-alkynyl imidazoles were prepared via the palladium(0)-catalyzed couplings of N-protected 4-haloimidazoles or 4-trimethylstannylimidazole **23** <96T13703>. This reaction process was used in the synthesis of fluorescent and biotinylated derivatives of the imidazole-containing substance, THI (**24**) <96TL6209>. N, N'-bisacylimidazolidines were developed into potential new acyl anion equivalents <96SL1109>. For example, the anion formed from **25** and sBuLi could be trapped with a variety of electrophiles. When E = CH_2Ph, TFA-hydrolysis of the imidazoline ring led to formation of phenylacetaldehyde which could be trapped *in situ* with 2,4-DNP.

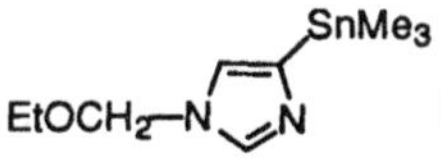
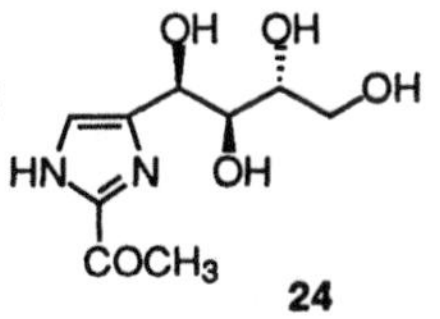
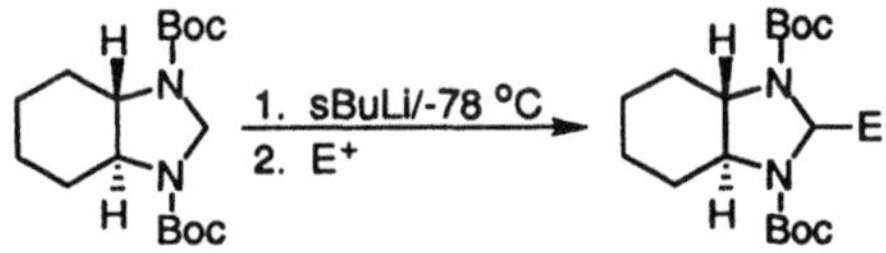

23 24 25

Imidazole and its derivatives continued to play an important role in asymmetric processes. Optically active pyrroloimidazoles **26** were prepared by the cycloaddition of homochiral imidazolium ylides with activated alkenes <96TL1707>. This reaction was used in the enantioselective preparation of pyrrolidines <96TL1711>. A review of the use of chiral imidazolidines in asymmetric synthesis was published <96PAC531> and the preparation and use of a new camphor-derived imidazolidinone-type auxiliary **27** was reported <96TL4565> <96TL6931>.

The development and use in peptide synthesis of the 1-adamantyloxymethyl protecting group for N^{π}-histidine **28** was reported <96JCS(P1)2139>. A procedure for the regiospecific alkylation of histidine and histamine at N-1(τ) via the corresponding tetrahydro oximidazo[1,5-*c*]pyrimidines **29** was also developed <96T5363>.

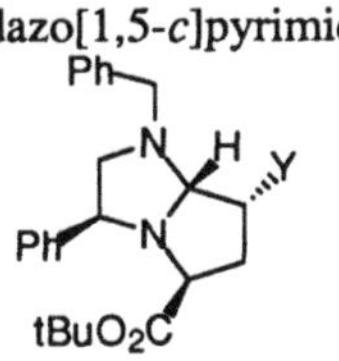
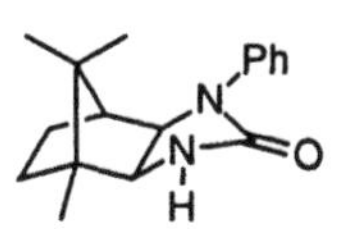
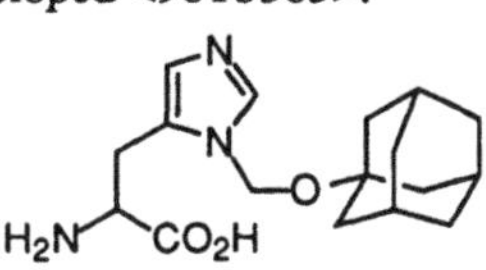
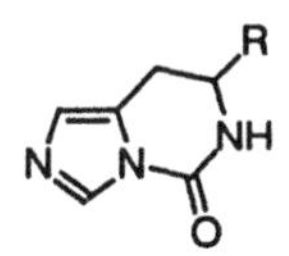

26 27 28 29

Four new procedures for preparing simple benzimidazole derivatives were reported. Benzimidazoles could be prepared by the condensation of 1,2-diaminobenzenes and orthoesters using KSF clay catalysis under microwave irradiation (MW) without solvent, or at reflux in toluene <96SC2895>. The same starting materials were converted to benzimidazolin-2-ones using a MW-induced condensation of urea in 1:10 N, N-dimethylacetamide (DMAC): diethylene glycol (DEG) <96JCR(S)92>. Other carbonyl and thiocarbonyl donors were also shown to work in this process <96JCR(S)94>. 2-Cyanobenzimidazoles were prepared from 1,2-diaminobenzenes and **30** <96TL4589> (e.g., R = Me, Ph, $PhCH_2$, etc.) in either a one or two-step process. Many complex N-substituted benzimidazolinones could be synthesized by the oxidative cyclization of 3-substituted 3-aryl-1-methoxy ureas like **31** <96TL2361>. Low-temperature (0 °C) ozonolysis of 5-methyl-2-mercaptoimidazole in $MeOH/CH_2Cl_2$ gave the sulfonic acid **32** (R = SO_3H) while at higher temperature (25 °C) the 2-methoxybenzimidazole (**32**; R = OMe) was formed <96SC3241>.

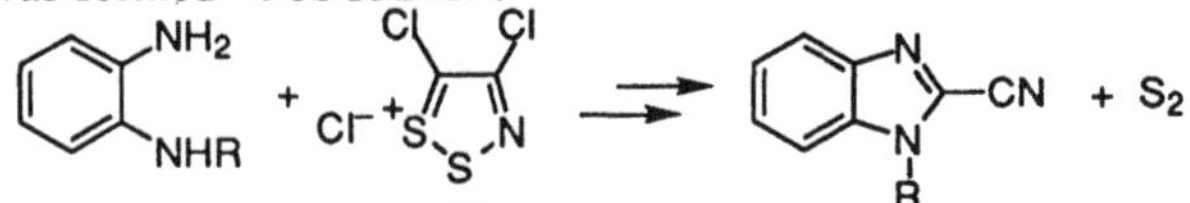

30

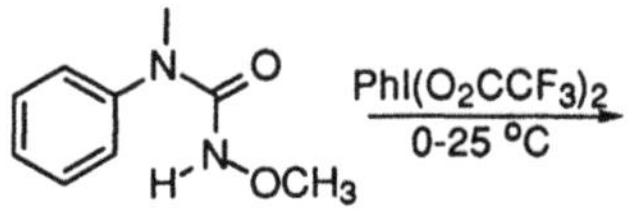

31

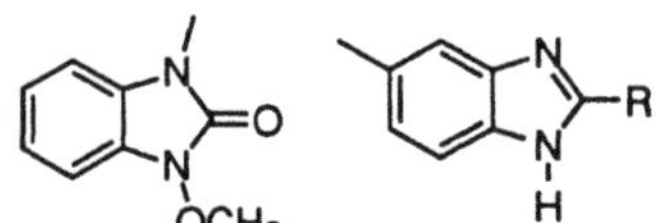

32

The photoaddition reaction of 2-phenylbenzimidazole with Michael acceptors was investigated <96JHC1031> as was the preparation and cycloaddition-reactivity of benzimidazole-2-carbonitrile oxide <96AJC199>. The nitration of 1-methylbenzimidazole was found to give only the 5- and 6-nitrated products as a mixture of isomers in 87% yield

<96BSB213>. 2-Vinylbenzimidazoles were prepared by β-elimination from dipolar ethyleneimidazolium inner salts <96H567><96CPB29>.

Several new approaches to ring fused imidazole derivatives were reported. For example, the imidazoline-2-thione **33** was converted into either the imidazo[2,1-*b*]thiazolone **34** or the imidazo[2,1-*b*][1,3]thiazinone **35** by PPA-mediated cyclization of the appropriate 2-sulfanylacid (R = *p*-chlorophenyl) <96JCR(S)322>. In like fashion, **36** could be transformed into 2-aroyl-3-methylthiazolo[3,2-a]benzimidazole derivatives by cyclization in acetic anhydride <96T10485>. The preparation and reactivity of 3-aminothiazolo[3,2-*a*]benzimidazole-2-carbonitrile (**38**) was also reported <96JCR(S)4>. 2-(Perfluoroalkyl)- and 2-(perfluoroaryl)-benzimidazoles were prepared by the oxidative-cyclization of perfluoroalkyl and aryl imidamides <96JOC3902>. Iodocyclization of 3-alkynyl-2-(substituted-amino)-1-imidazolin-4-ones **39** were found to give imidazo[1,2-*a*]imidazoles and/or imidazo[1,2-*a*]pyrimidines <96T6581> depending on the nature of the R group. The related iodocyclization of 3-alk-2-enyl-imidazolin-4-ones was also reported <96T2827>.

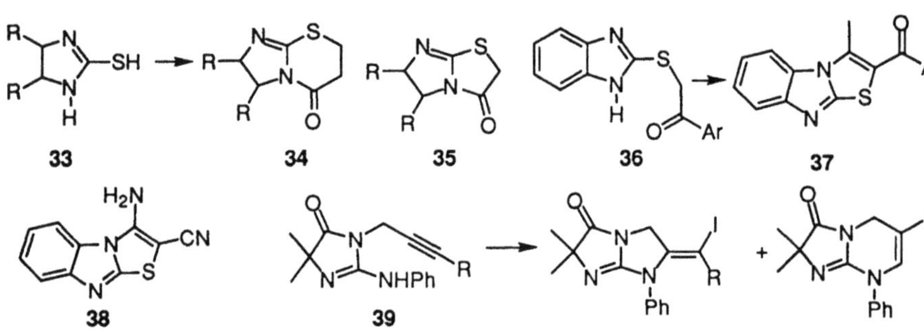

33 **34** **35** **36** **37** **38** **39**

Treatment of 5-amino-4-nitrosopyrazoles with DMF/$POCl_3$ gave 5-aminoimidazo[4,5-*c*]pyrazole derivatives **40** <96T7179> (R = CH_3, Ph). The imidazo[4,5-*d*]azepine ring system **41** was prepared by several methods <96PJC296>. Derivatives of diazacyclopenta[*c,d*]azulene **42** were also synthesized <96T9835>. A variety of imidazo[1,2-*a*]pyridines were prepared by the reaction of *p*-bromophenacyl bromide O-methyloxime (**43**) with substituted pyridines **44** <96S927> at RT in acetone followed by Et_3N/MeOH at 65 °C. The imidazolutidine moiety **46** of MK-996 was prepared from 2-amino-4,6-dimethylnicotinamide by a novel Hoffman rearrangement/cyclization <96H821>.

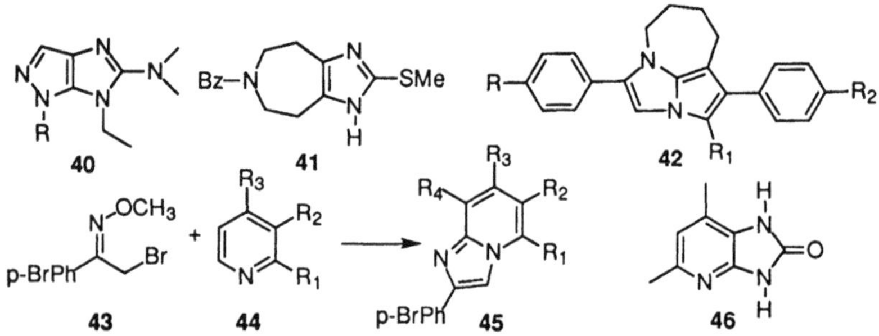

40 **41** **42** **43** **44** **45** **46**

Novel 1*H*-imidazo[1,2-*a*]indole-3-carboxylates **47** were prepared <96SC745>. Thermolyses of halogenated 4,5-dicyanoimidazole derivatives **48** (X = H, Y = F, Cl; X = I, Y = Cl, Br, I) at 100-290 °C led to formation of perhaps the ultimate fused-ring imidazole, hexacarbonitriletris(imidazo)triazene (HTT) <96JOC6666>.

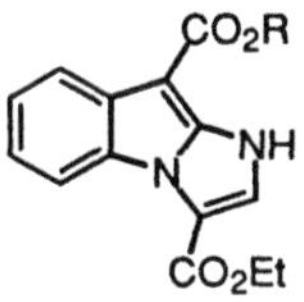

47

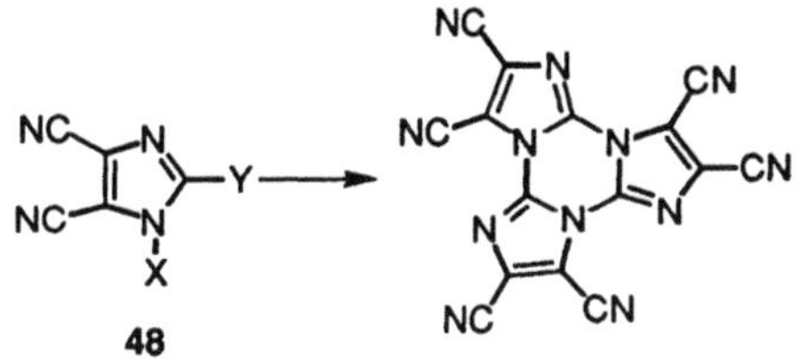

48 **HTT**

Solid-phase, parallel-synthesis approaches to the imidazole derivatives **49** <96TL937>, **50** and **51** <96TL835>, **52** <96TL751>, and **53** <96TL4865> were also reported in 1996.

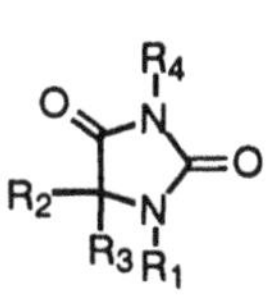

49

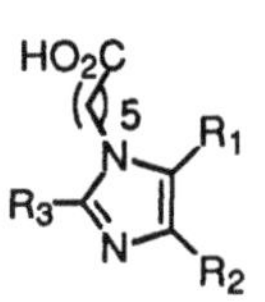

50

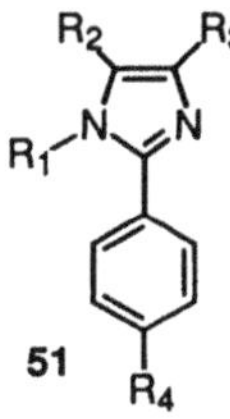

51

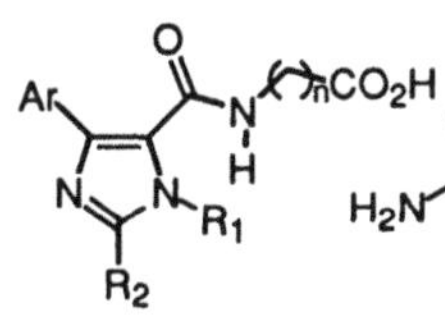

52

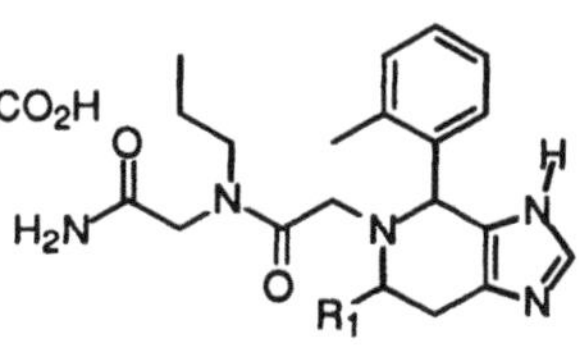

53

Three syntheses of tetraazafulvalenes derivatives were reported during the year. Two groups reported independent syntheses of **54** <96TL2357> <96AG(E)1011>. Deeply colored vinylogous tetraazafulvalenes **55** were also prepared <96S1302>. The imidazolium cyclophanes **56** were prepared, <96OPP345> and two methods for the synthesis of 1,3-dialkyl imidazolium cations **57** (Y = halide or $(CF_3SO_2)_2N$) <96IC1168> <96S697> were reported. Some of these imidazolium cations proved to be hydrophobic, highly conductive, ambient temperature molten salts.

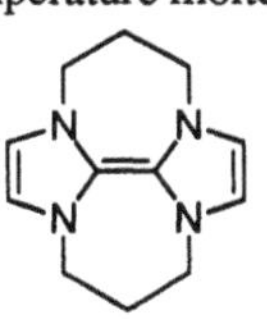

54

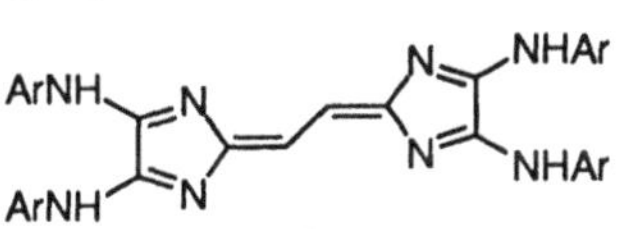

55

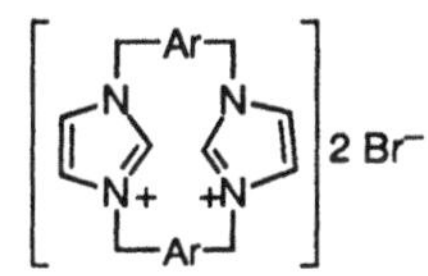

56

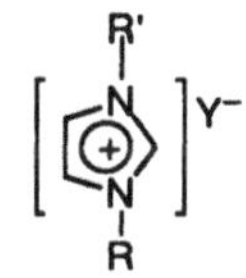

57

The amino acid **58** was used in the solid-phase synthesis of sequence-specific DNA binding polyamides containing N-methylimidazole and N-methylpyrrole amino acids <96JACS6141> and it was also reported that the imidazole-acridine conjugate **59** could effectively catalyze the cleavage of t-RNA <96TL4417>.

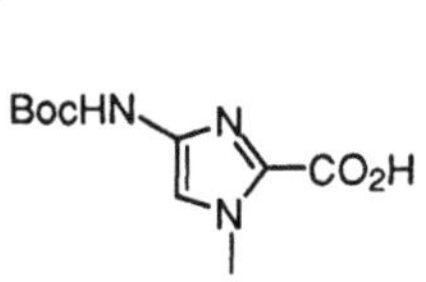

58

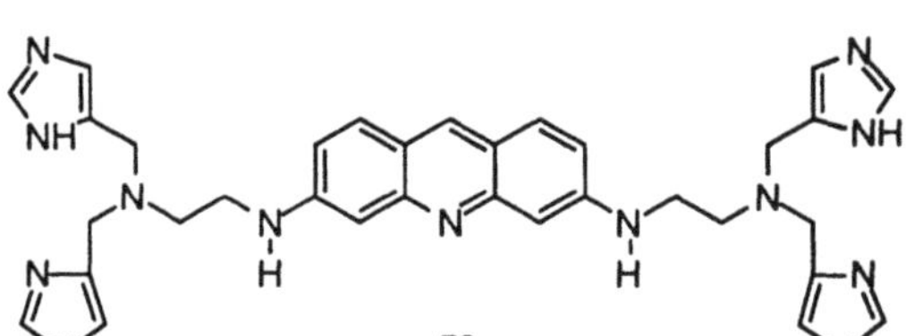

59

An imidazole-containing ligand **60** designed to position two metal ions was prepared <96SL734>. The novel tridentate ligand **61** was prepared in optically pure form <96CB59>. Imidazole-capped β-cyclodextrins were synthesized <96CC821> and the NMR behavior of related cyclodextrins was studied <96CC1943>. Imidazole-substituted porphyrins were prepared <96T9877> and the first example of a quadrapolar [14]metaazolophane **62** was also reported <96T15171>. The preparation of related cyclophanes was reported <96SL285> along with X-ray investigations of these novel compounds <96T15189>. The solid-state structures of six 4,5-disubstituted 2-benzimidazolones were determined and some were found to crystallize as molecular tapes <96JACS4018>. Reductive-amination of 2-methyl-nitroimidazoyl-1-yl-acetaldehyde with various diamines led to tetraimidazoyl tertiary diamines **63** <96TL6081>.

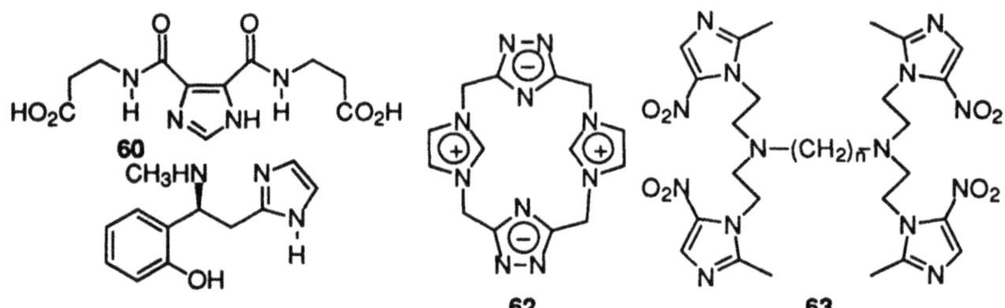

60 **61** **62** **63**

5.4.4 1,2,3-TRIAZOLES AND RING FUSED DERIVATIVES

Triazoles have been useful reagents in synthesis. Peptide coupling reagents include O-(benzotriazol-1-yl)-1,1,3,3-tetramethyluronium hexafluorophosphate (HBTU) and O-(7-azabenzotriazol-1-yl)-1,1,3,3-tetramethyluronium hexafluorophosphate (HATU). Analyses by X-ray techniques have demonstrated that these reagents are crystallized in their N-oxide form. These are typically represented as the O-bound uronium form in the literature. Semiempirical AM1 calculations indicate that, while O-alkylation is thermodynamically favored (as expected), N-alkylation is kinetically favored in solution <96JCR(S)302>. New methods for the synthesis of 1-hydroxytriazoles are of continued interest. Triazole 1-oxides, N-blocked at the 2- or 3- position can be functionalized by known methods. New unblocking procedures are reported <96ACS549>. The use of iodotrimethylsilane on benzyl blocked **1** leads to mixtures of the desired hydroxytriazoles as well as ring-substituted iodotriazoles. However, treatment of benzyl blocked **1** with concentrated HBr gives hydroxytriazoles **2** in excellent yields. For the less reactive 3-substituted derivatives **3**, the *p*-methoxybenzyl blocking group is required. Unblocking in concentrated sulfuric acid readily affords the hydroxytriazoles **2**.

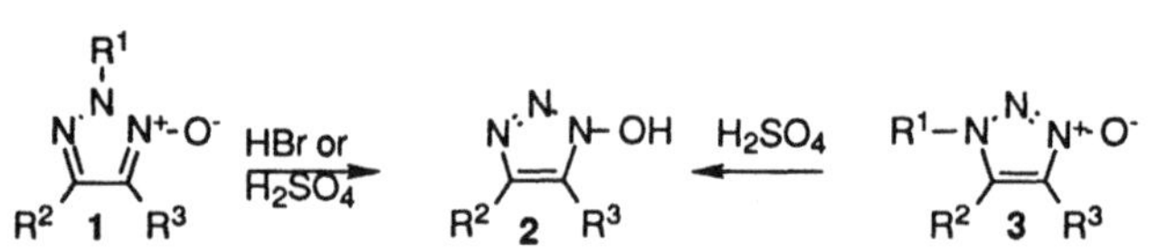

R^1 = Bn, *p*-MeOBn, R^2 = R^3 = H;
R^1 = Bn, R^2 = Me, R^3 = H;
R^1 = Bn, R^2 = H, R^3 = Cl

R^1 = *p*-MeOBn, R^2 = R^3 = H

In the benzotriazole area, there are continued references to their use in new synthetic methodology. The allyl 1-(benzotriazol-1-yl)alkyl ethers **4** and **8** undergo [2,3]-Wittig rearrangements to provide exclusively *E*-configuration homoallyl alcohols **5-7** and β,γ-unsaturated ketones **9** <96JOC4035>. This complements the known *Z*-selective methodologies. New methods using benzotriazoles are reported to synthesize α-aminoketones <96H273>, 5-acylaminooxazoles <95JHC1651>, dihydro- and α-hydroxytetrahydrofurans <96SC1385>, imidazoles, and pyrroles <95T13271>. In particular, 1,4-di- and

1,4,5-trisubstituted [2-(benzotriazol-1-yl)-methyl] pyrroles were prepared and elaborated into 1,2,4- and 1,2,4,5-substituted pyrroles **11** <96JOC1624>. The reaction of 1-propargylbenzotriazole **10** with α-bromoketones affords alkynyl oxiranes **10a**, which are cyclized to pyrroles **11** upon treatment with primary amines in refluxing isopropanol. Numerous examples are reported. In one interesting case treatment of **12**, R^1 = Ph, R^2 = H, R^3 = Bn, R^4 = Me, R^5 = Bt with NaCN gives **12**, R^1 = Ph, R^2 = CN, R^3 = Bn, R^4 = Me, R^5 = H via an $S_{N'}$ reaction. The readily available 3-(benzotriazol-yl)-1-ethoxyprop-1-ene **13** has been shown to be as useful three carbon synthon for $^{(-)}$CH=CHCHO, $^{(-)(-)}$C=CHCHO, $^{(-)(-)}$C=CHC$^{(-)}$=O, $^{(-)}$CH=CHC$^{(+)}$HOEt, and $^{(-)(-)}$C=CHC$^{(+)}$HOEt <95S1315>. Transformations to α,β-unsaturated aldehydes and ketones, furans, pyrroles, and allyl ethers are described. The valuable $^{(-)(-)}$C=CHC$^{(+)}$HOEt synthon is demonstrated in the conversion of **13** to **15**. Mild conditions are also described for the rearrangement of **14** to **16**.

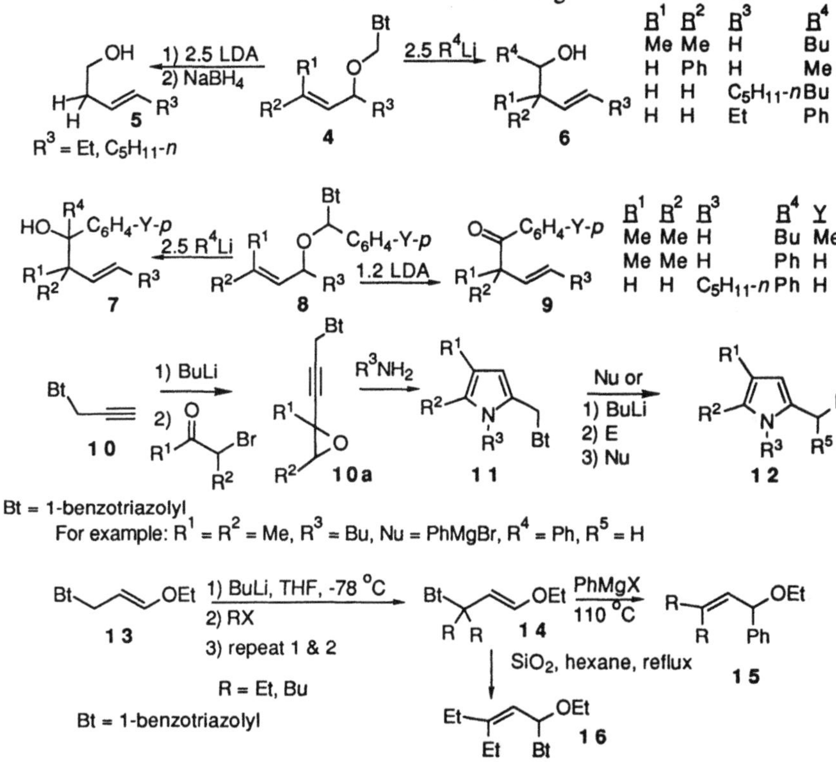

The reactions of 4-chloropyridines and quinolines **17** with benzotriazoles **18** in a modified Graebe-Ullman synthesis give excellent yields of γ-carbolines and their benzo-fused derivatives **20**. Excellent yields for preparation of the penultimate benzotriazole precursors **19** are reported as well. In the optimized 'one-pot' conditions, the combined neat substrates are heated with microwave irradiation (MW) for short (7-10 min) durations. The crude **19** is treated with $H_4P_2O_7$ and irradiated futher (4-6 min). The resultant γ-carbolines **20** were methylated to form the quaternary salts. These were tested and found to lack DNA intercalation properties <96JOC5587>.

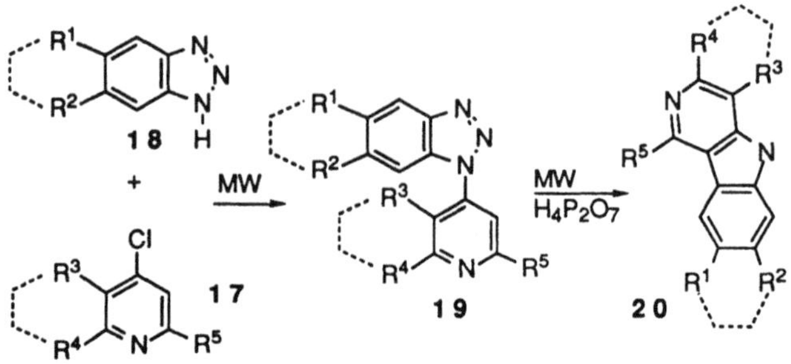

R^1	R^2	R^3	R^4	R^5
H	H	H	H	H
H	H	H	Me	Me
H	H	-(CH)4-		H
H	H	-(CH)4-		Me
-(CH)4-		H	H	H
-(CH)4-		-(CH)4-		H, Me

The addition of azides to alkynes is a well-known method to synthesize 1,2,3-triazoles; however, this approach suffers in that mixtures of regioisomers are typically formed. The reaction of nitro olefins **21** with azides, followed by the loss of HNO_2, selectively affords one regioisomer **22** in excellent yields <96BSB33>. Another triazole ring-forming reaction is the regiospecific preparation of 1-amino-5-fluoroalkyltriazoles **24** from the oxidation of bis-hydrazones **23** <96JFC45>. A new entry into 4-formyl-1,2,3-triazoles **27** is via the corresponding triazolylacylsilanes **26**. The reaction of acetylenic silylketone **25** with azides, provides triazoles **26** with ≥72% selectivity for the 1,4 isomers (in good isolated yields). Desilylation affords the formyl derivatives **27** <95TL9031>. The reaction of α-diazo, β-formyl carbonyl compounds **28** with various amines regiospecifically provides triazoles **29**. The yields are at best fair; however, the ease of synthesis and specificity of the reaction are notable <96HCA449>. A useful entry into the 5-acyl-2-aryl-1,2,3-triazole 1-oxide system **33** is via the bis-hydrazones of α-hydroxyiminodiones **32**. Despite the known conversion of **30** to **34** when treated with arylhydrazines, treatment of **30** with 2.2 equivalents of arylhydrazines at 0 °C affords bis-hydrazones **32**. Oxidation then provides **33** in generally good yields. In the case of *p*-tolylhydrazine the monohydrazones **31** are isolated and similarly oxidized to **33** <96JHC655>.

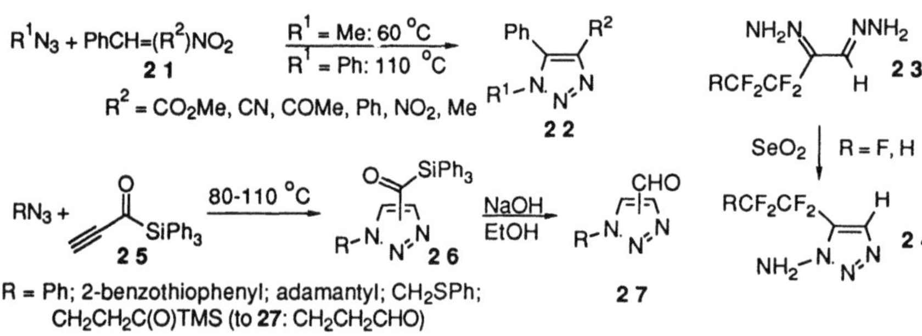

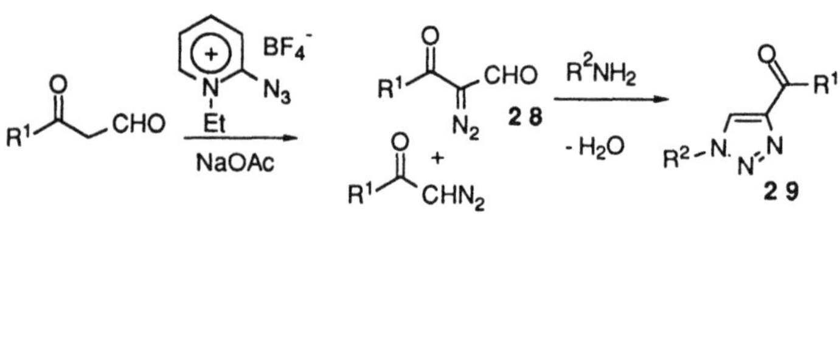

R^1	R^2
Ph	Me
Ph	4-Cl-Ph
Ph	2,4-Br2-Ph
H	OEt
H	Me

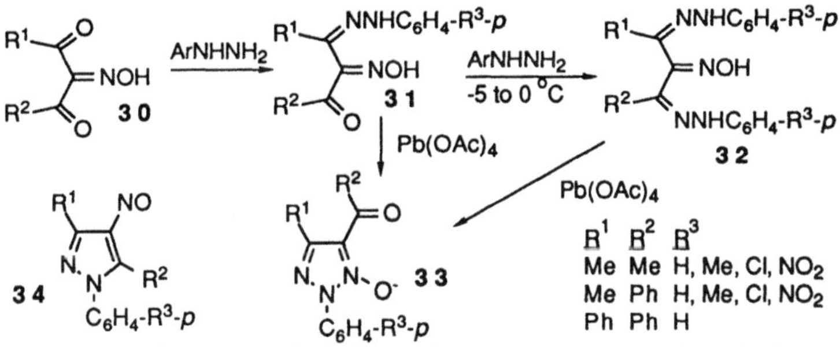

The 1-aminobenzotriazoles are desirable as benzyne precursors. The dianion of *N*-boc-1-aminobenzotriazole **34** reacts to introduce electrophiles at the 7 position to give **35** in excellent yields <96TL5615>. Pyrrolo[1,2-*c*][1,2,3]triazoles **37** are formed by trapping 1,2,3-triazolium-1-methanides **36** with dimethyl acetylenedicarboxylate (DMAD). These dipolarophiles are very reactive and prone to thermodynamically favored ring opening. Addition of DMAD followed by cesium fluoride results in fair yields of the fused heterocycles **37** <96JCS(P1)1617>. Bromine-lithium exchange of 1- or 2-methoxymethyl-4,5-dibromo-1,2,3-triazoles **38** and **39** each gives a single monoanion. These anions can be reacted with electrophiles to give **40** and **41**, respectively, in good yield. The second brominated site can subsequently also be lithiated, although only a single example (aqueous quenching) is reported <96JCS(P1)1341>. Electrophilic nitration of **42** unequivocally confirms the order of preferred electrophillic aromatic substitution as 2(8) > 4(10) >> 1(7) and 3(9) <95JOC6110>.

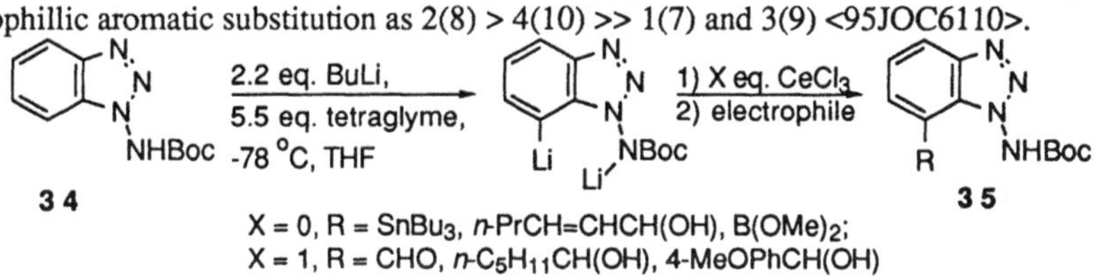

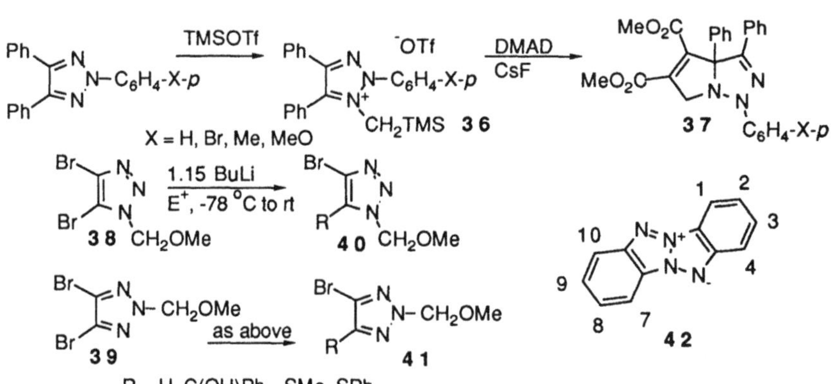

5.4.5 1,2,4-TRIAZOLES AND RING FUSED DERIVATIVES

The 1,2,4-triazole moiety was featured in the structures of several medicinal agents whose synthesis was reported during the year. Notable among these reports was the preparation of the antifungal agent Sch 42427 (**1**) <96TL611>, the synthesis of a metabolite (**2**) of MK-0462 <96SC1977>, and the preparation of a key intermediate (**3**) towards the synthesis of the highly active azole antifungals Sch 51048 and Sch 56592 <96TL5657>.

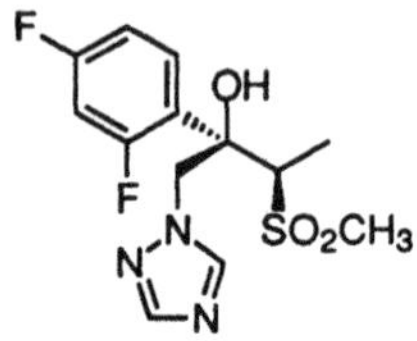

1

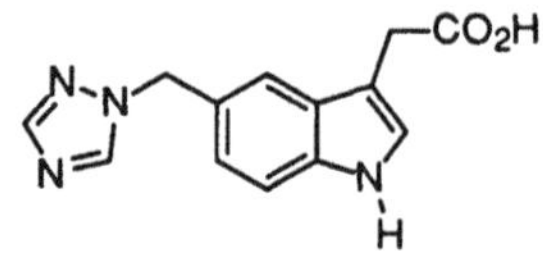

2

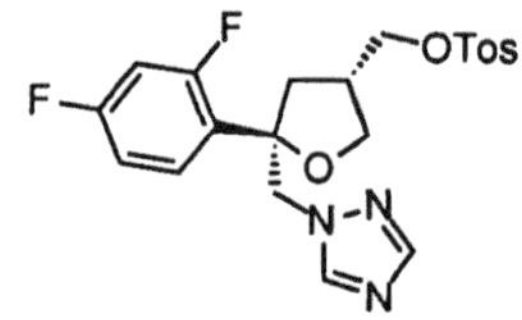

3

Simple 1,2,4-triazole derivatives played a key role in both the synthesis of functionalized triazoles and in asymmetric synthesis. 1-(α-Aminomethyl)-1,2,4-triazoles **4** could be converted into **5** by treatment with enol ethers <96SC357>. The novel C_2-symmetric triazole-containing chiral auxiliary (*S*,*S*)-4-amino-3,5-bis(1-hydroxyethyl)-1,2,4-triazole, SAT, (**6**) was prepared from (*S*)-lactic acid and hydrazine hydrate <96TA1621>. This chiral auxiliary was employed to mediate the diastereoselective 1,2-addition of Grignard reagents to the C=N bond of hydrazones. The diastereoselective-alkylation of enolates derived from ethyl ester **7** was mediated by a related auxiliary <96TA1631>.

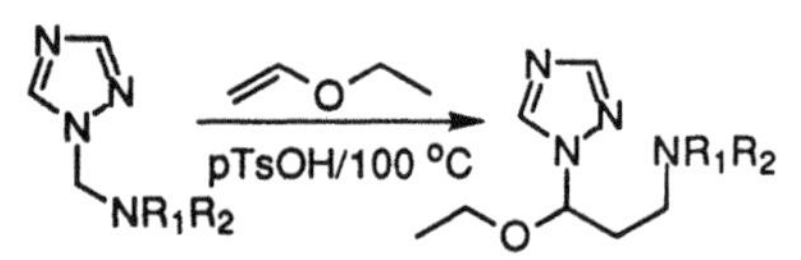

4 **5**

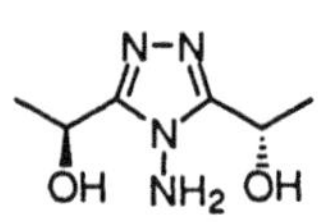

6

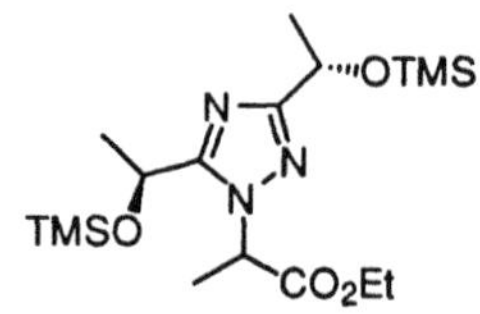

7

A few syntheses of the 1,2,4-triazole ring were reported in 1996. Treatment of the (1,1-dichloropropyl)azo compound **8** with $SbCl_5$ gave the chloro-substituted allenium salt **9**. This salt was found to undergo cycloaddition with MeCN to give a 76% yield of the 1,2,4-triazolium salt **10** <96S274>. Reaction of 3-[3-chlorophenyl]-1-[4-(3,4-methylenedioxybenzylidiene)aminophenyl]prop-2-enone with thiosemicarbazide in boiling AcOH was reported to give the triazole-derivative **11** <96SC3799>. Substituted 1,2,4-triazoline-5-thiones **12** were prepared in moderate to good yields by the ferric chloride-mediated oxidative-cyclization of some 2-methylsubstituted aldehyde semicarbazones <96JHC863>. The triazolidine **14** was prepared by the intramolecular-photocyclization of **13** <96LA575>.

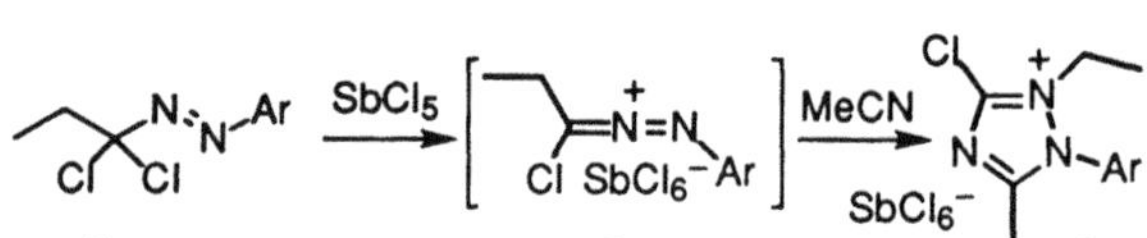

8 **9** **10**

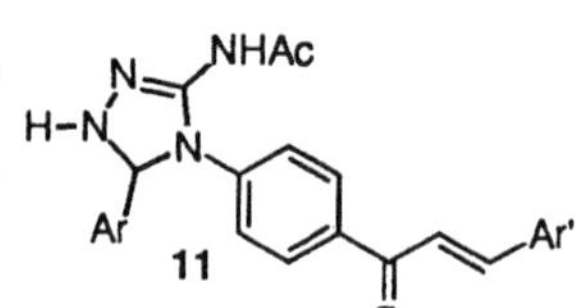

11

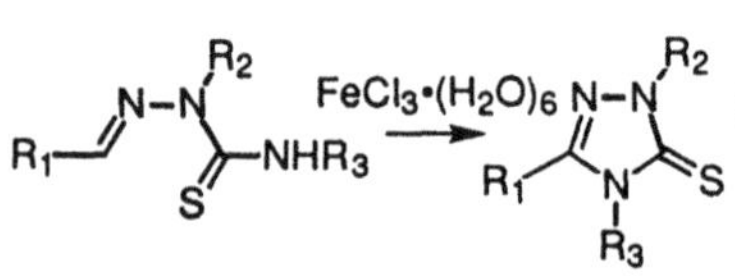

12

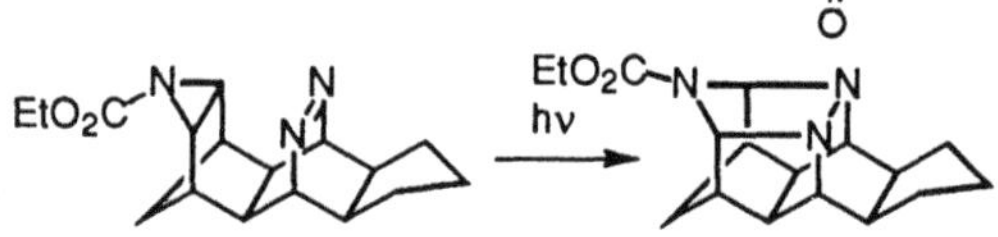

13 **14**

In contrast to the relatively limited number of non-fused 1,2,4-triazole syntheses that were reported in 1996, the preparation of several ring-fused 1,2,4-triazole-containing structures were published. For example, the first practical synthesis of fused[*a*]triazolo[1,4]benzodiazepine-5,11-diones **16** via the hydrazone **15** was reported (Y = H, Cl, Me; R = H, Me, Ph; X = O, S) <96JHC275>. Oxidative cyclization of N-heteroarylamidines allowed the preparation of

[1,2,4]triazolo[5,1-*b*]benzothiazoles **17** <96OPP362>. The sulfur-containing, fused-ring 1,2,4-triazoles **19** could be prepared by the cyclization of the thiourea-containing 1,2,4-triazole derivatives **18** in NaOH/MeOH <96JCR(S)388>. The oxadiazolo-triazole derivatives **21** were prepared by the $POCl_3$-mediated dehydration of **20** <96SC3827>. Microwave-induced cyclizations for the preparation of **23** from **22** were reported to be ~250 times faster than by classical methods <96JCR(S)254> (MTD = 5-methyl-1,3,4-thiadiazole). Finally, treatment with bromine converted the stable 3-methylmercapto-1,2,4-triazolium trifluoromethane sulfonates **24** directly into the desired imidazo[2,1-*c*]-1,2,4-triazoline-3-thiones **25** in moderate yields <96T791>.

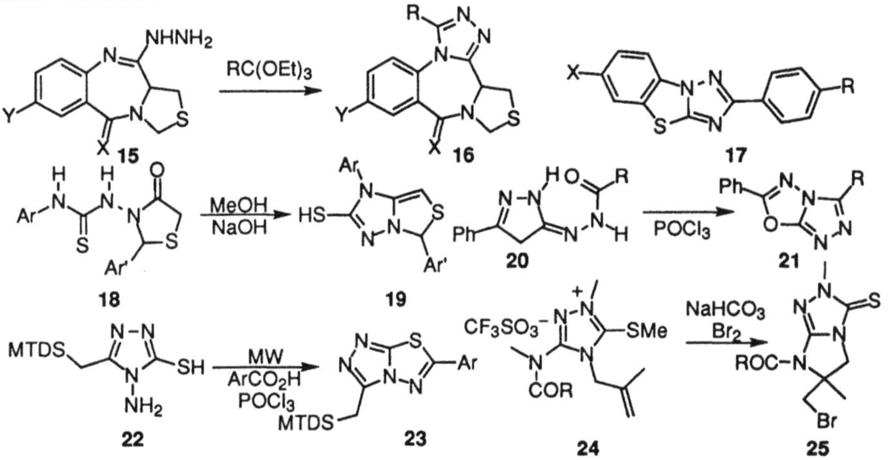

PTAD (4-phenyl-1,2,4-triazoline-3,5-dione) was employed as a novel dehydrogenating agent <96SC2587>. For example, the tetrahydrophthalazine **26** could be oxidized to the phthalazine **27** in excellent yield. The cycloaddition reaction of MTAD with cyclohexadiene formed the foundation for the preparation of **28** of high enantiomeric purity via a lipase-mediated desymmetrization process <96TL7951>.

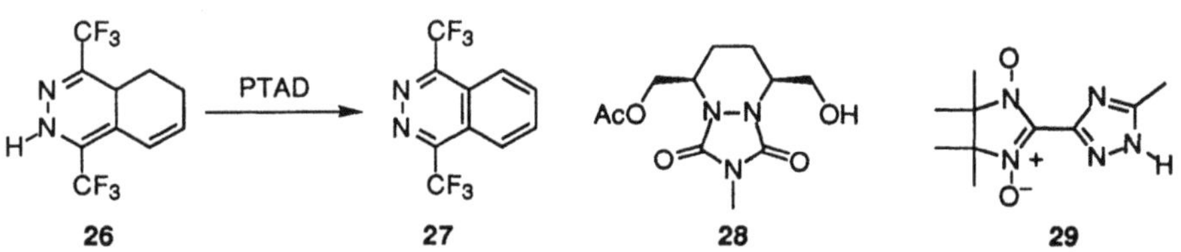

The preparation of novel triazole-containing 20-22 membered macrocyclic azacrown ether-thioethers was reported <96JCR(S)182> and the first selective synthetic method for the synthesis of dicyanotriazolehemiporhyrazines was published <96JOC6446>. 1,2,4-Triazole-containing polyimide beads were prepared and employed as Mo(VI) epoxidation catalyst supports. The 1,2,4-nitronyl nitroxide **29** was also synthesized and found to have remarkable magnetic properties <96AM60>.

5.4.6 TETRAZOLES AND RING FUSED DERIVATIVES

The tetrazole structure was featured in the structures of several medicinal agents whose synthesis was reported in 1996. Notable among these reports was the preparation of tetrazole-amide derivatives **1** of an ACAT inhibitor <96JMC2354>, the synthesis of substituted-tetrazole derivatives of the NK_1 receptor antagonist GR203040 (**2**; R = H) <96BML1015><96SL359>, and the preparation of tetrazole carboxylic acid bioisosteres of the folate-based thymidylate synthase inhibitor ICI 198583 <96BML631>. Glycosyl tetrazoles were employed effectively in the synthesis of the C-3 trisaccharide component of the antibiotic PI-080 <96JOC6>.

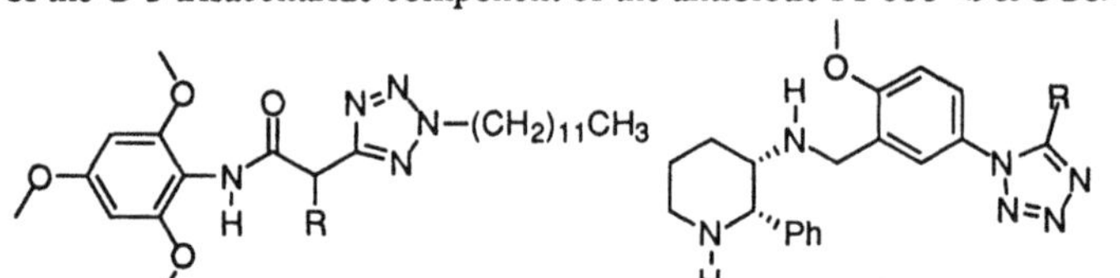

1 **2**

A few new methods for the preparation of simple tetrazole derivatives were reported in 1996. Tetrazole-substituted ureas **4** were prepared by the treatment of the weakly basic aminotetrazoles **3** with triphosgene followed by reaction of the putative carbamoyl chloride intermediate with a primary amine <96BML1753>. The alkylation of a variety of substituted tetrazoles **5** with primary and secondary alcohols using the Mitsunobo protocol was reported <96SC2687>. The 2-substituted tetrazole **6** was the major product of these reactions. In another study 5-methyltetrazole was N-tritylated and deprotonated using nBuLi <96TL3655>. The resulting anion was quenched with a wide range of electrophiles to give 5-substituted tetrazoles **7** in moderate yields. The deprotection of these tetrazoles was effected by treatment with HCl (g).

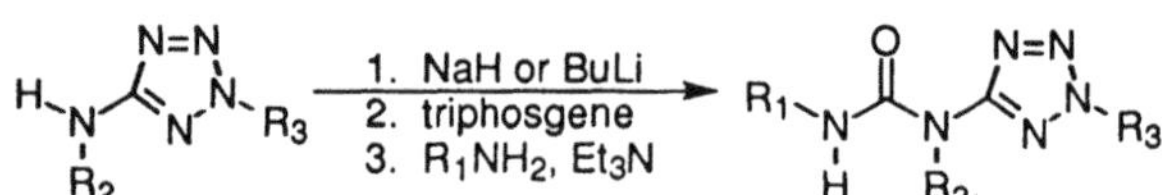

3 **4**

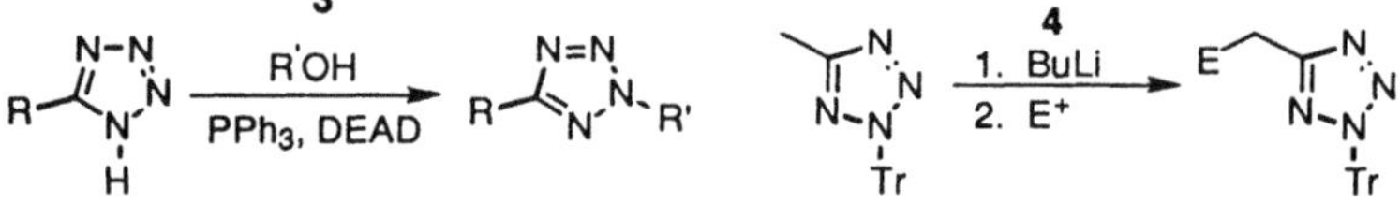

5 **6** **7**

A few notable methods for the preparation of simple ring fused tetrazoles were also reported during the year. The tetrazolobenzodiazepine **9** could be prepared in excellent yield by heating the azidomethylpyrrole **8** in xylene at reflux <96T10571>. The spiro tetrazole-containing compound **11** was produced by the CAN oxidation of the pyrrolidinone **10** in 60% yield <96T10169>. The reaction of the 3,5-dichloro-2H-1,4-oxazin-2-ones **12** with NaN_3 in DMF gave rise to the tetrazolo[5,1-*c*][1,4]oxazin-8-ones **13** in excellent yields <96T8813>.

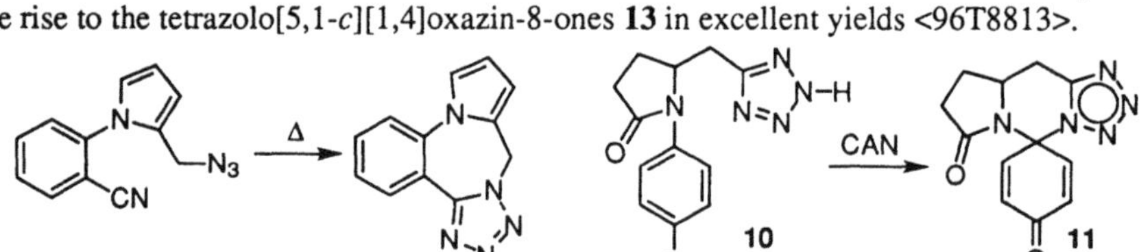

8 **9** **10** **11**

12 **13**

Several other studies concerning tetrazole-containing compounds were reported. The 1-bromo-glycosyl cyanide **15** was converted to the tetrazole derivative **16** in 72% yield by treatment with 2 eq LiN_3/DMF/3 d <96T9121>. Shorter reaction times led to formation of the 1-azido-pyranosyl cyanide. (Hydroxyphenyl) carbenes could be formed by thermal decomposition of the appropriate 5-substituted tetrazoles **17** <96JOC4462>. Ethoxycarbonylation of the 7-methylpyrrolotetrazolide ion gave the N1-substituted product **18** preferentially <96JOC5646>. The 1,3-dipolar cycloaddition reactions of **19** with alkenes was investigated <96JHC335> as was the formation and reactivity of the tetrazoyl(benzotriazol-1-yl)methane **20** <96JHC1107>.

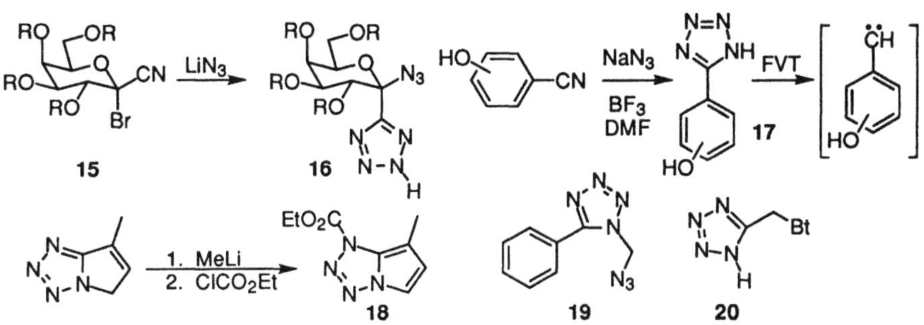

5.4.7 ACKNOWLEDGMENTS

The authors gratefully acknowledge Dr. Paul A. Burns (JRV) and Ms. Debra Sponholtz (MAW) for proofreading the manuscript and Ms. Jean Shein, '97 (MAW) for help in compiling the references.

5.4.8 REFERENCES

95AJC1587 C.M. Hartshorn, P.J. Steel, *Aust. J. Chem.* **1995**, *48*, 1587.

95CPB2123 M. Fujita, H. Egawa, M. Katoka, T. Miyamoto, J. Nakano, J.-i. Matsumoto, *Chem. Pharm. Bull.* **1995**, *43*, 2123.

95JHC1651 A.R. Katritzky, H. Wu, L. Xie, *J. Heterocycl. Chem.* **1995**, *32*, 1651.

95JMC3884 J.L. Kelley, R.G. Davis, E.W. McLean, R.C. Glen, F.E. Soroko, B.R. Cooper, *J. Med. Chem.* **1995**, *38*, 3884.

95JOC6110 G. Subramanian, J.H. Boyer, D. Buzatu, E.D. Stevens, M.L. Trudell, *J. Org. Chem.* **1995**, *60*, 6110.

95MC233 E.V. Tretyakov, S.F. Vasilevsky, *Mendeleev Commun.* **1995**, 233.

95NKK713 M. Morioka, M. Ohishi, M. Ohishi, H. Yoshida, T. Ogata, *Nippon Kagaku Kaishi* **1995**, 713.

95S1315 A.R. Katritzky, H. Wu, L. Xie, S. Rachwal, B. Rachwal, J. Jiang, G. Zhang, H. Lang, *Synthesis* **1995**, 1315.

95S1491 M.A.P. Martins, R. Freitag, A.F.C. Flores, N. Zanatta, *Synthesis* **1995**, 1491.

95T10941 K.R. Reddy, A. Roy, H. Ila, H. Junjappa, *Tetrahedron* **1995**, *51*, 10941.

95T13197 P. Hoffman, S. Huenig, L. Walz, K. Peters, H.G. von Schnering, *Tetrahedron* **1995**, *51*, 13197.

95T13271 A.R. Katritzky, L. Zhu, H. Lang, O. Denisko, Z. Wang, *Tetrahedron* **1995**, *51*, 13271.

95TL9031 A. Degl'Innocenti, P. Scafato, A. Capperucci, L. Bartoletti, A. Mordini, G. Reginato, *Tetrahedron Lett.* **1995**, *36*, 9031.

96ACS549 M. Begtrup, P. Vedsoe, *Acta Chem. Scand.* **1996**, *50*, 549.

96AM60 A. Lang, Y. Pei, L. Ouahab, O. Kahn, *Adv. Mater.* **1996**, *8*, 60.

96AG(E)1011	T.A. Taton, P. Chen, *Angew. Chem., Int. Ed. Engl.* **1996**, *35*, 1011.
96AJC199	D.J. Cundy, G.W. Simpson, *Aust. J. Chem.* **1996**, *49*, 199.
95AJC1587	C.M. Hartshorn, P.J. Steel, *Aust. J. Chem.* **1995**, *48*, 1587.
96BML185	Y. Wang, M.F.G. Stevens, *Bioorg. Med. Chem. Lett.* **1996**, *6*, 185.
96BML631	V. Bavetsias, A.L. Jackman, R. Kimbell, F.T. Boyle, G.M.F. Bisset, *Bioorg. Med. Chem. Lett.* **1996**, *6*, 631.
96BML833	J.W. Clitherow, P. Beswick, W.J. Irving, D.I.C. Scopes, J.C. Barnes, J. Clapham, J.D. Brown, D.J. Evans, A.G. Hayes, *Bioorg. Med. Chem. Lett.* **1996**, *6*, 833.
96BML1015	D.R. Armour, K.M.L. Chung, M. Congreve, B. Evans, S. Guntrip, T. Hubbard, C. Kay, D. Middlemiss, J.E. Mordaunt, N.A. Pegg, M.V. Vinader, P. Ward, S.P. Watson, *Bioorg. Med. Chem. Lett.* **1996**, *6*, 1015.
96BML1753	C.F. Purchase, II, A.D. White, M.K. Anderson, T.M.A. Bocan, R.F. Bousley, K.L. Hamelehle, R. Homan, B.R. Krause, P. Lee, et al., *Bioorg. Med. Chem. Lett.* **1996**, *6*, 1753.
96BSB33	J.C. Piet, G. Le Hetet, P. Cailleux, H. Benhaoua, R. Carrie, *Bull. Soc. Chim. Belg.* **1996**, *105*, 33.
96BSB213	M. Viktor, I. Dusan, *Bull. Soc. Chim. Belg.* **1996**, *105*, 213.
96CB59	A. Berkessel, M. Bolte, M. Frauenkron, T. Nowak, T. Schwenkreis, L. Seidel, A. Steinmetz, *Chem. Ber.* **1996**, *129*, 59.
96CB147	B. Wrackmeyer, J. Suess, W. Milius, *Chem. Ber.* **1996**, *129*, 147.
96CC821	D.-Q. Yuan, K. Ohta, K. Fujita, *Chem. Commun.* **1996**, 821.
96CC1217	T. Itoh, M. Miyazaki, H. Hasegawa, K. Nagata, A. Ohsawa, *Chem. Commun.* **1996**, 1217.
96CC1943	D.-Q. Yuan, K. Koga, M. Yamagushi, K. Fujita, *Chem. Commun.* **1996**, 1943.
96CPB29	E. Alcalde, M. Gisbert, L. Perez-Garcia, *Chem. Pharm. Bull.* **1996**, *44*, 29.
96CPB2123	M. Fujita, H. Egawa, M. Katoka, T. Miyamoto, J. Nakano, J.-i. Matsumoto, *Chem. Pharm. Bull.* **1995**, *43*, 2123.
96HCA61	J.H. Teles, J.-P. Melder, K. Ebel, R. Schneider, E. Gehrer, W. Harder, S. Brode, D. Enders, K. Breuer, G. Raabe, *Helv. Chim. Acta* **1996**, *79*, 61.
96HCA449	O. Sezer, K. Dabak, A. Akar, O. Anac, *Helv. Chim. Acta* **1996**, *79*, 449.
96HCA767	M. Reist, P.-A. Carrupt, B. Testa, S. Lehmann, J.J. Hansen, *Helv. Chim. Acta* **1996**, *79*, 767.
96H49	J.P. Zou, Z.E. Lu, L.H. Qiu, K.Q. Chen, *Heterocycles* **1996**, *43*, 49.
96H273	A.R. Katritzky, D. Cheng, R.P. Musgrave, *Heterocycles* **1996**, *42*, 273.
96H539	E. Perez, E. Sotelo, A. Loupy, R. Mocelo, M. Suarez, R. Perez, M. Autie, *Heterocycles* **1996**, *43*, 539.
96H567	E. Alcalde, M. Gisbert, L. Perez-Garcia, *Heterocycles* **1996**, *43*, 567.
96H821	C.H. Senanayake, L.E. Fredenburgh, R.A. Reamer, J. Liu, F.E. Roberts, G. Humphrey, A.A. Thompson, R.D. Larsen, T.R. Verhoeven, P.J. Reider, I. Shinkai, *Heterocycles* **1996**, *42*, 821.
96H1375	I. Kawasaki, N. Taguchi, Y. Yoneda, M. Yamashita, S. Ohta, *Heterocycles* **1996**, *43*, 1375.
96H1759	M. Morioka, M. Kato, H. Yoshida, T. Ogata, *Heterocycles* **1996**, *43*, 1759.
96IC1168	P. Bonhote, A.-P. Dias, N. Papageorgiou, K. Kalyanasundaram, M. Graetzel, *Inorg. Chem.* **1996**, *35*, 1168.
96JACS4018	K.E. Schwiebert, D.N. Chin, J.C. MacDonald, G.M. Whitesides, *J. Am. Chem. Soc.* **1996**, *118*, 4018.
96JACS6141	E.E. Baird, P.B. Dervan, *J. Am. Chem. Soc.* **1996**, *118*, 6141.
96JCR(S)4	A.E.-W.A.E.-H.O. Sarhan, H.A.H. El-Sherief, A.M. Mahamoud, *J. Chem. Res., Synop.* **1996**, 4.
96JCR(S)76	H. Uno, T. Kinoshita, K. Matsumoto, T. Murashima, T. Ogawa, N. Ono, *J. Chem. Res., Synop.* **1996**, 76.
96JCR(S)92	M.S. Khajavi, M. Hajihadi, R. Naderi, *J. Chem. Res., Synop.* **1996**, 92.
96JCR(S)94	M.S. Khajavi, M. Hajihadi, F. Nikpour, *J. Chem. Res., Synop.* **1996**, 94.
96JCR(S)182	A.H.M. Elwahy, A.A. Abbas, Y.A. Ibrahim, *J. Chem. Res., Synop.* **1996**, 182.
96JCR(S)244	H. Suzuki, N. Nonoyama, *J. Chem. Res., Synop.* **1996**, 244.
96JCR(S)254	M. Kidwai, P. Kumar, *J. Chem. Res., Synop.* **1996**, 254.
96JCR(S)302	J.M. Bofill, F. Albericio, *J. Chem. Res., Synop.* **1996**, 302.
96JCR(S)322	M.A.F. Sharaf, E.-E.H.M. Ezat, H.A.A. Hammouda, *J. Chem. Res., Synop.* **1996**, 322.
96JCR(S)388	R.H. Khan, R.K. Mathur, A.C. Ghosh, *J. Chem. Res., Synop.* **1996**, 388.

96JCS(P1)1341 B. Iddon, M. Nicholas, *J. Chem. Soc., Perkin Trans. 1* **1996**, 1341.
96JCS(P1)1545 U. Hanefeld, C.W. Rees, A.J.P. White, D.J. Williams, *J. Chem. Soc., Perkin Trans. 1* **1996**, 1545.
96JCS(P1)1617 R.N. Butler, P.D. McDonald, P. McArdle, D. Cunningham, *J. Chem. Soc., Perkin Trans. 1* **1996**, 1617.
96JCS(P1)2139 Y. Okada, J. Wang, T. Yamamoto, Y. Mu, T. Yokoi, *J. Chem. Soc., Perkin Trans. 1* **1996**, 2139.
96JFC45 G.G. Bargamov, M.D. Bargamova, *J. Fluorine Chem.* **1996**, *79*, 45.
96JHC41 C. Yamazaki, H. Arima, S. Udagawa, *J. Heterocycl. Chem.* **1996**, *33*, 41.
96JHC275 A.-C. Gillard, S. Rault, M. Boulouard, M. Robba, *J. Heterocycl. Chem.* **1996**, *33*, 275.
96JHC335 A.R. Katritzky, C.N. Fali, I.V. Shcherbakova, S.V. Verin, *J. Heterocycl. Chem.* **1996**, *33*, 335.
96JHC655 C.P. Hadjiantoniou-Maroulis, V. Ikonomou, E. Parisopoulou, *J. Heterocycl. Chem.* **1996**, *33*, 655.
96JHC671 A. Shafiee, K. Morteza-Semnani, A. Foroumadi, *J. Heterocycl. Chem.* **1996**, *33*, 671.
96JHC863 R. Noto, P.L. Meo, M. Gruttadauria, G. Werber, *J. Heterocycl. Chem.* **1996**, *33*, 863.
96JHC1031 R.H. Jones, C.A. Ramsden, H.L. Rose, *J. Heterocycl. Chem.* **1996**, *33*, 1031.
96JHC1107 A.R. Katritzky, D.A. Irina, I.A. Shcherbakova, J. Chem, S.A. Belyakov, *J. Heterocycl. Chem.* **1996**, *33*, 1107.
96JHC1345 P. Aulaskari, M. Ahlgren, J. Rouvinen, P. Vainiotalo, E. Pohjala, J. Vepsaelaeinen, *J. Heterocycl. Chem.* **1996**, *33*, 1345.
96JMC918 G.W. Rewcastle, B.D. Palmer, A.J. Bridges, H.D.H. Showalten, L. Sun, J. Nelson, A. McMichael, A.J. Kraker, D.W. Fry, W.A. Denny, *J. Med. Chem.* **1996**, *39*, 918.
96JMC2354 P.M. O'Brien, D.R. Sliskovic, J.A. Picard, H.T. Lee, C.F. Purchase, II, B.D. Roth, A.D. White, M. Anderson, M. S.B., T. Bocan, R. Bousley, K.L. Hamelehle, R. Homan, P. Lee, B.R. Krause, J.F. Reindel, R.L. Stanfield, D. Turluck, *J. Med. Chem.* **1996**, *39*, 2354.
96JOC6 A. Sobti, K. Kim, G.A. Sulikowski, *J. Org. Chem.* **1996**, *61*, 6.
96JOC428 N. Kise, K. Kashiwagi, M. Watanabe, J.-i. Yoshida, *J. Org. Chem.* **1996**, *61*, 428.
96JOC1624 A.R. Katritzky, J. Li, *J. Org. Chem.* **1996**, *61*, 1624.
96JOC2202 R. Bossio, S. Marcaccini, R. Pepino, T. Torroba, *J. Org. Chem.* **1996**, *61*, 2202.
96JOC2763 J. Ichikawa, M. Kobayashi, Y. Noda, N. Yokota, K. Amano, T. Minami, *J. Org. Chem.* **1996**, *61*, 2763.
96JOC3902 M. Kobayashi, K. Uneyama, *J. Org. Chem.* **1996**, *61*, 3902.
96JOC4035 A.R. Katritzky, H. Wu, L. Xie, *J. Org. Chem.* **1996**, *61*, 4035.
96JOC4462 A. Kumar, R. Narayanan, H. Shechter, *J. Org. Chem.* **1996**, *61*, 4462.
96JOC5204 Q. Dang, B.S. Brown, M.D. Erion, *J. Org. Chem.* **1996**, *61*, 5204.
96JOC5587 A. Molina, J.J. Vaquero, J.L. Garcia-Navio, J. Alvarez-Builla, B. de Pascual-Teresa, F. Gago, M.M. Rodrigo, M. Ballesteros, *J. Org. Chem.* **1996**, *61*, 5587.
96JOC5646 D. Moderhack, D. Decker, *J. Org. Chem.* **1996**, *61*, 5646.
96JOC6446 G. de la Torre, T. Torres, *J. Org. Chem.* **1996**, *61*, 6446.
96JOC6666 E.C. Coad, J. Kampf, P.G. Rasmussen, *J. Org. Chem.* **1996**, *61*, 6666.
96JOC6971 J. Alcazar, M. Begtrup, A. de la Hoz, *J. Org. Chem.* **1996**, *61*, 6971.
96JOC9009 M.A. de Heras, J.J. Vaquero, J.L. Garcia-Navio, J. Alvarez-Builla, *J. Org. Chem.* **1996**, *61*, 9009.
96LA575 S. Huenig, M. Schmitt, *Liebigs Ann.* **1996**, 575.
96LA683 V.J. Aran, M. Flores, P. Munoz, J.A. Paez, P. Sanchez-Verdu, M. Stud, *Liebigs Ann.* **1996**, 683.
96LA1581 J. Bohrisch, H. Faltz, M. Paetzel, J. Liebscher, *Liebigs Ann.* **1996**, 1581.
96MC98 S.F. Vasilevsky, T.A. Prikhod'ko, *Mendeleev Commun.* **1996**, 98.
96MC139 V. Vinogradov, I.L. Dalinger, B.I. Ugrak, S.A. Shevelev, *Mendeleev Commun.* **1996**, 139.
96OPP345 C.-H. Zhou, R.-G. Xie, H.-M. Zhao, *Org. Prep. Proced. Int.* **1996**, *28*, 345.
96OPP362 H.-M. Wang, L.-C. Chen, *Org. Prep. Proced. Int.* **1996**, *28*, 362.
96PJC296 M.A. Waly, S.B. Said, S.N. Ayyad, *Pol. J. Chem.* **1996**, *70*, 296.
96PAC531 A. Alexakis, P. Mangeney, N. Lensen, J.-P. Tranchier, R. Gosmini, S. Raussou, *Pure Appl. Chem.* **1996**, *68*, 531.
96SL285 E. Alcalde, M. Gisbert, *Synlett* **1996**, 285.
96SL359 M.S. Congreve, *Synlett* **1996**, 359.
96SL667 L.F. Tietze, A. Steinmetz, *Synlett* **1996**, 667.

96SL734 I. Mallik, S. Mallik, *Synlett* **1996**, 734.
96SL1109 I. Coldham, P.M.A. Houdayer, R.A. Judkins, D.R. Witty, *Synlett* **1996**, 1109.
96SC357 A.R. Katritzky, S. El-Zemity, H. Lang, E.A. Kadous, A.M. El-Shazly, *Synth. Commun.* **1996**, *26*, 357.
96SC569 H.H. Tso, Y.M. Chang, H. Tsay, *Synth. Commun.* **1996**, *26*, 569.
96SC745 I.T. Forbes, H.K.A. Morgan, M. Thompson, *Synth. Commun.* **1996**, *26*, 745.
96SC837 J.L. Gagnon, W.W. Zajac, Jr., *Synth. Commun.* **1996**, *26*, 837.
96SC1385 A.R. Katritzky, I.V. Shcherbakova, *Synth. Commun.* **1996**, *26*, 1385.
96SC1585 A.A. Cordi, T. Persigand, J.-P. Lecouve, *Synth. Commun.* **1996**, *26*, 1585.
96SC1977 C.-Y. Chen, D.R. Lieberman, L.J. Street, A.R. Guiblin, R.D. Larsen, T.R. Verhoeven, *Synth. Commun.* **1996**, *26*, 1977.
96SC2587 T. Klindert, G. Seitz, *Synth. Commun.* **1996**, *26*, 2587.
96SC2687 C.F. Purchase, II, A.D. White, *Synth. Comm.* **1996**, *26*, 2687.
96SC2895 D. Villemin, M. Hammadi, B. Martin, *Synth. Commun.* **1996**, *26*, 2895.
96SC3175 Z. Shi, H. Gu, L.-L. Xu, *Synth. Commun.* **1996**, *26*, 3175.
96SC3241 R. Saladino, C. Crestini, F. Occhionero, R. Nicoletti, *Synth. Commun.* **1996**, *26*, 3241.
96SC3799 M.R. Mahmoud, H.M.F. Madkour, *Synth. Commun.* **1996**, *26*, 3799.
96SC3827 G.A. El-Saraf, A.M. El-Sayed, *Synth. Commun.* **1996**, *26*, 3827.
96S274 Y. Guo, Q. Wang, J.C. Jochims, *Synthesis* **1996**, 274.
96S533 A. Arcadi, O.A. Attanasi, L. De Crescentini, E. Rossi, F. Serra-Zanetti, *Synthesis* **1996**, 533.
96S697 K.J. Harlow, A.F. Hill, T. Welton, *Synthesis* **1996**, 697.
96S927 V.A. Artyomov, A.M. Shestopalov, V.P. Litvinov, *Synthesis* **1996**, 927.
96S1076 G. Broggini, L. Garanti, G. Molteni, G. Zecchi, *Synthesis* **1996**, 1076.
96S1302 J. Brandenburg, C. Kapplinger, R. Beckert, *Synthesis* **1996**, 1302.
96T791 S. Ernst, S. Jelonek, J. Sieler, K. Schulz, *Tetrahedron* **1996**, *52*, 791.
96T901 B. Sun, K. Adachi, M. Noguchi, *Tetrahedron* **1996**, *52*, 901.
96T1177 N. Nakajima, M. Matsumoto, M. Kirihara, M. Hashimoto, T. Katoh, S. Terashima, *Tetrahedron* **1996**, *52*, 1177.
96T2827 M. Watanabe, H. Okada, T. Teshima, M. Noguchi, A. Kakehi, *Tetrahedron* **1996**, *52*, 2827.
96T4383 M.A. Plancquaert, M. Redon, Z. Janousek, H.G. Viehe, *Tetrahedron* **1996**, *52*, 4383.
96T5363 R. Jain, L.A. Cohen, *Tetrahedron* **1996**, *52*, 5363.
96T6581 M. Noguchi, H. Okada, M. Watanabe, K. Okuda, O. Nakamura, *Tetrahedron* **1996**, *52*, 6581.
96T7179 C.B. Vicentini, A.C. Veronese, M. Manfrini, M. Guarneri, *Tetrahedron* **1996**, *52*, 7179.
96T7939 E. Rossi, E. Pini, *Tetrahedron* **1996**, *52*, 7939.
96T8471 D. Barrett, H. Sasaki, T. Kinoshita, A. Fujikawa, K. Sakane, *Tetrahedron* **1996**, *52*, 8471.
96T8813 B.P. Medaer, K.J. Van Aken, G.J. Hoornaert, *Tetrahedron* **1996**, *52*, 8813.
96T9035 H. Faltz, J. Bohrisch, W. Wohlauf, M. Paetzel, P.G. Jones, J. Liebscher, *Tetrahedron* **1996**, *52*, 9035.
96T9121 L. Somsák, E. Sós, Z. Györgydeák, J.-P. Praly, G. Descotes, *Tetrahedron* **1996**, *52*, 9121.
96T9237 A. Diaz-Ortiz, J.R. Carrillo, E. Diez-Barra, A. de la Hoz, M.J. Gomez-Escalonilla, A. Moreno, F. Langa, *Tetrahedron* **1996**, *52*, 9237.
96T9541 J. Suwinski, P. Wagner, E.M. Holt, *Tetrahedron* **1996**, *52*, 9541.
96T9835 V.A. Kovtunenko, K.G. Nazarenko, A.M. Demchenko, *Tetrahedron* **1996**, *52*, 9835.
96T9877 L.R. Milgrom, P.J.F. Dempsey, G. Yahioglu, *Tetrahedron* **1996**, *52*, 9877.
96T10169 L.T. Giang, J. Fetter, K. Lempert, M. Kajtár-Peredy, A. Gömöry, *Tetrahedron* **1996**, *52*, 10169.
96T10485 A.E.-W.A.O. Sarhan, H.A.H. El-Sherief, A.M. Mahmoud, *Tetrahedron* **1996**, *52*, 10485.
96T10557 J.-M. Vatele, S. Hanessian, *Tetrahedron* **1996**, *52*, 10557.
96T10571 D. Korakas, A. Kimbaris, G. Varvounis, *Tetrahedron* **1996**, *52*, 10571.
96T11075 P. Cornago, C. Escolastico, M.M.D. Santa, R.M. Claramunt, C.C. Fernandez, F.C. Foces, J.P. Fayet, J. Elguero, *Tetrahedron* **1996**, *52*, 11075.
96T13703 M.D. Cliff, S.G. Pyne, *Tetrahedron* **1996**, *52*, 13703.
96T14297 M.A. de Heras, J.J. Vaquero, J.L. Garcia-Navio, J. Alvarez-Builla, *Tetrahedron* **1996**, *52*, 14297.
96T14905 J. Suwinski, W. Szczepankiewicz, E.M. Holt, *Tetrahedron* **1996**, *52*, 14905.
96T15171 E. Alcalde, M. Alemany, M. Gisbert, *Tetrahedron* **1996**, *52*, 15171.
96T15189 E. Alcalde, M. Gisbert, C. Alvarez-Rua, S. Garcia-Granda, *Tetrahedron* **1996**, *52*, 15189.

96TL187 S. Shuto, K. Haramuishi, A. Matsuda, *Tetrahedron Lett.* **1996**, *37*, 187.

96TL611 D. Gala, D.J. DiBenedetto, J.E. Clark, B.L. Murphy, D.P. Schumacher, M. Steinman, *Tetrahedron Lett.* **1996**, *37*, 611.

96TL751 C. Zhang, E.J. Moran, T.F. Woiwode, K.M. Short, A.M.M. Mjalli, *Tetrahedron Lett.* **1996**, *37*, 751.

96TL835 S. Sarshar, D. Siev, A.M.M. Mjalli, *Tetrahedron Lett.* **1996**, *37*, 835.

96TL937 B.A. Dressman, L.A. Spangle, S.W. Kaldor, *Tetrahedron Lett.* **1996**, *37*, 937.

96TL1707 R.C.F. Jones, K.J. Howard, J.S. Snaith, *Tetrahedron Lett.* **1996**, *37*, 1707.

96TL1717 R.C.F. Jones, K.J. Howard, J.S. Snaith, *Tetrahedron Lett.* **1996**, *37*, 1717.

96TL1829 H. Yamanaka, T. Takekawa, K. Morita, T. Ishihara, J.T. Gupton, *Tetrahedron Lett.* **1996**, *37*, 1829.

96TL1901 C.A. Ramsden, B.J. Sargent, C.D. Wallett, *Tetrahedron Lett.* **1996**, *37*, 1901.

96TL2357 Z. Shi, V. Goulle, R.P. Thummel, *Tetrahedron Lett.* **1996**, *37*, 2357.

96TL2361 A.G. Romero, W.H. Darlington, E.J. Jacobsen, J.W. Mickelson, *Tetrahedron Lett.* **1996**, *37*, 2361.

96TL2369 H. Kang, W. Fenical, *Tetrahedron Lett.* **1996**, *37*, 2369.

96TL3655 B.E. Huff, M.E. LeTourneau, M.A. Staszak, J.A. Ward, *Tetrahedron Lett.* **1996**, *37*, 3655.

96TL3915 C.W.G. Fishwick, R.J. Foster, R.E. Carr, *Tetrahedron Lett.* **1996**, *37*, 3915.

96TL4323 V.V. Khau, M.J. Martinelli, *Tetrahedron Lett.* **1996**, *37*, 4323.

96TL4417 A. Lorente, J.F. Espinosa, M. Fernandez-Saiz, J.-M. Lehn, W.D. Wilson, Y.Y. Zhong, *Tetrahedron Lett.* **1996**, *37*, 4417.

96TL4423 A. Horvath, *Tetrahedron Lett.* **1996**, *37*, 4423.

96TL4565 C. Palomo, M. Oiarbide, A. Gonzalez, J.M. Garcia, F. Berree, *Tetrahedron Lett.* **1996**, *37*, 4565.

96TL4589 O.A. Rakitin, C.W. Rees, O.G. Vlasova, *Tetrahedron Lett.* **1996**, *37*, 4589.

96TL4767 M. Periasamy, M.R. Reddy, J.V.B. Kanth, *Tetrahedron Lett.* **1996**, *37*, 4767.

96TL4865 S.M. Hutchins, K.T. Chapman, *Tetrahedron Lett.* **1996**, *37*, 4865.

96TL4969 T. Hayashi, E. Kishi, V.A. Soloshonok, Y. Uozumi, *Tetrahedron Lett.* **1996**, *37*, 4969.

96TL5039 A. Allen, J.P. Anselme, *Tetrahedron Lett.* **1996**, *37*, 5039.

96TL5265 M.C. Rezende, E.L. Dall'Oglio, C. Zucco, *Tetrahedron Lett.* **1996**, *37*, 5265.

96TL5503 S. Achab, *Tetrahedron Lett.* **1996**, *37*, 5503.

96TL5615 S.K.Y. Li, D.W. Knight, P.B. Little, *Tetrahedron Lett.* **1996**, *37*, 5615.

96TL5657 A.K. Saksena, V.M. Girijavallabhan, H. Wang, Y.-T. Liu, R.E. Pike, A.K. Ganguly, *Tetrahedron Lett.* **1996**, *37*, 5657.

96TL6081 L.-X. Yang, K.G. Hofer, *Tetrahedron Lett.* **1996**, *37*, 6081.

96TL6209 A.T. Ung, S.G. Pyne, *Tetrahedron Lett.* **1996**, *37*, 6209.

96TL6931 C. Palomo, M. Oiarbide, A. Gonzalez, J.M. Garcia, F. Berree, A. Linden, *Tetrahedron Lett.* **1996**, *37*, 6931.

96TL7249 R. Bohlmann, P. Strehlke, *Tetrahedron Lett.* **1996**, *37*, 7249.

96TL7951 S. Grabowski, H. Prinzbach, *Tetrahedron Lett.* **1996**, *37*, 7951.

96TL8121 Y.-Z. Xu, K. Yakushijin, D.A. Horne, *Tetrahedron Lett.* **1996**, *37*, 8121.

96TL9353 P. Molina, P.M. Fresenda, S. Garcia-Zafra, *Tetrahedron Lett.* **1996**, *37*, 9353.

96TA1621 A.R. Katritzky, S.R. El-Zemity, P. Leeming, C.M. Hartshorn, P.J. Steel, *Tetrahedron: Asymmetry* **1996**, *7*, 1621.

96TA1631 A.R. Katritzky, J. Wang, P. Leeming, *Tetrahedron: Asymmetry* **1996**, *7*, 1631.

96TA1641 A. Gutcait, K.-C. Wang, H.-W. Liu, J.-W. Chern, *Tetrahedron: Asymmetry* **1996**, *7*, 1641.

96TA1717 M. Barz, E. Herdtweck, W.R. Thiel, *Tetrahedron: Asymmetry* **1996**, *7*, 1717.

Chapter 5.5

Five-Membered Ring Systems : With N & S (Se) Atoms

Paul A. Bradley and David J. Wilkins
Knoll Pharmaceuticals, Research Department, Nottingham, England

5.5.1 ISOTHIAZOLES

The cycloaddition reactions of both arylalkyl and aryl azides with the isothiazole 1,1-dioxide (**1**) were found to be highly regioselective, leading to triazines (**2**). In the reaction with arylalkyl azides in refluxing benzene these triazolines (**2**) could be isolated and on heating at higher temperatures resulted in the elimination of nitrogen to give aziridines (**3**). However, reaction of (**1**) with aryl azides gave aziridines (**3**) directly as the major product; a satisfactory yield of the triazoline (**2**; R = Ph) could be isolated only after a seventy five day reaction at room temperature <96T7183>.

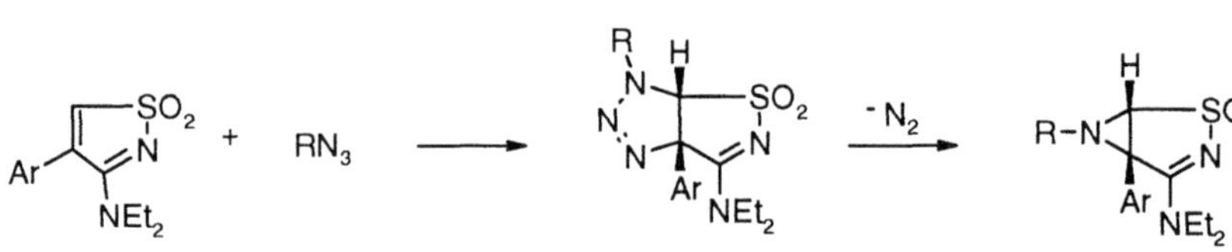

1 2 3

Ar= p-MeOC$_6$H$_4$; R= Ph, p-MeOC$_6$H$_4$, p-NO$_2$C$_6$H$_4$, Bn and Ph(CH$_2$)$_2$

Heating the aziridines (**3**; R = Bn, Ph, p-MeOC$_6$H$_4$, p-NO$_2$C$_6$H$_4$; Ar = p-MeOC$_6$H$_4$) at or slightly above their melting points generally gave the corresponding thiadiazine dioxides (**5**) as the major product; the thiazetes (**4**) and pyrazoles (**6**) were also isolated in lower yield. Prolonged heating of a mixture of these three compounds led to the disappearance of (**5**); compounds (**4**) and (**6**) being the resulting thermally stable products. The results of these thermolysis reactions led to the formulation of a mechanistic model for these transformations which involved an equilibrium between (**3**) and the open-chain intermediates A, B and C and the bicyclic intermediate D. In the case of the *N*-benzyl derivative (**3**; R = Bn), thermolysis led to a mixture of the thiadiazine (**5**; R = Bn) and the pyrimidine (**7**) whose mechanism of formation was proposed <96T7183>.

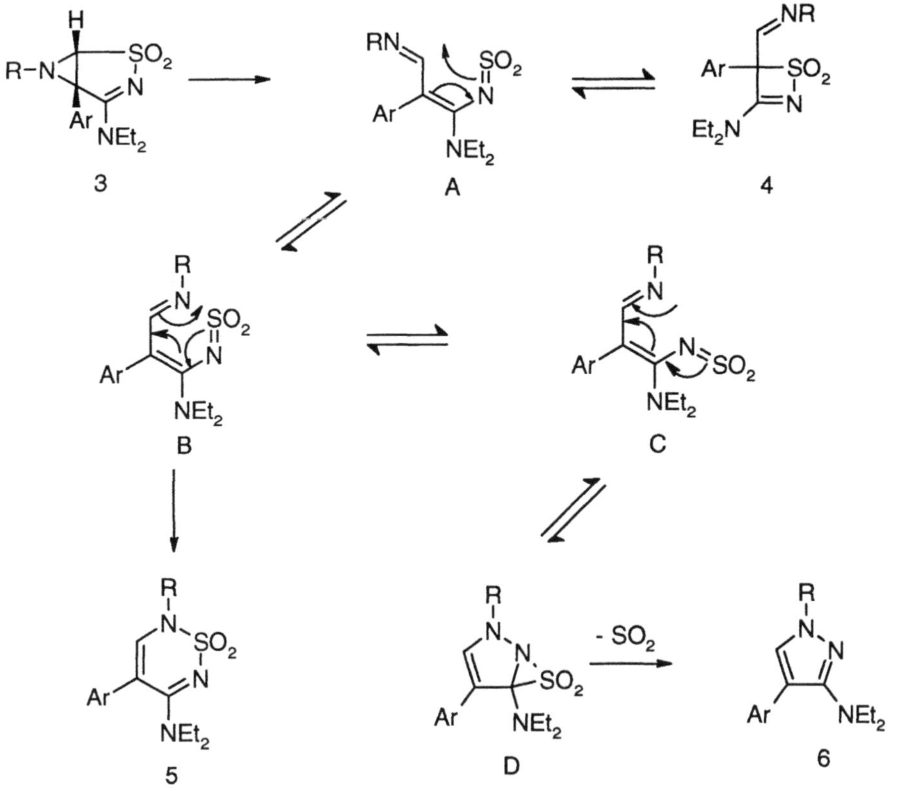

Ar = p-MeOC$_6$H$_4$

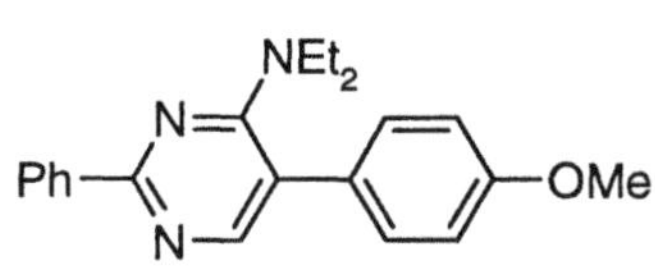

7

A variety of interesting ring expansions and annelations of 1,2-benzisothiazoles have recently been reported giving rise to new heterocyclic ring systems. The first of these involved the initial condensation of the lithium derivative (**8**) with aromatic aldehydes, followed by addition of HCl to give the isothiazole (**9**). When (**9**) was dissolved in trifluoroacetic acid it underwent an intermolecular Friedel-Crafts alkylation reaction to give the tricycle (**10**). The ring expanded thiazepine (**13**) was produced in the reaction of (**9**; R = OMe) with 1-diethylamino-1-propyne in acetonitrile. This may have arisen by 1,2-addition to the isothiazole (**9**; R = OMe) and then subsequent electrocyclic ring opening- ring expansion of the intermediate (**12**). Reaction of the alkyne in ether, however, led to the formation of the dihydropyridine (**11**) which was envisaged to arise by an initial 1,4-addition to the isothiazole (**9**; R = OMe) or by a 1,3-sigmatropic shift in the intermediate (**12**) <96T3339>.

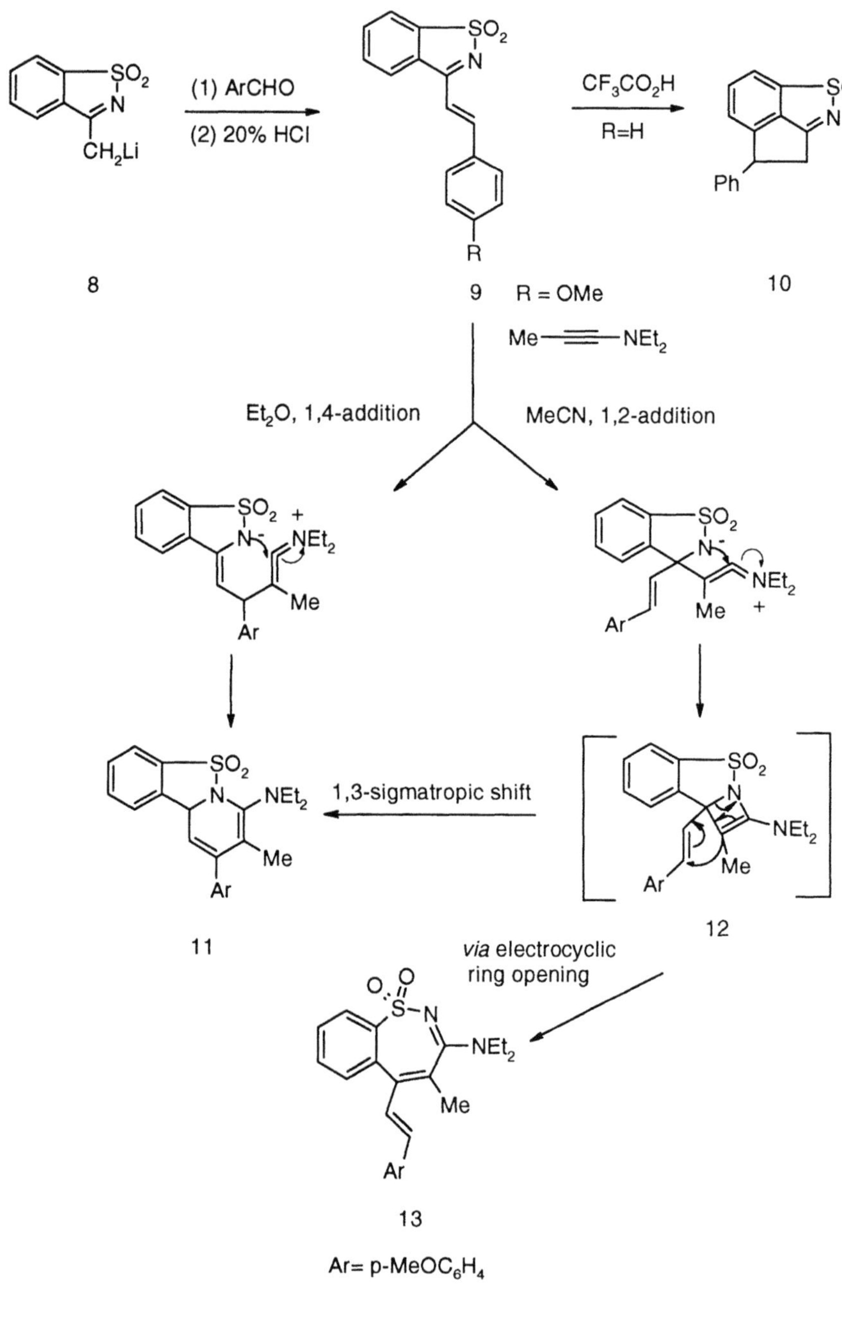

Two new pyridone derivatives (**14**) and (**15**) have been prepared by cycloaddition of saccharin pseudochloride (**16**; R = Cl) with Danishefsky's diene and by treatment of (**16**; R = Me) with cinnamoyl chloride. The synthesis of two more ring expanded derivatives (**17**) and (**18**) *via* cycloaddition to benzisothiazoles was also described <96T3339>.

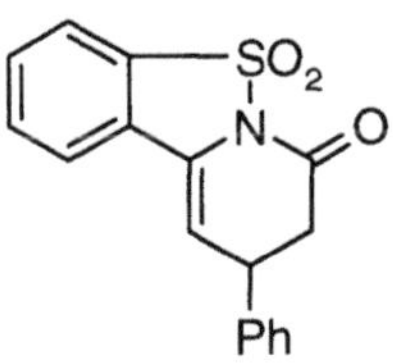

14

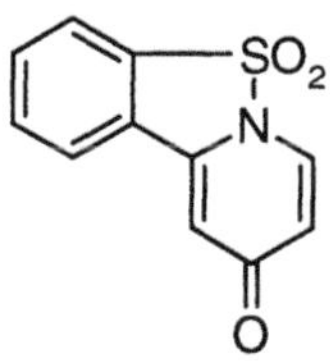

15

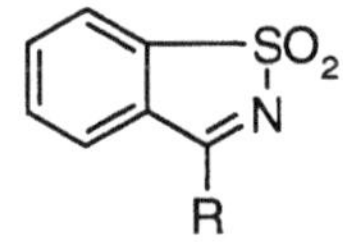

16

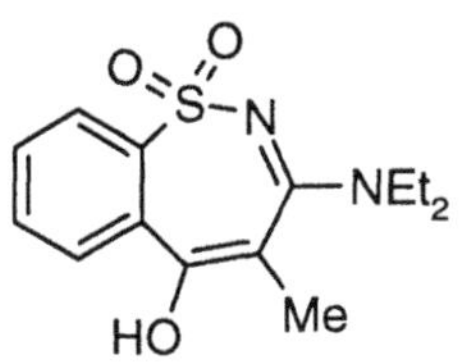

17

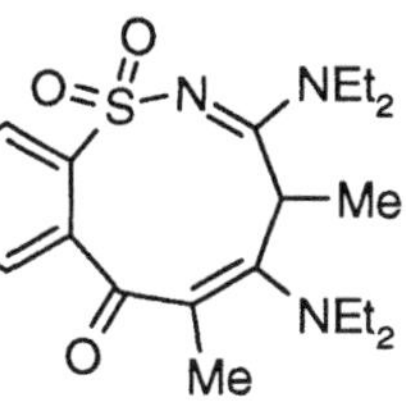

18

Isothiazol-3(2*H*)-ones showed moderate dipolarphilicity towards nitrile oxides, leading to isoxazoles after breakdown of an initial cycloadduct. When the isothiazolone ring contains a 5-carbonyl substituent, the isothiazolone ring acts as an electron-withdrawing group, which activates the 5-carbonyl substituent on the ring for cycloaddition with mesitonitrile giving preferential reaction at the carbonyl bond before reaction at the ring double bond, thus affording mono and bis adducts <96JHC713>.

Treatment of the sulpholene (**19**) with hydroxylamine in refluxing ethanol has led to the formation of isothiazol-3-sulpholenes (**21**) which presumably progressed *via* the oxime intermediate (**20**). Subsequent heating of (**21**) in toluene at 185°C in a sealed tube led to the generation of the diene (**22**) which could be trapped with *N*-phenylmaleimide and with DMAD, thus providing a new route to 1,2-benzisothiazole derivatives <96TL4189>.

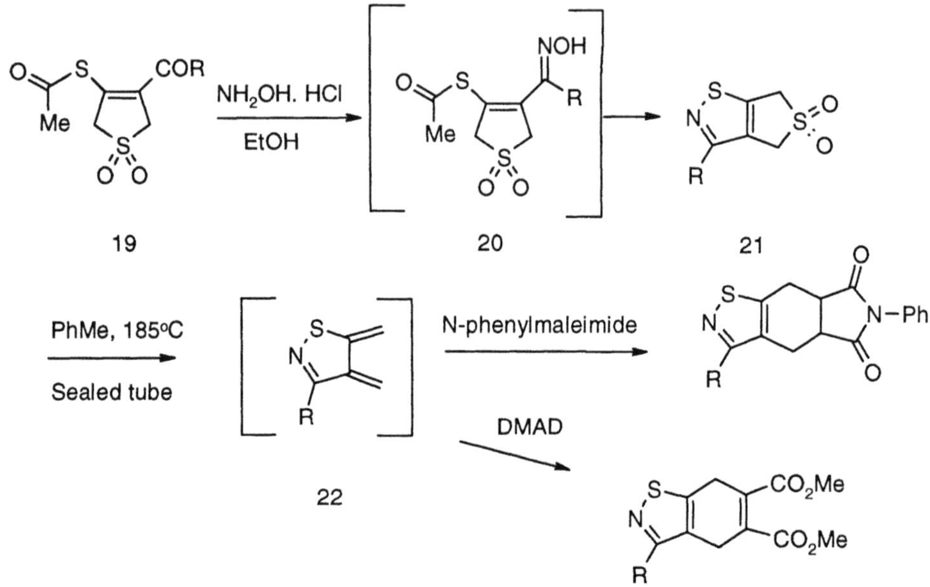

19 20 21 22

The first synthesis of stable 3-hydroperoxy-sultams (**24**) which are a new class of sultam with oxidising properties, was reported. The synthesis involved oxidation of the isothiazolium salts (**23**) with hydrogen peroxide in acetic acid. Reduction of (**24**) with aqueous sodium bisulphite afforded the corresponding novel 3-hydroxysultams whereas thermolysis in ethanol resulted in the elimination of water to give 3-ketone derivatives, which are versatile as dieneophiles <96T783>.

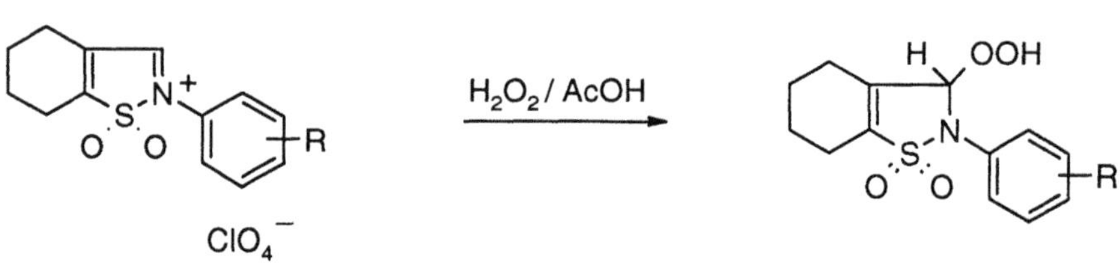

23 24

R= H, 2-Me, 2,6-$(Me)_2$, 4-Br, 4-OMe

N-Aryl-substituted isothiazolium salts (**25**) which contain an active 7-methylene group react under basic conditions to give a mixture of *cis* and *trans* spirocyclic isothiazolium salts (**26**). The *cis* isomers were stable, but the *trans* isomers on treatment with more base gave thianthrene derivatives <96JPR424>.

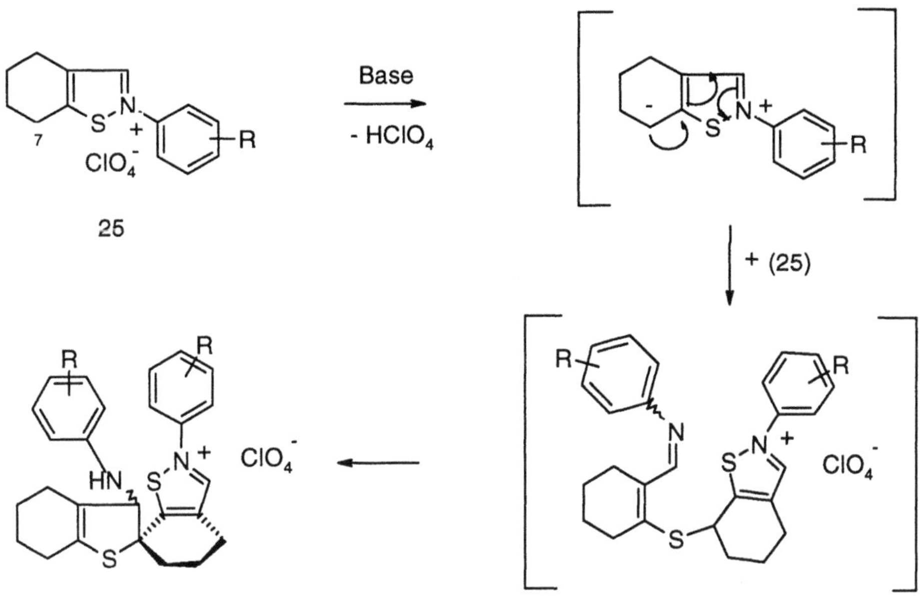

25

26

R= H, 4-Me, 4-OMe, 4-Cl, 4-SO_2Me, 3-Me, 3-OMe, 3-Cl

1-(Saccharin-1-yl)pyridinium-3-oxide underwent cycloaddition reactions across the 2,6-position of the pyridine ring with various dipolarophiles affording 2-oxo-8-azabicyclo[3.2.1]oct-3-enes <96MI169>.

The addition of trithiazyl trichloride $(NSCl)_3$ to 2,5-disubstituted furans and to *N*-2,5-trisubstituted pyrroles has led to the formation of isothiazole derivatives <96JHC1419>.

3-Chloro-4-cyano-2-methyl-5-phenylisothiazolium fluorosulphonate (**28**) which was prepared by methylation of (**27**) with $MeSO_3F$, underwent condensation reactions with lithium salts of substituted cyclopentadiene derivatives at -70°C to give 4-(*N*-methylamino)thialene-3-carbonitriles (**29**). The reactions were thought to proceed *via* initial nucleophilic attack of the cyclopentadiene at the sulphur atom of the isothiazole ring which gave an open-chain intermediate. Subsequent cyclisation and HCl elimination then gave the thialene compounds (**29**) [see <96MI638>].

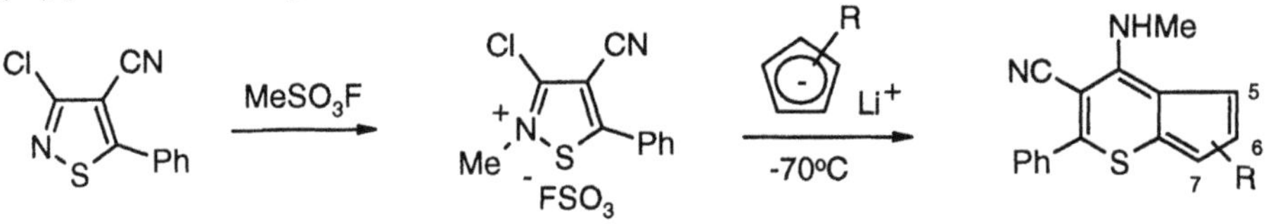

27 28 29

R = H, 6-^{t}Bu, 5-^{t}Bu and 5,7-(^{t}Bu)$_2$

The mechanism for the synthesis of 3-aminopyrroles by ring transformation-desulphurisation of substituted 2-methylisothiazolium salts was investigated and evidence for the intermediacy of 3-alkylideneaminothioacrylamides (**30**) and 2*H*-1,3-thiazines (**31**) obtained <96JCS(P1)2239>.

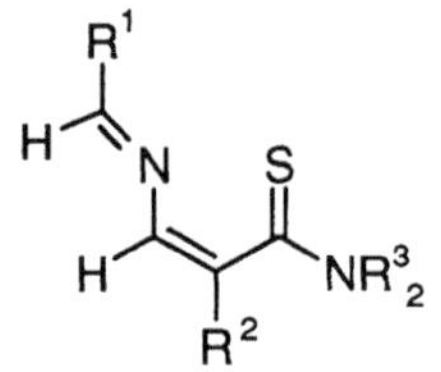

30

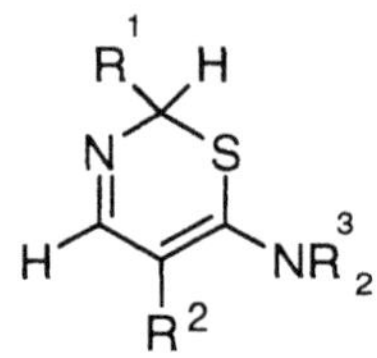

31

5.5.2 THIAZOLES

Thioamides react under mild conditions with conjugated azoalkenes (**32**) to give thiazolines (**33**) that exhibit hydrazono hydrazino tautomerism. X-ray diffraction studies on (**33**) showed the compound existed as the hydrazono tautomer <95S1397>.

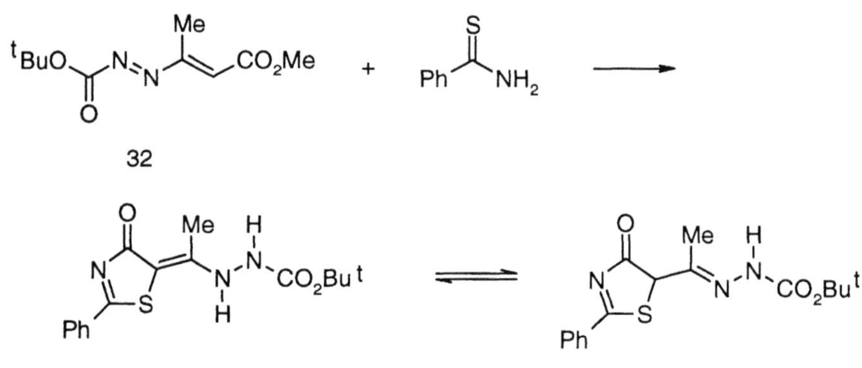

32

33

5-Arylmethylene-2-thioxo-4-thiazolidinones (**34**) react with phosphonium ylides to give dihydrofuro[2,3-*d*]thiazol-2(3*H*)-ones (**35**) in refluxing ethyl acetate, while performing the reaction in refluxing toluene led to the pyrone derivative (**36**); both of these products result from an initial 1,4-addition to the exocyclic double bond <95T11411>.

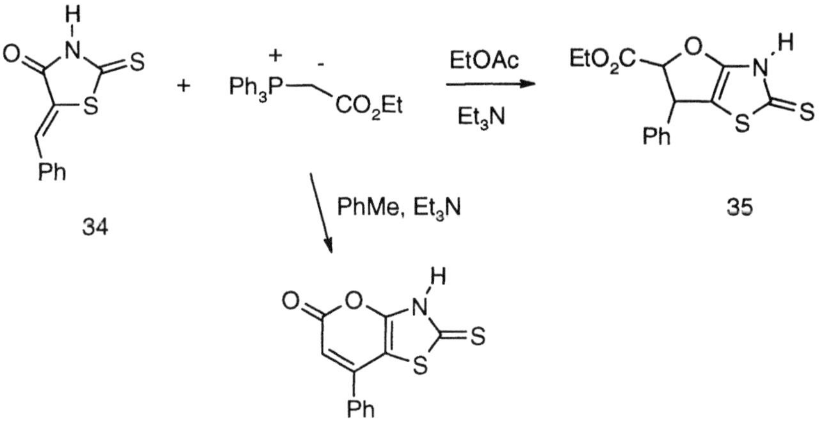

34 35 36

The condensation reaction between the triisobutylaluminium complex of cysteamine HCl and a variety of carboxylic esters affords thiazolines in good yields. When this method is applied to chiral α-aminoesters, α-aminothiazoline's are obtained in high optical purity <96TL2935>.

A variety of thiazolines and thiazoles were synthesised by the condensation reaction of thiocyanates and thioureas with α-haloketimines, these imines are a new type of reactive bielectrophilic reagent <96JHC1179>.

The reaction of troponephenylhydrazone with carbon disulphide afforded the bicyclic thiazole (**37**) in quantitative yield. *N*-methoxytroponimine when treated with phenylisothiocyanate afforded a mixture of cycloheptatriene derivatives (**38a**) and (**38b**). Both of these reactions proceed *via* an [8+2] cycloaddition <95H1675>.

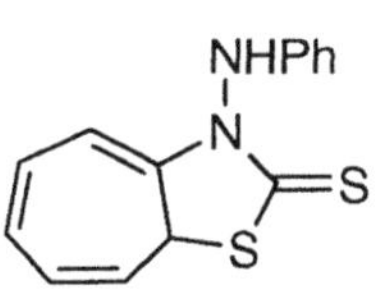

37

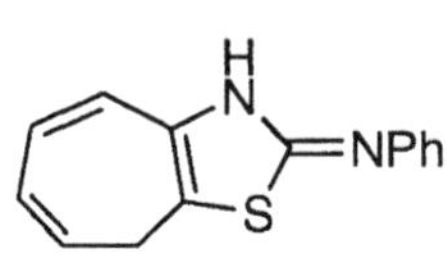

38a

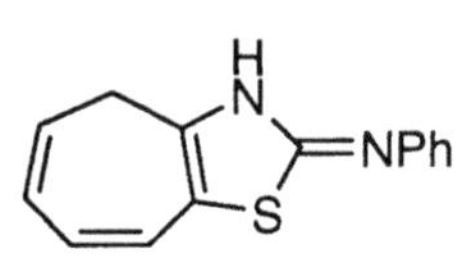

38b

Dialkyl esters of cystine (**39**) and lanthionine (**40**) undergo a surprising thermolysis reaction at between 25 °C and 80 °C to afford *cis* and *trans* methyl 2-methylthiazolidine-2,4-dicarboxylates (**43**) in protic solvents. A two stage process is proposed for this transformation. An initial β-elimination reaction gives the thiol (**41**) and the enamine (**42**). Thiol addition to the imine tautomer of (**42**) is then followed by loss of ammonia and an intramolecular cyclisation to give (**43**) <96CC843>.

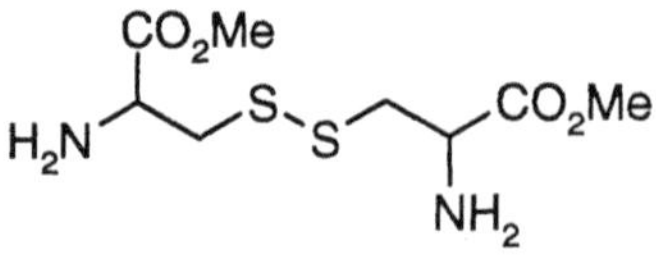

39

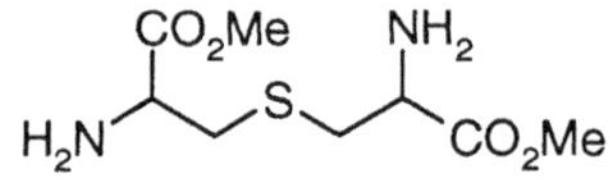

40

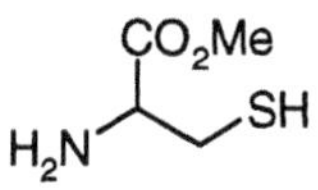

41

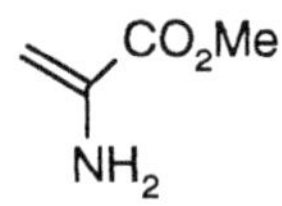

42

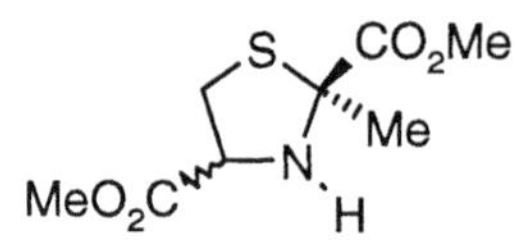

43

A direct conversion of the oxazoline (**44**) to the thiazoline (**45**) can be achieved by the thiolysis of the oxazoline (**44**) with H_2S in methanol in the presence of triethylamine followed by cyclodehydration with Burgess reagent. This method is essentially free from racemisation and has been used in the transformation of peptide substrates <95TL6395>.

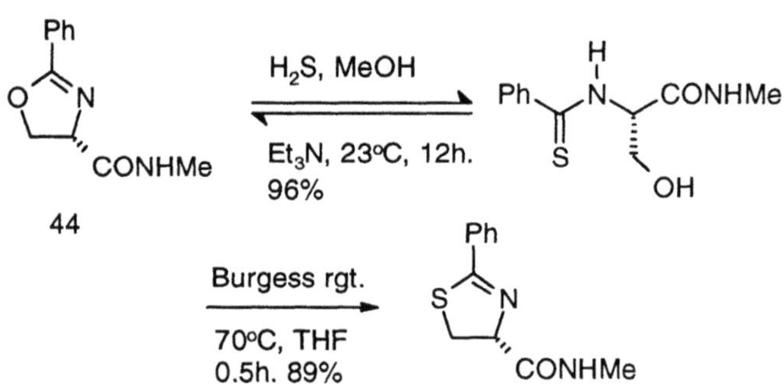

44

45

Phenyliodonium ylids of cyclic dicarbonyl compounds (**46**) react with thiourea to form the thiouronium ylid (**47**) which on heating is converted into the fused thiazole (**48**), this method is applicable to subtituted thioureas provided they have at least one free amino group. This reaction can be considered to be a modification of the Hantzsch thiazole synthesis <96JHC575>.

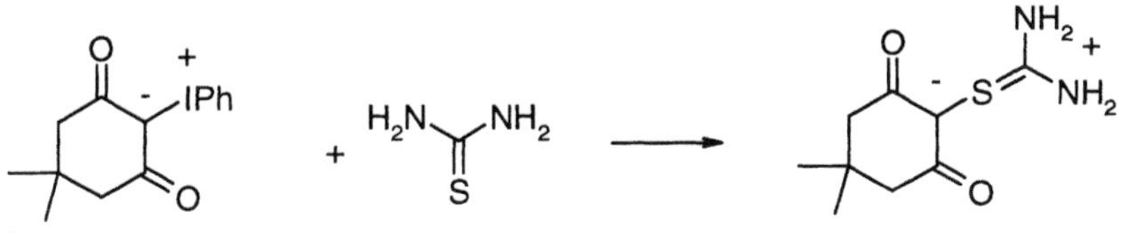

46 47

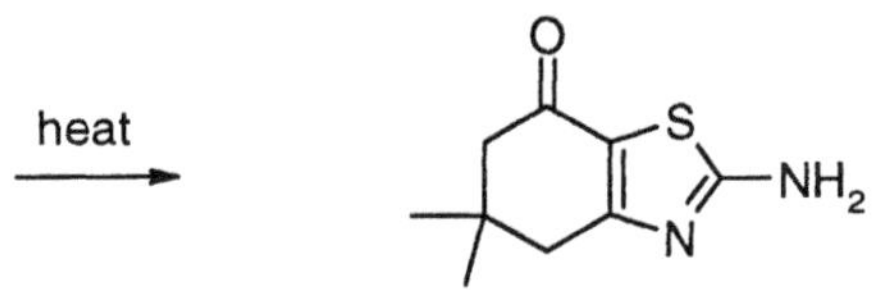

48

Another modification of the Hantzsch thiazole synthesis afforded C-4 thiazolylmethyl phosphonium salts (**49**). These ylids could then undergo Wittig condensations to furnish a wide variety of 2,4-disubstituted thiazoles <96TL983>.

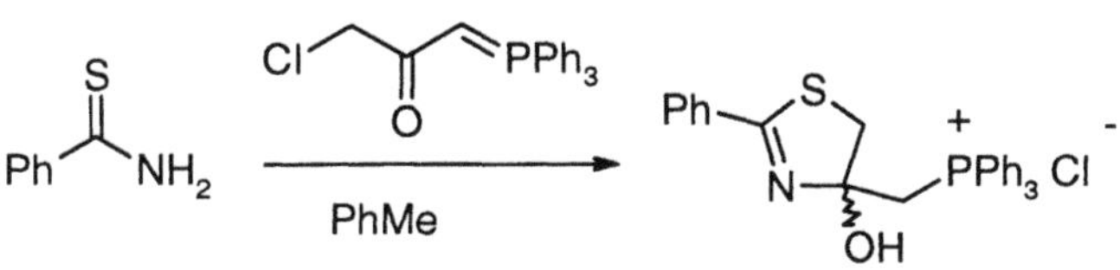

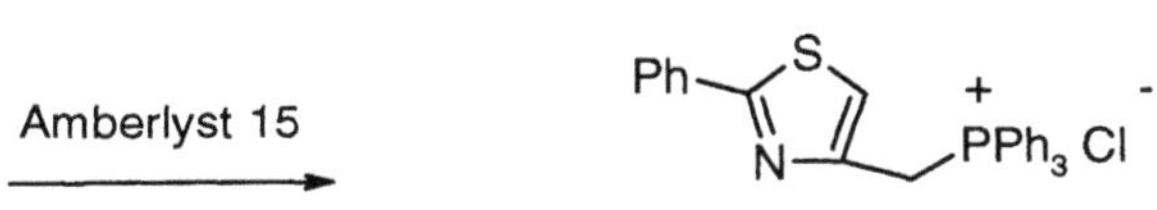

49

The carbanion of 2,3-dimethylthiazolidine-4-one reacted with nitroarenes to give either a ring opened product (**50**) *via* a VNS (vicarious nucleophilic substitution) reaction or a product resulting from oxidative nucleophilic substitution of hydrogen (**51**). Ring opening VNS reactions with 5-membered *S*-heterocycles are limited to those heterocycles which show some conformational flexibility <96TL983>.

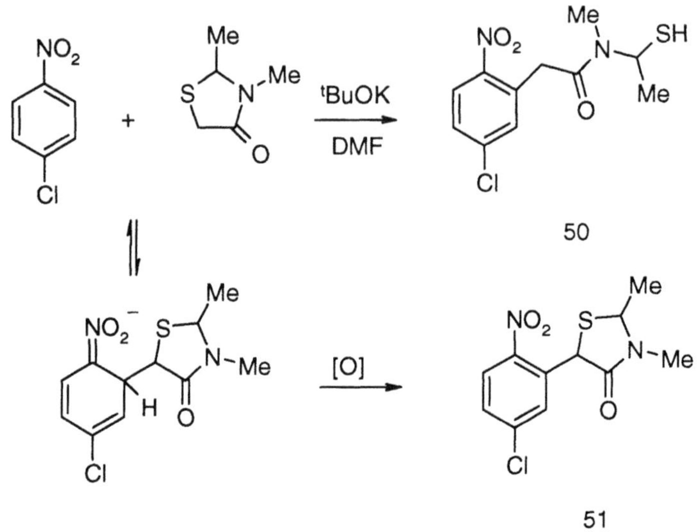

50

51

1,3-Dipolar cycloaddition reactions of thioisomunchnones (1,3-thiazolium-4-olates) have not been as extensively studied as those of munchnones (1,3-oxazolium-5-olates) despite offering rapid access to novel heterocyclic compounds. The cycloaddition of the thioisomunchnone (**52**) with trans-β-nitrostyrene results in the formation of two diastereoisomeric 4,5-dihydrothiophenes (**53**) and (**54**) *via* transient cycloadducts. These cycloadducts then undergo rearrangement under the reaction conditions <96JOC3738>.

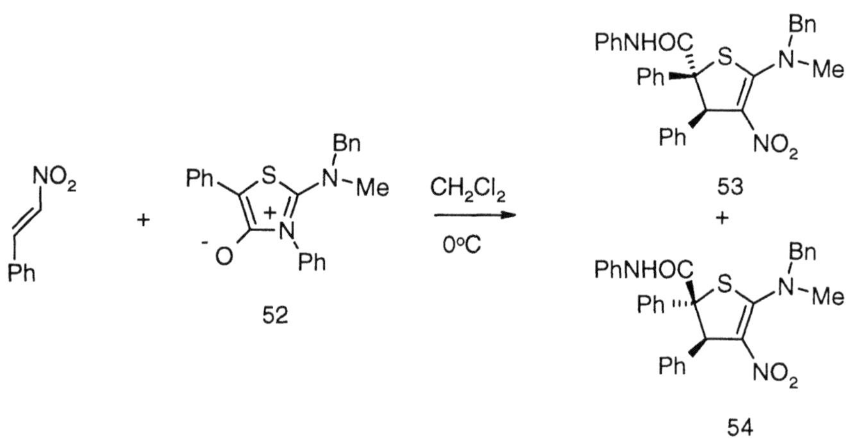

52

53

54

Dithiolane isocyanate iminium methylides (**55**), are a new type of azomethine methylide derived 1,3-dipole, and undergo efficient and regioselective cycloaddition to thiocarbonyls to yield predominantly thiazolidine-2-thiones (**56**) <96TL711>.

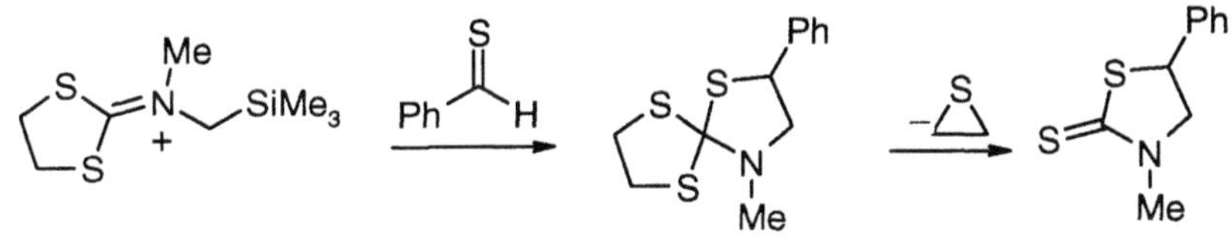

55 56

Stille cross coupling reactions usually proceed under mild neutral conditions. 2-Substituted thiazolines can be obtained by the cross coupling reaction of 2-bromothiazolines with various tributylstannyl compounds. Previous attempts at a palladium mediated coupling of 2-trimethylstannylthiazoline led to only decomposition of the substrate <96TL4857>.

2-Trichloromethyl substituted thiazolidines have been prepared from chloral. Reaction of chloral with anilines afforded the corresponding imines which were then treated *in situ* with thioglycolic acid to give a series of 2-trichloromethyl substituted thiazolidinones <96HC227>.

A convenient procedure for the solution phase preparation of a 2-aminothiazole combinatorial library has been reported. The Hantzch synthesis of 2-aminothiazoles has been adapted to allow the ready solution phase preparation of libraries of discrete 2-aminothiazoles <96BMC1409>.

3-Thiazolines (**57**) can be phosphonylated in a stereospecific manner by diastereomerically pure phosphites (**58**) to give 4-thiazolidinylphosphonates (**59**) *via* a Pudovik reaction <96SC1903>.

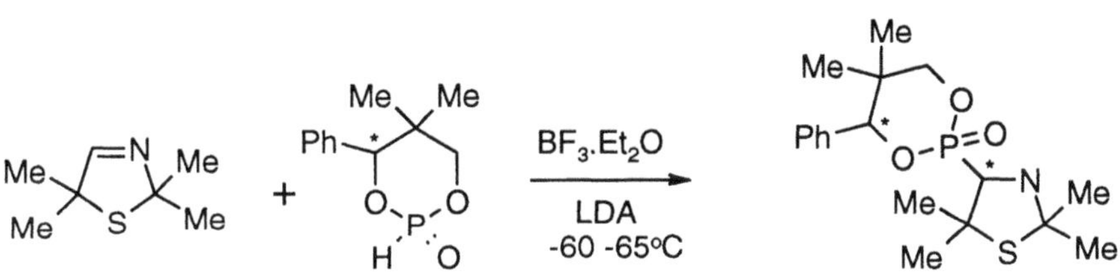

57 58 59

In contrast to *N*-disubstituted 3-hydroxyanilines which react with Vilsmeier reagent to give *N*-disubstituted salicylaldehydes, their heteroanalogous *N*-disubstituted 2-amino-4-hydroxythiazoles (**60**) react with Vilsmeier reagent to give *N*-disubstituted 2-amino-4-chlorothiazole-5-aldehydes (**61**) <96JPR51>.

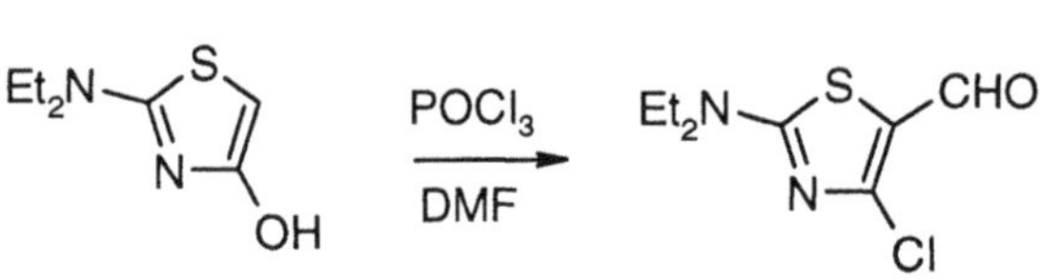

60 61

The benzthiazole (**62**), an example of a stabilised 1-azabuta-1,3-diene, undergoes Inverse type Diels-Alder reactions with electron-rich dienophiles under extremely mild conditions.

When (**62**) was treated with ethyl vinyl ether the cycloadduct (**63**) is afforded. If (**62**) is reacted with electron donating dienophiles such as allyl alcohols, transesterification and intramolecular cycloaddition occurs in the presence of a catalytic amount of distannoxane catalyst to give *cis*-fused polycyclic systems such as (**64**) <96T733>.

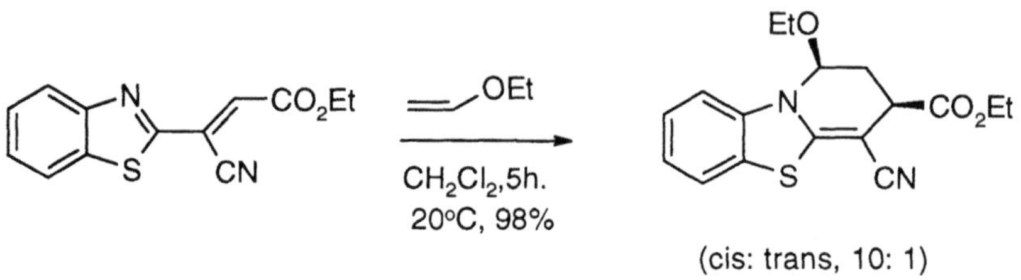

62

63

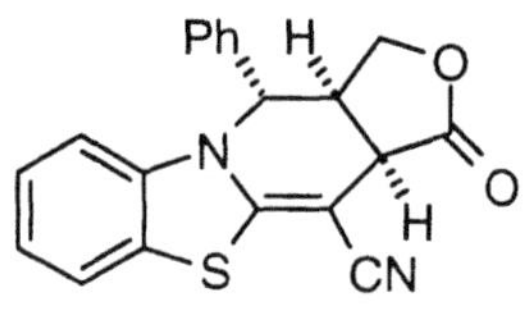

64

A study of the mechanism of the reaction of 2-silylthiazole (**65**) with formaldehyde has concluded that the reaction occurs *via* the initial fast formation of an *N*-(silyloxymethyl)thiazolium-2-ylide (**66**) followed by a rate determining second addition of formaldehyde to give (**67**). This is followed by a fast 1,6-silyl migration and loss of a molecule of formaldehyde to give the final product (**68**) <96JOC1922>.

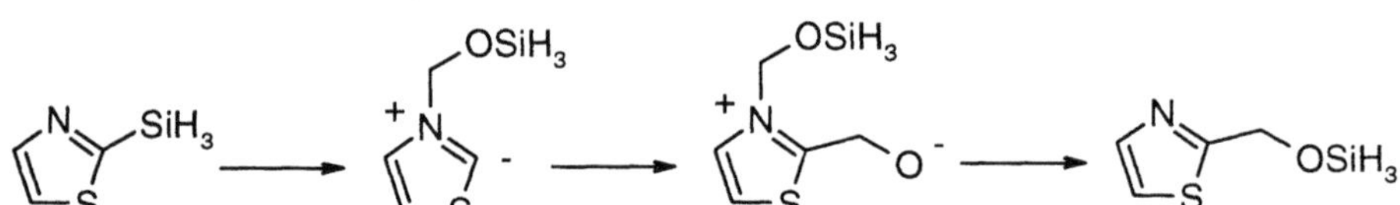

65 66 67 68

The reaction of 2-bromo-5-nitrothiazole with weakly basic secondary aliphatic amines gave the expected 2-amino products. The isomeric 5-bromo-2-nitrothiazole with such amines gave mixtures of the expected 5-amino products along with 2-aminated 5-nitrothiazole rearrangement products. A mechanism was proposed which involves the slow thermal isomerisation of the 5-bromo-2-nitrothiazole to the much more reactive 2-bromo isomer which competes, in the case of relatively weak amine nucleophiles, with direct but slow displacement of the 5-bromo group to form the normal displacement product <96JHC1191>.

Bis(aminoalkyl)bithiazoles are useful as DNA cleavage agents. Bleomycin contains a 2,4′-bithiazole moiety which plays an important role in the interaction with double stranded DNA during the cleavage reaction. The 2,2′-bis(aminomethyl)-4,4′-bithiazole (**70**) has been synthesised by the condensation of 1,4-dibromobutane-2,3-dione with Boc-glycinethioamide

(**69**). In the presence of Cu (II), the bithiazole (**70**) is a simple DNA cleaving agent, the first example of a 4,4′-bithiazole having this property <96CPB1761>.

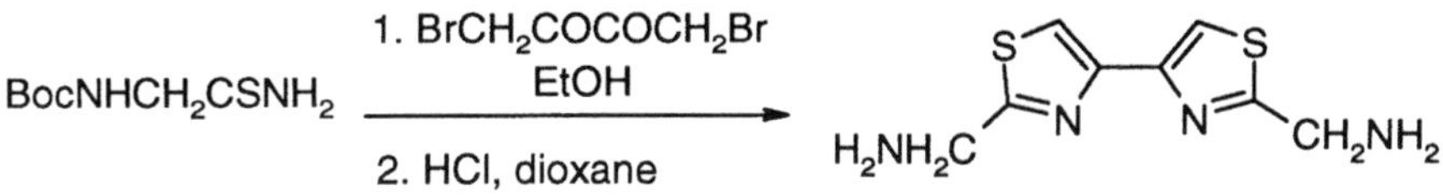

69 70

5.5.3 THIADIAZOLES

5.5.3.1 1,2,3-Thiadiazoles

A range of 4-substituted 1,3-dithiole-2-thiones (**71**) and 2,6-substituted 1,4-dithiafulvalenes (**73**) were synthesised from 4-substituted 1,2,3-thiadiazoles (**72**). Reaction of (**72**) with NaH in a mixture of CS_2 and acetonitrile led to the formation of (**71**), whereas absence of CS_2 gave fulvalenes (**73**). This route was found to be very efficient for the preparation of 4-formyl-1,3-dithiole-2-thione (**71**; R = CHO), which was previously difficult to prepare, and thus allowed the synthesis of the novel 2,6(7)-bisformyltetrathiafulvalene (**74**) <96T3171>.

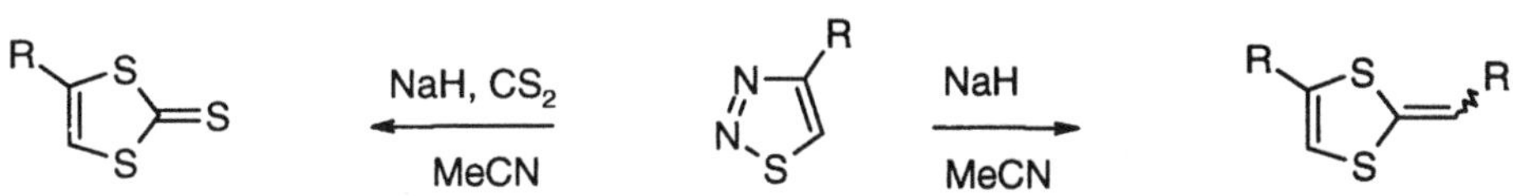

71 72 73

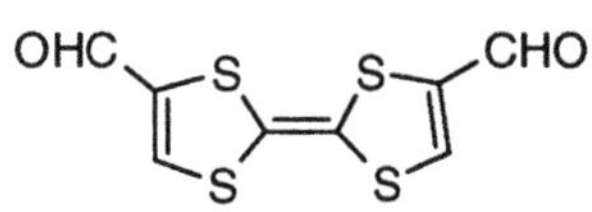

74

4-(1,4-Dithiafulvenyl) substituted 1,2,3-thiadiazoles (**75**) were readily obtained from 4-formyl-1,2,3-thiadiazole (**72**; R=CHO) by a Wittig reaction. Treatment of (**75**) with NaH in CS_2 and acetonitrile produced new fulvenes (**77**), whereas the absence of CS_2 gave extended tetrathiafulvalenes (**76**) <96T3171>.

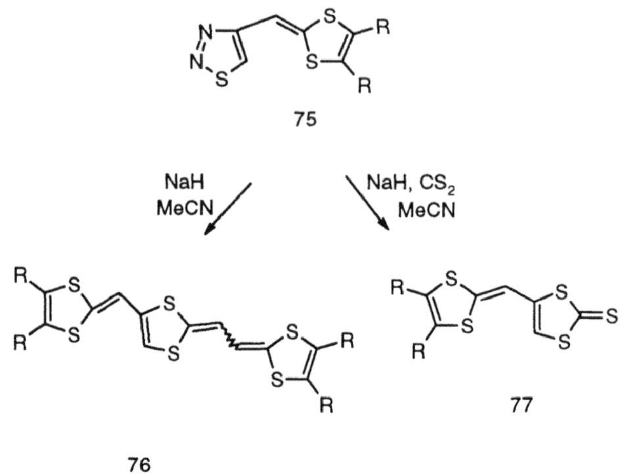

75

76

77

The literature synthesis of the racemic 2-azidonitrile (**80**) by diazotisation of the 5-amino-1,2,3-thiadiazole (**78**) could not be repeated; the 2-chloronitrile (**79**) was the only product obtained from the reaction <96TA607>.

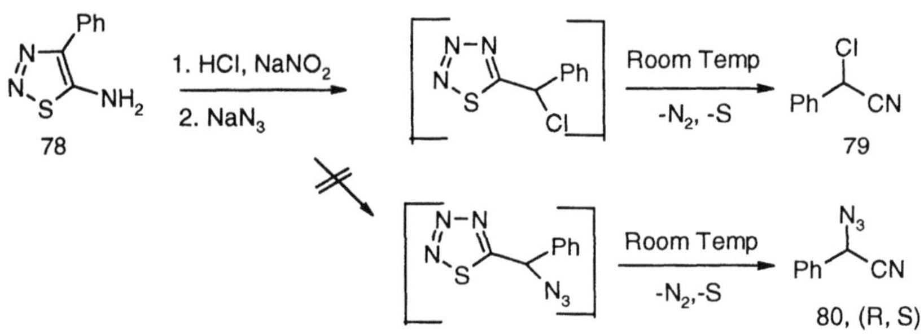

78 79 80, (R, S)

The novel 6a-λ^4-thia-1,2,4,6-tetraazapentalenes (**82**; R = CO_2Et) was prepared by treatment of the known 1,2,3-thiadiazolino salt (**81**) with *N*-methylbenzimidoyl chloride in pyridine. An alternative approach to the synthesis of tetraazapentalenes (**82**) *via* 1,2,4-thiadiazoles will be discussed in section 5.5.3.2 <96BSB335>.

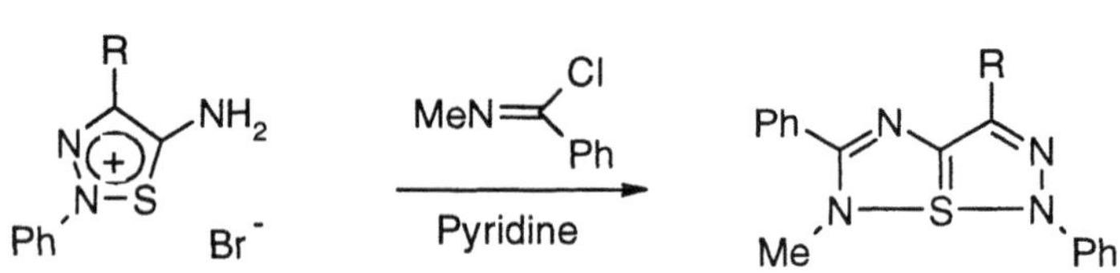

81 82

5.5.3.2 1,2,4-Thiadiazoles

Treatment of the thiadiazole (**83**) with phenylhydrazine gave the hydrazone (**84**) which on methylation with Meerwein's reagent gave the thiapentalene (**82**). Methylation occurred exclusively at N-2 due to hydrogen bonding of the hydrazono group with N-4 in (**84**). X-ray analysis of (**82**; R = CO_2Et; whose preparation was described in the preceding section) indicated that the 6a-λ^4-thia-1,2,4,6-tetraazapentalenes (**82**; R = CO_2Et and Ph) could be represented by the dual canonical forms (**85a**) and (**85b**) with a preference for structure (**85a**) <96BSB335>.

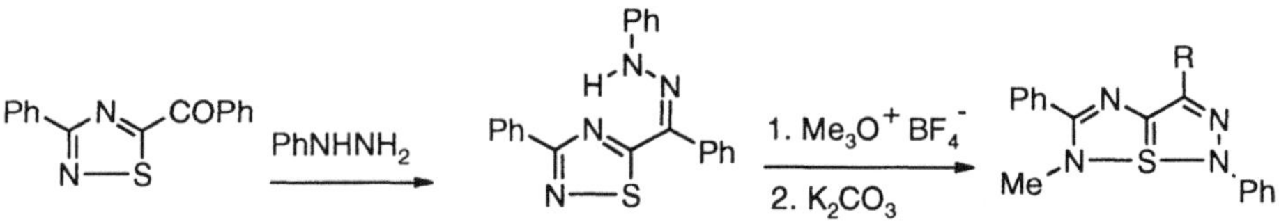

83 84 (82; R=Ph)

85a 85b

R = CO_2Et

The first *N*-oxides of the 1,2,4-thiadiazole ring system have been reported and were prepared by condensation of benzamidoximes (**86**) with 4,5-dichloro-1,2,3-dithiazolium chloride (**87**). [15]*N*-labelling showed the compounds to be 4-oxides (**88**) and a mechanism was proposed for their formation. Alkyl amidoximes and arylamidoximes with electron-withdrawing substituents did not give *N*-oxides, but only the dithiazolone (**89**) and the dithiazolthione (**90**) <96CC1273>.

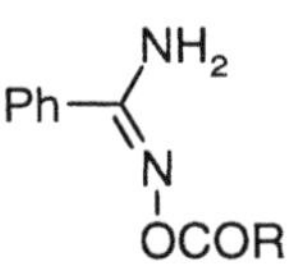

86

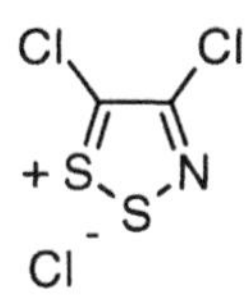

87

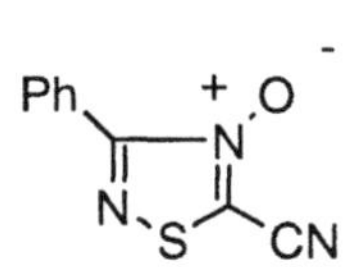

88

R= Me, Ph, NHMe and 4-ClC_6H_4NH

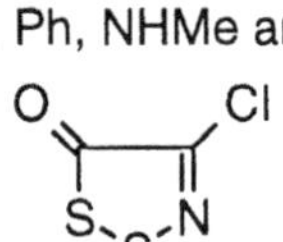

89

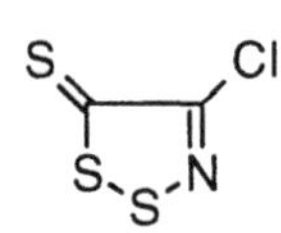

90

Reaction of equimolar amounts of the thiocarbamate (**91**) with (chlorocarbonyl)sulphenyl chloride gave 1,2,4-dithiazoline-5-one (**92**) and the 1,2,4-thiadiazole (**93**); the relative amounts of (**92**) and (**93**) being very dependent on the solvent used in the reaction. The mechanism of formation of both (**92**) and (**93**) was discussed <96JOC6639>.

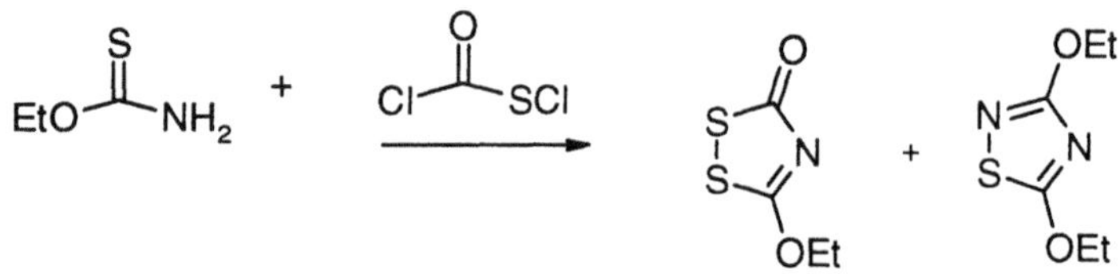

91 92 93

5.5.3.3 1,2,5-Thiadiazoles

An alternative synthesis of 3,4-disubstituted 1,2,5-thiadiazoles was developed because existing methods were unsatisfactory. This involved the coupling reaction of 4-substituted 3-halogeno and 3-trifluoromethylsulphonyl 1,2,5-thiadiazoles (**94**) with arylstannanes in the presence of palladium catalysts to give 4-substituted 3-aryl derivatives (**95**) in good yields. To avoid ring opening of the thiadiazole nucleus, couplings were performed with tributylarylstannanes in toluene at 120°C under nitrogen <96H2435>.

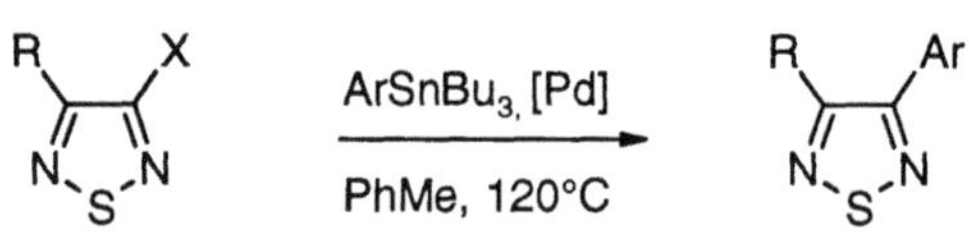

94 95

R = Ph, 2,6-$Cl_2C_6H_3$, ^tBu; X = Cl, Br, OTf; Ar = Ph, 4-ClC_6H_4, 4-MeC_6H_4, 4-$MeOC_6H_4$;

[Pd] = $Pd(PPh_3)_4$, $PdCl_2(PPh_3)_2$

Trithiazyl chloride $(NSCl)_3$ has proved to be a versatile molecule for the synthesis of the 1,2,5-thiadiazole ring system. Reaction with alkenes and alkynes gave 1,2,5-thiadiazoles (**97**) in generally good yield. Treatment of α,β-unsubstituted pyrroles (for example, 1,2,3-triphenylpyrrole) with $(NSCl)_3$ led to addition across the unsubstituted double bond, giving thiadiazoles (**96**) in good yield. When pyrroles with no substituents on carbon were used as substrates, addition across both double bonds occurred, affording tricycles (**100**). [None of the bicyclic compound (**98**) could be seen or isolated]. These pyrrole-trimer reaction pathways were discussed in detail. Finally, reaction of $(NSCl)_3$ with E,E-1,4-diphenylbuta-1,3-diene in CCl_4 gave a complex mixture, from which five sulphur-nitrogen containing heterocycles were isolated; the major component being the bis-1,2,5-thiadiazole (**99**) <96JHC1419>.

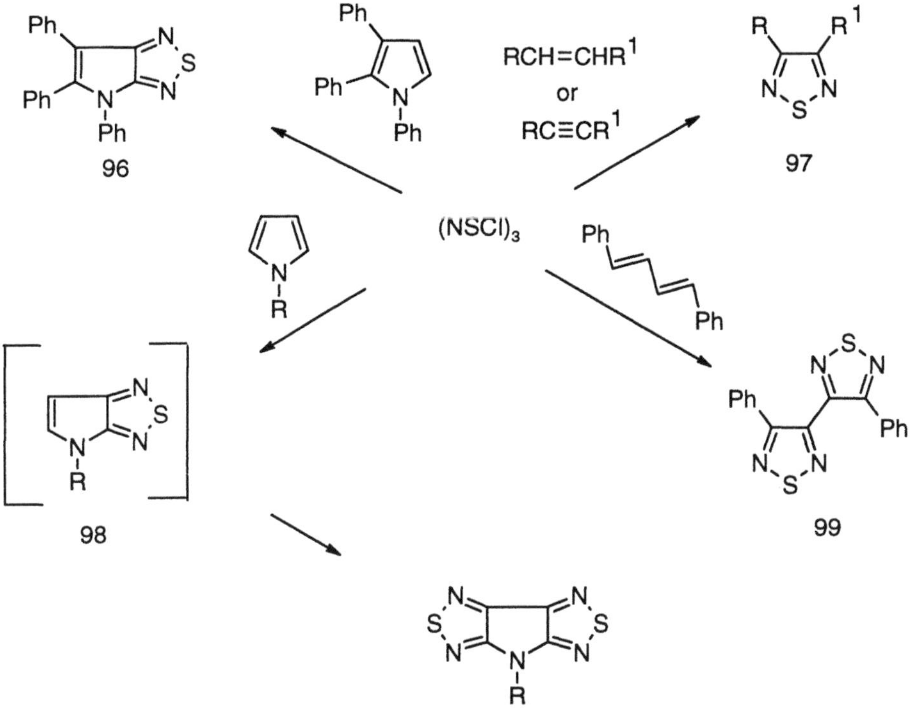

5.5.3.4 1,3,4-Thiadiazoles

The [1,3,4]thiadiazolo[3,2-*a*]pyrimidines (**102**) were prepared by condensation of 2-amino-1,3,4-thiadiazoles (**101**) with 2,4-pentadione in the presence of formic acid-phosphorus pentoxide. This method was higher yielding than the methanesulphonic acid-phosphorus pentoxide mediated synthesis. Treatment of 5-imino-6*H*-[1,3,4]-thiadiazolo[3,2-*a*]pyrimidin-7-ones (**104**) with both methanesulphonic acid-phosphorus pentoxide and with formic acid-phosphorus pentoxide furnished 7-amino-5-ones (**105**). *N*-Formylation of the 5-imino group in (**104**) did not occur with formic acid and so reaction with triethyl orthoformate was examined, and led to the unexpected formation of isocyanates (**103**) <96JHC1367>.

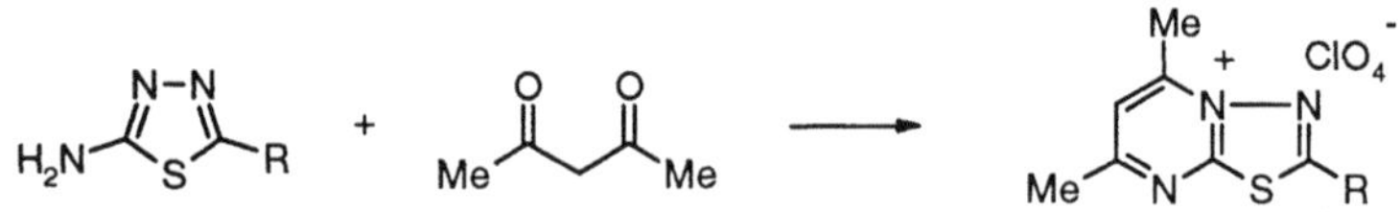

101

R= H, Me,Ph

102

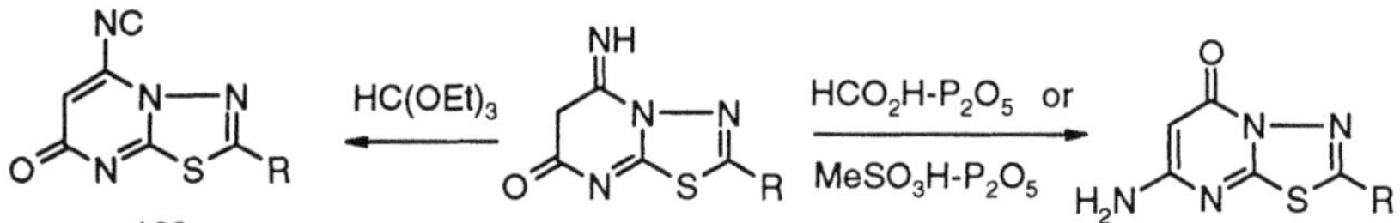

103 104 105

Reaction of pyridinium thiocyanatoacetamides (**106**) with a strong base (e.g potassium t-butoxide) in ethanol gave mesoionic *N*-[2-(1,3,4-thiadiazolo[3,2-*a*]pyridino)]acetamidates (**107**) or (**108**) whose structures were confirmed by the X-ray analysis of (**107**; R = Me). Possible mechanisms for the formation of the mesoionic derivatives were discussed <96BCJ1769>.

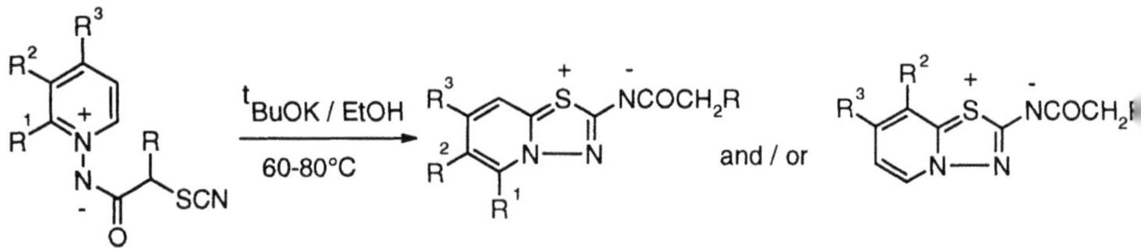

106 107 108

Equimolar amounts of aromatic aldehydes, thioglycolic acid and thionohydrazides in sulphuric acid at room temperature afforded 2-methylthio-5-aryl-5*H*-thiazolo[4,3-*b*]-1,3,4-thiadiazoles in a "one pot" procedure <96HC243>.

Reaction of the pyridine-2-thiol (**109**) with ketones and with triethyl orthoformate has led to *N,S*-acetals (**112**) and 1,3,4-thiadiazoles (**110**) which on methylation and subsequent ring opening gave two new pyridine derivatives (**113**) and (**111**) <96JPR516>.

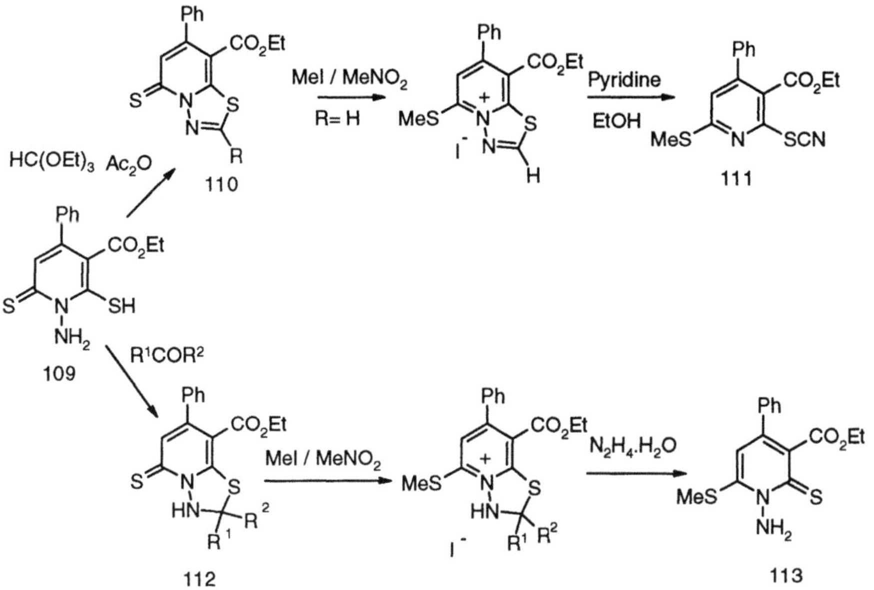

5.5.4 SELENAZOLES AND SELENADIAZOLES

The oxidation of the semicarbazone (**114**) with selenium dioxide in glacial acetic acid afforded the 1,2,3-selenadiazole (**115**) <96PS7>. This method has been used to prepare a number of 1,2,3-selenadiazoles <96PJC1143, 96LA239> and <96PS155>.

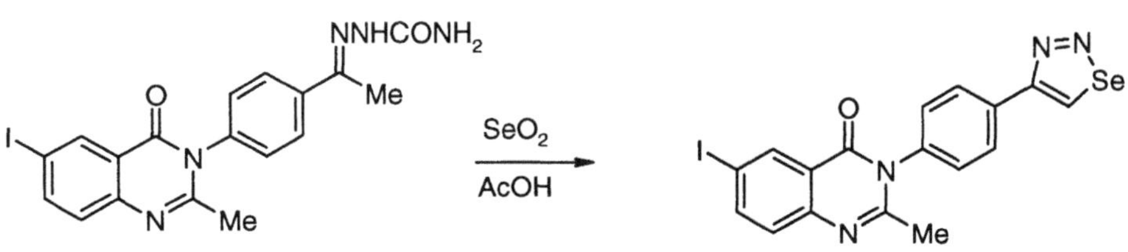

114 115

2,1,3-Benzoselenadiazole (bsd) has been used as a ligand in the ruthenium complex [RuClH-(CO)(bsd)(PPh_3)$_2$]. This complex was used to catalyse the transalkynylation and catalytic demercuration of bis(alkynyl)mercurials <96CC1059>.

Nitroarenes react with ethyl isocyanate in the presence of DBU to give pyrroles or pyridine-*N*-oxides depending on the structure of the starting nitro compounds. Two novel heterocyclic ring systems (**116**) and (**117**) containing a 1,2,5-selenadiazole ring have been synthesised starting from 4- and 5-nitro-2,1,3-benzoselenadiazole respectively. 4-Nitro-2,1,3-benzoselenadiazole was prepared by the condensing *o*-phenylenediamine with selenium dioxide followed by nitration. 5-Nitro-2,1,3-benzoselenadiazole was prepared by condensing 4-nitrophenylene-1,2-diamine with selenium dioxide <96JCS(P1)1403>.

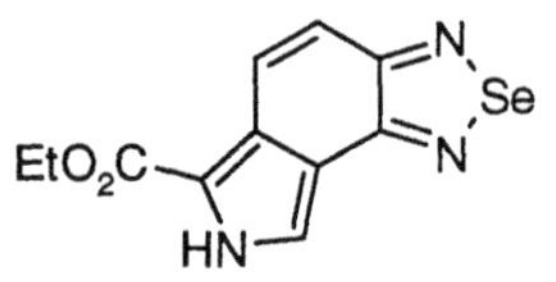

116

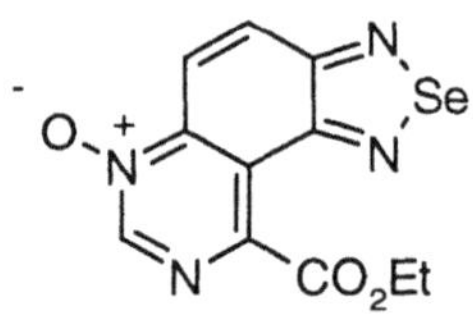

117

5.5.5 REFERENCES

95H1675 K. Ito, Y. Hara, R. Sakakibara and K. Saito, *Heterocycles*, **1995**, 41(8), 1675.

95S1397 O. A. Attanasi, L. De Cresentini, E. Foresti, R. Galarini, S. Santeusanio and F. Serra-Zanetti, *Synthesis*, **1995**, 1397.

95T11411 W. M. Abdou, El Sayed, M. A. Yakout and N. A. F. Ganoub, *Tetrahedron*, **1995**, 51(42),11411.

95TL6395 P. Wipf, C. P. Miller, S. Venkatraman and P. C. Fritch, *Tetrahedron Lett.*, **1995**, 36(36), 6395.

96BCJ1769 A. Kakehi, S. Ito and Y. Hashimoto, *Bull.Chem.Soc.Jpn.*, **1996**, 69,1769.

96BMC1409 N. Bailey, A. W. Dean, D. B. Judd, D. Middlemiss, R. Storer and S. P. Watson, *Biorg. Med. Chem. Lett.*, **1996**, 6(12), 1409.

96BSB335 G. L' Abbe, P. Vossen, W. Dehaen and L. Van. Meervelt, *Bull. Soc. Chim. Belg*, **1996**, 105(6), 335.

96CC843 R. R. Hill and S. J. Robinson, *J. Chem. Soc., Chem. Commun.*, **1996**, 843.

96CC1059 R. B. Bedford, A. F. Hill, A. R. Thompsett, A. J. P. White and D. J. Williams, *J. Chem. Soc., Chem. Commun.*, **1996**, 1059.

96CC1273 O. A. Rakitin, C. W. Rees and O. G. Vlasova, *J. Chem. Soc., Chem. Commun.*, **1996**, 1273

96CPB1761 H. Sasaki, S. Takanori, K. Yamamoto and Y. Nakamoto, *Chem. Pharm. Bull.*, **1996**, 44, 1761.

96H2435 Y. Hanasaki, *Heterocycles*, **1996**,43(11), 2435.

96HC243 S. Sh. Shukurov, M. A. Kukaniev, A. M. Alibaeva and B. M. Bobogaribov, *Chem. Heterocycl. Cpds.*, **1996**, 32(2),243.

96JCS(P1)1403 T. Murashima, K. Fujita, K. Ono, T. Ogawa, H. Uno and N. Ono, *J. Chem. Soc. Perkin Trans1*, **1996**, 1403.

96JCS(P1)2239 A. Rolfs, P. G Jones and J. Liebscher, *J.Chem.Soc.,Perkin Trans 1*, **1996**, 2239

96JHC575 H. Kamproudi, S. Spyroudis and P. Tarantili, *J. Heterocycl. Chem.*, **1996**, 33, 575.

96JHC731 E. Coutouli-Argyropoulou and C. Anastasopoulos, *J.Heterocycl. Chem.*, **1996**, 33,731.

96JHC1191 H. H. Lee, B. D. Palmer, M. Boyd and W. A. Denny, *J. Heterocycl. Chem.*, **1996**, 33, 1191.

96JHC1179 N. De Kimpe, W. De Cock, M. Keppens, D. De Smaele and A. Meszaros, *J. Heterocycl. Chem.*, **1996**, 33, 1179.

96JHC1201 N. S.Cho, C. S. Ra, D. Y. Ra, J. S. Song and S. K.Kang, *J.Heterocycl. Chem.*, **1996**, 33,1201.

96JHC1367 K. Takenaka and T. Tsuji, *J.Heterocycl.Chem.*, **1996**, 33,1367.

96JHC1419 X. G. Duan, X. L. Duan, C. W. Rees and T. Y. Yue, *J.Heterocycl. Chem.*, **1996**, 33,1419.

96JOC1922 Y-D. Wu, J. K. Lee, K. N. Houk and A. Dondoni, *J. Org. Chem.*, **1996**, *61*, 1922.

96JOC3738 M. Avalos, R. Babiano, A. Cabanillas, P. Cintas, F. J. Higes, J. L. Jimenez and J. C. Palacios, *J. Org. Chem.*, **1996**, *61*, 3738.

96JOC6639 L. Chen, T. R Thompson, R. P. Hammer and G. Barany, *J.Org.Chem.*, **1996**,61,6639.

96JPR51 J. E. Israel, R. Flaig and H. Hartmann, *J. Prakt. Chem.*, **1996**, 338, 51.

96JPR424 B. Schulze, B. Friedrich, S. Wagner and P. Fuhrmann, *J.Prakt.Chem.*, **1996**, 338,424.

96JPR516 M. Rehwald, H. Schafer, K. Gewald and M.Gruner, *J.Prakt.Chem.*, **1996,** 338,516.

96LA239 D. Prim, D. Joseph and G. Kirsch, *Liebigs Ann.* Chem., **1996**, 239.

96MI169 S. A. El-Abbady, S. M. Agami and W. A. M. Mokbel, *Heterocyclic. Commun.*,**1996**, 2,169.

96MI227 R. Issac, J. Tierney, L. M. Mascavage, A. Findeisen and J. Kilburn, *Heterocyclic Commun.*, **1996**, 2, 227.

96MI638 K. Hartke and C. Ashry, *Pharmazie*, **1996**, 51(9),638.

96PJC1143 T. I. El-Emary, *Polish J. Chem.*, **1996**, 70, 1143.

96PS7 M. M. Ghorab, S. G. Abdel-Hamide and M. M. Abou Zeid, *Phosphorus, Sulfur and Silicon*, **1996**, 112, 7.

96PS155 A. H. Mandour, T. H. El-Shihl, A. Abdel-Latif Nehad and Z. E. El-Bazza, *Phosphorus, Sulfur and Silicon*, **1996**, 113, 155.

96SC1903 H. Groger and J. Martens, *Synth. Commun.*, **1996**, 26(10), 1903.

96TA607 F. Effenberger, A. Kremser and U. Stelzer, *Tetrahedron Asymmetry, 1996*, 7(2),607.

96TL711 C. W. G. Fishwick and R. G. Foster, *Tetrahedron Lett.*, **1996**, 37(5), 711.

96TL983 D. R. Williams, D. A. Brooks, J. L. Moore and A. O. Stewart, *Tetrahedron Lett.*, **1996**, 37(7), 983.

96TL2935 C. A. Busacca, Y. Dong and E. M. Spinelli, *Tetrahedron Lett.*, **1996**, 37(17), 2935.

96TL4189 H. H. Tso and M. Chandrasekharan, *Tetrahedron Lett.*, **1996**, 37(24),4189.

96TL4857 W. D. Schmitz and D. Tomo, *Tetrahedron Lett.*, **1996**, 37(28), 4857.

96T733 M. Sakamoto, M. Nagano, Y. Suzuki, K. Satoh and O. Tamura, *Tetrahedron*, **1996**, 52(3), 733.

96T783 B. Schulze, S. Kirrbach, K. Illgen and P. Fuhrmann, *Tetrahedron*, **1996**, 52(3),783.

96T3171 R. P. Clausen and J. Becher, *Tetrahedron*, **1996**, 52(9),3171.

96T3189 M. Makosza, M. Sypniewski and T. Glinka, *Tetrahedron*, **1996**, 52(9), 3189.

96T3339 R. A. Abramovitch, I. Shinkai, B. J. Mavunkel, K. M. More, S. O'Connor, G. H. Ooi, W. T. Pennington, P. C. Srinivasan and J. S. Stowers, *Tetrahedron*, **1996**, 52(9),3339.

96T7183 F. Clerici, F. Galletti and D. Pocar, *Tetrahedron*, **1996**, 52(20),7183.

Chapter 5.6

Five-Membered Ring Systems: With O & S (Se, Te) Atoms

R. Alan Aitken and Lawrence Hill
University of St. Andrews, UK

5.6.1 1,3-DIOXOLES AND DIOXOLANES

New catalysts for the reaction of carbonyl compounds with 1,2-diols to form 1,3-dioxolanes include scandium triflate <96SL839> and *N*-benzoylhydrazinium salts <96MI362> and metal-catalysed reaction of thioketones with 1,2-diols also gives 1,3-dioxolanes <96H(43)851>. The anions of nitroalkanes add to *o*-quinones to give 1,3-dioxoles <96JCS(P2)1429>, and carbonyl ylides, generated either from α-iodosilyl ethers and SmI_2 <96JA3533> or from α-chloromethoxysilanes and CsF <96SL234>, add to carbonyl compounds to afford 1,3-dioxolanes. The carboxylation of propylene oxide to give 4-methyl-1,3-dioxolan-2-one may be efficiently catalysed by an aluminium phthalocyanin <96BCJ2885>, the I^- form of an ion exchange resin <96MI513>, or a polyethylene glycol/potassium iodide complex <96MI701>. Titanium catalysts are effective for the reaction of 1,3-dioxolan-2-one with methanol to give dimethyl carbonate <96CC2281> and reaction of chiral styrene epoxide with acetone to give the 2,2-dimethyl-1,3-dioxolane with clean inversion of configuration <95JAP07247280>, while phase-transfer catalysts allow the carboxylation of phenyl glycidyl ether to give 4-phenoxymethyl-1,3-dioxolan-2-one to proceed with 1 atm. of CO_2 <96MI26>.

A detailed spectroscopic and theoretical study of the conformation of dioxolanes **1** has appeared <96T8275>, and a theoretical study has shown that the anomeric effect explains the non-planarity of 1,3-dioxole <96JA9850>. The tetraalkynyldioxolanone **2** has been prepared and its structure and reactivity studied <96HCA634>. Both enantiomers of the chiral glycolic acid equivalent **3** can be prepared from D-mannitol <96HCA1696>, and lipase-mediated kinetic

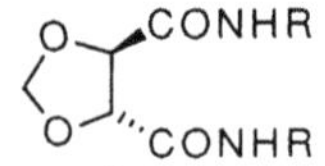

1

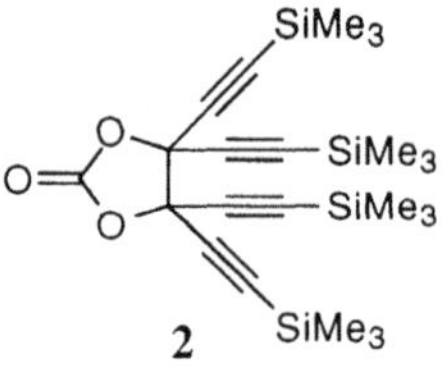

2

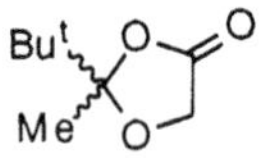

3

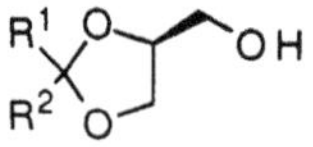

4

5

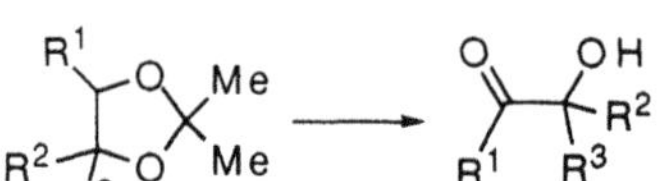

6 **7**

resolution has been used to obtain chiral 4-hydroxymethyldioxolanes **4** <96TA3037>. Regioselective ring opening of the bis-dioxolane **5** provides access to synthetically useful chiral 1,2-diols <96SL53>.

Cleavage of 2-substituted-1,3-dioxolanes to give carbonyl compounds may be achieved using Ph_3P/CBr_4 <96CC341> or $CpTiCl_3$ <96MI184>. Oxidative ring-opening of dioxolanes **6** to give the α-hydroxyketones **7** can be achieved using dimethyldioxirane <96TL115>, and the photosensitised cleavage of **8** has been examined <96T4911>. Selective cleavage of **9** to give **10** is possible using $BH_3{\bullet}Me_2S$ followed by $BF_3{\bullet}Et_2O$ <96SL231>. Dioxolane **11** acts as a chemically stable equivalent of chiral $PhCH_2CH(OH)CHO$ <96SC3453>, and the mandelic acid-derived dioxolanone **12** undergoes highly selective ring-opening by Grignard reagents <96TL1421>. Spiro dioxolane–sulfoxides such as **13** have been used to achieve desymmetrisation of cyclic *cis*-1,2-diols <96TA29>. Mercury photosensitised oxidative dimerisation of 1,3-dioxolane has been reported <96TL6853> and photolysis of 4-chloromethyl-2-dimethylamino-1,3-dioxolane in hex-1-ene gives both the 2-hexyl product and the 2,2'-bis(dioxolane) <95MI142>. 2,2-Diisopropyl-1,3-dioxole undergoes a stereoselective Paterno-Büchi reaction with methyl trimethylpyruvate to give mainly **14** <96TL1195>, and palladium catalysed reaction of 1,3-dioxoles with ArBr and $Bu^n{}_3SnPh$ affords **15** <95SL1225>. The absolute stereochemistry of the various isomers of **16** obtained from diphenylnitrone cycloaddition to 2-methylene-4-phenyl-1,3-dioxolane has been determined <96MRC(34)52> and their reactivity examined <96JCS(P1)259>.

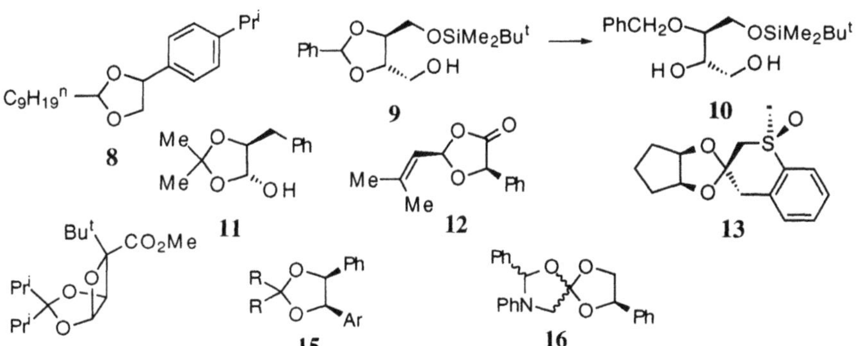

8 **9** **10** **11** **12** **13** **14** **15** **16**

The stereoselectivity of conjugate addition and cyclopropanation of the chiral nitrovinyldioxolanes **17** can be effectively controlled <96TL6307>, and good selectivity is observed in the ultrasound-promoted cycloaddition of nitrile oxides to alkenyldioxolanes **18** <95MI877, 95JOC7701>. Asymmetric Simmons-Smith cyclopropanation of **19** proceeds with

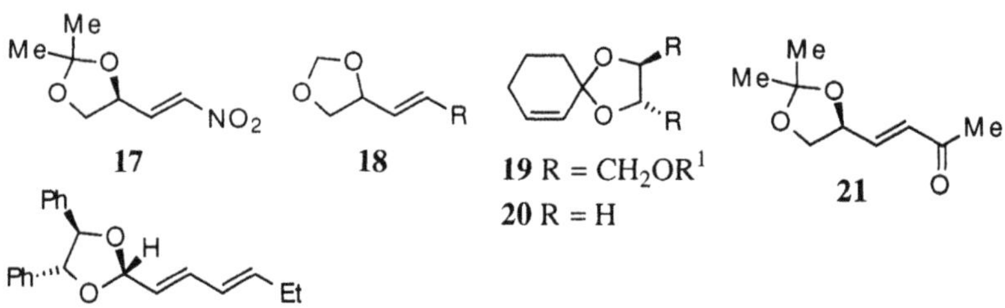

17 **18** **19** R = CH_2OR^1 **20** R = H **21** **22**

high selectivity <96JOC3906> and copper-catalysed cyclopropanation of **20** with dimethyl diazomalonate proceeds cleanly, in contrast to the same reaction of **18**, where competing side-reactions are observed <95HCA2036>. The regio- and stereoselectivity of organometallic addition to **21** has been examined <95T12843> and, in the presence of $BF_3{\bullet}Et_2O$, PhCu undergoes conjugate addition to both the β- and δ-positions of **22** with opening of the dioxolane ring <96OM1957>.

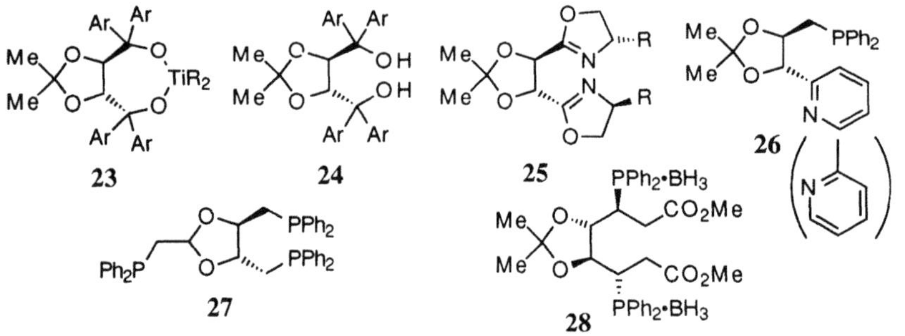

23 **24** **25** **26** **27** **28**

Various titanium 'TADDOL' compounds **23** have been used as effective catalysts for asymmetric cyclopropanation of allylic alcohols <95JA11367>, iodocyclisation <95TL9333> and Diels-Alder reactions <96LA63>, and polymer- and dendrimer-bound analogues have also been used as chiral catalysts for a variety of reactions <96HCA1710>. The diols **24** have also been used to mediate other asymmetric reactions such as the addition of organocerium reagents to aldehydes <96TL2675> and the zirconium-mediated Meerwein-Ponndorf-Verley reduction <96RTC140>. The diols **24** and related dioxolanediols have been used for resolution of a wide variety of chiral compounds by means of crystalline host-guest complexes <95CL809, 95JAP07242653, 95JAP07285955, 96JAP0803076, 96JAP0859539>. The dioxolane-based bis-oxazolines **25** are effective ligands for copper catalysed asymmetric cyclopropanation and aziridination of double bonds <96SL677, 96TL4073> and dioxolane-based phosphines such as **26** <96TA885>, **27** <96SL267>, and **28** <96TL4713> have been assessed as chiral ligands for transition metal catalysed asymmetric reactions.

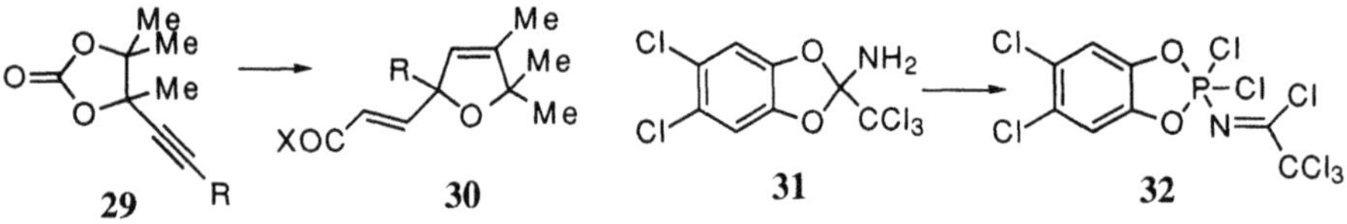

29 **30** **31** **32**

Palladium catalysed reaction of **29** with acrylic acid derivatives $CH_2{=}CHCOX$ proceeds with decarboxylation to give the dihydrofurans **30** <96CC919> and $ZnCl_2$ mediated reaction of 1,3-dioxolanes with ketenes results in insertion into the 1,2-bond to afford 1,4-dioxepin-5-ones <96AG(E)1970>. A mechanism has been suggested for the unexpected reaction of **31** with PCl_5 to give **32** <95ZOB1054>.

The dioxolanones **33** <95JAP07291959> and **34** <95JAP07285960> are both reported to be useful as solvents, while **35** has been used as an X-ray contrast medium <96MIP19487>. Carboxydioxolanes such as **36** are useful for controlled release of volatile aldehyde pheromones <96JCR(S)274> and pharmaceutically active amines R_2NH can be administered in

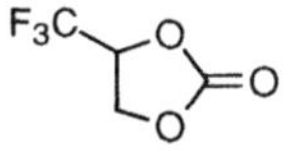

33

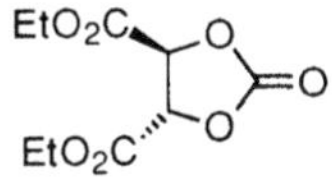

34

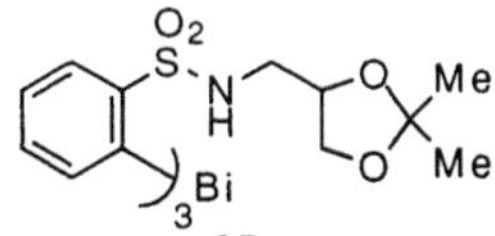

35

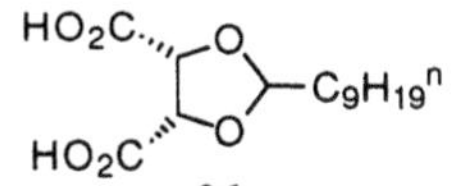

36

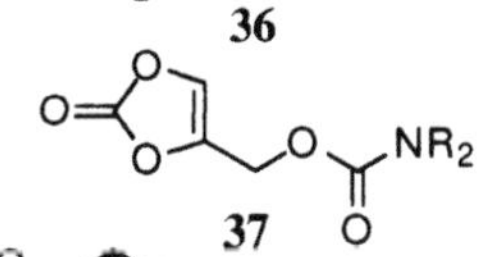

37

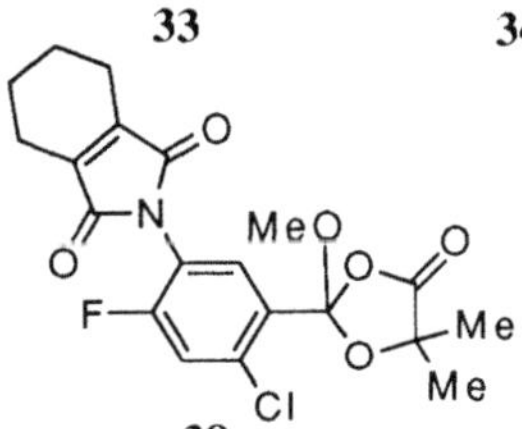

38

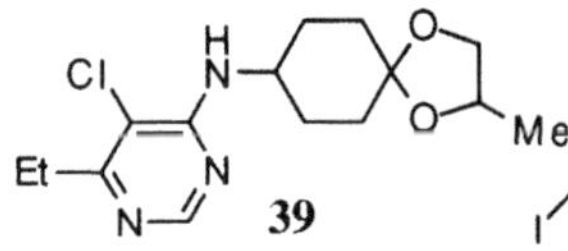

39

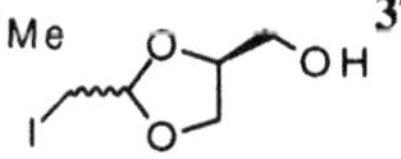

40

the form of pro-drugs **37** <95USP5466811>. Agrochemical applications include herbicides such as **38** <96GEP19538472> and pesticides such as **39** <96GEP4436509>, and a simple synthesis of the mucolytic drug domiodol **40** from D-mannitol has been described <95MI623>.

5.6.2 1,3-DITHIOLES AND DITHIOLANES

A reliable procedure for the preparation of **41**, an important intermediate for TTF synthesis has appeared <96OS270> and **42** has been prepared for the first time <96IZV775>. Thiocarbonyl ylides generated from 1,3,4-thiadiazolines undergo cycloaddition to the C=S bond of sulfines to give dithiolane *S*-oxides such as **43** formed from diphenylsulfine <95PJC1649, 96HCA31>. A range of silicon and tin-substituted 1,3-benzodithioles have been prepared and their oxidation potentials measured <96CL171>. A mechanism has been proposed for the unexpected formation of **44** from air oxidation of allyl dithiobenzoate <95SUL67> and a range of cinnamyl dithiocarboxylates, $RC(S)SCH_2CH{=}CHPh$, react with tetracyanoethylene to give **45** <95IZV1804>. Propargylic dithiocarbonates, $RC{\equiv}CCH_2SC(S)OMe$, exist in equilibrium with the zwitterionic dithioles **46** which may be trapped by conjugate addition <96CC743>. Some evidence has been obtained for the formation of dithiolanes **47** in the reaction of $Bu^n{}_3P{\bullet}CS_2$ with maleic anhydride and *N*-phenylmaleimide <95ZOB1101> and the synthesis of the phosphonoketene dithioacetals **48** has been described <96JOC8132>. X-ray structures have been reported for **49** <96MI473> and **50** <96MI103>.

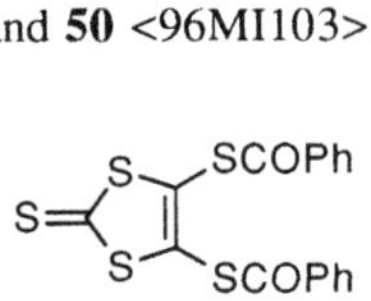

41

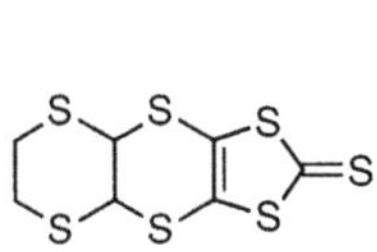

42

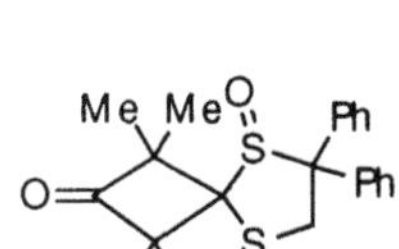

43

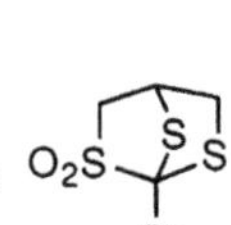

44

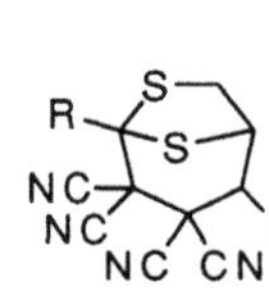

45

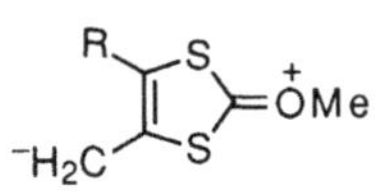

46

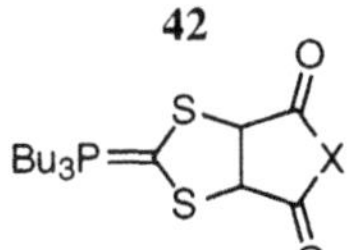

47 X = O, NPh

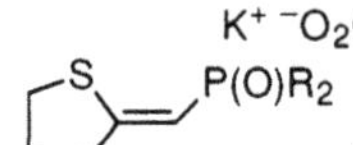

48

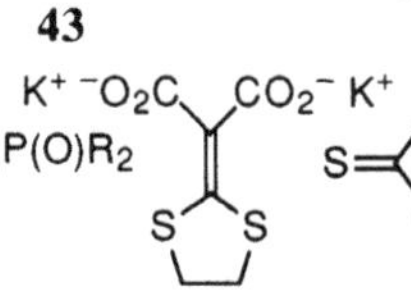

49

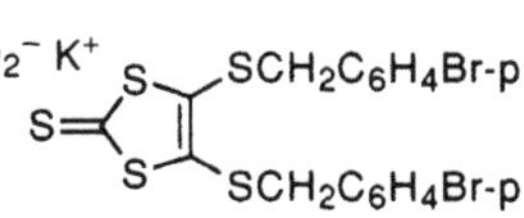

50

2,2-Disubstituted-1,3-dithiolanes can be converted into the corresponding carbonyl compounds using oxone and wet alumina <96SL767> and to the geminal difluorides using fluorine and iodine in acetonitrile <96MIP03357>. Nickel catalysed ring opening of dithiolanes with Grignard reagents to give alkenes has been reviewed <96PAC(68)105,96SL201> and the reaction can also be performed with geminal bimetallic Mg/Zn reagents <96OM3099>. Generation of a radical at the 2-position of 1,3-dithiolanes in the presence of $(Me_3Si)_3SiH$ results in ring opening <96SL237> and such radicals have also been used in cyclisation reactions <96T9713>. Reaction of 1,3-dithiolane-2-thiones with epoxides in the presence of HBF_4 gives the dithiolan-2-ones and thiiranes by way of a spiro dithiolane-oxathiolane intermediate <96JCS(P1)289> and the reaction of 1,3-dithiole-2-thione with chlorosulfonyl isocyanate has been examined <96MI169>. Treatment of 1,3-benzodithiole-2-thione with benzyne followed by HCl gives the stable salt **51** whose reactivity has been examined and **52** is similarly formed from the 2-selenone <96CC205,96BCJ2349>. The first 1,3-thiaselenolo [5,4-*c*]quinolines **53** have been reported <96PJC54>. Stabilising O–X interactions are observed in a series of nitrosomethylenedithioles **54** <96JCS(P2)2367> and their diselenole analogues **55** <96JOC2877> due to the charged form shown. The dithiolium-4-olate **56** undergoes $4\pi + 6\pi$ cycloaddition with a variety of fulvenes <96CC1011>. A variety of metal complexed dithiole systems have been prepared including **57** <95ICA(238)57>, **58** <95ICA(239)117>, **59** and **60** <95MI1581> and **61** <96OM1966>.

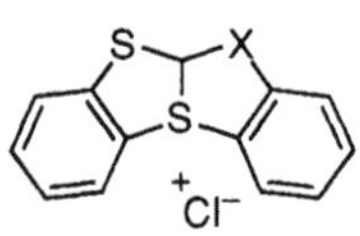
51 X = S
52 X = Se

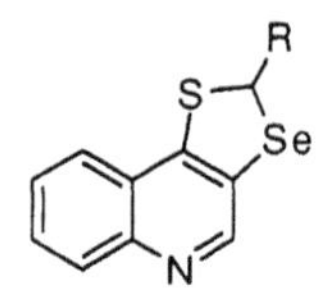
53

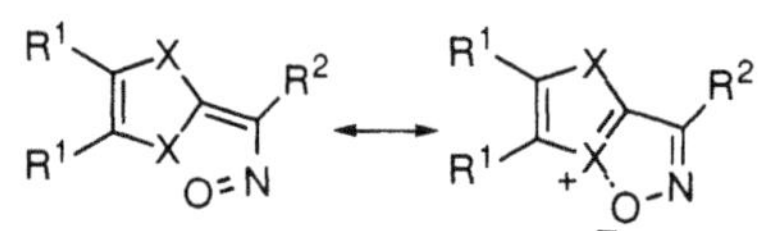
54 X = S
55 X = Se

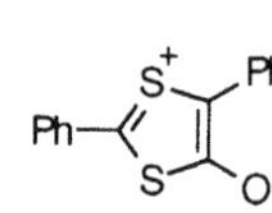
56

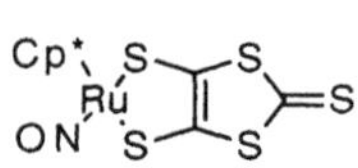
57

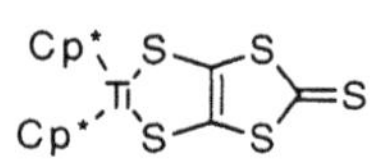
58

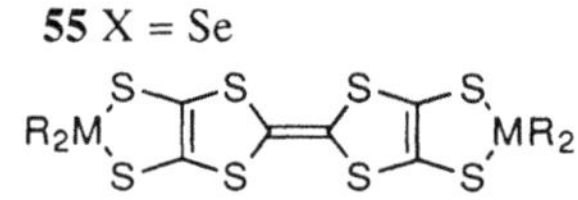
59 M = Ti, R = Cp–Pr^i
60 M = Ni, $R_2 = Ph_2P(CH_2)_3PPh_2$

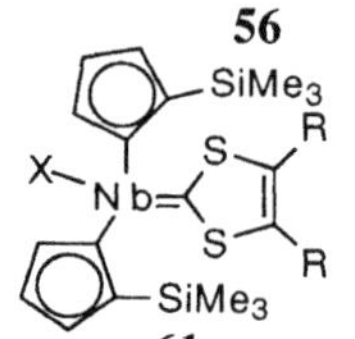
61

The high level of activity in the synthesis of tetrathiafulvalenes (TTFs) and related systems has continued and several major reviews of the area have appeared <95AHC(62)249, 95MI1481, 96SR(18)1>. A mechanistic study on the P(III) mediated coupling of 1,3-dithiole-

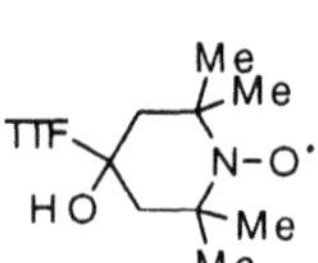
62

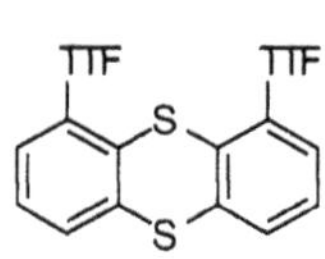
63

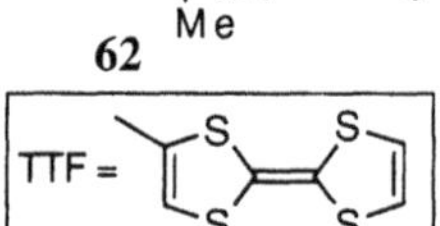

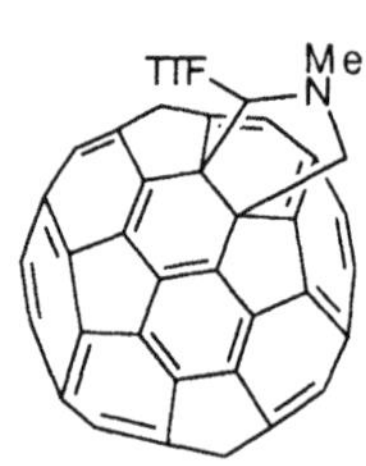
64

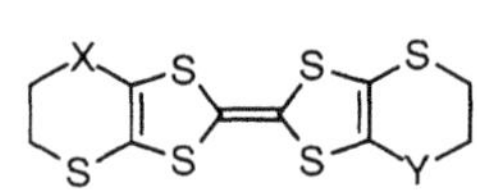
65 X = Y = S
66 X = S, Y = O
67 X = Y = O

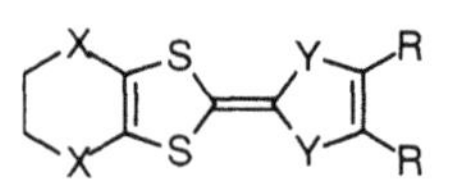
68 X = Se, Y = S
69 X = S, Y = Se

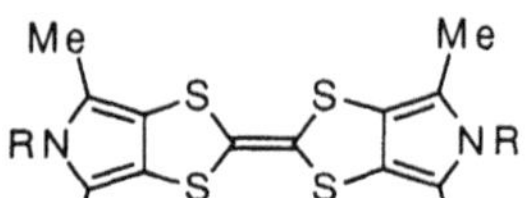
70

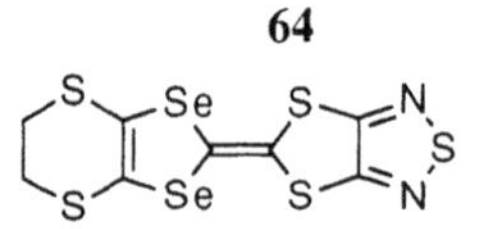
71

2-thiones to give TTFs has appeared <96JPR523>, an efficient synthesis of tetratellurafulvalene has been described <96JOC7006> and further applications of TTF-mediated radical cyclisation have been described <96CC737, 96CC739>. New monosubstituted TTF derivatives include the radical **62** <95MI153>, **63** <96T4745>, the C_{60}-bound compound **64** which forms semiconducting charge transfer complexes <96TL5979> and a phthalocyanin with eight TTF units attached <96AM63>. Langmuir-Blodgett films of tetra(methylthio)-TTF with $C_{60}Br_6$ and $C_{60}Br_{24}$ have been reported <96MI571>, the preparation and X-ray structure of tetra-2-thienyl-TTF have been described <96CC2423> and TTFs have been obtained by coupling of **50** and its dichloro and diiodo analogues with loss of sulfur <96MI597>.

An improved method for synthesis of BEDT-TTF **65** has been described <95PS(106)145> and new results on its mono- and di-oxygen analogues, **66** and **67**, have appeared <96AM807, 96S198>. New selenium-containing analogues of EDT-TTF include **68** <96SM(78)89, 96T11063> and **69** <96CC1955>. The reaction of pyrimidine-fused TTFs with iodine has been examined <96KGS123> and **70** <96JOC8117> and **71** <96CC521> have also been prepared. Conducting radical cation salts of **72** <95CL1069> and **73** <96CC2517, 96JOC3987> with AuI_2^- as counterion have been reported and compounds of structure **74** have been prepared <96JOC3650>. Quinone-containing TTF analogues such as **75** have been prepared <96TL2503, 96ZN(B)901> and the structure and properties of the zwitterionic compound **76** have been described <96CEJ1275>.

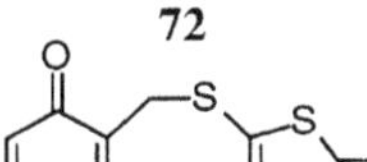

72

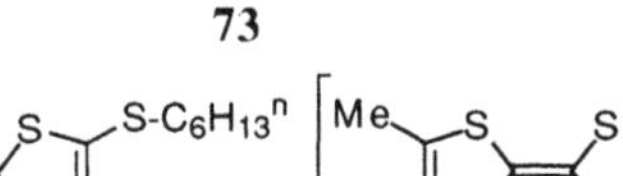

73

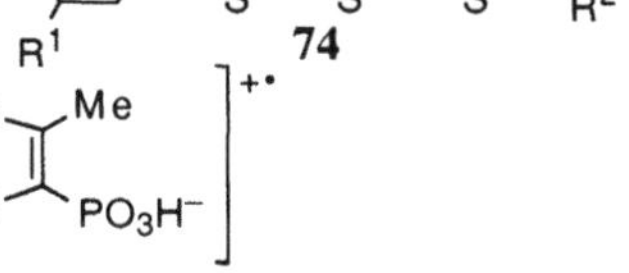

74

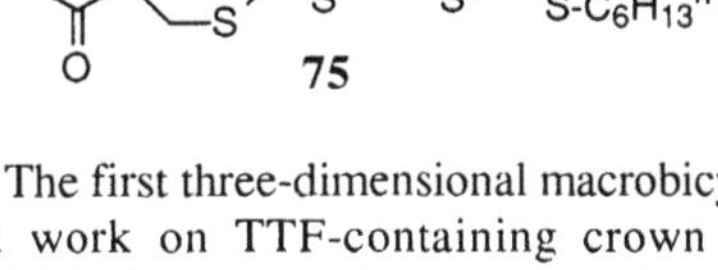

75

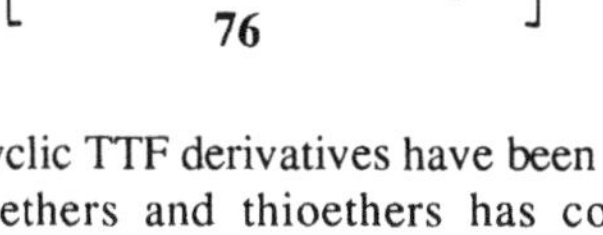

76

The first three-dimensional macrobicyclic TTF derivatives have been obtained <96CC615> and work on TTF-containing crown ethers and thioethers has continued <96LA551, 96JCS(P1)1995> with structures of this type being used to obtain the first TTF-containing

77

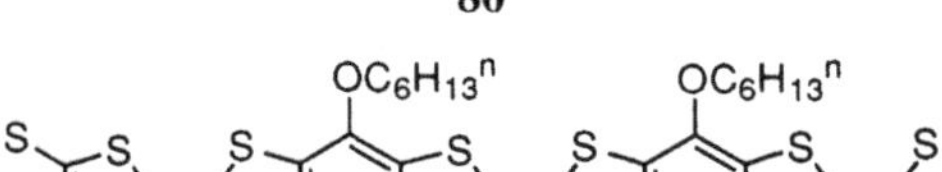

80

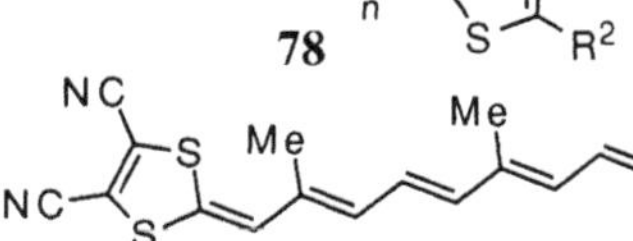

78

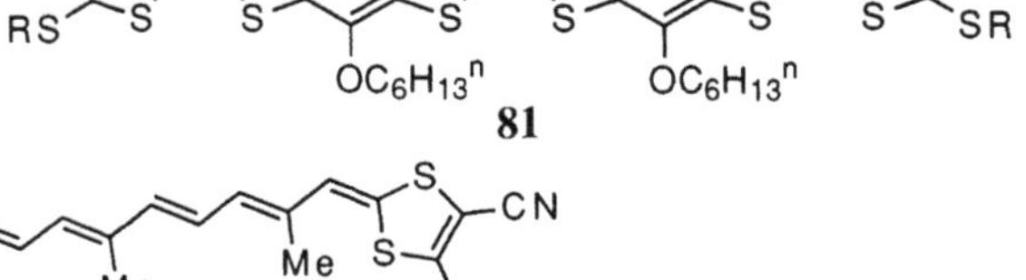

81

79

catenanes and rotaxanes <96CC639, 96CEJ624>. The preparation and electrochemistry of mercury and nickel tetrathio-TTF complexes has been reported <96CC1363, 96JCS(D)823>.

Work on TTF homologues has continued with compounds such as **77** <96CC363> and a variety of structures **78** <96CM1182, 96BSF301> being prepared. Carotenoid TTF analogues such as **79** which may be considered as molecular wires have been obtained <96HCA1497>. Other extended analogues with aromatic and heteroaromatic spacer groups have also been reported <95MRCS94, 96AM804, 96CM2291, 96CC2021>. The tris-fused TTFs **80** have been examined <96CL43> and benzobis(TTF)s <95JA9995> and higher analogues such as **81** <95MI1059> have also been described.

5.6.3 1,3-OXATHIOLES AND OXATHIOLANES

A fluorinated ion-exchange resin containing sulfonic acid groups is effective as a catalyst for reaction of carbonyl compounds with 2-hydroxythiols to give 1,3-oxathiolanes <96JAP07247283>. A mechanism has been proposed for the formation of **82** from the electrochemical reduction of $Ph_2C(Br)COBr$ in the presence of H_2S <96T1259>. Thiocarbonyl ylides, generated from the corresponding 1,3,4-thiadiazolines, undergo cycloaddition to pyruvates to afford products such as **83** <96HCA1537>. Treatment of hindered thiones such as adamantane-2-thione or 2,2,4,4-tetramethyl-3-thioxocyclobutanone with dimethyl fumarate and phenyl azide results in a complex sequence of reactions to give oxathiolanes such as **84** <96PJC595, 96HCA1305>. Reaction of 2-mercaptoquinoline with $Me_2C(OH){-}C{\equiv}C{-}CN$ takes an unusual course to afford **85** <95KGS1694>. The biological activity of the griseofulvin analogues **86** has been examined <96AP361> and antiviral activity has been claimed for **87** <95MIP29176>.

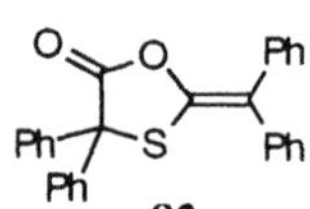

82

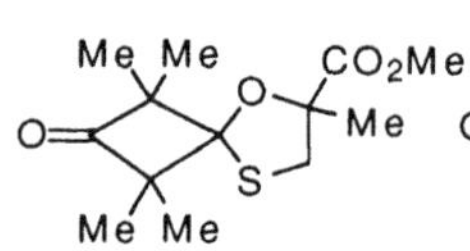

83

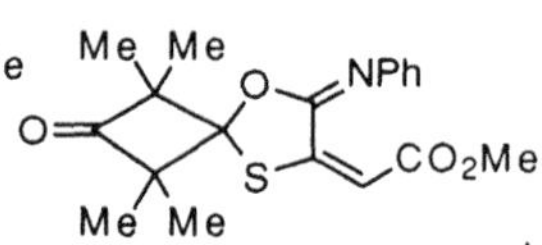

84

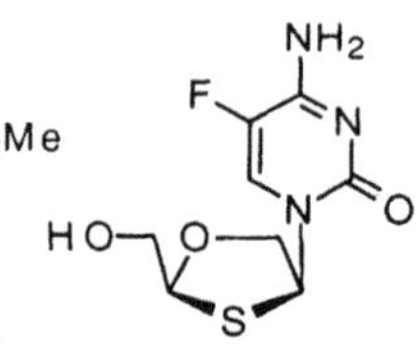

87

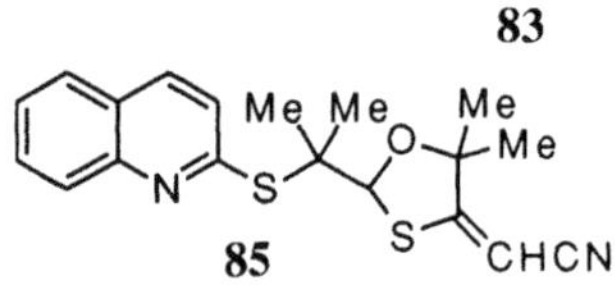

85

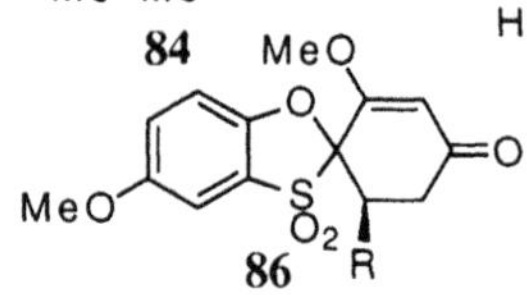

86

5.6.4 1,2-DIOXOLANES

The ring-opening of bicyclic 1,2-dioxolanes with vinyllithium and vinyl magnesium bromide has been examined <96TL6635>, photolysis of 3,3-bis(*p*-methoxyphenyl)-1,2-dioxolanes results in ring-opening to give β-hydroxy- and β-aryloxyketones <96CC2407> and a range of 3-hydroperoxy-1,2-dioxolanes are reported to be good oxidants for the formation of sulfoxides, *N*-oxides and epoxides <96JHC1399>.

5.6.5 1,2-DITHIOLES AND DITHIOLANES

A review covering all aspects of the chemistry of 1,2-dithiol-3-ones and -3-thiones has appeared <95SR(16)173> and new methods for the synthesis of the latter include reaction of dialkyl malonates with P_2S_5 and sulfur <96TL2137> and treatment of 3-aminoacrylonitriles with sulfur and either P_2S_5 or Lawesson's reagent <96SUL235>. Reaction of **88** with Na_2S in air affords the 1,2-dithiole **89** <95LA2011> and the dark blue "pseudoazulene" **91** is formed by treatment of **90** with S_2Cl_2 <96CC427>. The Pummerer reaction of **92** with triflic anhydride gives the triflate of the dication **93** <96CC311>. A review of *peri*-dichalcogenoarenes has appeared <96YGK752>.

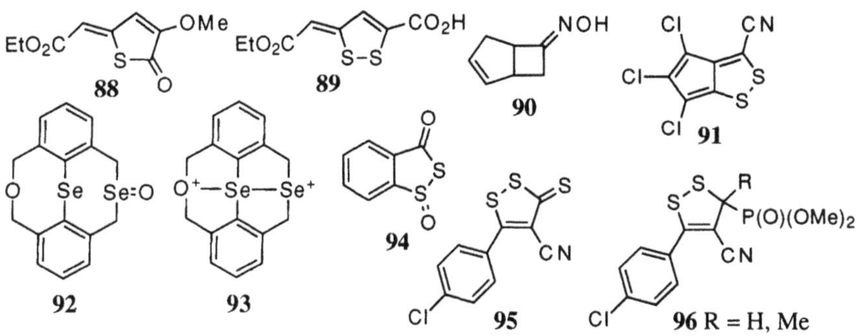

88 **89** **90** **91** **92** **93** **94** **95** **96** R = H, Me

The reaction of 1,2-dithiolanes with 2- and 4-picolyllithium has been examined <96PS(112)101> and the reactions of thioanhydrides such as **94** with both thiols <95JOC3964> and amines <96TL5337> have been reported. Treatment of 1,2-dithiolium salts with lithium or thallium cyclopentadienide results in formation of a variety of bi-, tri- and tetracyclic products <96LA109>. Reaction of **95** with trimethyl phosphite gives some of the desired coupling product but also the phosphonates **96** <96PS(109)557>.

5.6.6 1,2-OXATHIOLES AND OXATHIOLANES

A review on the synthesis and properties of sultines includes material on 1,2-oxathiolane 2-oxides <96UK156>. The reactivity of *o*-sulfobenzoic anhydride (2,1-benzoxathiol-3-one 1,1-dioxide) towards anilines has been examined <95ZOR548>. Terpene-derived chiral sultine **97** reacts selectively with Grignard reagents to give the sulfoxides **98** <96S603>. Reaction of the related selenium compound **99** with a lithium amide proceeds by replacement of the chlorine with an amino group and this intermediate then undergoes a 2,3-sigmatropic rearrangement to afford an allylic amine <96JOC2932>. The preparation and chemistry of benzoxatelluroles

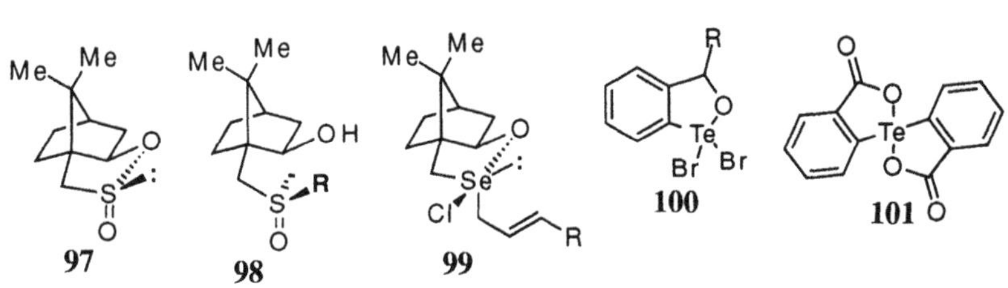

97 **98** **99** **100** **101**

such as **100** has been reported <96T3365> and the structure and reactivity of spiro compounds such as **101** has been studied in detail <95HAC481, 96CL365, 96CL859>.

5.6.7 THREE AND FOUR HETEROATOMS

Ozonolysis of enol ethers <96JCS(P1)871> or enol esters <96MI264> in the presence of an acyl cyanide to trap the intermediate carbonyl oxide affords 3-cyano-1,2,4-trioxolanes and stable ozonides such as **102** have been prepared as potential anti-malaria agents <96JCS(P1)1101>. Monomers containing a cyclic sulfite such as **103** have found an application in the manufacture of contact lenses <96JAP08165288> and cyclic sulfates have been used as the basis for a method of selective inversion of hydroxy groups in chiral 3-amino-1,2-diols <96JOC7162>.

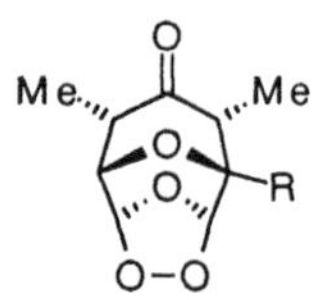

102

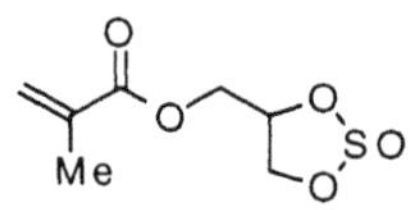

103

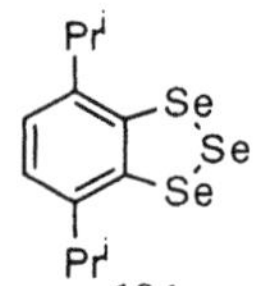

104

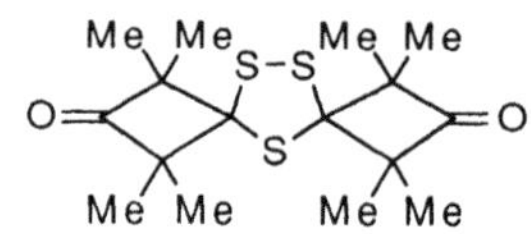

105

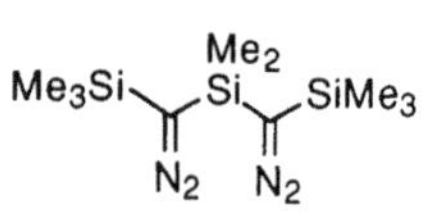

106

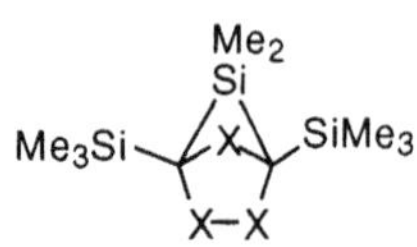

107 X = S, Se

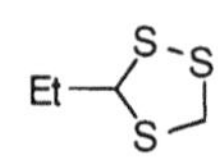

108

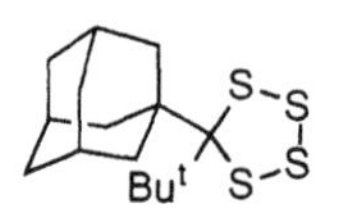

109

Stable 1,2,3-naphtho- and phenanthro-trithioles have been prepared <96CL757> and further studies on 1,2,3-benzotriselenoles have appeared <96H(43)1843>, with examples such as **104** finding application as electrochromic materials <95JAP07304766>. Reaction of hindered thiones such as adamantane-2-thione and 3-thioxo-2,2,4,4-tetramethylcyclobutanone with azides takes a rather complex course but thiocarbonyl sulfides are generated which react with the starting thione to afford 1,2,4-trithiolanes such as **105** among other products <96PJC437, 96PJC880>. Reaction of the silyldiazo compound **106** with sulfur or selenium affords the silicon bridged 1,2,4-trithiolane and triselenolane **107** <96G147>. Simple trithiolanes such as **108**, which is readily prepared from propanesulfenyl chloride, Na_2S and DMF, have an application as flavour compounds <96JAP08116914>. The first stable crystalline tetrathiolane **109** has been prepared and its X-ray structure determined <96CC2681>.

5.6.8 REFERENCES

95AHC(62)249 J. Garin, *Adv. Heterocycl. Chem.*, **1995**, *62*, 249.

95CL809 F. Toda and H. Miyamoto, *Chem. Lett.*, **1995**, 809.

95CL1069 J.-i. Yamada, S. Takasaki, M. Kobayashi, H. Anzai, N. Tajima, M. Tamura, Y. Nishio and K. Kajita, *Chem. Lett.*, **1995**, 1069.

95HAC481 Y. Takaguchi and N. Furukawa, *Heteroatom Chem.*, **1995**, *6*, 481 [*Chem. Abstr.*, **1996**, *124*, 261185].

95HCA2036 O. Sezer, A. Daut and O. Anaç, *Helv. Chim. Acta*, **1995**, *78*, 2036.

95ICA(238)57 H. Shen, R.A. Senter, S.G. Bott and M.G. Richmond, *Inorg. Chim. Acta*, **1995**, *238*, 57.

95ICA(239)117 F. Guyon, M. Fourmigué, P. Audebert and J. Amaudrut, *Inorg. Chim. Acta*, **1995**, *239*, 117.

95IZV1804 I.V. Magedov, S.Yu. Shapakin, A.S. Batsanov, Yu.T. Struchkov and V.N. Drozd, *Izv. Akad. Nauk, Ser. Khim.*, **1995**, 1804 [*Chem. Abstr.*, **1996**, *124*, 288914].

95JA9995 K. Lahlil, A. Moradpour, C. Bowlas, F. Menou, P. Cassoux, J. Bonvoisin, J.-P. Launay, G. Dive and D. Dehareng, *J. Am. Chem. Soc.*, **1995**, *117*, 9995.

95JA11367 A.B. Charette and C. Brochu, *J. Am. Chem. Soc.*, **1995**, *117*, 11367.

95JAP07242653 F. Toda, *Jpn. Pat.* 07 242 653 (1995) [*Chem. Abstr.*, **1996**, *124*, 146152].

95JAP07247280 M. Mukoyama, T. Takai, T. Nagata and T. Yamada, *Jpn. Pat.* 07 247 280 (1995) [*Chem. Abstr.*, **1996**, *124*, 146135].

95JAP07285955 F. Toda, *Jpn. Pat.* 07 285 955 (1995) [*Chem. Abstr.*, **1996**, *124*, 316971].

95JAP07285960 J. Takuma and I. Kawakami, *Jpn. Pat.* 07 285 960 (1995) [*Chem. Abstr.*, **1996**, *124*, 202229].

95JAP07291959 M. Tojo, A. Kato and M. Ikeda, *Jpn. Pat.* 07 291 959 (1995) [*Chem. Abstr.*, **1996**, *124*, 176074].

95JAP07304766 K. Sato, T. Kikuchi, A. Sasaki and S. Ogawa, *Jpn. Pat.* 07 304 766 (1995) [*Chem. Abstr.*, **1996**, *124*, 202608].

95JOC3964 S.J. Behroozi, W.K. Kim and K.S. Gates, *J. Org. Chem.*, **1995**, *60*, 3964.

95JOC7701 T.-J. Lu, J.-F. Yang and L.-J. Sheu, *J. Org. Chem.*, **1995**, *60*, 7701.

95KGS1694 L.V. Andriyankova, A.G. Malkina and B.A. Trofimov, *Khim. Geterotsikl. Soedin.*, **1995**, 1694 [*Chem. Abstr.*, **1996**, *125*, 33518].

95LA2011 H.-D. Stachel and K. Zeitler, *Liebigs Ann. Chem.*, **1995**, 2011.

95MI142 D.K. Kurbanov, T. Khodzhalyev, K. Patyshakuliev, A. Taganlyev and K. Khekimov, *Izv. Akad. Nauk Turkm.*, **1995**, 142 [*Chem. Abstr.*, **1996**, *125*, 114526].

95MI153 S. Nakatsuji, N. Akashi, K. Suzuki, T. Enoki, N. Kinoshita and H. Anzai, *Mol. Cryst. Liq. Cryst. Sci. Technol., Sect. A*, **1995**, *268*, 153.

95MI623 P. Ferraboschi, P. Grisenti and E. Santaniello, *Chirality*, **1995**, 623 [*Chem. Abstr.*, **1996**, *124*, 232293].

95MI877 T.-J. Lu and L.-J. Sheu, *J. Chin. Chem. Soc. (Taipei)*, **1995**, 877 [*Chem. Abstr.*, **1996**, *124*, 175908].

95MI1059 Y.-J. Shen, C.-Z. Dong and C.-Y. Ni, *Gaodeng Xuexiao Huaxue Xuebao*, **1995**, 1059 [*Chem. Abstr.*, **1996**, *124*, 175892].

95MI1481 M.R. Bryce, *J. Mater. Chem.*, **1995**, *5*, 1481 [*Chem. Abstr.*, **1996**, *124*, 86834].

95MI1581 R.D. McCullough, J.A. Belot, J. Seth, A.L. Rheingold, G.P.A. Yap and D.O. Cowan, *J. Mater. Chem.*, **1995**, *5*, 1581 [*Chem. Abstr.*, **1996**, *124*, 117480].

95MIP29176 T.S. Mansour and H. Jin, *PCT Int. Appl.* WO 29 176 (1995) [*Chem. Abstr.*, **1996**, *124*, 176137].

95MRCS94 M. Scholz, G. Gescheidt, U. Schöberl and J. Daub, *Magn. Reson. Chem.*, **1995**, *33*, S94 [*Chem. Abstr.*, **1996**, *124*, 145263].

95PJC1649 G. Mloston and H. Heimgartner, *Pol. J. Chem.*, **1995**, *69*, 1649 [*Chem. Abstr.*, **1996**, *124*, 289307].

95PS(106)145 S.-G. Liu, P.-J. Wu, Y.-Q. Liu and D.-B. Zhu, *Phosphorus Sulfur Silicon Relat. Elem.*, **1995**, *106*, 145 [*Chem. Abstr.*, **1996**, *124*, 175890].

95SL1225 H. Oda, K. Hamataka, K. Fugami, M. Kosugi and T. Migita, *Synlett*, **1995**, 1225.

95SR(16)173 C.T. Pedersen, *Sulfur Reports*, **1995**, *16*, 173 [*Chem. Abstr.*, **1996**, *125*, 221634].

95SUL67 V.N. Drozd, S.Yu. Shapakin, I.V. Magedov, D.S. Yufit and Yu.T. Struchkov, *Sulfur Lett.*, **1995**, *18*, 67 [*Chem. Abstr.*, **1996**, *124*, 289373].

95T12843 J. Leonard, S. Mohialdin, D. Reed, G. Ryan and P.A. Swain, *Tetrahedron*, **1995**, *51*, 12843.

95TL9333	T. Inoue, O. Kitagawa, O. Ochiai, M. Shiro and T. Taguchi, *Tetrahedron Lett.*, **1995**, *36*, 9333.
95USP5466811	J. Alexander, *US Pat.* 5 466 811 (1995) [*Chem. Abstr.*, **1996**, *124*, 176148].
95ZOB1054	V.I. Boiko, L.I. Samarai and V.V. Pirozhenko, *Zh. Obshch. Khim.*, **1995**, *65*, 1054 [*Chem. Abstr.*, **1996**, *124*, 117451].
95ZOB1101	Yu.G. Shtyrlin, N.I. Tyryshkin, G.G. Iskhakova, G.G. Gavrilov, V.D. Kiselev and A.I. Konovalov, *Zh. Obshch. Khim.*, **1995**, *65*, 1101 [*Chem. Abstr.*, **1996**, *124*, 146288].
95ZOR548	L.V. Kuritsyn, A.I. Sadovnikov and G.Yu. Babikova, *Zh. Org. Khim.*, **1995**, *31*, 548 [*Chem. Abstr.*, **1996**, *124*, 260136].
96AG(E)1970	J. Mulzer, D. Trauner and J.W. Bats, *Angew. Chem., Int. Ed. Engl.*, **1996**, *35*, 1970.
96AM63	M.A. Blower, M.R. Bryce and W. Davenport, *Adv. Mater. (Weinheim, Ger.)*, **1996**, *8*, 63.
96AM804	Y. Misaki, T. Sasaki, T. Ohta, H. Fujiwara and T. Yamabe, *Adv. Mater. (Weinheim, Ger.)*, **1996**, *8*, 804.
96AM807	M. Moge, J. Hellberg, K.W. Törnroos and J.-U. von Schütz, *Adv. Mater. (Weinheim, Ger.)*, **1996**, *8*, 807.
96AP361	M. Friedrich, W. Meichle, H. Bernhard, G. Rihs and H.-H. Otto, *Arch. Pharm. (Weinheim, Ger.)*, **1996**, *329*, 361.
96BCJ2349	J. Nakayama, A. Kimata, H. Taniguchi and F. Takahashi, *Bull. Chem. Soc. Jpn.*, **1996**, *69*, 2349.
96BCJ2885	K. Kasuga, T. Kato, N. Kabata and M. Handa, *Bull. Chem. Soc. Jpn.*, **1996**, *69*, 2885.
96BSF301	T.-T. Nguyen, Y. Gouriou, M. Sallé, P. Frère, M. Jubault, A. Gorgues, L. Toupet and A. Riou, *Bull. Soc. Chim. Fr.*, **1996**, *133*, 301.
96CC205	J. Nakayama, A. Kimata, H. Taniguchi and F. Takahashi, *Chem. Commun.*, **1996**, 205.
96CC311	H. Fujihara, T. Nakahodo and N. Furukawa, *Chem. Commun.*, **1996**, 311.
96CC341	C. Johnstone, W.J. Kerr and J.S. Scott, *Chem. Commun.*, **1996**, 341.
96CC363	Y. Misaki, H. Fujiwara, T. Yamabe, T. Mori, H. Mori and S. Tanaka, *Chem. Commun.*, **1996**, 363.
96CC427	O.A. Rakitin, C.W. Rees and T. Torroba, *Chem. Commun.*, **1996**, 427.
96CC521	T. Naito, H. Kobayashi, A. Kobayashi and A.E. Underhill, *Chem. Commun.*, **1996**, 521.
96CC615	P. Blanchard, N. Svenstrup and J. Becher, *Chem. Commun.*, **1996**, 615.
96CC639	Z.-T. Li and J. Becher, *Chem. Commun.*, **1996**, 639.
96CC737	J.A. Murphy, F. Rasheed, S.J. Roome and N. Lewis, *Chem. Commun.*, **1996**, 737.
96CC739	R.J. Fletcher, D.E. Hibbs, M. Hursthouse, C. Lampard, J.A. Murphy and S.J. Roome, *Chem. Commun.*, **1996**, 739.
96CC743	M. Poelart, W. Roger and S.Z. Zard, *Chem. Commun.*, **1996**, 743.
96CC919	C. Darcel, C. Bruneau, M. Albert and P.H. Dixneuf, *Chem. Commun.*, **1996**, 919.
96CC1011	H. Kato, T. Kobayashi, M. Ciobanu, H. Iga, A. Akutsu and A. Kakehi, *Chem. Commun.*, **1996**, 1011.
96CC1363	N. Le Narvor, N. Robertson, T. Weyland, J.D. Kilburn, A.E. Underhill, M. Webster, N. Svenstrup and J. Becher, *Chem. Commun.*, **1996**, 1363.
96CC1955	J.-i. Yamada, S. Satoki, H. Anzai, K. Hagiya, M. Tamura, Y. Nishio, K. Kajita, E. Watanabe, M. Konno, T. Sato, H. Nishikawa and K. Kikuchi, *Chem. Commun.*, **1996**, 1955.
96CC2021	Y. Yamashita, M. Tomura and K. Imaeda, *Chem. Commun.*, **1996**, 2021.
96CC2281	T. Tatsumi, Y. Watanabe and K.A. Koyano, *Chem. Commun.*, **1996**, 2281.
96CC2407	M. Kamata, Y. Nishikata and M. Kato, *Chem. Commun.*, **1996**, 2407.
96CC2423	A. Charlton, A.E. Underhill, G. Williams, M. Kalaji, P.J. Murphy, D.E. Hibbs, M.B. Hursthouse and K.M.A. Malik, *Chem. Commun.*, **1996**, 2423.

96CC2517 J.-i. Yamada, S. Mishima, H. Anzai, M. Tamura, Y. Nishio, K. Kajita, T. Sato, H. Nishikawa, I. Ikemoto and K. Kikuchi, *Chem. Commun.*, **1996**, 2517.

96CC2681 A. Ishii, J. Yinan, Y. Sugihara and J. Nakayama, *Chem. Commun.*, **1996**, 2681.

96CEJ624 Z.-T. Li, P.C. Stein, J. Becher, D. Jensen, P. Mørk and N. Svenstrup, *Chem. Eur. J.*, **1996**, *2*, 624.

96CEJ1275 A. Dolbecq, M. Fourmigué, F.C. Krebs, P. Batail, E. Canadell, R. Clérac and C. Coulon, *Chem. Eur. J.*, **1996**, *2*, 1275.

96CL43 H. Nishikawa, S. Kawauchi, Y. Misaki and T. Yamabe, *Chem. Lett.*, **1996**, 43.

96CL171 K. Nishiwaki and J.-i. Yoshida, *Chem. Lett.*, **1996**, 171.

96CL365 Y. Takaguchi and N. Furukawa, *Chem. Lett.*, **1996**, 365.

96CL757 S. Ogawa, S. Nobuta, R. Nakayama, Y. Kawai, S. Niizuma and R. Sato, *Chem. Lett.*, **1996**, 757.

96CL859 Y. Takaguchi and N. Furukawa, *Chem. Lett.*, **1996**, 859.

96CM1182 M.R. Bryce, A.J. Moore, B.K. Tanner, R. Whitehead, W. Clegg, F. Gerson, A. Lamprecht and S. Pfenninger, *Chem. Mater.*, **1996**, *8*, 1182.

96CM2291 A. Benahmed-Gasmi, P. Frere, E.H. Elandaloussi, J. Roncali, J. Orduna, J. Garin, M. Jubault, A. Riou and A. Gorgues, *Chem. Mater.*, **1996**, *8*, 2291.

96G147 Y. Kabe, T. Watanabe and W. Ando, *Gazz. Chim. Ital.*, **1996**, *126*, 147 [*Chem. Abstr.*, **1996**, *125*, 10982].

96GEP4436509 W. Schaper, R. Preus, P. Braun, M. Kern, W. Knauf, B. Sachse, U. Sanft, A. Waltersdorfer, W. Bonin *et al.*, *Ger. Pat.* 4 436 509 (1996) [*Chem. Abstr.*, **1996**, *125*, 33670].

96GEP19538472 J. Wenger, *Ger. Pat.* 19 538 472 (1996) [*Chem. Abstr.*, **1996**, *125*, 33621].

96H(43)851 I. Shibuya, E. Katoh, Y. Gama, A. Oishi, Y. Taguchi and T. Tsuchiya, *Heterocycles*, **1996**, *43*, 851.

96H(43)1843 S. Ogawa, T. Ohmiya, T. Kikuchi, Y. Kawai, S. Niizuma and R. Sato, *Heterocycles*, **1996**, *43*, 1843.

96HCA31 G. Mloston, A. Linden and H. Heimgartner, *Helv. Chim. Acta*, **1996**, *79*, 31.

96HCA634 R.R. Tykwinski, F. Diederich, V. Gramlich and P. Seiler, *Helv. Chim. Acta*, **1996**, *79*, 634.

96HCA1305 G. Mloston, J. Romanski, A. Linden and H. Heimgartner, *Helv. Chim. Acta*, **1996**, *79*, 1305.

96HCA1497 G. Märkl, A. Pöll, N.G. Aschenbrenner, C. Schmaus, T. Troll, P. Kreitmeir, H. Nöth and M. Schmidt, *Helv. Chim. Acta*, **1996**, *79*, 1497.

96HCA1537 G. Mloston, T. Gendek and H. Heimgartner, *Helv. Chim. Acta*, **1996**, *79*, 1537.

96HCA1696 P. Renaud and S. Abazi, *Helv. Chim. Acta*, **1996**, *79*, 1696.

96HCA1710 D. Seebach, R.E. Marti and T. Hintermann, *Helv. Chim. Acta*, **1996**, *79*, 1710.

96IZV775 R.V. Pisarev, V.V. Kalashnikov, A.I. Kotov and E.B. Yagubskii, *Izv. Akad. Nauk, Ser. Khim.*, **1996**, 775 [*Chem. Abstr.*, **1996**, *125*, 275769].

96JA3533 M. Hojo, H. Aihara and A. Hosomi, *J. Am. Chem. Soc.*, **1996**, *118*, 3533.

96JA9850 D. Suárez, T.L. Sordo and J.A. Sordo, *J. Am. Chem. Soc.*, **1996**, *118*, 9850.

96JAP07247283 N. Wakao, Y. Hino and R. Ishikawa, *Jpn. Pat.* 07 247 283 (1996) [*Chem. Abstr.*, **1996**, *124*, 146165].

96JAP0803076 F. Toda, *Jpn. Pat.* 08 03 076 (1996) [*Chem. Abstr.*, **1996**, *124*, 342245].

96JAP0859539 F. Toda, *Jpn. Pat.* 08 59 539 (1996) [*Chem. Abstr.*, **1996**, *125*, 57985].

96JAP08116914 H. Tamura and Y. Kurobayashi, *Jpn. Pat.* 08 116 914 (1996) [*Chem. Abstr.*, **1996**, *125*, 86652].

96JAP08165288 K. Watanabe and T. Matsura, *Jpn. Pat.* 08 165 288 (1996) [*Chem. Abstr.*, **1996**, *125*, 221549].

96JCR(S)274 P. Gaviña, N.L. Lavernia, R. Mestres and E. Muñoz, *J. Chem. Res. (S)*, **1996**, 274.

96JCS(D)823 N. Le Narvor, N. Robertson, E. Wallace, J.D. Kilburn, A.E. Underhill, P.N. Bartlett and M. Webster, *J. Chem. Soc., Dalton Trans.*, **1996**, 823.

96JCS(P1)259 A. Díaz-Ortiz, E. Díez-Barra, A. de la Hoz, P. Prieto and A. Moreno, *J. Chem. Soc., Perkin Trans. 1*, **1996**, 259.

96JCS(P1)289 M. Barbero, I. Degani, S. Dughera, R. Fochi and L. Piscopo, *J. Chem. Soc., Perkin Trans. 1*, **1996**, 289.

96JCS(P1)871 H. Kuwabara, Y. Ushigo and M. Nojima, *J. Chem. Soc., Perkin Trans. 1*, **1996**, 871.

96JCS(P1)1101 L.C. de Almeida Barbosa, D. Cutler, J. Mann, M.J. Crabbe, G. Kirby and D.C. Warhurst; *J. Chem. Soc., Perkin Trans. 1*, **1996**, 1101.

96JCS(P1)1995 M. Wagner, D. Madsen, J. Markussen, S. Larsen, K. Schaumburg, K.-H. Lubert, J. Becher and R.-M. Olk, *J. Chem. Soc., Perkin Trans. 1*, **1996**, 1995.

96JCS(P2)1429 S. Itoh, J. Maruta and S. Fukuzumi, *J. Chem. Soc., Perkin Trans. 2*, **1996**, 1429.

96JCS(P2)2367 M.R. Bryce, M.A. Chalton, A.S. Batsanov, C.W. Lehmann and J.A.K. Howard, *J. Chem. Soc., Perkin Trans. 2*, **1996**, 2367.

96JHC1399 A. Baumstark, Y.-X. Chen and A. Rodriguez, *J. Heterocycl. Chem.*, **1996**, *33*, 1399.

96JOC2877 A. Chesney, M.R. Bryce, M.A. Chalton, A.S. Batsanov, J.A.K. Howard, J.-M. Fabre, L. Binet and S. Chakroune, *J. Org. Chem.*, **1996**, *61*, 2877.

96JOC2932 N. Kurose, T. Takahashi and T. Koizumi, *J. Org. Chem.*, **1996**, *61*, 2932.

96JOC3650 Y. Misaki, H. Fujiwara and T. Yamabe, *J. Org. Chem.*, **1996**, *61*, 3650.

96JOC3906 S.-M. Yeh, L.-H. Huang and T.-Y. Luh, *J. Org. Chem.*, **1996**, *61*, 3906.

96JOC3987 J.-i. Yamada, S. Satoki, S. Mishima, N. Akashi, K. Takahashi, N. Masuda, Y. Nishimoto, S. Takasaki and H. Anzai, *J. Org. Chem.*, **1996**, *61*, 3987.

96JOC7006 D.E. Herr, M.D. Mays, R.D. McCullough, A.B. Bailey and D.O. Cowan, *J. Org. Chem.*, **1996**, *61*, 7006.

96JOC7162 S.J. Kemp, J. Bao and S.F. Pedersen, *J. Org. Chem.*, **1996**, *61*, 7162.

96JOC8117 K. Zong, W. Chen, M.P. Cava and R.D. Rogers, *J. Org. Chem.*, **1996**, *61*, 8117.

96JOC8132 T. Minami, T. Okauchi, H. Matsuki, M. Nakamura, J. Ichikawa and M. Ishida, *J. Org. Chem.*, **1996**, *61*, 8132.

96JPR523 E. Fanghänel, L. Van Hinh, G. Schukat and A. Herrmann, *J. Prakt. Chem./Chem.-Ztg.*, **1996**, *338*, 523.

96KGS123 O.Ya. Neiland, V.Zh. Tilika, A.A. Supe and A.S. Edzhinya, *Khim. Geterotsikl. Soedin.*, **1996**, 123 [*Chem. Abstr.*, **1996**, *125*, 195560].

96LA63 L.F. Tietze, C. Ott and U. Frey, *Liebigs Ann. Chem.*, **1996**, 63.

96LA109 K. Hartke and X.-p. Popp, *Liebigs Ann. Chem.*, **1996**, 109.

96LA551 M. Wagner, S. Zeltner and R.-M. Olk, *Liebigs Ann. Chem.*, **1996**, 551.

96MI26 D.-W. Park, J.-Y. Moon, J.-G. Yang, S.-H. Park and J.-K. Lee, *Kongop Hwahak*, **1996**, *7*, 26 [*Chem. Abstr.*, **1996**, *124*, 343163].

96MI103 F. Qi, *Chin. Chem. Lett.*, **1996**, 103 [*Chem. Abstr.*, **1996**, *124*, 343165].

96MI169 H. Pajouhesh and M. Mahkam, *J. Sci., Islamic Repub. Iran*, **1996**, *7*, 169 [*Chem. Abstr.*, **1996**, *125*, 328618].

96MI184 S. Yan, N. Chen, J. Li and Y. Zhang, *Hecheng Huaxue*, **1996**, *4*, 184 [*Chem. Abstr.*, **1996**, *125*, 221750].

96MI264 T.S. Huh, M.K. Lee, M.J. Kim and K. Griesaum, *Erdoel, Erdgas, Kohle*, **1996**, *112*, 264 [*Chem. Abstr.*, **1996**, *125*, 142644].

96MI362 S.-B. Lee, H. Jung and K.W. Lee, *Bull. Korean Chem. Soc.*, **1996**, *17*, 362 [*Chem. Abstr.*, **1996**, *124*, 343205].

96MI473 K.M. Kim, O.-S. Jung, Y.S. Sohn and M.-J. Jun, *Bull. Korean Chem. Soc.*, **1996**, *17*, 473 [*Chem. Abstr.*, **1996**, *125*, 85924].

96MI513 H. Zhu, L.-B. Chen and Y.-Y. Jiang, *Chin. Chem. Lett.*, **1996**, *7*, 513 [*Chem. Abstr.*, **1996**, *125*, 195474].

96MI571 Y. Xiao, *Chin. Sci. Bull.*, **1996**, *41*, 571 [*Chem. Abstr.*, **1996**, *125*, 195475].

96MI597 H.Q. Li and Z.Q. Yao, *Chin. Chem. Lett.*, **1996**, *7*, 597 [*Chem. Abstr.*, **1996**, *125*, 275700].

96MI701 H. Zhu, L.-B. Chen and Y.-Y. Jiang, *Polym. Adv. Technol.*, **1996**, *7*, 701 [*Chem. Abstr.*, **1996**, *125*, 328552].

96MIP03357 R.D. Chambers and G. Sandford, *PCT Int. Appl.* WO 03 357 (1996) [*Chem. Abstr.*, **1996**, *125*, 9849].

96MIP19487 H. Suzuki, K. Tanikawa, K. Miyaji and N. Suzuki, *PCT Int. Appl.* WO 19 487 (1996) [*Chem. Abstr.*, **1996**, *125*, 143015].

96MRC(34)52 A. Moreno, A. Díaz-Ortiz, E. Díez-Barra, A. de la Hoz, F. Langa, P. Prieto and T.D.W. Claridge, *Magn. Reson. Chem.*, **1996**, *34*, 52 [*Chem. Abstr.*, **1996**, *124*, 231739].

96OM1957 H. Rakotoarisoa, R.G. Perez, P. Mangeney and A. Alexakis; *Organometallics*, **1996**, *15*, 1957.

96OM1966 A. Antiñolo, I. del Hierro, M. Fajardo, S. Garcia-Yuste, A. Otero, O. Blacque, M.M. Kubicki and J. Amaudrut, *Organometallics*, **1996**, *15*, 1966.

96OM3099 H.-R. Tseng and T.-Y. Luh, *Organometallics*, **1996**, *15*, 3099.

96OS270 T.K. Hansen, J. Becher, T. Jørgensen, K. Sukumar Varma, R. Khedekar and M.P. Cava, *Org. Synth.*, **1996**, *73*, 270.

96PAC(68)105 T.-Y. Luh, *Pure Appl. Chem.*, **1996**, *68*, 105 [*Chem. Abstr.*, **1996**, *124*, 289292].

96PJC54 A. Maslankiewicz, L. Skrzypek and A. Niedbala, *Pol. J. Chem.*, **1996**, *70*, 54 [*Chem. Abstr.*, **1996**, *125*, 10976].

96PJC437 G. Mloston, J. Romanski and H. Heimgartner, *Pol. J. Chem.*, **1996**, *70*, 437 [*Chem. Abstr.*, **1996**, *125*, 10707].

96PJC595 G. Mloston, J. Romanski, A. Linden and H. Heimgartner, *Pol. J. Chem.*, **1996**, *70*, 595 [*Chem. Abstr.*, **1996**, *125*, 86562].

96PJC880 G. Mloston, J. Romanski, A. Linden and H. Heimgartner, *Pol. J. Chem.*, **1996**, *70*, 880 [*Chem. Abstr.*, **1996**, *125*, 195526].

96PS(109)557 W.M. Abdou, I.T. Hennawy and O.E. Khoshnich; *Phosphorus Sulfur Silicon Relat. Elem.*, **1996**, *109–110*, 557 [*Chem. Abstr.*, **1996**, *125*, 328914].

96PS(112)101 M. Tazaki, S. Okai, T. Hieda, S. Nagahama and M. Takagi, *Phosphorus Sulfur Silicon Relat. Elem.*, **1996**, *112*, 101 [*Chem. Abstr.*, **1996**, *125*, 195539].

96RTC140 K. Krohn and B. Knauer, *Recl. Trav. Chim. Pays-Bas*, **1996**, *115*, 140.

96S198 J. Hellberg and M. Moge, *Synthesis*, **1996**, 198.

96S603 R. Kawecki and Z. Urbanczyk-Lipkowska, *Synthesis*, **1996**, 603.

96SC3453 H.-O. Kim, D. Friedrich, E. Huber and N.P. Peet, *Synth. Commun.*, **1996**, *26*, 3453.

96SL53 T.-M. Yuan, Y.-T. Hsieh, S.-M. Yeh, J.-J. Shyue and T.-Y. Luh, *Synlett*, **1996**, 53.

96SL201 T.-Y. Luh, *Synlett*, **1996**, 201.

96SL231 S. Saito, A. Kuroda, K. Tanaka and R. Kimura, *Synlett*, **1996**, 231.

96SL234 M. Hojo, N. Ishibashi and A. Hosomi, *Synlett*, **1996**, 234.

96SL237 M. Lesage and P. Arya, *Synlett*, **1996**, 237.

96SL267 J. Holz, A. Kless and A. Börner, *Synlett*, **1996**, 267.

96SL677 A.M. Harm, J.G. Knight and G. Stemp, *Synlett*, **1996**, 677.

96SL767 P. Ceccherelli, M. Curini, M.C. Marcotullio, F. Epifano and O. Rosati, *Synlett*, **1996**, 767.

96SL839 K. Ishihara, Y. Karumi, M. Kubota and H. Yamamoto, *Synlett*, **1996**, 839.

96SM(78)89 J.M. Fabre, S. Chakroune, A. Javidan, M. Calas, A. Souizi and L. Ouahab, *Synth. Met.*, **1996**, *78*, 89 [*Chem. Abstr.*, **1996**, *125*, 58413].

96SR(18)1	G. Schukat and E. Fanghänel; *Sulfur Reports*, **1996**, *18*, 1 [*Chem. Abstr.*, **1996**, *125*, 86525].
96SUL235	A. Corsaro, U. Chiacchio, G. Gumina and V. Pistara, *Sulfur Lett.*, **1996**, *19*, 235 [*Chem. Abstr.*, 1996, **125**, 328550].
96T1259	J.I. Lozano and F. Barba, *Tetrahedron*, **1996**, *52*, 1259.
96T3365	I.D. Sadekov, A.A. Maksimenko and V.I. Minkin, *Tetrahedron*, **1996**, *52*, 3365.
96T4745	J.M. Lovell, R.L. Beddoes and J.A. Joule, *Tetrahedron*, **1996**, *52*, 4745.
96T4911	P. Gaviña, N.L. Lavernia, R. Mestres and M.A. Miranda, *Tetrahedron*, **1996**, *52*, 4911.
96T8275	C. Alemán, A. Martinez de Ilarduya, E. Giralt and S. Muñoz-Guerra, *Tetrahedron*, **1996**, *52*, 8275.
96T9713	A. Nishida, N. Kawahara, M. Nishida and O. Yonemitsu, *Tetrahedron*, **1996**, *52*, 9713.
96T11063	J. Garín, J. Orduna, M. Savirón, M.R. Bryce, A.J. Moore and V. Morisson, *Tetrahedron*, **1996**, *52*, 11063.
96TA29	N. Maezaki, M. Soejima, A. Sakamoto, I. Sakamoto, Y. Matsumori, T. Tanaka, T. Ishida, Y. In and C. Iwata, *Tetrahedron Asymmetry*, **1996**, *7*, 29.
96TA885	G. Chelucci, M.A. Cabras, C. Botteghi, C. Basoli and M. Marchetti, *Tetrahedron Asymmetry*, **1996**, *7*, 885.
96TA3037	E. Väntinnen and L. T. Kanerva, *Tetrahedron Asymmetry*, **1996**, *7*, 3037.
96TL115	R. Curci, L. D'Accolti, A. Dinoi, C. Fusco and A. Rosa, *Tetrahedron Lett.*, **1996**, *37*, 115.
96TL1195	S. Buhr, A.G. Griesbeck, J. Lex, J. Mattay and J. Schröer, *Tetrahedron Lett.*, **1996**, *37*, 1195.
96TL1421	B. Heckmann, C. Mioskowski, R.K. Bhatt and J.R. Falck, *Tetrahedron Lett.*, **1996**, *37*, 1421.
96TL2137	M. L. Aimar and R. H. Rossi, *Tetrahedron Lett.*, 1996, **37**, 2137.
96TL2503	J. L. Segura, N. Martín, C. Seoane and M. Hanack, *Tetrahedron Lett.*, 1996, **37**, 2503.
96TL2675	N. Greeves, J. E. Pease, M. C. Bowden and S. M. Brown, *Tetrahedron Lett.*, 1996, **37**, 2675.
96TL4073	A. V. Bedekar and P. G. Andersson, *Tetrahedron Lett.*, 1996, **37**, 4073.
96TL4713	P. Pellon, C. Le Goaster and L. Toupet, *Tetrahedron Lett.*, **1996**, *37*, 4713.
96TL5337	W. Kim, J. Dannaldson and K.S. Gates, *Tetrahedron Lett.*, **1996**, *37*, 5337.
96TL5979	N. Martin, L. Sanchez, C. Seoane, R. Andreu, J. Garin and J. Orduna, *Tetrahedron Lett.*, **1996**, *37*, 5979.
96TL6307	G. Galley, J. Hübner, S. Anklam, P.G. Jones and M. Pätzel, *Tetrahedron Lett.*, **1996**, *37*, 6307.
96TL6635	M.K. Schwaebe and R.D. Little, *Tetrahedron Lett.*, **1996**, *37*, 6635.
96TL6853	O. Genkinger and J. Bargon, *Tetrahedron Lett.*, **1996**, *37*, 6853.
96UK156	O.B. Bondarenko, L.G. Saginova and N.V. Zyk, *Usp. Khim.*, **1996**, *65*, 156 [*Chem. Abstr.*, **1996**, *125*, 33501].
96YGK752	T. Otsubo, Y. Aso and K. Takimiya, *Yuki Gosei Kagaku Kyokaishi*, **1996**, *54*, 752 [*Chem. Abstr.*, **1996**, *125*, 247641].
96ZN(B)901	T. Kniess and R. Mayer, *Z. Naturforsch., Teil B*, **1996**, *51*, 901 [*Chem. Abstr.*, **1996**, *125*, 167929].

Chapter 5.7

Five-Membered Ring Systems with O & N Atoms

G. V. Boyd
The Hebrew University, Jerusalem, Israel

5.7.1 ISOXAZOLES

Enamino ketones **1** react with hydroxylamine hydrochloride to give 4,5-diarylisoxazoles **2** <96JOC5435>. Biohydrogenation of the nitropropene nitriles **3** (R = H, Me or Ar) with baker's yeast affords good yields of the aminoisoxazoles **4** <96SL695>. Hydroxylamine hydrochloride transforms ethoxydihydropyrans into (cyanoethyl)isoxazoles, e.g. **5**→**6** <96JHC383>. An efficient synthesis of the cholinergic channel activator ABT-418 **7**, which is being evaluated for the treatment of Alzheimer's disease, from *N*-methylproline methyl ester has been described <96JOC356>.

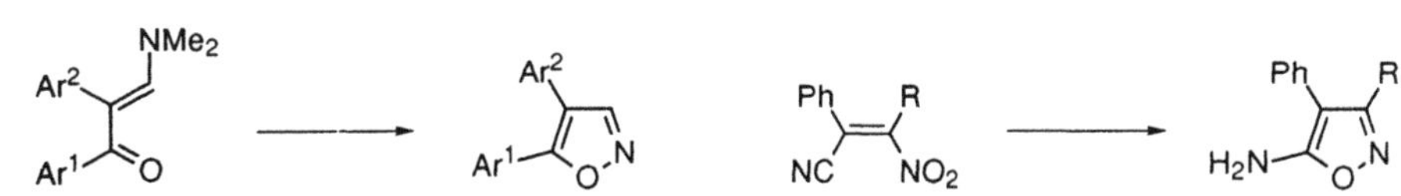

1 **2** **3** **4**

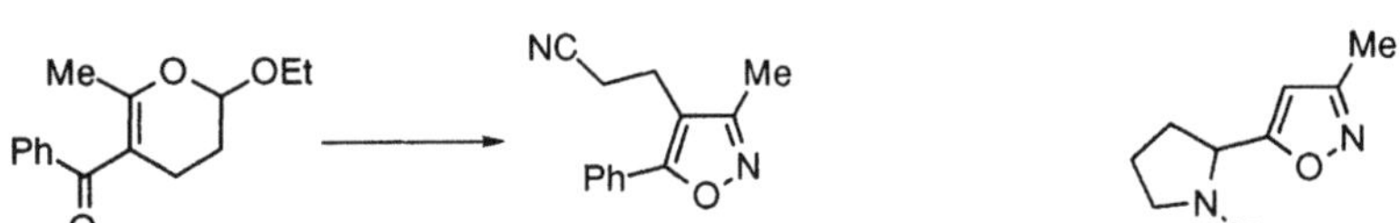

5 **6** **7**

A one-pot synthesis of 3-amino-1,2-benzisoxazoles from *o*-fluorobenzonitrile and acetohydroxamic acid is shown in Scheme 1 <96TL2885>. The amide oxime **8** (R = 4-pyridyl) cyclises to the benzisoxazole **9** under the influence of potassium t-butoxide <96TL995>. Heating the azidoisoxazole aldehyde **10** results in the formation of the isoxazoloisoxazole **11** with extrusion of nitrogen <95JHC1189>. An isoxazolopyrimidine is produced when the diester **12** is heated in refluxing xylene in the presence of molecular sieves. The reaction is thought to proceed by way of the ketene shown in Scheme 2 <95CC2457>.

The isoxazolopyrylium olate **14** is generated by the action of rhodium(II) acetate on the diazo compound **13**; in the presence of dimethyl acetylenedicarboxylate the cycloadduct **15** is isolated <96JCS1035>.

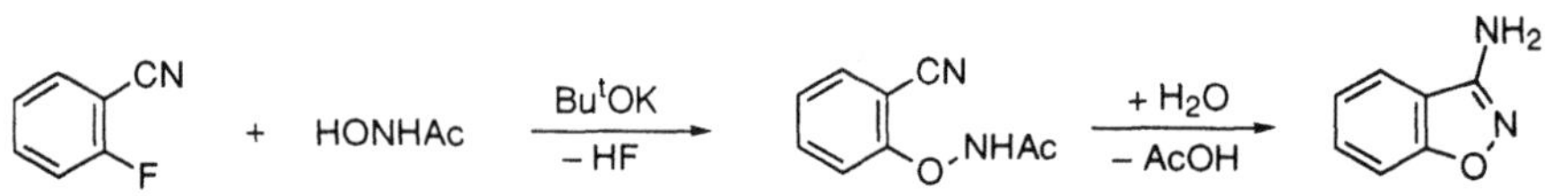

Scheme 1

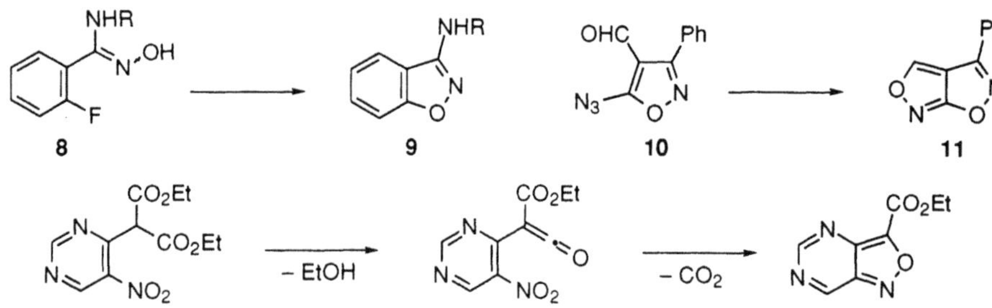

8 9 10 11

12

Scheme 2

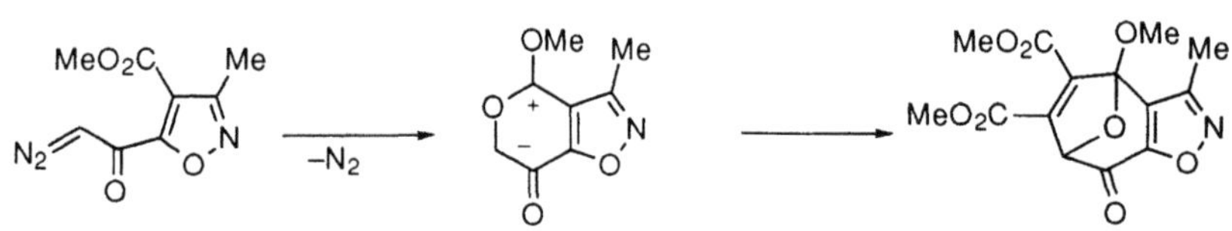

13 14 15

5.7.2 ISOXAZOLINES

The photochemical cyclisation of β,γ-unsaturated ketoximes to 2-isoxazolines, e.g., **16→17**, has been reported <95RTC514>. 2-Isoxazolines are obtained from alkenes and primary nitroalkanes in the presence of ammonium cerium nitrate and formic acid <95MI399>. Treatment of certain 1,3-diketones with a nitrating mixture generates acyl nitrile oxides, which can be trapped *in situ* as dipolar cycloadducts (see Scheme 3) <96SC3401>.

16 17

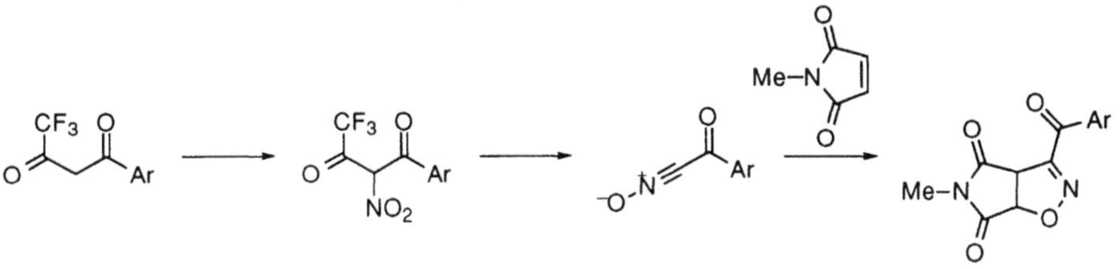

Scheme 3

The chemistry of 5-hetero substituted-4-methylene-2-isoxazolines **18** has been reviewed <95SL1208>. Symmetrically linked bis(isoxazolines), e.g., **21**, are obtained by a one-pot reaction of an oxime with a dipole-generating component **19** and a dipole-trapping component **20** <96TL4597>. Several 1,3-dipolar cycloadducts of nitrile oxides to C_{60} fullerene have been described <96T5043>. The reaction of C_{60} fullerene with cyanogen bis(*N*-oxide), ONCCNO, gives the bis(fullereno[1,2-*d*]isoxazolin)-3-yl **22** <96TL4137>. Treatment of C_{60} fullerene with *N*-(trimethylsilyloxy)nitrone, followed by toluene-*p*-sulfonic acid, gives the fused isoxazoline **23** (Scheme 4) <96SL815>. A mixture of three isomeric monoadducts of C_{70} fullerene to 2,4,6-trimethoxybenzonitrile oxide has been obtained <96LA1609>.

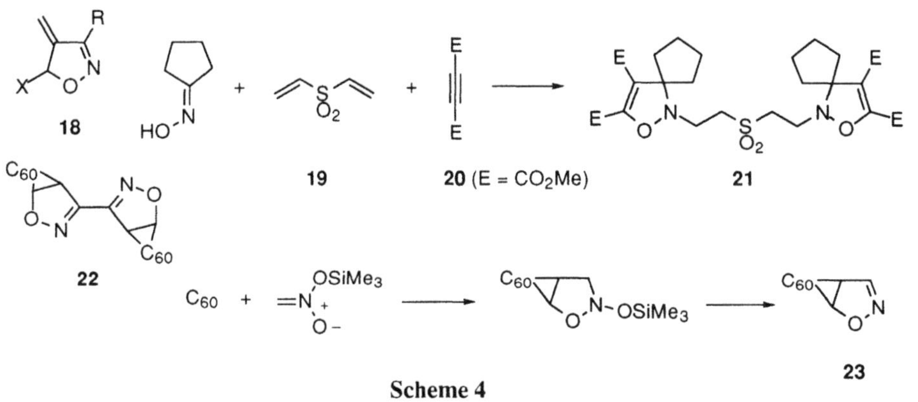

Scheme 4

The base-catalysed reaction of α-bromo-α,β-unsaturated ketones with aliphatic nitro compounds leads to 2-isoxazoline *N*-oxides by tandem conjugate addition–ring closure (Scheme 5) <95JOC6624>. *N*-Acyl-3-isoxazolin-5-ones are transformed into oxazoles by photolysis or by flash vacuum pyrolysis (Scheme 6) <96TL675>.

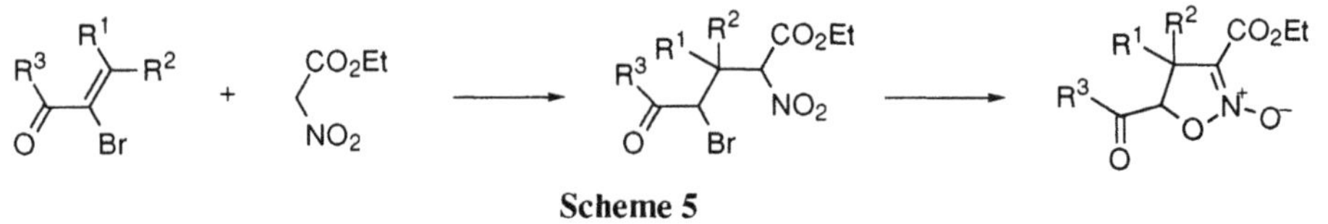

Scheme 5

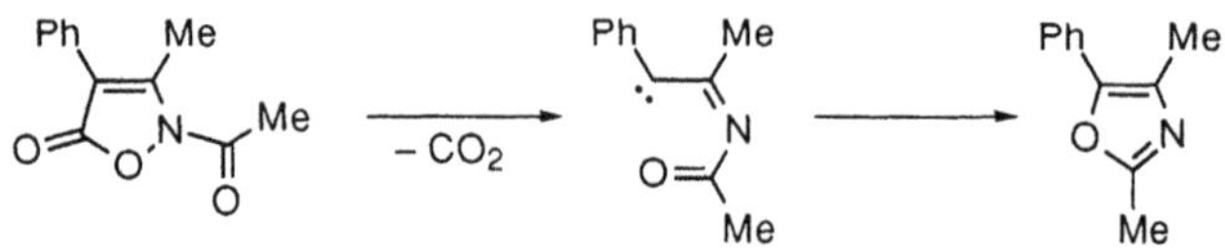

Scheme 6

5.7.3 ISOXAZOLIDINES

Treatment of cyclobutanone with *N*-methylhydroxylamine generates a nitrone, which can be trapped as the 1,3-dipolar cycloadduct **24** in the presence of styrene <96T9187>. Addition of ethyl trifluoroacetoacetate, which reacts in the enol form **25**, to the nitrone **26** under microwave irradiation and in the absence of a solvent furnishes the isoxazolidine **27** in 76% yield <95JFC(75)215>. The nitrone **28** undergoes a spontaneous intramolecular cycloaddtion to give the enantionerically pure 3-oxa-2,7-diazabicyclo[3.3.0]octane **29** <95T10497>. The chiral nitrone complex **30** cyclises to the diastereomerically pure fused isoxazolidine **31**, which can be decomplexed with air and sunlight <95TA1711>.

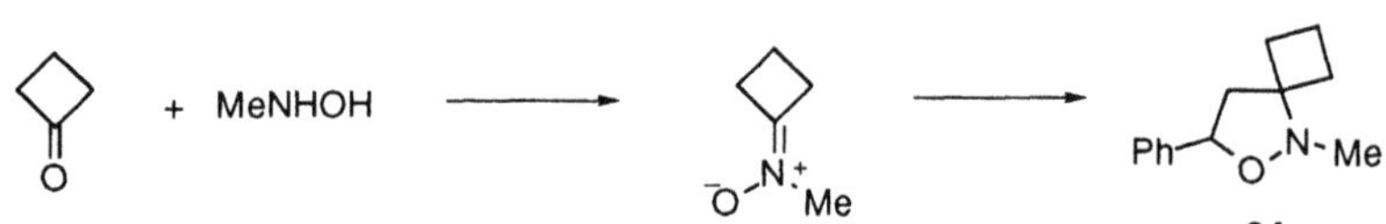

24

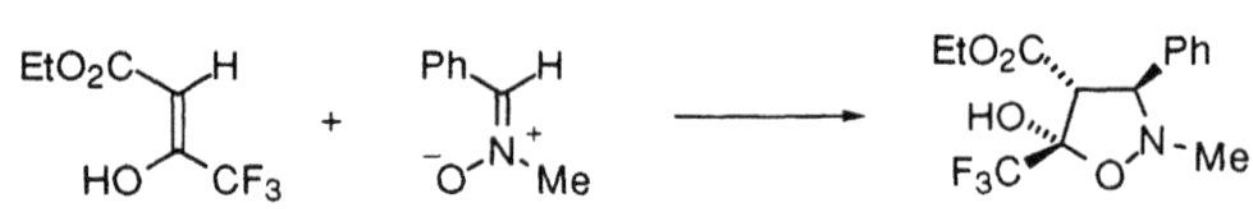

25 **26** **27**

28 **29**

30 **31**

5.7.4 OXAZOLES

Phenacyl benzoate condenses with acetamide in boiling xylene under boron trifluoride etherate catalysis to give a high yield of 2,4-diphenyloxazole in a general synthesis of diaryloxazoles (Scheme 7) <96T10131>.

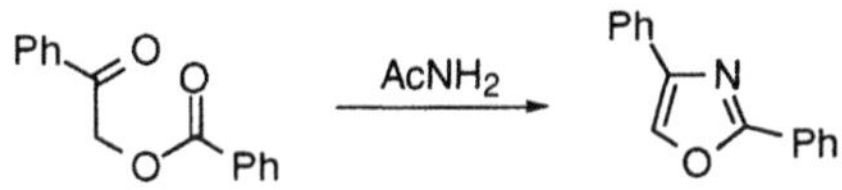

Scheme 7

Treatment of *N*-benzoyl-L-alanine with oxalyl chloride, followed by methanolic triethylamine, yields methyl 4-methyl-2-phenyloxazole-5-carboxylate **32** <95CC2335>. α-Keto imidoyl chlorides, obtained from acyl chlorides and ethyl isocyanoacetate, cyclise to 5-ethoxyoxazoles by the action of triethylamine (e.g., Scheme 8) <96SC1149>. The azetidinone **33** is converted into the oxazole **34** when heated with sodium azide and titanium chloride in acetonitrile <95JHC1409>. Another unusual reaction is the cyclisation of compound **35** to the oxazole **36** on sequential treatment with trifluoroacetic anhydride and methanol <95JFC(75)221>.

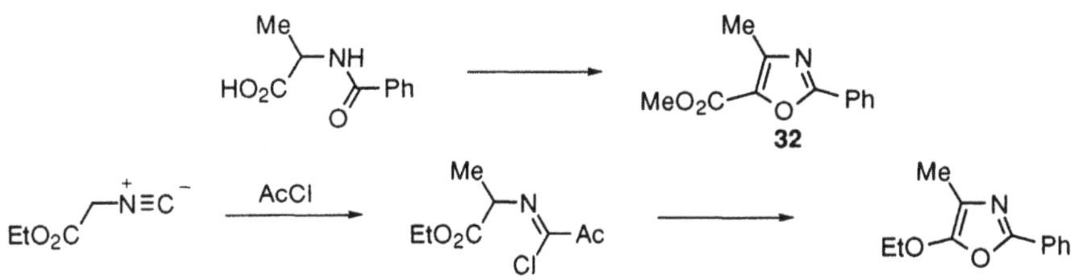

Scheme 8

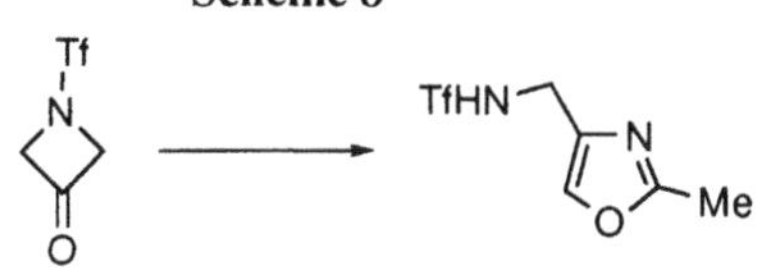

33 **34**

35 **36**

A synthesis of 5-(aroylamino)-2-aryloxazoles **39** is outlined in Scheme 9. Heating the glycol **37** (Bt = benzotriazol-1-yl), prepared from glyoxal and benzotriazole, with an amide in the presence of an ion exchange resin yields the acylated diamine **38**, which cyclises by the action of sodium hydride in DMF <95JHC1651>.

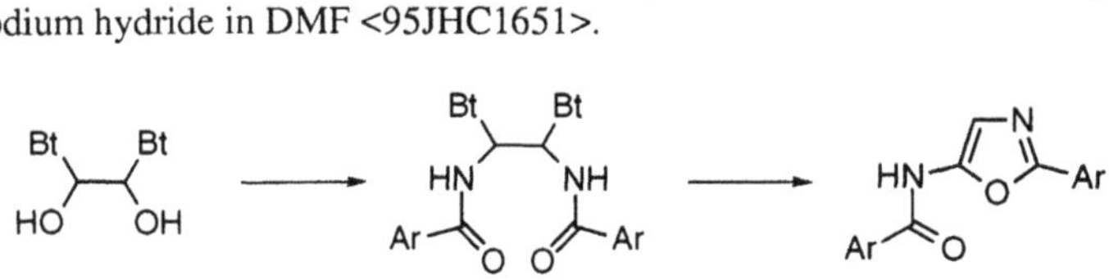

37 **38** **39**

Scheme 9

The synthesis of 2-substituted oxazoles from 2-lithiooxazoles is impeded by the tendency of the latter to undergo ring-opening <Scheme 10>. It has been shown that this can be overcome by locking the electron pair of the nitrogen atom in place by complexation with a Lewis acid. Thus treatment of oxazole or 5-phenyloxazole with THF–borane, followed by lithiation and addition of an electrophile, affords the desired products in good yields <96JOC5192>. Sequential treatment of 5-phenyloxazole with butyllithium, zinc chloride, copper(I) iodide and aroyl chlorides gives 2-aroyl-5-phenylaxazoles in about 70% yield <95TL9453>.

Scheme 10

Benzoxazoles are produced in high yield from α-acylphenol oximes by a Beckmann rearrangement using zeolite catalysts <95SC3315>. The reaction of the *o*-benzoquinone **40** with aromatic aldehyde oximes produces the benzoxazoles **41** <95ZOR1060>. The fused oxazolium salts **43** (R^1 = Me, Et, Pr^i, or Ph; R^2 = Me or Pr^i) are formed from tropone and nitrilium hexachloroantimonates **42** <96JPR598>.

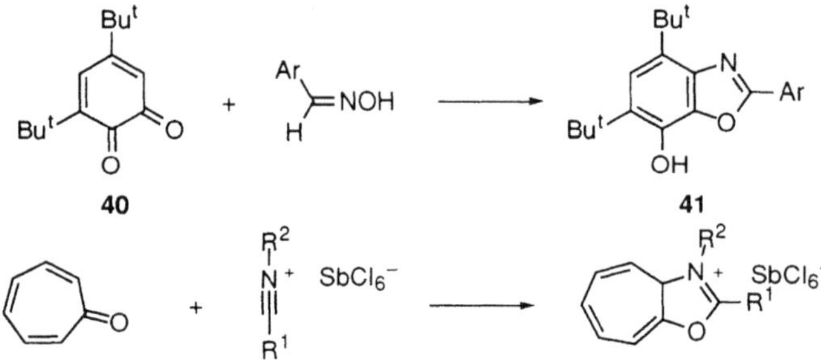

5.7.5 OXAZOLINES

Heating α,α,α-tris(hydroxymethyl)methylamine with a carboxylic acid RCO_2H in a domestic microwave oven for less than 5 min affords 80–95% of the 2-oxazoline **44** <96SL245>. Similarly, microwave irradiation of mixtures of β-amino alcohols and aryl cyanides in the presence of zinc chloride leads to oxazolines (e.g., Scheme 11) <96SC1335>. Treatmsnt of the azirine **45** with aldehydes RCHO (acetaldehyde, benzaldehyde, cinnamaldehyde, furfuraldehyde *etc.*) in the presence of 1,4-diazabicyclo[2.2.2]octane results in ring-expansion to the 3-oxazolines **46** <96JOC3749>. The Ritter reaction of malononitrile with (1*S*, 2*R*)-indanediol **47** in dichloromethane in the presence of trifluoromethanesulfonic acid affords the chiral bis(oxazoline) **48**; analogous products are obtained from succinonitrile and glutaronitrile <96TL813>. The copper complex of the optically active ligand **49** catalyses the enantioselective allylic oxidation of cycloalkenes by t-butyl perbenzoate to yield the esters

50 (n = 1–3) <95SL1245>. Diethylaluminium cyanide promotes the conjugate addition of cyanide to the enantiomerically pure α,β-unsaturated 2-oxazoline **51** to afford **52** diastereoselectively <96SL51>. The chiral bis(oxazolines) **53** (R = Bu^t or Ph) catalyse the cyclopropanation of styrene with ethyl diazoacetate to give the (1*S*, 2*S*)-isomer **54** enantioselectively; the same catalysts promote the formation of the aziridine (*R*)-**55** from styrene and the imine PhI=NTs <916TL4073, 96TL6189, 96SL677>. Almost complete diastereoselectivity was observed in the free-radical cyclisation of the oxazolinone **56** to the pyrrolooxazolidinone **57** by means of tributyltin hydride in the presence of AIBN <96JCS465>.

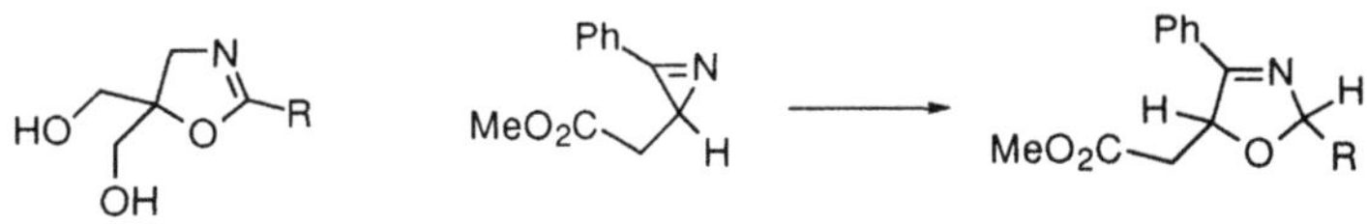

44 **45** **46**

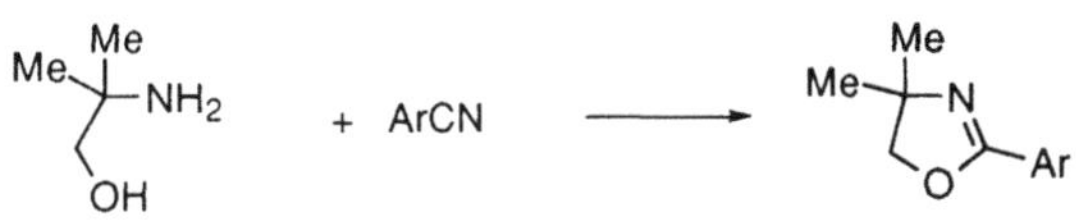

Scheme 11

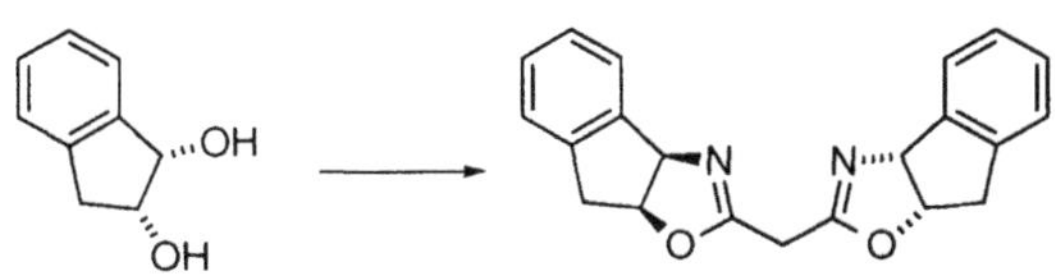

47 **48**

49 **50**

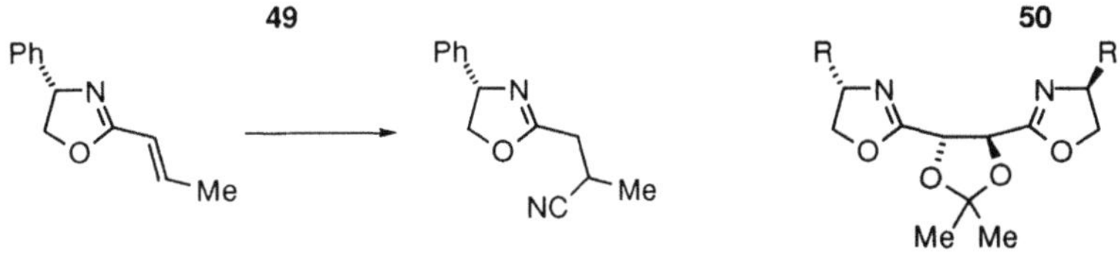

51 **52** **53**

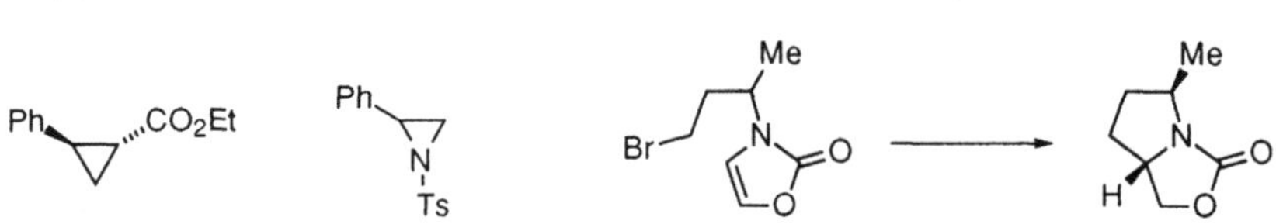

54 **55** **56** **57**

Cathodic reduction of mixtures of benzil imines **58** and *N*-arylcarbonimidoyl dichlorides **59** produces 3,4,5-triaryl-2-(arylimino)-4-oxazolines **60** <95T10375>. Carbodiimides **61** containing ester groups in the α-position cyclise in the presence of tetrabutylammonium fluoride to give 2-arylamino-4-benzylidene-2-oxazolin-5-ones **62** <96S690>. The Diels-Alder addition of cyclopentadiene to the exocyclic double bond of the oxazolinone **63,** catalysed by silica gel/zinc chloride, results mainly in the *exo*-product **64**; by contrast, silica gel/montmorillonite clays promote the formation of the *endo*-adduct **65** <95T9217>. Diels-Alder reactions of the chiral oxazolinone **66** exhibit diastereofacial selectivity: 2,3-dimethyl-1,3-butadiene, for instance, yields the cycloadduct **67** of high optical purity <95T8923>. Chiral 2-bromo-2-oxazolines couple with organostannanes $RSnMe_3$ (R = MeC=C, PhC=C, BuCH=CH or 2-furyl) in the presence of palladium catalysts to yield the corresponding oxazolines (Scheme 12) <96TL1747>. An oxidative rearrangement of 2-substituted 2-oxazolines **68** (R^1, R^2 = H, alkyl or aryl) is shown in Scheme 13 <96JOC2044>.

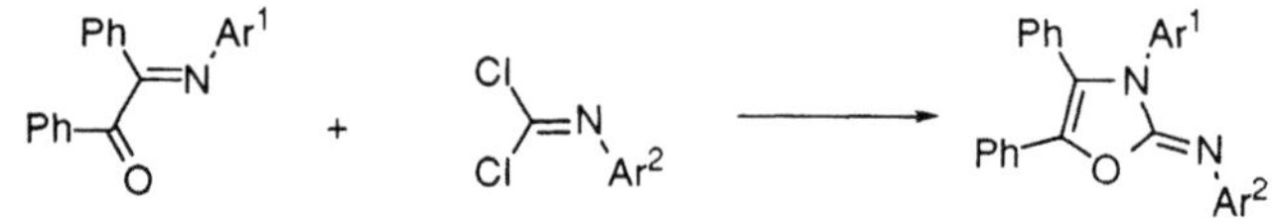

58 **59** **60**

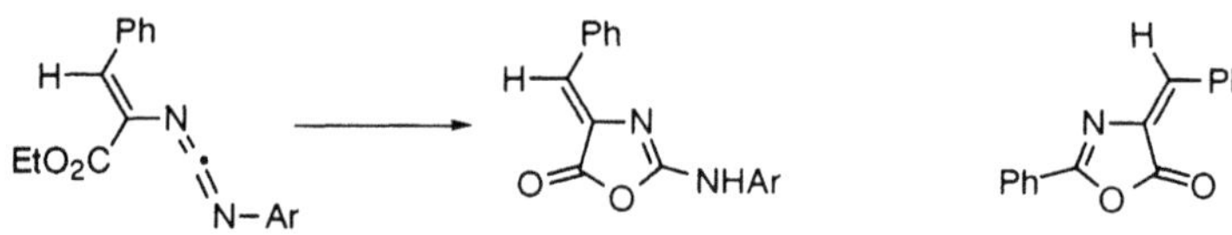

61 **62** **63**

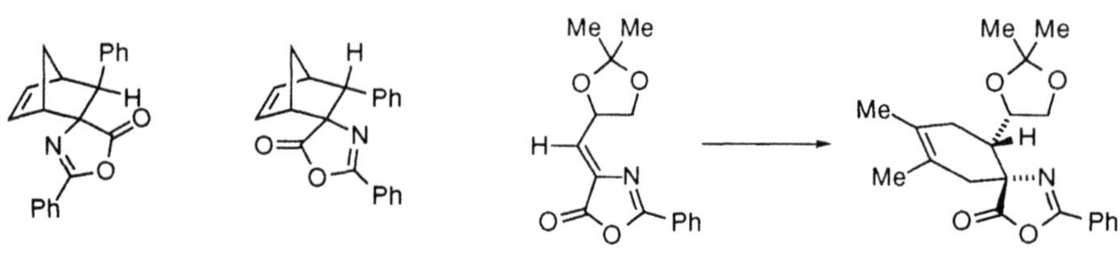

64 **65** **66** **67**

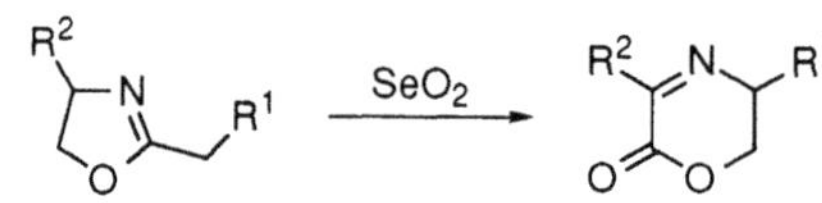

68

Scheme 12 **Scheme 13**

5.7.6 OXAZOLIDINES

The "one-pot domino reaction" of *N*-benzylaniline with benzaldehyde in refluxing toluene results in a mixture of oxazolidines *via* a transient azomethine ylide (Scheme 14) <96S367>. The 2-benzoyloxazolidine **69** rearranges spontaneously to the oxazine **70** <96JHC1271>. The ring-closure of derivatives **71** (R = H or Me) of (*R*)-phenylglycinol to oxazolidin-2-ones

72 is favoured by *N*-methyl substitution (the "Thorpe-Ingold effect") <95BSF808>. *N*-Alkyloxazolidin-2-ones are obtained by the action of di-t-butyl dicarbonate/sodium iodide on aziridines (e.g., Scheme 15) <96T2097>. Treatment of the ylide **73** with benzaldehyde in the presence of caesium fluoride gives a transient adduct **74**, which loses thiirane to afford the oxazolidine thione **75** <96TL711>. β-Amino alcohols react with carbon disulfide under mild basic conditions to yield oxazolidine-2-thiones **76** <95JOC6604>. Oxazolidin-2-one couples with carboxylic acids in the presence of Mukaiyama's reagent (2-chloro-1-methylpyridinium iodide/triethylamine) to yield the acyl derivatives **77** (R = alkyl, aryl, CH_2=CH, PhCH=CH, or PhC=C) <96SC261>. The enolate dianion of the chiral ester **78**, generated by adding two equivalents of sodium hexamethyldisilazide, reacts with alkyl halides in the presence of HMPA to give *anti*-products **79** with high diastereoselectivity <96TL5723>.

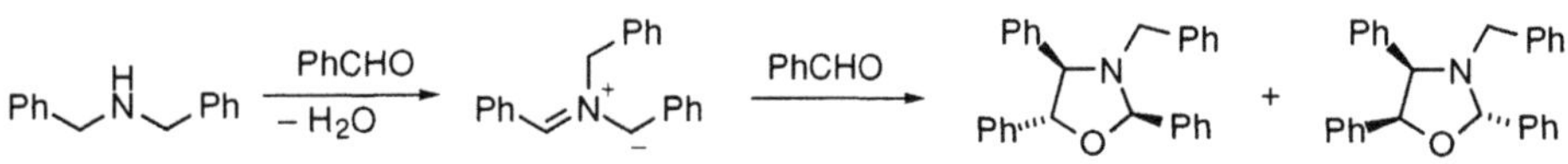

Scheme 14

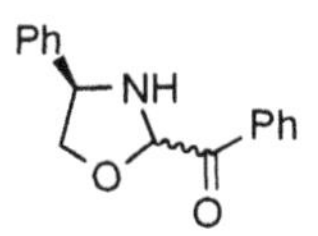

69

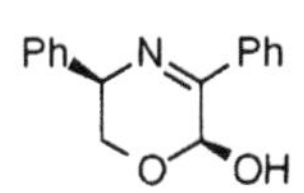

70

71 **72**

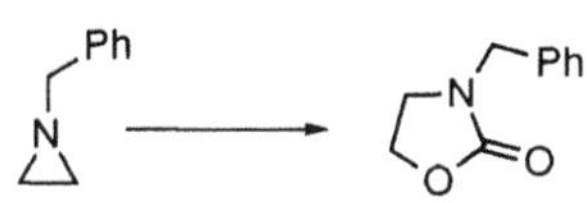

Scheme 15

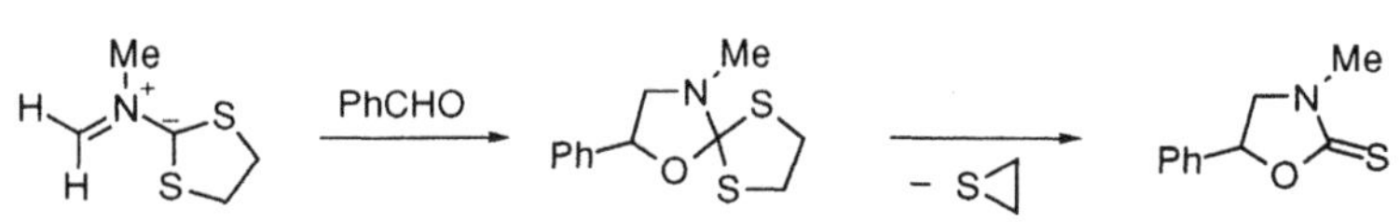

73 **74** **75**

76

77

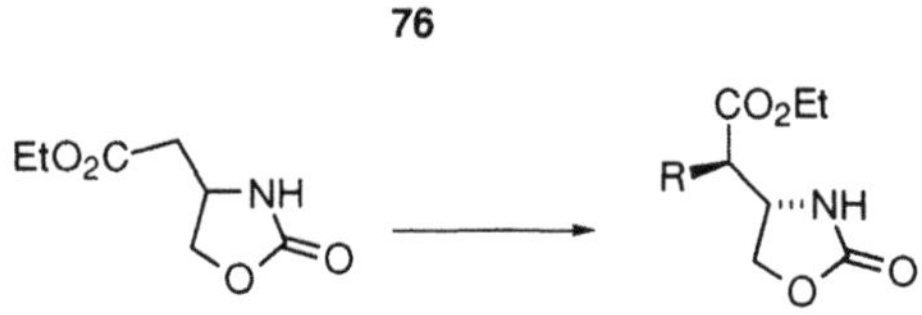

78 **79**

The base-catalysed reaction of prop-2-ynylamines with carbon dioxide results in 5-methyleneoxazolidin-2-ones (e.g. **80**) <96CC1699>. The fused oxazolidinone **82** is formed by the joint action of *N*-phenyltrifluoromethanesulfonimide and potassium t-butoxide on the hydroxylamine derivative **81** <95TL7539>.

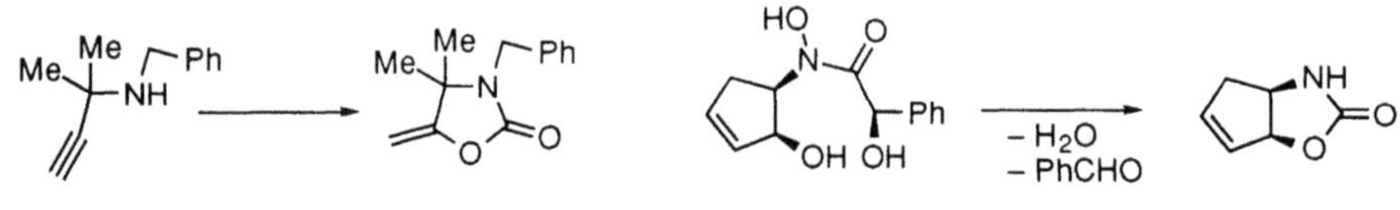

80 **81** **82**

Triethylborane-mediated reactions of lithium enolates derived from chiral *N*-acyloxazolidin-2-ones with ethyl difluoroiodoacetate give products with better than 86% diastereomeric excess (Scheme 16) <96CPB1314>. The lithium enolate **83** condenses with the ester **84** to yield the chiral *N*-[(*E*)-4-methoxycarbonyl-4-pentenoyl]oxazolidin-2-one **85** diastereoselectively <95TA1503>. The oxazolidinone **86,** prepared from oxazolidin-2-one and maleic anhydride in the presence of triethylamine, forms the fumaroid ester **87** on treatment with oxalyl chloride, followed by methanol <95SL1025>. Diels-Alder reactions of the fused oxazolidinone **88**, derived from 1-aminoindan-2-ol, with 1,3-dienes give adducts with good diastereo- and *endo*-selectivity, e.g., **89** with isoprene <95TL7619>. The asymmetric 1,3-dipolar cycloaddition of the nitrone **90** to 3-crotonoyloxazolidin-2-one **91** in the presence of phosphine–palladium complexes proceeds in high yield to give the isoxazolidine **92** in up to 91% enantiomeric excess <96TL5947>.

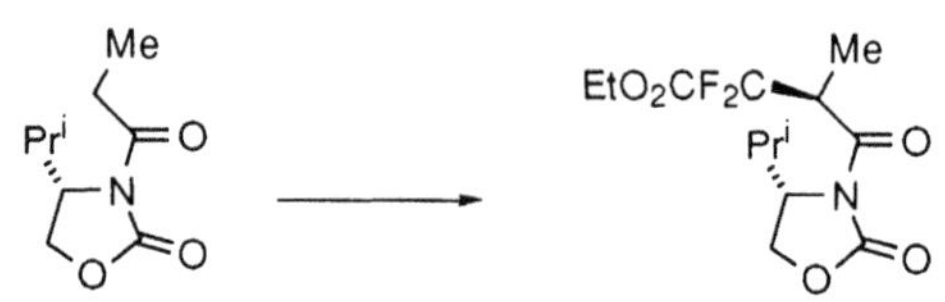

Scheme 16

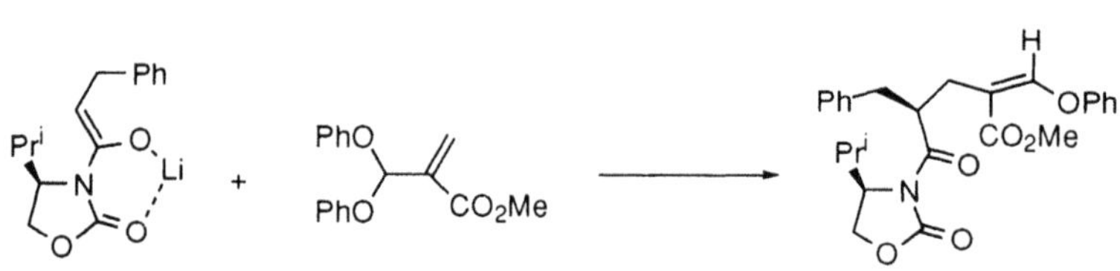

83 **84** **85**

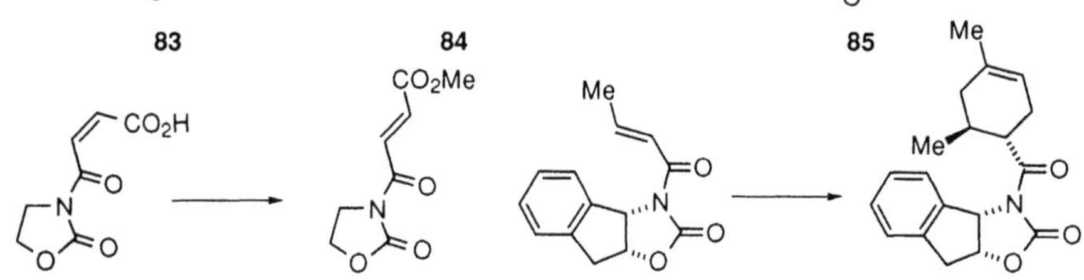

86 **87** **88** **89**

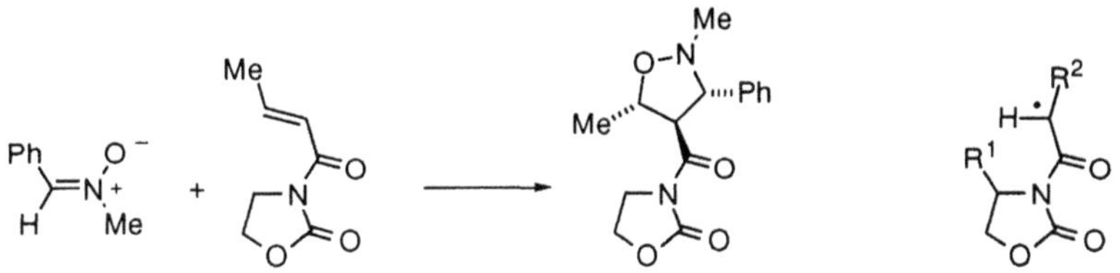

90 **91** **92** **93**

Several examples of stereocontrol of radical reactions of *N*-acyloxazolidin-2-ones have been reported by Sibi *et al.* The rotamer **93** is attainable when a Lewis acid chelates the carbonyl groups and steric interaction orients the group R^2 *syn* to the exocyclic carbonyl group. Thus treatment of the bromo derivative **94** with allyltributyltin and triethylborane/oxygen in the presence of magnesium bromide or scandium(III) triflate gives **95** diastereospecifically <96AG(E)190>. Similarly, isopropyl radicals, generated from 2-iodopropane and tributyltin hydride/triethylborane/oxygen, react with the oxazolidinones **96** to afford **97**, Maximum yield and diastereoselectivity are obtained in the presence of ytterbium(III) triflate <95JA10779, 96JOC6090>. The selective intramolecular 5-*exo* radical cyclisation of the *N*-enoyloxazolidinone **98** to the bicyclic heterocycle **99**, induced by tributyltin hydride/AIBN, occurs with high stereoselectivity <96JA3063>. The (*S*)-cycloadduct **101** is obtained from cyclopentadiene and 3-acryloyloxazolidin-2-one **100** in the presence of catalytic amounts of magnesium perchlorate and (*R*)-2,2'-isopropylidenebis(4-phenyl-2-oxazolidine) **53** (R = Ph); if water is added the (*R*)-enantiomer is formed <96TL3027>. (4*S*)-4-Phenyl-*N*-methacryloyloxazolidin-2-one reacts with thioacetic acid in the absence of catalysts to give the adduct **102** with 94% diastereoselectivity <95TA1633>. Superoxide anion, produced by the electroreduction of oxygen in dipolar aprotic solvents such as acetonitrile or DMF, activates carbon dioxide to give a carboxylating agent which converts *N*-benzylchloroacetamide into the oxazolidine-2,4-dione **103** <96JOC380>.

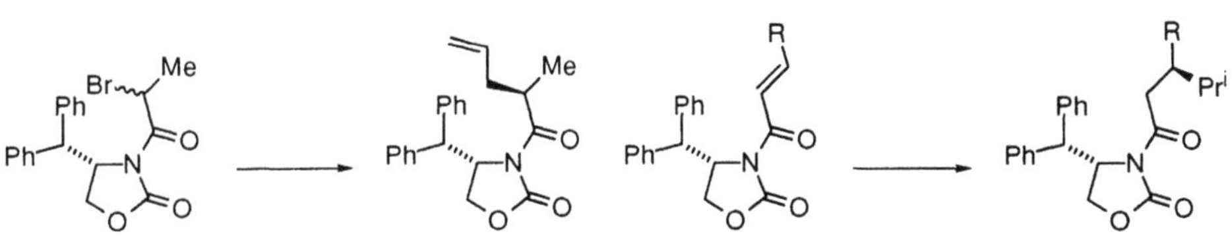

94 **95** **96** **97**

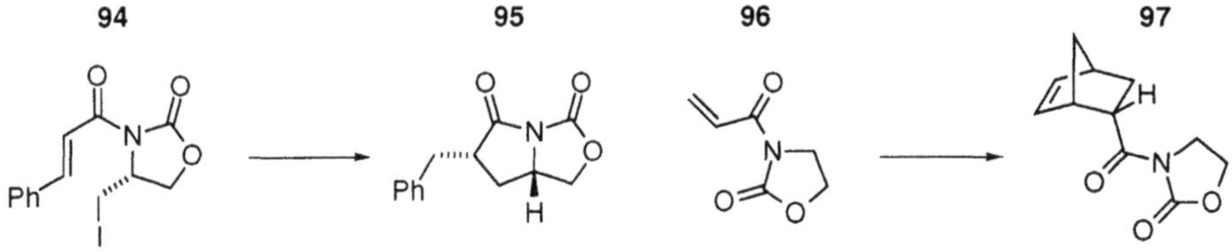

98 **99** **100** **101**

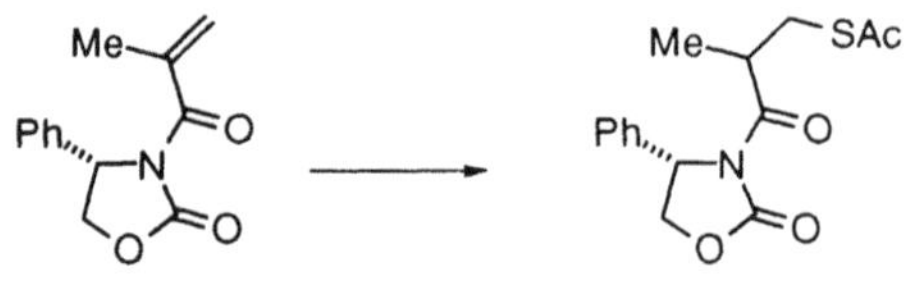

102

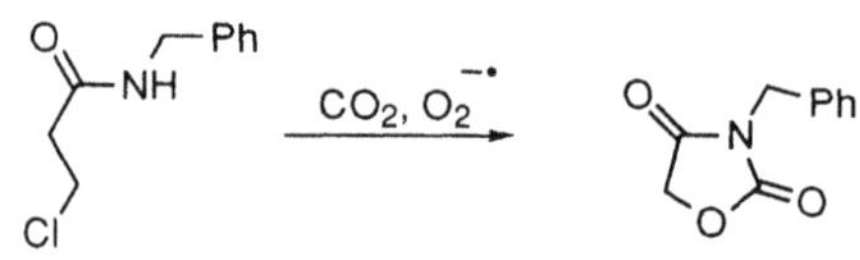

103

5.7.7 OXADIAZOLES

Compounds of type **104,** in which the olefinic double bond and the azoxy group are held rigidly parallel, undergo a thermal acid-catalysed intramolecular [3 + 2] cycloaddition reaction to yield the first stable derivatives **105** of 1,2,3-oxadiazoline <95LA1801>. The "tandem 1,3-dipolar cycloaddition" of sydnones to cycloocta-1,5-diene results in tricyclic compounds (Scheme 17) <96JHC719>. Sydnones **106** (R = Ar or benzyl) are chlorinated at C-4 by means of $PhICl_2$ <96SC1441> and acetylated by acetic anhydride in the presence of the solid clay catalyst Montmorillonite K-10 <96SC2757>. A review of the photochemistry of 1,2,4- and 1,2,5--oxadiazoles has appeared <95H(41)2095>. Diaminoglyoxime **107** (from glyoxime and hydroxylamine hydrochloride) reacts with aqueous potassium hydroxide in a stainless steel autoclave to give diaminofurazan **108** <95JHC1405>. The diaminoazofurazan **109** is converted into the betaine **110** by the action of thionyl chloride <96KGS253>. The synthesis of the furazan macrocycle **111** has been described <96JOC1510>.

104 **105**

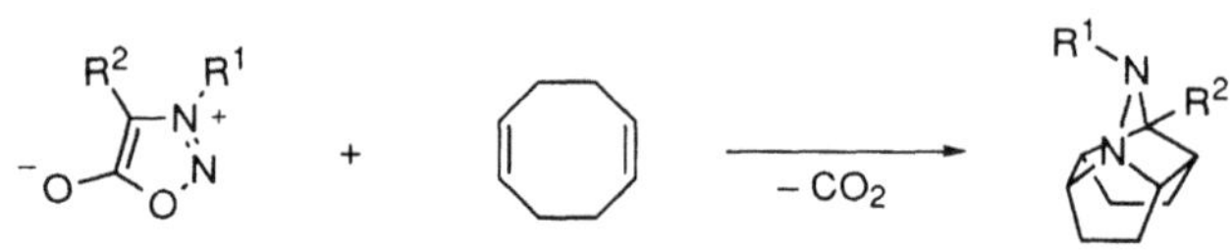

Scheme 17

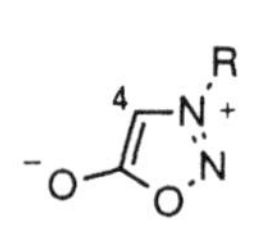

106

107 **108**

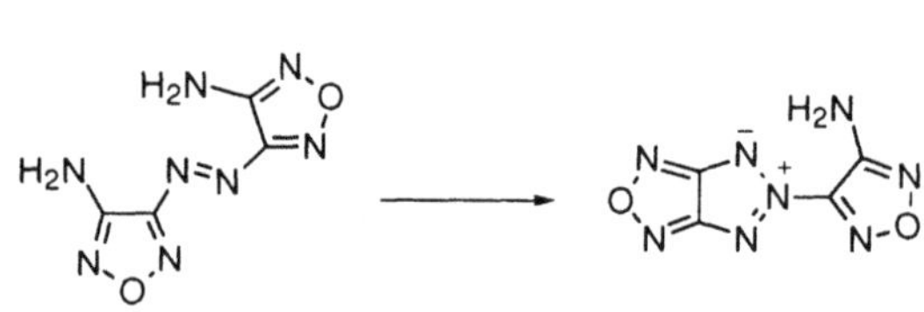

109 **110**

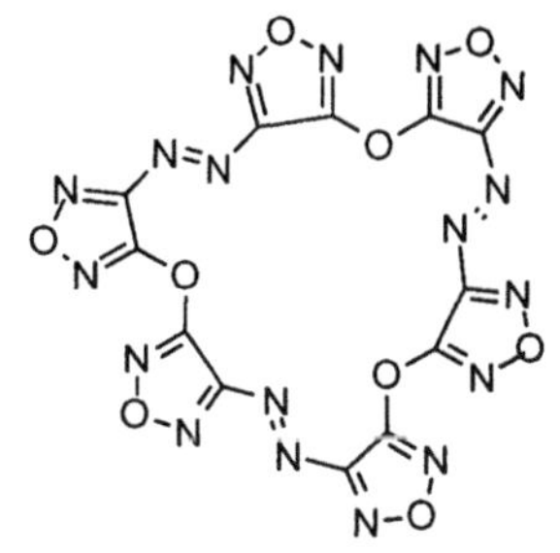

111

The 1,3,4-oxadiazole **113** is formed from the azo compound **112** by the action of triphenylphosphine <96SL652>. A general synthesis of 1,3,4-oxadiazolines consists in boiling an acylhydrazone with an acid anhydride (e.g., Scheme 18) <95JHC1647>. 2-Alkoxy-2-amino-1,3,4-oxadiazolines are sources of alkoxy(amino)carbenes; the spiro compound **114,** for instance, decomposes in boiling benzene to nitrogen, acetone and the carbene **115,** which was trapped as the phenyl ether **116** in the presence of phenol <96JA4214>.

112 **113**

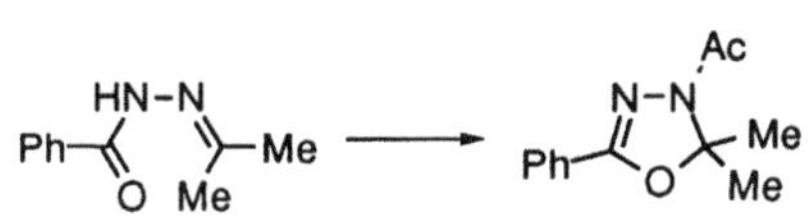

Scheme 18

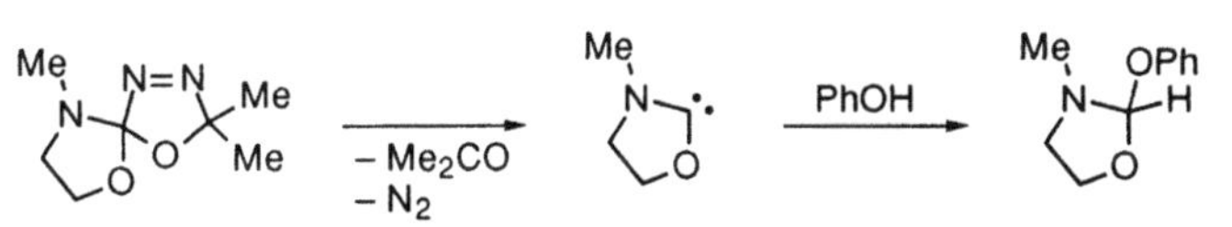

114 **115** **116**

5.7.8 REFERENCES

95BSF808 C. Agami, F. Couty, L. Hamon and O. Venier, *Bull. Soc. Chim. Fr.*, 1995, **132**, 808.

95CC2335 T. Cynkowski, G. Cynkowska, P. Ashton and P. A. Crooks, *J. Chem. Soc, Chem. Commun.*, 1995, 2335.

95CC2457 K. J. Duffy and G. Tennant, *J. Chem. Soc, Chem. Commun.*, 1995, 2457.

95H(41)2095 N. Vivona and S. Buscemi, *Heterocycles*, 1995, **41**, 2095.

95JA10779 M. P. Sibi, C. P. Jasperse and J. Ji, *J. Am Chem. Soc.*, 1995, **117**, 10779.

95JFC(74)221 B. G. Jones, S. K. Branch, M. D. Threadgill and D. E. V. Wilman, *J. Fluorine Chem.*, 1995, **74**, 221.

95JFC(75)215 A. Loupy, A. Petit and D. Bonnet-Delpon, J. *Fluorine Chem.*, 1995, **75**, 215.

5JHC1189 D. J. Anderson and C. R. Muchmore, *J. Heterocycl. Chem.*, 1995, **32**, 1189.

95JHC1405 A. Gunasekaran, T. Jayachandran, J. H. Boyer and M. L. Trudell, ., 1995, **32,** 1405.

95JHC1409 A. P. Marchand, D. Rajagopal and S. G. Bott, *J. Heterocycl. Chem.*, **32**, 1409.

95JHC1647 M. J. Hearn and P.-Y. Chanyaputhipong, *J. Heterocycl. Chem.*, 1995, **32,** 1647.

95JHC1651	A. R. Katritzky, H. Wong and L. Xie, *J. Heterocycl. Chem.*, 1995, **32,** 1651.
95JOC6604	D. Delaunay, L. Toupet and M. LeCorre, *J. Org. Chem.*, 1995, **60**, 6604.
95JOC6624	C. Galli, E. Marotta, P. Righi and G. Rosini, *J. Org. Chem.*, 1995, **60**, 6624.
95LA18O1	S. Hünig and M. Schmitt, *Liebigs Ann. Chem.*, 1995, 1801.
95MI399	T. Sugiyama, *Appl. Organomet. Chem.*, 1995, **9**, 399.
95RTC514	D. Armesto, A. Ramos, M.J. Ortiz, M.J. Mancheno and B.P. Mayoral, *Recl. Trav. Chim. Pays-Bas*, 1995, **114**, 514.
95SC3315	B.M. Bhawal, S.P. Mayabhate, A.P. Likhite and A.R.A.S. Deshmukh, *Synth. Commun.*, 1995, **25**, 3315.
95SL1025	J. Knol and B.L. Feringa, *Synlett*, 1995, 1025.
95SLI208	G. Broggini, C. La Rosa and G. Zecchi, *Synlett*, 1995, 1208.
95SL1245	K. Kawasaki, S. Tsumura and T. Katsuki, *Synlett*, 1995, 1245.
95T8923	E. Buñuel, C. Cativiela and M.D. Diaz-de-Villegas, *Tetrahedron*, 1995, **51**, 8923.
95T9217	C. Cativiela, J.I. Garcia, J.A. Mayoral, E. Pires and R. Brown, *Tetrahedron*, 1995, **51**, 9217.
95T10375	A. Guirado, A. Zapata and P.G. Jones, *Tetrahedron*, 1995, **51**, 10375.
95T10497	H.G. Aurich, C. Gentes and K. Harms, *Tetrahedron*, 1995, **51**, 10497
95TA1503	H. Kanno and K. Osanai, *Tetrabedron Asymmetry*, 1995, **6**, 1503.
95TA1633	T.-C. Tseng and M.-J. Wu, *Tetrabedron Asymmetry*, 1995, **6**, 1633.
95TA1711	C. Baldoli, P. Del Buttero, E. Licandro, S. Maiorana and A. Papagni, *Tetrabedron Asymmetry*, 1995, **6**, 1711.
95TL7539	J. P. Muxworthy, J. A. Wilkinson and G. Procter, *Tetrahedron Lett.*, 1995, **36**, 7539.
95TL7619	I.W. Davies, C. H. Senanayake, L. Castonguay, R. D. Larsen, T.R. Verhoeven and P. J. Reider, *Tetrahedron Lett.*, 1995, **36**, 7619.
95TL9453	N. K. Harn, G. J. Gramer and B. A. Anderson, *Tetrahedron Lett.*, 1995, **36**, 9453.
95ZOR1060	V. N. Komissarov, *Zh. Org. Khim.*, 1995, **31**, 1060 (*Chem. Abstr.*, 1996, **124**, 289319).
96AG(E)190	M. P. Sibi and J. Ji, *Angew. Chem., Int. Ed. Engl.*, 1996, **35**, 190.
96CC1699	M. Costa, G.P. Chiusoli and M. Rizzardi, *J. Chem. Soc, Chem. Commun.*, 1996, 1699.
96CPB1314	K. Iseki, D. Asada, M. Takahashi, T. Nagai and Y. Kobayashi, *Chem. Pharm. Bull.*, 1996, **44**, 1314.
96JA3063	M. P. Sibi and J. Jianguo, *J. Am Chem. Soc.*, 1996, **118**, 3063.
96JA4214	P. Couture, J. K.Tarlouw and J. Warkentin, *J. Am Chem. Soc.*, 1996, **118**, 4214.
96JCS465	Y. Yuasa, J. Ando and S. Shibuya, *J. Chem. Soc., Perkin Trans. 1*, 1996, 465.
96JCS1035	C. Plüg and W. Friedrichsen, *J. Chem. Soc., Perkin Trans. 1*, 1996, 1035.
96JHC383	M. Yamauchi, S. Akayama, T. Watanabe, K. Okamura and T. Date, *J. Heterocycl. Chem.*, 1996, **33,** 383.
96JHC719	G. W. Gribble and B. H. Hirth, *J. Heterocycl. Chem.*, 1996, **33**, 719.
96JHCI271	T. Sheradsky and E. R. Silcoff, *J. Heterocycl. Chem.*, 1996, **33**, 1271.
96JOC356	S.J. Wittenberger, *J. Org. Chem.*, 1996, **61**, 356.
96JOC380	M. A. Casadi, S. Cesa, F. M. Moracci, A. Incsi and M. Feroci, *J. Org. Chem.*, 1996, **61**, 380.
96JOC1510	A. B.Sheremetev, V. O.Kulagina and E. I. Ivanova, *J. Org. Chem.*, 1996, **61**, 1510.
96JOC2044	C. M. Shafer and T. F. Molinski, *J. Org. Chem.*, 1996, **61**, 2044.
96JOC3749	M. C. M. Sá and A. Kascheres, *J. Org. Chem.*, 1996, **61**, 3749.
96JOC5192	E. Vedejs and S. D. Monahan, *J. Org. Chem.*, 1996, **61**, 5192.
96JOC5435	E. Dominguez, E. lbeas, E. Martinez de Marigorta, J. K. Palacios and R. SanMartin, *J. Org. Chem.*, 1996, **61**, 5435.
96JOC6090	M. P. Sibi and J. Ji, *J. Org. Chem.*, 1996, **61**, 6090.

96JPR598 R. Abu-El-Halawa, W. Wirschun, A. H. Moustafa and I. C. Jochims, *J. Prakt. Chem. /Chem. -Ztg.*, 1996, **338**, 598.
96KGS253 V. E. Eman, M. S. Sukhanov, O. V. Lebedev, L. V. Batog and L. I. Khmelnitskii, *Khim. Geterotsikl. Soedin.*, 1996, 253 (*Chem. Abstr.*, 1996, **125**, 167882).
96LA1609 H. Irngartinger, C. M. Koehler, G. Baum and D. Fenske, *Liebigs Ann. Chem.*, 1996, 1609.
96S367 C. Wittland, M. Arend and N. Risch, *Synthesis*, 1996, 367.
96S690 P. Molina, E. Aller, M.Écija and A. Lorenzo, *Synthesis, 1996*, 690.
96SC261 J. Knol and B. L. Feringa, *Synth. Commun.*, 1996, **26**, 261.
96SC1149 W. -S. Huang, Y. -X. Zhang and C. -Y. Yuan, *Synth. Commun.*, 1996, **26**, 1149.
96SC1335 D. S. Clarke and R. Wood, *Synth. Commun.*, 1996, **26**, 1335.
96SC1441 S. Ito and K. Turnbull, *Synth. Commun.*, 1996, **26**, 1441.
96SC2757 K. Turnbull and J. C. George, *Synth. Commun.*, 1996, **26**, 2757.
96SC3401 S. Barrett, P. D. Bentley and T. R. Perrior, *Synth. Commun.*, 1996, **26**, 3401.
96SL51 W. Dahuron and N. Langlois, *Synlett*, 1996, 51.
96SL245 A. L. Marrero-Terrero and A. Loupy, *Synlett*, 1996, 245,
96SL652 J. Kosmrlj, M. Kocevar and S. Polanc, *Synlett*, 1996, 652.
96SL677 A. M. Harm, J. G. Knight and G. Stemp, *Synlett*, 1996, 677.
96SL695 A. Navarro-Ocaña, M. Jiménez-Estrada, M. B. González-Paredes and E. Bárzana, *Synlett*, 1996, 695.
96SL815 M. Ohno, A. Yashiro and S. Eguchi, *Synlett*, 1996, 815.
96T2097 J. Sepúlveda-Arques, T. Armero-Alarte, A. Acero-Alarcón, E. Zaballos-Garcia, B. Y. Solesio and J. E. Carrera, *Tetrahedron,* 1996, **52**, 2097.
96T5043 M. S. Meier and N. Poplawska, *Tetrahedron*, 1996, **52**, 5043.
96T9187 A. Goti, *Tetrahedron*, 1996, **52**, 9187.
96TI0131 W. Huang, J. Pei, B. Chen, W. Pei and X. Ye, *Tetrahedron*, 1996, **52**, 10131,
96TL675 K. H. Ang, R. H. Prager, J. A. Smith, B. Weber and C. M. Williams, *Tetrahedron Lett.*, 1996, **37**, 675.
96TL711 C. W. G. Fishwick, R. J. Foster and R. E. Carr, *Tetrahedron Lett.*, 1996, **37**, 711.
96TL813 I. W. Davies, C. H. Senanayake, R. D. Larsen, T. R. Verhoeven and P. J. Reider, *Tetrahedron Lett.*, 1996, **37**, 813.
96TL995 D. M. Fink and B. E. Kurys, *Tetrahedron Lett.*, 1996, **37**, 995.
96TLI747 A. I. Meyers and K.A. Novachek, *Tetrahedron Lett.*, 1996, **37**, 1747.
96TL2885 M. G. Palermo, *Tetrahedron Lett.*, 1996, **37**, 2885.
96TL3027 G. Desimoni, G. Faita and P. P. Righetti, *Tetrahedron Lett.*, 1996, **37**, 3027.
96TL4073 A. V. Bedekar and P. G. Andersson, *Tetrahedron Lett.*, 1996, **37**, 4073.
96TL4137 H. Irngartinger and A. Weber, *Tetrahedron Lett.*, 1996, **37**, 4137.
96TL4597 M. T. McKiernan and F. Heaney, *Tetrahedron Lett.*, 1996, **37**, 4597.
96TL5723 D.-C. Ha, K.-E. Kil, K.-S. Choi and H.-S. Park, *Tetrahedron Lett.*, 1996, **37**, 5723.
96TL5947 K. Hori, H. Kodama, T. Ohta and I. Furukawa, *Tetrahedron Lett.*, 1996, **37**, 5947.
96TL6189 A. M. Harm, J. G. Knight and G. Stemp, *Tetrahedron Lett.*, 1996, **37**, 6189.

Chapter 6.1

Six-Membered Ring Systems: Pyridine and Benzo Derivatives

Daniel L. Comins
North Carolina State University, Raleigh, NC, USA

Sean O'Connor
Alliant Techsystems, Magna, UT, USA

6.1.1 INTRODUCTION

The chemistry of pyridine and benzo derivatives continues to attract considerable attention. Many synthetic derivatives of these heterocycles are present in therapeutic agents, herbicides and fungicides. The synthetic potential of pyridine and its derivatives as building blocks for the construction of complex molecules continues to act as a stimulator to spark imagination and creativity in the areas of synthetic organic chemistry and material science. This review covers selected recent advances in the field and is not an exhaustive coverage of the literature. Several pertinent reviews appeared during 1996 <96CHEC-II(5)1, 96CHEC-II(5)37,96CHEC-II(5)91,96CHEC-II(5)135,96CHEC-II(5)167,96CHEC-II(5)245, 96COS(3)259, 96AHC(65)1, 96AHC(65)39, 96AHC(65)93>.

6.1.2 PYRIDINES

6.1.2.1 Preparation of Pyridines

Reaction of vinyliminophosphoranes with aldehydes gives betaines which undergo either intra- or intermolecular cyclization to give pyridines **1** or dihydropyridines **2** in moderate yields <95TL(36)8283, 96JOC(61)8094>.

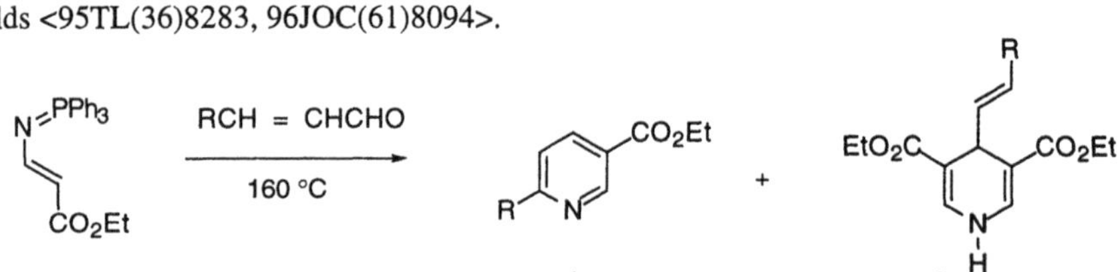

1 **2**

The aza-Wittig reaction of phosphazenes **3** with α,β-unsaturated aldehydes gives 3-azatrienes, which on heating are converted to pyridines **4** <96TL(37)6379>.

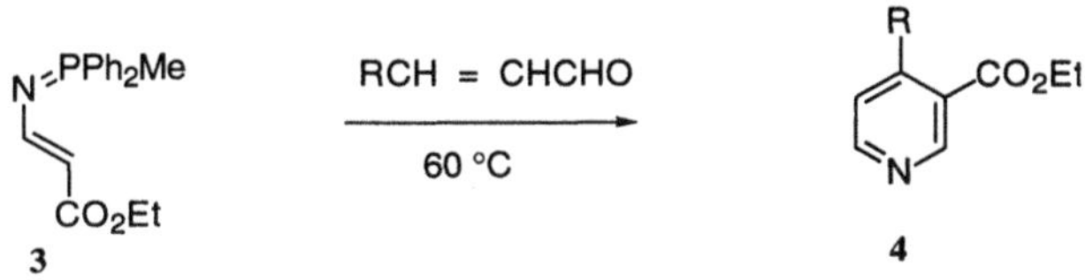

An efficient procedure for the synthesis of 2,4,6-trisubstituted and 2,3,4,6-tetrasubstituted pyridines **5** and **6** involves the one-pot reaction of *in situ* generated α,β-unsaturated imines with carbanions <95TL(36)9297>.

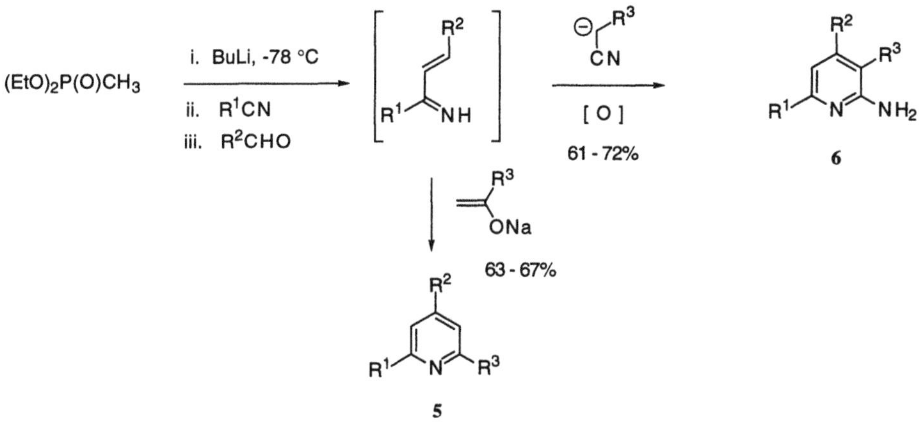

The regioselective halogenation of pyridinium-*N*-(2′-pyridyl)aminide **7** with *N*-halosuccinimides, combined with a reduction of the N-N bond, provides a convenient preparation of 5-halo- and 3,5-dihalo-2-aminopyridines **8** <95T(51)8649>.

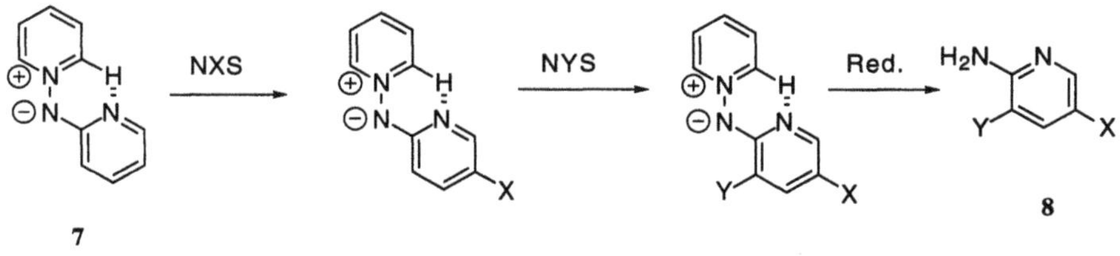

The base-catalyzed Michael addition of 2-chlorocyanoacetate to α,β-unsaturated ketones or aldehydes affords 5-oxopentenenitrile derivatives. In the presence of anhydrous HCl, these compounds cyclize to yield 2-chloro-3-pyridinecarboxylates. The process is highly regiospecific and useful in the synthesis of 2,3-disubstituted pyridines <95T(51)13177>.

The palladium(0)-catalysed coupling of 2-bromocyclopentene-1-carboxaldehyde *N*,*N*-dimethylhydrazone with vinylzinc-, 2-furyl-, or 2-thienylzinc halides provides a route to the corresponding dimethylhydrazones, which cyclize thermally to pyridines, e.g. **9** <95T(51)9119>.

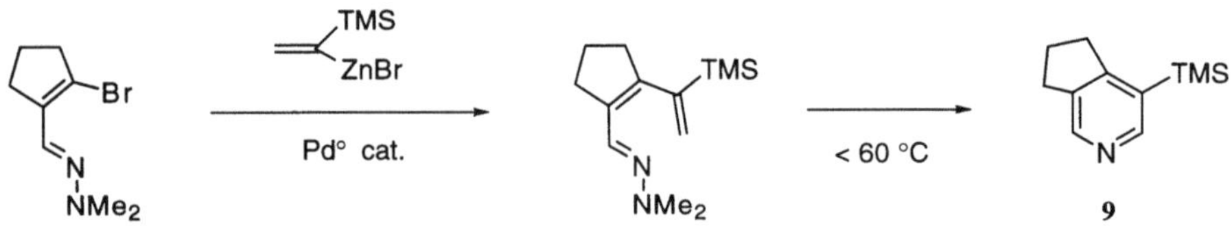

9

Trichloro-1,2,4-triazine reacts with cycloheptene and cyclododecene to give 2,6-dichloropyridine derivatives (e.g., **10**) via an intermediate dihydropyridine formed by Diels-Alder addition and loss of N_2 <96JCS(P1)519>.

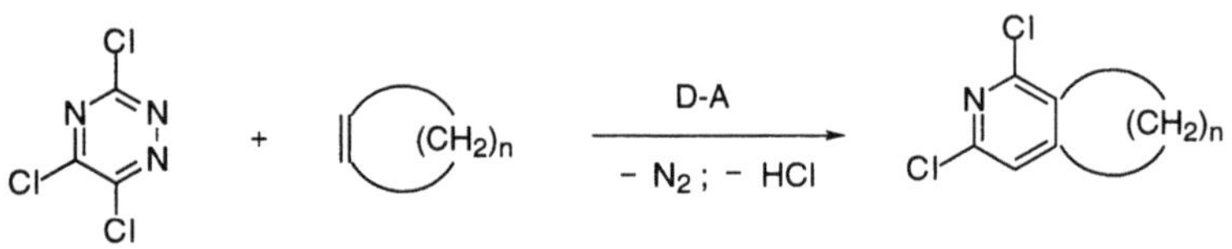

10 n = 5, 10

A useful synthesis of polysubstituted pyridines (i.e., **11**) is based on the regioselective addition of lithiated β-enaminophosphonates to unsaturated carbonyl compounds. These pyridines can also be obtained via a one-pot reaction from metalated phosphonates and sequential addition of nitriles and unsaturated carbonyl compounds <96TL(37)4577>.

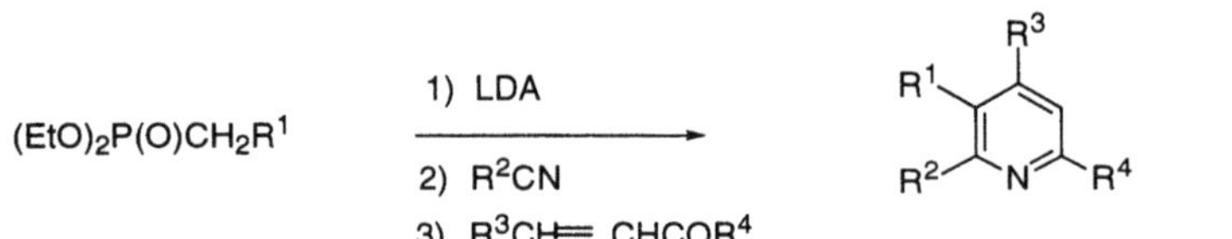

11 55 - 91%

Trimethylsilylketene and acyl isocyanates generate 4-trimethylsiloxy-1,3-oxazin-6-ones **12** *in situ*, which smoothly react with the enamines of cycloalkanones to give bicyclic 2-pyridones **13** <96TL(37)4977>. The heterocycles **12** also undergo the Diels-Alder reaction with dimethyl acetylenedicarboxylate or methyl propiolate to furnish substituted 2-pyridones <96TL(37)4973>.

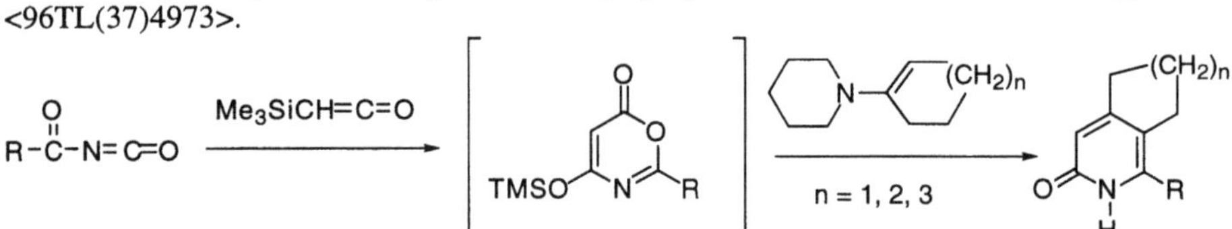

12 **13**

The cycloadducts formed from the Diels-Alder reaction of 3-amino-5-chloro-2(1*H*)-pyrazinones with methyl acrylate in toluene are subject to two alternative modes of ring transformation yielding either methyl 6-cyano-1,2-dihydro-2-oxo-4-pyridinecarboxylates or the corresponding 3-amino-6-cyano-1,2,5,6-tetrahydro-2-oxo-4-pyridinecarboxylates. From the latter compounds, 3-amino-2-pyridones can be generated through subsequent loss of HCN <96JOC(61)304>. Synthesis of 3-spirocyclopropane-4-pyridone and furo[2,3-*c*]pyridine derivatives can be achieved by the thermal rearrangement of nitrone and nitrile oxide cycloadducts of bicyclopropylidene <96JOC(61)1665>.

The mechanism of formation of pyridines from α,β-unsaturated nitriles and active cyano compounds has been investigated. These processes proceed through a Michael adduct which undergoes a regioselective cyclization to the corresponding pyridine <96H(43)33>.

Reaction of 1,3-dicarbonyl compounds with *N,N*-dimethylformamide dimethyl acetal followed by malonamide in the presence of sodium hydride gives 5,6-disubstituted 1,2-dihydro-2-oxopyridine-3-carboxamides, whereas reaction of the intermediate enamines with cyanothioacetamide or cyanoacetamide in the presence of piperidine provides 2-thioxopyridine-3-carboxamides and 4,5-disubstituted 1,2-dihydro-2-oxopyridine-3-carboxamides, respectively <95S923>. β-Enaminonitriles **14** react with β-ketoesters and alkyl malonates, in the presence of stoichiometric amounts of tin(IV) chloride, to afford 4-aminopyridines **15** and 4-amino-2-pyridones **16** <95T(51)12277>.

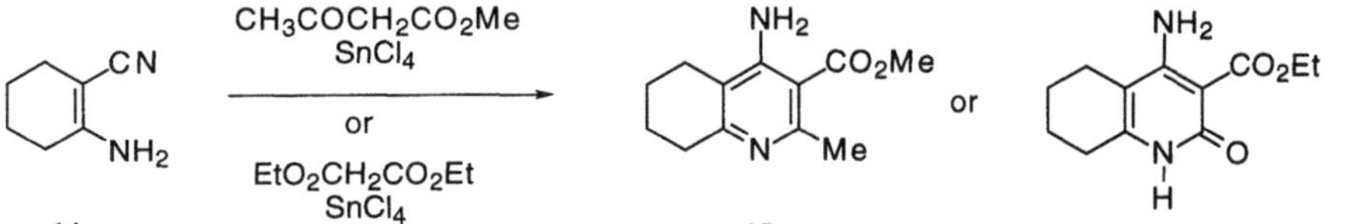

14 **15** **16**

Halo-substituted phthalazines react with two equivalents of ynamines to give penta-substituted pyridines **17** through N-N bond cleavage of the pyridazine ring <96H(43)199>.

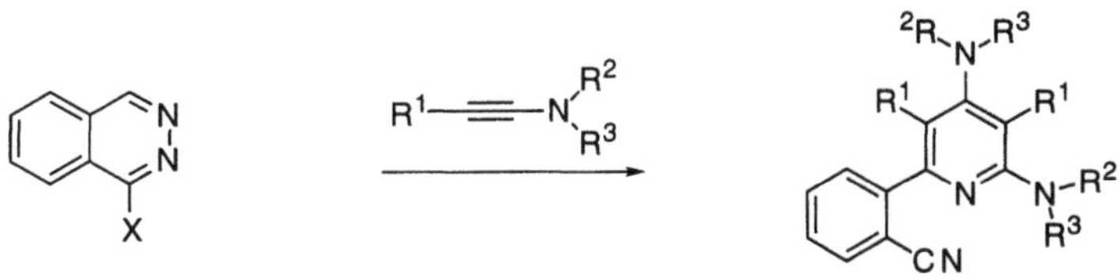

X = Cl, Br

17

When 2,4,6-triarylpyrylium perchlorates and the 4,4′-dimethylbispyridinium dibromides **18** were heated in the presence of triethylamine/acetic acid in boiling ethanol, a double pyrylium ring transformation occurred to give high yields of the 4,4′-bis(2,4,6-triarylphenyl)bispyridinium salts **19** <96JHC(33)783>.

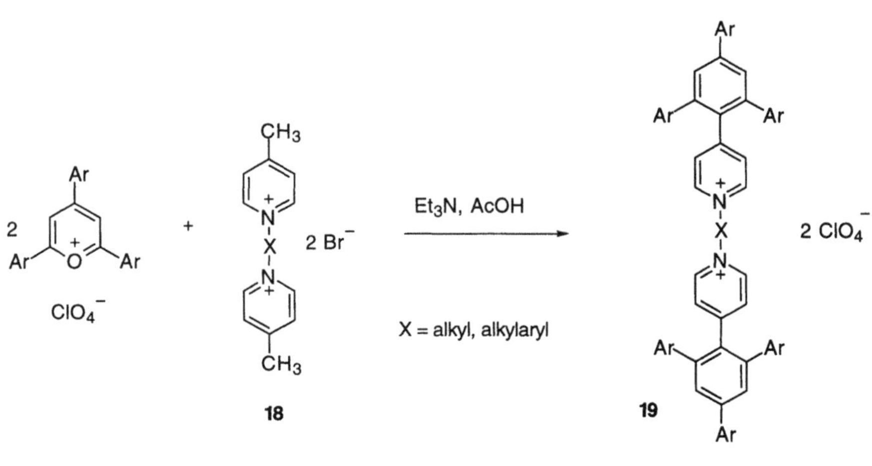

The reaction of 2-phenyl-1,1-dicyanoallyl anion **20** with *N*-methyl-2,2-dimethylpropionitrilium ion in acetonitrile solution resulted in formation of enamine **21**, which on heating undergoes a ring closure to pyridine **22** <96SCA(50)623>.

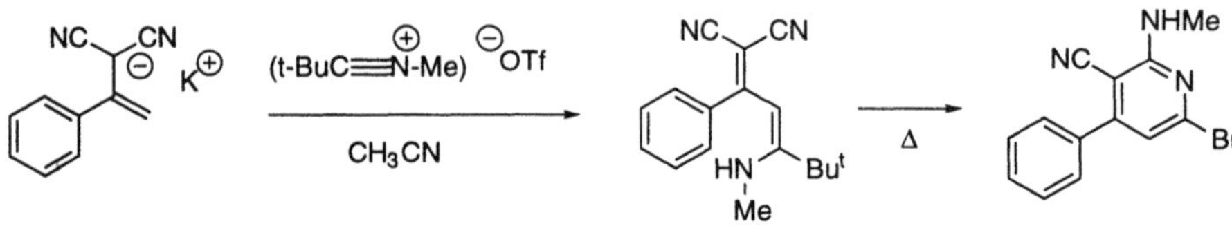

20 **21** **22**

Treatment of triphenylpyrylium perchlorate with α-methylheterocycles, (e.g., **23**), yields fused pyridinium derivatives (e.g., **24**) and low yields of aryl derivatives **25** <96MI99>.

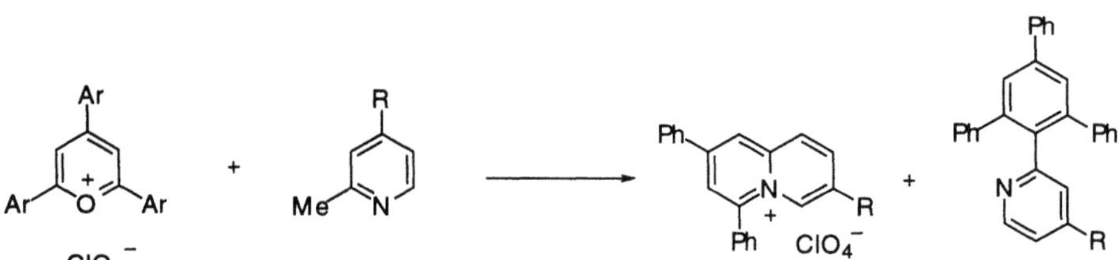

23 **24** **25**

The scope and mechanism of the isomerization of arylamines to methyl-substituted aromatic heterocycles have been studied. Aniline, toluidines, naphthylamines and *m*-phenylenediamine all gave the corresponding ortho-methyl-substituted aza-aromatics when exposed to high NH_3 pressure and elevated temperature in the presence of acid catalysts, e.g., zeolites. The yields of pyridines formed by this process range from low to moderate <95JC(155)268>.

Substituted 2-aza-1-(dimethylamino)-3-(methylthio)-1,3-dienes react readily with a variety of electron-deficient dienophiles to give pyridine derivatives (e.g., **26**) <96T(52)10095>.

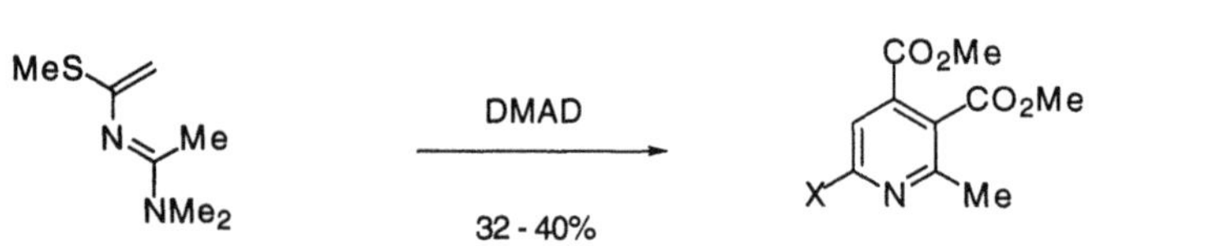

26 X = SMe, NMe_2

A review of the synthesis of chalcogenobispyridines (oxy-, seleno-, telluro-, and thiobispyridines) has appeared <95JHC(32)1671>. Several of the chalcogenobispyridines have displayed a wide range of biological and industrial applications.

Hoornaert and coworkers have continued to develop the preparation of polysubstituted 2-pyridinecarboxylic acid derivatives **27** from the reaction of various functionalized 2-oxa-5-axabicyclo[2.2.2]oct-5-en-3-ones **28** with nucleophiles <96T(52)2591, 96T(52)6997, 96T(52)12529>.

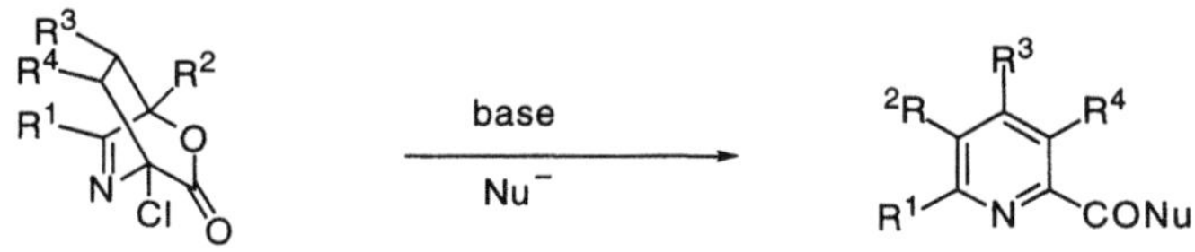

27 **28**

Various 4-(arylthio)pyridine-2-(1*H*)-ones, substituted in their 3-, 5-, and 6-positions, have been synthesized starting from ethyl 2-ethylaminocrotonate and diethyl malonate. Biological studies revealed that some of them show potent HIV-1 specific reverse transcriptase inhibitory properties <95JMC(38)4679>. Beckmann rearrangement of 4,5,6-trisubstituted 3-acetylpyridine-2-one oximes gives exclusively 4,5,6-trisubstituted 3-acetylaminopyridin-2-ones in satisfactory yields <96H(42)35>. The Hantzsch condensation continues to be an effective way to prepare novel 1,4-dihydropyridines related to the drug nifedipine <96JHC(33)157, 96TL(37)4177>. A new solid-phase synthesis has been developed for combinatorial library synthesis of nifedipine derivatives <96JOC(61)924>.

6.1.2.2 Reactions of Pyridines

An interesting palladium-mediated amination of halopyridines to form 2-, 3-, or 4-substituted aminopyridines under mild conditions has been reported <96JOC(61)7240>. A carbene intermediate is proposed as a reactive species in the base-induced reaction of benzoyl chloride salts of pyridine *N*-oxides with carbonitriles to give 2-pyridyldiacylamines as the main products <96TL(37)69>. A good yield of 4-iodopyridine can be obtained from 4-aminopyridine by using a low temperature diazotization-Sandmeyer reaction. The 4-iodopyridine was efficiently converted to the air-stable pinacol ester of 4-pyridylboronic acid <96SC(26)3543>. An improved large-scale preparation of 4-iodopicolinic acid via methyl 4-chloropicolinate has been developed <96SC(26)2017>.

Unsymmetrical quaterpyridines were prepared by Pd(0)-catalyzed cross-coupling of a pyridyl borane or stannane to a chlorinated bipyridine <95T(51)11393>. Reaction of trimethylstannylpyridines with bromopyridines, or bromopyridine *N*-oxides, in the presence of Pd(0) provided all nine pyridinylpyridine *N*-oxides in satisfactory yields. Similarly, nicotelline and 2,2′:6;2″-terpyridine were produced in good yields <96H(42)189>. An efficient synthesis of 3,5-diarylpyridines using a Ni(II)-catalyzed coupling of various arylmagnesium bromides with dihalo pyridines has been reported <95BMC(5)2143>. Various substituted 2-alkoxypyridines were converted into the corresponding 2-chloropyridines **29** under Vilsmeier-Haack conditions <96JCR(S)194>.

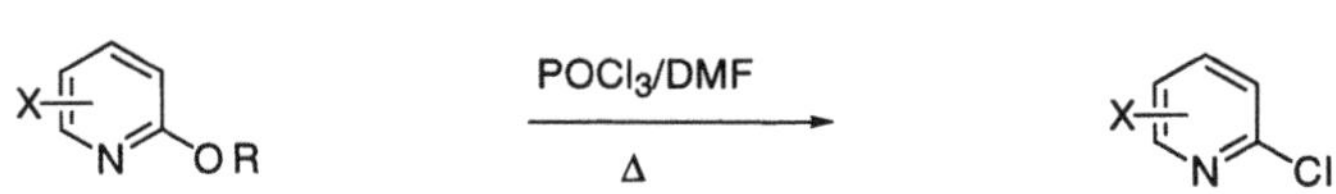

R = Me, Et, Bn
X = H, NH_2, Me, Cl, Br, CO_2Me, NO_2

29

Wakefield and Varvounis have continued their study of abnormal nucleophilic substitution in 3-trichloromethylpyridines <95T(51)12791>. For example, morpholine and 3-

trichloromethylpyridine in acetonitrile under reflux gave the aldehyde **30**. The analogous reaction in THF at room temperature provided the amide **31**.

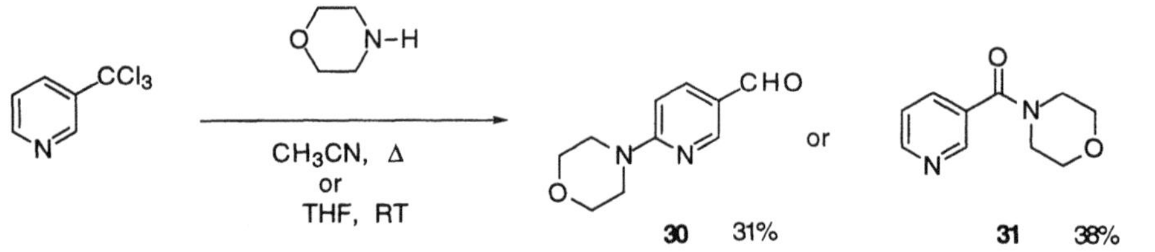

Several substituted fluoropyridines were prepared in high yield from the corresponding aminopyridines using an improved deaminative fluorination procedure <96T(52)23>. Base has been shown to have a remarkable effect on acceleration of the rate of Suzuki coupling of sterically bulky boronic acids with halopyridines in non-aqueous solvent <96TL(37)1043>. A procedure amenable to the preparation of bulk quantities of 3-mercaptopicolinic acid from 3-bromo-2-cyanopyridine has been described <96TL(37)3617>. While 2-chloropyridine reacts poorly with deactivated sulfanions derived from heterocyclic thiols, 2-chloropyridines carrying a trifluoromethyl group at C-3 or C-5 react efficiently yielding unsymmetrical sulfides <95BSF(132)1053>.

The synthesis of 4-substituted pyridines via 1,4-addition of Grignard reagents to pyridinecarboxamides has been studied. After addition of Grignard reagents to pyridinecarboxamides **32**, oxidation of the dihydropyridine intermediates with NCS or oxygen provides the substituted pyridines **33** in good yields <95T(51)9531>.

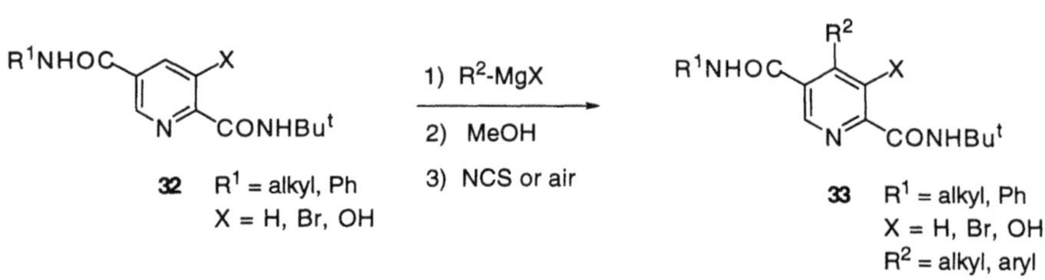

A variation of an approach developed by Meyers was used to prepare nifedipine-type 1,4-dihydropyridines **35** from pyridine **34** using an oxazoline-directed aryllithium 1,4-addition reaction <96H(43)2425>.

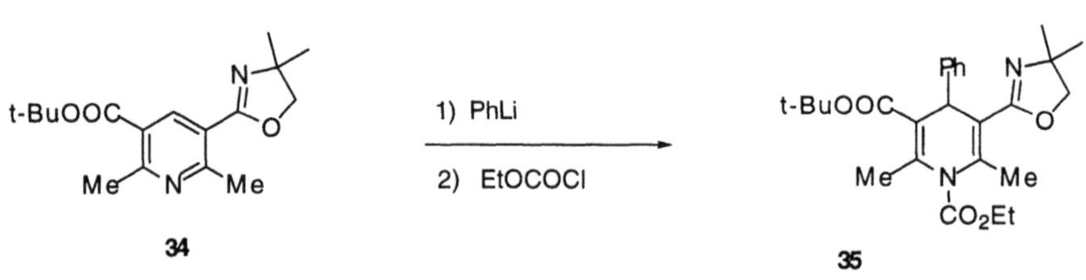

Directed metallation continues to be developed as a convenient method for regiospecific substitution of pyridines. A mild and general procedure for the preparation of structurally diverse 4-alkyl-2-aminopyridines **37** involves the lithiation/alkylation of aminopyridine derivative **36** <96JOC(61)4810>.

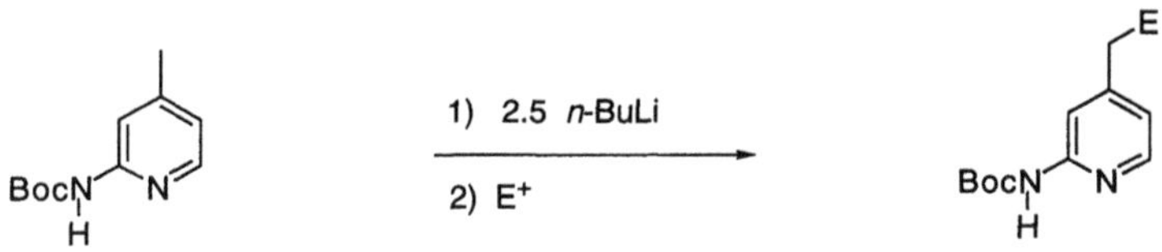

36 37

Lithiation of 3,5-dibromopyridine with LDA and subsequent reaction with electrophiles provide 4-alkyl-3,5-dibromopyridines in high yield <96TL(37)2565>. The synthesis of aza-anthraquinones **39** via metallation of the pyridine ring of **38** was reported by Epsztajn <96T(52)11025>.

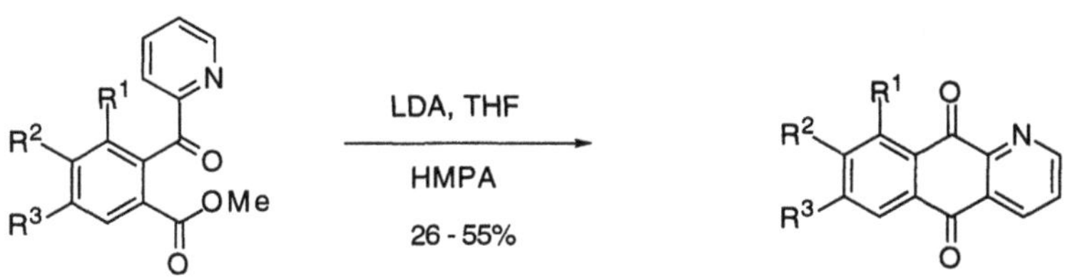

38 39

N-Allyl-2(1*H*)-pyridones **41** are obtained by a palladium(II)-catalyzed Claisen rearrangement of 2-(allyloxy)pyridines **40** <96TL(37)2829>.

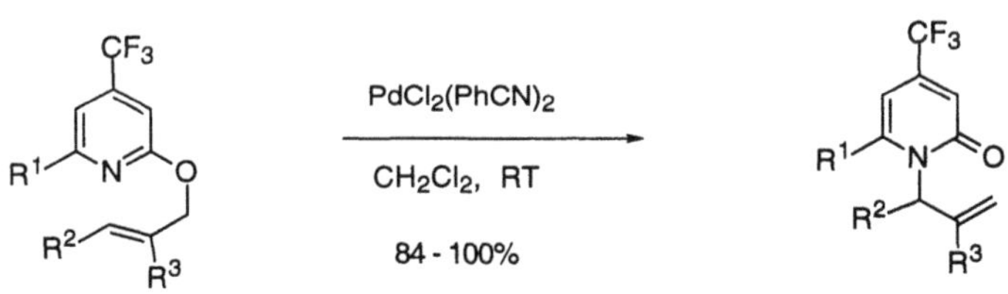

40 41

The hypervalent iodine oxidation of (pyridylalkyl)trimethylsilanes to the corresponding alcohols and esters has been reported <96H(43)1151>. Oxidation of 2-(phenylethynyl)pyridine with H_2O_2/AcOH followed by ROH/Na_2CO_3 provides 6-alkoxy-2-phenacylpyridines **42** via an intermediate *N*-oxide <96H(43)1179>.

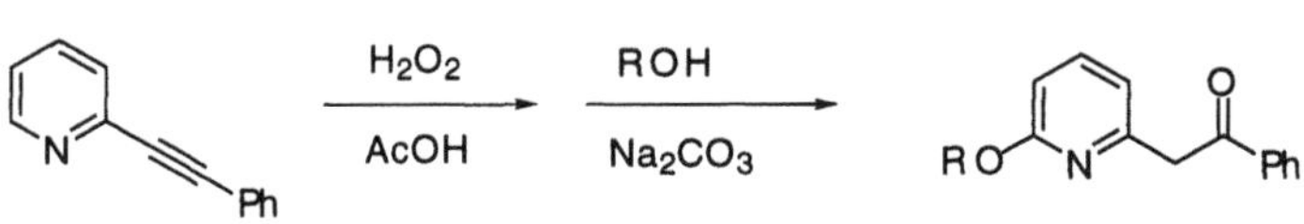

42 8 - 62%

Hoornaert has studied Diels-Alder reactions of pyridine *o*-quinodimethane analogs generated from functionalized *o*-bis(chloromethyl)pyridines <96T(52)11889>. The photochemical cycloaddition of 2-alkoxy-3-cyano-4,6-dimethylpyridine with methacrylonitrile gives a bicyclic azetine, 6-alkoxy-3,5-dicyano-2,5,8-trimethyl-7-azabicyclo[4.2.0]octa-2,7-diene, in moderate yield <96CC1349>. Regiospecific hydroxylation of 3-(methylaminomethyl)pyridine to 5-(methylaminomethyl)-2-(1*H*)-pyridone by *Arthrobacter ureafaciens* has been reported <96MI173>.

The direct bromination of 2-hydroxypyridine is regioselective at C-3 whereas chlorination is not. Quéguiner has developed a halide exchange reaction. Heating 3-bromo-2-hydroxypyridine in pyridine hydrochloride at reflux gives 3-chloro-2-hydroxypyridine in good yield <96TL(37)6695>. The directed aminomethylation of 3-hydroxy-2(1*H*)-pyridones and 3-hydroxy-4(1*H*)-pyridones has been studied <96T(52)1835>.

The C-5 position of *N*-acyl-2-alkyl-2,3-dihydro-4-pyridones can be substituted by halogenation followed by a cross-coupling or carbonylation reaction <95TL(36)9141>. Selective reductions of 1-methyl-4-phenyl-2-pyridone were investigated. Treatment with BH_3 in THF unexpectedly led to the formation of 4-phenylpyridine in 98% yield. $LiAlH_4$ and Red-Al® in THF gave varying amounts of 3,6-dihydro-1-methyl-4-phenyl-2-pyridone and 1-methyl-4-phenyl-1,2,3,6-tetrahydropyridine. L-Selectride® in THF gave exclusively the 1,4-reduction product **43** <96JOC(61)309>.

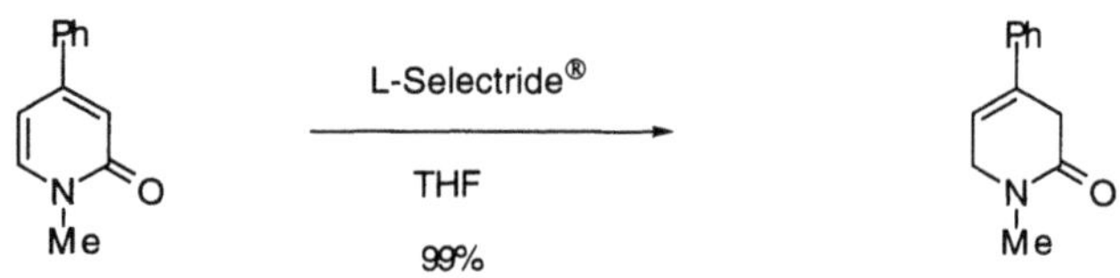

43

The first synthesis of 3-mercapto-2(1*H*)-pyridone has been achieved starting from the thiazolopyridine **44** <96JOC(61)662>.

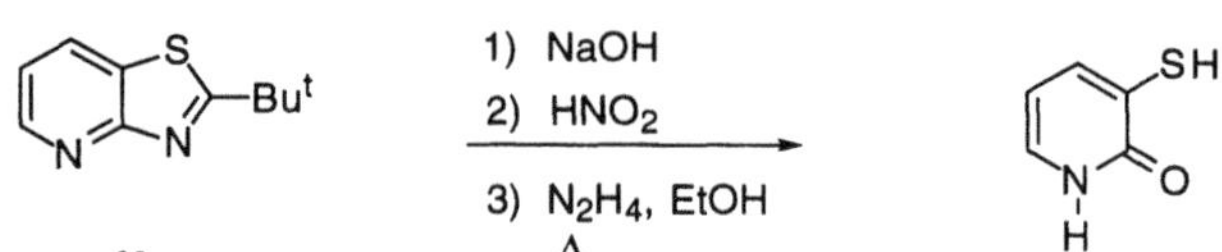

44

Introduction of an oxygen substituent at C-6 of a Hantzsch-type dihydropyridine having a sulfinyl group at C-5 affects the reduction of ketones with respect to both reactivity and stereoselectivity <96CC2535>.

6.1.2.3 Pyridiniums

A study on the effectiveness of various cyclohexyl-based chiral auxiliaries in the asymmetric addition of Grignard reagents to 1-acylpyridinium salts **45** has been published <96TL(37)3807>.

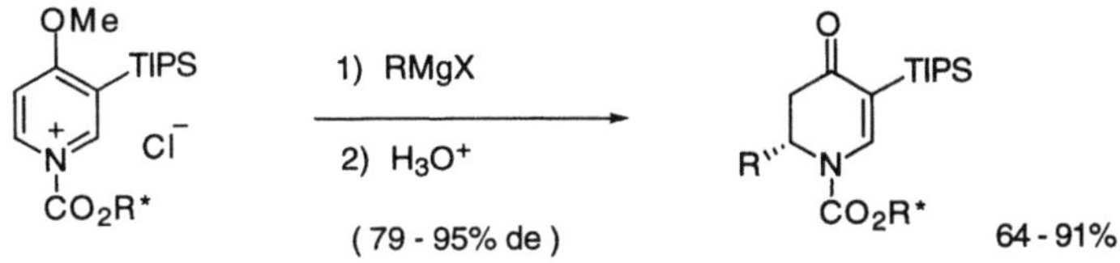

45

It has been shown that liquid SO_2 was not necessary as solvent for the direct nitration of pyridine and substituted pyridines with dinitrogen pentoxide. Several of the standard organic solvents tested gave good yields of 3-nitropyridines even if the reaction medium contained only a minor amount of sulfur dioxide <96SCA(50)556>. A general approach to imidazo[1,2-*a*]pyridines by ferricyanide oxidation of quaternary pyridinium salts has been reported <96CCC(61)126>. Simple pyridines are converted in significant amounts into 1-methyl-, 1-ethyl-, 1-propyl- and 1-pentylpiperidines on reaction with formic acid at high temperatures <96JPR(338)220>.

Two-electron reduction of 4,4′-(1,3-propanediyl)bis-(*N*-methylpyridinium) dibromide with sodium amalgam affords bis-1,4-dihydropyridine **46** <96CL151>.

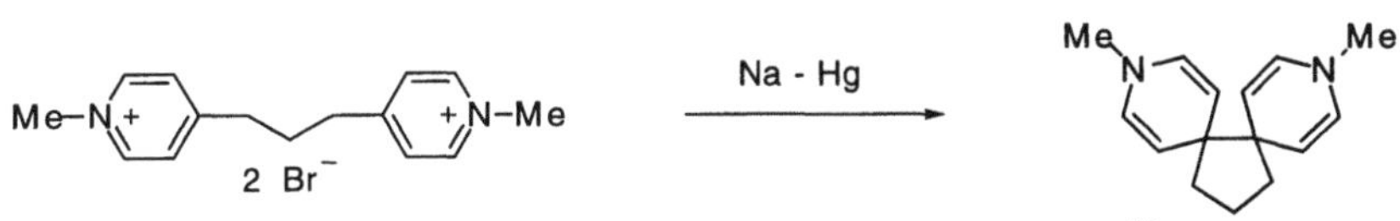

46

A series of new *N*-(pyridylmethyl)azinium salts were synthesized from 2,4,6-triphenylpyrylium tetrafluoroborate <95T(51)12425>.

6.1.3 QUINOLINES

6.1.3.1 Preparation of Quinolines

Palladium-catalyzed reaction of alkyne **47** with a variety of aryl and vinyl halides afforded alkenes **48** in good yield. Cyclization to quinolines **49** was performed by treating **48** with TsOH in EtOH <96T(52)10225>.

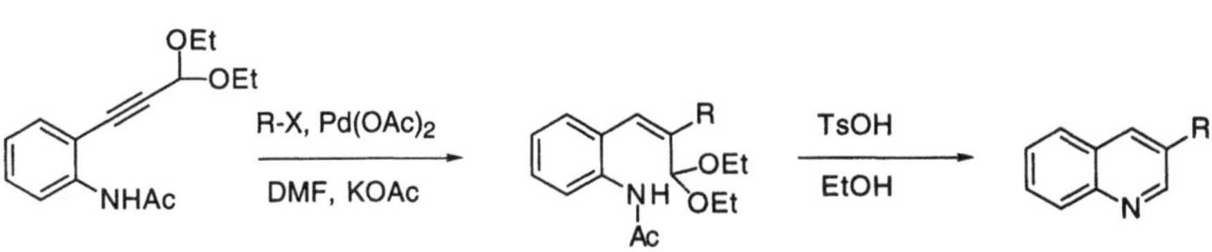

47 **48** **49**

Cyclohexanones have been converted to 8-chloroquinolines through a series of reactions involving imination, α-alkylation with *N,N*-disilyl protected ω-bromoamines, transimination, α-chlorination of the resulting bicyclic imines, dehydrochlorination and dehydrogenation <96T(52)3705>. A short, high yielding one-pot synthesis of acenaptho(1,2-*b*)benzoquinolines has been described, which involves a thermal cyclization of enaminoimine hydrochlorides from 2-chloro-1-formylacenaphthylene and naphthylamine <96SC(26)3959>.

The [4+2] cycloaddition reaction of *N*-arylaldimines with vinyl ethers is effectively catalyzed by ytterbium(III) triflate to give quinoline derivatives (e.g., **50**) in good yield <95S801>.

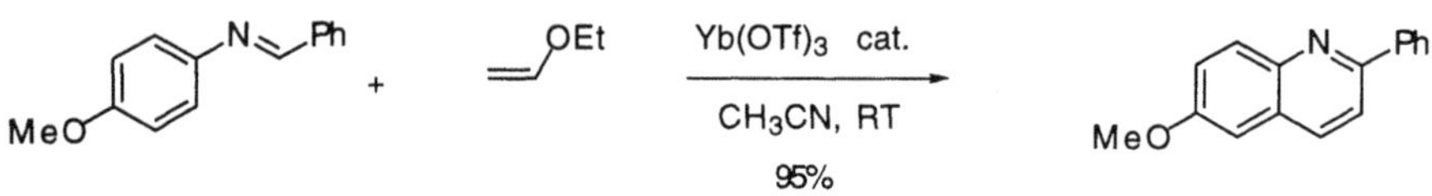

50

The irradiation of haloakenimines **51** leads to the formation of 3-haloquinolines **52** <96JOC(61)7195>.

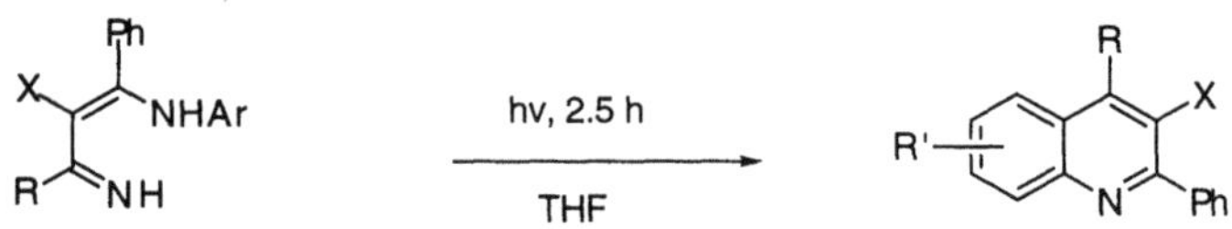

51 **52** X = Cl, Br, I

A facile preparation of 4-(heptafluoropropyl)quinoline **53** occurred on treatment of *ortho*-(nonafluorobutyl)aniline with the lithium enolate of acetaldehyde <96TL(37)4655>.

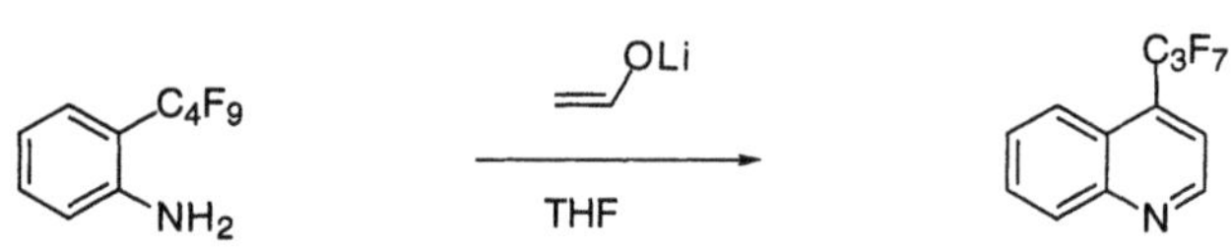

55%

53

A direct preparation of selectively protected derivatives of 3-hydroxyquinoline-2-carboxylates **55** was found through use of a modified Friedlander condensation employing the readily accessible *O*-methyloxime **54** <95JOC(60)7369>.

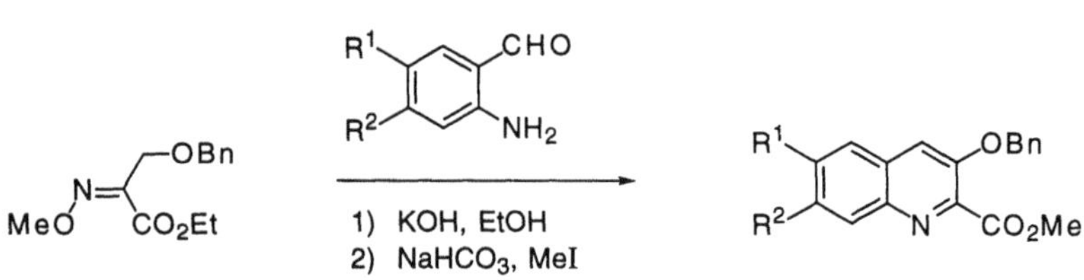

54 **55**

52-70%

Lithiation of diarylbenzotriazol-1-ylmethanes **56** followed by addition of copper(I) iodide gave 6-arylphenanthridine derivatives **57** in moderate yields. According to the substituent effect, it was suggested that the reactions proceed via radical intermediates <96JHC(33)607>.

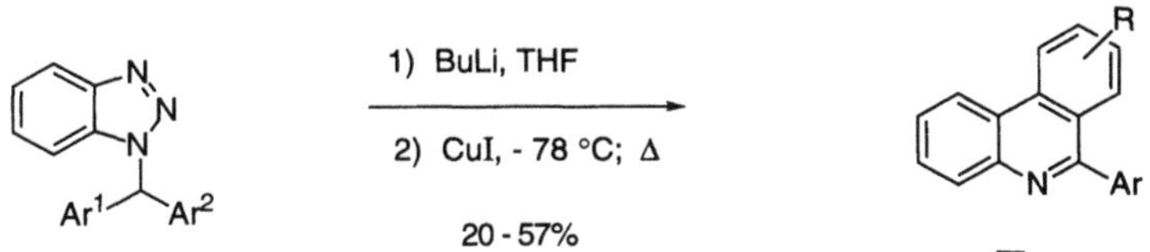

56 **57**

N-Methylformanilide reacts with various electron-rich alkenes in $POCl_3$ solution to give *N*-methylquinolinium salts generally in good yield. In addition, the reaction of *N*-methylformanilide with amides in $POCl_3$ gives 4-quinolones on alkaline workup <95T(51)12869>. The first example of a library of quinolone antibacterial agents prepared by solid phase synthesis has been described <96TL(37)4815>.

Reaction of 3-benzamido-2,5-dioxo-5,6,7,8-tetrahydro-(2*H*)-1-benzopyrans **58** with nitrogen-containing nucleophiles (amines and hydrazines) provides the corresponding quinolin-2-(1*H*)-ones **59** <96H(43)809>.

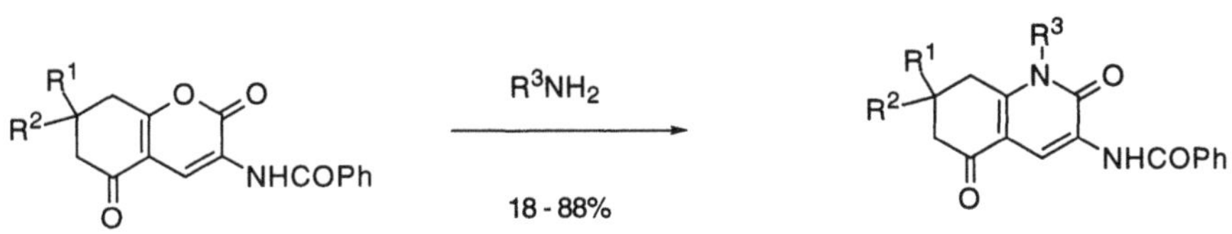

58 **59**

Substituted phenacyl anthranilates were prepared by reaction of sodium anthranilate with substituted phenacyl halides in DMF. The phenacyl esters were cyclized in polyphosphoric acid or by heating to give the respective substituted 2-aryl-3-hydroxyquinoline-4(1*H*)-ones in high yield <95CCC(60)1357>.

The 5,8-dimethoxy-2(1*H*)-quinolinones **60** were obtained in moderate yield via an intramolecular Wittig reaction <95LA1895>.

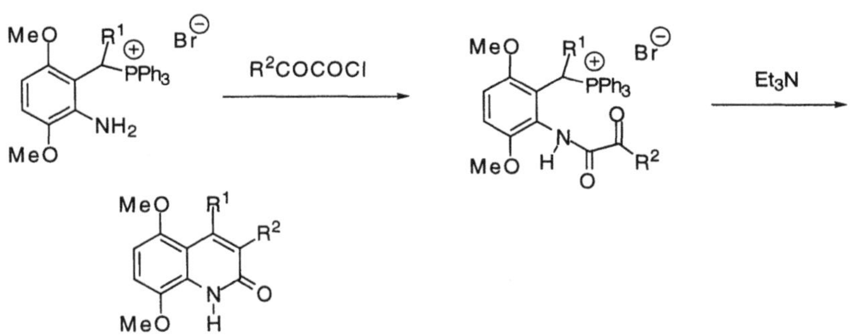

60

Substituted 1,2,3,4-tetrahydroquinolines (e.g., **61**) are formed with high regio- and stereoselectivity in high yield by intermolecular [4+2] cycloadditions of cationic 2-aza-butadienes and various dienophiles <95CC2137, 96SL34>.

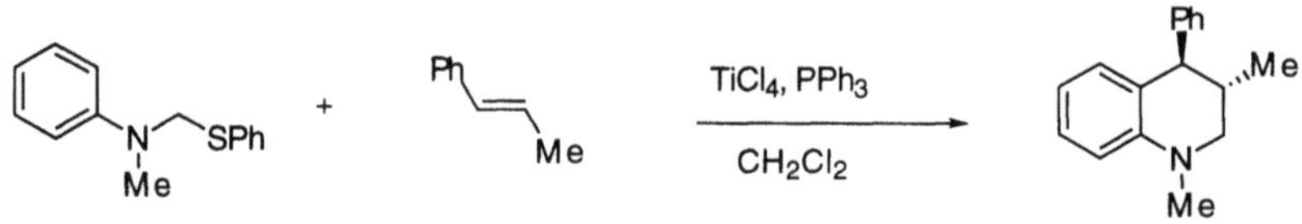

61

A convenient synthesis of benzo[*g*]quinoline via the condensation of β-naphthylamine with malondialdehyde followed by cyclization in polyphosphoric acid was reported <95H(41)2221>.

The intramolecular Diels-Alder reaction of 1-amino substituted isobenzofuran **62** leads to hydrobenzo[*h*]quinolines **63** <95TL(36)8581>.

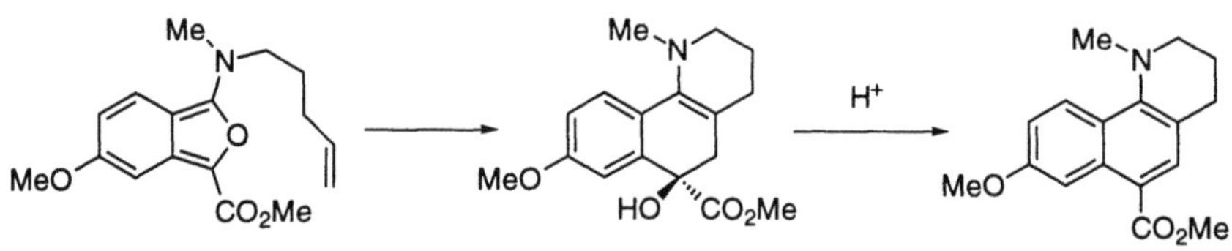

62 **63**

N-(1,1-Dialkylpropargyl)anilines can be cyclized to 2,2-disubstituted 1,2-dihydroquinolines **64** by refluxing in toluene containing cuprous chloride <95TL(36)7721>.

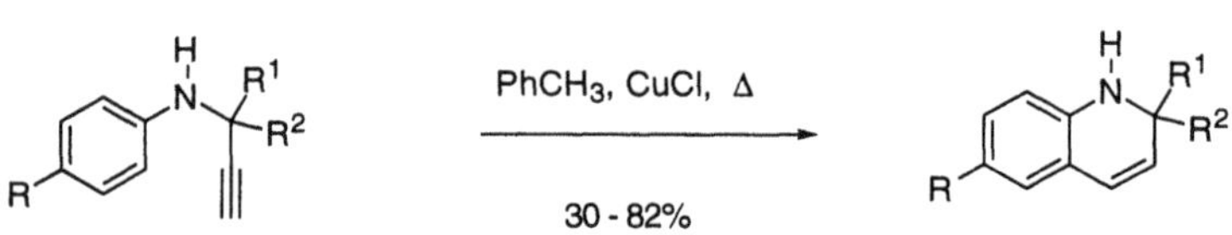

64

Various 1-alkyl-4-(benzotriazol-1-yl)-1,2,3,4-tetrahydroquinolines have been prepared by condensation of *N*-alkylaniline with two equivalents of an aldehyde and one equivalent of benzotriazole <95JOC(60)7631>. Quinolones **66** were simply prepared in good yield by heating a mixture of the appropriate vinylogous amide **65** and $NaHCO_3$ in the presence of a catalytic amount of palladium(II) acetate and triphenylphosphine in DMF under a carbon monoxide atmosphere <96CC2253>.

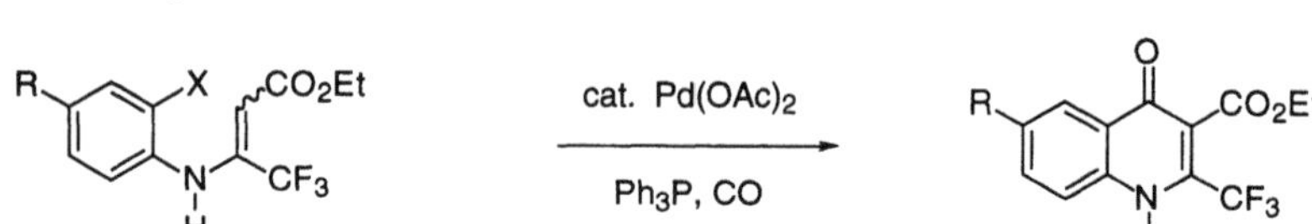

65 **66**

6.1.3.2 Reactions of Quinolines

A *cine*-substitution occurred on the addition of triethylamine to a mixture of 2-quinolone **67** and a 1,3-dicarbonyl compound to give **68** <96BCJ(69)1377>.

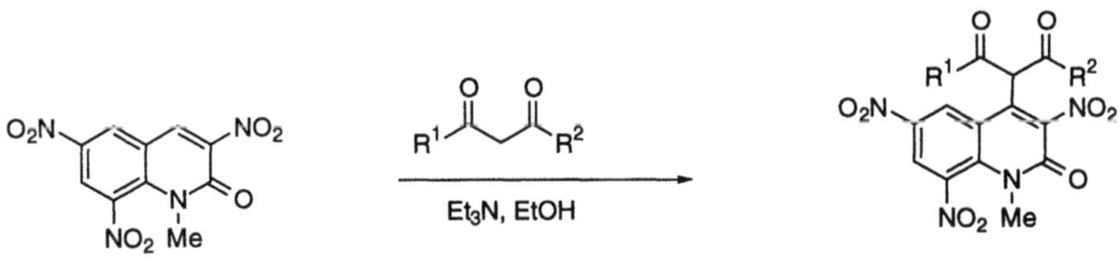

67 **68**

3-Nitro-, 4-nitro-, 7-nitro-, 5,7-dinitro-, and 6,8-dinitroquinoline react with the carbanion of chloromethyl phenyl sulfone to give products of substitution of hydrogen at positions 4; 3; 8; 6, 8; and 5,7; respectively <96LA641>. The base-induced reaction of benzoyl chloride salts of quinoline *N*-oxides with carbonitriles to give 2-quinolyldiacylamines as the main products has been reported <96TL(37)69>.

The *ortho*-directing effect of the carbonyl group in the regioselective lithiation of 2(1*H*)-quinolinone has been studied by Avendaño <95S1362>.

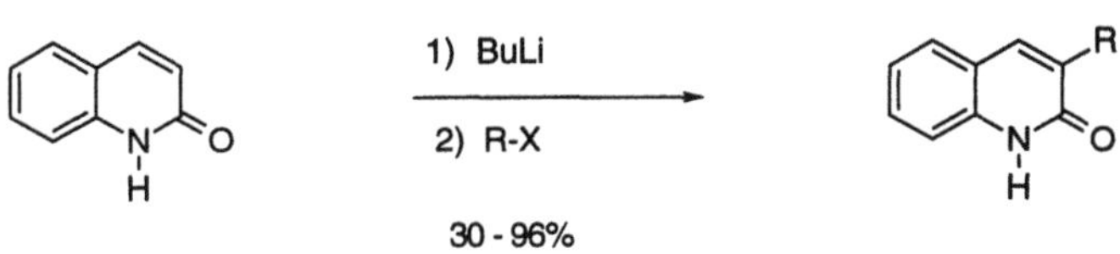

A methoxy group at the 5- or 6-position of 2(1*H*)-quinolinones is compatible with the regioselective lithiation/electrophilic substitution imposed by the *ortho*-directing effect of the quinolinone lithium salt. The coordination effect of a methoxy group at the 8-position changes the reaction course, precluding the *ortho*-directed metallation and effecting a conjugate addition at the 4-position (e.g., **69**) <96H(43)817>.

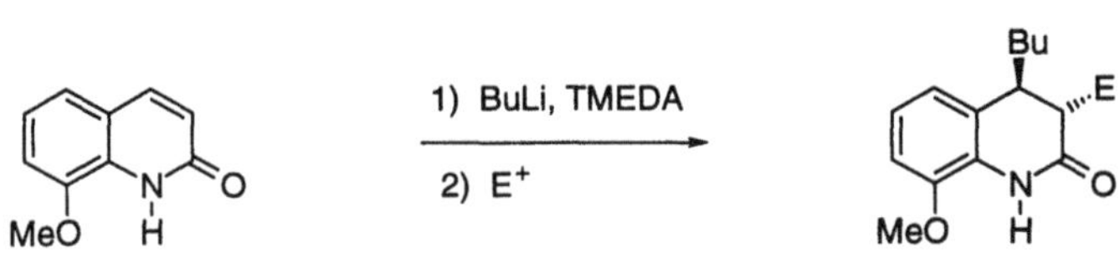

69

6.1.4 ISOQUINOLINES

6.1.4.1 Preparation of Isoquinolines

Condensation of 2-cyanomethyl benzaldehydes with amines carried out in the presence of a catalytic amount of trifluoroacetic acid affords 3-aminoisoquinolines (e.g.,**70**) <95T(51)12439>.

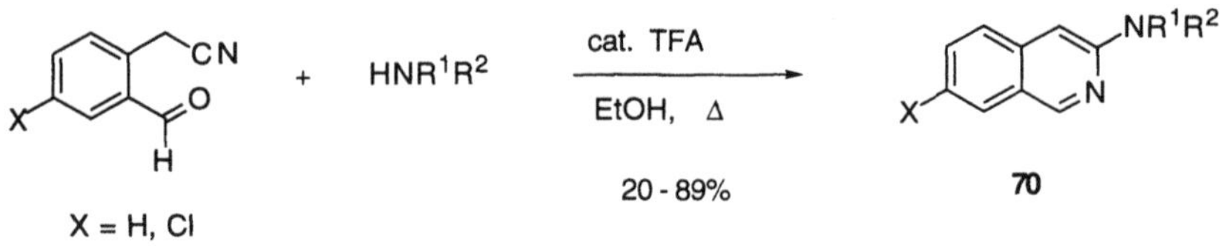

Cyclization of allene **71** using a catalyst system comprising 10 mol % $Pd(OAc)_2$, 20 mol % PPh_3 and K_2CO_3 in the presence of amine **72** affords isoquinolone **73** <95CC1903>.

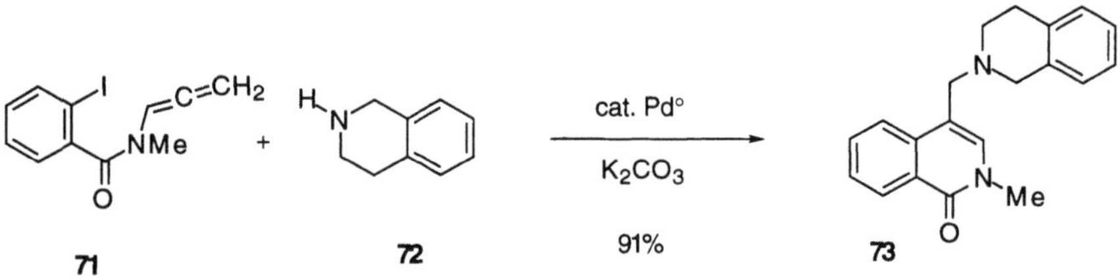

Irradiation of substituted α-dehydrophenylalanine **74** in acetonitrile gives isoquinoline **75** as a major product <96TL(37)5917>.

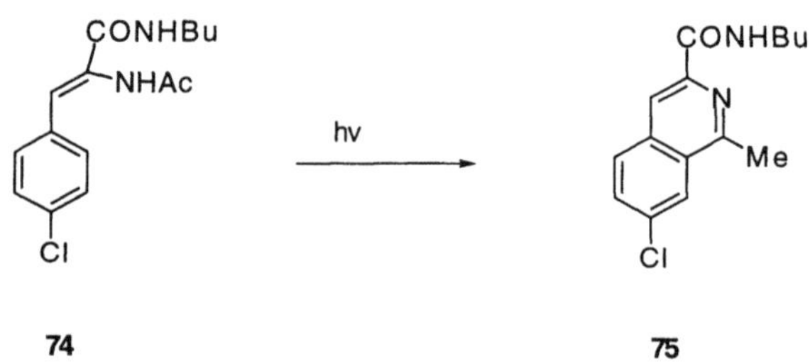

Esters of 2-(2-methylphenyl)hydrazinecarboxylic acids were metallated with excess LDA, and the resulting polyanion intermediates were condensed with aromatic esters followed by acid cyclization to give 3-substituted 1(2*H*)-isoquinolones **76** <96SC(26)1763>.

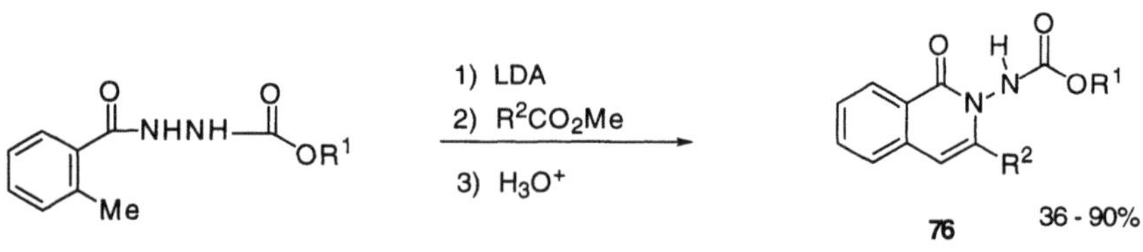

A new synthetic route to 2-alkyl-4-aryl-1(2*H*)-isoquinolones **77** involves the base promoted cyclization of phosphorylated *o*-aroylbenzamides <96T(52)4433>.

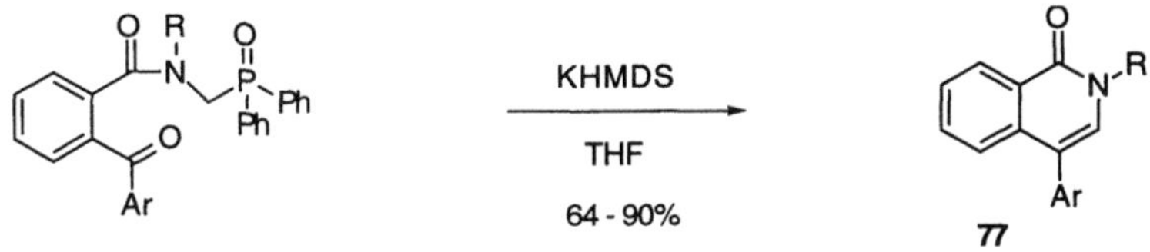

77

An asymmetric synthesis of 1-aryltetrahydroisoquinolines **79** from chiral amide **78** was reported <96TL(37)4369>. Optically active *cis*- or *trans*-1,3-disubstituted tetrahydroisoquinolines can be prepared by a modification of this procedure.

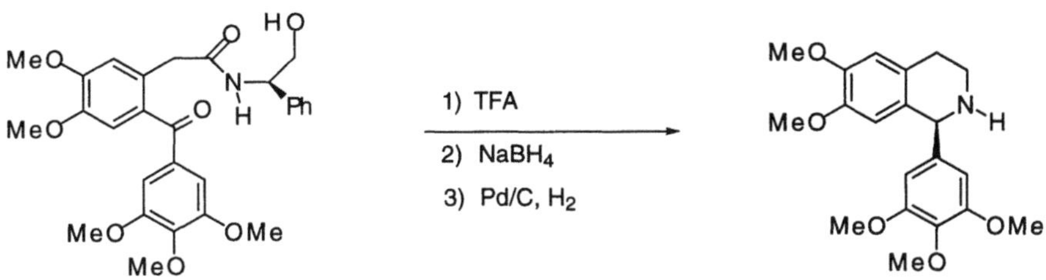

78 79

A novel asymmetric synthesis of 1-aryl-1,2,3,4-tetrahydroisoquinolines is based on the addition of chiral phenylacetaldehyde acetals to acylimines <95TL(36)8003>.

The Bischler-Napieralski and Pictet-Spengler reactions continue to produce new dihydro- and tetrahydroisoquinolines <96TL(37)5453, 95T(51)12159, 96H(43)1605>.

Intramolecular Diels-Alder reaction of substituted furans has been investigated as a route to the synthesis of isoquinoline alkaloids. Tetrahydroisoquinoline **81** was prepared from furan **80** in 40% yield <95JCS(P1)2393>.

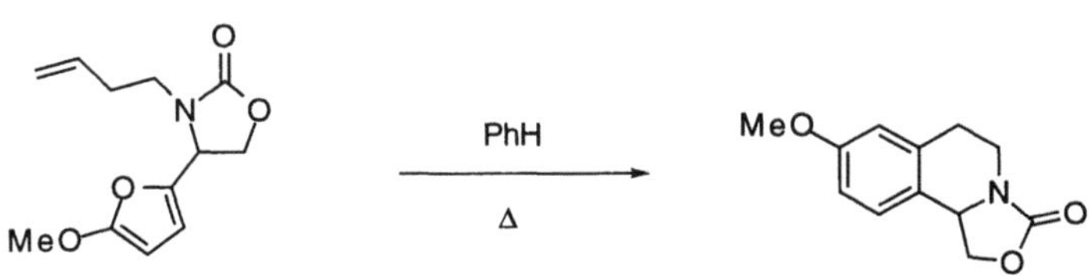

80 81

Wender reported a nickel(0)-catalyzed intramolecular [4+2] cycloaddition as a mild, efficient, and practical method for the assembly of hydroisoquinoline derivatives (e.g., **82**) <96JOC(61)824>.

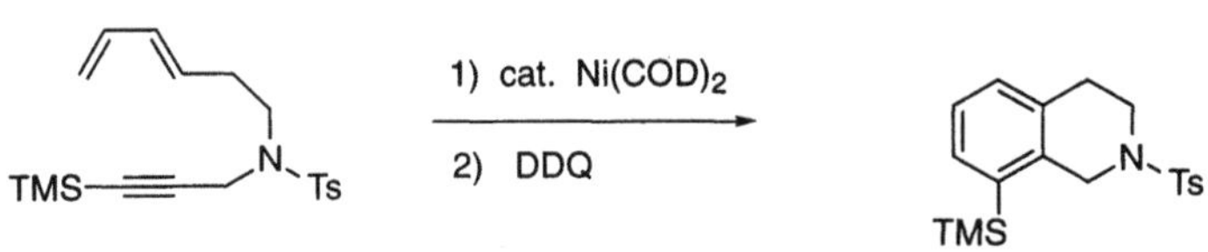

82

The vinylogous Pummerer reaction of an amido-substituted sulfoxide has been used for the preparation of tetrahydroisoquinoline **83** <95JOC(60)7082>.

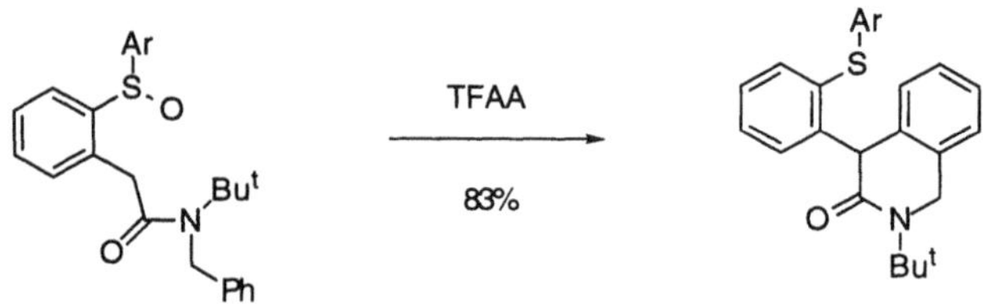

83

The reaction of electron deficient *N*-benzenesulfonyl-β-phenethylamines **84** with ethyl chloro(methylthio)acetate gives tetrahydroisoquinoline-1-carboxylates **85** in high yields <96H(42)141>.

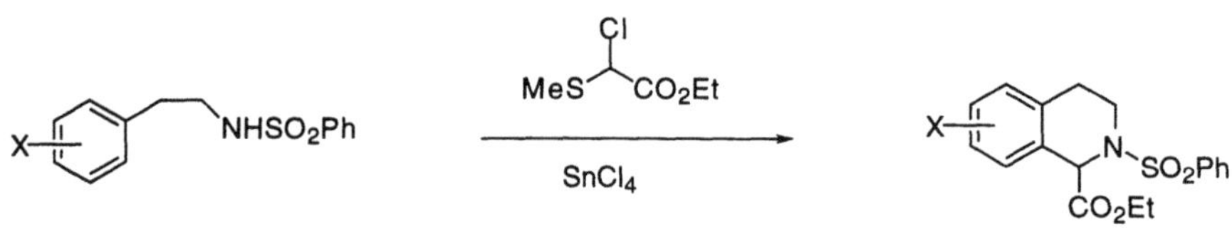

84 44 - 91% **85**

The acid-catalyzed cyclization of *p*-toluenesulfonamides of *N*-benzylaminoacetaldehyde usually yields 1,2-dihydroisoquinolines. However, cyclization of intermediates bearing a substitutent α to the carbonyl group affords 3-substituted 2-*p*-toluenesulfonyl tetrahydroisquinolin-4-ol derivatives, capable of further transformation into the related 1,2,3,4-tetrahydroisoquinolines <95CJC(73)1348>.

Tetrahydroisoquinolines have been synthesized on the Merrifield resin in good yields and high purity via the Bischler-Napieralski approach <95TL(36)7709>.

6.1.4.2 Reactions of Isoquinolines

A direct preparation of 1-nitroisoquinolines from the corresponding isoquinolines was accomplished with potassium nitrite and acetic anhydride in DMSO <96JCS(P1)1777>. This method has preparative merit since 1-nitroisoquinolines are not readily accessible by other methods.

An efficient synthesis of 2-methyl-1-(2-oxoalkyl)-1,2-dihydroisoquinolines **86** via addition of ketone enolates to isoquinolinium methiodide has been reported <96JOC(61)4830>.

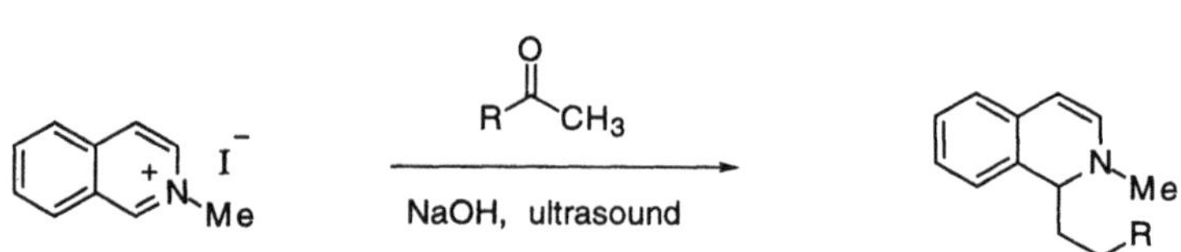

72 - 80% **86**

While 3-arylisoquinolin-1(2*H*)-ones **87** are obtained by air oxidation of 3,4-dihydroisoquinolinium salts, the use of DDQ in dioxane results in a selective dehydrogenation to the corresponding *N*-substituted isoquinolinium salts **88** <95T(51)12721>.

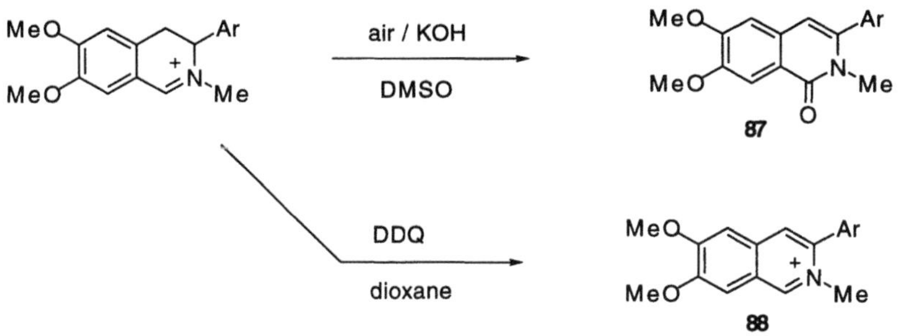

Nucleophilic addition reactions of allylic tin reagents to chiral 3-substituted 3,4-dihydroisoquinolines **89** activated by acyl chlorides afford trans 1,3-disubstituted 1,2,3,4-tetrahydroisoquinolines **90** stereoselectively <95CL1003>.

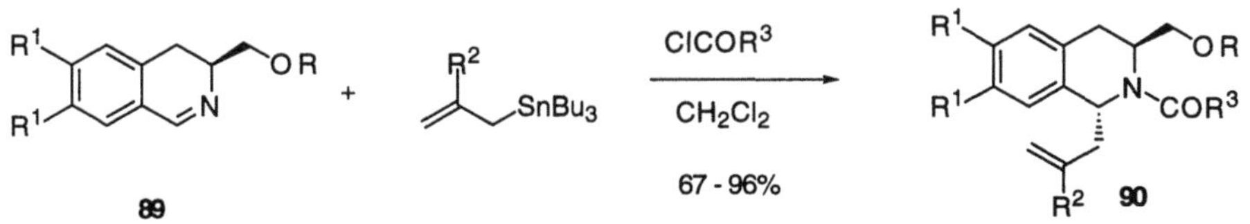

The intermolecular α-amidoarylation of salts **91** containing a nitro group at C-7 has been studied <96SC(26)127>.

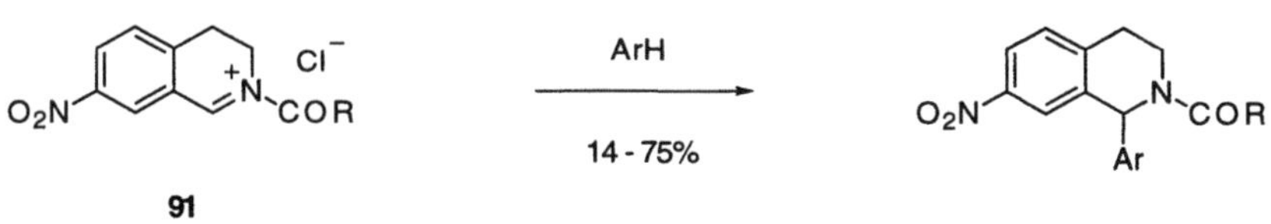

A short synthesis of 6-cyano-1,2,3,4-tetrahydroisoquinoline used an improved method of aryl triflate cyanation that employs zinc(II) cyanide as the cyanide source <95SC(25)3255>.

Alkylation of the lactam **92** via its enolate has been studied and shown to be highly stereoselective. The 4-substituted 1,4-dihydroisoquinoline-3-one **93** was obtained in high yield with greater than 97% de <96T(A)(7)417>.

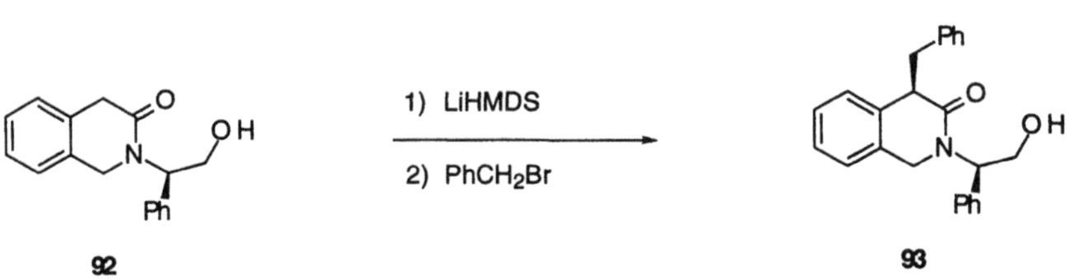

6.1.5 ACRIDINES

A highly diastereoselective synthesis of substituted 1,2,3,4,4a,9,9a,10-octahydroacridines **95** with five stereogenic centers has been achieved by domino imine condensation/intramolecular polar [4π + 2π] cycloaddition of anilines and ω-unsaturated aldehydes **94** <96JPR(338)468>.

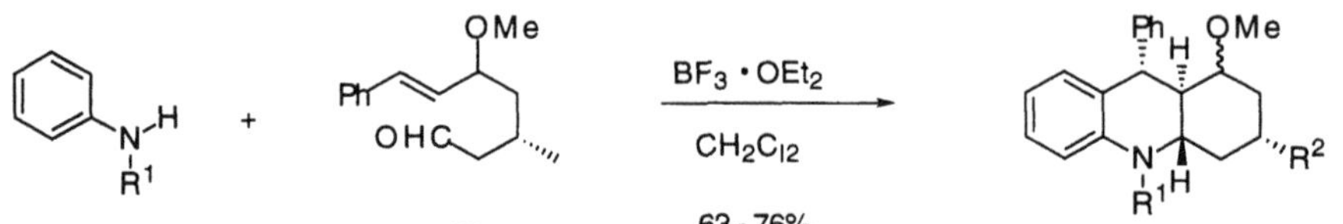

94

95

A systematic study on the addition of alkyllithiums to polycyclic monoazaarenes has appeared. The dibenzacridine **97** was prepared by addition of *n*-butyllithium to **96** followed by aromatization with DDQ in refluxing benzene <96JOC(61)1863>.

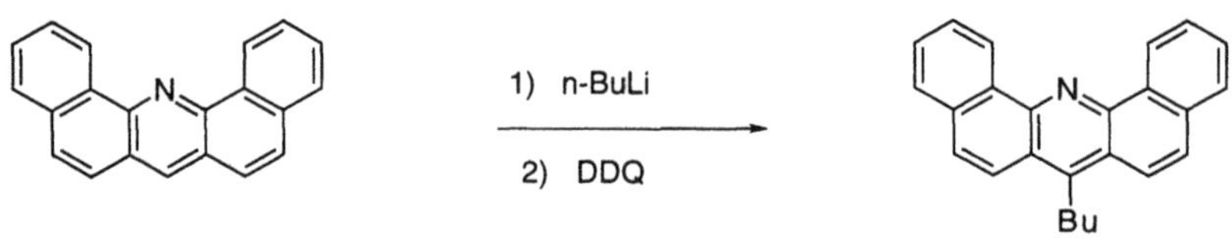

96

97

6.1.6 PIPERIDINES

6.1.6.1 Synthesis of Piperidines

The stereochemical features of piperidine synthesis via oxidative Mannich cyclizations of vinyl- and allylsilane-containing α-silylamines and α-silylamides have been studied <96TL(37)571>. A mild preparation of 2,6-piperidinediones from the reaction of substituted glutaric acids and alkyloxyamines in the presence of *N*-ethyl-*N*-dimethylaminopropylcarbodiimide hydrochloride was reported <96H(43)1923>. Studies have appeared on the reverse-Cope reaction for the preparation of various piperidine derivatives <95JOC(60)5795, 96TL(37)6417>. Investigations continue on the synthesis of piperidines using intramolecular Schmidt reactions of alkyl azides with ketones <95JA(117)10449> and tertiary alcohols <96T(52)12039>.

The *N*-alkylation of a range of 2-benzoyl-3-alkylaziridines **98** with *tert*-butyl bromoacetate followed by Wittig olefination gave directly piperidine derivatives **99** <95JCS(P1)2739>.

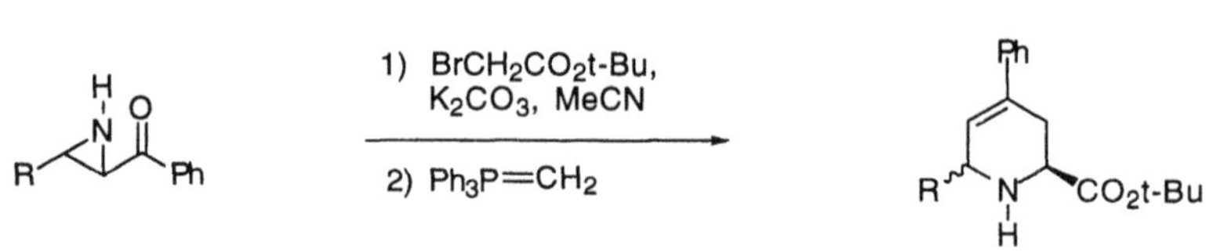

98

99

Cyclization of ketoester **100** occurs readily to give the 2,6-disubstituted piperidine **101**. The geometry about the carbon-carbon double bond of the Michael acceptor in **100** was shown to have a significant impact on the diastereoselectivity of the cyclization process <96JCS(P1)967>.

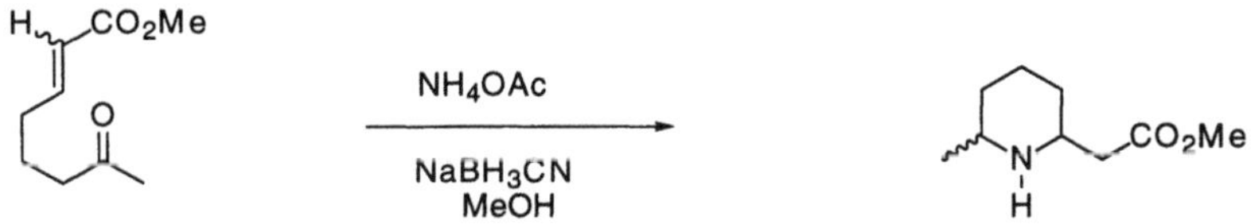

100 73 - 86% **101**

Termination of cyclic carbopalladation of alkynes via carbonylative lactamization can be achieved more satisfactorily with alkenyl or aryl halides containing an ω-carboxamido or ω-sulfonamido group than with those containing an ω-amino group. The method appears to be satisfactory for the preparation of certain piperidines (e.g., **102**) <96T(52)11529>.

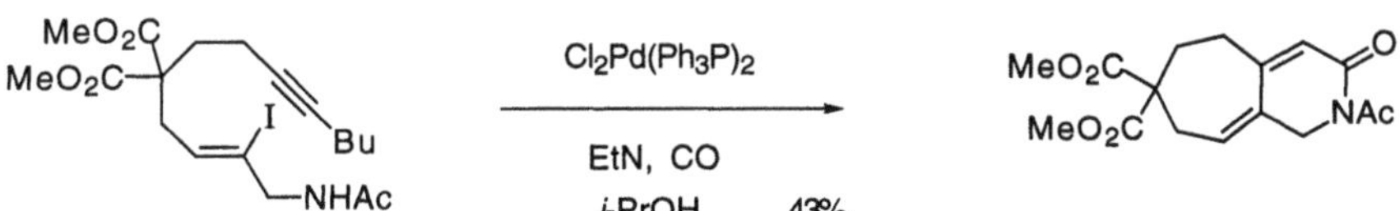

102

Palladium-catalyzed cyclization of *N*-alkenyl-2-alkynlamides occurred smothly in the presence of $CuCl_2$ and LiCl affording valerolactams **103** in moderate yield <96T(52)10945>.

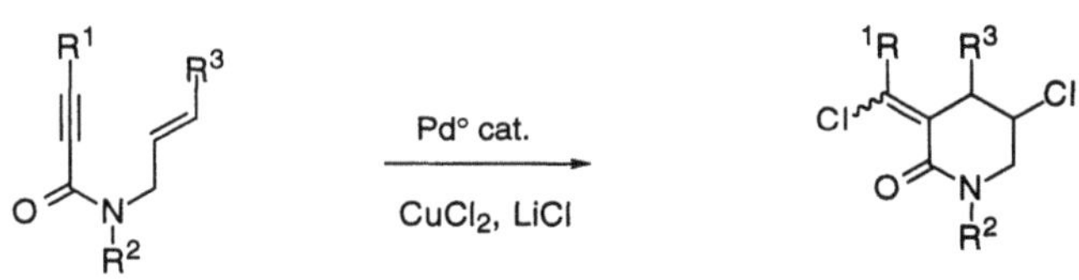

103

Neier and coworkers have shown that piperidine diones (e.g., **105**) can be prepared stereoselectively using a cascade Diels-Alder/acylation reaction of ketene acetal **104** <96T(52)11643>.

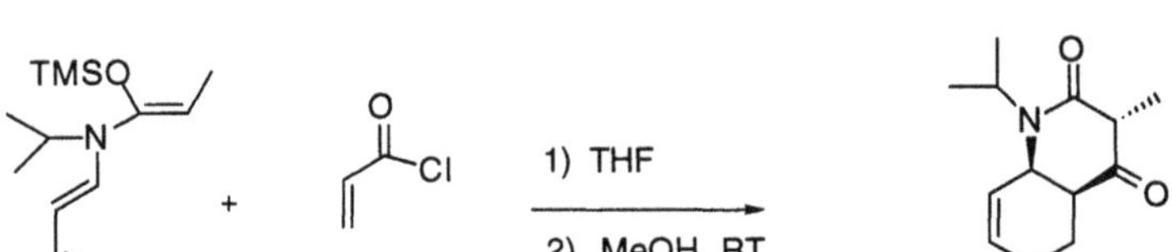

104 60% **105**

Several new 5,6-dihydropyridine-2(1*H*)-thiones were prepared by cyclization of the appropriate *N*-3-oxoalkylthioamides <96TL(37)5203>. Various 6-phenyl-3,6-dihydro-2*H*-thiopyran-4-amines were converted to 1-phenyl-5,6-dihydropyridine-2(1*H*)-pyridones, which were alkylated and hydrolyzed to give 6-methylthio-1-phenyl-2,3-dihydro-4(1*H*)-pyridones <95M(126)1367>. A solid phase synthesis of some pipecolic acid derivatives using an intramolecular Pauson-Khand cyclization was reported <96TL(37)3433>. A chiral bicyclic oxazolidine prepared from (*R*)-phenylglycinol was used in a synthesis of (2*S*,3*S*)-3-hydroxpipecolic acid <96TL(37)4001>.

[Ni(cyclam)](ClO_4)$_2$-catalyzed electroreduction of olefinic bromide **106** produced piperidone **107** in moderate yield <96JOC(61)677>.

106 **107**

Radical cyclization of dihydropyridones **108** provided piperidine derivatives **109** containing a trifluoromethyl group at the bridgehead position adjacent to nitrogen <96JOC(61)8826>.

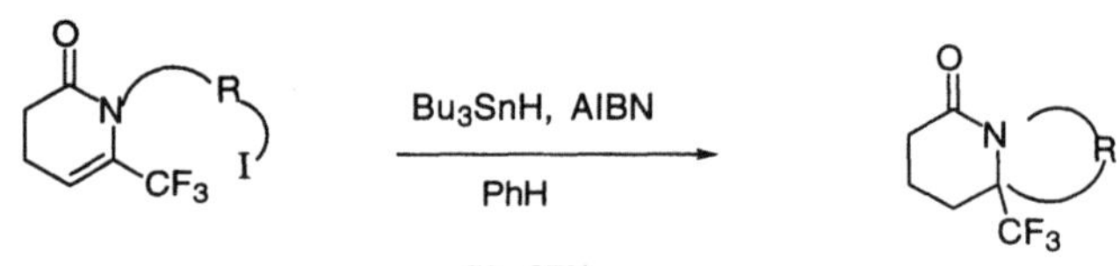

108

109 R = alkyl, aryl, alkenyl

Overman has reported that aprotic chloride- and bromide-promoted alkyne-iminium ion cyclizations to give piperidine derivatives occur under milder conditions than direct cyclizations of the corresponding secondary amine with formaldehyde and a halide salt <96H(42)549>. Treatment of 5-alkynylamines with 0.5-1.2 equiv of *n*-butyllithium brought about a facile anionic cyclization to give the corresponding piperidines having an exo double bond in high yields <96H(42)385>. A new method for the preparation of 4-oxo-2,2,6,6-tetramethylpiperidine in good yield has been developed <96SC(26)3565>.

6.1.6.2 Reactions of Piperidines

A study of the controlled double-bond migration in palladium-catalyzed intramolecular arylation of enamidines to give spiro-compounds containing the α-aryl piperidine ring has been reported by Hallberg and coworkers <96JOC(61)7147>. Intramolecular Heck reactions of *N*-acyl-2,3-dihydro-4-pyridones provide polycyclic heterocycles **110** with high regio- and stereoselectivity <96TL(37)793>.

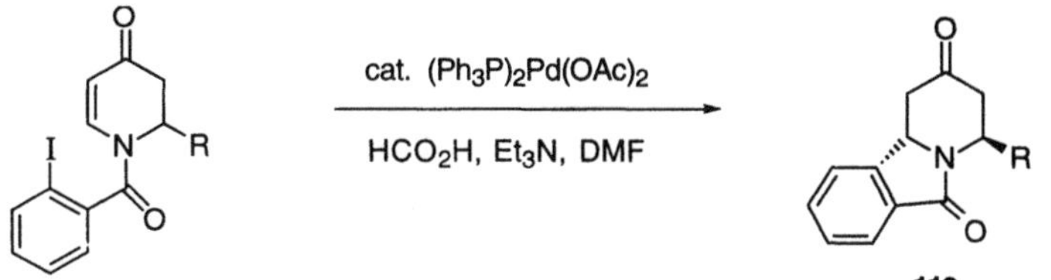

110

The oxidation of silyl enol ethers **111** with palladium(II) acetate is a convenient method for the preparation of synthetically useful 2,6-disubstituted 2,3-dihydro-4-pyridones **112** <95TL(36)9449>.

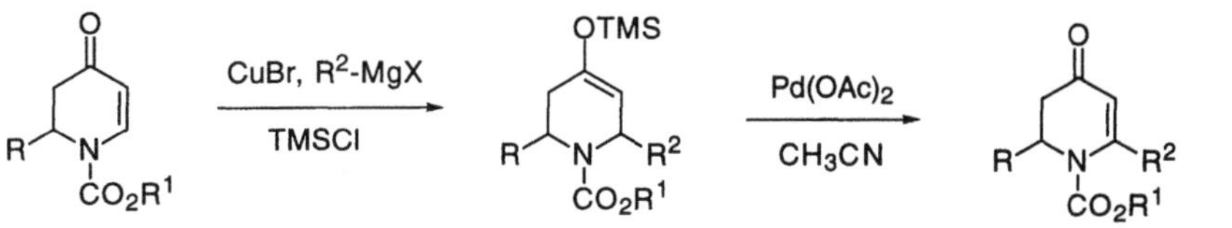

111 **112** 63-79%

Tetrahydropyridine-4-carboxylic acid **113** and related 2- and/or 6-methylated analogs were synthesized using the methoxycarbonylation of their corresponding vinyl triflates <96H(43)2131>.

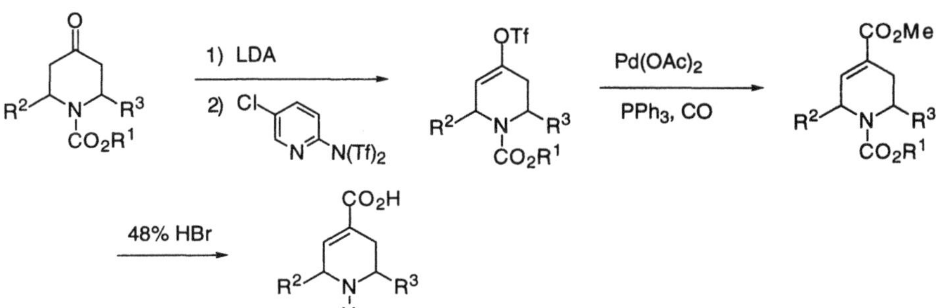

113

A method for regioselective introduction of a bis(methoxycarbonyl)-methyl group into the 4-position of the piperidine skeleton was explored, and this method was applied to the preparation of *cis*- and *trans*-2,4-disubstituted piperidines (e.g., **114**) starting from 2-piperidinecarboxylic acid <96TL(37)5715>.

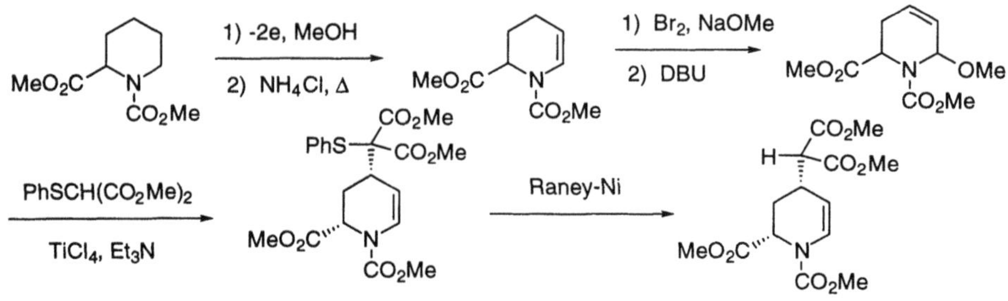

114

A general synthesis of spiro-piperidinyl heterocycles **116** was obtained from intermediate **115** via an efficient radical reaction <96TL(37)5233>.

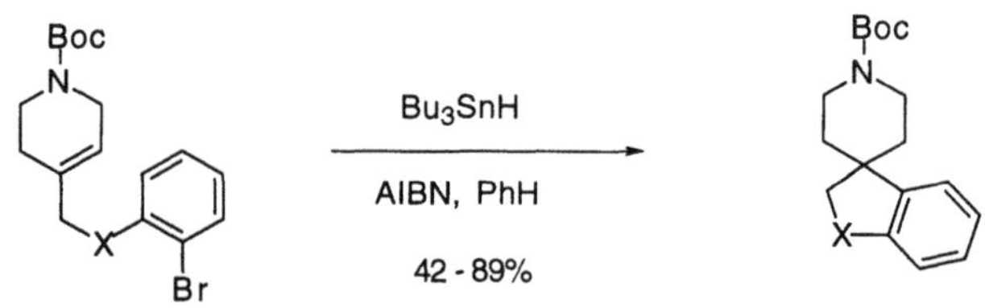

115 X = S, O, NH, NAc

116

Diastereoselective alkylation of lactams derived from *R*-(-)-phenylglycinol is an efficient method for the preparation of various substituted piperidines <96T(52)7719, 96T(52)7727, 96TL(37)849>. The asymmetric synthesis of a series of 2-(1-aminoalkyl)piperidines using (-)-2-cyano-6-phenyloxazolopiperidine has been described <96JOC(61)6700>.

The *N*-benzoyl ester **117** undergoes dihalocarbene addition to yield the cycloadduct **118** <96JCS(P1)2553>. Attempted ring-expansions of **118** were unsuccessful.

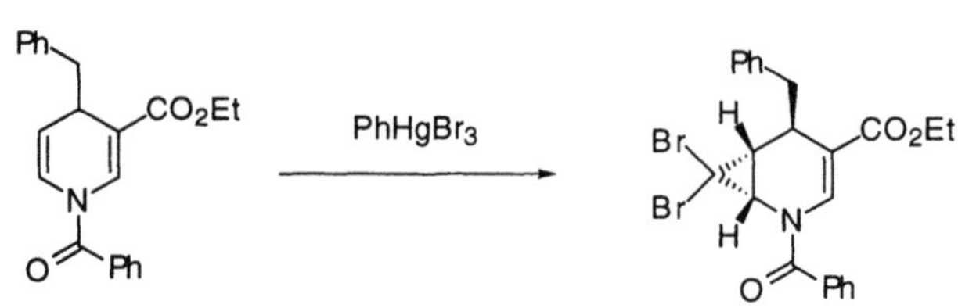

117

118

6.1.7 REFERENCES

95BMC(5)2143	J. Tagat, S. W. McCombie, B. E. Barton, J. Jackson and J. Shortall, *Bioorg. Med. Chem. Lett.* **1995**, *5*, 2143.
95BSF(132)1053	M. Chbani, J.-P. Bouillon, J. Chastanet, M. Soufiaoui, R. Beugelmans, *Bull. Soc. Chim. Fr.* **1995**, *132*, 1053.
95CC1903	R. Grigg, V. Sridharan and L.-H. Xu, *J. Chem. Soc., Chem. Comm.* **1995**, 1903.
95CC2137	U. Beifuss and S. Ledderhose, *J. Chem. Soc., Chem. Commun.* **1995**, 2137.
95CCC(60)1357	P. Hradil and J. Jirman, *Collect. Czech. Chem. Commun.* **1995**, *60*, 1357.
95CJC(73)1348	V. L. Ponzo and T. S. Kaufman, *Can. J. Chem.* **1995**, *73*, 1348.
95CL1003	B. Hatano, Y. Haraguchi, S. Kozima and R. Yamaguchi, *Chem. Lett.* **1995**, 1003.
95H(41)2221	I. Takeuchi, M. Ushida, Y. Hamada, T. Yuzuri, H. Suezawa and M. Hirota, *Heterocycles*, **1995**, *41*, 2221.
95JA(117)10449	G. L. Milligan, C. J. Mossman and J. Aubé, *J. Am. Chem. Soc.* **1995**, *117*, 10449.
95JC(155)268	T. Stamm, H. W. Kouwenhoven, D. Seebach and R. Prins, *J. Catal.* **1995**, *155*, 268.
95JCS(P1)2393	T. Hudlicky, G. Butora, S. P. Fearnley, A. G. Gum, P. J. Persichini III, M. R. Stabile and J. S. Merola, *J. Chem. Soc., Perkin Trans. 1* **1995**, 2393.
95JCS(P1)2739	I. Coldham, A. J. Collis, R. J. Mould and R. E. Rathmell, *J. Chem. Soc., Perkin Trans. 1* **1995**, 2739.
95JHC(32)1671	S.J. Dunne, L. A. Summers and E. I. von Nagy-Felsobuki, *J. Heterocycl. Chem.* **1995**, *32*, 1671.
95JMC(38)4679	V. Dollé, E. Fan, C. H. Nguyen, A.-M. Aubertin, A. Kirn, M. L. Andreola, G. Jamieson, L. Tarrago-Litvak and E. Bisagni, *J. Med. Chem.* **1995**, *38*, 4679.
95JOC(60)5795	E. Ciganek, J. M. Read, Jr. and J. C. Calabrese, *J. Org. Chem.* **1995**, *60*, 5795.
95JOC(60)7082	J. T. Kuethe, J. E. Cochran and A. Padwa, *J. Org. Chem.* **1995**, *60*, 7082.
95JOC(60)7369	D. L. Boger and J.-H. Chen, *J. Org. Chem.* **1995**, *60*, 7369.
95JOC(60)7631	A. R. Katritzky, B. Rachwal and S. Rachwal, *J. Org. Chem.* **1995**, *60*, 7631.
95LA1895	P. Ferrer, C. Avendaño and M. Söllhuber, *Liebigs Ann.* **1995**, 1895.
95M(126)1367	R. Weis, U. di Vora, W. Seebacher and K. Schweiger, *Monatsh. Chem.* **1995**, *126*, 1367.
95S801	Y. Makioka, T. Shindo, Y. Taniguchi, K. Takaki and Y. Fujiwara, *Synthesis* **1995**, 801.
95S923	F. A. Abu-Shanab, F. M. Aly and B. J. Wakefield, *Synthesis* **1995**, 923.
95S1362	M. Fernández, E. de la Cuesta, C. Avendaño, *Synthesis* **1995**, 1362.
95SC(25)3255	H. G. Selnick, G. R. Smith and A. J. Tebben, *Synth. Commun.* **1995**, *25*, 3255.
95T(51)8649	C. Burgos, F. Delgado, J. L. Garcia-Navio, M. L. Izquierdo and J. Alvarez-Builla, *Tetrahedron* **1995**, *51*, 8649.
95T(51)9119	T.L. Gilchrist, P. D. Kemmitt and A. L. Germain, *Tetrahedron* **1995**, *51*, 9119.
95T(51)9531	W. Schlecker, A. Huth and E. Ottow, *Tetrahedron* **1995**, *51*, 9531.
95T(51)11393	J. A. Zoltewicz and M. P. Cruskie, Jr., *Tetrahedron* **1995**, *51*, 11393.
95T(51)12159	N. Sotomayor, E. Dominguez and E. Lete, *Tetrahedron* **1995**, *51*, 12159.
95T(51)12277	A. C. Veronese, R. Callegari and C. F. Morelli, *Tetrahedron* **1995**, *51*, 12277.
95T(51)12425	J. Agejas, A. M. Cuadro, M. Pastor, J. J. Vaguero, J. L. Garcia-Navio and J. Alvarez-Builla, *Tetrahedron* **1995**, *51*, 12425.
95T(51)12439	T. Zdrojewski and A. Jonczyk, *Tetrahedron* **1995**, *51*, 12439.
95T(51)12721	N. Sotomayor, E. Dominguez and E. Lete, *Tetrahedron* **1995**, *51*, 12721.
95T(51)12791	D. Cartwright, J. R. Ferguson, T. Giannopoulos, G. Varvounis and B. J. Wakefield, *Tetrahedron* **1995**, *51*, 12791.
95T(51)12869	O. Meth-Cohn and D. L. Taylor, *Tetrahedron* **1995**, *51*, 12869.
95T(51)13177	T. Y. Zhang, J. R. Stout, J. G. Keay, E. F. V. Scriven, J. E. Toomey and G. L. Goe, *Tetrahedron* **1995**, *51*, 13177.
95TL(36)7709	W.D.F. Meutermans and P. F. Alewood, *Tet. Lett.* **1995**, *36*, 7709.
95TL(36)7721	N. M. Williamson, D. R. March and A. D. Ward, *Tet. Lett.* **1995**, *36*, 7721.
95TL(36)8003	B. Wüsch and S. Nerdinger, *Tet. Lett.* **1995**, *36*, 8003.
95TL(36)8283	P. Molina, A. Pastor and M. J. Vilaplana, *Tet. Lett.* **1995**, *36*, 8283.
95TL(36)8581	O. Peters and W. Friedrichsen, *Tet. Lett.* **1995**, *36*, 8581.
95TL(36)9141	D. L. Comins, S. P. Joseph and X. Chen, *Tet. Lett.* **1995**, *36*, 9141.
95TL(36)9297	A. S. Kiselyov, *Tet. Lett.* **1995**, *36*, 9297.
95TL(36)9449	D. L. Comins, S. P. Joseph and D. D. Peters, *Tet. Lett.* **1995**, *36*, 9449.

96ACS(50)556 B. Arnestad, J. M. Bakke, I. Hegbom and E. Ranes, *Acta. Chem. Scand.* **1996**, *50*, 556.
96ACS(50)623 P. Kolsaker, H. Karlsen and C. Romming, *Acta Chem. Scand.* **1996**, *50*, 623.
96AHC(65)1 O. Meth-Cohn; *Adv. Heterocycl. Chem.*, *65*, 1.
96AHC(65)39 M. Bohle and J. Liebscher; *Adv. Heterocycl. Chem.*, *65*, 39.
96AHC(65)93 A. F. Khlebnikov, M. S. Novikov and R. R. Kostikov; *Adv. Heterocycl. Chem.*, *65*, 93.
96BCJ(69)1377 N. Nishiwaki, A. Tanaka, M. Uchida, Y. Tohda and M. Ariga, *Bull. Chem. Soc. Jpn.* **1996**, *69*, 1377.
96CC1349 M. Sakamoto, T. Sano, M. Takahashi, K. Yamaguchi, T. Fujita and S. Watanabe, *J. Chem. Soc., Chem. Commun.* **1996**, 1349.
96CC2253 E. J. Latham and S. P. Stanforth, *J. Chem. Soc. Chem. Commun.* **1996**, 2253.
96CC2535 K. Miyashita, M. Nishimoto, H. Murafuji, A. Murakami, S. Obika, Y. In, T. Ishida and T. Imanishi, *J. Chem. Soc., Chem. Commun.* **1996**, 2535.
96CCC(61)126 R. Kubik, S. Böhm, I. Ruppertová and J. Kuthan, *Collect. Czech. Chem. Commun.* **1996**, *61*, 126.
96CHEC-II(5)1 C. D. Johnson; *Comp. Heterocycl. Chem.*, 2nd ed., **1996**, *5*, 1.
96CHEC-II(5)37 D. L. Comins and S. P. Joseph; *Comp. Heterocycl. Chem.*, 2nd ed., **1996**, *5*, 37.
96CHEC-II(5)91 N. Dennis; *Comp. Heterocycl. Chem.*, 2nd ed., **1996**, *5*, 91.
96CHEC-II(5)135 M. Lounasmaa and A. Tolvanen; *Comp. Heterocycl. Chem.*, 2nd ed., **1996**, *5*, 135.
96CHEC-II(5)167 G. Jones; *Comp. Heterocycl. Chem.*, 2nd ed., **1996**, *5*, 167.
96CHEC-II(5)245 M. Balasubramanian and J. G. Keay; *Comp. Heterocycl. Chem.*, 2nd ed., **1996**, *5*, 245.
96CL151 T. Muramatsu, A. Toyota, Y. Ikegami, S. Onodera and H. Hagiwara, *Chem. Lett.* **1996**, 151.
96COS(3)259 T. Harrison; *Comp. Org. Synth.* **1996**, *3*, 259.
96H(42)35 Y. Yamamoto and Y. Ogawa, *Heterocycles* **1996**, *42*, 35.
96H(42)141 H. Kohno and Y. Sekine, *Heterocycles* **1996**, *42*, 141.
96H(42)189 Y. Yamamoto, T. Tanaka, M. Yagi, and M. Inamoto, *Heterocycles* **1996**, *42*, 189.
96H(42)385 M. Tokuda, H. Fujita, M. Nitta, and H. Suginome, *Heterocycles* **1996**, *42*, 385.
96H(42)549 Y. Murata and L. E. Overman, *Heterocycles* **1996**, *42*, 549.
96H(43)33 A. Lorente, J. L. Alvarez-Barbas, J. L. Soto, P. Gómez-Sal and A. Manzanero, *Heterocycles* **1996**, *43*, 33.
96H(43)199 K. Iwamoto, S. Suzuki, E. Oishi, A. Miyashita and T. Higashino, *Heterocycles* **1996**, *43*, 199.
96H(43)809 P. Trebse, S. Polanc, M. Kocevar and T. Solmajer, *Heterocycles* **1996**, *43*, 809.
96H(43)817 T. Moreno, M. Fernández, E. de la Cuesta and C. Avendaño, *Heterocycles* **1996**, *43*, 817.
96H(43)1151 I. P. Andrews, N. J. Lewis, A. McKillop and A. S. Wells, *Heterocycles* **1996**, *43*, 1151.
96H(43)1179 N. Nishiwaki, M. Ariga, M. Komatsu and Y. Ohshiro, *Heterocycles* **1996**, *43*, 1179.
96H(43)1605 F. Fülöp, J. Tari, G. Bernáth and P. Sohár, *Heterocycles* **1996**, *43*, 1605.
96H(43)1923 J. Zhu, P. H. Chuong, P. Lemoine, A. Tomas and H. Galons, *Heterocycles* **1996**, *43*, 1923.
96H(43)2131 M. Rohr, S. Chayer, F. Garrido, A. Mann, M. Taddei and C.-G. Wermuth, *Heterocycles* **1996**, *43*, 2131.
96H(43)2425 C.-Y. Cheng, J.-Y. Chen and M.-J. Lee, *Heterocycles* **1996**, *43*, 2425.
96JCR(S)194 L.-L. Lai, P.-Y. Lin, J.-S. Wang, J. R. Hwu, M.-J. Shiao and S.-C. Tsay, *J. Chem. Res. (S)* **1996**, 194.
96JCS(P1)519 M.G. Barlow, L. Sibous, N. N. E. Suliman and A. E. Tipping, *J. Chem. Soc., Perkin Trans. 1* **1996**, 519.
96JCS(P1)967 M. G. Banwell, C. T. Bui, H. T. T. Pham and G. W. Simpson, *J. Chem. Soc., Perkin Trans. 1* **1996**, 967.
96JCS(P1)1777 W. Baik, S. Yun, J. U. Rhee and G. A. Russell, *J. Chem. Soc., Perkin Trans. 1* **1996**, 1777.
96JCS(P1)2553 K. J. McCullough, J. MacTavish, G. R. Proctor and J. Redpath, *J. Chem. Soc., Perkin Trans. 1* **1996**, 2553.
96JHC(33)157 N. Iqbal and E. E. Knaus, *J. Heterocycl. Chem.* **1996**, *33*, 157.
96JHC(33)607 A. R. Katritzky and B. Yang, *J. Heterocycl. Chem.* **1996**, *33*, 607.
96JHC(33)783 T. Zimmermann and K. Schmidt, *J. Heterocycl. Chem.* **1996**, *33*, 783.
96JOC(61)304 S. M. Vandenberghe, K. J. Buysens, L. Meerpoel, P. K. Loosen, S. M. Toppet and G. J. Hoornaert, *J. Org. Chem.* **1996**, *61*, 304.

96JOC(61)309 S. Mabic and N. Castagnoli, Jr., *J. Org. Chem.* **1996**, *61*, 309.
96JOC(61)662 K. Smith, D. Anderson and I. Matthews, *J. Org. Chem.* **1996**, *61*, 662.
96JOC(61)677 M. Ihara, A. Katsumata, F. Setsu, Y. Tokunagu and K. Fukumoto, *J. Org. Chem.* **1996**, *61*, 677.
96JOC(61)824 P.A. Wender and T. E. Smith, *J. Org. Chem.* **1996**, *61*, 824.
96JOC(61)924 M. F. Gordeev, D. V. Patel and E. M. Gordon, *J. Org. Chem.* **1996**, *61*, 924.
96JOC(61)1665 A. Goti, B. Anichini, A. Brandi, S. Kozhushkov, C. Gratkowski and A. de Meijere, *J. Org. Chem.* **1996**, *61*, 1665.
96JOC(61)1863 J. K. Ray, B. C. Roy and G. K. Kar, *J. Org. Chem.* **1996**, *61*, 1863.
96JOC(61)4810 N. C. Ihle and A. E. Krause, *J. Org. Chem.* **1996**, *61*, 4810.
96JOC(61)4830 F. Diaba, I. Lewis, M. Grignon-Dubois and S. Navarre, *J. Org. Chem.* **1996**, *61*, 4830.
96JOC(61)6700 O. Froelich, P. Desos, M. Bonin, J.-C. Quirion and H.-P. Husson, *J. Org. Chem.* **1996**, *61*, 6700.
96JOC(61)7147 L. Ripa and A. Hallberg, *J. Org. Chem.* **1996**, *61*, 7147.
96JOC(61)7195 P. J. Campos, C.-Q. Tan, M. A. Rodriguez and E. Anon, *J. Org. Chem.* **1996**, *61*, 7195.
96JOC(61)7240 S. Wagaw and S. L. Buchwald, *J. Org. Chem.* **1996**, *61*, 7240.
96JOC(61)8094 P. Molina, A. Pastor and M. J. Vilaplana, *J. Org. Chem.* **1996**, *61*, 8094.
96JOC(61)8826 T. Okano, T. Sakaida and S. Eguchi, *J. Org. Chem.* **1996**, *61*, 8826.
96JPR(338)220 A. R. Katritzky, R. L. Parris, E. S. Ignatchenko, M. Balasubramanian, and R. A. Barcock, *J. Prakt. Chem.* **1996**, *338*, 220.
96JPR(338)468 U. Beifuss, H. Gehm, A. Herde, and S. Ledderhose, *J. Prakt. Chem.* **1996**, *338*, 468.
96LA641 M. Grzegozek, M. Wozniak and H. C. van der Plas, *Liebigs Ann.* **1996**, 641.
96MI99 A. I. Pyschev, V. V. Krasnikov, A. E. Zibert, D. E. Tosyniyan and S. V. Verin, *Mendeleev Commun.* **1996**, 99.
96MI173 T. Ishikawa, K. Maeda, K. Hayakawa and T. Kajima, *J. Mol. Catal.* **1996**, 173.
96SC(26)127 A. P. Venkov and S. Statkova-Abeghe, *Synth. Commun.* **1996**, *26*, 127.
96SC(26)1763 M. U. Koller, A. C. Church, C. L. Griffith, M. A. Hines, R. J. Lachicotte, R. A. Taylor and C. F. Beam, *Synth. Commun.* **1996**, *26*, 1763.
96SC(26)2017 O. Lohse, *Synth. Commun.* **1996**, *26*, 2017.
96SC(26)3543 C. Coudret, *Synth. Commun.* **1996**, *26*, 3543.
96SC(26)3565 A. Wu, W. Yang and X. Pan, *Synth. Commun.* **1996**, *26*, 3565.
96SC(26)3959 J. K. Ray, G. K. Kar and M. K. Haldar, *Synth. Commun.* **1996**, *26*, 3959.
96SL34 U. Beifuss, O. Kunz, S. Ledderhose, M. Taraschewski, and C. Tonko, *Synlett* **1996**, 34.
96TA(7)417 N. Philippe, V. Levacher, G. Dupas, J. Duflos, G. Quéguiner and J. Bourguignon, *Tetrahedron: Asymm.* **1996**, *7*, 417.
96T(52)23 N. Yoneda and T. Fukuhara, *Tetrahedron* **1996**, *52*, 23.
96T(52)1835 M. K. Patel, R. Fox and P. D. Taylor, *Tetrahedron* **1996**, *52*, 1835.
96T(52)2591 K. J. Dubois, C. C. Fannes, F. Compernolle and G. J. Hoornaert, *Tetrahedron* **1996**, *52*, 2591.
96T(52)3705 N. De Kimpe and M. Keppens, *Tetrahedron* **1996**, *52*, 3705.
96T(52)4433 A. Couture, E. Deniau, P. Grandclaudon and P. Woisel, *Tetrahedron* **1996**, *52*, 4433.
96T(52)6997 K. J. Dubois and G. J. Hoornaert, *Tetrahedron* **1996**, *52*, 6997.
96T(52)7719 L. Micouin, M. Bonin, M.-P. Cherrier, A. Mazurier, A. Tomas, J.-C. Quirion and H.-P. Husson, *Tetrahedron* **1996**, *52*, 7719.
96T(52)7727 R. Rodriguez, M. A. Estiarte, A. Diez and M. Rubiralta, A. Colell, C. Garcia-Ruiz and J. C. Fernández-Checa, *Tetrahedron* **1996**, *52*, 7727.
96T(52)10095 G. Morel, E. Marchand, J.-P. Pradère, L. Toupet and S. Sinbandhit, *Tetrahedron* **1996**, *52*, 10095.
96T(52)10225 S. Cacchi, G. Fabrizi, F. Marinelli, L. Moro and P. Pace, *Tetrahedron* **1996**, *52*, 10225.
96T(52)10945 H. Jiang, S. Ma, G. Zhu and X. Lu, *Tetrahedron* **1996**, *52*, 10945.
96T(52)11025 J. Epsztajn, A. Jozwiak, J. K. Krysiak and D. Lucka, *Tetrahedron* **1996**, *52*, 11025.
96T(52)11529 C. Copéret, S. Ma, T. Sugihara and E. Negishi, *Tetrahedron* **1996**, *52*, 11529.
96T(52)11643 A. Franz, P.-Y. Eschler, M. Tharin and R. Neier, *Tetrahedron* **1996**, *52*, 11643.
96T(52)11889 P. R. Carly, S. L. Cappelle, F. Compernolle and G. J. Hoornaert, *Tetrahedron* **1996**, *52*, 11889.
96T(52)12039 W. H. Pearson and B. M. Gallagher, *Tetrahedron* **1996**, *52*, 12039.

96T(52)12529	K. J. Dubois, C. C. Fannes, S. M. Toppet and G. J. Hoornaert, *Tetrahedron* **1996**, *52*, 12529.
96TL(37)69	Y. Tagawa, J. Tanaka, K. Hama, Y. Goto and M. Hamana, *Tet. Lett.* **1996**, *37*, 69.
96TL(37)571	S.-K. Khim, X. Wu and P. S. Mariano, *Tet. Lett.* **1996**, *37*, 571.
96TL(37)793	D. L. Comins, S. P. Joseph and Y. Zhang, *Tet. Lett.* **1996**, *37*, 793.
96TL(37)849	L. Micouin, J.-C. Quirion and H.-P. Husson, *Tet. Lett.* **1996**, *37*, 849.
96TL(37)1043	H. Zhang and K. S. Chan, *Tet. Lett.* **1996**, *37*, 1043.
96TL(37)2565	Y. G. Gu and E. K. Bayburt, *Tet. Lett.* **1996**, *37*, 2565.
96TL(37)2829	A. C. S. Reddy, B. Narsaiah and R. V. Venkataratnam, *Tet. Lett.* **1996**, *37*, 2829.
96TL(37)3433	G. L. Bolton, *Tet. Lett.* **1996**, *37*, 3433.
96TL(37)3617	K. A. Memoli, *Tet. Lett.* **1996**, *37*, 3617.
96TL(37)3807	D. L. Comins and L. Guerra-Weltzien, *Tet. Lett.* **1996**, *37*, 3807.
96TL(37)4001	C. Agami, F. Couty and H. Mathieu, *Tet. Lett.* **1996**, *37*, 4001.
96TL(37)4177	I. Katsuyama, K. Funabiki, M. Matsui, H. Muramatsu and K. Shibata, *Tet. Lett.* **1996**, *37*, 4177.
96TL(37)4369	G. Gosmann, D. Guillaume and H.-P. Husson, *Tet. Lett.* **1996**, *37*, 4369.
96TL(37)4577	F. Palacios, A. M. O. de Retana and J. Oyarzabal, *Tet. Lett.* **1996**, *37*, 4577.
96TL(37)4655	L. Strekowski, S.-Y. Lin, H. Lee and J. C. Mason, *Tet. Lett.* **1996**, *37*, 4655.
96TL(37)4815	A. A. MacDonald, S. H. DeWitt, E. M. Hogan and R. Ramage, *Tet. Lett.* **1996**, *37*, 4815.
96TL(37)4973	K. Takaoka, T. Aoyama and T. Shioiri, *Tet. Lett.* **1996**, *37*, 4973.
96TL(37)4977	K. Takaoka, T. Aoyama and T. Shioiri, *Tet. Lett.* **1996**, *37*, 4977.
96TL(37)5203	A. S. Fissyuk, M. A. Vorontsova and D. V. Temnikov, *Tet. Lett.* **1996**, *37*, 5203.
96TL(37)5233	M.-H. Chen and J. A. Abraham, *Tet. Lett.* **1996**, *37*, 5233.
96TL(37)5453	G. E. Stokker, *Tet. Lett.* **1996**, *37*, 5453.
96TL(37)5715	Y. Matsumura, Y. Yoshimoto, C. Horikawa, T. Maki and M. Watanabe, *Tet. Lett.* **1996**, *37*, 5715.
96TL(37)5917	K. Kubo, S. Yaegashi, K. Susaki, T. Sakurai and H. Inoue, *Tet. Lett.* **1996**, *37*, 5917.
96TL(37)6379	F. Palacios and G. Rubiales, *Tet. Lett.* **1996**, *37*, 6379.
96TL(37)6417	M. P. Coogan and D. W. Knight, *Tet. Lett.* **1996**, *37*, 6417.
96TL(37)6695	F. Mongin, O. Mongin, F. Trécourt, A. Godard and G. Quéguiner, *Tet Lett.* **1996**, *37*, 6695.

Chapter 6.2

Six-Membered Ring Systems: Diazines and Benzo Derivatives

Michael P. Groziak
SRI International, Menlo Park, CA, USA

6.2.1 INTRODUCTION

Isolated and benzo-fused diazine rings are key structural elements in many natural and synthetic compounds of current interest. This contribution relates highlights from many of the studies on the diazines pyridazine, pyrimidine, pyrazine, and their benzo-fused derivatives cinnoline, phthalazine, quinazoline, quinoxaline, and phenazine published in English in the journal literature during 1996, as covered by *Chem. Abstr.* through volume 126, issue 5.

6.2.2 REVIEWS AND GENERAL STUDIES

Potentially tautomeric methyl/methylidene derivatives of pyridine and diazines were the subject of a review focusing on hydrogen isotope exchanges, alkylations, and acylations <96KGS816>. Alkylation of various azines and diazines has afforded a set of congeners of pyridazomycin (**1**), an antimicrobial natural product <96PHA76>. The synthesis and configurational assignment of a collection of aryl diazinyl ketoximes <96H151> was followed by a ^{13}C NMR stereochemical assignment <96MI429> and then a biological SAR study <96JMC4058> of ketoxime ether diazine analogs of the antithrombic agent ridogrel (**2**). Pyrimidines and pyridazines were among the compounds obtained from cyclization reactions involving ethyl 2-amino-4,5,6,7-tetrahydrobenzo[*b*]thiophene-3-carboxylate <96JCR(S)440>.

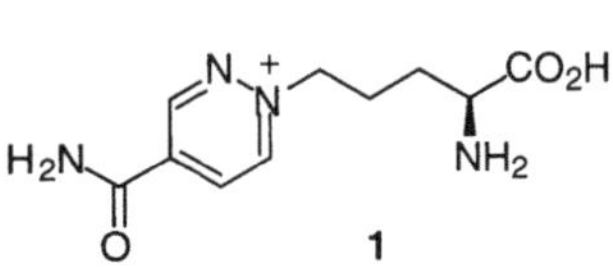

1

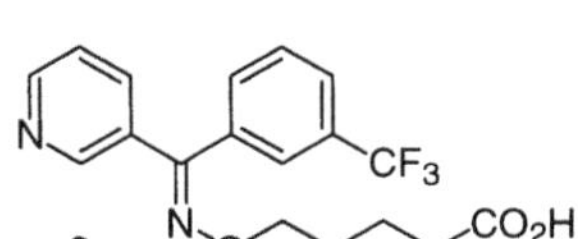

2

Aspects of benzonitrile oxide cycloaddition to all three parent diazines have been investigated <96T6421>, and the stereochemistry of resultant biscycloadducts determined <96JCR(S)220>.

Ortho-directed lithiation followed by reaction with TsN_3 and reduction has led to an improved, general synthesis of aminodiazines **3a-c** <96S838>. In some cases, tetrazolo-fused diazines could be isolated as intermediates.

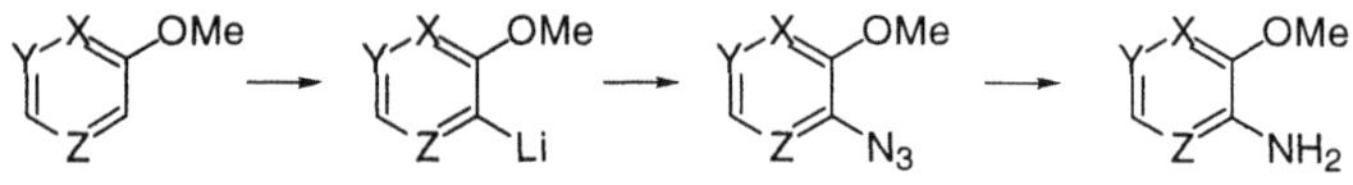

3a, X=N, Y=CH, Z=N; **b**, X=CH, Y=Z=N; **c**, X=Y=N, Z=CH

6.2.3 PYRIDAZINES AND BENZODERIVATIVES

X-ray crystal structures of the 4-methyl, 5-methyl, 6-methyl, and 2-(*p*-tolyl) derivatives of pyridazin-3(2*H*)-one have been determined <96AX(C)2622>. An *N*-ethylation of 3-(2-pyrrolyl)pyridazine has been shown to give the 1-alkyl isomer <96AX(C)1002>. The structure of a triphenylboroxin derivative was found to contain two 1,2-dihydro-2,3,1-benzodiazaborine (boracinnoline) units <96AX(C)2826>. Structures of unsolvated and water-included 6-chloro-4-(*N*,*N*-dimethylaminemethylidenohydrazine)-2-phenyl-3(2*H*)-pyridazinone were determined <96MI251>. Two of six benzodiazepine receptor-active 3-acylaminomethyl, 2,6-disubstituted imidazo[1,2-*b*]pyridazines were also examined by X-ray crystallography <96AJC451>, as were 2-(4-methanesulfonylphenyl)-6-phenylhexahydro-4-pyridazinol <96JHC213>, four 5,6-dihydrobenzo[*c*]cinnolines of an electrochemical study <96JA5020>, and one "face-to-face" benzo/pyridazino potential precursor to an azapagodane (azadodecahedrane) <96TL4491>.

6.2.3.1 Syntheses

Inverse-electron-demand Diels-Alder reactions continue to be popular ones for preparing pyridazines. For instance, unusual bicyclic pyridazine-containing endoperoxides were accessed from dimethyl 1,2,4,5-tetrazine-3,6-dicarboxylate and unsaturated bicyclic endoperoxides <96TL921>, new pyridazino-fused heterocycles were derived from cyclic ketene-*S*,*N*-acetals <96H1967>, and even polymer-supported 1,2,4,5-tetrazines have given rise to pyridazines **4** <96TL8151>. Routes to pyrimido[4',5':4,5]thieno[2,3-*c*]pyridazines <96M739, 96M537>, benzothiazole-containing pyrazolo[3,4-*d*]pyridazines <96JCR(S)416>, spiro- and tricyclic pyridazino[4,5-*c*]pyridazinones <96H1613>, 3-(coumarin-3-yl)pyridazines <96HAC137>, pyrido[2',3':4,5]thieno[2,3-*c*]pyridazines <96H1073>, 2,6-diarylhexahydro-4-pyridazinols **5** <96JHC213>, and "inward-facing" 3,6-di(2-pyridyl)pyridazine ligands <96TL2825> have been developed.

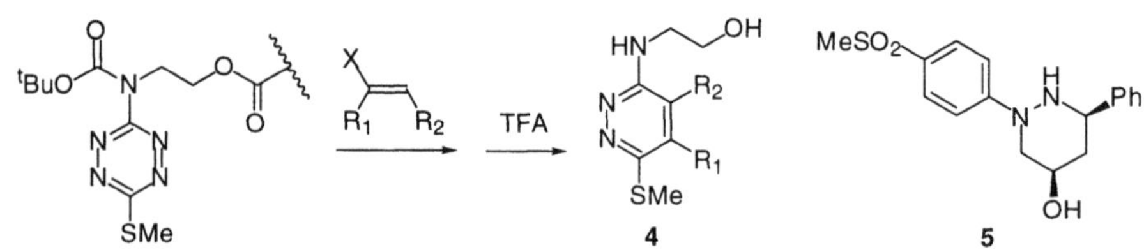

4 **5**

The reaction of α-oxophenylhydrazones with 2-amino-1,1,3-tricyanopropene and diethyl 1-cyano-2-aminopropene-1,3-dicarboxylate has been shown to afford polyfunctionally substituted pyridazines and cinnolines <96JCR(S)434>. A study of haloazodienes has led to a new, general synthesis of substituted pyridazines <96JOC8921>. By this approach, dihalohydrazones **6** are treated with Hünig's base, and the haloazodiene that forms *in situ* is condensed with an electron rich olefin to give tetrahydropyridazines **7**. Upon treatment with base, these latter compounds readily afford pyridazines **8**. An improvement in the synthesis of pyridazinediones has been noted when microwave irradiation is employed <96TL4145>. Various *N*- and *O*-protected 3-(4-pyridazinyl)isoserines have been shown to be useful 1,2-diazine-containing synthetic building blocks <96H1057>, and the synthesis of a podophyllotoxin A-ring pyridazine analog has been accomplished <96T14235>.

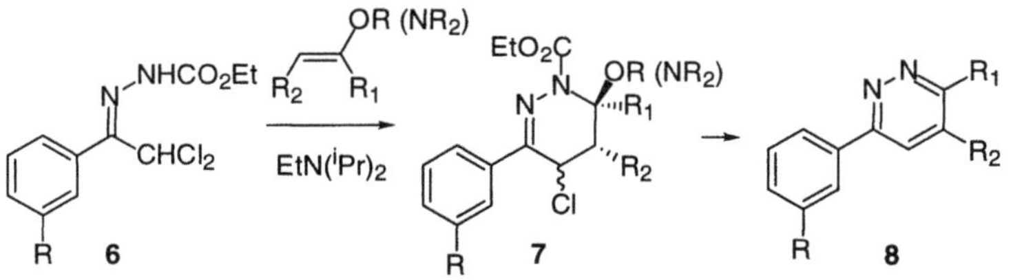

6 7 8

6.2.3.2 Reactions

In a combined theoretical and experimental study, the [3+2] dipolar cycloaddition of 3-(4-halophenyl)pyridazinium ylides to acrylate and propiolate esters led to eight new pyrrolopyridazines <96T8853>. An improvement in the regioselectivity of metalation of 3-chloro-6-methoxypyridazine has been achieved by the use of very hindered bases <96T10417>. The reactivity of 5-amino-7-substituted thieno[3,4-*d*]pyridazinones <96T11915> and 4-substituted-7-aminothieno[3,4-*d*]pyridazines <96PJC589> toward condensation with electron-poor olefins and acetylenes has been studied, as have the ring transformative reactions of halo-substituted fused pyridazines with ynamines <96H199>. The base-mediated protodesilylation of (2*H*)-pyridazinon-5-yl silylmethyl sulfides was studied <96MI603>, and the alkylative retro-ene reaction in 4,5-dichloro-1-hydroxymethylpyridazin-6-ones **9** was shown to constitute a synthesis of 4,5-dichloro-1-(ω-phthalimido- and saccharin-2'-ylalkyl)pyridazin-6-ones **10** <96JHC615>, and a convenient palladium-catalyzed methoxycarbonylation of pyridazin-3-yl triflates was developed <96H1459>. A mechanistic evaluation of C-10 halogenation in 1-substituted quinoxalino[2,3-*c*]cinnolines has been conducted <96JCS(P1)1699>.

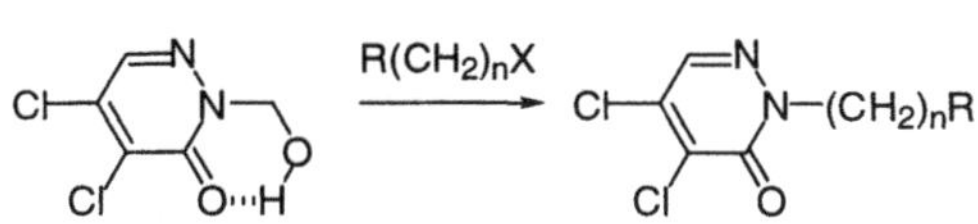

9 10

6.2.3.3 Applications

Biological SAR studies covered pyridazines and 1,2,4-triazolo[4,3-*b*]pyridazines <96MI35, 96MI47> as antimicrobials, 2-arylimidazo[1,2-*b*]pyridazines <96AJC443> and 1,2,3,4-tetra- and 1,2,3,4,5,6-hexahydropyrido[3,4-*d*]pyridazines <96F431> as CNS agents, 6-[4-(benzylamino)-7-quinazolinyl]-4,5-dihydro-5-methyl-3(2*H*)-pyridazinones as cardiotonics <96JMC297>, ω-sulfamoylalkyloxyimidazo[1,2-*b*]pyridazines as antiasthmatics <96CPB122>,

6-phenyl-3(2*H*)-pyridazinones as antinociceptives <96EJM65>, imidazo[1,2-*b*]pyridazines as benzodiazepine receptor ligands <96AJC435, 96EJM651>, tricyclic pyridazinones as aldose reductase inhibitors <96JMC4396>, oxazolo[3',2':1,2]pyrrolo[3,4-*d*]pyridazines and imidazolo[1',2':1,2]pyrrolo[3,4-*d*]pyridazines as potential non-nucleoside HIV-1 reverse transcriptase inhibitors <96AP403>, trazodone-like 4-phenyl-6-aryl-2-[3-(4-arylpiperazin-1-yl)propyl]pyridazin-3-ones as analgesics <96CPB980>, fused thiazolo[4,5-*d*]pyridazines as antibacterials <96MI433>, 5,6-dihydrobenzo[*f*]cinnolin-2(3*H*)ones as antihypertensives and antiaggregating benzo[*h*]cinnolinones <96F653>, and a cinnoline derived from a β-carboline as an antibacterial and antifungal <96H1887>.

6.2.4 PYRIMIDINES AND BENZODERIVATIVES

X-ray crystal structure determinations of 6-(3-methylbenzyl)-2-(2-methylpropyl)thio-4(3*H*)-pyrimidinone <96AX(C)2115>, 5,5-diethyl-1-methyl-1*H*,2*H*,3*H*,5*H*-pyrimidine-2-spiro-9'-fluorene-4,6-dione <96AX(C)2081>, 2-chloro-5-*p*-toluoylamino- and *p*-chlorobenzoylamino-*N*-*p*-tolyl-4-*p*-tolylamino-6-pyrimidinecarboxamide <96AX(C)2106>, 2-bromomethyl-3-(2-ethylphenyl)-4(3*H*)-quinazolinone <96AX(C)1703>, 4-(2-diethylaminovinyl)pyrimidine/TfOH 1:1 and 1:2 adducts <96JHC439>, 7-ethoxycarbonylmethyl-5-methyl-1,2,4-triazolo[1,5-*a*]pyrimidine <96AX(C)1031>, [1,2-(*Z*)-bis(methoxycarbonyl)vinyl]-amino-5-dimethylamino-2-methoxy-3-methyl-4(3*H*)-pyrimidinone <96AX(C)254>, 4-benzyl-2-(dimethylamino)-pyrimidine-5-carboxylic acid and its methyl ester <96JMC3671>, and several tetraazathiapentalenes fused with pyrimidine rings like **11** <96JA6355> have been conducted. Crystallographic analysis of the alkaloids (+)-vasicine (**12**, X=H_2) and (-)-vasicinone (**12**, X=O) described in section **6.2.4.1** were also performed. X-ray diffraction and NMR spectroscopy were combined in a mechanistic analysis of the transformation of 6-ethoxycarbonyl-7-(2-dimethylaminovinyl)pyrazolo[1,5-*a*]pyrimidines into 7-oxo-6-[pyrazol-3'(5')-yl]-4,7-dihydropyrazolo[1,5-*a*]pyrimidines <96JCS(P2)1147>. Other NMR investigations included a ROESY and NOESY analysis of the solution structures presented by 6-formyluracil nucleosides <96MI1041>, an analysis of restricted rotation in 2-substituted 2,3-dihydro-3-*o*-tolyl and -*o*-(chlorophenyl)-4(1*H*)-quinazolinones <96JHC1067>, and a study of benzoxazine, benzothiazine and quinazoline nitrate esters <96MRC958>. A conformational study of 4-(2-furyl)-2-(methylamino)pyrimidine <96JHC1207> and a quantum-chemical one of functionalized 1*H*-pyrimidines <96MI27> were performed.

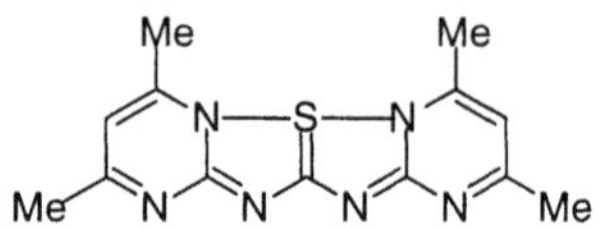

11

6.2.4.1 Syntheses

Many unusual pyrimidine-based compounds were investigated last year. Among these were steroidal pyrimidines <96SC3511>, *t*-butylcyclopentane-fused pyrimidin-4-ones <96H625>, pyrimidine derivatives of cyanimidodithiocarbonates <96M725>, pyrimidine *ortho*-quinodimethanes <96T1735>, bi(pyrimidiniumolates) <96T5475>, fused pyrimidine mono-*N*-

oxides <96H389>, bisguanidinoalkanes and guanidinoalkanes *N*- or *N'*-substituted with pyrimidines, <96AJC573>, phosphonium ylide-betaines of pyrimidines <96T8835>, and linear tricyclic analogs of adenine <96JHC319>. A reversal of the absolute stereochemistry of the pyrrolo[2,1-*b*]quinazoline alkaloids vasicine (**12**, X=H_2), vasicinone (**12**, X=O), vasicinol, and vasicinolone has been done <96TA25>. Asymmetric synthesis of (*S*)-(+)- and (*R*)-(-)-3-[[4-(2-methoxyphenyl)piperazin-1-yl]methyl]-5-methylthio-2,3-dihydroimidazo[1,2-*c*]quinazolines **13** <96TA1641> and of (*S*)-(-)-2-(1-hydroxyethyl)quinazolin-4(3*H*)-one (chrysogenine, a fungal metabolite) <96JCR(S)306> have been accomplished. The solid phase synthesis of 1,3-disubstituted (1*H*,3*H*)-quinazoline-2,4-diones <96BMCL1483>, pyrido[2,3-*d*]pyrimidines **14** <96TL4643>, and 3-substituted quinazoline-2,4-diones **15** <96TL7031> have been reported. Many of these are suitable for combinatorial applications.

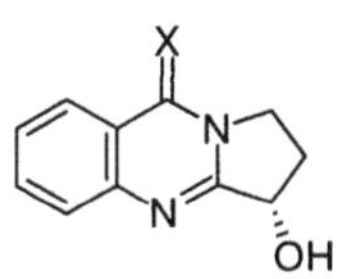

12

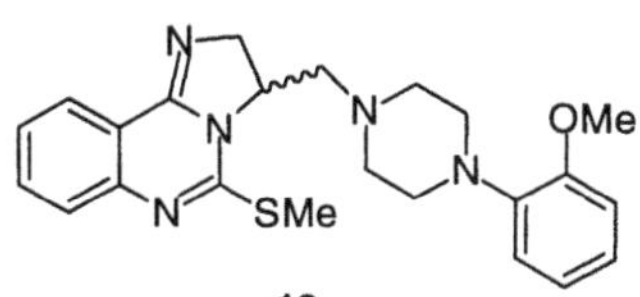

13

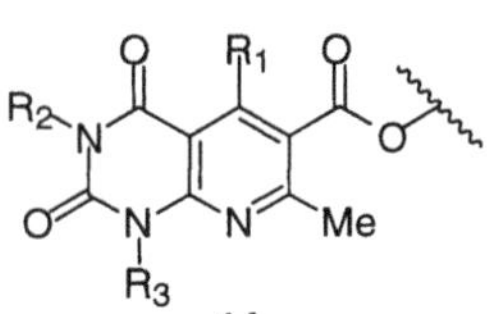

14

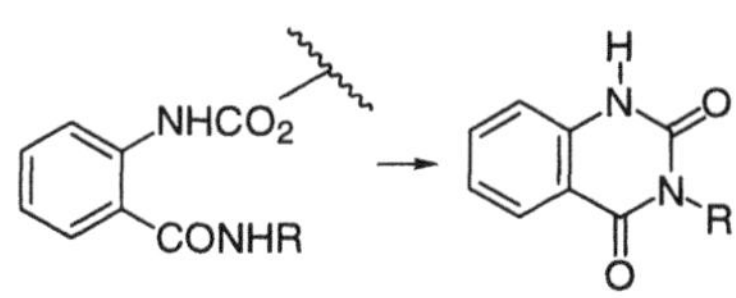

15

General syntheses of quinazolinones <96SC475>, pyrimidines <96JPR451, 96IJC(B)144>, 2,4-disubstituted 6-fluoroalkylpyrimidin-4(3*H*)-ones and pyrimidines **16** <96S997>, heterocycle substituted and 2,4-difunctionalized pyrimidines <96T7973>, and new 3-substituted fervenulins **17** <96JHC949> were reported. Those of more elaborated ring systems include pyrrolo[2,3-*d*]-, thieno[2,3-*d*]-, isoxazolo[5,4-*d*]- and 1,2,3-triazolo[4,5-*d*]pyrimidinones and-quinazolones <96H691>, furo[3,2-*e*][1,2,4]triazolo[4,3-*c*]pyrimidines and furo[3,2-*e*]tetrazolo[1,5-*c*]pyrimidines <96JHC659>, 1,3,4-oxa/thiadiazolo[3,2-*a*]pyrimidin-5-ones <96IJHC59>, 2-thioxo-4(3*H*)-quinazolinones <96IJHC315>, 11*H*-pyrido[2,1-*b*]quinazolin-11-ones <96SC3869>, 3,4-dihydroquinazolines and quinazolinethiones <96SC3167>, polyfunctionally substituted thieno[3,2-*d*]pyrimidines <96M955>, pyrrolo[1,2-*c*]pyrimidines **18** from pyrrole-2-carboxaldehydes **19** <96TL4263>, pyrido[2,3-d:6,5-d']dipyrimidines <96IJC(B)852>, 4-substituted 1-(hydroxyalkyl)-1*H*-pyrazolo[3,4-*d*]pyrimidines <96T2271>, 2-amino-1,5-dihydro-4*H*-pyrrolo[3,2-*d*]-pyrimidin-4-one (9-deazaguanine) <96TL4339>, pyrazolo[3,4-*d*]pyrimidines and pyrazolo[3,4-*c*]pyrazoles <96IJC(B)848>, 2,3-dihydro-4-oxoisoxazolo[5,4-*d*]pyrimidin-6-thiones <96IJC(B)478>, and 2,3-dihydroimidazo[1,2-*c*]pyrimidines <96SC453>.

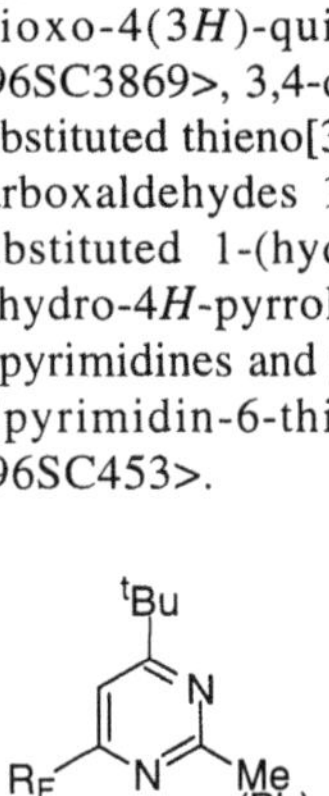

16

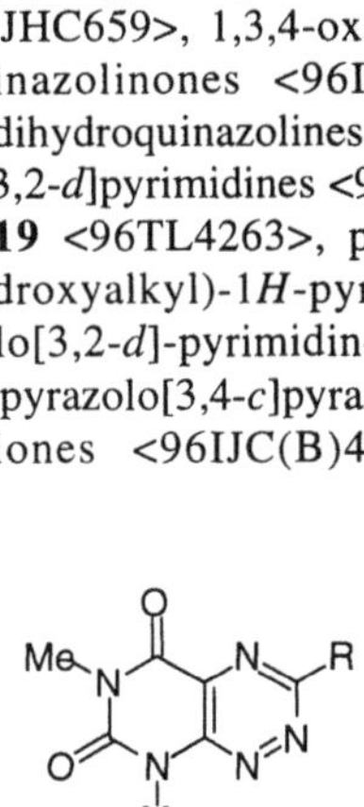

17

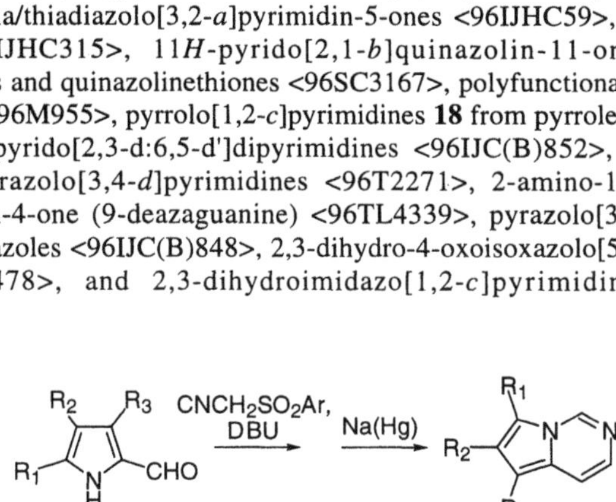

18 **19**

One-pot syntheses of 1,3-diaryl-6-ethoxycarbonyl-1,2,3,4-tetrahydro-4,7-dioxo-2-thioxo-7*H*-pyrano[2,3-*d*]pyrimidines <96IJC(B)1319> and of hexahydro-1,2,4-triazolo[4,3-*a*]pyrimidine-3,7-diones <96IJC(B)567> were reported. Revision of a reported structure of an indolo[1,2-*b*]indazole has led to an access to the indazolo[2,3-*a*]quinazolines <96CPB1099>. Entry to the new benzimidazo[1,2-*a*]pyrazolo[1,5-*c*]quinazoline ring system has occurred <96MC15>, as has that to 1*H*-pyrrolo[3,2-*g*]quinazoline **20** and its [2,3-*f*] angular isomer **21** <96JOC1155>. A Fischer-indole approach to pyrrolo[2,3-*d*]pyrimidines was reported <96H323>, the pyrrolo[2,3-*d*]pyrimidine marine alkaloid rigidin was synthesized <96JCS(P1)459>, and 5-substituted 2-aminopyrrolo[2,3-*d*]pyrimidin-4-ones were prepared <96JHC767>. The ring-contracted pyrrolo[2,3-*d*]pyrimidine-based analog **22** of 5,10-dideaza-5,6,7,8-tetrahydrofolic acid has been synthesized <96JOC1261>. Synthesis of pyrano[2,3-*d*]pyrimidines <96IJC(B)1205, 96IJC(B)673, 96IJC(B)248> has received attention, as has that of thieno[2,3-*d*]pyrimidines <96HAC29, 96IJC(B)715, 96H349> and pyrido[2,3-*d*]pyrimidines <96CCC901, 96JHC45, 96M917, 96JCS(P1)1999, 96IJC(B)1208>.

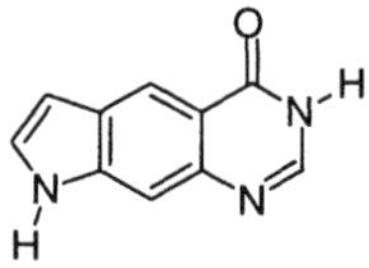

20

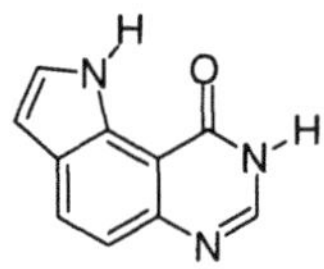

21

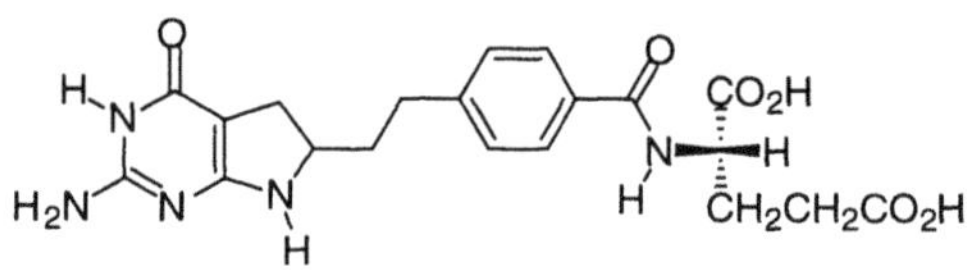

22

2*H*-Imidazo- and 2,3-dihydropyrimido[1,2-*c*]quinazolines were obtained by quinazoline ring annelation <96CCC957>. A [1,3,4]thiadiazolo[3,2-*a*]pyrimidine synthesis utilizing the presence of HCO_2H was reported <96JHC1367>, and pyrimidines **23** were obtained from 3-trifluoromethylsulfonyloxypropeniminium triflates and nitriles <96JHC439>, from 2-aza-1,3-dienes with electron-releasing substituents at the 1,3-positions <96T10095>, and by reaction of acyl isothiocyanates with nucleophiles <96PS15>. 3,4-Dihydro-2*H*-pyrido[1,2-*a*]pyrimidines **24** were synthesized from 2-aminopyridines <96TL2615>. New 5-methylene-6*H*-pyrimidine-2,4-dione thymine analogs **25** were obtained by *N,N'*-cyclization of carbodiimides with 2-(bromomethyl)acrylic acid <96JHC1259>. Directed lithiation of 2-substituted 5-aminopyridines has led to 6-substituted pyrido[3,4-*d*]pyrimidin-4(3*H*)-ones <96JCS(P1)2221>. Dimerization of 2-aminobenzoic acids under Vilsmeier conditions has led to 4-(3*H*)-quinazolinones **26** <96TL5015>. A partial synthesis of a major metabolite of the potent bronchodilator 7,8,9,10-tetrahydroazepino[2,1-*b*]-quinazolin-12(6*H*)-one was conducted <96IJC(B)345>.

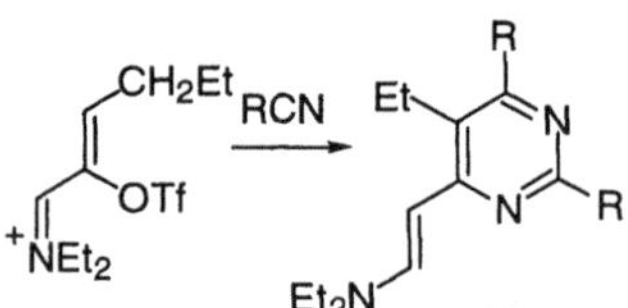

23

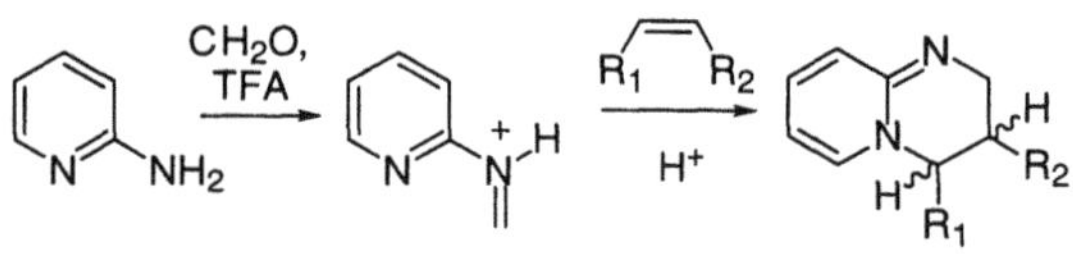

24

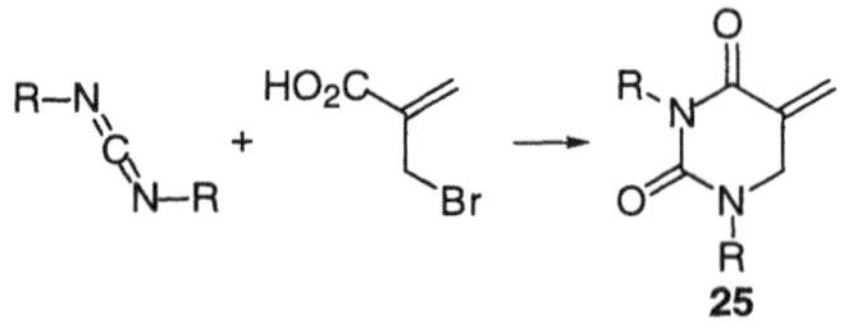

25

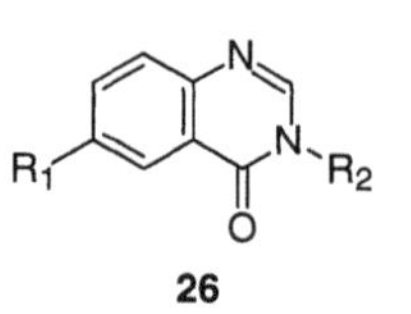

26

6.2.4.2 Reactions

The kinetics of Ce(IV) oxidation of pyrimidines has been assessed <96MI713>. Pyrido[1,2-*a*]pyrazines **27** (below) have been transformed into imidazo[1,2-*a*]pyridines, imidazo[1,2-*a*]pyrimidines and 2-oxa-6a,10c-diazaaceanthrylenes <96JHC639>. Nucleophilic and electrophilic reagent interaction with 4,6-diaryl-2(1*H*)-pyrimidine-2-thiones has been assessed <96RRC109>, as has the Pd-catalyzed allylic coupling of 1,2,3-triazolo[4,5-*d*]pyrimidines (8-azapurines) <96JOC6199>. The reactions of 3-amino-2(1*H*)-thioxo-4(3*H*)-quinazolinone with chloroacetic acids has been examined <96PJC197>. The ring closure reaction of 5-hydroxypyrido[2,3-*d*]pyrimidine-2,4,7-triones to benzo[*b*]pyrimido[4,5-*h*]1,6-naphthyridine-1,3,6-triones was studied <96JPR151>. In studies of intramolecular 1,3-dipolar cycloadditions, the hydrazone-azomethine imine isomerization <96T901> and facile oxime-nitrone isomerization <96T887> at the periphery of pyrido[1,2-*a*]pyrimidines were examined. Fusion reactions of *N*-heterocycles to thiopyrano[4',3':4,5]thieno[2,3-*d*]pyrimidines were also studied <96CCC147>.

Factors that affect the hydrolysis of 5-monoalkyl substituted barbituric acids have been examined <96PJC570> and the photocycloaddition of mono- and dithiobarbiturates with an *N*-side chain alkenyl group was shown to afford fused pyrimidines <96H117>. An oxidative and thermal ring opening in 2-oxoisoxazolo[2,3-*a*]pyrimidines **28** was found to afford pyrimidin-2-yl phenyl ketones **29** (R_1=H, Me, R_2=H, Me, Ph) <96JOC3212>. Methylthiomethyl (MTM) protection for the N1 of pyrrolo[2,3-*d*]pyrimidine-2,4-diones has been developed <96TL759>. Mild removal in the presence of BOM, MOM, and PNPE protecting groups involves treating, for instance, **30** with SO_2Cl_2 followed by SiO_2 in aqueous THF. The *N*-glucosides of certain 6-substituted 3-amino-2-aryl-4(3*H*)-quinazolinones were found susceptible to Amadori rearrangement <96IJC(B)147>. In a study of tetraazathiapentalenes fused with pyrimidine rings, the nature of hypervalent N-S-N bond was probed by restricting the internal rotation of the pyrimidine ring <96JA6355>. The lithiation of 3-(acylamino)-2-unsubstituted-, 3-(acylamino)-2-ethyl-, and 3-(acylamino)-2-propyl-4(3*H*)-quinazolinones (**31**, R_1=H, Et, Pr, respectively) <96JOC647> and of 2-alkyl-3-amino- and 2-alkyl-3-(methylamino)-4(3*H*)-quinazolinones (**32**, R_2=H and Me, respectively) <6JOC656> were investigated.

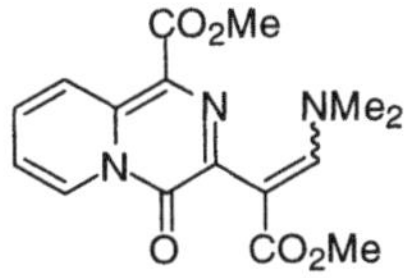

27

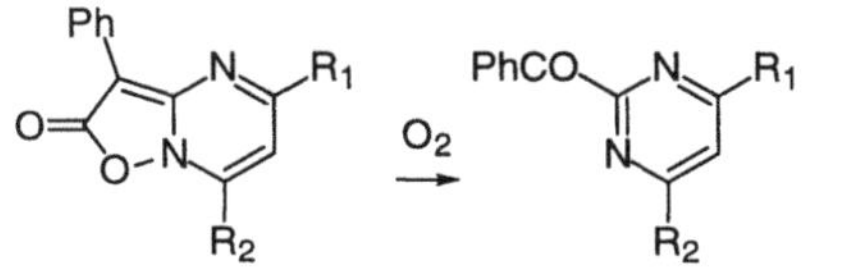

28 **29**

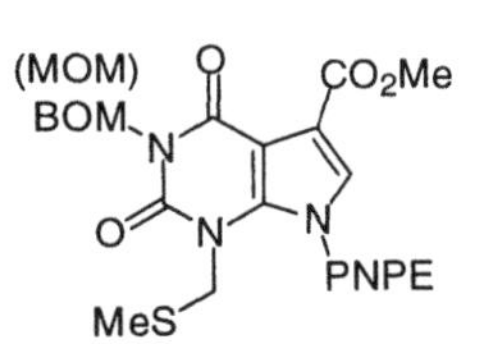

30

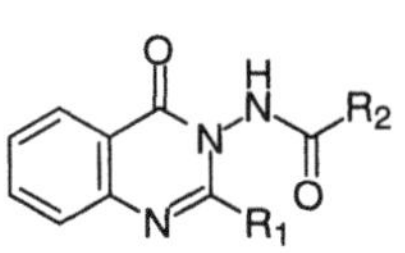

31

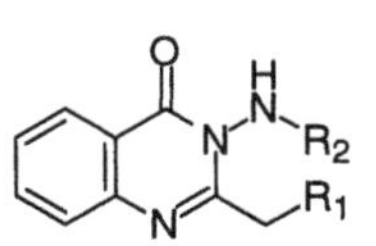

32

6.2.4.3 Applications

3-(Acetoxyamino)quinazolinones found continued use as aziridination agents. α,β-Unsaturated esters bearing allylic hydroxy groups <96JCS(P1)1951> and alkenes <96TL5179> were suitable substrates, the latter reacting in diastereoselective fashion when a chiral 1'-(*t*-butyldimethylsilyloxy)ethyl substituent was present on the 3,4-dihydroquinazolin-4-one reagent framework, as in **33**. Enantiomerically pure β-amino acids are now obtainable from 2-*t*-butyl-1-carbomethoxy-2,3-dihydro-4(1*H*)-pyrimidinone <96OS201>. 1-Hydroxy-4,6-dimethyl-2(1*H*)-pyrimidinone has been used to develop a new benzyloxycarbonylating agent <96HC141>. Synthetic utilities of the readily prepared 2-(trichloromethyl)pyrimidines **34** (R_1=H, Me, or Ph, R_2=CO_2Me, H, or Ph, R_3=H or alkoxy) have been developed <96JOC2470>.

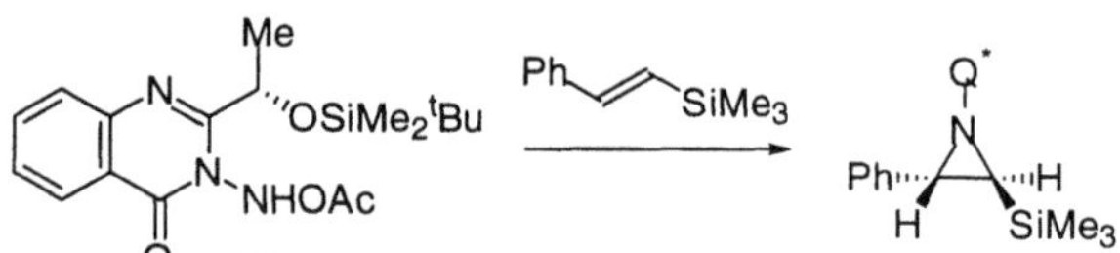

33

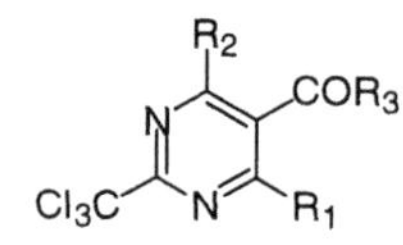

34

Pyrimidines and related condensed ring systems continue to be attractive as anticancer agents <96JHC229>. A nonlinear QSAR study of DHFR inhibition by pyrimidines was conducted <96JMC3526>, and 4-(2,6-diaminopyrimidin-5-yl)alkyl-L-glutamic acids <96BMCL473>, quinazolines <96JMC73>, and 5-deaza pyrido[2,3-*d*]pyrimidines <96JMC1438> were examined as antifolates. Subjects of other biological SAR studies include those of thiazolo[4,5-*d*]pyrimidines as antimicrobials, antivirals, and antitumor agents <96PHA927, 96MI28>, pyrrolo[3,2-*f*]quinazolines as antifolate antimicrobials and anticancer agents <96JMC892>, thiazolo[2,3-*b*]pyrazolo[4,3-*c*]pyrimidin-5(3*H*)-ones <96IJHC193>, pyrazolo[1,5-*a*]pyrido-[3,4-*e*]pyrimidin-6-ones <96F451>, and quinazolin-4-ones <96IJHC25> as antimicrobials, and 2-aminopyrimidines and 2(1*H*)-pyrimidinones <96MI91>, 3-pyrimidine(thiol)-substituted cephalosporins <96MI15>, and pyrrolo[2,3-*d*]pyrimidines and pyrido[2,3-*d*]pyrimidines <96BMC593> as antibacterials. Also examined were pyrano[2,3-*d*]pyrimidines as antibacterials and antifungals <96IJC(B)742>, pyrrolo[2,3-*d*]pyrimidines as antiinfluenza agents <96BMCL565> and antifolates <96JMC4563>, 1,3,4-thiadiazolo[3,2-*a*]pyrimidines and imidazo[2,1-*b*]-1,3,4-thiadiazoles as leishmanicides <96IJC(B)238>, and pyrrolo[2,3-*d*]pyrimidin-4-amines as molluscicidals <96JCR(S)127>.

Many pyrimidine-based heterocycles were examined as potential A1 and/or A2α adenosine antagonists. Among these were 1,2,3-triazolo[5,4-*e*]1,2,4-triazolo[1,5-*c*]pyrimidines <96F297>, pyrazolo[4,3-*e*]-1,2,4-triazolo[1,5-*c*]pyrimidines <96JMC1164>, pyrrolo[2,3-*d*]pyrimidines and pyrimido[4,5-*b*]indoles <96JMC2482>, pyrazolo[3,4-*d*]pyrimidines <96JMC4156, 96BMCL357>, and 1*H*-pyrazolo[4,3-*d*]pyrimidin-7(6*H*)-one and 5*H*-pyrazolo[4,3-*d*]-1,2,3-triazin-4(3*H*)-one <96AF365>. Imidazo[1,2-*a*]pyrimidines are of interest as CNS agents <96MI209>, 2,4-diamino-5-arylpyrimidines as antimalarials <96AJC647>, quinazolines as antirheumatics <96JMC5176>, 2- and 4-(heteroaryl)pyrimidines as antiulcer agents <96CPB1700>, 6-chloro-2,3-dihydro-4(1*H*)-quinazolinones as antiemetics <96AF911>, and pyrimidines as xanthine oxidase inhibitors <96JMC2529>. Thieno[2,3-*d*]-pyrimidin-4(3*H*)-ones <96AF273, 96AF981> and 2-arylquinazolines and 4(3*H*)-quinazolinones <96JMC1433> are potential antihyperlipemics, imidazo[1,5-*a*]pyrimidinyl- and pyrazolo[1,5-*a*]pyrimidinyl-1,2,4-oxadiazoles are potential 5-HT3 antagonists <96MI37>, and pyrazolo[3,4-*d*]pyrimidines are potential antihypertensives <96PHA540>. Pyrimidinyl-salicylic and -thiosalicylic acids have been prepared as herbicidals <96MI293>.

Studies of pyrido[*d*]pyrimidines <96JMC1823>, 3,4-dihydro-1*H*,6*H*-[1,4]oxazino-[3,4-*b*]quinazolin-6-ones <96BMC547>, and fused tricyclic quinazolines <96JMC918> as tyrosine kinase inhibitors, pyrazolo[4,3-*d*]pyrimidin-7-ones as 3α-hydroxy steroid dehydrogenase inhibitors <96PHA983>, phenylaminopyrimidines as protein kinase C <96AP371> and PDGF-receptor autophosphorylation inhibitors <96BMCL1221>, 4(1*H*)-quinazolinones as cholecystokinin receptor ligands <96JHC1163, 96F333>, pyrazolo[1,5-*c*]quinazolines <96JMC2915> and 1,2,3-triazolo[1,5-*a*]quinazolines <96F131> as benzodiazepine receptor ligands, and pyrimido-pyrimidines as 5-lipoxygenase inhibitors <96MI61> have been reported. 1,6-Diphenylpyrazolo[3,4-*d*]thiazolo[3,2-*a*]-4*H*-pyrimidin-4-one shows NK-2 antagonist properties <96BMCL59>, and lipophilic quinazoline inhibitors of thymidylate synthase have been developed <96JMC695, 96JMC904>. Biological evaluations of 2,4-dioxo-1,2,3,4,5,6,7,8-octahydropyrimido[4,5-*d*]pyrimidines <96MI39>, of heterocycles containing the 2-phenyl-6-iodo-quinazolinyl-4-oxy moiety <96IJHC189>, of pyrimidines containing sulfur <96IJC(B)388>, of 4'-(R)-hydroxy-5'-(*S*)-hydroxymethyl-tetrahydrofuranyl purines and pyrimidines <96BMC609>, and of *N*-[5*H*-[1]benzopyrano[4,3-*d*]pyrimidin-2-yl]-*N*-methylglycinamides <96F137> have been conducted. The effect of a 1,6-dihydro-2-substituted-6-oxo-5-pyrimidinecarboxylic acid on EtOH-induced microvascular injury was assessed <96AF779>.

6.2.5 PYRAZINES AND BENZODERIVATIVES

X-ray crystal structure determinations of 1,4-dihydro-2,3-quinoxalinediones <96AX(B)487>, divalent metal 2,3-pyrazinedicarboxylates <96CJC433>, a 3,4-pyrazino-3',4'-ethylenedithio-2,2',5,5'-tetrathiafulvalenium salt <96AX(C)159>, methyl 3-(triphenylphosphoranylidene-amino)pyrazine-2-carboxylate <96JCS(P1)247>, and 1-(benzo[*a*]phenazin-5-yl)-2-acetyloxy-naphthalene <96AX(C)222> have been conducted. NMR investigations include a ^{1}H and ^{13}C NMR study of imidazo[1,2-*a*]pyrazines <96MRC409>, an ^{17}O and ^{14}N NMR one of quinoxaline-2(1*H*),3(4*H*)-diones and *N,N'*-substituted oxamides <96JHC643, 96TL3191>, a ^{13}C NMR one of quinoxaline spirans and carboxyureides <96H1873>, and a ^{15}N NMR one of pyrazine, methylpyrazines, and their *N*-oxides <96MRC567>. An electrochemical and ESR study of 2,7-disubstituted phenazines <96CPB1448>, a pulse radiolysis study of reduction free radicals of imidazo[1,2-*a*]quinoxaline *N*-oxides **35**, hypoxia-selective cytotoxic antitumor drugs <96JA5648>, and a photochemical one of 1-benzyloxy-2(1)-pyrazinones <96H883> have been conducted.

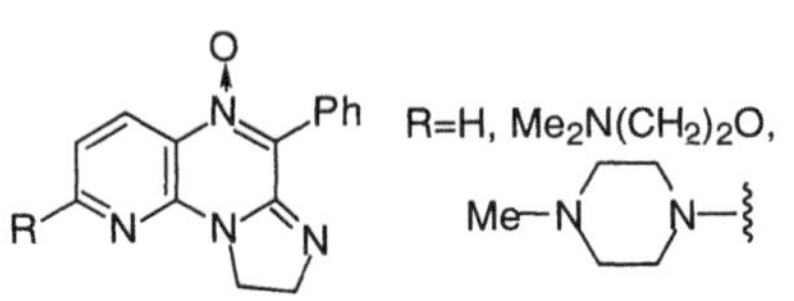

35

6.2.5.1 Syntheses

Studies of nonsymmetrical dimeric steroid-pyrazine marine alkaloids <96AC(E)611> have led to the total synthesis of dihydrocephalostatin 1 and to the first biologically active cephalostatin analogs <96JA10672, 96AC(E)1572>. A stereoselective route to 6-oxoperhydropyrrolo[1,2-*a*]pyrazines was developed <96T13991>. A Suzuki coupling approach using **36** (prepared in 4 steps from chloropyrazine) gave access to pyrazines related to coelenterazine <96SL509>. 2,3,4,4a,5,6-Hexahydro-1*H*-pyrazino[1,2-*a*]quinoline was prepared by a [3+2] cycloaddition reaction <96TL7343>. 5-Alkyl-2-pyrazinecarboxamide, 5-alkyl-2-pyrazinecarbonitrile and 5-alkyl-2-acetylpyrazine synthetic intermediates for antiinflammatory agents were developed <96CCC1093>. Quinoxalines **37** have been accessed by thermal rearrangement of benzimidazoles **38** <96TL3355> and by Pd-catalyzed oligomerization of 1,2-diisocyanoarenes <96H597>. An improved preparation of 4,7-phenanthrolino-5,6:5',6'-pyrazine was developed <96SC2197>. 6-Alkylidenated 3,6-dihydropyrazin-2(1*H*)-ones obtained by the action of MeO^- on 6-bromomethylpyrazin-2(1*H*)-ones were converted into piperazine-2,5-diones and pyrazin-2(1*H*)-ones <96JCS(P1)231>. An intramolecular aza-Wittig reaction has led to pyrazino[2,3-*e*][1,4]diazepines <96TL81>, and transamination of 1-(2-aminophenyl)-2-iminomethylpyrroles was shown to give pyrrolo[1,2-*a*]quinoxalines <96T10751>. Construction of the 2,3-dihydrofuro[2,3-*b*]quinoxaline skeleton was reported<96HC325>, and 2-(2-nitrophenoxy)quinoxaline was prepared and its surfactant-mediated basic hydrolysis studied <96T11665>. A naphtho[2,3-*f*]quinoxaline-7,12-dione was obtained from mitoxantrone <96SC3929>, and quinoxaline 1,4-dioxides **39** and their isomers from 4,5(6,7)-dimethylbenzofuroxan **40** <96JHC1057>. Indolo[2,3-*b*]quinoxaline-based ketonic Mannich bases have been prepared <96MI10>.

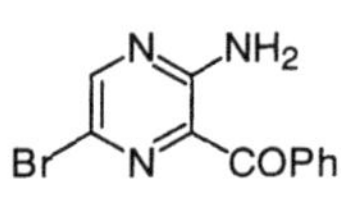

36

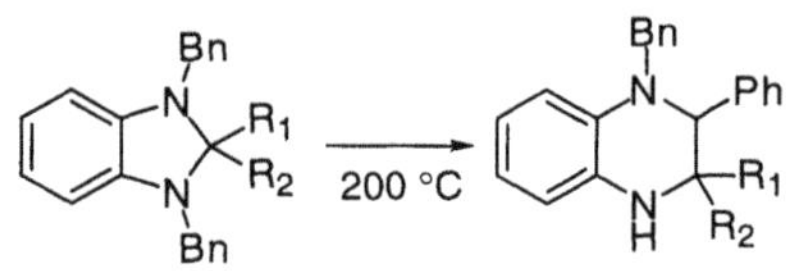

37 **38**

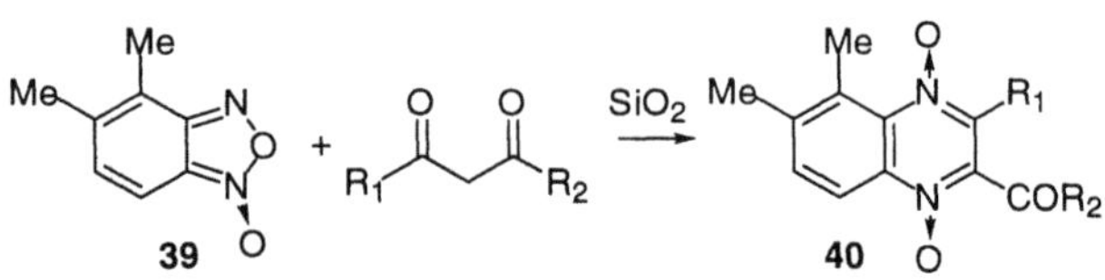

39 **40**

Some other synthetic accomplishments during 1996 include new approaches to 2,3-disubstituted quinoxalines <96PHA428> and to quinoxalines and related polycyclics <96JCS(P1)2443>. Intermediate nitrileimines **41** were found to be viable precursors to pyrazolo[1,5-*a*]quinoxalines **42** <96S1076>. The 1,5-dihydropyridazino[3,4-*b*]quinoxalines **43** and 2-(pyrazol-4-yl)quinoxalines <96JHC757>, perfluoroalkyl- and perfluoroalkylether-substituted quinoxalines <96JFC27>, 2-phenylimidazo[1,2-*a*]-pyrazine-3-acetates <96M947>, and 1-benzyloxy-2(1*H*)-pyrazinones <96H883> also received attention.

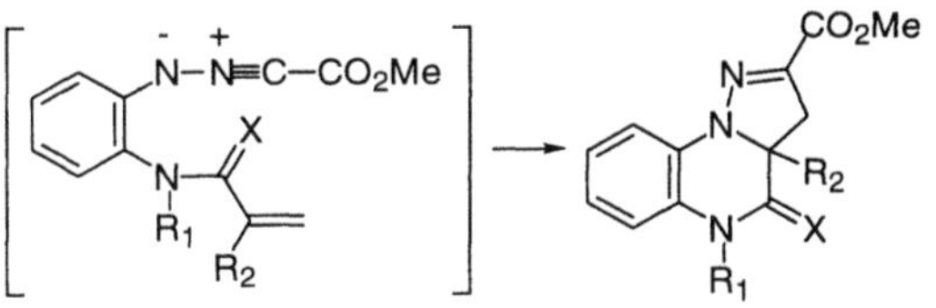

41 42

R_1=Ph, R_2=H, X=H_2; R_1=Ph, R_2=Me, X=H_2; R_1=Ac, R_2=H, X=H_2; R_1=Me, R_2=H, X=O

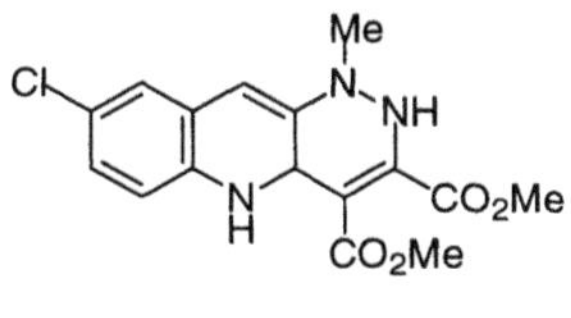

43

6.2.5.2 Reactions

Azomethine ylides **44** of pyrrolo[1,2-*a*]pyrazines have been shown to undergo 1,3-dipolar cycloaddition with suitable dipolarophiles to give dipyrrolo[1,2-*a*]pyrazines **45**, pyrazolo[1,5-*a*]pyrrolo[2,1-*c*]pyrazines **46**, and heterobetaines **477** <96JOC4655>. Pyrazines with alkenyl side chains like **48** undergo intramolecular Diels-Alder reaction in TFA to give bridged adducts <96TL8205>. A retro-Diels Alder regenerating starting material occurs in hot $PhNO_2$. In 2(1*H*)-pyrazinones this has led to pyrrolo[3,4-*b*]- and [3,4-*c*]pyridin(on)es and related 1,7-naphthyridinones and 2,7-naphthyridines <96T9161>. The alkylation of chloropyrazine *N*-oxides by Ni-catalyzed cross coupling with dialkylzincs has been achieved <96JHC1047>, the reductive metalation of 2,3,5,6-tetraphenylpyrazine has led to 1,2-dihydro-1,4-diazines <96RTC377>, and tetra-2,3-quinoxalinoporphyrazine has been angularly annelated <96CJC508>. Olefin cycloadditive ring transformation of 3-amino-5-chloro-2(1*H*)-pyrazinones **49** has been shown to afford methyl 6-cyano-1,2-dihydro-2-oxo-4-pyridinecarboxylates **50** or the corresponding 3-(diethylamino)-1,2-dihydro-2-oxo-4-pyridinecarboxylates **51** depending upon the substitution pattern <96JOC304>.

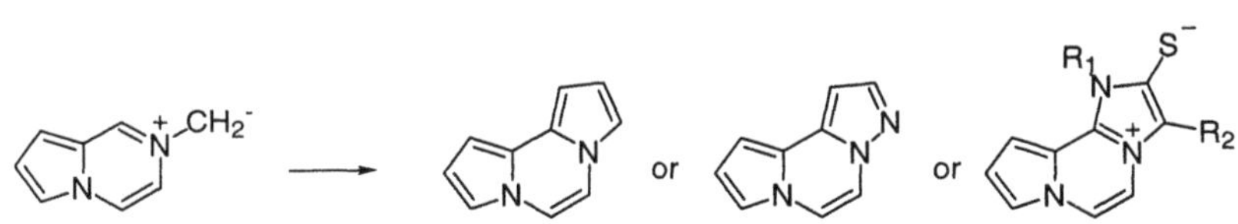

44 45 46 47

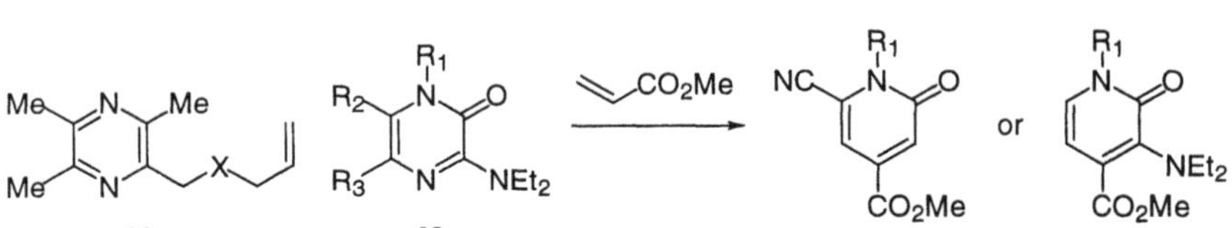

48 49 50 51

6.2.5.3 Applications

Electroluminescence applications of a poly(phenylquinoxaline) have been described <96SM105>. Biological investigations of pyrazinecarboxamides and -carbothioamides as antituberculotics <96CCC1109, 96CCC1102> and tri- and tetrasubstituted pyrazines as ant pheromones <96JCS(P1)2345> have been conducted. Quinoxalines are of interest as antispasmodics <96MI199>, 5-HT3 receptor antagonists <96AF401>, inhibitors of PGF receptor tyrosine kinase <96JMC2170>, streptonigrin analogs <96JHC447>, and antifolates <96F559>. Other interests include imidazo[1,5-*a*]quinoxaline amides, carbamates, and ureas as agonists of the $GABA_A$/benzodiazepine receptor complex <96JMC158>, imidazo[1,5-*a*]quinoxalin-4-ones and imidazo[1,5-*a*]quinoxaline ureas as $GABA_A$/benzodiazepine receptor ligands <96JMC3820>, 1,4-dihydro-6,7-quinoxaline-2,3-diones as α-amino-3-hydroxy-5-methyl-4-isoxazolepropionate receptor agonists and antagonists <96JMC4430>, and 1-substituted [1]benzothieno[2,3-*b*]pyrazin-2(1*H*)-ones as antagonists of KCl-induced contractions <96EJM607>. SAR studies of 2,3(1*H*,4*H*)-quinoxalinediones as AMPA receptor antagonists <96JMC3971> and 6,7-disubstituted pyrido[2,3-*b*]pyrazine-2,3(1*H*,4*H*)-diones as α-amino-3-hydroxy-5-methylisoxazole-4-propionate receptor antagonists <96JMC1331> have been conducted. A QSAR one of pyrazinoate esters as antimycobacterials have been done <96JMC3394>. Finally, 6-amino-3-methylpyrazin-2(1*H*)-one now has been employed as a non-standard oligonucleotide residue to base-pair with 5-aza-7-deazaisoguanine <96HCA1863, 96HCA1881>.

6.2.6 REFERENCES

<96AC(E)611> A. Ganesan, *Angew. Chem., Int. Ed. Engl.* **1996**, *35*, 611.

<96AC(E)1572> M. Droegemueller, R. Jautelat, E. Winterfeldt, *Angew. Chem., Int. Ed. Engl.* **1996**, *35*, 1572.

<96AF273> C.J. Shishoo, K.S. Jain, I.S. Rathod, B. Thakkar, S.B. Brahmbhatt, T.P. Gandhi, R. Bangaru, R.K. Goyal, *Arzneim. Forsch.* **1996**, *46*, 273.

<96AF365> P.G. Baraldi, B. Cacciari, A. Dalpiaz, S. Dionisotti, C. Zocchi, M.J. Pineda de las Infantas, G. Spalluto, K. Varani, *Arzneim. Forsch.* **1996**, *46*, 365.

<96AF401> B. Lasheras, A. Berjon, R. Montanes, J. Roca, G. Romero, M.J. Ramirez, J. Del Rio, *Arzneim. Forsch.* **1996**, *46*, 401.

<96AF779> Y. Ishizuka, T. Kamisaki, H. Okamoto, M. Kawashima, M. Sato, *Arzneim. Forsch.* **1996**, *46*, 779.

<96AF911> C. Baldazzi, M. Barbanti, R. Basaglia, A. Benelli, A. Bertolini, S. Piani, *Arzneim. Forsch.* **1996**, *46*, 911.

<96AF981> A.K. Gadad, S.G. Kapsi, R.I. Anegundi, S.R. Pattan, C.S. Mahajanshetti, C.J. Shishoo, *Arzneim. Forsch.* **1996**, *46*, 981.

<96AJC435> P. Matyus, G.B. Barlin, P.W. Harrison, M.G. Wong, L.P. Davies, *Aust. J. Chem.* **1996**, *49*, 435.

<96AJC443> G.B. Barlin, L.P. Davies, S.J. Ireland, *Aust. J. Chem.* **1996**, *49*, 443.

<96AJC451> G.B. Barlin, L.P. Davies, P.W. Harrison, S.J. Ireland, A.C. Willis, *Aust. J. Chem.* **1996**, *49*, 451.

<96AJC573> C.B. Elmes, G. Holan, G.T. Wernert, D.A. Winkler, *Aust. J. Chem.* **1996**, *49*, 573.

<96AJC647> G.B. Barlin, B. Kotecka, K.H. Rieckmann, *Aust. J. Chem.* **1996**, *49*, 647.

<96AP371> J. Zimmermann, G. Caravatti, H. Mett, T. Meyer, M. Mueller, N.B. Lydon, D. Fabbro, *Arch. Pharm.* **1996**, *329*, 371.

<96AP403> B. Barth, M. Dierich, G. Heinisch, B. Matuszczak, K. Mereiter, J. Soder, H. Stoiber, *Arch. Pharm.* **1996**, *329*, 403.

<96AX(B)487> M. Kubicki, T.W. Kindopp, M.V. Capparelli, P.W. Codding, *Acta Crystallogr., Sect. B: Struct. Sci., B52* **1996**, *3*, 487.

<96AX(C)159> U. Geiser, J.A. Schlueter, A.M. Kini, C.A. Achenbach, A.S. Komosa, J.M. Williams, *Acta Crystallogr., C52* **1996**, *1*, 159.

<96AX(C)222> V. McKee, M. McCann, N.C. Connaughton, M.G. Kennedy, *Acta Crystallogr., C52* **1996**, *1*, 222.

<96AX(C)254> J.N. Low, G. Ferguson, J. Cobo, M. Nogueras, A. Sanchez, *Acta Crystallogr., C52* **1996**, *1*, 254.

<96AX(C)1002> R.A. Jones, A.K. Powell, A.P. Whitmore, *Acta Crystallogr., C52* **1996**, *4*, 1002.

<96AX(C)1031> M. Fettouhi, A. Boukhari, B. El Otmani, E.M. Essassi, *Acta Crystallogr., C52* **1996**, *4*, 1031.

<96AX(C)1703> S. E.-S. Barakat, H.J. Lindner, *Acta Crystallogr., C52* **1996**, *7*, 1703.

<96AX(C)2081> Y. Gong, P.D. Robinson, M.J. Bausch, *Acta Crystallogr., C52* **1996**, *8*, 2081.

<96AX(C)2106> J. Mazurek, T. Lis, R. Jasztold-Howorko, *Acta Crystallogr., C52* **1996**, *8*, 2106.

<96AX(C)2115> A. Ettorre, A. Mai, M. Artico, S. Massa, A. De Montis, P. La Colla, *Acta Crystallogr., C52* **1996**, *8*, 2115.

<96AX(C)2622> A.J. Blake, H. McNab, *Acta Crystallogr., C52* **1996**, *10*, 2622.

<96AX(C)2826> P.D. Robinson, M.P. Groziak, L. Yi, *Acta Crystallogr., C52* **1996**, *8*, 2826.

<96BMC547> L. Orfi, J. Koekoesi, G. Szasz, I. Koevesdi, M. Mak, I. Teplan, G. Keri, *Bioorg. Med. Chem.* **1996**, *4*, 547.

<96BMC593> L.F. Kuyper, J.M. Garvey, D.P. Baccanari, J.N. Champness, D.K. Stammers, C.R. Beddell, *Bioorg. Med. Chem.* **1996**, *4*, 593.

<96BMC609> H.-W. Yu, L.-R. Zhang, J.-C. Zhou, L.-T. Ma, L.-H. Zhang, *Bioorg. Med. Chem.* **1996**, *4*, 609.

<96BMCL59> S. Guccione, M. Modica, J. Longmore, G. Uccello Barretta, A. Santagati, M. Santagati, F. Russo, *Bioorg. Med. Chem. Lett.* **1996**, *6*, 59.

<96BMCL357> S.-A. Poulsen, R.J. Quinn, *Bioorg. Med. Chem. Lett.* **1996**, *6*, 357.

<96BMCL473> L.S. Gossett, L.L. Habeck, S.B. Gates, S.L. Andis, J.F. Worzalla, R.M. Schultz, L.G. Mendelsohn, W. Kohler, M. Ratnam, et al., *Bioorg. Med. Chem. Lett.* **1996**, *6*, 473.

<96BMCL565> M. Sznaidman, E.A. Meade, L.M. Beauchamp, S. Russell, M. Tisdale, *Bioorg. Med. Chem. Lett.* **1996**, *6*, 565.

<96BMCL1221> J. Zimmermann, E. Buchdunger, H. Mett, T. Meyer, N.B. Lydon, P. Traxler, *Bioorg. Med. Chem. Lett.* **1996**, *6*, 1221.

<96BMCL1483> A.L. Smith, C.G. Thomson, P.D. Leeson, *Bioorg. Med. Chem. Lett.* **1996**, *6*, 1483.

<96CCC147> E.K. Ahmed, J. Froehlich, F. Saute,r *Collect. Czech. Chem. Commun.* **1996**, *61*, 147.

<96CCC901> J.I. Borrell, J. Teixido, B. Martinez-Teipel, B. Serra, J.L. Matallana, M. Costa, X. Batllori, *Collect. Czech. Chem. Commun.* **1996**, *61*, 901.

<96CCC957> K. Spirkova, S. Stankovsky, *Collect. Czech. Chem. Commun.* **1996**, *61*, 957.

<96CCC1093> V. Opletalova, A. Patel, M. Boulton, A. Dundrova, E. Lacinova, M. Prevorova, M. Appeltauerova, M. Coufalova, *Collect. Czech. Chem. Commun.* **1996**, *61*, 1093.

<96CCC1102> M. Dolezal, J. Hartl, A. Lycka, V. Buchta, Z. Odlerova, *Collect. Czech. Chem. Commun.* **1996**, *61*, 1102.

<96CCC1109> J. Hartl, M. Dolezal, J. Krinkova, A. Lycka, Z. Odlerova, *Collect. Czech. Chem. Commun.* **1996**, *61*, 1109.

<96CJC433> L. Mao, S.J. Rettig, R.C. Thompson, J. Trotter, S. Xia, *Can. J. Chem.* **1996**, *74*, 433.

<96CJC508> S.V. Kudrevich, M.G. Galpern, E.A. Luk'yanets, J.E. van Lier, *Can. J. Chem.* **1996**, *74*, 508.

<96CPB122> M. Kuwahara, Y. Kawano, H. Shimazu, Y. Ashida, A. Miyake, *Chem. Pharm. Bull.* **1996**, *44*, 122.

<96CPB980> F. Rohet, C. Rubat, P. Coudert, E. Albuisson, J. Couquelet, *Chem. Pharm. Bull.* **1996**, *44*, 980.

<96CPB1099> H. Katayama, O. Kato, K. Kaneko, *Chem. Pharm. Bull.* **1996**, *44*, 1099.

<96CPB1448> T. Michida, H. Sayo, *Chem. Pharm. Bull.* **1996**, *44*, 1448.

<96CPB1700> M. Ikeda, K. Maruyama, Y. Nobuhara, T. Yamada, S. Okabe, *Chem. Pharm. Bull.* **1996**, *44*, 1700.

<96EJM65> V. Dal Piaz, M.P. Giovannoni, G. Ciciani, D. Barlocco, G. Giardina, G. Petrone, G.D. Clarke, *Eur. J. Med. Chem.* **1996**, *31*, 65.

<96EJM607> F. Guerrera, L. Salerno, M.C. Sarva, M.A. Siracusa, A. Rescifina, R. Scalia, A. Bianchi, *Eur. J. Med. Chem.* **1996**, *31*, 607.

<96EJM651> P.W. Harrison, G.B. Barlin, L.P. Davies, S.J. Ireland, P. Matyus, M.G. Wong, *Eur. J. Med. Chem.* **1996**, *31*, 651.

<96F131> G. Biagi, I. Giorgi, O. Livi, V. Scartoni, S. Velo, A. Lucacchini, G. Senatore, B. De Santis, A. Martinelli, *Farmaco* **1996**, *51*, 131.

<96F137> O. Bruno, S. Schenone, A. Ranise, F. Bondavalli, M. D'Amico, W. Filippelli, L. Berrino, F. Rossi, *Farmaco* **1996**, *51*, 137.

<96F297> P.G. Baraldi, B. Cacciari, G. Spalluto, M.J. Pineda De Las Infants, C. Zocchi, S. Ferrara, S. Dionisotti, *Farmaco* **1996**, *51*, 297.

<96F333> A. Varnavas, L. Lassiani, E. Luxich, M. Zacchigna, E. Boccu, *Farmaco* **1996**, *51*, 333.

<96F431> H. Sladowska, J. Potoczek, G. Rajtar, M. Sieklucka-Dziuba, M. Mlynarczyk, Z. Kleinrok, *Farmaco* **1996**, *51*, 431.

<96F451> F. Bruni, S. Selleri, A. Costanzo, G. Guerrini, M.L. Casilli, C. Sacco, R. Donato, *Farmaco* **1996**, *51*, 451.

<96F559> M. Loriga, M. Fiore, P. Sanna, G. Paglietti, *Farmaco* **1996**, *51*, 559.

<96F653> G.A. Pinna, M.M. Curzu, P. Fraghi, E. Gavini, M. D'Amico, *Farmaco* **1996**, *51*, 653.

<96H117> H. Takechi, M. Machida, *Heterocycles* **1996**, *42*, 117.

<96H151> G. Heinisch, W. Holzer, T. Langer, P. Lukavsky, *Heterocycles* **1996**, *43*, 151.

<96H199> K.-i. Iwamoto, S. Suzuki, E. Oishi, A. Miyashita, T. Higashino, *Heterocycles* **1996**, *43*, 199.

<96H323> E.C. Taylor, B. Hu, *Heterocycles* **1996**, *43*, 323.

<96H349> E.C. Taylor, H.H. Patel, G. Sabitha, R. Chaudhari, *Heterocycles* **1996**, *43*, 349.

<96H389> S. Ostrowski, *Heterocycles* **1996**, *43*, 389.

<96H597> Y. Ito, Y. Kojima, M. Suginome, M. Murakami, *Heterocycles* **1996**, *42*, 597.

<96H625> Z. Szakonyi, F. Fulop, G. Bernath, P. Sohar, *Heterocycles* **1996**, *42*, 625.

<96H691> A. Miyashita, K. Fujimoto, T. Okada, T. Higashino, *Heterocycles* **1996**, *42*, 691.

<96H883> J. Ohkanda, T. Kumasaka, A. Takasu, T. Hasegawa, A. Katoh, *Heterocycles* **1996**, *43*, 883.

<96H1057> G. Heinisch, T. Langer, J. Tonnel, K. Mereiter, K. Wurst, *Heterocycles* **1996**, *43*, 1057.

<96H1073> M.C. Veiga, J.M. Quintela, C. Peinador, *Heterocycles* **1996**, *43*, 1073.

<96H1459> M. Rohr, D. Toussaint, S. Chayer, A. Mann, J. Suffert, C.-G. Wermuth, *Heterocycles* **1996**, *43*, 1459.

<96H1613> T. Yamasaki, Y. Yoshihara, Y. Okamotro, T. Okawara, M. Furukawa, *Heterocycles* **1996**, *43*, 1613.

<96H1873> P.J. Zeegers, M.J. Thompson, *Heterocycles* **1996**, *43*, 1873.

<96H1887> S. Anjum, T. Sarfraz, Y. Ahmad, R. Atta ur, *Heterocycles* **1996**, *43*, 1887.

<96H1967> S. Pippich, H. Bartsch, *Heterocycles* **1996**, *43*, 1967.

<96HAC29> Z. E.-S. Kandeel, *Heteroat. Chem.* **1996**, *7*, 29.

<96HAC137> S.I. Aziz, *Heteroat. Chem.* **1996**, *7*, 137.

<96HC141> A. Katoh, S. Kondoh, J. Ohkanda, *Heterocycl. Commun.* **1996**, *2*, 141.

<96HC325> E.S. El Ashry, H.A. Hamid, Y. El Kilany, *Heterocycl. Commun.* **1996**, *2*, 325.

<96HC417> M. Noguchi, B. Sun, M. Gotoh, K.-i. Tokunaga, S. Nishimura, *Heterocycl. Commun.* **1996**, *2*, 417.

<96HCA1863> J. Voegel, S.A. Benner, *Helv. Chim. Acta* **1996**, *79*, 1863.

<96HCA1881> J. Voegel, S.A. Benner, *Helv. Chim. Acta* **1996**, *79*, 1881.

<96IJC(B)144> A.A. Hataba, M.G. Assy, R.M. Fikry, *Indian J. Chem.* **1996**, *35B*, 144.

<96IJC(B)147> M.A. Saleh, M.A. Abdo, M.F. Abdel-Megeed, G.A. El-Hiti, *Indian J. Chem.* **1996**, *35B*, 147.

<96IJC(B)238> V.J. Ram, N. Haque, *Indian J. Chem.* **1996**, *35B*, 238.

<96IJC(B)248> V.K. Ahluwalia, R. Sahay, R. Kumar, *Indian J. Chem.* **1996**, *35B*, 248.

<96IJC(B)345> S.C. Sharma, U. Zutshi, O.P. Gupta, K.L. Dhar, C. K., Atal *Indian J. Chem.* **1996**, *35B*, 345.

<96IJC(B)388> M.S. Amine, S.A. Nassar, M.A. El-Hashash, S.A. Essawy, A.A. Hashish, *Indian J. Chem.* **1996**, *35B*, 388.

<96IJC(B)478> N.A. Devi, L.W. Singh, *Indian J. Chem.* **1996**, *35B*, 478.

<96IJC(B)567> A.R. Chowdhury, S. Sharma, A.P. Bhaduri, *Indian J. Chem.* **1996**, *35B*, 567.

<96IJC(B)673> V.K. Ahluwalia, U. Das, R. Kumar, *Indian J. Chem.* **1996**, *35B*, 673.

<96IJC(B)715> V.K. Ahluwalia, A. Dahiya, M. Bala, *Indian J. Chem.* **1996**, *35B*, 715.

<96IJC(B)742> V.K. Akluwalia, M. Bala, *Indian J. Chem.* **1996**, *35B*, 742.

<96IJC(B)848> V.K. Ahluwalia, A. Dahiya, M. Bala, *Indian J. Chem.* **1996**, *35B*, 848.

<96IJC(B)852> V.K. Ahluwalia, U. Das, R. Kumar, *Indian J. Chem.* **1996**, *35B*, 852.

<96IJC(B)1205> V.K. Ahluwalia, P. Sharma, R. Aggarwal, *Indian J. Chem.* **1996**, *35B*, 1205.

<96IJC(B)1208> V.K. Ahluwalia, A. Dahiya, *Indian J. Chem.* **1996**, *35B*, 1208.

<96IJC(B)1319> V.K. Akluwalia, S. Dudeja, R. Sahay, R. Kumar, *Indian J. Chem.* **1996**, *35B*, 1319.

<96IJHC25> M.A. Aziza, M.W. Nassar, S.G. Abdel-Hamide, A.E. El-Hakim, A.S. El-Azab, *Indian J. Heterocycl. Chem.* **1996**, *6*, 25.

<96IJHC59> M.M. Dutta, J.C.S. Kataky, *Indian J. Heterocycl. Chem.* **1996**, *6*, 59.

<96IJHC189> S.G. Abdel-Hamide, A.E. El-Hakim, R.M. Abdel-Rahman, *Indian J. Heterocycl. Chem.* **1996**, *5*, 189.

<96IJHC193> S.C. Bennur, M.B. Talawar, U.V. Laddi, Y.S. Somannavar, V.V. Badiger, H.M. Virupakshaiah, *Indian J. Heterocycl. Chem.* **1996**, *5*, 193.

<96IJHC315> R. Lakhan, R.K. Banerjee, *Indian J. Heterocycl. Chem.* **1996**, *5*, 315.

<96JA5020> M. Dietrich, J. Heinze, C. Krieger, F.A. Neugebauer, *J. Am. Chem. Soc.* **1996**, *118*, 5020.

<96JA5648> K.I. Priyadarsini, M.F. Dennis, M.A. Naylor, M. R.L. Stratford, P. Wardman, *J. Am. Chem. Soc.* **1996**, *118*, 5648.

<96JA6355> K. Ohkata, M. Ohsugi, K. Yamamoto, M. Ohsawa, K. Akiba, *J. Am. Chem. Soc.* **1996**, *118*, 6355.

<96JA10672> C. Guo, S. Bhandaru, P.L. Fuchs, M.R. Boyd, *J. Am. Chem. Soc.* **1996**, *118*, 10672.

<96JCR(S)127> W.M. Basyouni, K.A.M. El-Bayouki, M.M. El-Sayed, H. Hosni, *J. Chem. Res., Synop* **1996**, *3*, 127.

<96JCR(S)220> G. Grassi, F. Risitano, G. Bruno, F. Nicolo, *J. Chem. Res., Synop* **1996**, *5*, 220.

<96JCR(S)306> D.K. Maiti, P.P. Ghoshdastidar, P.K. Bhattacharya, *J. Chem. Res., Synop* **1996**, *6*, 306.

<96JCR(S)416> A.M. Farag, K.M. Dawood, Z.E. Kandeel, *J. Chem. Res., Synop* **1996**, *9*, 416.

<96JCR(S)434> I.A. El-Sakka, *J. Chem. Res., Synop* **1996**, *9*, 434.

<96JCR(S)440> H.F. Zohdi, W.W. Wardakhan, S.H. Doss, R.M. Mohareb, *J. Chem. Res., Synop* **1996**, *10*, 440.

<96JCS(P1)231> K.J. Buysens, D.M. Vandenberghe, S.M. Toppet, G.J. Hoornaert, *J. Chem. Soc., Perkin Trans. 1* **1996**, 231.

<96JCS(P1)247> T. Okawa, S. Eguichi, A. Kakehi, *J. Chem. Soc., Perkin Trans. 1* **1996**, 247.

<96JCS(P1)459> T. Sakamoto, Y. Kondo, S. Sato, H. Yamanaka, *J. Chem. Soc., Perkin Trans. 1* **1996**, 459.

<96JCS(P1)1699> I.W. Harvey, D.M. Smith, C.R. White, *J. Chem. Soc., Perkin Trans. 1* **1996**, 1699.

<96JCS(P1)1951> R.S. Atkinson, P.J. Williams, *J. Chem. Soc., Perkin Trans. 1* **1996**, 1951.

<96JCS(P1)1999> B. Baruah, D. Prajapati, J.S. Sandhu, A.C. Ghosh, *J. Chem. Soc., Perkin Trans. 1* **1999**, 1999.

<96JCS(P1)2221> G.W. Rewcastle, W.A. Denny, R.T. Winters, N.L. Colbry, H.D.H. Showalter, *J. Chem. Soc., Perkin Trans. 1* **1996**, 2221.

<96JCS(P1)2345> N. Sato, T. Matsuura, *J. Chem. Soc., Perkin Trans. 1* **1996**, 2345.

<96JCS(P1)2443> A.R. Ahmad, L.K. Mehta, J. Parrick, *J. Chem. Soc., Perkin Trans. 1* **1996**, 2443.

<96JCS(P2)1147> S. Chimichi, F. Bruni, B. Cosimelli, S. Selleri, G. Valle, *J. Chem. Soc., Perkin Trans. 2* **1996**, 1147.

<96JFC27> K.J.L. Paciorek, S.R. Masuda, W.H. Lin, J.H. Nakahara, *J. Fluorine Chem* **1996**, *78*, 27.

<96JHC45> R. Rodriguez, M. Suarez, E. Ochoa, A. Morales, L. Gonzalez, N. Martin, M. Quinteiro, C. Seoane, J.L. Soto, *J. Heterocycl. Chem.* **1996**, *33*, 45.

<96JHC213> R. Bakthavatchalam, E. Ciganek, J.C. Calabrese, *J. Heterocycl. Chem.* **1996**, *33*, 213.

<96JHC229> E.-S.A.M. Badawey, *J. Heterocycl. Chem.* **1996**, *33*, 229.

<96JHC319> P.A. Harris, W. Pendergast, *J. Heterocycl. Chem.* **1996**, *33*, 319.

<96JHC439> R. Rahm, G. Maas, *J. Heterocycl. Chem.* **1996**, *33*, 439.

<96JHC447> K.V. Rao, C.P. Rock, *J. Heterocycl. Chem.* **1996**, *33*, 447.

<96JHC615> S.-K. Kim, S.-D. Cho, J.-K. Moon, Y.-J. Yoon, *J. Heterocycl. Chem.* **1996**, *33*, 615.

<96JHC639> P. Kolar, A. Pizzioli, M. Tisler, *J. Heterocycl. Chem.* **1996**, *33*, 639.

<96JHC643> I.P. Gerothanassis, G. Varvounis, *J. Heterocycl. Chem.* **1996**, *33*, 643.

<96JHC659> C.C. Lockhart, J.W. Sowell, Sr., *J. Heterocycl. Chem.* **1996**, *33*, 659.

<96JHC757> Y. Kurasawa, A. Takano, K. Kato, A. Takada, H.S. Kim, Y. Okamoto, *J. Heterocycl. Chem.* **1996**, *33*, 757.

<96JHC767> G.C. Hoops, J. Park, G.A. Garcia, L.B. Townsend, *J. Heterocycl. Chem.* **1996**, *33*, 767.

<96JHC949> S. Werner-Simon, W. Pfleiderer, *J. Heterocycl. Chem.* **1996**, *33*, 949.

<96JHC1047> N. Sato, T. Matsuura, *J. Heterocycl. Chem.* **1996**, *33*, 1047.

<96JHC1057> T. Takabatake, T. Miyazawa, M. Hasegawa, *J. Heterocycl. Chem.* **1996**, *33*, 1057.

<96JHC1067> R. Noto, M. Gruttadauria, P.L. Meo, A. Pace, *J. Heterocycl. Chem.* **1996**, *33*, 1067.

<96JHC1163> G. Pentassuglia, B. Bertani, D. Donati, A. Ursini, *J. Heterocycl. Chem.* **1996**, *33*, 1163.

<96JHC1207> J.L. Mokrosz, A.J. Bojarski, D.B. Harden, L. Strekowski, *J. Heterocycl. Chem.* **1996**, *33*, 1207.

<96JHC1259> J.M. Anglada, T. Campos, F. Camps, J.M. Moreto, L. Pages, *J. Heterocycl. Chem.* **1996**, *33*, 1259.

<96JHC1367> K. Takenaka, T. Tsuji, *J. Heterocycl. Chem.* **1996**, *33*, 1367.

<96JMC73> V. Bavetsias, A.L. Jackman, R. Kimbell, W. Gibson, F.T. Boyle, G. M.F. Bisset, *J. Med. Chem.* **1996**, *39*, 73.

<96JMC158> E.J. Jacobsen, R.E. TenBrink, L.S. Stelzer, K.L. Belonga, D.B. Carter, H.K. Im, W.B. Im, V.H. Sethy, A.H. Tang, et al., *J. Med. Chem.* **1996**, *39*, 158.

<96JMC297> Y. Nomoto, H. Takai, T. Ohno, K. Nagashima, K. Yao, K. Yamada, K. Kubo, M. Ichimura, A. Mihara, H. Kase, *J. Med. Chem.* **1996**, *39*, 297.

<96JMC695> L.F. Hennequin, F.T. Boyle, J.M. Wardleworth, P.R. Marsham, R. Kimbell, A.L. Jackman, *J. Med. Chem.* **1996**, *39*, 695.

<96JMC892> L.F. Kuyper, D.P. Baccanari, M.L. Jones, R.N. Hunter, R.L. Tansik, S.S. Joyner, C.M. Boytos, S.K. Rudolph, V. Knick, et al., *J. Med. Chem.* **1996**, *39*, 892.

<96JMC904> T.R. Jones, M.D. Varney, S.E. Webber, K.K. Lewis, G.P. Marzoni, C.L. Palmer, V. Kathardekar, K.M. Welsh, S. Webber, et al., *J. Med. Chem.* **1996**, *39*, 904.

<96JMC918> G.W. Rewcastle, B.D. Palmer, A.J. Bridges, H.D.H. Showalten, L. Sun, J. Nelson, A. McMichael, A.J. Kraker, D.W. Fry, W.A. Denny, *J. Med. Chem.* **1996**, *39*, 918.

<96JMC1164> P.G. Baraldi, B. Cacciari, G. Spalluto, M.J. Pineda de Villatoro, C. Zocchi, S. Dionisotti, E. Ongini, *J. Med. Chem.* **1996**, *39*, 1164.

<96JMC1331> J. Ohmori, H. Kubota, M. Shimizu-Sasamata, M. Okada, S. Sakamoto, *J. Med. Chem.* **1996**, *39*, 1331.

<96JMC1433> Y. Kurogi, Y. Inoue, K. Tsutsumi, S. Nakamura, K. Nagao, H. Yoshitsugu, Y. Tsuda, *J. Med. Chem.* **1996**, *39*, 1433.

<96JMC1438> A. Gangjee, A. Vasudevan, S.F. Queener, R.L. Kisliuk, *J. Med. Chem.* **1996**, *39*, 1438.

<96JMC1823> G.W. Rewcastle, B.D. Palmer, A.M. Thompson, A.J. Bridges, D.R. Cody, H. Zhou, D.W. Fry, A. McMichael, A.J. Kraker, W.A. Denny, *J. Med. Chem.* **1996**, *39*, 1823.

<96JMC2170> A. Gazit, H. App, G. McMahon, J. Chen, A. Levitzki, F.D. Bohmer, *J. Med. Chem.* **1996**, *39*, 2170.

<96JMC2482> C.E. Mueller, U. Geis, B. Grahner, W. Lanzner, K. Eger, *J. Med. Chem.* **1996**, *39*, 2482.

<96JMC2529> G. Biagi, A. Costantini, L. Costantino, I. Giorgi, O. Livi, P. Pecorari, M. Rinaldi, V. Scartoni, *J. Med. Chem.* **1996**, *39*, 2529.

<96JMC2915> V. Colotta, D. Catarzi, F. Varano, G. Filacchioni, L. Cecchi, A. Galli, C. Costagli, *J. Med. Chem.* **1996**, *39*, 2915.

<96JMC3394> K.E. Bergmann, M.H. Cynamon, J.T. Welch, *J. Med. Chem.* **1996**, *39*, 3394.

<96JMC3526> J.D. Hirst, *J. Med. Chem.* **1996**, *39*, 3526.

<96JMC3671> P. Dorigo, D. Fraccarollo, G. Santostasi, I. Maragno, M. Floreani, P.A. Borea, L. Mosti, L. Sansebastiano, P. Fossa, et al., *J. Med. Chem.* **1996**, *39*, 3671.

<96JMC3820> E.J. Jacobsen, L.S. Stelzer, K.L. Belonga, D.B. Carter, W.B. Im, V.H. Sethy, A.H. Tang, P.F. VonVoigtlander, J.D. Petke, *J. Med. Chem.* **1996**, *39*, 3820.

<96JMC3971> J. Ohmori, M. Shimizu-Sasamata, M. Okada, S. Sakamoto, *J. Med. Chem.* **1996**, *39*, 3971.

<96JMC4058> G. Heinisch, W. Holzer, F. Kunz, T. Langer, P. Lukavsky, C. Pechlaner, H. Weissenberger, *J. Med. Chem.* **1996**, *39*, 4058.

<96JMC4156> S.-A. Poulsen, R.J. Quinn, *J. Med. Chem.* **1996**, *39*, 4156.

<96JMC4396> L. Costantino, G. Rastelli, K. Vescovini, G. Cignarella, P. Vianello, A. Del Corso, M. Cappiello, U. Mura, D. Barlocco, *J. Med. Chem.* **1996**, *39*, 4396.

<96JMC4430> G. Sun, N.J. Uretsky, L.J. Wallace, G. Shams, D.M. Weinstein, D.D. Miller, *J. Med. Chem.* **1996**, *39*, 4430.

<96JMC4563> A. Gangjee, F. Mavandadi, R.L. Kisliuk, J.J. McGuire, S.F. Queener, *J. Med. Chem.* **1996**, *39*, 4563.

<96JMC5176> A. Baba, N. Kawamura, H. Makino, Y. Ohta, S. Taketomi, T. Sohda, *J. Med. Chem.* **1996**, *39*, 5176.

<96JOC304> S.M. Vandenberghe, K.J. Buysens, L. Meerpoel, P.K. Loosen, S.M. Toppet, G.J. Hoornaert, *J. Org. Chem.* **1996**, *61*, 304.

<96JOC647> K. Smith, G.A. El-Hiti, M.F. Abdel-Megeed, M.A. Abdo, *J. Org. Chem.* **1996**, *61*, 647.

<96JOC656> K. Smith, G.A. El-Hiti, M.F. Abdel-Megeed, M.A. Abdo, *J. Org. Chem.* **1996**, *61*, 656.

<96JOC1155> H.D.H. Showalter, L. Sun, A.D. Sercel, R.T. Winters, W.A. Denny, B.D. Palmer, *J. Org. Chem.* **1996**, *61*, 1155.

<96JOC1261> E.C. Taylor, W.B. Young, C. Spanka, *J. Org. Chem.* **1996**, *61*, 1261.

<96JOC2470> A. Guzman, M. Romero, F.X. Talamas, R. Villena, R. Greenhouse, J.M. Muchowski, *J. Org. Chem.* **1996**, *61*, 2470.

<96JOC3212> G. Zvilichovsky, V. Gurvich, *J. Org. Chem.* **1996**, *61*, 3212.

<96JOC4655> J.M. Mínguez, M.I. Castellote, J.J. Vaquero, J.L. Garcia-Navio, J. Alvarez-Builla, O. Castano, J.L. Andrés, *J. Org. Chem.* **1996**, *61*, 4655.

<96JOC6199> M.J. Konkel, R. Vince, *J. Org. Chem.* **1996**, *61*, 6199.

<96JOC8921> M.S. South, T.L. Jakuboski, M.D. Westmeyer, D.R. Dukesherer, *J. Org. Chem.* **1996**, *61*, 8921.

<96JPR151> A. F.A. Khattab, T. Dang Van, W. Stadlbauer, *J. Prakt. Chem./Chem. Ztg.* **1996**, *338*, 151.

<96JPR451> S. Brandl, R. Gompper, K. Polborn, *J. Prakt. Chem./Chem. Ztg.* **1996**, *338*, 451.

<96KGS816> O.A. Zagulyaeva, I.V. Oleinik, *Khim. Geterotsikl. Soedin.* **1996**, *6*, 816.

<96M537> J.M. Quintela, M.C. Veiga, R. Alvarez-Sarandes, C. Peinador, *Monatsh. Chem.* **1996**, *127*, 537.

<96M725> R. Foldenyi, *Monatsh. Chem.* **1996**, *127*, 725.

<96M739> J.M. Quintela, M.C. Veiga, S. Conde, C. Peinador, *Monatsh. Chem.* **1996**, *127*, 739.

<96M917> A.F. Khattab, T. Kappe, *Monatsh. Chem.* **1996**, *127*, 917.

<96M947> L. Avallone, M.G. Rimoli, E. Abignente, *Monatsh. Chem.* **1996**, *127*, 947.

<96M955> S.M. Sherif, *Monatsh. Chem.* **1996**, *127*, 955.

<96MC15> G.N. Lipunova, G.A. Mokrushina, E.B. Granovskaya, O.M. Chasovskikh, V.N. Charushin, *Mendeleev Commun* **1996**, *1*, 15.

<96MI10> M.A. Azam, N. Ahmed, S. Srinivas, B. Suresh, *J. Inst. Chem.* **1996**, *68*, 10.

<96MI15> Y.-Z. Kim, J.-C. Lim, J.-H. Yeo, C.-S. Bang, S.-S. Kim, T.H. Lee, S.H. Oh, Y.-C. Moon, C.-S. Lee, *Korean J. Med. Chem.* **1996**, *6*, 15.

<96MI27> I. Yildirim, M. Tezcan, Y. Guzel, E. Saripinar, Y. Akcamur, *Turk. J. Chem.* **1996**, *20*, 27.

<96MI28> M.A. El-Sherbeny, F.A. Badria, S.A. Kheira, *Med. Chem. Res.* **1996**, *6*, 28.

<96MI35> A.A. Moenes, F.E. Goda, A.S. Tantawy, S.M. Kheira, A.-K.M. Ismaiel, *Alexandria J. Pharm. Sci.* **1996**, *10*, 35.

<96MI37> J.-W. Chern, C.-C. Lee, S.-L. Lin, C.-S. Tung, *Chin. Pharm. J.* **1996**, *48*, 37.

<96MI39> H. Sladowska, M. Sieklucka-Dziuba, G. Rajtar, R. Wydro, Z. Kleinrok, *Acta Pol. Pharm.* **1996**, *53*, 39.

<96MI47> A.A. Moenes, F.E. Goda, A.S. Tantawy, S.M. Kheira, A.K.M. Ismaiel, *Alexandria J. Pharm. Sci.* **1996**, *10*, 47.

<96MI61> A. Basha, J.D. Ratajczyk, R.D. Dyer, P. Young, G.W. Carter, C.D.W. Brooks, *Med. Chem. Res.* **1996**, *6*, 61.

<96MI91> V.M. Barot, *Asian J. Chem.* **1996**, *8*, 91.

<96MI199> I. Isikdag, U. Ucucu, K. Benkli, N. Gundogdu, Y. Ozturk, S. Aydin, B. Ergun, *Boll. Chim. Farm.* **1996**, *135*, 199.

<96MI209> D. Matosiuk, T. Tkaczynski, J. Stefanczyk, *Acta Pol. Pharm.* **1996**, *53*, 209.

<96MI251> A. Katrusiak, A. Katrusiak, *J. Mol. Struct.* **1996**, *374*, 251.

<96MI293> Y. Nezu, N. Wada, Y. Saitoh, S. Takahashi, T. Miyazawa, *Nippon Noyaku Gakkaishi* **1996**, *21*, 293.

<96MI429> G. Heinisch, W. Holzer, T. Langer, P. Lukavsky, *Sci. Pharm.* **1996**, *64*, 429.

<96MI433> M. Makki, H.M. Faidallah, *J. Chin. Chem. Soc.* **1996**, *43*, 433.

<96MI603> Z.J. He, Z.M. Li, *Chin. Chem. Lett.* **1996**, *7*, 603.

<96MI713> S. Lakshmi, R. Renganathan, *Int. J. Chem. Kinet.* **1996**, *28*, 713.

<96MI1041> M.P. Groziak, R. Lin, W.C. Stevens, L.L. Wotring, L.B. Townsend, J. Balzarini, M. Witvrouw, E. De Clercq, *Nucleosides Nucleotides* **1996**, *15*, 1041.

<96MRC409> L. Avallone, P. De Caprariis, A. Galeone, M.G. Rimoli, *Magn. Reson. Chem.* **1996**, *34*, 409.

<96MRC567> C. Sakuma, M. Maeda, K. Tabei, A. Ohta, *Magn. Reson. Chem.* **1996**, *34*, 567.

<96MRC958> G. Bertolini, F. Ferrario, S. Pravet-Toni, A. Sala, *Magn. Reson. Chem.* **1996**, *34*, 958.

<96OS201> F.J. Lakner, K.S. Chu, G.R. Negrete, J.P. Konopelski, *Org. Synth.* **1996**, *73*, 201.

<96PHA76> J. Easmon, G. Heinisch, W. Holzer, B. Matuszczak, *Pharmazie* **1996**, *51*, 76.

<96PHA428> B. Zaleska, B. Bialas, *Pharmazie* **1996**, *51*, 428.

<96PHA540> S.A. El-Feky, Z.K. Abd El-Samii, *Pharmazie* **1996**, *51*, 540.

<96PHA927> S.M. Rida, N.S. Habib, E.A.M. Badawey, H.T.Y. Fahmy, H.A. Ghozlan, *Pharmazie* **1996**, *51*, 927.

<96PHA983> O. Migliara, S. Plescia, D. Schillaci, *Pharmazie* **1996**, *51*, 983.

<96PJC197> W. Nawrocka, B. Sztuba, *Pol. J. Chem.* **1996**, *70*, 197.

<96PJC570> H.J. Barton, L. Matusik, J. Bojarski, *Pol. J. Chem.* **1996**, *70*, 570.

<96PJC589> A.M. Hussein, A.A. Atalla, I.S. Abdel-Hafez, M.H. Elnagdi, *Pol. J. Chem.* **1996**, *70*, 589.

<96PS15> M.G. Assy, *Phosphorus, Sulfur Silicon Relat. Elem.* **1996**, *108*, 15.

<96RRC109> F. M.A. Soliman, M.A. El-Hashash, L. Souka, A.S. Salman, *Rev. Roum. Chim.* **1996**, *41*, 109.

<96RTC377> S. Kaban, N. Ocal, *Recl. Trav. Chim. Pays Bas* **1996**, *115*, 377.

<96S838> N. Plé, A. Turck, K. Couture, G. Quéguiner, *Synthesis* **1996**, *7*, 838.

<96S997> H.P. Guan, Q.S. Hu, C.M. Hu, *Synthesis* **1996**, *8*, 997.

<96S1076> G. Broggini, L. Garanti, G. Molteni, G. Zecchi, *Synthesis* **1996**, *9*, 1076.

<96SC453> M. Rahmouni, A. Derdour, J.P. Bazureau, J. Hamelin, *Synth. Commun.* **1996**, *26*, 453.

<96SC475> Y. Cheng, M.-X. Wang, W.-X. Gan, Z.-T. Huang, *Synth. Commun.* **1996**, *26*, 475.

<96SC2197> S. Imor, R.J. Morgan, S. Wang, O. Morgan, A.D. Baker, *Synth. Commun.* **1996**, *26*, 2197.

<96SC3167> M.L. El Efrit, B. Hajjem, H. Zantour, B. Baccar, *Synth. Commun.* **1996**, *26*, 3167.

<96SC3511> K.R. Rapole, A.H. Siddiqui, B. Dayal, A.K. Batta, S.J. Rao, P. Kumar, G. Salen, *Synth. Commun.* **1996**, *26*, 3511.

<96SC3869> R.F. Pellon, R. Carrasco, L. Rodes, *Synth. Commun.* **1996**, *26*, 3869.

<96SC3929> P. Chang, *Synth. Commun.* **1996**, *26*, 3929.

<96SL509> K. Jones, M. Keenan, F. Hibbert, *SynLett* **1996**, *6*, 509.

<96SM105> D. O'Brien, A. Bleyer, D.D. C Bradley, S. Meng, *Synth. Met.* **1996**, *76*, 105.

<96T887> M. Gotoh, B. Sun, K. Hirayama, M. Noguchi, *Tetrahedron* **1996**, *52*, 887.

<96T901> B. Sun, K. Adachi, M. Noguchi *Tetrahedron* **1996**, *52*, 901.

<96T1735> A.C. Tome, J.A.S. Cavaleiro, R.C. Storr, *Tetrahedron* **1996**, *52*, 1735.

<96T2271> B. Zacharie, T.P. Connolly, R. Rej, G. Attardo, C.L. Penney, *Tetrahedron* **1996**, *52*, 2271.

<96T5475> P. Laackmann, W. Fredrichsen, *Tetrahedron* **1996**, *52*, 5475.

<96T6421> A. Corsaro, G. Perrini, V. Pistara, *Tetrahedron* **1996**, *52*, 6421.

<96T7973> A. Garcia Martinez, A. Herrera Fernandez, F. Moreno Jimenez, P.J. Munoz Martinez, C. Alonso Martin, L.R. Subramanian, *Tetrahedron* **1996**, *52*, 7973.

<96T8835> L. Van Meergvelt, O.B. Smolii, N. Mischchenko, D.B. Shakhnin, E.A. Romanenko, B.S. Drach, *Tetrahedron* **1996**, *52*, 8835.

<96T8853> I.I. Mangalagiu, I.I. Druta, M.A. Constantinescu, I.V. Humelnicu, M.C. Petrovanu, *Tetrahedron* **1996**, *52*, 8853.

<96T9161> K.J. Buysens, D.M. Vandenberghe, G.J. Hoornaert, *Tetrahedron* **1996**, *52*, 9161.

<96T10095> G. Morel, E. Marchand, J.-P. Pradere, L. Toupet, S. Sinbandhit, *Tetrahedron* **1996**, *52*, 10095.

<96T10417> L. Mojovic, A. Turck, N. Plé, M. Dorsy, B. Ndzi, G. Quéguiner, *Tetrahedron* **1996**, *52*, 10417.

<96T10751> D. Korakas, A. Kimbaris, G. Varvounis, *Tetrahedron* **1996**, *52*, 10751.

<96T11665> A. Cuenca, A. Strubinger, *Tetrahedron* **1996**, *52*, 11665.

<96T11915> F. Al-Omran, M. M.A. Khalik, H. Al-Awadhi, M.H. Elnagdi, *Tetrahedron* **1915**, *52*, 11915.

<96T13991> M. Martin-Martinez, R. Herranz, M.T. Garcia-Lopez, R. Gonzalez-Muniz, *Tetrahedron* **1996**, *52*, 13991.

<96T14235> E. Bertounesque, T. Imbert, C. Monneret, *Tetrahedron* **1996**, *52*, 14235.

<96TA25> B.S. Joshi, M.G. Newton, D.W. Lee, A.D. Barber, S.W. Pelletier, *Tetrahedron: Asymmetry* **1996**, *7*, 25.

<96TA1641> A. Gutcait, K.-C. Wang, H.-W. Liu, J.-W. Chern, *Tetrahedron: Asymmetry* **1996**, *7*, 1641.

<96TL81> T. Okawa, S. Eguchi, *Tetrahedron Lett.* **1996**, *37*, 81.

<96TL759> E.D. Edstrom, X. Feng, S. Tumkevicius, *Tetrahedron Lett.* **1996**, *37*, 759.

<96TL921> M. Balci, N. Saracoglu, A. Menzek, *Tetrahedron Lett.* **1996**, *37*, 921.

<96TL2615> J.M. Mellor, G.D. Merriman, H. Rataj, G. T.L. Reid, P. Laackmann, W. Fredrichsen, *Tetrahedron Lett.* **1996**, *52*, 2615.

<96TL2825> D.N. Butler, P.M. Tepperman, R.A. Gau, R.N. Warrener, *Tetrahedron Lett.* **1996**, *37*, 2825.

<96TL3191> I.P. Gerothanassis, J. Cobb, A. Kimbaris, J.A.S. Smith, G. Varvounis, *Tetrahedron Lett.* **1996**, *37*, 3191.

<96TL3355> G.M. Reddy, P.L. Prasunamba, P. S.N. Reddy, *Tetrahedron Lett.* **1996**, *37*, 3355.

<96TL4145> L. Bourel, A. Tartar, P. Melnyk, *Tetrahedron Lett.* **1996**, *37*, 4145.

<96TL4263> J.M. Mínguez, J.J. Vaquero, J.L. Garcia-Navio, J. Alvarez-Builla, *Tetrahedron Lett.* **1996**, *37*, 4263.

<96TL4339> A.J. Elliott, J.A. Montgomery, D.A. Walsh, *Tetrahedron Lett.* **1996**, *37*, 4339.

<96TL4491> T. Mathew, M. Keller, D. Hunkler, H. Prinzbach, *Tetrahedron Lett.* **1996**, *37*, 4491.

<96TL4643> M.F. Gordeev, D.V. Patel, J. Wu, E.M. Gordon, *Tetrahedron Lett.* **1996**, *37*, 4643.

<96TL5015> V.J. Majo, P.T. Perumal, *Tetrahedron Lett.* **1996**, *37*, 5015.

<96TL5179> R.S. Atkinson, M.P. Coogan, I.S.T. Lochrie, *Tetrahedron Lett.* **1996**, *37*, 5179.

<96TL7031> L. Gouilleux, J.-A. Fehrentz, F. Winternitz, J. Martinez, *Tetrahedron Lett.* **1996**, *37*, 7031.

<96TL7343> R.C. Bernotas, G. Adams, *Tetrahedron Lett.* **1996**, *37*, 7343.

<96TL8151> J.S. Panek, B. Zhu, *Tetrahedron Lett.* **1996**, *37*, 8151.

<96TL8205> B. Chen, C.-Y. Yang, D.-Y. Ye, *Tetrahedron Lett.* **1996**, *37*, 8205.

Chapter 6.3

Triazines, Tetrazines and Fused Ring Polyaza Systems

Derek T. Hurst
Kingston University, Kingston upon Thames, UK.

6.3.1 INTRODUCTION

This year, and the latter part of 1995, has seen the usual large number of references to the topics covered by this chapter. In December the second edition of "Comprehensive Heterocyclic Chemistry" was published which provides a substantial coverage of most aspects of heterocyclic chemistry, including the material of this chapter, for the last 10 years.

6.3.2 SYNTHESIS

6.3.2.1 Triazines

A convenient synthesis of trisubstituted 1,3,5-triazines uses acylamidines **1** and amidines <95JOC8428>, whilst 1,3,5-triazin-2(1*H*)-ones **3** are obtained from *N*-acyl-*N*′-carbamoyl-*S*-methylisothioureas **2** <96H(43)839>.

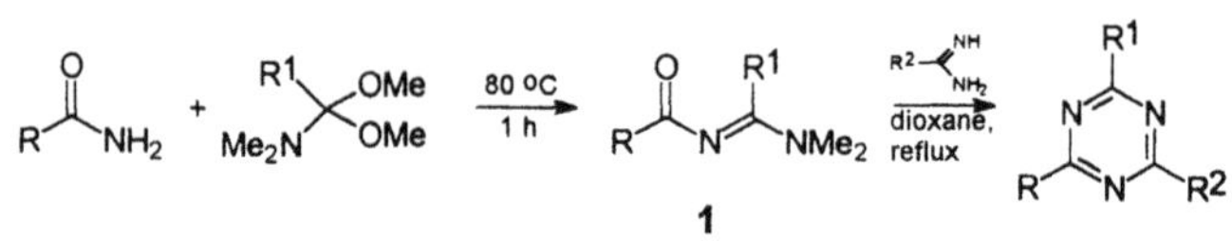

R= var. Ar; R1= H, Me; R2= Me, Me2N and others (14 examples)

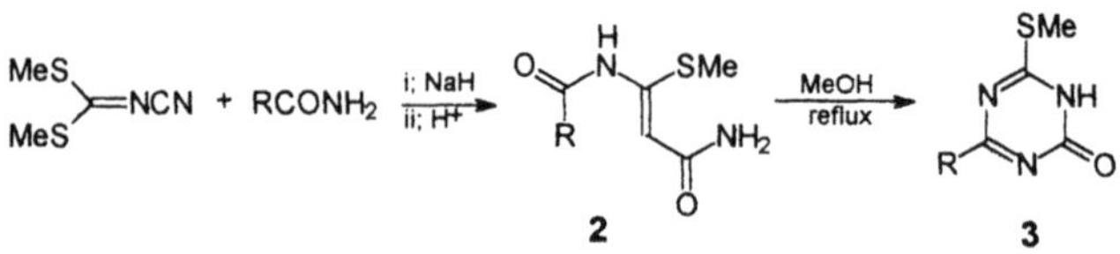

Regioselective cyclocondensation of *N*-(1-chlorobenzyl) benzimidoylchlorides with thioureas yields 1,3,5-triazine-2(1*H*)-thiones **4** but if the reaction is carried out in the presence of triethylamine thiadiazines **5** are obtained <95ZOB1246; 96CA(124)317108>.

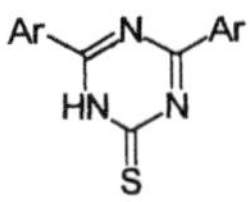

4

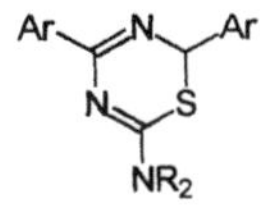

5 Ar= Ph, *p*FC$_6$H$_4$; R= H, Ph

The condensation of nitrilimines with α-amino esters gives 1,4-dihydro-1,2,4-triazin-6-ones **6** <94MI01; 96CA(124)55907>.

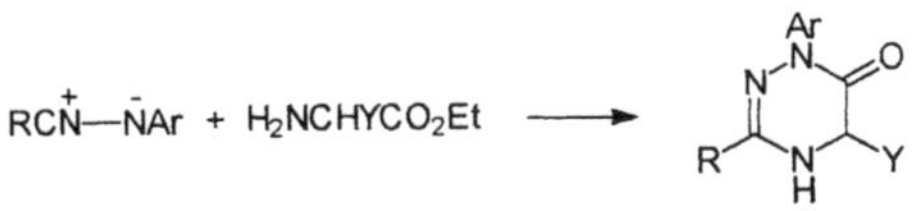

6

R= PhNH, 2-naphthyl, OEt, Me, Ph, 2-thienyl; Ar= *p*MeC$_6$H$_4$, *p*ClC$_6$H$_4$; Y= H, Bu

An unusual synthesis of the 1,6-dihydro-1,2,3-triazine system **8** involves the Bamford-Stevens reaction of cis-aziridinylketone tosylhydrazones **7** in the presence of a "slight excess" of sodium hydride or sodium ethoxide. If a large excess of sodium ethoxide is used then isopropylamino-3,5-diarylpyrazoles are formed <96H(43)1759>.

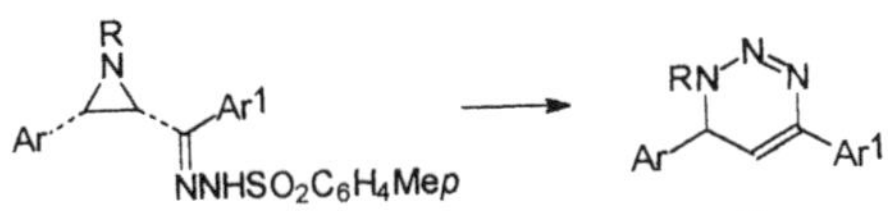

7 Ar, Ar1= Ph, var. Cl, MeO, NO$_2$C$_6$H$_4$ **8**

6.3.2.2 Tetrazines

Nitrilimines react with hydrazones of aliphatic aldehydes and ketones to yield addition products **9** which cyclise when treated with palladium charcoal at room temperature to give 1,6-dihydro-*s*-tetrazines **10** <96JCR(S)174>.

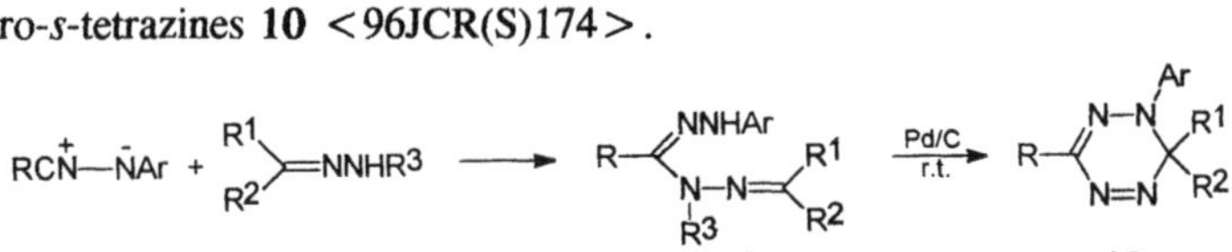

9 **10**

R= CO$_2$Me, Ac; R1, R2= H, Me, Bu and others; Ar= Ph and others

Benzo-1,2,3,4-tetrazine 1-*N*-oxides **13** can be obtained by intramolecular cyclisation of 2-(*t*-butylazoxy)phenyldiazonium tetrafluoroborates **11** which leads to 2-(*t*-butyl)benzo-1,2,3,4-tetrazinium 4-*N*-oxides **12** *via* N to N migration of the *t*-butyl group followed by elimination of the *t*-butyl group. From DMSO solutions containing 1% water 2-alkyl-6-oxo-2,6-dihydrobenzo-1,2,3,4-tetrazine 4-*N*-oxides **14** are obtained (Scheme 1)<96MC22>.

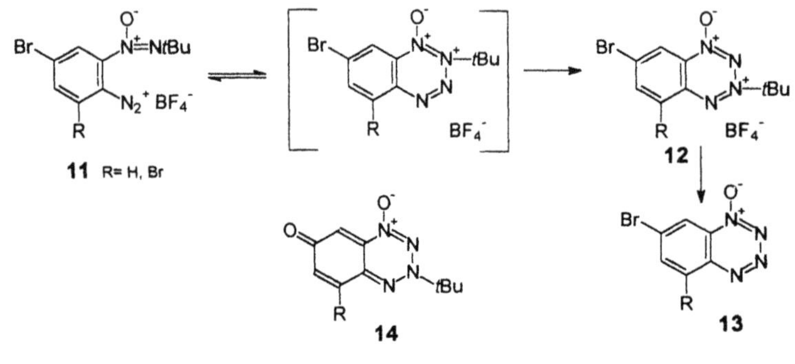

Scheme 1

The polyazapolysulfur ring system 1,3,2,4,6-dithiatriazine **15** is obtained by reacting 1-aryl-2,2,2-trifluoroethanone oximes with tetrasulfur tetranitride in refluxing toluene. However, the yield is only moderate <96JHC295>.

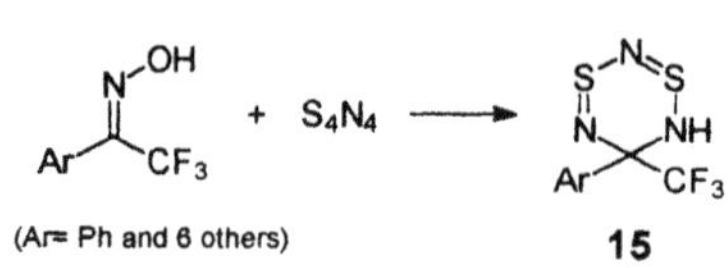

6.3.2.3 Purines and related compounds

A general synthesis of pyrrolo[3,4-*d*]pyrimidinones **17** is provided by the chromous ion mediated reductive cyclisation of the pyrrolidinones **16** with amidines. The suggested mechanism is that shown in Scheme 2 <95JOC7687>.

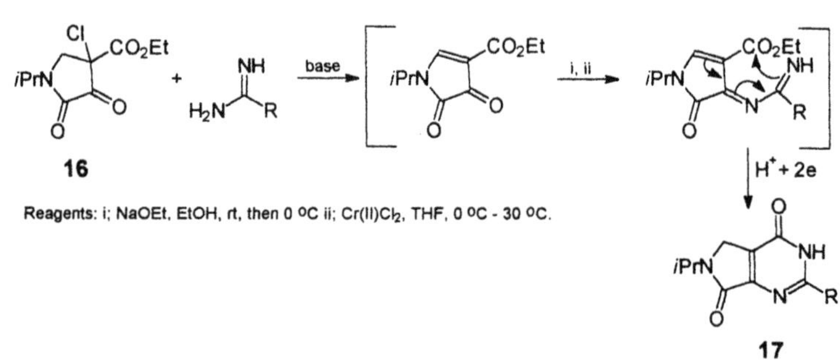

Scheme 2

A new entry to the 2,3-dihydroimidazo[1,2-*c*]pyrimidine system **18** is the reaction of *N*-acylimidates with imidazoline ketene aminals under focussed microwave irradiation <96SC453>.

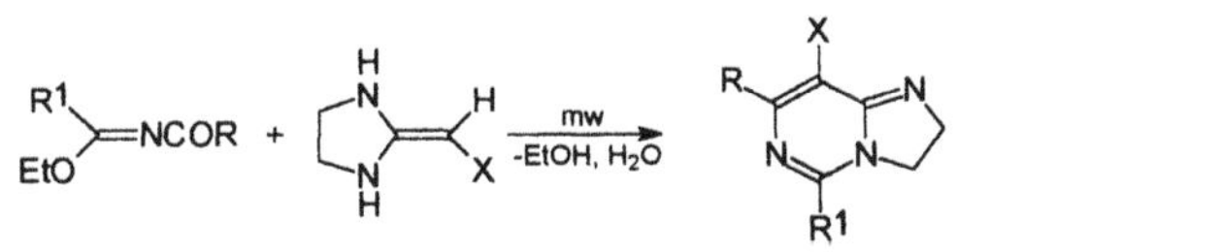

18 X= CN, CO_2Et; R, R1= Me, Et

The synthesis of pyrrolo[3,4-*c*]pyridazines **20** is achieved by intramolecular aza-Wittig reactions of the phosphazines **19** <95JHC1457>.

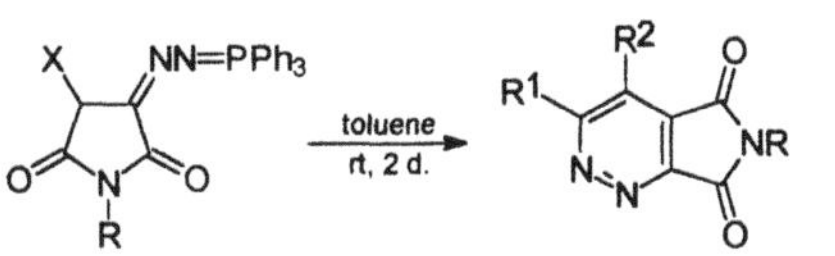

19 X= O= or $R1O_2CCH$=CH-   **20** R= H, Me; R1= H, Me, CO_2Et; R2= Me, OMe, OtBu, CH_2CO_2Et

Cyclocondensation of 3-trifluoroacetyl substituted lactams with cyclic 1,3-bisnucleophiles gives pyrimido[1,2-*a*]benzimidazoles **21** and 1,2,4-triazolo[4,3-*a*]pyridines **22**. The use of amidines yields pyrrolopyrimidines <95JCS(P1)2907; 94H(37)915>.

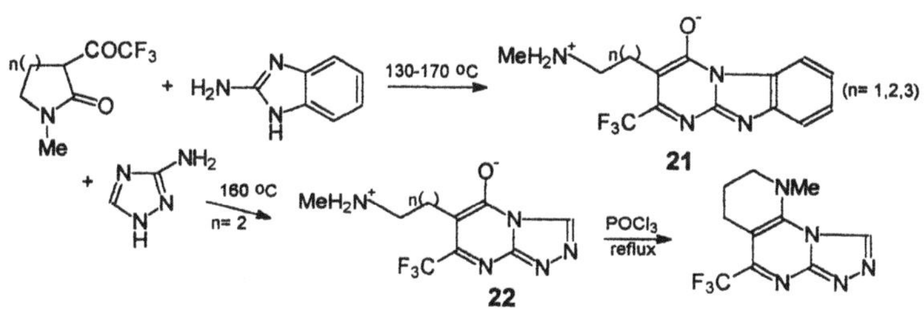

Imidazo[4,5-*d*]1,2,3-triazines and pyrazolo[3,4-*d*]-1,2,3-triazines have been synthesised as analogues of the potent anticonvulsant BW78U79 **23** by the reactions shown in Scheme 3. Similar reactions have been used to obtain imidazopyridazines <95JHC1417, 1423>.

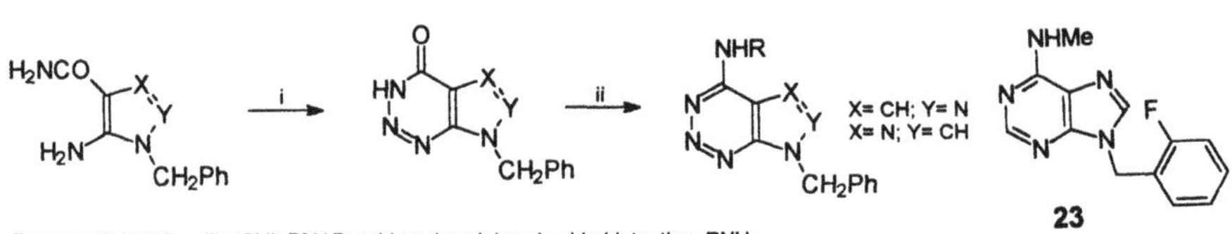

Reagents:i; $NaNO_2$, dil.HCl ii; DMAP, pchlorophenylphosphochloridate, then RNH_2

Scheme 3

The antitumour drug temozolomide **24** (R= Me) has been attracting attention and two new routes have been developed. One starts from 5-aminoimidazole-4-carboxamide, whilst the other starts from 5-diazoimidazole-4-carboxamide which reacts with trimethylsilyl isocyanate to give 8-carbamoylimidazo[5,1-*d*]1,2,3,5-tetrazin-4(3*H*)one **24** (R= H).

The dicarbamoylaminoimidazole **25** failed in a cyclisation to yield **24** (R= H) but gave 8-carbamoylaminoimidazo[1,5-*a*]-*s*-triazin-4(3*H*)one **26** instead <95JCS(P1)2783; 96MI01; 96CA(124)232405>.

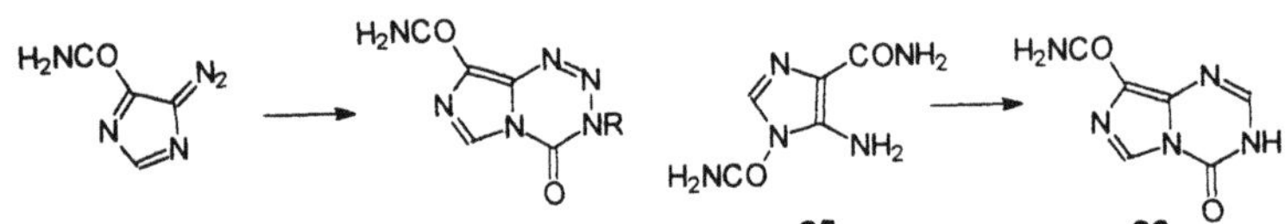

24 R= H, Me **25** **26**

A one-step synthesis of pyrazolo[3,4-*d*]pyrimidines **27** is the inverse electron demand cycloaddition of 2,4,6-tris(ethoxycarbonyl)-1,3,5-triazine and 5-aminopyrazoles involving loss of ammonia, then EtO_2CCN <96JOC5204>.

R= H, Me, Et and others **27**

1,1-Diacetylcyclopropane reacts with 3-amino-1,2,4-triazole in acetic acid (either aqueous or glacial) to give triazolopyrimidines **28** <96IZV1322>, whilst a selection of fused-ring pyrimidines, for example the 1,2,3-triazolo[4,5-*d*]pyrimidinones, **29** has been conveniently prepared using amino heteroarenecarboxamides and esters <96H(42)691>.

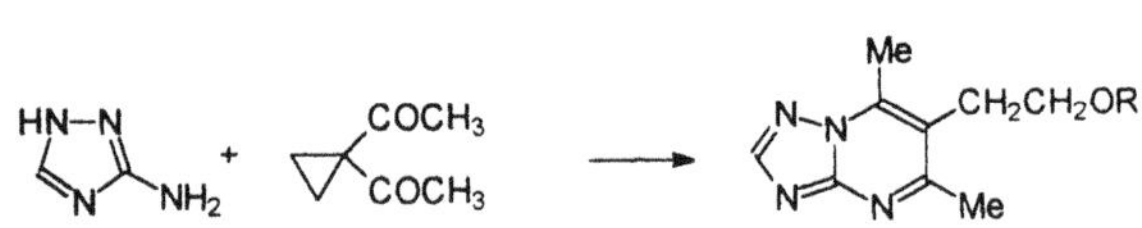

28 R= H or Ac

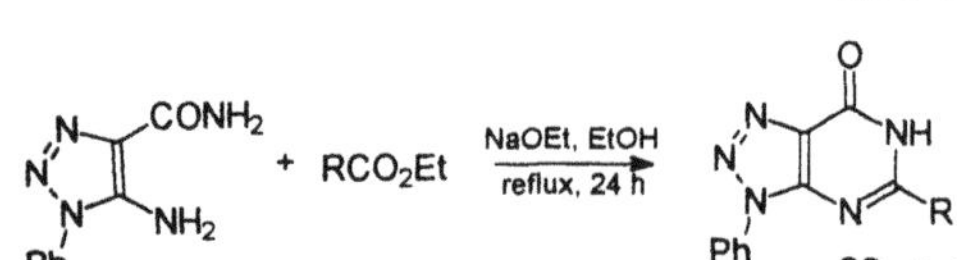

29 R= H, Me, Et, *etc.*

1,2,4-Triazolo[4,3-*a*]quinazolin-5-ones **30** are obtained from hydrazonoyl halides with 2-mercapto-4(3*H*)quinazolinone (Scheme **4**) <95H(41)1999>.

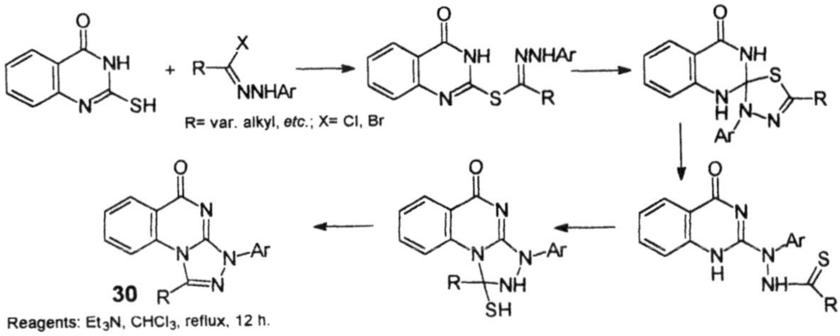

Scheme 4

An attempt to prepare triazolothiadiazepines **31** by reacting bromoalkenecarbonyl substituted thiophenes with *N*-aminotriazole thiols failed and triazolo[3,4-*b*]-1,3,4-thiadiazines **32** were produced <95IJCB939>.

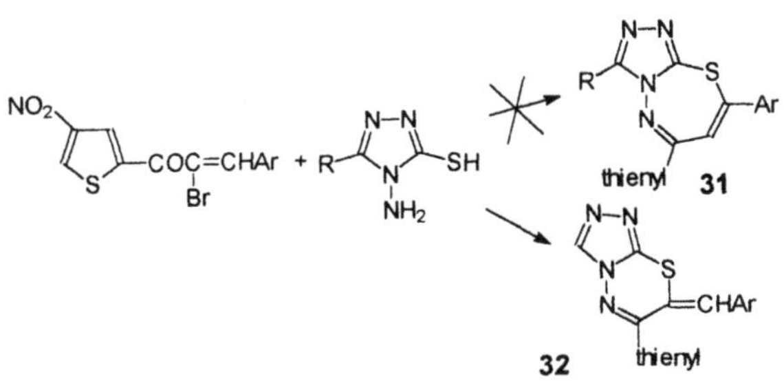

6.3.2.4 Pteridines and related compounds

Methyl 3-(triphenylphosphoranylideneamino)pyrazine-2-carboxylate **33** has been synthesised and used to obtain pteridines **34**, **35** by reaction with isocyanates followed by adding alcohols or amines (Scheme 5) <96JCS(P1)247>.

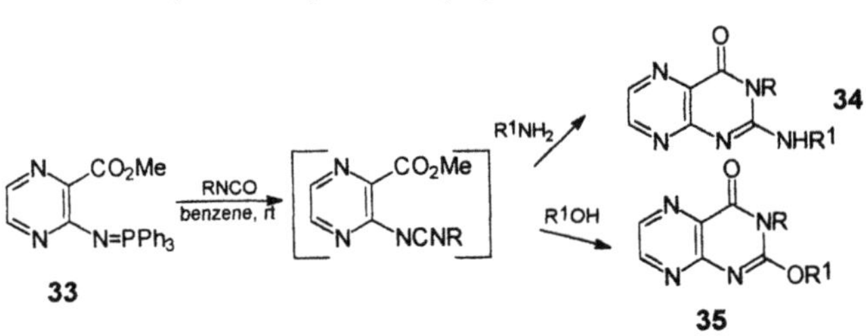

Scheme 5

6-Amino-2-dimethylamino-5-nitrosopyrimidin-4(3*H*)-one **36** reacts with dimethyl phenacylsulfonium bromides when refluxed in pyridine to give pteridinediones **37** rather than the isomeric 5-*N*-oxides **38** <96H(43)437>.

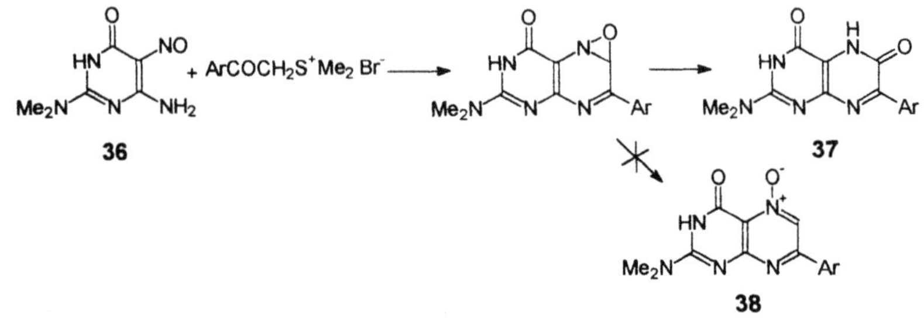

2,4,5-Triaminopyrimidin-6(1*H*)-one **39** reacts with a range of fluorocarbonyls to give fluoropterins (Scheme **6**). The fluoropterins are even more insoluble than the parent compounds and are also stable to nucleophiles and bases in dilute aqueous solution <96T13017>.

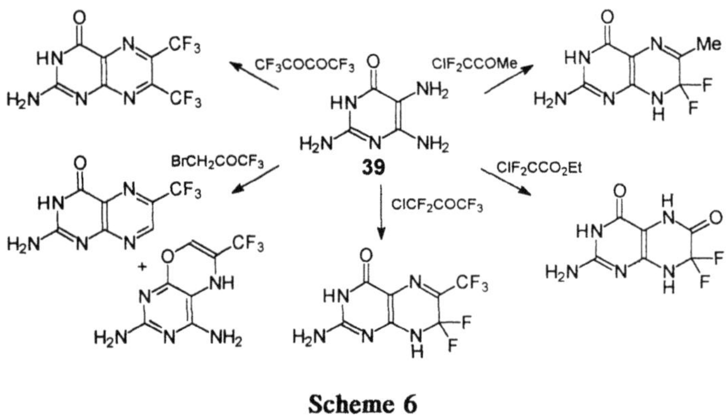

Scheme 6

6.3.2.5 Miscellaneous ring systems

An effective synthesis of the pyrido[1',2':1,2]imidazo[5,4-*c*]isoquinoline system **41** uses the 3-aceta (or benza)mido-2-phenylimidazo[1,2-*a*]pyridine **40** and phosphorus oxychloride. An alternative approach using the carbodiimide **42**, obtained from the corresponding iminophosphorane and an isocyanate, is also successful <95H(41)2019>.

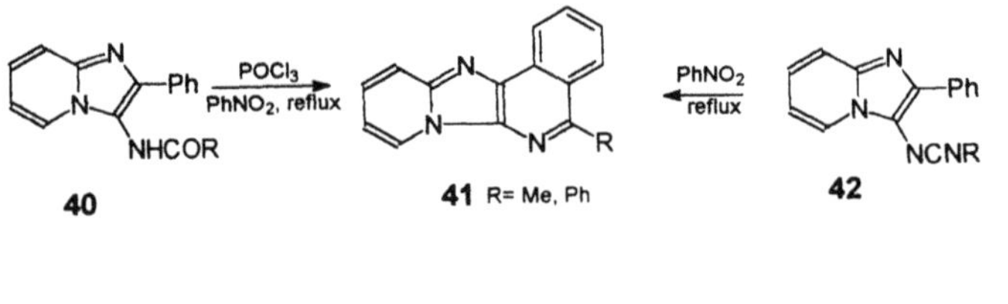

A series of 1,3,4-thiadiazolo[2,3-*c*]-1,2,4-triazino[5,6-*b*]indoles **44** has been prepared by cyclisation of **43** for use as fungicides <95IJCB1010>.

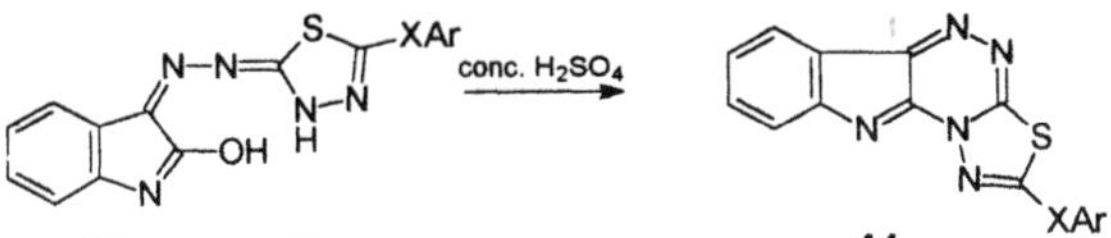

43 X= CH_2O, NH **44**

Several new 2,4-disubstituted pyrrolo[2,1-*f*][1,2,4]triazines, which can be further elaborated, and pyrrolo[5,1-*c*]pyrimido[4,5-*e*][1,2,4]triazines have been made by the reaction of *N,N*-dimethyl-dichloromethyleniminium chloride with 1-aminopyrrole-2-carbonitrile and ethyl 4-amino-3-cyanopyrazolo[5,1-*c*][1,2,4]triazine-8-carboxylate respectively (Scheme **7**) <96T3037>.

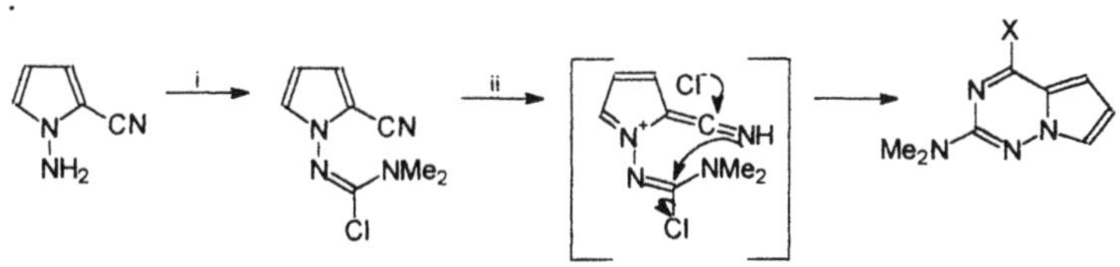

Reagents: i; $CCl_2N^+Me_2Cl^-$, $(CH_2Cl)_2$, reflux 1h. ii; HCl(g), rt.

Scheme 7

The use of heterocyclic 1,2,3-amino, cyano, methylthio compounds with DMAD or DEAD in dimethyl sulfoxide in the presence of potassium carbonate yields polycyclic products, for example those shown in Scheme **8** <96H(42)53>.

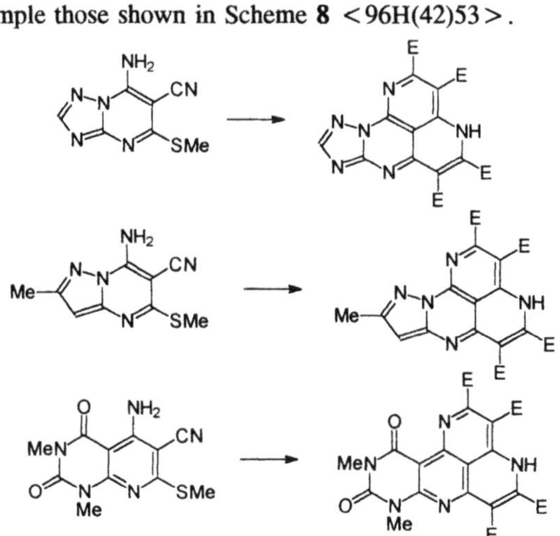

Reagents: i; $CCl_2N^+Me_2Cl^-$, THF. ii; HCl(g)

E= CO_2Me, CO_2Et

Scheme 8

Reaction of thienopyrimidinediamines **45** with formaldehyde in the presence of hydrochloric acid gives 3,4-dihydro-5*H*-1-thia-3,5,6,8-tetraazaacenaphthalenes **46**. The use of triethyl orthoformate gives the 3,4-dehydro products <95LA1703; 96KGS103; 96CA(125)167905>.

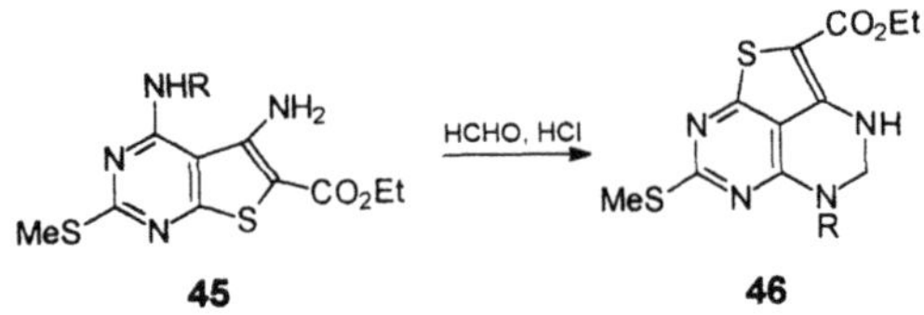

Cyclocondensation of the pyrimidinecarboxaldehyde **47** with benzimidazolethiones **48** gives 5*H*-benzimidazo[2',1':2,3][1,3]thiazino[6,5-*d*]pyrimidine **49** and the new heterocyclic system benzimidazo[2,1-*b*]pyrimido[5,4-*f*][1,3,4]thiadiazepine **50** <96KGS427; 96CA(125) 167911>.

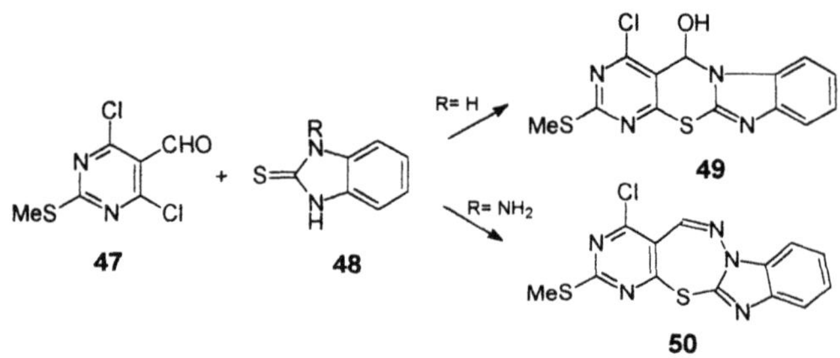

The pyrazoles **51** react with triethyl orthoformate and hydrazine to give the pyrazolopyrimidines **52** which react further with acetic anhydride and benzoyl chloride to give pyrazolotriazolopyrimidines **53** and with α-ketohydrazonoyl halides to give pyrazolopyrimidotriazines **54** <95MI01; 96CA(124)86924>.

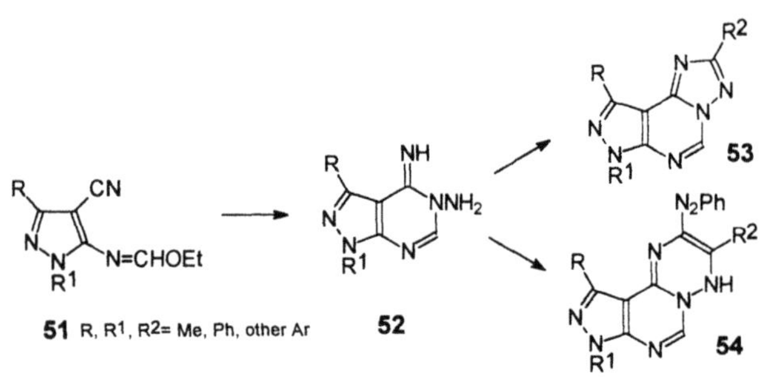

Iminothiodiazolopyrimidinones **55** and 2,4-pentanediones condense in polyphosphoric acid to yield 9-oxo-1,2,3,4-thiadiazolo[3,2-*a*]pyrido[3,2-*e*]pyrimidines **56** <95IZV2037; 96CA(124) 289413>.

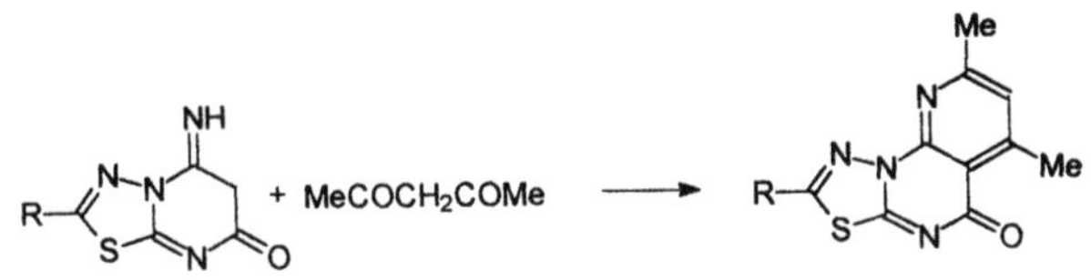

55 **56**

Novel thieno[2',3':3,4]pyrazolo[1,5-*a*]pyrimidines and triazines **58** can be prepared by reacting the thienopyrazoles **57** with 1,3-biselectrophiles <95MI02; 96CA(124)176031>.

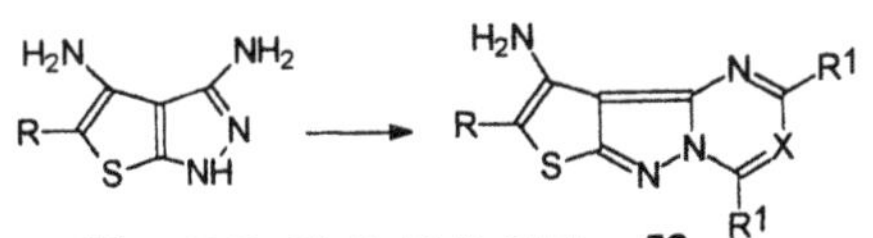

57 R= CO_2Me; R1= Me, Ph; X= CH, N **58**

4,6-Dihydroxypyrimidine reacts with aldehydes to yield bispyrimidinediols which cyclise in acetic acid/acetic anhydride to give the pyranopyrimidines **59** <95ZOB1161; 96CA(124) 146052>. These products react with phosphorus oxychloride to give the corresponding chloro derivatives which undergo ready substitution with nucleophiles <96ZOB824; 96CA(125)328655>.

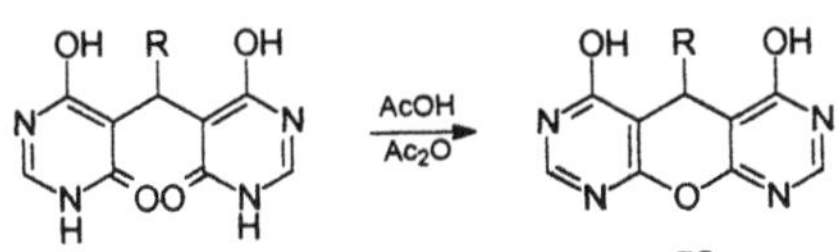

R= H, Me, Pr, *etc.* **59**

A development of the reaction described above for the synthesis of pyrrolo[2,1-*f*][1,2,4]triazines using 1-aminopyrrole-2-carbonitrile and *N*,*N*-dimethyldichloromethyl-iminium chloride utilising ethyl 4-amino-3-cyanopyrazolo[5,1-*c*][1,2,4]triazine-8-carboxylate **60** yields pyrazolo[5,1-*c*]pyrimido[4,5-*e*][1,2,4]triazines **61** <96T3037>.

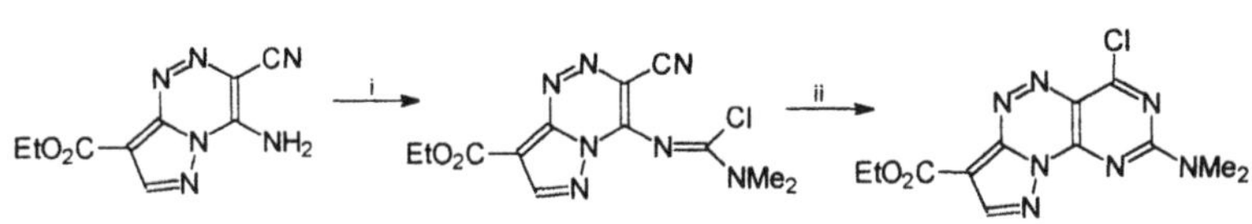

60 **61**

Reagents: i; $CCl_2N^+Me_2Cl^-$, THF. ii; HCl(g)

The pyrimidines **62** undergo cyclisation on refluxing in dioxane to yield not only the pyrazolopyrimidines **63**, but the novel pyrazolo[3',4':4,5]pyrido[2,3-*d*]pyrimidines **64** by an intramolecular 1,3-dipolar cycloaddition reaction (Scheme **9**)<96JCS(P1)1999>.

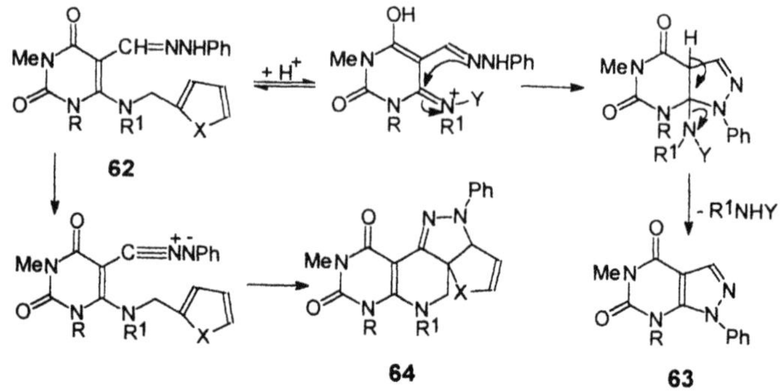

R= H, Me; X= O,S; R1= Me, Ar

Scheme 9

The imidazo[4,5-*e*][1,3]diazepinone system **66**(R^1= H), a seven membered ring analogue of guanine, has been synthesised from the imidazoles **65**. The acycloguanine analogue **66**(R^1= $CH_2OCH_2CH_2OH$; R= H) was also synthesised from the appropriate imidazole starting material <96JCS(P1) 2257>.

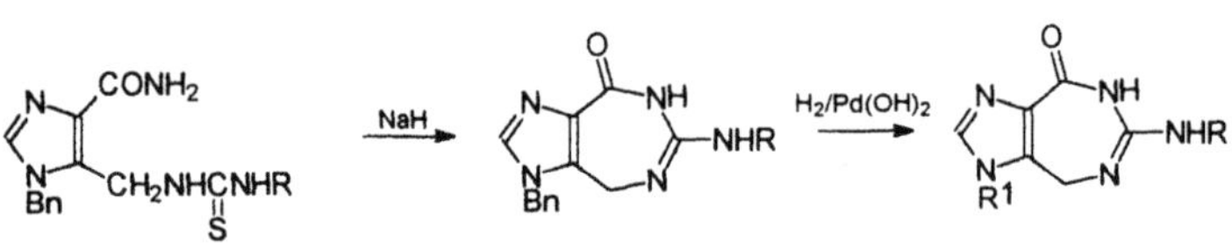

65 (R= Me, Bn, CPh_3) **66**

Routes have been developed to furo[3,2-*e*][1,2,4]triazolo[4,3-*c*]pyrimidines **67** and furo[3,2-*e*]tetrazolo[1,5-*c*]pyrimidines **68** which are shown in Scheme **10** <96JHC659>.

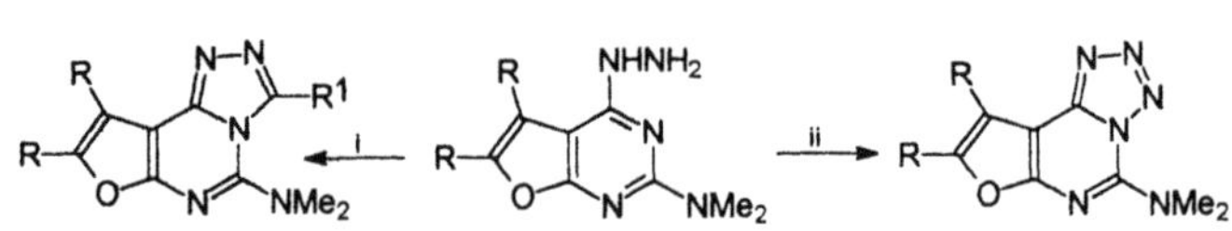

67 **68**

Reagents: i; $HC(OMe)_3$ or $MeC(OEt)_3$, 70 °C, 24 h. ii; $NaNO_2$, AcOH, rt, 24 h.

Scheme 10

The synthesis of the novel condensed ring system 4,5-dihydro[1,2,3]triazolo[5,1-*f*][1,2,4]triazine **69** has been achieved by the method shown in Scheme **11** <96JHC599>.

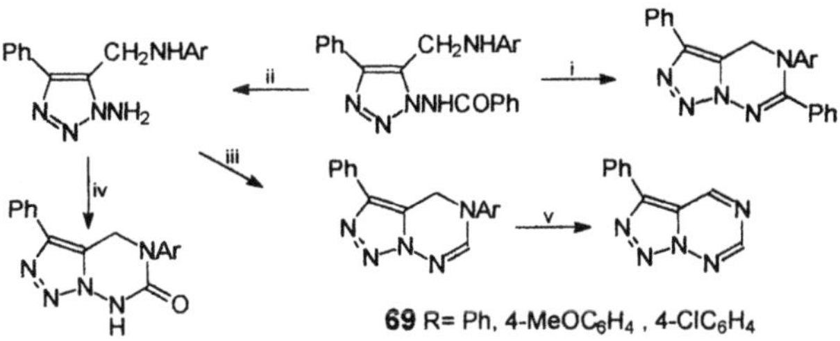

69 R= Ph, 4-$MeOC_6H_4$, 4-ClC_6H_4

Reagents: i; PCl_5, C_6H_6, reflux; ii: HCl/EtOH, reflux; iii; $HC(OEt)_3$, TsOH, r.t. iv: $COCl_2/NEt_3$, toluene, r.t. v; CAN

Scheme 11

The hydrazine **70** reacts with triethyl orthoformate, sodium nitrite in acetic acid, or pyruvic acid, to yield 1,2,4-triazolo[4,3-*c*]-, tetrazolo[1,5-*c*]- and 1,2,4-triazino[5,6-*c*]pyrano [4',3':4,5]pyrrolo[3,2-*e*]pyrimidine derivatives **71**(X= CH), **71**(X= N) and **72** respectively <95KGS700; 96CA(124)176023>.

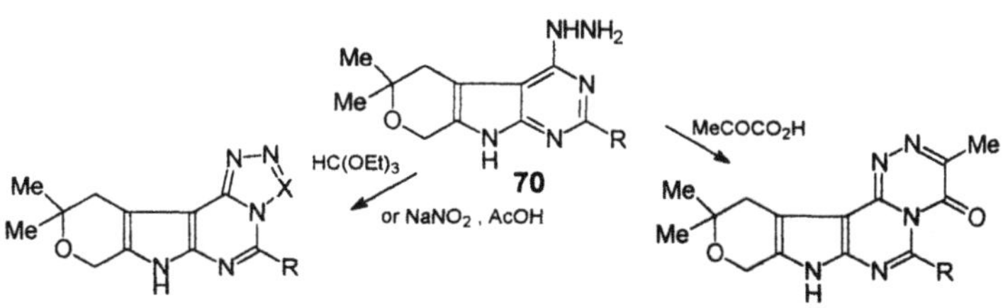

71 R= H, SMe; X= CH, N

72

The diaminopyridodipyrimidinedione **73** has been synthesised as a cytosine-like Tecton designed for self assembly into a helical superstructure. Self recognition occurs in the solid state <96MI02; 96CA(125)221771>.

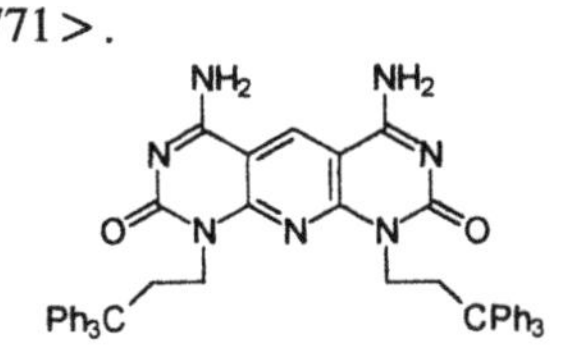

73

6.3.3 REACTIONS

6.3.3.1 Triazines

The reaction of 5,6-diphenyl-3-cyano-1,2,4-triazine with hydrazine gives the amidrazone which, with 1,2-diketones, yields 3,3'-bistriazinyls **74** <95MI03; 96CA(124)86945>.

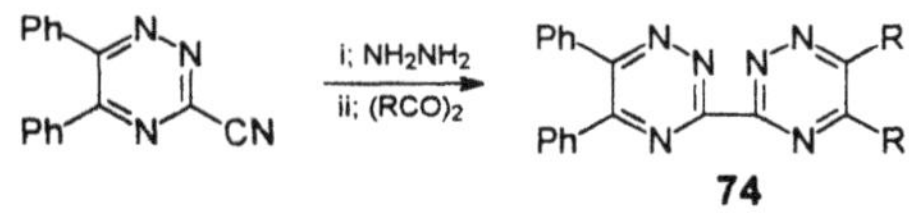

74

Some 1,2,4-triazines having a 1,2,4-triazinone substituent, and some 1,2,4-triazolo[4,3-*b*]-1,2,4-triazinones have been shown to have *in vitro* anti-HIV and anticancer activity <95MI04; 96CA(124)86961>.

2,4,6-Tris(trinitromethyl)-1,3,5-triazine reacts with nucleophiles such as alcohols and amines to give products in which the trinitromethyl groups are displaced in turn to yield mono, di, then trisubstituted compounds. Some of the products which have been obtained are shown in Scheme **12** <95CHE(31)596>.

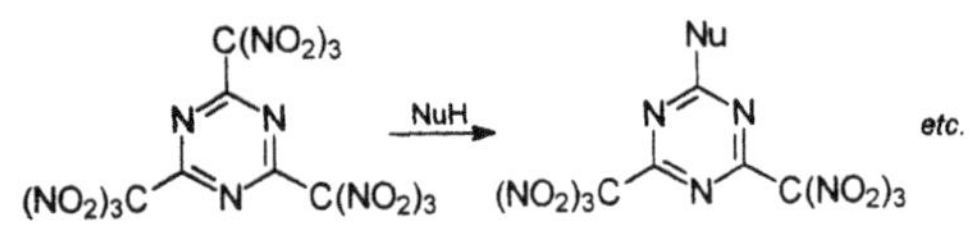

NuH= ROH, NH_3, RNH_2, R_2NH; R= Me, Et

Scheme 12

The inverse electron demand reactions of 2,4,6-tris(ethoxycarbonyl)-1,3,5-triazine and 5-aminopyrazoles to provide a one-step synthesis of pyrazolo[3,4-*d*]pyrimidines **27** has been referred to above <96JOC5204>.

1,3,5-Triazine itself reacts with the 2,3-dihydro-1,4-diazepinium salt **75** (which acts as a nucleophile) to give the pyrimido[3,4-*d*][1,4]diazepinium salt **76** <96ZN(B)421.

75 **76**

Reagents: piperidine, MeOH, rt, 14 d.

A novel route to functionalised 3-aminopyridazines **77** is the reaction of 3-chloro-6-phenyl-1,2,4-triazine with *C*-nucleophiles. The mechanism proposed for this reaction is shown in Scheme **13** <96TL5795>.

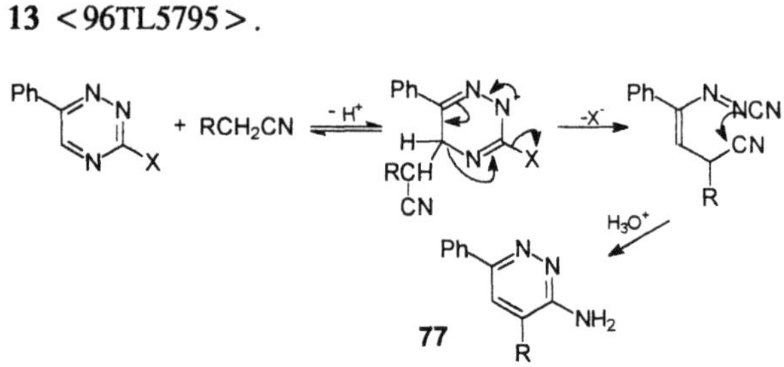

77

Scheme 13

5,6-Diamino-3-methylthio-1,2,4-triazine **78** reacts with phenacyl bromide to give the enamine **79** and no cyclic product. However, the cyclisation is successful with bromoacetone to give the pyrazinotriazine **80**. The cyclisation is also successful using 5,6-diaminotriazin-2(1*H*)-one <96H(43)1007>.

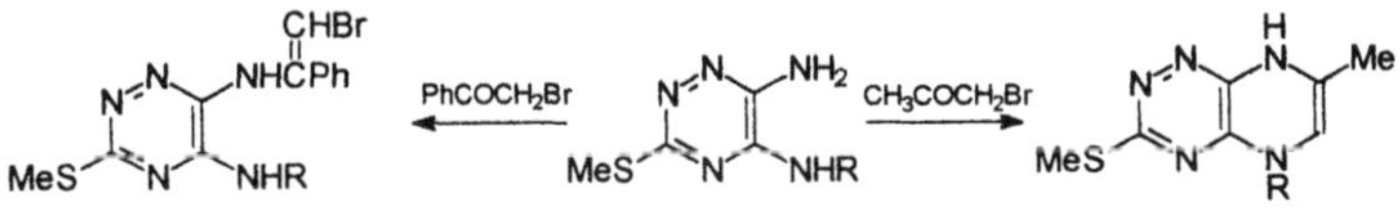

79 **78** **80** R= H, Me

3-Thioxo-1,2,4-triazin-5-ones and hydrazonyl halides react together in the presence of triethylamine in dichloromethane to yield 1,2,4-triazolo[4,3-*b*][1,2,4]-triazin-5-ones **81** <95MI05; 96CA(124)317107>.

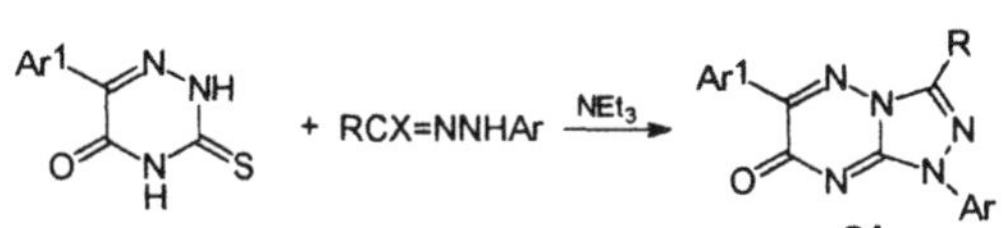

81

Another cyclisation involving 4-amino-6-methyl-3-thioxo-1,2,4-triazin-5-one **82** is the reaction with propargyl bromide followed by palladium(II) catalysis to yield a [1,2,4] triazino[3,4-*b*][1,3,4]thiadiazine **83**, or with phenacyl bromide to give **84** <96MI03; 96CA(125)195597; 96MI04; 96CA(125)247767>.

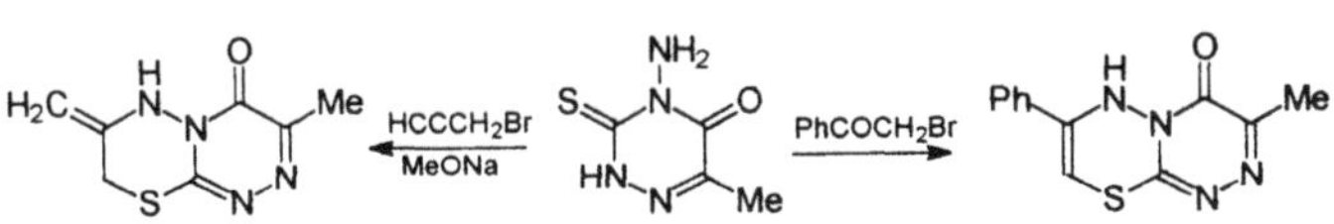

83 **82** **84**

3-Propargylmercapto-6-methyl-1,2,4-triazin-5(2*H*)-one **85** cyclises to the thiazolo[3,2-*b*] [1,2,4]triazin-7-one **86** which can be brominated, and subsequently aminated, yielding the products **87** <95MI06; 96CA(124)176024>.

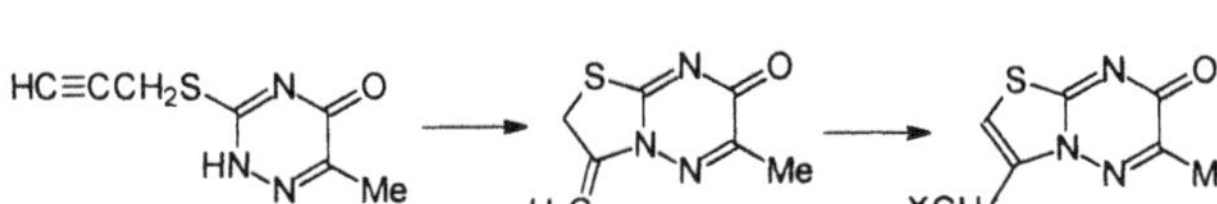

85 **86** **87** X= Br, R₂N

The oxazolotriazinone **88** undergoes ring transformation when treated with ammonia or with primary amines to give imidazo[1,2-*b*][1,2,4]triazines **89**. Similar treatment with hydrazine yields triazino[4,3-*b*][1,2,4]triazines **90** <95MI07; 96CA(125)202176>.

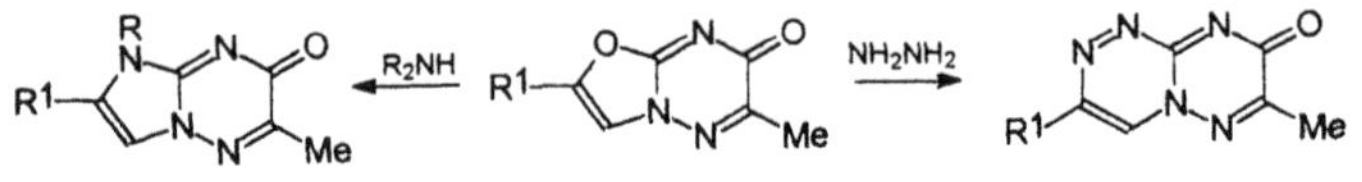

89 **88** **90**

6-Phenyl-1,2,4-triazine 4-oxide **91** reacts with 1,3-dimethyluracil-6-hydrazones **92** in DMF in the presence of triethylamine to give pyrazolo[3,4-*d*]pyrimidines **93** (Scheme **14**) <95MC229>.

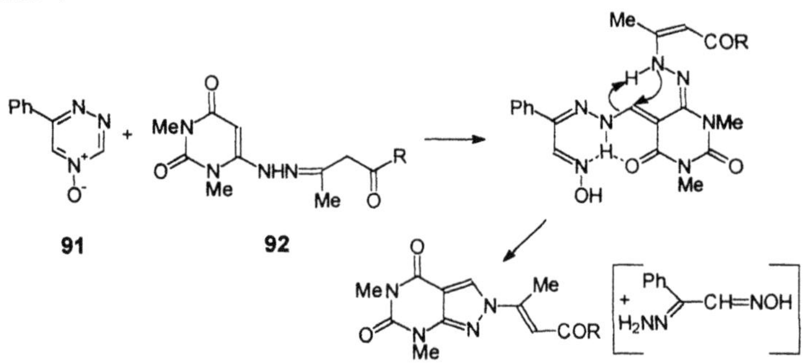

91 **92** **93**

Scheme 14

1-Alkyl-3-morpholino-5-phenyl-1,2,4-triazinium iodides **94** undergo very easy de-quaternisation when treated with triethylamine in ethanol or acetone at room temperature. The reaction probably proceeds by a radical mechanism through the radical intermediate **95** <95MC104>.

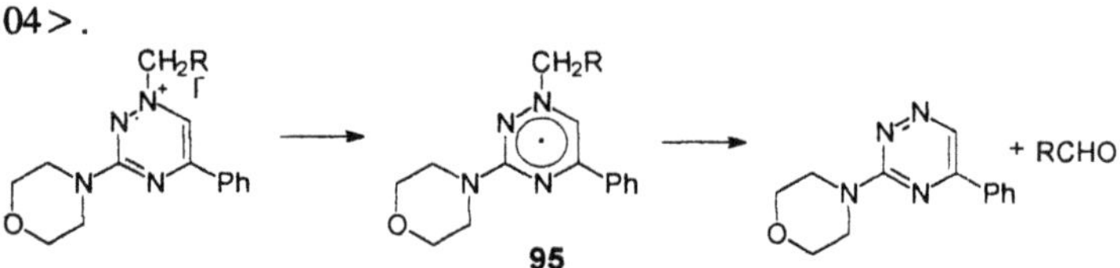

94 **95**

6.3.3.2 Tetrazines

3-Aryl-1,2,4,5-tetrazines are oxidised by methyl(trifluoromethyl)dioxirane to their previously unknown *N*-oxides **96**. NMR studies have shown that *N*-1 is oxidised regioselectively <96T2377>.

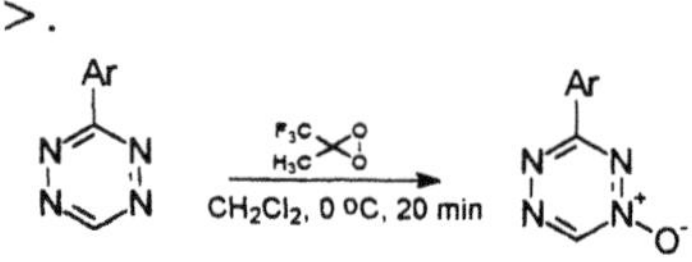

96

The 3,6-bis(2-arylethenyl)-1,2,4,5-tetrazines **97** have been synthesised and have been shown to have liquid crystal properties <95JPR(337)641>.

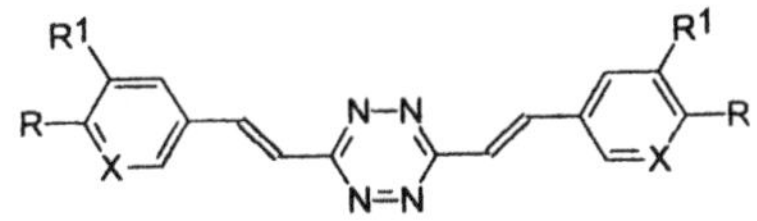

97

Spectroscopic studies have been carried out on a number of benzo-1,2,3,4-tetrazine 1,3-di-*N*-oxides **98** and furazanotetrazine 1,3-di-*N*-oxide **99** to investigate their characteristic vibration frequencies and electronic parameters <95MC100>.

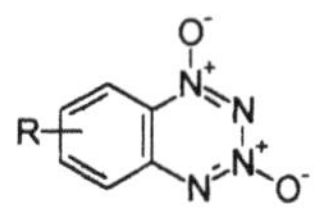

98

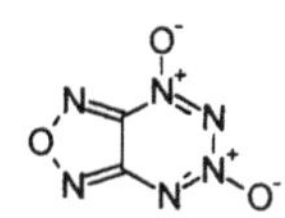

99

6.3.3.3 Purines and related compounds

The pyrrolopyrimidines **100** react with chloroamines to yield cyclic products **101** which are being investigated for antitumour activity <95MI08; 96CA(124)117234>.

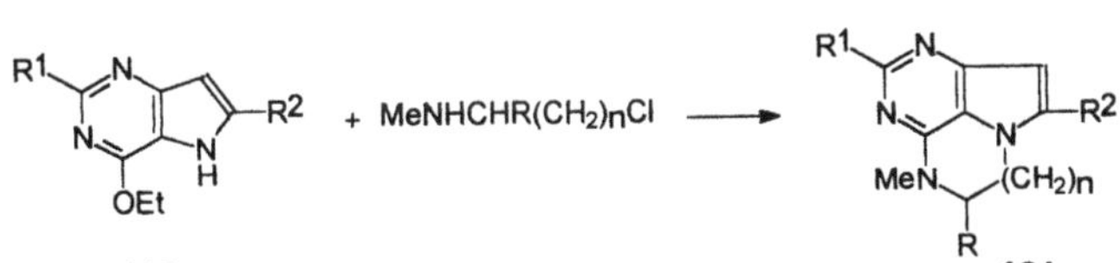

100 **101**

Electrophilic amination of adenine using H_2NOSO_3H in alkaline media gives 1-, 3-, 7- and 9-aminoadenine in a 1:1:3:1 ratio. This differs from methylation using dimethyl sulfate. Amination using dinitrophenoxyamine in DMF gives mainly 1-aminoadenine <96MI05; 96CA(124)289077>.

1,9-, 1,3-, 1,7- And 3,7-dibenzylpurin-2-one forms stable adducts when reacted with Grignard reagents. The reaction takes place in the purine 6- or 8-position and substantial differences are found in the regiochemical outcome of the reaction with the different isomers. The 6-position is the most susceptible to attack but the 3,7-dibenzyl isomer preferentially adds at the 8-position, although it has been reported that the 3,7-dimethyl compound prefers to react with Grignard reagents at the 6-position <94S203>. Rearomatisation is normally carried out by DDQ or manganese dioxide. An example is shown in Scheme **15** <96T12979>.

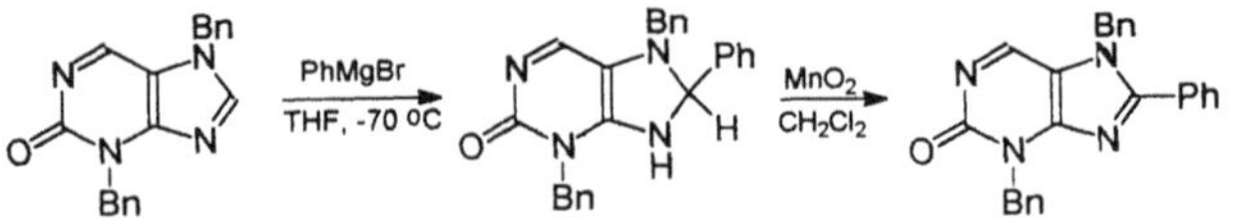

Scheme 15

A convenient synthesis of 6-cyanopurines from *N*-THP protected 6-chloropurines is that shown in Scheme **16**. This method should be applicable to other chloro nitrogen heterocycles <95CCC1386>.

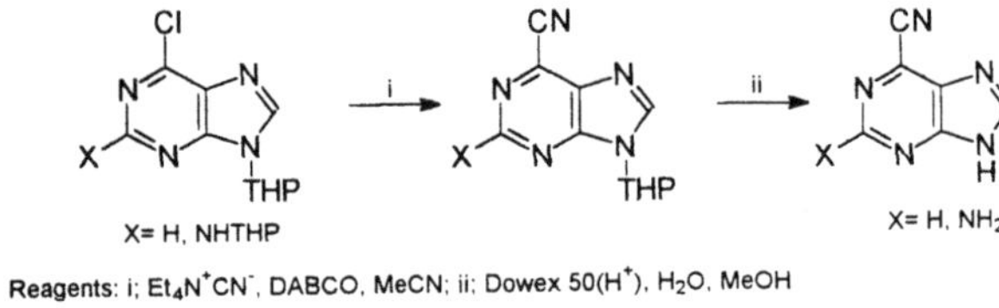

Scheme 16

9-Benzyl-6-chloropurine reacts with tetrakis(triphenylphosphine)palladium in DCE to give, not only the 6-purinylpalladium(II) complex **102**, but a dinuclear complex **103**. Using Stille coupling ($RSnBu_3$) only the 6-substituted purine is obtained <96ACS462>.

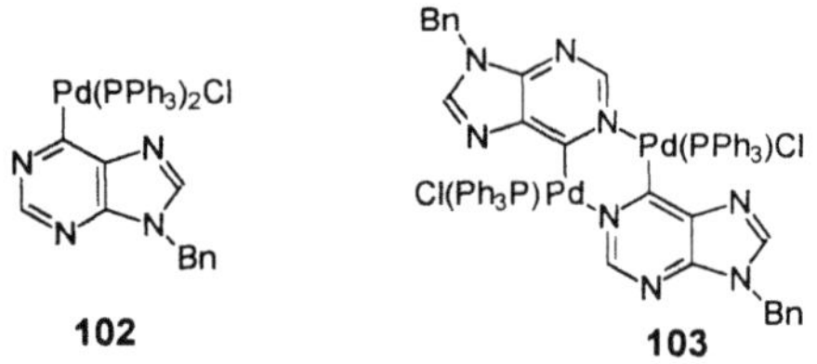

102 **103**

6.3.3.4 Pteridines and related compounds

Folic acid models have been synthesised either from the bromomethylpteridine **104** or from the oxadiazine compound **105** (Scheme 17) <96JHC341>.

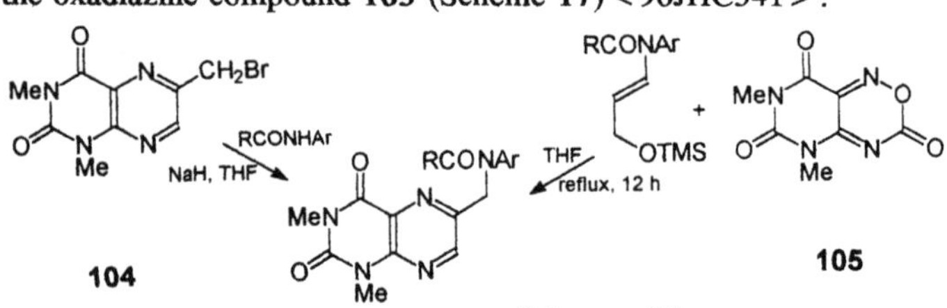

104 **105**

Scheme 17

The synthesis of 6-azidomethyl-5,6,7,8-tetrahydropterin **108** has been carried out from **106** *via* the intermediate **107** using the Mitsunobu reaction with diphenylphosphoryl azide followed by deprotection <95MI09; 96CA(124)232123>.

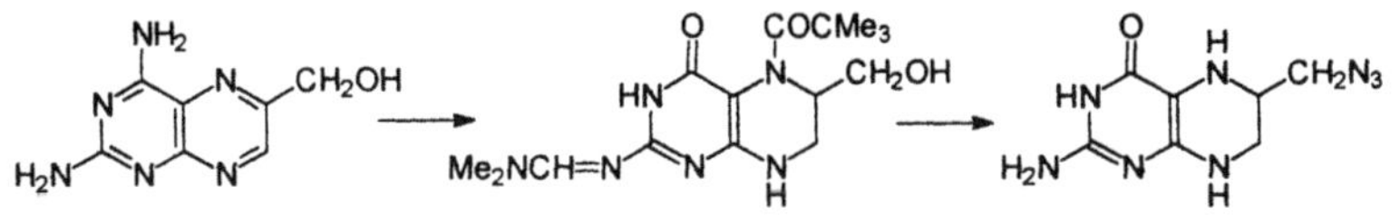

106 **107** **108**

An unprecedented ring contraction of the 7-azapteridine system has been observed when fervenulone **109** reacts with *t*-butyl bromoacetate using potassium carbonate/acetonitrile to yield **110** and the *O*-alkylate in a 7:1 ratio. The mechanism proposed for this reaction is shown in Scheme **18**. The use of sodium hydride/DMF yields the *N*-alkylated product **111** <95JOC7063>.

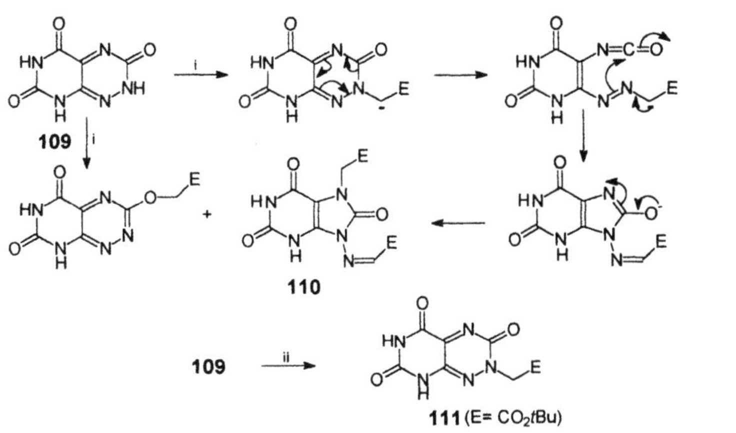

Reagents: i; $BrCH_2CO_2tBu$ (2.2 equiv.), K_2CO_3 (2.1 equiv.), MeCN, 90 °C; ii; NaH, DMF, rt then $BrCH_2CO_2tBu$ (1.2 equiv.), DMF, 85 °C.

Scheme 18

6.3.3.5 Miscellaneous ring systems

The pyrimido[4',5':4,5]thieno[2,3-*c*]pyridazines **113** (X= OH), easily obtained from the thienopyridazines **112**, are readily converted to the corresponding chloro compounds using phosphoryl chloride which undergo facile nucleophilic substitution with a variety of nucleophiles <96M537>.

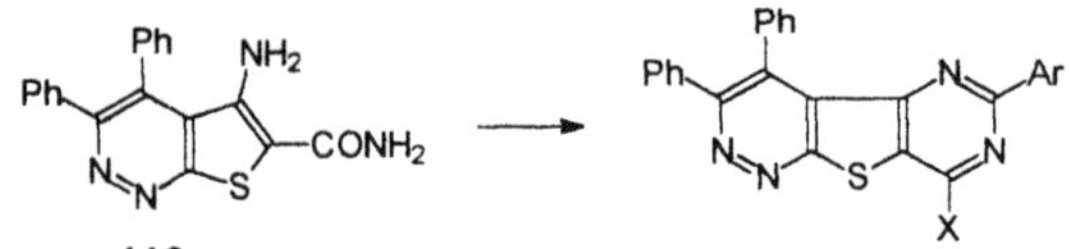

112

113 X= OH, Cl, OEt, $NHNH_2$, NHR

A series of annulated purines **114-6** have been synthesised as potential inhibitors of xanthine oxidase but, in general, they showed poor activity and the simple pyrimidines **117** were more effective *in vitro* <96MI06; 96CA(125)86586; 96JMC2529>.

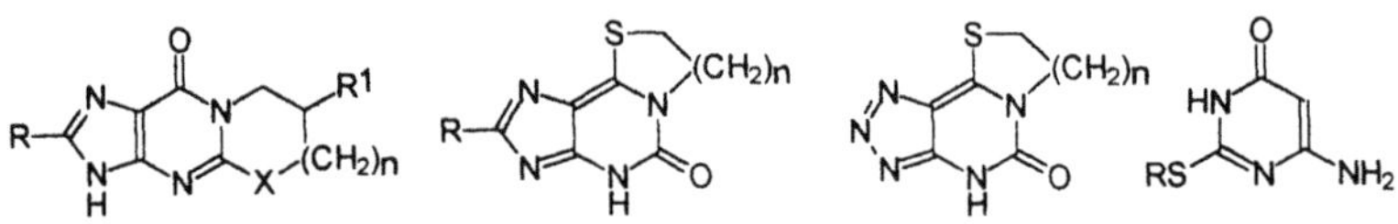

114 **115** **116** **117**

R= H, OH, SH, Ph; R1= H, Me; n= 0, 1, 2

A study of the halogenation of polyaza heterocycles is in progress. In the case of quinoxalino[2,3-*c*]cinnolines it has been found that using HCl or HBr in chloroform halogenation occurs at *C*-10, with protonation at *N*-12, and subsequent oxidative aromatisation (Scheme **19**) <96JCS(P1)1699>.

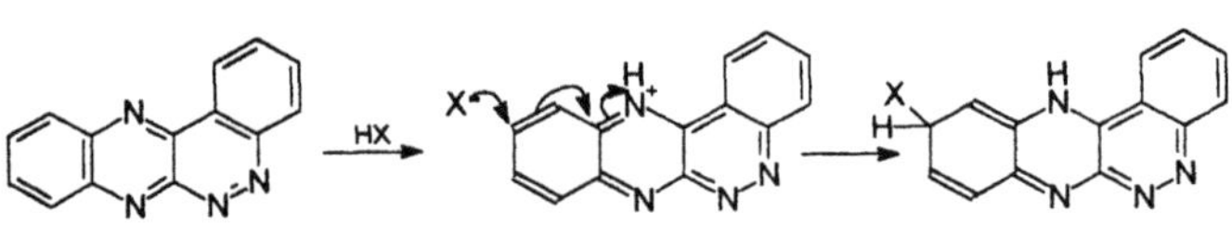

Scheme 19

References

94H(37)915 J.-P.Bouillon, V.Bouillon, C.Wynants, Z.Janousek and H.G.Viehl, *Heterocycles* **1994**, *37*, 915.

94MI01 M.S.Algharib, *Zagazig J.Pharm.Sci.* **1994**, *3*, 156; *Chem.Abstr.* **1996**, *124*, 55907.

94S203 T.Iwamura, Y.Okamoto, M.Yokomoto, H.Shimizu, M.Hori and T.Kataoka, *Synthesis* **1994**, 203.

95CCC1386 M.Hocek and A.Holy, *Collect.Czech.Chem.Commun.* **1995**, *60*, 1386.

95CHE596 A.V.Shastin, T.I.Goddovikova, S.P.Golova, L.I.Khmel'nitskii and B.L.Korsunski, *Chem. Heterocycl.Compd.(Engl.Transl.)* **1995**, *31*, 596.

95H(41)1999 H.A.Abdelhadi, T.A.Abdallah and H.M.Hassaneen, *Heterocycles* **1995**, *41*, 1999.

95H(41)2019 O.Chavignon, M.Raihane, P.Deplat, J.L.Chabard, A.Gueiffier, Y.Blache, G.Dauphin and J.C.Teulade, *Heterocycles* **1995**, *41*, 2019.

95IJCB939 B.Kalluraya, A.D'Souza and B.S.Holla, *Indian J.Chem.Sec.B* **1995**, *34*, 939.

95IJCB1010 S.Tiwara, N.Tiwara, T.Agrawal and N.H.Khan, *Indian J.Chem.,Sec.B* **1995**, *34*, 1010.

95IZV2037 S.Sh.Shukurov, M.A.Kukaniev and M.I.Nasyrov, *Izv.Akad.Nauk SSSR,Ser.Khim.* **1995**, 2037; *Chem.Abstr.* **1996**, *124*, 289413.

95JCS(P1)2783 Y.Wang, M.F.G.Stevens, W.T.Thompson and B.P.Shutts, *J.Chem.Soc.,Perkin Trans.1* **1995**, 2783.

95JCS(P1)2907 J.-P.Bouillon, Z.Janousek, H.G.Viehl, B.Tinant and J.-P.Declerq, *J.Chem.Soc.,Perkin Trans.1* **1995**, 2907.

95JHC1417 J.L.Kelley, D.C.Wilson, V.L.Styles, F.E.Soroko and B.R.Cooper, *J.Heterocycl.Chem.* **1995**, *32*, 1417.

95JHC1423 J.L.Kelley, D.C.Wilson, V.L.Styles, F.E.Soroko and B.R.Cooper, *J.Heterocycl.Chem.* **1995**, *32*, 1423.

95JHC1457 H.Poschenreider and H.-D.Stachel, *J.Heterocycl.Chem.* **1995**, *32*, 1457.

95JOC7063 M.M.Mehotra, D.D.Sternbach, R.D.Rutkowske and P.L.Feldman, *J.Org.Chem.* **1995**, *60*, 7063.

95JOC7687 J.B.Campbell and J.W.Firov, *J.Org.Chem.* **1995**, *60*, 7687.

95JOC8428 C.Chen, R.Dagnino Jr. and J.R.McCarthy, *J.Org.Chem.* **1995**, *60*, 8428.

95JPR641 T.Lifka and H.Meier, *J.Prakt.Chem./Chem.-Ztg.* **1995**, *337*, 641; *Chem.Abstr.* **1996**, *124*, 176038.

95KGS700 E.G.Paronikyan and A.S.Noravyan, *Khim.Geterotsikl.Soedin.* **1995**, 700; *Chem.Abstr.* **1996**, *124*, 176023.

95LA1703 S.Tumkevicius, *Liebigs Ann.Chem.* **1995**, 1703.

95MC100 K.I.Rezchikova, A.M.Churakov, V.Shylapochnikov and V.A.Tartakovskii, *Mendeleev Commun.* **1995**, 100.

95MC104 O.N.Chupakin, B.V.Rudakov, P.McDermott, S.G.Alexeev, V.N.Charushin and F.Hegarty, *Mendeleev Commun.* **1995**, 104.

95MC229 Y.A.Azev, H.Neunhoffer, S.Foro, H.J.Lindner and S.V.Shorshnev, *Mendeleev Commun.* **1995**, 229.

95MI01 A.A.Fahmi and M.S.Algharib, *Zagazig J.Pharm.Sci.* **1995**, *4*, 272; *Chem.Abstr.* **1996**, *124*, 86924.

95MI02 D.Briel, *Pharmazie* **1995**, *50*, 675; *Chem.Abstr.* **1996**, *124*, 176031.

95MI03 J.-K.Lee, S.-N.Kim and S.-G.Lee, *J.Korean Chem.Soc.* **1995**, *39*, 755; *Chem.Abstr.* **1996**, *124*, 86945.

95MI04 A.M.Abdel-Halim, Z.El-Gendy and R.M.Abdel-Rahman, *Pharmazie* **1995**, *50*, 726.

95MI05 A.K.Mansour, N.M.Elwan, H.A.Abdelhadi, T.A.Abdallah and H.M.Hassaneen, *Sulfur Lett.* **1995**, *18*, 105; *Chem.Abstr.* **1996**, *124*, 317107.

95MI06 M.M.Heravi and M.Shafaie, *Indian J.Heterocycl.Chem.* **1995**, *5*, 79; *Chem.Abstr.* **1996**, *124*, 176024.

95MI07 M.M.Heravi, M.Bakavoli and Z.Sadjada Hashemi, *Iran J.Chem./Chem.Eng.* **1995**, *14*, 41; *Chem.Abstr.* **1996**, *124*, 202176.

95MI08 R.G.Glushkov, O.S.Sizova, G.A.Modnikova, A.S.Sokolova and V.A.Chernov, *Khim.-Pharm.Zh.* **1995**, *29*, 19; *Chem.Abstr.* **1996**, *124*, 117234.

95MI09 G.Heizmann, V.Groehn, B.Almaas, J.Haavik, T.Flatmark and W.Pfleiderer, *Pteridines* **1995**, *6*, 153; *Chem.Abstr.* **1996**, *124*, 232123.

95ZOB1161 A.V.Moskvin, N.M.Petrova, I.I.Polkovnikov, S.P.Saenchuk and B.V.Ivin,*Zh. Obshch.Khim.* **1995**,*31*,1161; *Chem.Abstr.* **1996**,*124*,146052.
95ZOB1246 V.V.Kiselev, V.S.Zyabrev, E.A.Romanenko, A.V.Kharchenko and B.S.Drach,*Zh. Obshch. Khim.* **1995**,*31*,1246; *Chem.Abstr.* **1996**,*124*,317108.
96ACS462 L.-L. Gundersen,*Acta Chem.Scand.* **1996**,*52*,462.
96H(42)53 Y.Tominga and N.Yoshioka,*Heterocycles* **1996**,*42*,53.
96H(42)691 A.Miyashita, K.Fujimoto, T.Okada and T.Higashino,*Heterocycles* **1996**,*42*,691.
96H(43)1007 C.-C.Tzeng, K.-H.Lee, Y.-L.Chen and T.-C.Wang,*Heterocycles* **1996**,*43*,1007.
96H(43)437 E.C.Taylor, M.Takahashi and N.Kobayashi,*Heterocycles* **1996**,*43*,437.
96H(43)839 S.Kohra, K.Ueda and Y.Tominga,*Heterocycles* **1996**,*43*,839.
96H(43)1759 M.Morioka, M.Kato, H.Yoshida and T.Ogata,*Heterocycles* **1996**,*43*,1759.
96IZV1322 M.M.Vartanyan, T.Yu.Soloveva, O.L.Eliseev and M.E.Panina,*Izv.Akad.Nauk SSSR,Ser.Khim.* **1996**,1322; *Chem.Abstr.* **1996**,*125*,195557.
96JCR(S)174 A.Q.Hussein,*J.Chem.Res.(Synop.)* **1996**,174.
96JCS(P1)247 T.Okawa, S.Eguchi and A.Kakehi,*J.Chem.Soc.,Perkin Trans.1* **1996**,247.
96JCS(P1)1699 I.W.Harvey, D.M.Smith and C.R.White,*J.Chem.Soc.,2erkin Trans.1* **1996**,1699.
96JCS(P1)1999 B.Baruah, D.Prajapati, J.S.Sandhu and A.C.Ghosh,*J.Chem.Soc.,Perkin Trans.1* **1996**,1999.
96JCS(P1)2257 P.K.Bridson, H.Huang and X.Lin,*J.Chem.Soc.,Perkin Trans.1* **1996**,2257.
96JHC295 K.-J.Kim, H.-S.Li and K.Kim, *J.Heterocycl.Chem.* **1996**,*33*,295.
96JHC341 M.Igarashi, T.Kambe and M.Tada,*J,Heterocycl.Chem.* **1996**,*33*,341.
96JHC599 E.Laskos, P.S.Lianis and N.A.Rodios,*J.Heterocycl.Chem.* **1996**,*33*,599.
96JHC659 C.C.Lockhart and J.W.Sewell,*J.Heterocycl.Chem.* **1996**,*33*,659.
96JMC2529 G.Biagi, A.Costantini, L.Costantino, I.Girgio, O.Livi, P.Pecorari, M.Rinaldi and V.Scartoni, *J.Med.Chem.* **1995**,*39*,2529.
96JOC5204 Q.Dang, B.S.Brown and M.D.Erion,*J.Org.Chem.* **1996**,*61*,5204.
96KGS103 S.Tumkevicius,*Khim.Geterotsiikl.Soedin.* **1996**,103; *Chem.Abstr.* **1996**,*125*,167905.
96KGS427 A.Brukstus, T.Sadanskas and S.Tumkevicius,*KhimGeterotsikl.Soedin.* **1996**,427; *Chem.Abstr.* **1996**,*125*,167911.
96M537 J.M.Quintela, M.C.Veiga, R.Alvarez-Sarandes and C.Peinador,*Monatsh.Chem.* **1996**,*127*,537.
96MC22 A.M.Churakov, O.Yu.Smirnov, Y.A.Strelenko, S.L.Ioffe, V.A.Tartakovski, Y.T.Struchkov, F.M.Dolgushin and A.I.Yanovsky,*Mendeleev Commun.* **1996**,22.
96MI01 Y.Wang and M.F.G.Stevens,*Biorg.Med.Chem.Lett.* **1996**,*6*,185; *Chem.Abstr.* **1996**,*124*, 232405.
96MI02 P.M.Petersen, W.Wu, E.E.Fenlon, S.Kim and S.C.Zimmerman,*Biorg.Med.Chem.* **1996**,*4*, 1107; *Chem.Abstr.* **1996**,*125*,221771.
96MI03 M.M.Heravi and P.Khosrofar,*J.Sci.Islamic Repub.Iran* **1996**,*7*,86; *Chem.Abstr.* **1996**,*125*, 195597.
96MI04 M.M.Heravi, M.Shafaie, M.Bakavoli, M.M.Sadaghi and A.R.Koshdast,*Orient.J.Chem.* **1996**,*12*,43; *Chem.Abstr.* **1996**,*125*,247767.
96MI05 T.Saga, T.Kaiya, S.Asano and K.Kohda,*Nucleosides and Nucleotides* **1996**,*15*,219; *Chem. Abstr.* **1996**,*124*,289077.
96MI06 G.Biagi, I.Giorgi, O.Livi and V.Scartoni,*Farmaco* **1996**,*51*,301; *Chem.Abstr.* **1996**,*125*, 86586.
96SC453 M.Rahmouni, A.Derdour, J.P.Bazureau and J.Hamelin,*Synth.Commun.* **1996**,*26*,453.
96T2377 W.Adam, C.van Barneveld and D.Golsch,*Tetrahedron* **1996**,*52*,2377.
96T3037 J.M.Quintella, M.J.Moreira and C.Peinador,*Tetrahedron* **1996**,*52*,3037.
96T12979 G.Andresen, L.-L.Gundersen and F.Rise,*Tetrahedron* **1996**,*52*,12979.
96T13017 C.Dunn, C.L.Gibson and C.J.Suckling,*Tetrahedron* **1966**,*52*,13017.
96TL5795 A.Rykowski and E.Wolinska,*Tetrahedron Lett.* **1996**,*37*,5795.
96ZN(B)421 H.Moehrle and W.U.von der Lieck-Waldheim,*Z.Naturforsch.,Teil B* **1996**,*51*,421.
96ZOB824 A.V.Moskvin, N.M.Petrova, K.A.Krasnov and B.A.Ivin,*Zh.Obshch.Khim.* **1966**,*32*, 824; *Chem.Abstr.* **1996**,*125*,328655.

Chapter 6.4

Six-Membered Ring Systems: With O and/or S Atoms

John D. Hepworth and B. Mark Heron
University of Central Lancashire, Preston, UK

Introduction

Total syntheses of macrocycles containing 6-membered *O*-heterocyclic units reported during 1996 include those of swinholide A <96CEJ847, 96JA3059>, brevetoxins <96AG(E)589, 96TL6365>, (+)-milbemycin D <96JA7513> and bafilomycin <96TL1073>. Other natural products which have been synthesised include ambruticin <96SYN297>, (+)-herboxidiene A <96SYN652>, staurosporine <96JA2825, 96JA10656>, a methylated mycalamide <96JCS(P1)1797> siccanin <96JA5146>, (-)-invictolide <96T12177>, phenoxan <96JOC4853, 96TL2997>, (+)-camptothecin <96AG(E)1692>, dihydrokawain-5-ol <96JOC9103>, strobilurin E <96TL7955>, sordinin <96TL3741> and zaragozic acid A <96JOC9115>.

Several reviews contain material relevant to this chapter, including those on saturated oxygen heterocycles <96COS229>, saturated and unsaturated lactones <96COS295>, rapamycin <96COS345>, dibenzodioxins <96RCR27>, radical cyclisation reactions <96OR(48)301>, cycloadditions of *o*-benzoquinones <96SL1143>, the addition of isocyanates to vinyl ethers <96CC2689>, domino reactions <96CRV115>, cascade processes of metallo carbenoids <96CRV223>, the use of (π-allyl)tricarbonyliron lactone complexes in synthesis <96CRV423> and potassium channel modulators <96SYN307>.

The ingenuity of heterocyclic chemists in their search for new compounds stretches the imagination when reports of the isolation of *O*-heterocycles from possum urine <96AJC1> and lemming droppings <95TL5847> are digested. The discovery of polyphenol glycosides with a new flavonoid skeleton in Bordeaux wine <96TL7739> is much more tasteful, but who knows what we will make of them!

6.4.1 HETEROCYCLES CONTAINING ONE OXYGEN ATOM

6.4.1.1 Pyrans

2H-Pyran-2-iminium salts (**2**) are formed by the acid-catalysed cyclisation of the pentadienal (**1**), itself accessible from the addition of HCl to 5-dimethylaminopenta-2-en-4-ynal, a push-pull enyne <96HCA192>.

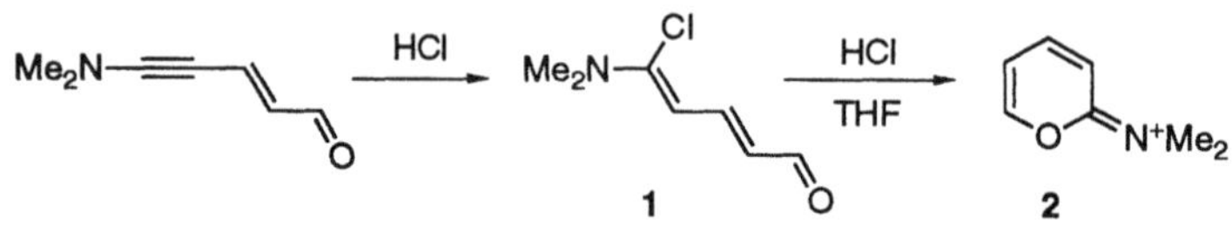

Hetero-fused 4*H*-pyrans result from the cyclisation of 1,5-hydroxynitriles <96BSF229> and 1,5-diketones <96HA567>; the latter route can be adapted to give the corresponding thiopyrans.

Dimedone and related compounds behave as carbon nucleophiles towards oxazines and oxazolidines leading to the octahydroxanthene derivative (**3**). In the presence of stoichiometric quantities of reactive *C*-nucleophiles, such as substituted acetonitriles, 2,3-disubstituted partially reduced 4*H*-1-benzopyrans result <96T14273>.

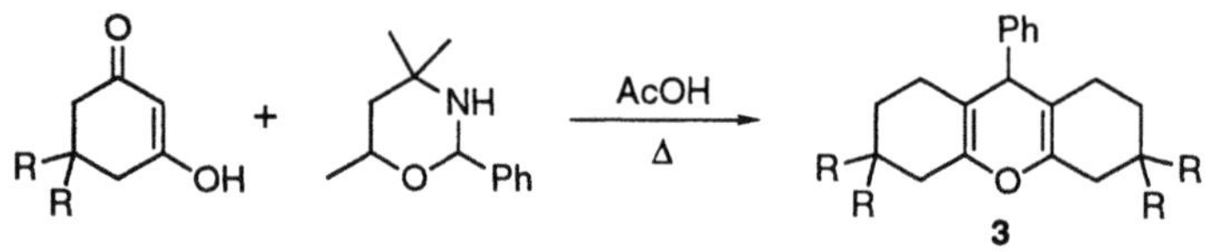

The hetero-Diels-Alder (HDA) reaction involving aldehydes is catalysed by Pd(II) complexes, such that even non-activated dienes give satisfactory yields of 5,6-dihydro-2*H*-pyrans <96TL6351>. Chiral aluminium complexes are effective in controlling both the chemo- and enantio- selectivity of the reaction of alkyl glyoxylates with dienes containing an allylic C-H bond; high yields of the dihydropyran result with enantiomeric excesses of up to 97% <96CC2373>.

The extent of diastereoselectivity observed in the reaction of 1-(1-phenylalkoxy)buta-1,3-dienes with indantrione and alloxane is associated with the steric requirements of the alkoxy function in the chiral auxiliary <96SYN105>.

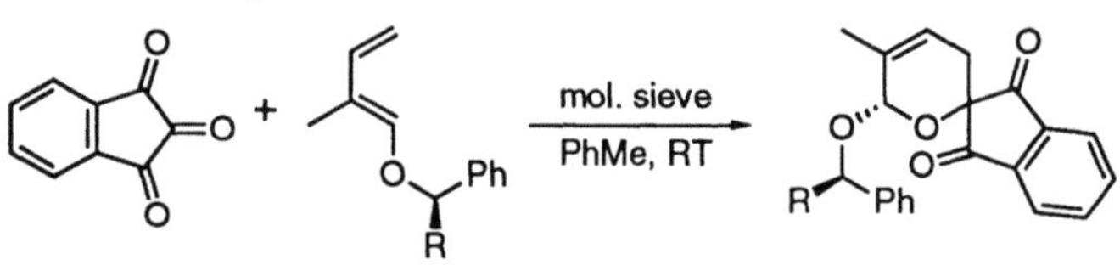

Both (*E*)-1-phenylsulfonyl and (*S*)-(+)-3-*p*-tolylsulfinyl -alk-3-en-2-ones can exhibit high diastereoselectivity in their reactions with vinyl ethers and styrenes, with the dienophile having a dominant influence on the stereochemical outcome <96T1205, 96TL3687>. Indol-2-ylideneacetic acid esters can act as both dienophile and heterodiene in cycloaddition reactions; in the latter case pyrano[3,2-*b*]indoles are formed <96SYN519>.

Cyclisation of 1-aryl-6-hydroxyhex-2-ynyl carbonates to the 2-arylidene-5,6-dihydropyran catalysed by Pd(0) is stereospecific, giving only the (*Z*)-isomer <96SL553>, whilst alk-1-yn-5-ols form dihydropyranylidene carbenes under the influence of the tungsten pentacarbonyl-THF complex. The carbenes are a source of 2-stannyldihydropyrans <96TL4675>.

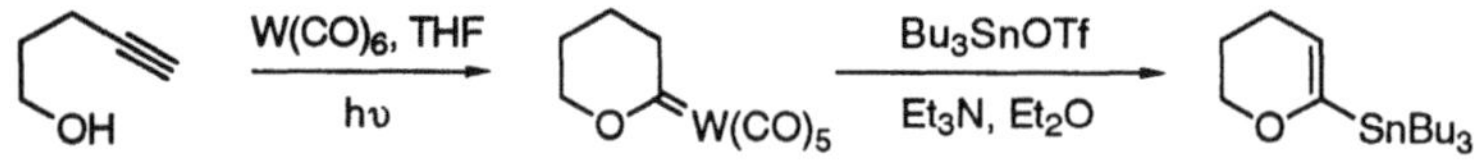

Both the Tebbe and Petasis reagents, $Cp_2CH_2ClAlMe_2$ and Cp_2TiMe_2, effect the direct conversion of alkenic esters to dihydropyrans. This olefin metathesis has been successfully applied to the synthesis of complex polyether frameworks <96JA1565, 96JA10335>.

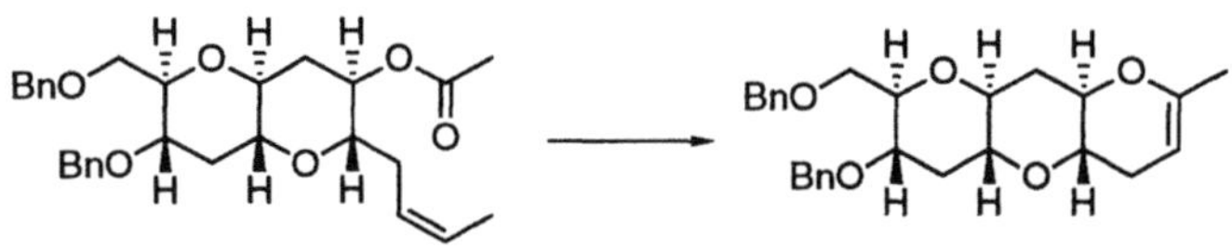

Ring expansion of tetrahydrofurans to dihydropyrans results when their 2-*N*-aziridinyl imines are heated <96CC909> and when their 2-ω-alkyl bromides are treated with Ag_2O in a nucleophilic acidic solvent <96JCS(P1)413>. Alkyl carbenes and bicyclic oxonium ion intermediates are invoked, respectively.

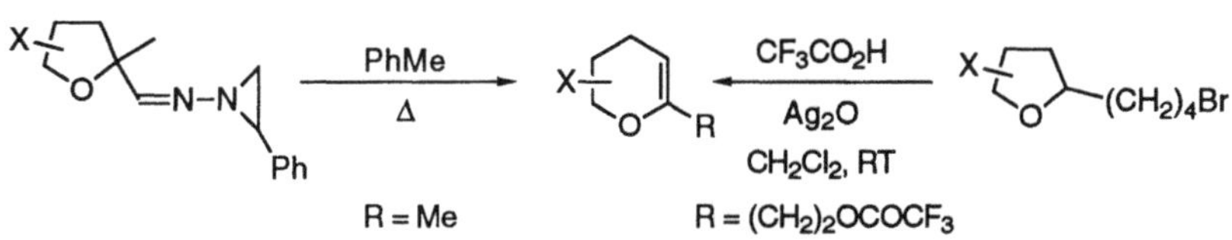

Interest continues in 2,2'-bisdihydropyrans. Chiral 2-hydroxymethyl-3,4-dihydropyrans have been obtained from the racemic acetate by hydrolysis by porcine pancreatic lipase and converted to the halomethyl and phenylthiomethyl derivatives <96SL787>. Both groups are of value as protecting and resolving reagents for 1,2-diols <96SL789, 96SL791>. The copper catalysed Stille coupling of pyran derived enol triflates and stannyl enol ethers has been optimised to give the cross coupled bisdihydropyrans and shown to be applicable to the formation of complex polyether systems <96AG(E)889>.

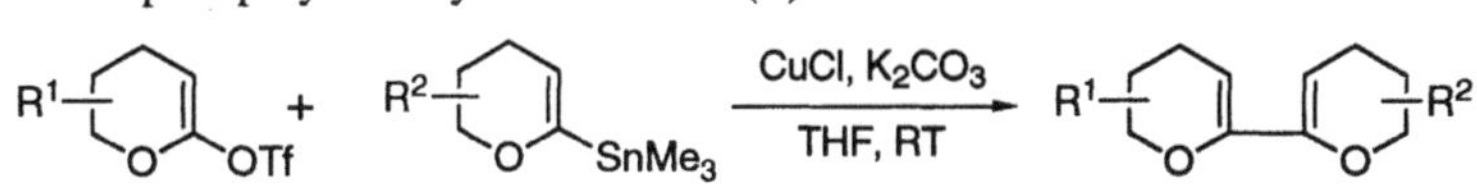

3,4-Dihydropyrans can be converted into 4-cyanoethylisoxazoles and piperid-2-ones by reaction with hydroxylamine and arylsulfonyl isocyanates, respectively <96JHC383, 96JOC2865>. Reaction of $(PhIO)_n$/$TMSN_3$ with dihydropyrans is a preliminary to the preparation of various aminopyrans <96TL303> and 3-aminotetrahydropyrans are formed stereoselectively from the Lewis-acid catalysed intramolecular cyclisation of hydrazones derived from a γ-alkoxyallylstannane <96CC841>.

Tetrahydropyrans have been obtained with high stereoselectivity by the cyclisation of substituted hexenols <96JCS(P1)967, 96JCS(P2)2407>, hexenals <96TL3059>, 1,5-diols <96TL2463>, epoxy alcohols <96SYN647, 96TL6173>, alkenyl bromoalkyl ethers <96JOC677> and iodoketones <96JOC6673>. Oxaadamantane has been synthesised using this methodology <96TL5673>, as has the tricyclic furanochroman framework of phomactin <96TL275>.

The photochemical cyclisation of 2-aryl-3-cycloalkenyloxychromones results in the formation of a spiropyran unit in a tetracyclic array <96TL8913>.

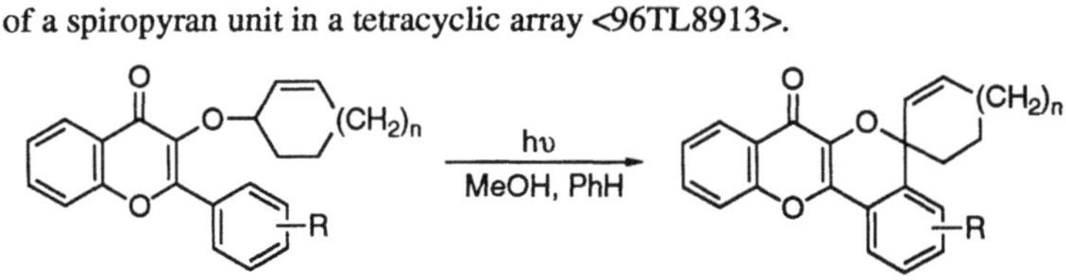

A ring opening - ring closure sequence yields tetrahydropyrans from 1,3-dioxanes involving the Petasis reagent <96TL141> and the Prins reaction <96TL8679>, whilst the rearrangment and ring expansion of tetrahydrofurans occurs stereoselectively under the influence of zinc or silver acetates <96TL213>. Extrusion of SO_2 from an oxathiane involving a Ramberg-Bäcklund olefination leads to a *trans*-fused pyrano[3,2-*b*]pyran <96TL2865>. The reductive elimination of the sulfonyl group from 1,6-dioxaspiro[4.5]decan-10-ones is accompanied by the cleavage of the spiroketal ring and leads to a mixture of *cis*- and *trans*- 1,6-dioxadecalins <96TL3179>.

Stereocontrol in the formation of spiroketals has been achieved in the alkylation of 2-(benzenesulfonyl)pyrans with allylsilanes <96JOC7860> and using a double carbonyl cyclisation strategy <96SL1065>. Spirocyclisation of protected dihydroxydiketones yields *cis*- and *trans*- 1,7,9-trioxadispiro[5.1.5.3]hexadecanes; the latter isomer is the thermodynamically more stable <96TL5461>.

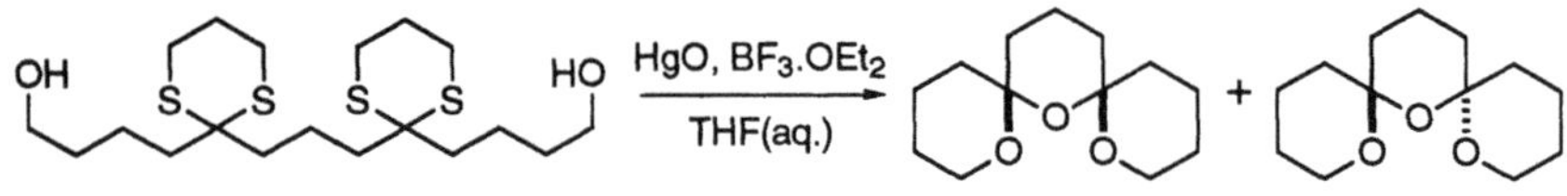

Infrared and Raman spectra of 4*H*-pyran and 1,4-dioxin have been analysed <96JST255> and interconversion barriers for dihydropyrans and some dioxins have been determined <96BKS7>.

6.4.1.2 Benzopyrans (Chromenes)

The cyclisation of naphthyl propargyl ethers occurs efficiently under microwave irradiation leading to naphthopyrans, but naphthofurans are formed in the presence of base <96JCR(S)338>. The thermal rearrangement of naphthyl 3-trimethylsilylprop-2-ynyl ethers yields the 4-trimethylsilyl derivatives of naphthopyrans <96H(43)751>.

Interest in enantioselective epoxidation continues and 2,2-dimethylchromenes appear to be particularly suitable substrates for the evaluation of the catalytic system <96JCS(P1)1757, 96SL1079, 96TL3895>.

The 3,4-didehydro-2*H*-1-benzopyran (**4**), generated when 3,4-dibromochromenes are treated with organometallic reagents, reacts with dienes to give the dibenzo[*b,d*]pyran system <96TL1313>.

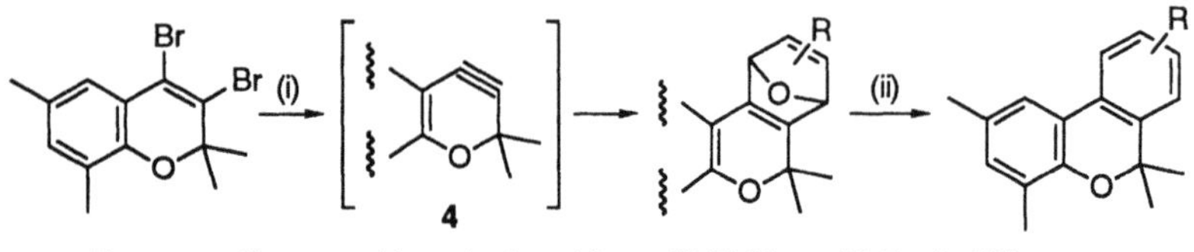

Reagents: (i) excess Mg, substituted furan, THF, N_2, Δ; (ii) Zn, AcOH, warm.

6.4.1.3 Dihydrobenzopyrans (Chromans)

A review of the photogeneration of quinone methides includes examples of the synthesis of chromans and related compounds <96CJC465>. These reactive intermediates, generated by chemical and electrochemical oxidation, are involved in the synthesis of euglobals, chromans and spirochromans isolated from *Eucalyptus* sp. <96CC1763, 96JCS(P1)1435>. A synthesis of robustadials, isolated from the same species and also of interest because of their pharmacological activity, is based on chroman-4-one chemistry <96ACS132>.

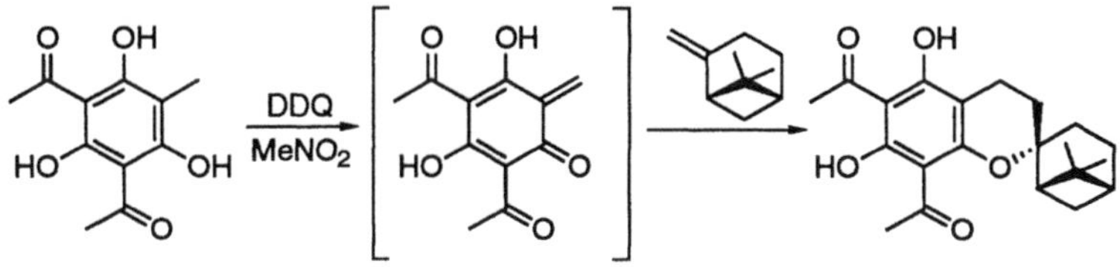

An intramolecular variant of this cycloaddition process is combined with a Knoevenagel reaction in a total synthesis of the insectan leporin A, a pyrano[3,2-*c*]pyridine derivative <96JOC2839>.

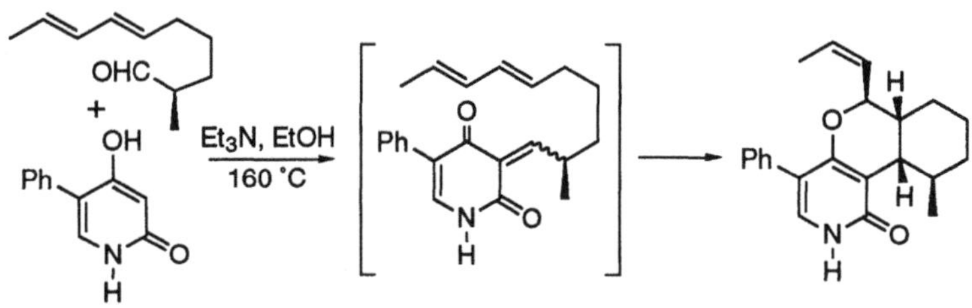

A neat stereoselective synthesis of *trans*-fused tetrahydropyrano[3,2-*c*][1]benzopyrans involves treating salicylaldehydes with alk-4-en-1-ols and triethyl orthoformate. The selectivity is attributed to steric repulsion in the *endo* transition state, the precursor of the *cis*-fused compound <96CL889>.

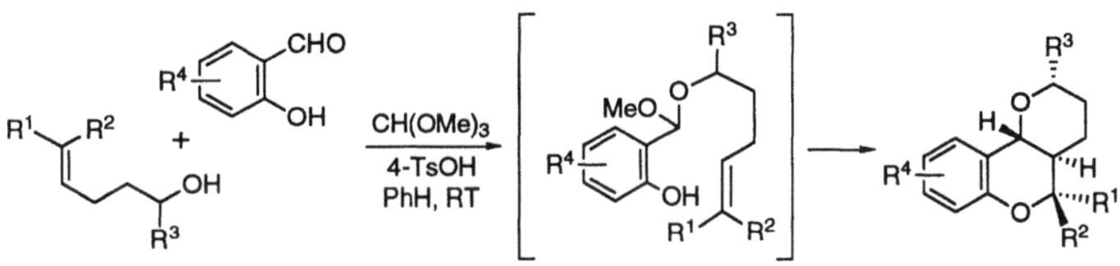

Chromans have been obtained by the Pd-catalysed intramolecular cross coupling reaction involving aryl halides containing an *ortho*-hydroxyalkyl substituent <96JA10333>.

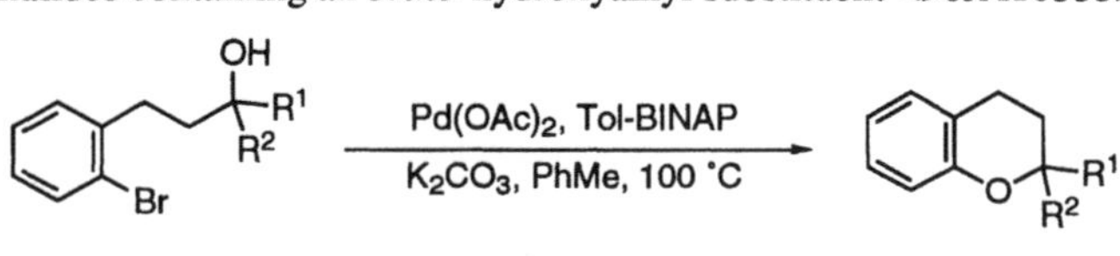

Enediynes are of interest because of their ability to cleave DNA. The 10-membered oxadiyne (**5**) cycloaromatises to the isochroman (**6**) <96TL2433> and a biradical intermediate is proposed to account for the formation of an isochroman from an acyclic enediyne <96TL5397>.

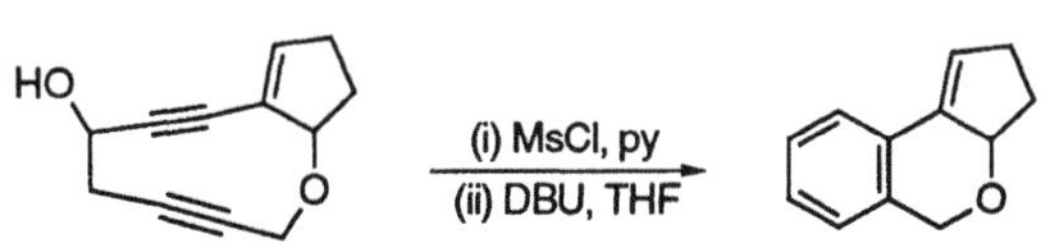

5 **6**

Benzopyrano[4,3-*c*]isoxazolidine derivatives feature as chiral auxiliaries in the Horner-Emmons synthesis of axially chiral cyclohexylidene compounds <96TL1077> and in a synthesis of the marine polypropionate (+)-siphonarienone <96TL1081>. The isoxazolidines are also a source of chroman-based 1,3-aminoalcohols <96SYN1280>. The complex 4-aminochroman-3-ol (**7**) derived from a 3,4-epoxychroman, undergoes an acid promoted 1,4-oxygen to nitrogen migration <96TL4447>.

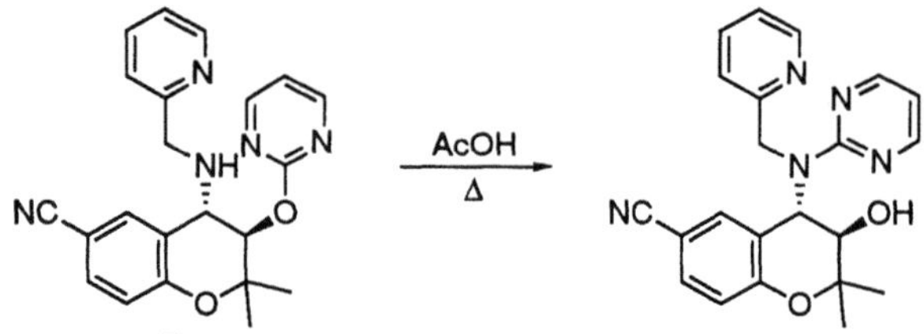

7

Cyclobuta[*b*]chroman-4-ols, derived from chromones by a [2+2] photocycloaddition to ethylene, are prone to acid-catalysed rearrangements. Elaboration of the parent system prior to rearrangement has enabled the marine sesquiterpene filiformin <96JOC4391>, the benzo-1,3-dioxan nucleus of averufin <96JOC9164> and cyclobuta[*b*][1]benzoxepin-8,9-diones <96CC1965> to be synthesised.

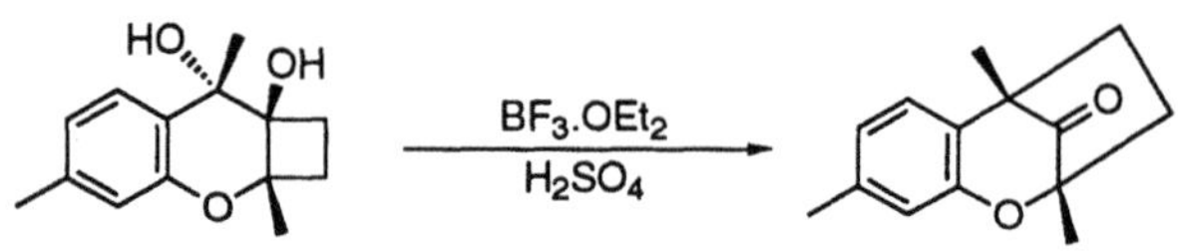

Substituted 4-aryl-1,3-dioxolanes are ring opened by $TiCl_4$ and at low temperatures subsequently recyclise *ortho* to the activating group on the aryl ring to give 4-hydroxyisochromans. A further stereoselective isomerisation to benzofurans occurs at higher temperatures <96JCS(P1)2241>.

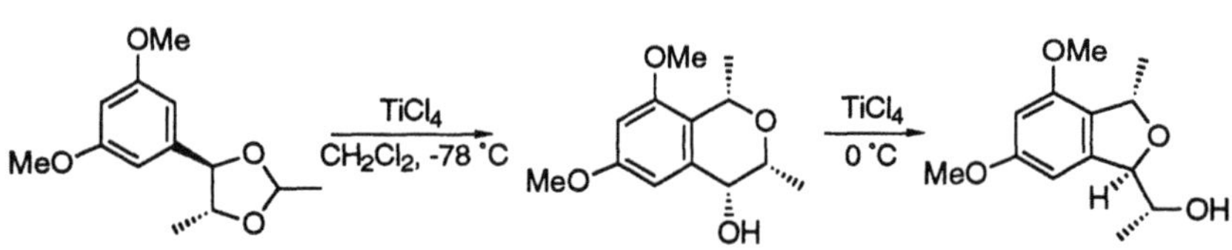

1-Trifluoromethylbenzocyclobutenols yield isochroman-1-ols on treatment with aromatic aldehydes and LTMP and hence serve as laterally-lithiated 2-methyltrifluoromethyl-acetophenone <96SL57>.

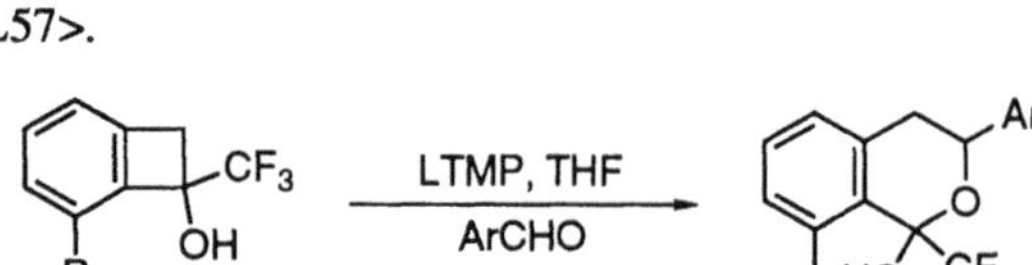

A number of azabenzisochromanquinone antibiotics have been synthesised by the HDA reaction between isochromanquinones and 1-aza-1,3-dienes <96LA1385>.

6.4.1.4 Pyranones

Stannyl derivatives of 2*H*-pyran-2-one, accessible from bromopyranones by cross coupling with organotin reagents, themselves take part in Pd(0)-catalysed cross coupling with enol triflates. This methodology offers a new approach to steroidal pyran-2-ones <96JOC6693>.

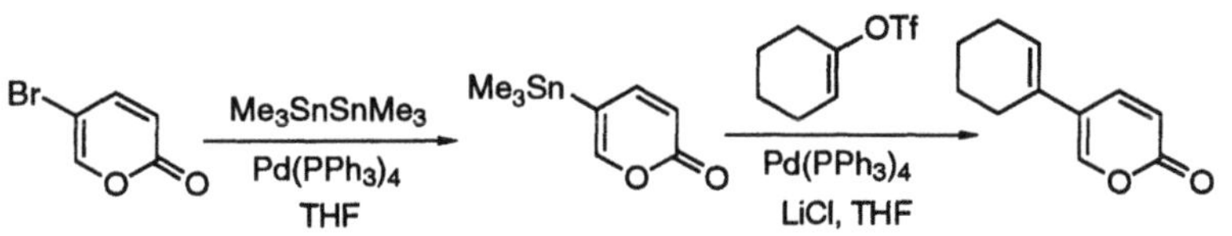

Pyran-4-ones are formed when acyl ketenes derived from dioxofurans and dioxinones react with vinyl ethers. Intermediate products are 1,3,5-triketones and reduced pyranones <96CPB956, 96H(43)2457, 96TL6499>.

Both 2- and 4- pyranones are versatile intermediates in synthesis. Interest has been maintained in the Diels-Alder reaction with electron rich dienophiles and conditions necessary to achieve stereocontrol have been reported <96JOC671, 96PAC113, 96H(43)745>. Fulveneketene acetal reacts with pyran-2-one in a [6+4] cycloaddition to give 4-hydroxyazulenes <96CC937>. An intramolecular Diels-Alder cycloaddition involving an alkyne is part of a synthesis of the lycorine alkaloids <96JOC1650>. Pyran-2-ones with a tethered furan substituent at C-3 undergo a crossed [4+4] photocycloaddition yielding polycyclic cyclooctanoid systems <96SL1173>.

3-Substituted 6-aryl-4-methylthiopyran-2-ones may be converted into terphenyls or 4,6-diarylpyran derivatives through the carbanion induced ring-opening reaction with acetophenones <96TL93>.

Both the mode and site of nucleophilic attack of pyran-2-ones by alkoxides is influenced by the counter ion such that a variety of products can be obtained. In particular, chelation effects play a significant role <96JCS(P1)2715>.

An intramolecular lactonisation features as the final stage in an enantiospecific assembly of the pentacyclic quassinoid framework <96CC2369>.

Complete diastereoselection is observed in the HDA reaction of Danishefsky's diene with *o*-substituted benzaldehyde chromium tricarbonyl complexes. Decomplexation is facile and good yields of 2-aryl-2,3-dihydropyran-4-ones result <96SL258>. *Cis*-2,3-disubstituted pyranones are accessible from the Lewis-acid catalysed HDA reaction between (triisopropylsilyloxy) dienes and aldehydes and dehydrogenation of the resulting dihydropyrans <96JOC7600>.

Formation of a 6-hydroxydihydropyran-3-one by the oxidative rearrangement of a furan followed by its conversion to a pyrylium ylide forms part of a synthesis of the taxane skeleton <96T14081>.

Dihydropyran-4-ones are a source of phenols *via* an intramolecular [2+2] photocycloaddition reaction and a Lewis-acid catalysed cleavage of the cyclobutane moiety <96TL1663>.

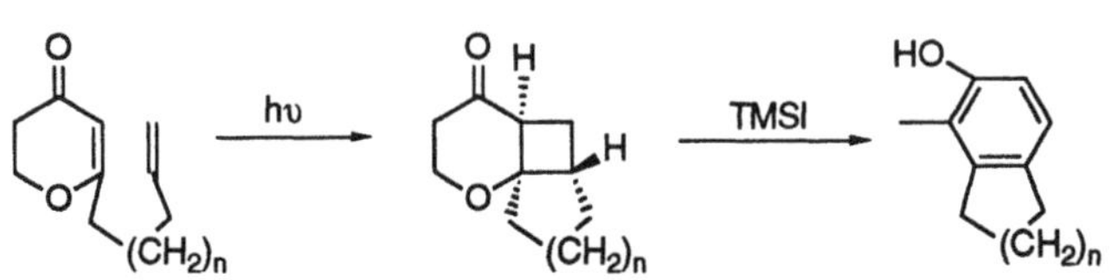

Conjugate addition of azide ion to dihydropyran-2,5-diones affords the 3-amino derivative <96SL341>, whilst reaction with bisnucleophiles provides a route to piperazines, thiazines and diazepines <96JHC703>.

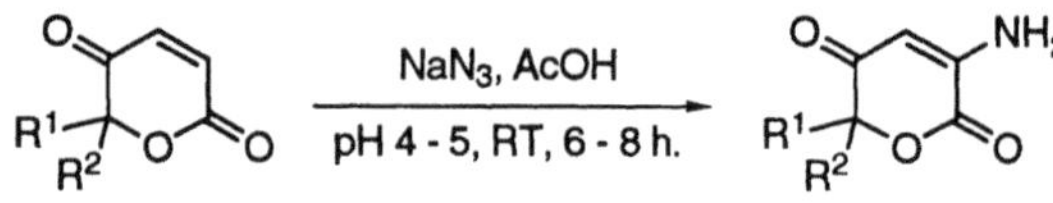

6.4.1.5 Coumarins

Alkynoates react with electron rich phenols to give coumarins with good regioselectivity in the presence of formic acid and a Pd-catalyst. Yields are good even in instances where the Pechmann synthesis is reported to be unreliable <96JA6305>.

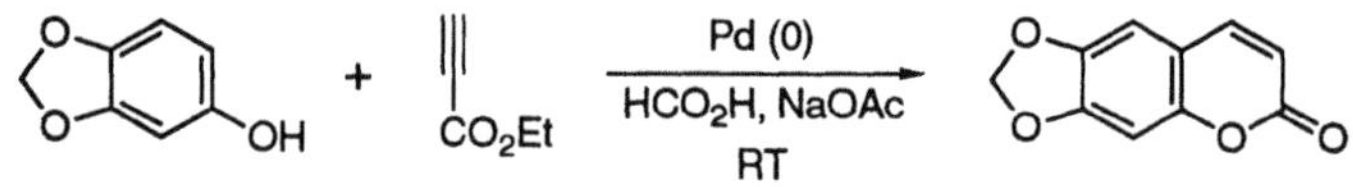

The Suzuki cross coupling reaction features in a synthesis of 4-arylcoumarins from the 4-halogeno derivatives <96JCS(P1)2591> and in a route to the thieno[3,4-*c*]coumarin (**8**) from which coumarin 3,4-quinodimethane (**9**) can be thermally generated <96T3117>.

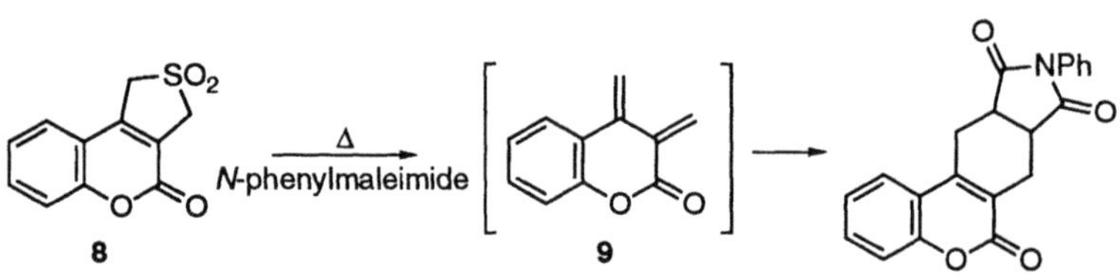

8 **9**

The Knoevenagel reaction between *o*-hydroxyaryl aldehydes and ketones and substituted acetonitriles affords high yields of 3-substituted coumarins in aqueous alkaline media <96H(43)1257>, whilst 4-hydroxycoumarins have been elaborated to pyrano [3,2-*c*]benzopyran-5-ones by reaction with aromatic aldehydes and malononitrile <96P148>. The imine (**10**) resulting from the complex reaction of *o*-hydroxyacetophenone with malononitrile undergoes a 1,5-tautomeric shift in solution <96JCS(P1)1067>.

10

An improved route to fluorinated 4-hydroxycoumarins has been reported, based on a facile decarboxylation-deacetylation of their 3-(3-oxopropanoic acid) derivatives <96TL1551>. The reaction of methyl salicylates with triphenylphosphoranylidene ketene, $Ph_3P{=}C{=}C{=}O$, affords 4-methoxycoumarins <96JCS(P1)2799> and the formation of coumarin 3-phosphonates from salicylaldehydes and phosphonoacetates, $EtO_2CCH_2P(O)(OR)_2$, has been investigated <96T12597>.

Asymmetric syntheses of warfarin <96TL8321> and the axially chiral bicoumarin, isokotanin A <96TL3015> have been reported. The former is based on a Rh-catalysed asymmetric hydrogenation of a 3-(α,β-unsaturated ketone) substituted coumarin, whilst the key steps of the latter are an asymmetric Ullmann coupling and a selective demethylation. The stereochemistry of the fused dihydrocoumarin resulting from Li/NH_3 reduction of

3-methoxy-7,8,9,10-tetrahydro-6*H*-dibenzo[*b,d*]pyran-6-one has been established as *cis* <96AJC719>.

The cyclopenta[*c*]coumarin derivative (**12**), which occurs with sesquiterpenes in liverworts, has been synthesised in both racemic and enantiomeric forms by lactonisation and further manipulation of the cyclopentenylbenzene derivative (**11**) <96SYN863>. Liverworts have been used to illustrate the value of direct nmr analysis of $CDCl_3$ plant extracts <96CC2187>.

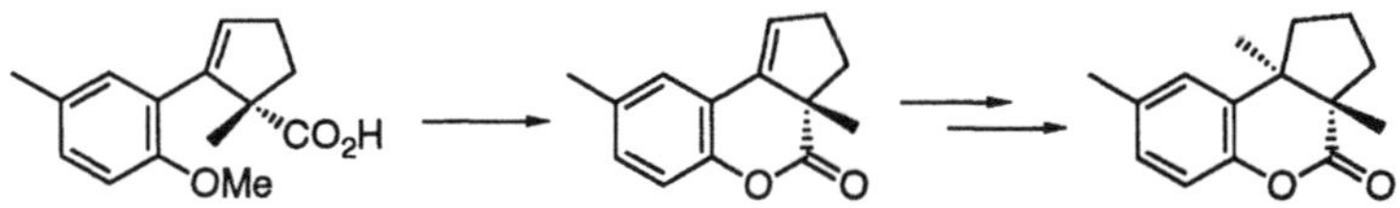

11 **12**

Irradiation of mixtures of 2-methylbenzoyl cyanide and benzoyl cyanide results in the exclusive formation of the mixed cycloadduct, from which elimination of HCN leads to isocoumarins <96CJC221>.

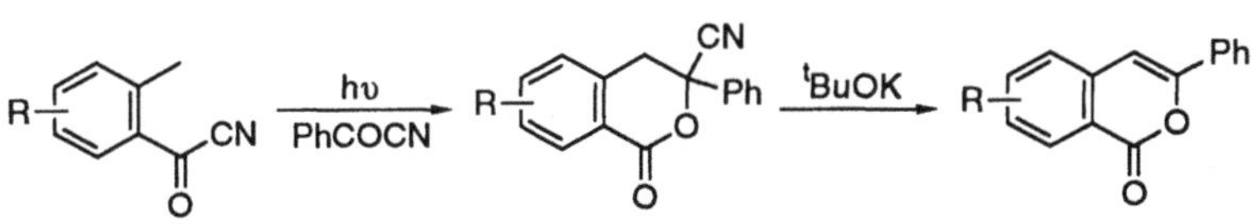

The 1-imino-1*H*-2-benzopyran (**13**) ring opens on treatment with nitrogen nucleophiles. Different modes of recyclisation are possible and (**13**) is therefore a source of a variety of *N*-heterocycles <96AJC485>.

Asymmetric dihydroxylation procedures feature in enantioselective syntheses of naturally occurring 3,4-dihydroisocoumarins <96JOC4190, 96JOC5371, 96TL8053>. An efficient route to dihydrocoumarins from *o*-tolylacetic acids involves oxidation of an isochroman using Al_2O_3-supported $KMnO_4$. Sequential Birch reduction of the dihydroisocoumarins and alkylation occurs with high diastereoselectivity. The resulting products provide fused lactones on treatment with trifluoromethanesulfonic acid <96JOC5631>. The Birch reduction alone of 8-phenyl derivatives also exhibits diastereoselectivity, with the stereocontrol at C-8 being influenced by the nature of the alcohol present in the reaction mixture <96TL6511>.

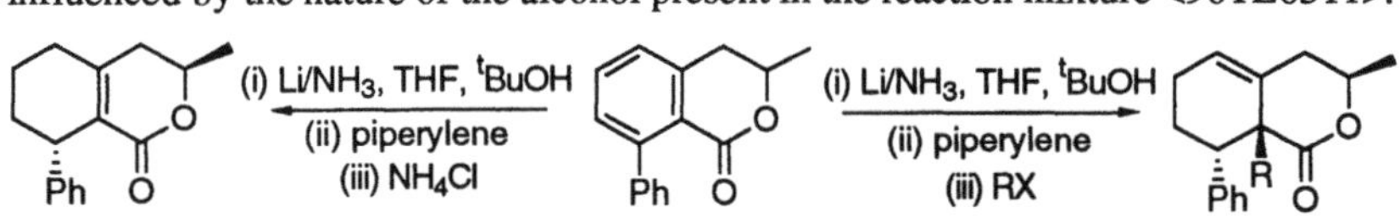

Both inter- and intra- molecular Diels-Alder reactions of 2-benzopyran-3-ones occur with high *endo*-selectivity and have been used to synthesise (-)-podophyllotoxin (**14**) and 4a-substituted *cis*-BC fused hexahydrophenenthrenes (**15**), respectively <96JCS(P1)151, 96JCS(P1)705>.

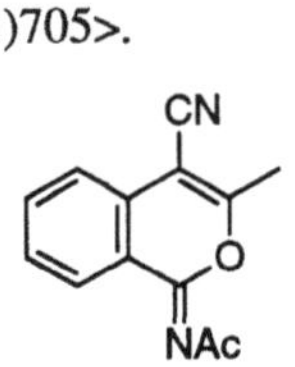

13

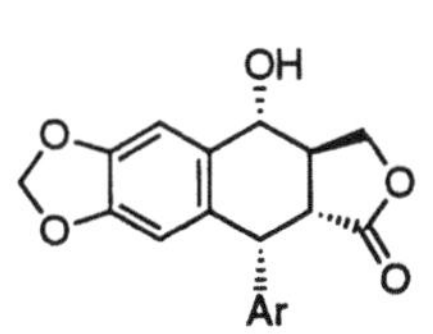

14

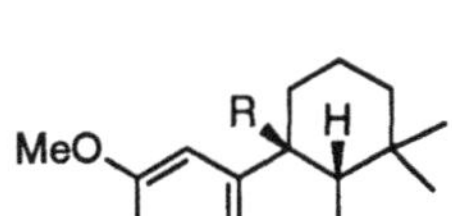

15

6.4.1.6 Chromones

Cyclisation of *o*-hydroxyphenyl ethynyl ketones under basic conditions is known to produce benzopyran-4-ones and benzofuranones by 6-*endo-dig* and 5-*exo-dig* processes, respectively. However, both cyclisations are reversible in aprotic media thereby generating anions, of which that derived from the pyranone is rapidly and irreversibly protonated and hence selective formation of the chromone results <96T9427>.

The mechanisms of the cyclisation of 2'-hydroxychalcone derivatives which can lead to flavanones, flavones and aurones have been reviewed <95MI1> and the formation of 3-hydroxy- chromanones and -flavanones from 1-(2-hydroxyphenyl)-2-propen-1-ones *via* the epoxide has been optimised <96JOC5375>.

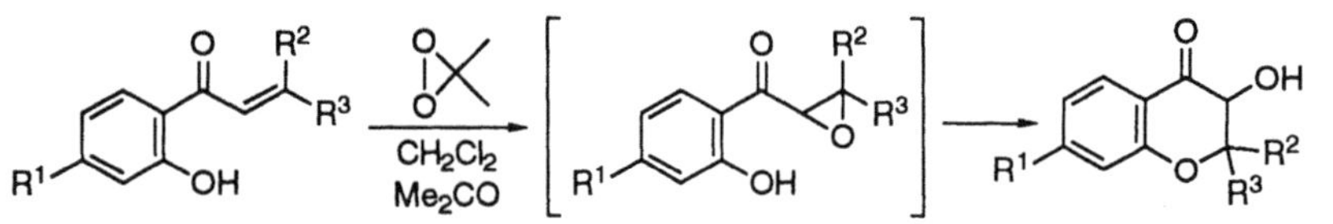

4-(2-Formylphenoxy)but-2-enoates, available from salicyclaldehydes and 4-bromo-crotonates, undergo an intramolecular Stetter reaction, which, in the presence of a chiral triazolium salt, affords chroman-4-ones with good enantiomeric excesses <96HCA1899>.

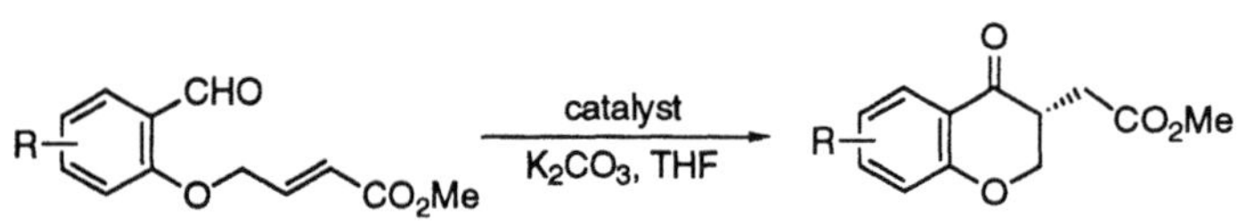

2-Monosubstituted 3-bromochroman-4-ones are converted into chromones on photolysis. For certain 2,2-disubstituted derivatives, dehydrohalogenation is accompanied by a 1,2-shift of a 2-substituent to afford a 2,3-disubstituted chromone <96H(43)339>.

6-Oxaestrone has been synthesised from 7-methoxychroman-4-one <96JCS(P1)841>.

Formation of the complex chroman-3-one (**16**) by a Dieckmann reaction and its conversion to the 4-diazo derivative are signifcant steps in the synthesis of the pentacyclic system (**17**) <96TL5243>.

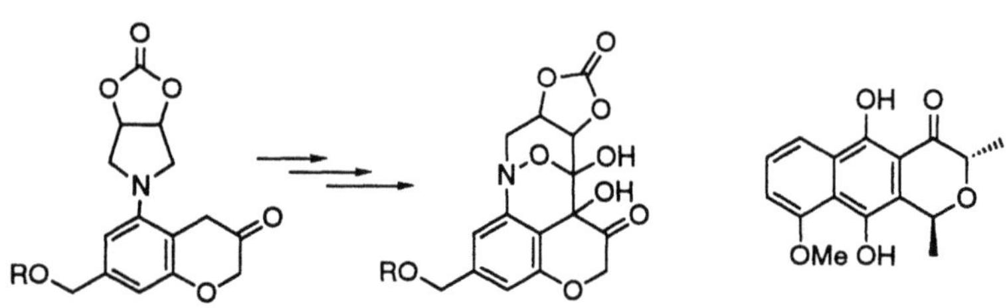

16 **17** **18**

Two syntheses of hongconin (**18**), a naturally occurring isochroman-4-one which exhibits antianginal activity, have been described. One utilises the annulation of phthalide unit to optically pure dihydropyran-3-ones <96JOC455>, whilst a similar Michael addition to the bicyclic pyranone levoglucosenone and subsequent enolate methylation are essentials of the second route <96JOC459>.

6.4.1.7 Flavonoids

The antioxidant properties of flavonoids are attributable to the ring whose radical has the lower reduction potential. Conjugation between the 2-aryl and the fused benzene rings is very inefficient <96JCS(P2)2497>.

The first enantioselective synthesis of *cis*- and *trans*- 3-hydroxyflavanones is based on the Lewis-acid-catalysed reaction of phenylmethanethiol with chalcone epoxides <96CC2747>. Further support for the intermediacy of epoxides in the Algar-Flynn-Oyamada flavone synthesis has been provided by the isolation of epoxides in the corresponding preparation of 3-hydroxy-2-phenylquinol-4-ones <96JCS(P2)269>.

(*E*)-3-Benzylideneflavanones are converted into 3-(α-hydroxybenzyl)flavones on treatment with NBS and dibenzoyl peroxide <96TL8001>.

6.4.1.8 Xanthones

Dilithiated diphenyl ether reacts with heteroaryl esters to give 9-substituted xanthen-9-ols, offering an alternative approach to pharmaceuticals containing a 9-xanthylidene unit <96TL5073>.

A further example of the utility of the cyclobuta[*b*]chroman system in synthesis (see 6.4.1.3) is provided by the conversion of the cyclobutenedione derivatives (**19**) into substituted xanthones. Compounds (**19**) are obtained in high yield from salicylaldehydes and squarate esters and their reaction with alkenyl, aryl and heteroaryl Li compounds is both facile and high yielding <96JA12473>.

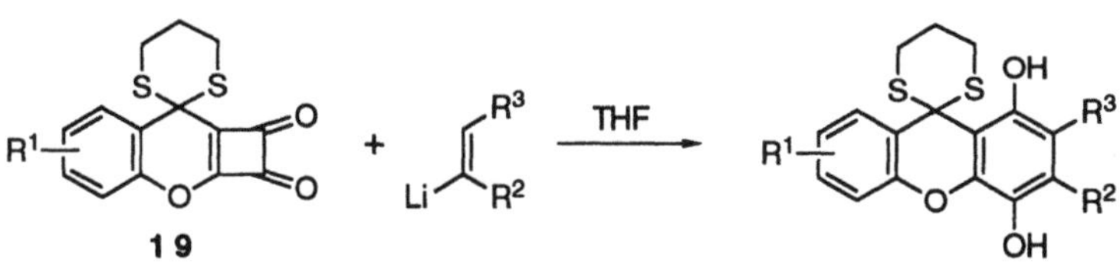

19

The xanthene framework is formed when salicylaldehyde reacts with alkenylcyclohexenes in the presence of a bentonite clay. Two intramolecular heterocyclisations are involved <96TL6181>.

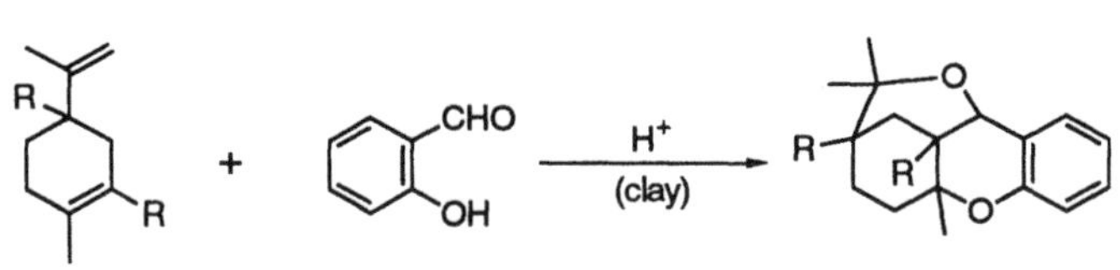

The xanthene unit of the dibenzopyranoazepine alkaloid, clavizepine, has been obtained from a dibenzoxepinediol by a pinacol rearrangement <96JOC5818>.

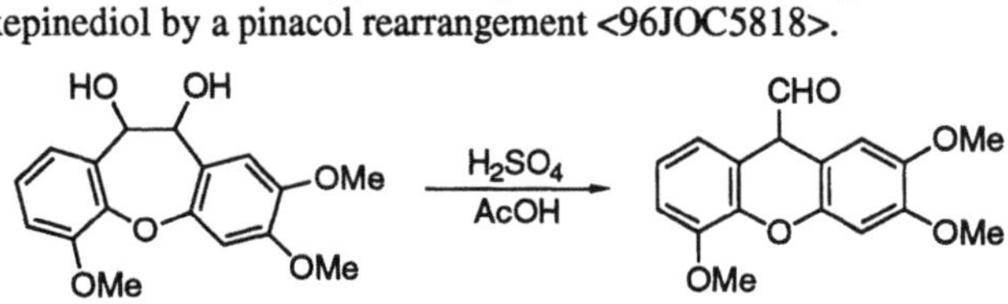

Benzo[*k,l*]xanthenes result from the chromium-mediated benzannulation of the carbene complex (**20**) <96CC895>, whilst the photooxidative cyclisation of (*E*)-2-styrylchromones (**21**) leads to benzo[*a*]xanthones <96HCM251>.

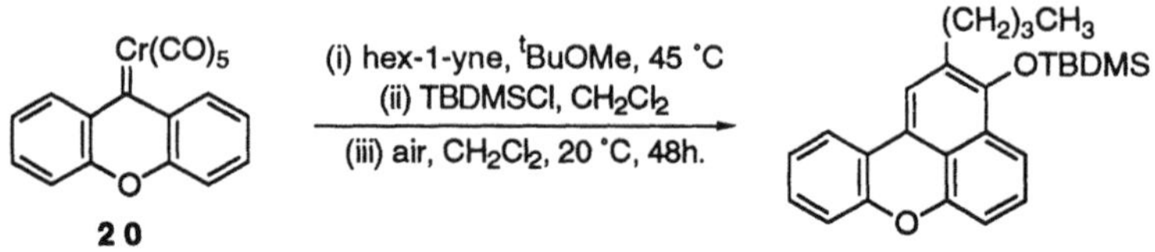

20

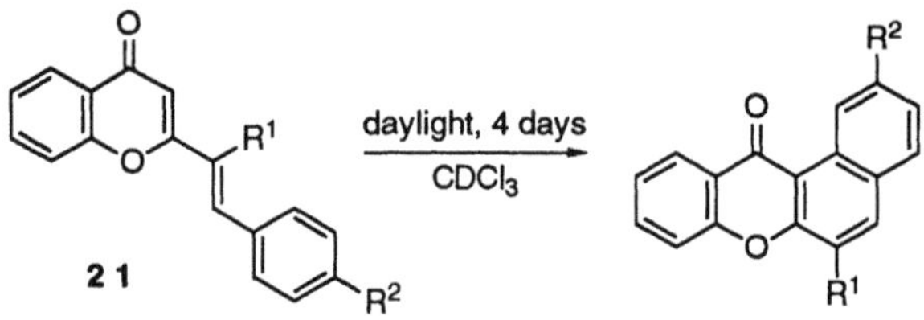

21

Dealkylation of 1-methoxyxanthone occurs on reaction with $BeCl_2$, but neither the 2- nor 3- isomers are affected <96T13623>.

Polymer supported xanthene derivatives have been used in the solid phase synthesis of 1-aminophosphinic acids, $RCH(NH_2)PH(O)OH$, <96TL1647> and of *C*-terminal peptide amides <96JOC6326>. Xanthene units also feature in crown ethers <96JCS(P2)2091>, calixarenes <96JOC5670> and in a flexible template for a β-sheet nucleator <96JOC7408>.

6.4.1.9 Pyrylium Salts

β-Cyclomanganated chalcones couple with alkynes to give pyranyl complexes (**22**) which on treatment with I_2 yield pyrylium salts <96JOM(507)103>.

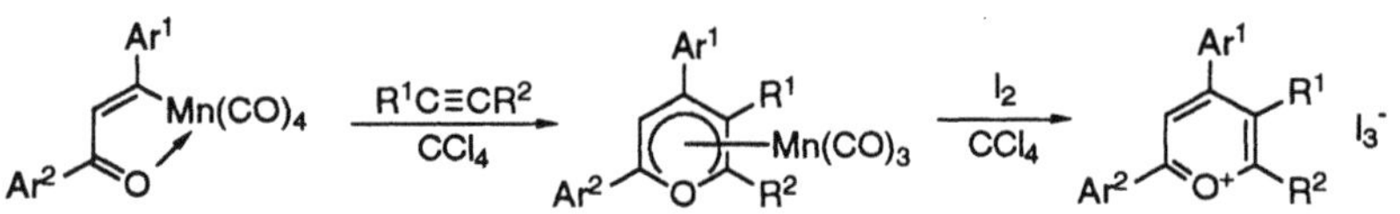

22

Quinolizinium and other fused pyridinium salts are formed when α-methylheterocycles react with 2,4,6-triphenylpyrylium, which thus behaves as a C3-synthon <96MC99>. Pyrylium salts also feature in a stereocontrolled route to conjugated dienynes which has led to a synthesis of Carduusyne A, a marine metabolic fatty acid <96TL1913> and in the formation of pyridinium containing crown ethers <96LA959>.

6.4.2 HETEROCYCLES CONTAINING ONE SULFUR ATOM

6.4.2.1 Thiopyrans and analogues

The reaction of 1-thiabuta-1,3-dienes with di-(-)-menthyl fumarate exhibits high *endo*-selectivity when carried out thermally or when catalysed by Lewis acids. In both cases, the preferred product is the 3,4-*cis*-dihydrothiopyran (**23**) which is a source of optically pure diols <96JCS(P1)1897>.

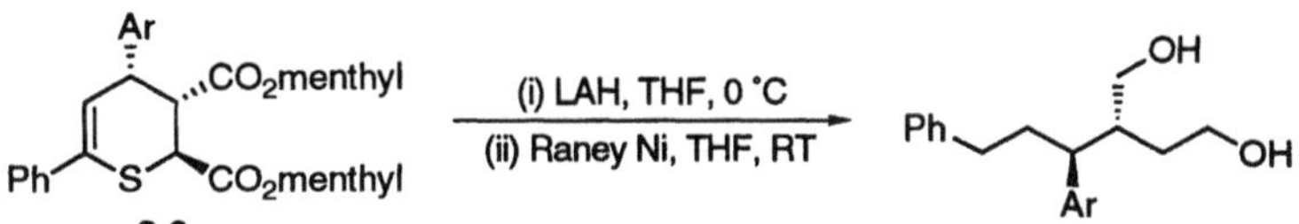

23

However, the homochiral 2-*N*-(*R*)-(-)-α-[(2-naphthyl)ethylamino]-4-phenyl-1-thiabuta-1,3-diene reacts with dienophiles when activated by acetyl chloride to give the *exo* product, exemplified by the cyclopenta[*b*]thiopyran derivative (**24**) <96TL123>.

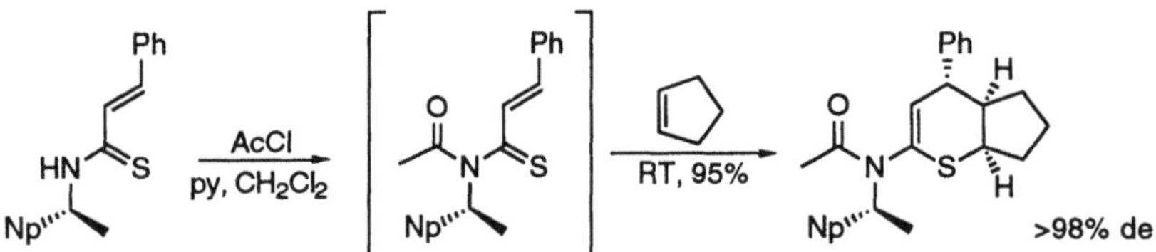

24

2*H*-Thiopyran-2-thiones result from the reaction of alkynic dienophiles with the disulfide (**25**) <96BCJ2091> and 2-*tert*.-butylfulvenes afford cyclopenta[*c*]thiopyrans, pseudo-azulenes, on reaction with the mesoionic compound (**26**), diphenyldithioliumolate, invoving a [4+6] cycloaddition <96CC1011>.

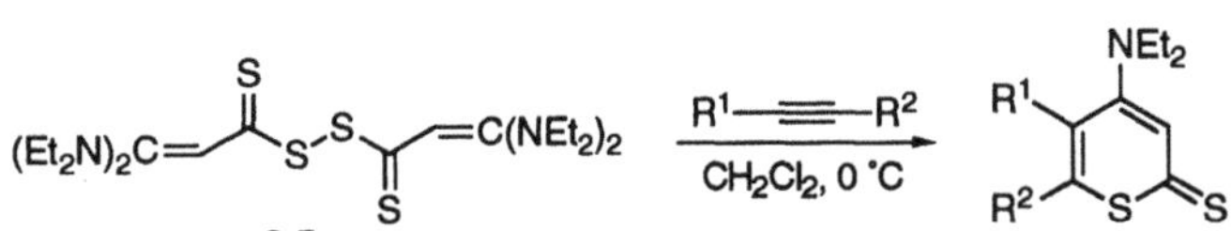

25

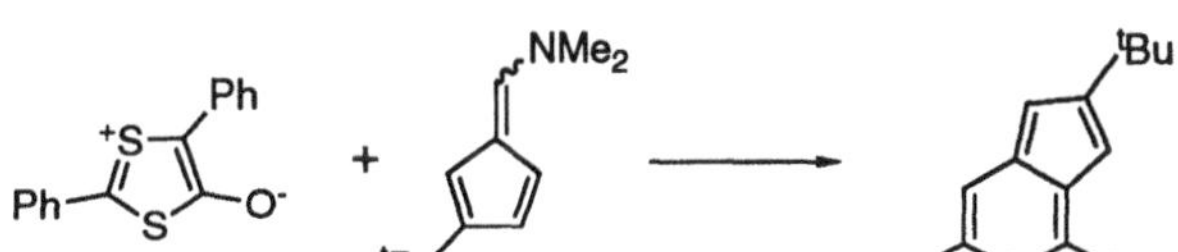

26

α-Thioxothioamides can behave as dithiabutadienes and thiabutadienes in their reaction with alkynic dienophiles leading to 1,4-dithiins, which spontaneously extrude sulfur to form thiophenes, and fused thiopyrans, respectively <96BSF903>. The photolytic conversion of 1,2-dithiins to thiophenes has been shown to proceed *via* a 2,6-dithiacyclo[3.1.0.]hexene <96JA4719>. 4-Dimethylamino-5,6-dihydro-2*H*-thiopyran-2-thiones have been converted into 3-aminothiophene derivatives <96M1027>.

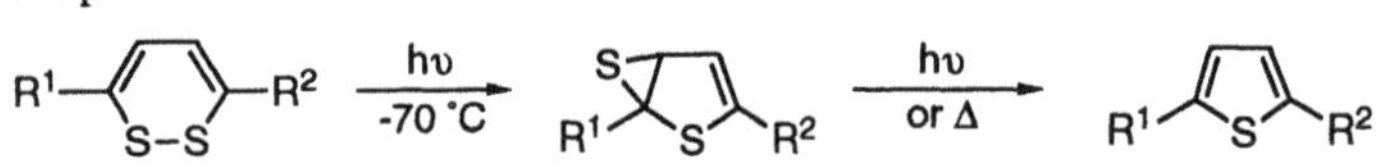

The first examples of intramolecular diene transmissive HDA reactions have been reported. Thus, the divinyl thioketones, derived from the ketones by treatment with Lawesson's reagent, spontaneously cyclise to the thiopyran derivative (**27**) and react further as shown <96CC811>.

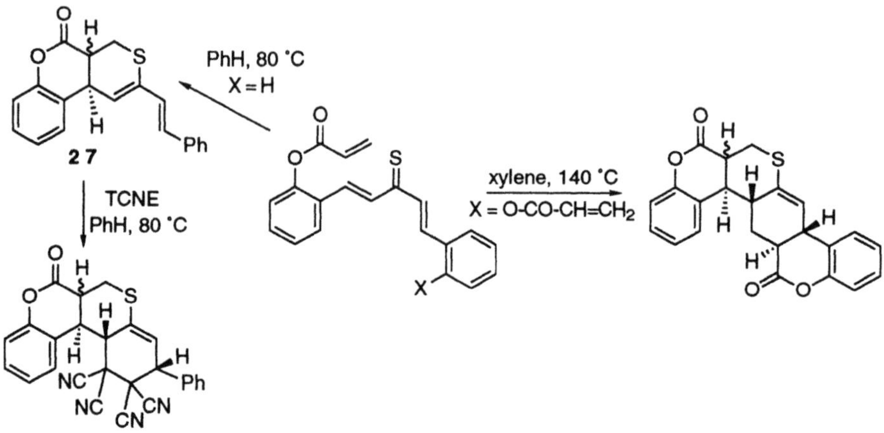

Similarly, the macrocyclic α,β-unsaturated thioketone (**28**) undergoes a stereoselective transannular HDA reaction < 96SL72>.

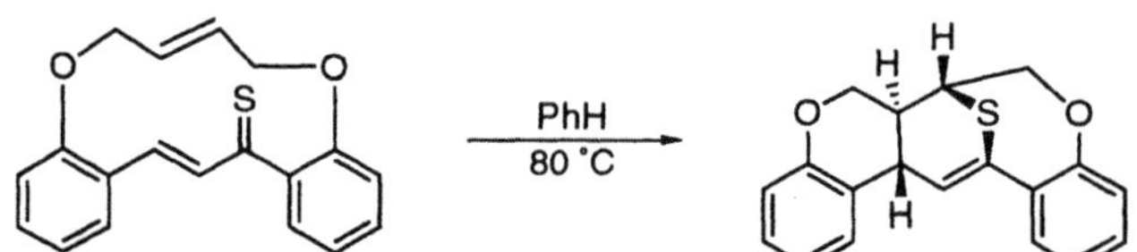

3,6-Dihydro-2*H*-thiopyrans, derived from dimethylbuta-1,3-dienes, $Na_2S_2O_3.5H_2O$ and various activated alkyl halides, ring contract on treatment with a strong base leading to vinyl cyclopropanes and cyclopentenes <96JOC4725>.

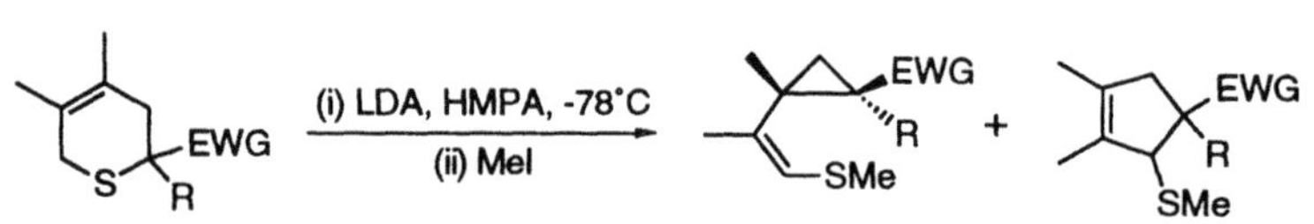

Intramolecular Diels-Alder reactions of 4-(triisopropysilyloxy)-2*H*-thiopyrans incorporating a dienophilic side chain in the ring are only successful when the alkene is activated by an electron withdrawing group. The length of the tether and its location strongly influence both the viability of the cycloaddition and its stereoselectivity <96CJC1418>.

4-Aryl-1,2-dithiolium salts are a source of either dithiatricyclodecadienes (**29**) or cyclopenta[*b*]thiopyrans (**30**) depending on whether the initial ring opening resulting from reaction with a metal cyclopentadienide is followed by an intramolecular cycloaddition or a condensation <96LA109>.

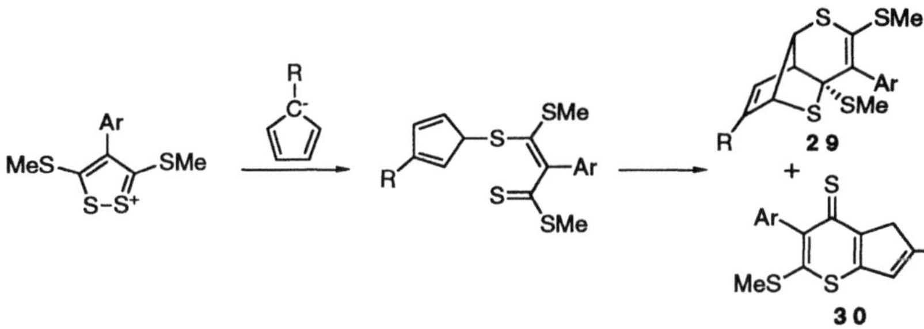

A dibenzo[*b,d*]thiopyran is formed when thiobenzophenones react with benzyne, confirming that the [4+2] cycloaddition is preferred to the [2+2] mode <96TL8883>.

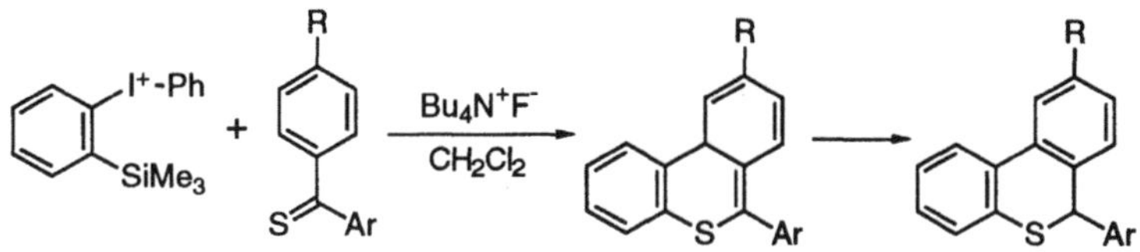

The photocycloaddition of alkenes to 2*H*,8*H*-thiopyrano[3,2-*g*]benzothiopyran-2,8-dione and the analogous thiopyranobenzopyran and pyranobenzopyran can be controlled by selection of the wavelength of the irradiating light. At wavelengths >395 nm the mono-adducts are formed in a *cis-trans* ratio of 4:1, whilst at shorter wavelengths bis-adducts are formed. It is therefore possible to form unsymmetrical bis-adducts <96LA291>.

Benzothiopyrylium salts undergo a polar [4+2⁺] cycloaddition to give 1,10a-dihydro-4*H*-4a-thioniaphenanthrene salts, which afford 2-substituted 2*H*-1-benzothiopyrans on treatment with nucleophiles <96JCS(P1)2227>.

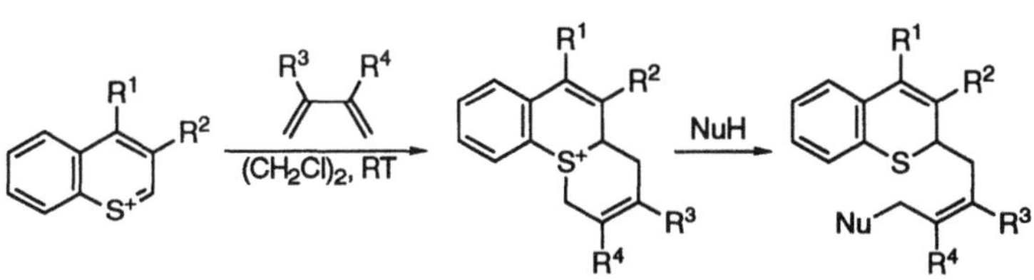

This behaviour is similar to that reported for the 2-isomers <94JCS(P1)3129>, but it is now shown that 2-benzothiopyrylium salts can behave as electron deficient dienes affording tricyclic molecules such as (**31**) and (**32**) by a [4⁺+2] cycloaddition <96CC2185>.

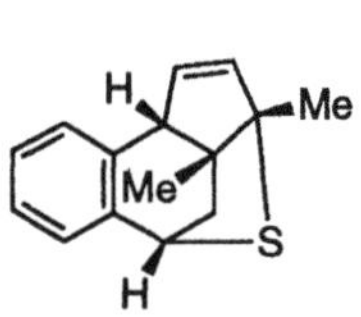

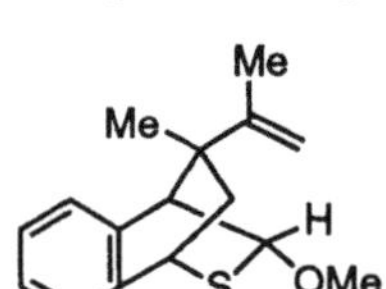

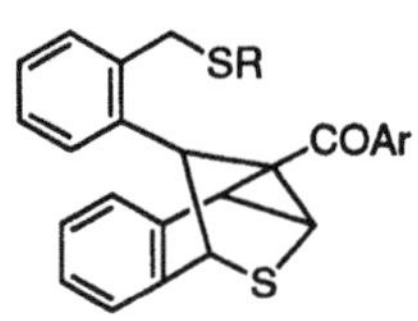

31 **32** **33**

In the presence of a base, 3-acyl derivatives of 2-benzothiopyrylium salts form a stabilised ylide which undergoes a cycloaddition reaction with unchanged starting material, affording the complex benzothiopyran (**33**) <96CC1659>.

The formation of the *cis*-annulated hexahydrothioxanthene (**34**) from thiophenol and hept-6-enals may also result from a $[4^{+}+2]$ cycloaddition, although the minor stereoisomers probably arise from a cationic non-concerted cyclisation <96SL396>.

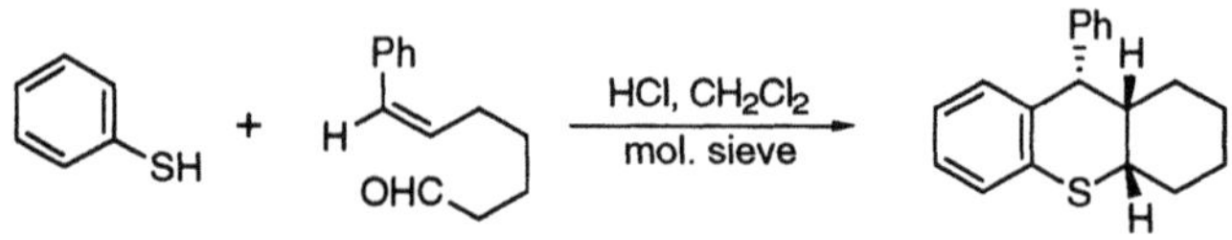

34

Trifluoromethyl derivatives of tetrahydrothiopyran-4-ols result from the Michael addition of hydrosulfide to α,β-unsaturated trifluoromethyl ketones <96JOC1986> and tetrahydrothiopyran-4-ones arise from the reaction of H_2S with 1,5-diaryl-2-chloropenta-1,4-dien-3-ones <96PS(108)93>.

The synthesis of substituted 3,4-dihydrothiopyrans by the reaction of 2-amino-4,5-dihydrothiophene-3-carbonitriles with ethyl diazoacetoacetate involves rearrangement of initially formed 1,4-oxathiocines <96LA725>.

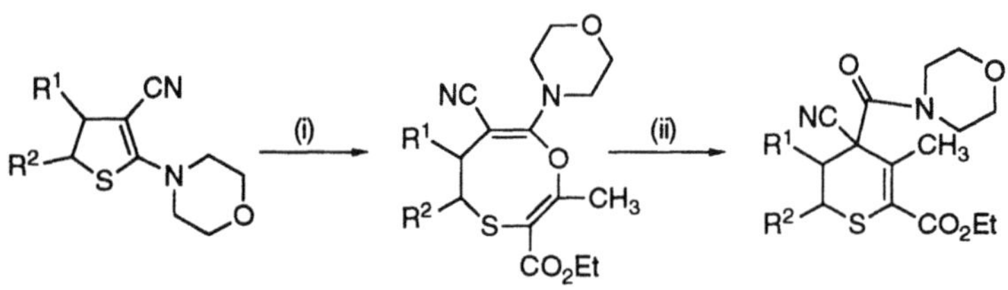

Reagents: (i) $CH_3COC(N_2)CO_2Et$, $Rh(OAc)_2$, C_6H_5F, 70 °C; (ii) 140 °C.

The electrochemical reduction of 4*H*-thiopyrans bearing four electron-withdrawing substituents leads to 5,6-dihydro-2*H*-thiopyrans. Four diastereoisomers are produced, their relative proportions depending on the electrolytic conditions. Their conformations have been established using the vinylic proton as an nmr probe and confirmed in some instances by X-ray analysis <96JCS(P2)2623>.

Radical promoted reactions feature in a synthesis of 3-substituted derivatives of 2,3-dihydro- and tetrahydro- thiopyran-4-ones from the 3-methylene compounds <96SL261> and in the formation of 2-methyltetrahydroselenopyran from a selenoalkyl (phenyltelluro)formate <96JOC5754>.

Cyclisation of the allylic alcohols derived from the condensation of 2-*tert*.butylthiobenzaldehyde and methylketones provides a new route to substituted 2*H*-1-benzothiopyrans <96TL5077>.

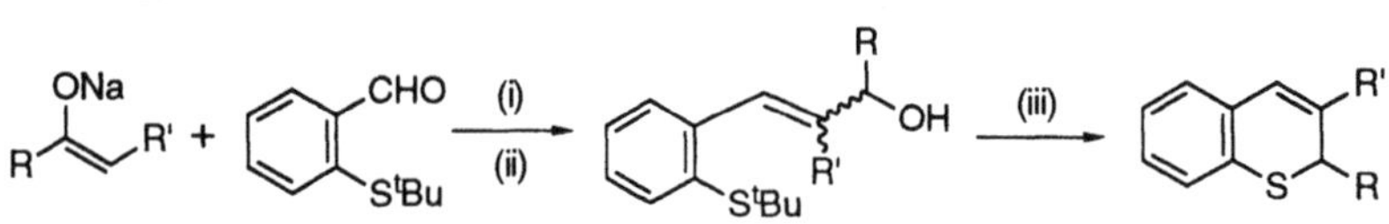

Reagents: (i) THF, -78 °C to RT; (ii) $LiAlH_4$, THF, 0 °C; (iii) TFA, CH_2Cl_2, 0 °C

Benzo[*c*]thiophene is a source of 3-substituted derivatives of 3,4-dihydro-1*H*-2-benzothiopyrans *via* ring opening to the dianion and subsequent reaction with electrophiles <96JOC1859>.

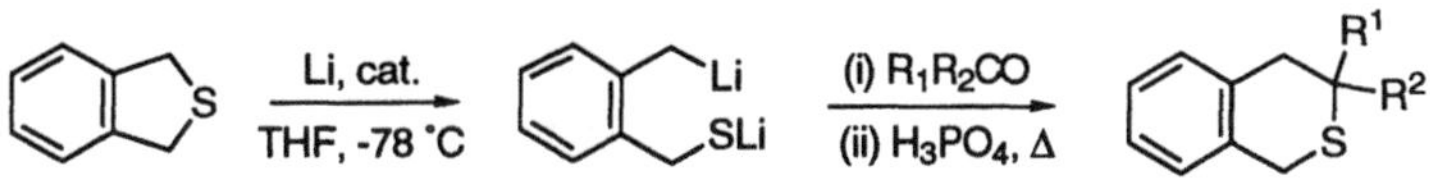

6.4.3 HETEROCYCLES CONTAINING TWO OR MORE OXYGEN ATOMS

6.4.3.1 Dioxins

In the presence of Mn(OAc)3, acylacetonitriles react with alkenes at room temperature to form 1,2-dioxan-3-ols. At higher temperatures dihydrofurans are produced. The initial step is the generation of acylcyanomethyl radicals <96TL4949>.

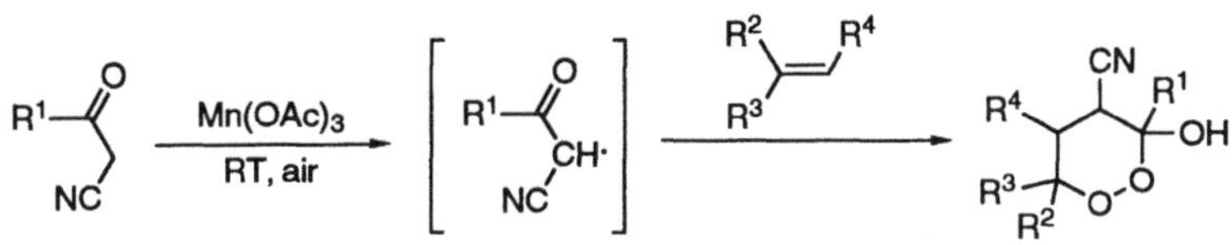

The endoperoxy hydroperoxide (**36**) results from the hydroperoxide (**35**) by sequential peroxy radical generation, 6-*exo trig* cyclisation and oxygen trapping <96SL349>.

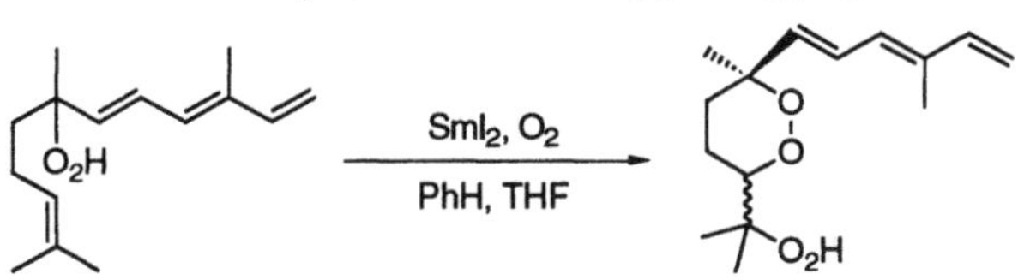

35 **36**

Addition of an organolithium and Grignard reagent across the peroxy bridge of endoperoxides gives *cis*-cyclohex-2-en-1,4-diols. Alkyl, cycloalkyl, alkenyl and aryl moieties can be transferred to oxygen <96TL6635>.

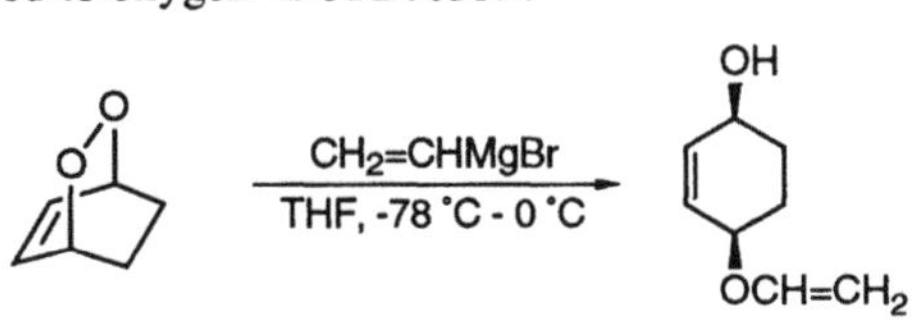

The photooxygenation of chiral 1,2-dihydronaphthalene-2-carboxylic acids leads to a mixture of the diendoperoxide (**37**) and hydroperoxide (**38**) arising from a double [4+2] cycloaddition and an ene reaction, respectively <96CC2585>.

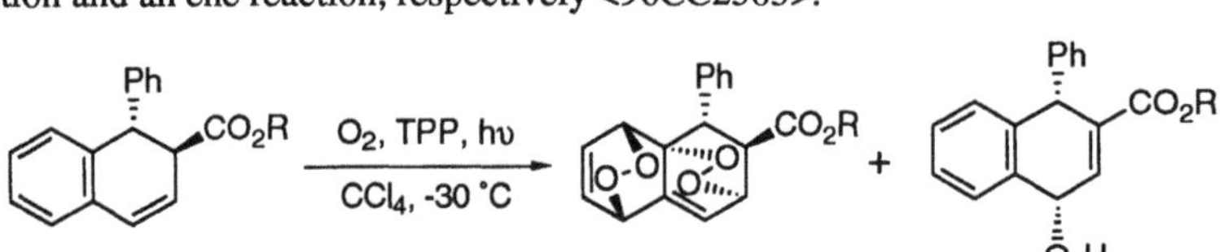

37 **38**

When aryl acrylates and phenyl salicylates react with aliphatic aldehydes in the presence of DABCO, the normal Baylis-Hillman product (**39**) often reacts further to give the acetal (**40**) <96TL1715, 96TL3755>.

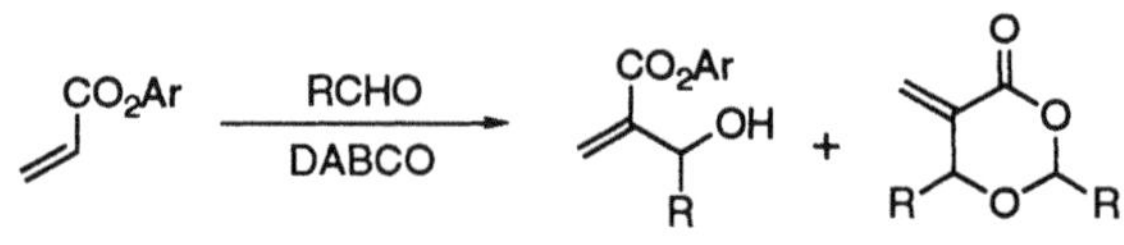

39 **40**

Diastereomerically pure 1,3-dioxanes are formed when optically pure 1-aryl-2,2-dimethylpropan-1,3-diols react with phenylglyoxals; only ketalisation is observed <96RTC407>.

1,3-Dioxan-4-ones gave the equatorial acetate stereoselectively on sequential reaction with DIBALH and acetic anhydride. The cationic species which result from treatment of the acetate with a Lewis acid reacts readily with nucleophiles <96JOC8317>. The yields and diastereoselectivity of such coupling processes are strongly dependent on the reaction conditions and the structure of the dioxane <96SL536>.

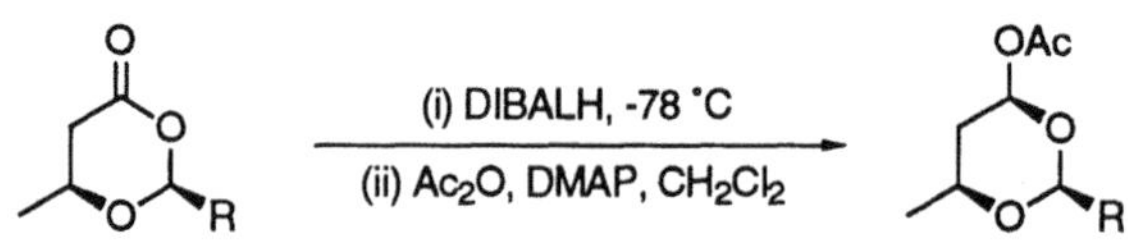

The boron enolates derived from (*S*)-4-silylated 2,2-dimethyl-1,3-dioxan-5-one undergo *anti* diastereoselective aldol reactions which provide access to protected oxopolyols of high stereochemical integrity <96SYN1095>.

The stereochemistry of the vinyl ether is retained during its reaction with the α-peroxy lactone (**41**) which leads to a 1,4-dioxan-2-one <96JOC8432>.

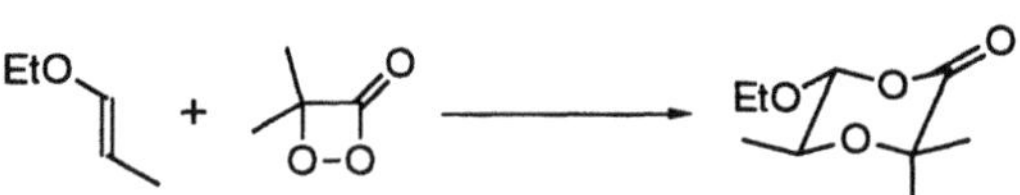

41

Only *cis*-disubstituted and trisubstituted alkenes yield 1,4-dioxan-2-ones by way of a cycloaddition reaction when oxidised by dimethyl α-peroxy lactone. An open 1,6-dipolar intermediate is postulated, involving stereoelectronic control <96JA4778>.

trans-2,3-Disubstituted benzo-1,4-dioxanes result when the ethers derived from 1,2-dihydroxybenzenes and the epoxide (**42**) are converted to the imidate in a Vilsmeier reaction which subsequently undergoes an intramolecular nucleophilic displacement of DMF. This approach offers an alternative to the Mitsonobu reaction <96JCS(P1)2249>.

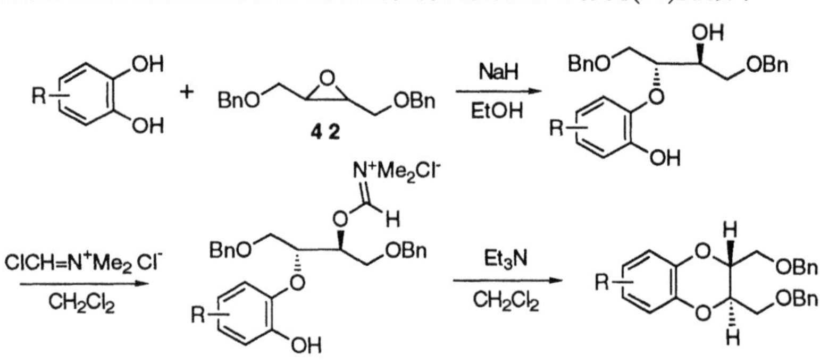

The Pd-catalysed coupling of aryl and heteroaryl halides with 2-hydroxyphenyl 2-propynyl ethers leads to (*Z*)-2-arylidene-1,4-benzodioxanes <96CC1067>.

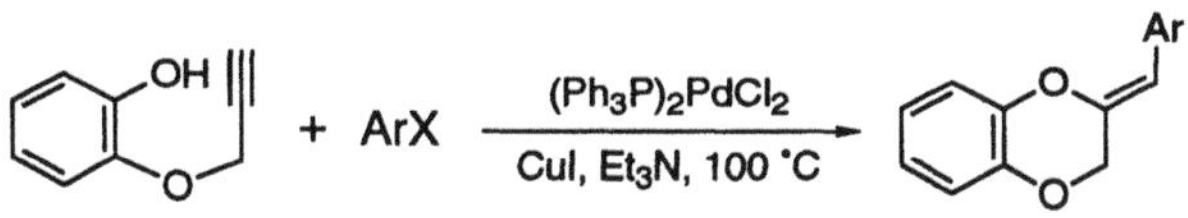

The reaction between *o*-quinones and electron rich dienes leads to benzodioxanes. It is proposed that an initial HDA followed by a [3,3] sigmatropic rearrangement account for the stereochemistry of the products <96JCS(P1)443>.

Cyclopenta[*b*]dioxanes (**44**) are accessible from the reaction of the dioxenylmolybdenum carbene complex (**43**) with enynes <96JOC159>, whilst an intramolecular and stereoselective cyclisation of (η^5-dienyl)tricarbonyliron(1+) cations affords chiral *trans*-2,3-disubstituted 1,4-dioxanes <96JOC1914>. 2,3-Dimethylidene-2,3-dihydro-1,4-benzodioxin is a precursor of the 3,8-dioxa-1*H*-cyclopropa[*b*]anthracene, which readily dimerises to dihydrotetraoxaheptacene (**45**) and the analogous heptaphene <96AJC533>.

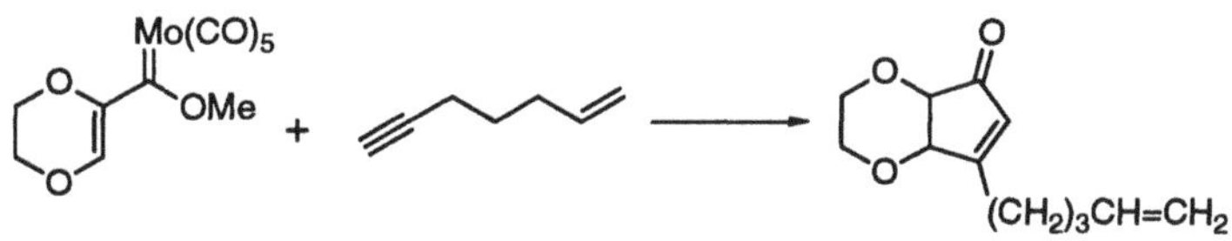

43 **44**

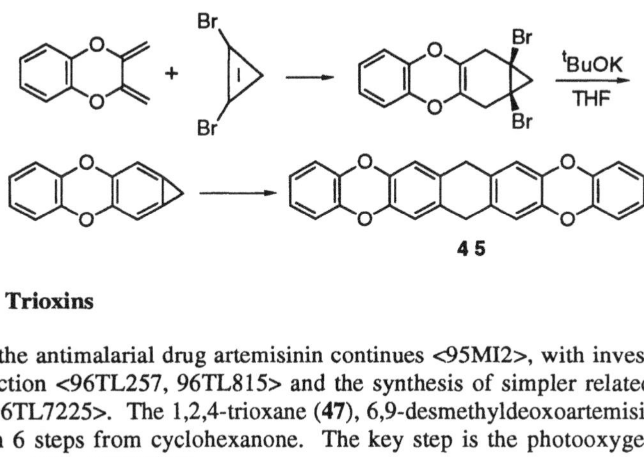

45

6.4.3.2 Trioxins

Interest in the antimalarial drug artemisinin continues <95MI2>, with investigations into its mode of action <96TL257, 96TL815> and the synthesis of simpler related compounds <96JA3537, 96TL7225>. The 1,2,4-trioxane (**47**), 6,9-desmethyldeoxoartemisinin, has been synthesised in 6 steps from cyclohexanone. The key step is the photooxygenation of the cyclic enol ether (**46**) <96BKS581>.

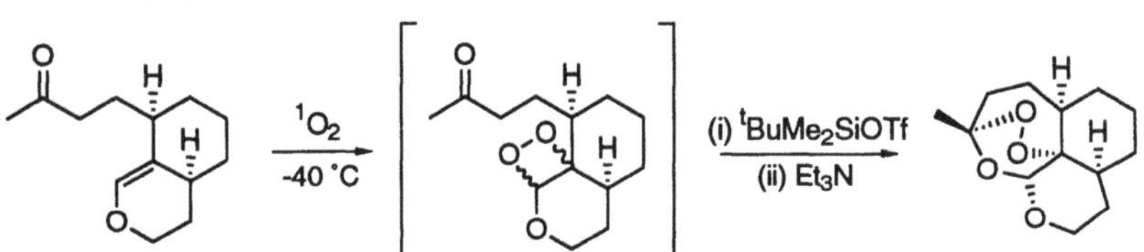

46 **47**

Both artemisinin and artemether undergo deoxygenation on treatment with zinc in AcOH <96HCA1475>, but Fe(II) salts rupture the peroxy linkage and lead to rearranged products

<96CC2213, 96HCA1475>. On the other hand, radicals derived from nBu_3SnH - AIBN leave the trioxane ring of 9-bromo-10-alkynoxydihydroartemisinin in tact whilst allowing the construction of a furan ring across the 9,10-bond <96JCS(P1)1015>.

6.4.4 HETEROCYCLES CONTAINING TWO OR MORE SULFUR ATOMS

6.4.4.1 Dithianes

On reaction with triflic anhydride, the mono oxides of 2,2'-bis(alkylthio)biphenyl are converted into a dithiadication (**48**) which spontaneously monodealkylates to give a thiasulfonium salt (**49**) <96TL667>.

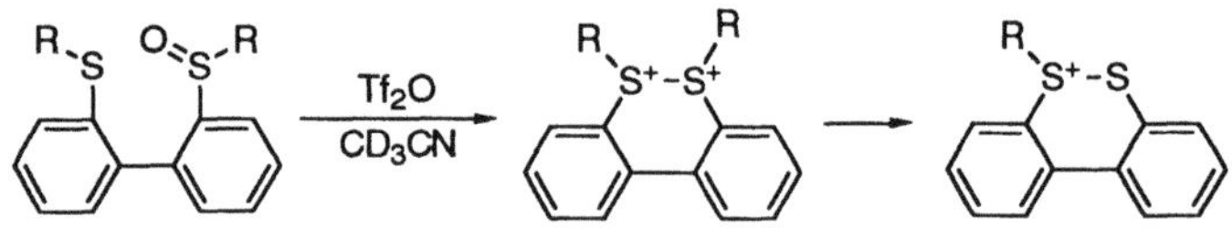

48 **49**

2-Pyridyl-1,3-dithianes (**50**) result from the reaction of picolyl lithium reagents with 1,2-dithiolanes in the presence of HMPT. An initial ring opening is followed by reaction at the carbanion site with a second mole of dithiolane <96PS(112)101>.

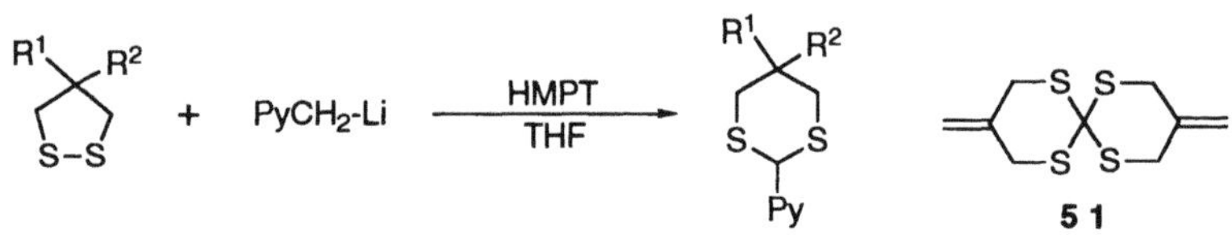

50 **51**

2-Methylenepropan-1,3-dithiol is converted into the tetrathiaspiro[5.5]undecane (**51**) on reaction with dichlorodiphenoxymethane <96AJC1261>.

A major aspect of 1,3-dithianes is their value in synthesis and several new applications are highlighted this year.

The Horner - Emmons reagent (**52**) is effective in the one carbon homologation of ketones possessing acidic α-hydrogen atoms <96SL875> and electron-deficient alkenes add to 2-phenylseleno-1,3-dithiane in a photo-initiated heteroatom stabilised radical atom transfer process, giving products of considerable synthetic potential <96TL2743>.

52

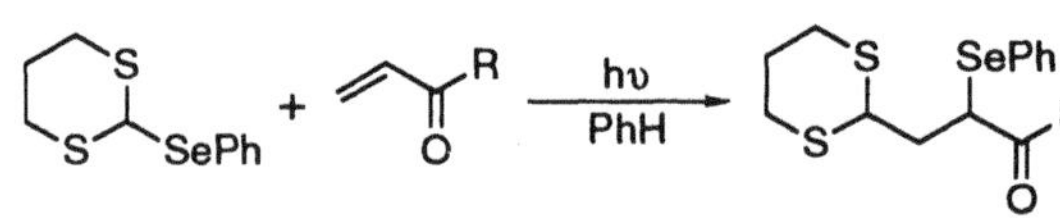

The radicals derived by hydrogen abstraction from 1,3-dithianes and 1,3-oxathianes undergo intramolecular addition to α,β-unsaturated esters and hence facilitate the synthesis of cycloalkanones <96T9713>.

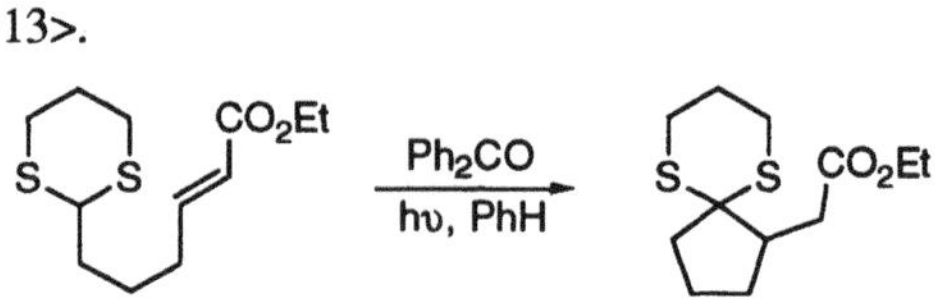

The 1,3-dithian-2-ylidene substituted carbene (**54**), accessible from the tosylhydrazone (**53**) by a Bamford - Stevens reaction, not only participates in cycloaddition reactions but is also a source of 4,8-dithiaspiro[2.5]oct-1-ene <96JCS(P1)2773>.

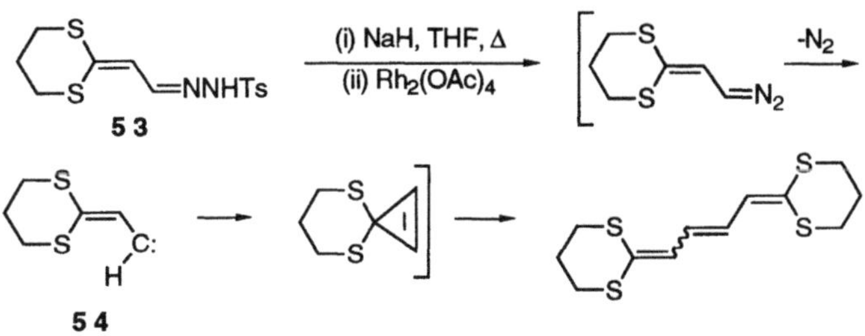

The metallation of 1,3-diselenanes is complex. When potassium diisopropylamide is used as base, deprotonation and alkylation affords the 2-equatorially substituted derivative <96TL2667>. However, with *tert.*butyllithium, Se-Li exchange is observed in preference to H-Li exchange in the reaction with 2-*ax*-methylseleno derivatives <96TL8015>. The reaction with nBuLi either forms the anion or cleaves a C-Se bond depending on the substituents present at the 2-, 4- and 6- positions <96TL8011>.

The regeneration of carbonyl compounds from 1,3-dithianes can be achieved using potassium hydrogen persulfate, Oxone™, supported on wet alumina <96SL767> and by periodic acid under non-aqueous conditions <96TL4331>. The deprotection of benzyl substituted 1,3-dithianes can be achieved using the one electron oxidant $[Fe(phen)_3](PF_6)_3$ <96SL315>.

The value of dithiane mono- and di- oxides as chiral acyl anion equivalents ensures a continued interest in their chemistry. Two groups have reported the enantioselective preparation of 2-substituted 1,3-dithiane 1-oxides using Sharpless methodology <96JCS(P1)1879, 96T2125>, whilst 1,3-dithiane 1-oxide has been obtained enantioselectively by whole cell bacterial oxidation <96TL6117>. The value of bacterial cyclohexanone monooxygenases in the enantiomeric oxidation of sulfides to sulfoxides has been reviewed <96CC2303>.

α-Hydroxyketones can be prepared enantioselectively using dithiane oxide methodology <96TL8289>.

Fused 1,4-dithiins are formed when the 1,8-diketones derived from dithiols and α-bromoketones are treated with Lawesson's reagent (LR) <96TL2821>.

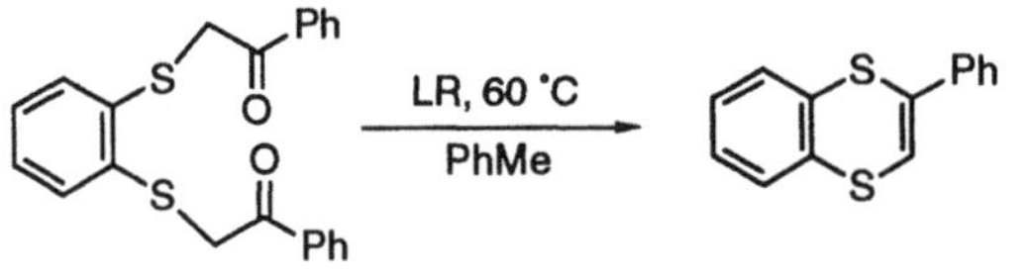

Both *o*- and *p*- substituted bis(methylthio)arylmethylium salts, derived by the *S*-methylation of substituted dithiobenzoates, dimerise following deprotonation to give 1,4-dithianes. Only with the former isomer is the quinone methide generated leading to a cyclooctene derivative <96LA1159>.

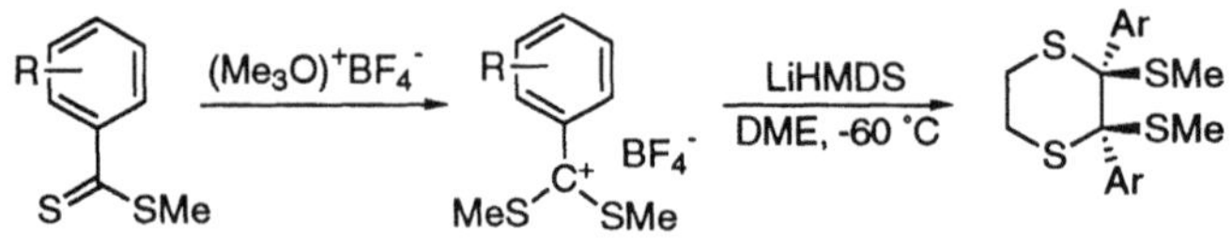

2-Chloro-1,4-benzodithiin 1,4-tetraoxide (**55**) can be used to introduce a benzene ring into a cyclic system, hence functioning as a benzyne equivalent. The sequence below is illustrative <96T14247>.

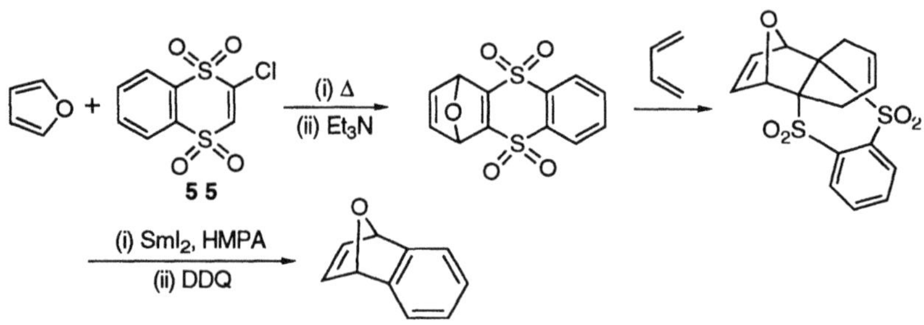

The introduction of substituents into the 1- and 2- positions of thianthrene can be achieved by deprotonation at C-1 and subsequent reaction with electrophiles and by formation of 2-lithiothianthrene from the corresponding bromo compound <96JCS(P1)2391>. Selective oxidation of both thianthrene and its 5-oxide have been described <96CEJ255, 96JCS(P1)2693>.

The 1,4,2-diselenazine (**56**), formed when a 1,3-diselenolium salt reacts with ammonia and iodine, readily fragments to a 1,2-diselenete or its valence isomer, a 1,2-diselenone. This species affords a 1,4-diselenin either by dimerisation or on trapping with DMAD <96CC2375>.

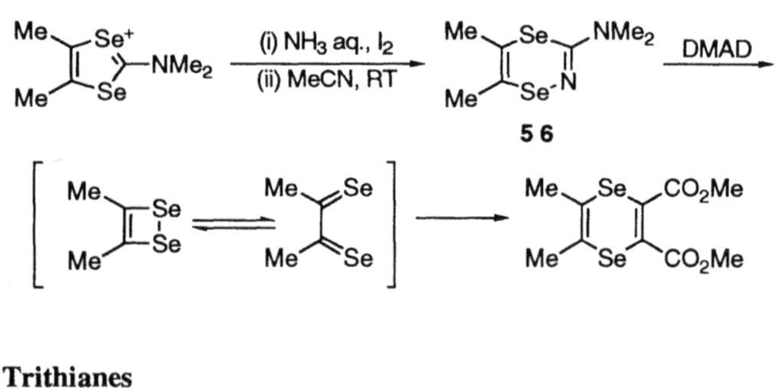

6.4.4.2 Trithianes

Reaction of the anion derived from the tosyl imide of 1,3,5-trithiane with alkyl iodides gives a mixture of the mono- and di- alkylated products, in which *anti* stereochemistry predominates. The analogous 1,3-dithiane derivative is only monoalkylated <96JCS(P1)313>.

The anion from 1,3,5-trithiane hexaoxide has been silylated and thence converted into a sulfoxonium ylide <96CB161>.

6.4.5 HETEROCYCLES CONTAINING BOTH OXYGEN AND SULFUR IN THE SAME RING

6.4.5.1 Oxathiines

An intramolecular cycloaddition reaction of the ethene sulfonate (**57**) occurs at high pressure and leads predominantly to the *cis* fused sultone (**58**). The minor product is the *trans* fused diastereoisomer <96JCS(P1)2297>.

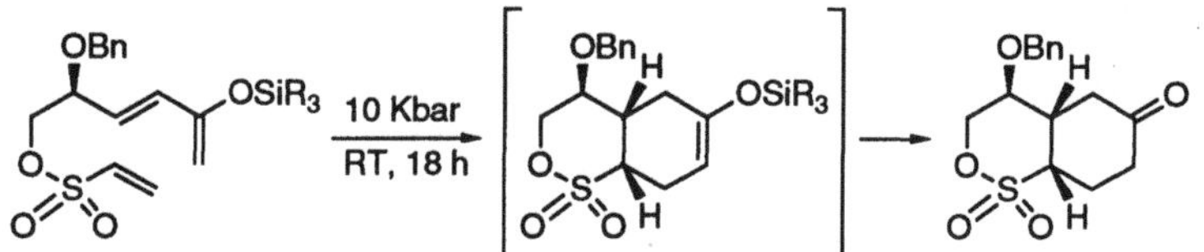

57 **58**

The desulfurisation of sultones results in the formation of an exocyclic alkene and can be achieved by an alkylation - elimination procedure <96TL3841>.

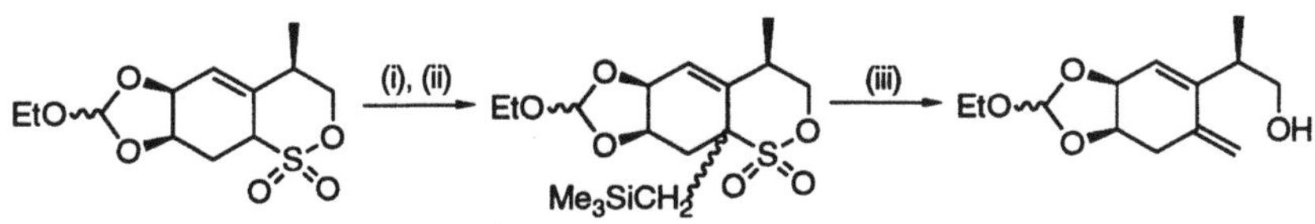

Reagents: (i) MeLi, THF, -78 °C; (ii) Me_3SiCH_2I, RT; (iii) $nBu_4N^+F^-$, THF, Δ.

4*H*-3,1-Benzoxathiin-2-ones (**59**), readily accessible from 2-mercaptobenzyl alcohols by treatment with trichloromethyl chloroformate, are a convenient source of 2*H*-benzo[*b*]thietes and hence are useful precursors of *o*-thiobenzoquinone methides. Thus extrusion of CO_2 occurs in boiling xylene and *in situ* reaction with dienophiles leads to benzothiopyrans <96SYN327>.

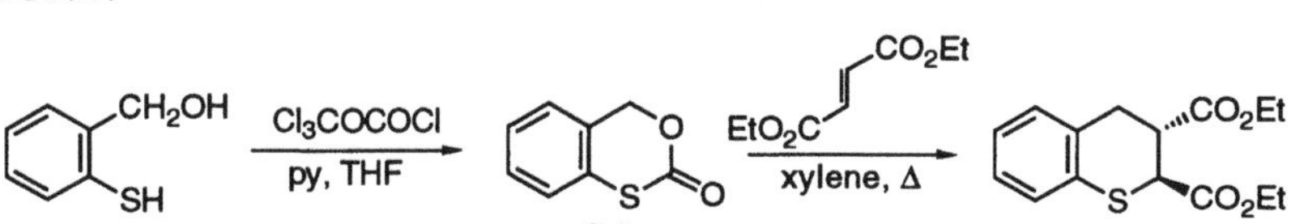

59

Three novel azaphenoxathiines have been obtained by reaction of 3-mercaptopyridin-2(1*H*)-one with activated chlorobenzenes and chloropyridines <96JOC662>.

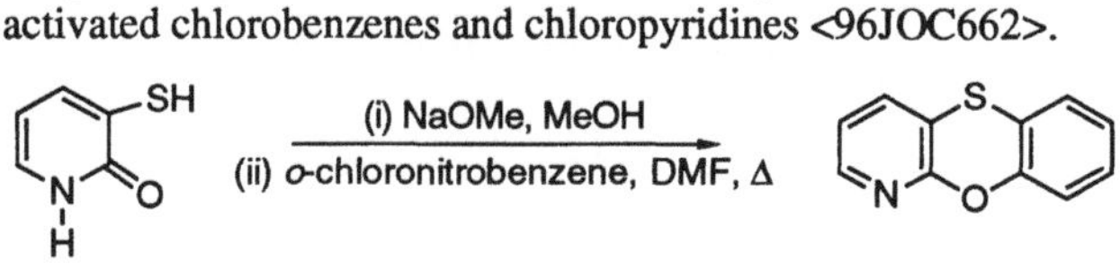

1,4-Oxathiin-3-carboxamides undergo a facile acid catalysed rearrangement to 1,4-thiazin-3-ones <96JOC3894>.

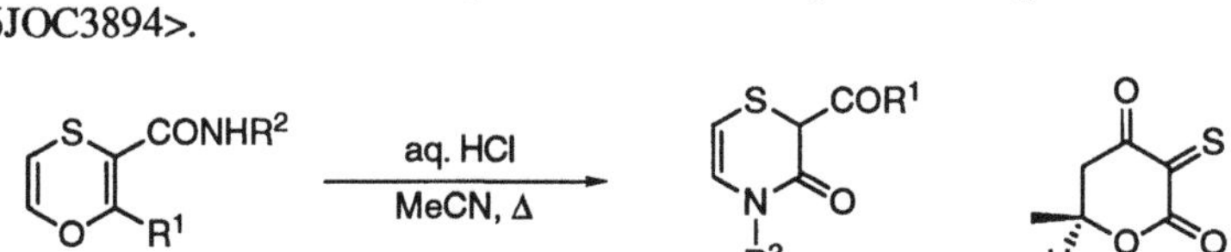

60

Various 1,4-oxathiins have been converted to the sulfoxides in a highly stereoselective reaction. The sulfoxides undergo a thermal retro-Diels-Alder reaction to α,α'-dioxosulfines which can behave both as electron deficient dienes and as dienophiles, affording different oxathiins and dihydrothiopyran 1-oxides, respectively <96T12233>. In like manner, *o*-thioquinone *S*-oxides have been generated and they too exhibit diene and dienophile behaviour <96T12247>.

1,4-Oxathiins derived from the cyclic diacylthione (**60**) and glycals by a cycloaddition reaction yield glycosides on desulfurization, offering an alternative approach to glycosyl transfer <96AG(E)777>.

REFERENCES

94JCS(P1)3129 H. Shimizu, S. Miyazaki, T. Kataoka, M. Hori, O. Muraoka, *J. Chem. Soc. Perkin Trans. 1*, **1994**, 3129.

95MI1 L. Main, '*Advances in Detailed Reaction Mechanisms,*' Ed. J. M. Coxon, Jai Press Inc., Greenwich, Connecticut, **1995**, vol. 4, p. 129.

95MI2 P. L. Olliaro, P. I. Trigg, *Bulletin of the World Health Organisation*, **1995**, *73*, 565.

95TL5847 H. Wang, J. B. Gloer, J. A. Scott, D. Malloch, *Tetrahedron Lett.*, **1995**, *36*, 5847.

96ACS132 I. R. Aukrust, L. Skattebøl, *Acta Chem. Scand.*, **1996**, *50*, 132.

96AG(E)589 K. C. Nicolaou, *Angew. Chem. Int. Ed. Engl.*, **1996**, *35*, 589.

96AG(E)777 G. Capozzi, A. Dios, R. W. Franck, A. Geer, C. Marzabadi, S. Menichetti, C. Nativi, M. Tamarez, *Angew. Chem. Int. Ed. Engl.*, **1996**, *35*, 777.

96AG(E)889 K. C. Nicolaou, M. Sato, N. D. Miller, J. L. Gunzner, J. Renaud, E. Untersteller, *Angew. Chem. Int. Ed. Engl.*, **1996**, *35*, 889.

96AG(E)1692 M. A. Ciufolini, F. Roschangar, *Angew. Chem. Int. Ed. Engl.*, **1996**, *35*, 1692.

96AJC1 R. M. Carman, A. C. Rayner, *Aust. J. Chem.*, **1996**, *49*, 1.

96AJC485 L. W. Deady, P. M. Loria, N. H. Quazi, *Aust. J. Chem.*, **1996**, *49*, 485.

96AJC533 M. J. Cooney and B. Halton, *Aust. J. Chem.*, **1996**, *49*, 533.

96AJC719 D. J. Collins, G. D. Fallon, A. Staffa, H. Tope, *Aust. J. Chem.*, **1996**, *49*, 719.

96AJC1261 M. K. Bromley, S. J. Gason, A. G. Jhingran, M. G. Looney, D. H. Solomon, *Aust. J. Chem.*, **1996**, *49*, 1261.

96BCJ2091 K. Akimoto, K. Masaki, J. Nakayama, *Bull. Chem. Soc. Jpn.*, **1996**, *69*, 2091.

96BKS7 J. Choo, S-N. Lee, K-H. Lee, *Bull. Korean Chem. Soc.*, **1996**, *17*, 7.

96BKS581 C. H. Oh, J. H. Kang, G. H. Posner, *Bull. Korean Chem. Soc.*, **1996**, *17*, 581.

96BSF229 S. Z. A. Sowellim, F. M. A. A. El-Taweel, A. G. A. Elagamey, *Bull. Soc. Chim. Fr.*, **1996**, *133*, 229.

96BSF903 E. Marchand, G. Morel, *Bull. Soc. Chim. Fr.*, **1996**, *133*, 903.

96CB161 W. Sundermeyer, A. Walch, *Chem. Ber.*, **1996**, *129*, 161.

9CC811 T. Saito, H. Kimura, K. Sakamaki, T. Karakasa, S. Moriyama, *J. Chem. Soc. Chem. Commun.*, **1996**, 811.

96CC841 I. Kadota, J-Y. Park, Y. Yamamoto, *J. Chem. Soc. Chem. Commun.*, **1996**, 841.

96CC895 K. H. Dötz, J. Pfeiffer, *J. Chem. Soc. Chem. Commun.*, **1996**, 895.

96CC909 S. Kim, J-Y. Yoon, C. M. Cho, *J. Chem. Soc. Chem. Commun.*, **1996**, 909.

96CC937 B-C. Hong, S-S. Sun, *J. Chem. Soc. Chem. Commun.*, **1996**, 937.

96CC1011 H. Kato, T. Kobayashi, M. Ciobanu, H. Iga, A. Akutsu, A. Kakehi, *J. Chem. Soc. Chem. Commun.*, **1996**, 1011.

96CC1067 C. Chowdhury, N. G. Kundu, *J. Chem. Soc. Chem. Commun.*, **1996**, 1067.

96CC1659 H. Shimizu, T. Yonezawa, T. Watanabe, K. Kobayashi, *J. Chem. Soc. Chem. Commun.*, **1996**, 1659.

96CC1763 K. Chiba, T. Arakawa, M. Tada, *J. Chem. Soc. Chem. Commun.*, **1996**, 1763.

96CC1965 B. C. Ranu, S. Bhar, A. Patra, N. P. Nayak, M. Mukherjee, *J. Chem. Soc. Chem. Commun.*, **1996**, 1965.

96CC2185 H. Shimizu, N. Araki, O. Muraoka, G. Tanabe, *J. Chem. Soc. Chem. Commun.*, **1996**, 2185.

96CC2187 D. S. Rycroft, *J. Chem. Soc. Chem. Commun.*, **1996**, 2187.

96CC2213 W-M. Wu, Z-J. Yao, Y-L. Wu, K. Jiang, Y-F. Wang, H-B. Chen, F. Shan, Y. Li, *J. Chem. Soc. Chem. Commun.*, **1996**, 2213.

96CC2303 S. Colonna, N. Gaggero, P. Pasta, G. Ottolina, *J. Chem. Soc. Chem. Commun.*, **1996**, 2303.

96CC2369 T. K. M. Shing, X. Y. Zhu, T. C. W. Mak, *J. Chem. Soc. Chem. Commun.*, **1996**, 2369.

96CC2373 A. Graven, M. Johannsen, K. A. Jørgensen, *J. Chem. Soc. Chem. Commun.*, **1996**, 2373.

96CC2375 S. Yoshida, M. R. Bryce, A. Chesney, *J. Chem. Soc. Chem. Commun.*, **1996**, 2375.

96CC2585 T. Linker, F. Rebien, G. Tóth, *J. Chem. Soc. Chem. Commun.*, **1996**, 2585.
96CC2689 M. Chmielewski, Z. Kaluza, B. Furman, *J. Chem. Soc. Chem. Commun.*, **1996**, 2689.
96CC2747 H. van Rensberg, P. S. van Heerden, B. C. B. Bezuidenhoudt, D. Ferreira, *J. Chem. Soc. Chem. Commun.*, **1996**, 2747.
96CEJ255 W. Adam, D. Golsch, F. C. Görth, *Chem. Eur. J.*, **1996**, *2*, 255.
96CEJ847 K. C. Nicolaou, A. P. Patron, K. Ajito, P. K. Richter, H. Khatuya, P. Bertinato, R. A. Miller, M. J. Tomaszewski, *Chem. Eur. J.*, **1996**, *2*, 847.
96CJC221 R. Connors, E. Tran, T. Durst, *Can. J. Chem.*, **1996**, *74*, 221.
96CJC465 P. Wan, B. Barker, L. Diao, M. Fischer, Y. Shi, C. Yang, *Can. J. Chem.*, **1996**, *74*, 465.
96CJC1418 D. E. Ward, T. E. Nixey, Y. Gai, M. J. Hrapchak, M. S. Abaee, *Can. J. Chem.*, **1996**, *74*, 1418.
96CL889 S. Inoue, M. Asami, K. Honda, H. Miyazaki, *Chem. Lett.*, **1996**, 889.
96COS229 C. J. Burns, D. S. Middleton, *Contemp. Org. Synth.*, **1996**, *3*, 229.
96COS295 I. Collins, *Contemp. Org. Synth.*, **1996**, *3*, 295.
96COS345 M. C. Norley, *Contemp. Org. Synth.*, **1996**, *3*, 345.
96CPB956 T. Saitoh, T. Oyama, K. Sakurai, Y. Niimura, M. Hinata, Y. Horiguchi, J. Toda, T. Sano, *Chem. Pharm. Bull.*, **1996**, *44*, 956.
96CRV115 L. F. Tietze, *Chem. Rev.*, **1996**, *96*, 115.
96CRV223 A. Padwa, D. M. Weingarten, *Chem. Rev.*, **1996**, *96*, 223.
96CRV423 S. V. Ley, L. R. Cox, G. Meek, *Chem. Rev.*, **1996**, *96*, 423.
96H(43)339 M. C. Jiménez, M. A. Miranda, R. Tormos, *Heterocycles*, **1996**, *43*, 339.
96H(43)745 P. W. Groundwater, D. E. Hibbs, M. B. Hursthouse, M. Nyerges, *Heterocycles*, **1996**, *43*, 745.
96H(43)751 M. R. Saidi, R. Zadmard, T. Saberi, *Heterocycles*, **1996**, *43*, 751.
96H(43)1257 G. Brufola, F. Fringuelli, O. Piermatti, F. Pizzo, *Heterocycles*, **1996**, *43*, 1257.
96H(43)2457 J. Toda, T. Saitoh, T. Oyama, Y. Horiguchi, T. Sano, *Heterocycles*, **1996**, *43*, 2457.
96HA567 M. S. Al-Thebeiti, M. F. El-Zohry, *Heteroatom Chem.*, **1996**, *6*, 567.
96HCA192 D. Berger, M. Neuenschwander, *Helv. Chim. Acta*, **1996**, *79*, 192.
96HCA1475 C. W. Jefford, M. G. H. Vicente, Y. Jacquier, F. Favarger, J. Mareda, P. Millasson-Schmidt, G. Brunner, U. Burger, *Helv. Chim. Acta*, **1996**, *79*, 1475.
96HCA1899 D. Enders, K. Breuer, J. Runsink, J. H. Teles, *Helv. Chim. Acta*, **1996**, *79*, 1899.
96HCM251 A. M. S. Silva, H. R. Tavares, J. A. S. Cavaleiro, *Heterocyclic Commun.*, **1996**, *2*, 251.
96JA1565 K. C. Nicolaou, M. H. D. Postema, C. F. Claiborne, *J. Am. Chem. Soc.*, **1996**, *118*, 1565.
96JA2825 J. T. Link, S. Raghavan, M. Gallant, S. J. Danishefsky, T. C. Chou, L. M. Ballas, *J. Am. Chem. Soc.*, **1996**, *118*, 2825.
96JA3059 K. C. Nicolaou, K. Ajito, A. P. Patron, H. Khatuya, P. K. Richter, P. Bertinato, *J. Am. Chem. Soc.*, **1996**, *118*, 3059.
96JA3537 G. H. Posner, S. B. Park, L. González, D. Wang, J. N. Cumming, D. Klinedinst, T. A. Shapiro, M. D. Bachi, *J. Am. Chem. Soc.*, **1996**, *118*, 3537.
96JA4719 E. Block, J. Page, J. P. Toscano, C.-X. Wang, X. Zhang, R. DeOrazio, C. Guo, R. S. Sheridan, G. H. N. Towers, *J. Am. Chem. Soc.*, **1996**, *118*, 4719.
96JA4778 W. Adam, L. Blancafort, *J. Am. Chem. Soc.*, **1996**, *118*, 4778.
96JA5146 B. M. Trost, F. J. Fleitz, W. J. Watkins, *J. Am. Chem. Soc.*, **1996**, *118*, 5146.
96JA6305 B. M. Trost, F. D. Toste, *J. Am. Chem. Soc.*, **1996**, *118*, 6305.
96JA7513 M. T. Crimmins, R. S. Al-Awar, I. M. Vallin, W. G. Hollis, Jr., R. O'Mahony, J. G. Lever, D. M. Bankaitis-Davis, *J. Am. Chem. Soc.*, **1996**, *118*, 7513.
96JA10333 M. Palucki, J. P. Wolfe, S. L. Buchwald, *J. Am. Chem. Soc.*, **1996**, *118*, 10333.
96JA10335 K. C. Nicolaou, M. H. D. Postema, E. W. Yue, A. Nadin, *J. Am. Chem. Soc.*, **1996**, *118*, 10335.
96JA10656 J. L. Wood, B. M. Stoltz, S. N. Goodman, *J. Am. Chem. Soc.*, **1996**, *118*, 10656.
96JA12473 L. Sun, L. S. Liebeskind, *J. Am. Chem. Soc.*, **1996**, *118*, 12473.
96JCR(S)338 F. M. Moghaddam, A. Sharifi, M. R. Saidi, *J. Chem. Res. (S)*, **1996**, 338.
96JCS(P1)151 E. J. Bush, D. W. Jones, *J. Chem. Soc. Perkin Trans. 1*, **1996**, 151.
96JCS(P1)313 G. Smith, T. J. Sparey, P. C. Taylor, *J. Chem. Soc. Perkin Trans. 1*, **1996**, 313.
96JCS(P1)413 T. Kamada, Ge-Qing, M. Abe, A. Oku, *J. Chem. Soc. Perkin Trans. 1*, **1996**, 413.
96JCS(P1)443 V. Nair, S. Kumar, *J. Chem. Soc. Perkin Trans. 1*, **1996**, 443.
96JCS(P1)705 D. W. Jones, F. M. Nongrum, *J. Chem. Soc. Perkin Trans. 1*, **1996**, 705.
96JCS(P1)841 Z. Cao, J. G. Liehr, *J. Chem. Soc. Perkin Trans. 1*, **1996**, 841.
96JCS(P1)967 M. G. Banwell, C. T. Bui, H. T. T. Pham, G. W. Simpson, *J. Chem. Soc. Perkin Trans. 1*, **1996**, 967.
96JCS(P1)1015 B. Venugopalan, P. J. Karnik, S. Shinde, *J. Chem. Soc. Perkin Trans. 1*, **1996**, 1015.
96JCS(P1)1067 C. N. O'Callaghan, T. B. H. McMurry, J. E. O'Brien, S. M. Draper, D. J. Wilcock, *J. Chem. Soc. Perkin Trans. 1*, **1996**, 1067.

96JCS(P1)1435	K. Chiba, J. Sonoyama, M. Tada, *J. Chem. Soc. Perkin Trans. 1*, **1996**, 1435.
96JCS(P1)1757	D. R. Boyd, N. D. Sharma, R. Boyle, T. A. Evans, J. F. Malone, K. M. McCombe, H. Dalton, J. Chima, *J. Chem. Soc. Perkin Trans. 1*, **1996**, 1757.
96JCS(P1)1797	P. Kocienski, P. Raubo, J. K. Davis, F. T. Boyle, D. E. Davies, A. Richter, *J. Chem. Soc. Perkin Trans. 1*, **1996**, 1797.
96JCS(P1)1879	Y. Watanabe, Y. Ono, S. Hayashi, Y. Ueno, T. Toru, *J. Chem. Soc. Perkin Trans. 1*, **1996**, 1879.
96JCS(P1)1897	T. Saito, H. Fujii, S. Hayashibe, T. Matsushita, H. Kato, K. Kobayashi, *J. Chem. Soc. Perkin Trans. 1*, **1996**, 1897.
96JCS(P1)2227	H. Shimizu, S. Miyazaki, T. Kataoka, *J. Chem. Soc. Perkin Trans. 1*, **1996**, 2227.
96JCS(P1)2241	R. G. F. Giles, R. W. Rickards, B. S. Senanayake, *J. Chem. Soc. Perkin Trans. 1*, **1996**, 2241.
96JCS(P1)2249	P. A. Procopiou, A. C. Brodie, M. J. Deal, D. F. Hayman, G. M. Smith, *J. Chem. Soc. Perkin Trans. 1*, **1996**, 2249.
96JCS(P1)2297	G. Galley, M. Pätzel, *J. Chem. Soc. Perkin Trans. 1*, **1996**, 2297.
96JCS(P1)2391	J. M. Lovell, J. A. Joule, *J. Chem. Soc. Perkin Trans. 1*, **1996**, 2391.
96JCS(P1)2591	G. M. Boland, D. M. X. Donnelly, J-P. Finet, M. D. Rea, *J. Chem. Soc. Perkin Trans. 1*, **1996**, 2591.
96JCS(P1)2693	M. Hirano, S. Yakabe, J. H. Clarke, T. Morimoto, *J. Chem. Soc. Perkin Trans. 1*, **1996**, 2693.
96JCS(P1)2715	L. Crombie, D. E. Games, A. W. G. James, *J. Chem. Soc. Perkin Trans. 1*, **1996**, 2715.
96JCS(P1)2773	H-G. Schwarz, E. Schaumann, *J. Chem. Soc. Perkin Trans. 1*, **1996**, 2773.
96JCS(P1)2799	J. Löffler, R. Schobert, *J. Chem. Soc. Perkin Trans. 1*, **1996**, 2799.
96JCS(P2)269	F. Gao, K. F. Johnson, J. B. Schlenoff, *J. Chem. Soc. Perkin Trans. 2*, **1996**, 269.
96JCS(P2)2091	R. S. Beddoes, B. G. Cox, O. S. Mills, N. J. Mooney, C. I. F. Watt, D. Kirkland, D. Martin, *J. Chem. Soc. Perkin Trans. 2*, **1996**, 2091.
96JCS(P2)2407	R. W. Hoffmann, B. C. Kahrs, J. Schiffer, J. Fleischhauer, *J. Chem. Soc. Perkin Trans. 2*, **1996**, 2407.
96JCS(P2)2497	S. V. Jovanovic, S. Steenken, Y. Hara, M. G. Simic, *J. Chem. Soc. Perkin Trans. 2*, **1996**, 2497.
96JCS(P2)2623	D. Rondeau, E. Raoult, A. Tallec, S. Sinbandhit, L. Toupet, A. Imberty, J. P. Pradère, *J. Chem. Soc. Perkin Trans. 2*, **1996**, 2623.
96JHC383	K. Okamura, T. Date, *J. Heterocycl. Chem.*, **1996**, *33*, 383.
96JHC703	C. D. Apostolopoulos, M. P. Georgiadis, E. A. Couladouros, *J. Heterocycl. Chem.*, **1996**, *33*, 703.
96JOC159	D. F. Harvey, E. M. Grezner, *J. Org. Chem.*, **1996**, *61*, 159.
96JOC455	P. P. Deshpande, K. N. Price, D. C. Baker, *J. Org. Chem.*, **1996**, *61*, 455.
96JOC459	J. S. Swenton, J. N. Freskos, P. Dalidowicz, M. L. Kerns, *J. Org. Chem.*, **1996**, *61*, 459.
96JOC662	K. Smith, D. Anderson, I. Matthews, *J. Org. Chem.*, **1996**, *61*, 662.
96JOC671	G. H. Posner, H. Dai, D. S. Bull, J-K. Lee, F. Eydoux, Y. Ishihara, W. Welsh, N. Pryor, S. Petr, Jr., *J. Org. Chem.*, **1996**, *61*, 671.
96JOC677	M. Ihara, A. Katsumata, F. Setsu, Y. Tokunaga, K. Fukumoto, *J. Org. Chem.*, **1996**, *61*, 677.
96JOC1650	D. Pérez, G. Burés, E. Guitián, L. Castedo, *J. Org. Chem.*, **1996**, *61*, 1650.
96JOC1859	J. Almena, F. Foubelo, M. Yus, *J. Org. Chem.*, **1996**, *61*, 1859.
96JOC1914	A. Braun, L. Toupet, J-P. Lellouche, *J. Org. Chem.*, **1996**, *61*, 1914.
96JOC1986	A. V. Sanin, V. G. Nenajdenko, V. S. Kuz'min, E. S. Balenkova, *J. Org. Chem.*, **1996**, *61*, 1986.
96JOC2839	B. B. Snider, Q. Lu, *J. Org. Chem.*, **1996**, *61*, 2839.
96JOC2865	E. Jao, P. B. Slifer, R. Lalancette, S. S. Hall, *J. Org. Chem.*, **1996**, *61*, 2865.
96JOC3894	H-G. Hahn, K. D. Nam, H. Mah, J. J. Lee, *J. Org. Chem.*, **1996**, *61*, 3894.
96JOC4190	P. Salvadori, S. Superchi, F. Minutolo, *J. Org. Chem.*, **1996**, *61*, 4190.
96JOC4391	A. Nath, J. Mal, R. V. Venkateswaran, *J. Org. Chem.*, **1996**, *61*, 4391.
96JOC4725	S. D. Larsen, P. V. Fisher, B. E. Libby, R. M. Jensen, S. A. Mizsak, W. Watt, W. R. Ronk, S. T. Hill, *J. Org. Chem.*, **1996**, *61*, 4725.
96JOC4853	D. Garey, M. Ramirez, S. Gonzales, A. Wertsching, S. Tith, K. Keefe, M. R. Peña, *J. Org. Chem.*, **1996**, *61*, 4853.
96JOC5371	A. Ramacciotti, R. Fiaschi, E. Napolitano, *J. Org. Chem.*, **1996**, *61*, 5371.
96JOC5375	T. Patonay, A. Lévai, C. Nemes, T. Timár, G. Tóth, W. Adam, *J. Org. Chem.*, **1996**, *61*, 5375
96JOC5631	A. G. Schultz, S. J. Kirincich, *J. Org. Chem.*, **1996**, *61*, 5631.
96JOC5670	O. Aleksiuk, S. E. Biali, *J. Org. Chem.*, **1996**, *61*, 5670.
96JOC5754	M. A. Lucas, C. H. Schiesser, *J. Org. Chem.*, **1996**, *61*, 5754.
96JOC5818	M. C. de la Fuente, L. Castedo, D. Dominguez, *J. Org. Chem.*, **1996**, *61*, 5818.
96JOC6326	Y. Han, S. L. Bontems, P. Hegyes, M. C. Munson, C. A. Minor, S. A. Kates, F. Albericio, G Barany, *J. Org. Chem.*, **1996**, *61*, 6326.
96JOC6673	D. Nicoletti, A. A. Ghini, G. Burton, *J. Org. Chem.*, **1996**, *61*, 6673.

96JOC6693 Z. Liu, J. Meinwald, *J. Org. Chem.*, **1996**, *61*, 6693.
96JOC7408 K. McWilliams, J. W. Kelly, *J. Org. Chem.*, **1996**, *61*, 7408.
96JOC7600 P. A. Evans, J. D. Nelson, *J. Org. Chem.*, **1996**, *61*, 7600.
96JOC7860 L. A. Paquette, J. Tae, *J. Org. Chem.*, **1996**, *61*, 7860.
96JOC8317 V. H. Dahanukar, S. D. Rychnovsky, *J. Org. Chem.*, **1996**, *61*, 8317.
96JOC8432 W. Adam, L. Blancafort, *J. Org. Chem.*, **1996**, *61*, 8432.
96JOC9103 R. W. Friesen, C. Vanderwal, *J. Org. Chem.*, **1996**, *61*, 9103.
96JOC9115 D. Stoermer, S. Caron, C. H. Heathcock, *J. Org. Chem.*, **1996**, *61*, 9115.
96JOC9164 J. Mal, A. Nath, R. V. Venkateswaran, *J. Org. Chem.*, **1996**, *61*, 9164.
96JOM(507)103 W. Tully, L. Main, B. K. Nicholson, *J. Organomet. Chem.*, **1996**, *507*, 103.
96JST255 J. Choo, K-H. Lee, J. Laane, *J. Mol. Struct.*, **1996**, *376*, 255.
96LA109 K. Hartke, X-P. Popp, *Leibigs Ann.*, **1996**, 109.
96LA291 C. P. Klaus, P. Margaretha, *Leibigs Ann.*, **1996**, 291.
96LA725 K. Yamagata, K. Akizuki, M. Yamazaki, *Leibigs Ann.*, **1996**, 725.
96LA959 M. Barboiu, C. T. Supuran, C. Luca, G. Popescu, C. Barboiu, *Leibigs Ann.*, **1996**, 959.
96LA1159 K. Hartke, N. Bien, W. Massa, S. Wocadlo, *Leibigs Ann.*, **1996**, 1159.
96LA1385 C. Tödter, H. Lackner, *Leibigs Ann.*, **1996**, 1385.
96M1027 H. Auer, R. Weis, K. Schweiger, *Monatsh. Chem.*, **1996**, *127*, 1027.
96MC99 A. I. Pyschev, V. V. Krasnikov, A. E. Zibert, D. E. Tosyniyan, S. V. Verin, *Mendeleev Commun.*, **1996**, 99.
96OR(48)301 B. Giese, B. Kopping, T. Göbel, J. Dickhaut, G. Thoma, K. J. Kulicke, F. Trach, *Org. React.*, **1996**, *48*, 301.
96P148 R. M. Shaker, *Pharmazie*, **1996**, *51*, 148.
96PS(108)93 M. El-Ghanam, *Phosphorus, Sulfur and Silicon*, **1996**, *108*, 93.
96PS(112)101 M. Tazaki, S. Okai,T. Hieda, S. Nagahama, *Phosphorus, Sulfur and Silicon*, **1996**, *112*, 101.
96PAC113 I. E. Markó, G. R. Evans, P. Seres, I. Chellé, Z. Janousek, *Pure Appl. Chem.*, **1996**, *68*, 113.
96RCR27 A. D. Kunzevich, V. F. Golovkov, V. R. Rembovsky, *Russ. Chem. Rev.*, 1996, *65*, 27.
96RTC407 H. Xianming, R. M. Kellogg, *Recl. Trav. Chim. Pays-Bas*, **1996**, *115*, 407.
96SL57 D. P. Becker, D. L. Flynn, *Synlett*, **1996**, 57.
96SL72 S. Moriyama, T. Karakasa, T. Inoue, K. Kurashima, S. Satsumabayashi, T. Saito, *Synlett*, **1996**, 72.
96SL258 C. Baldoli, P. Del Buttero, M. Di Ciolo, S. Maiorana, A. Papagni, *Synlett*, **1996**, 258.
96SL261 D. E. Ward, Y. Gai, Y. Lai, *Synlett*, **1996**, 261.
96SL315 M. Schmittel, M. Levis, *Synlett*, **1996**, 315.
96SL341 E. A. Couladouros, C. D. Apostolpoulos, *Synlett*, **1996**, 341.
96SL349 S. Fielder, D. D. Rowan, M. S. Sherburn, *Synlett*, **1996**, 349.
96SL396 U. Beifuss, H. Gehm, M. Taraschewski, *Synlett*, **1996**, 396.
96SL536 G-J. Boons, R. Eveson, S. Smith, T. Stauch, *Synlett*, **1996**, 536.
96SL553 C. Fournier-Nguefack, P. Lhoste, D. Sinou, *Synlett*, **1996**, 553.
96SL767 P. Ceccherelli, M. Curini, M. C. Marcotullio, F. Epifano, O. Rosati, *Synlett*, **1996**, 767.
96SL787 S. V. Ley, S. Mio, B. Meseguer, *Synlett*, **1996**, 787.
96SL789 S. V. Ley, S. Mio, *Synlett*, **1996**, 789.
96SL791 S. V. Ley, S. Mio, B. Meseguer, *Synlett*, **1996**, 791.
96SL875 D. Mink, G. Deslongchamps, *Synlett*, **1996**, 875.
96SL1065 K. T. Mead, R. Zemribo, *Synlett*, **1996**, 1065.
96SL1079 T. Hashihayata, Y. Ito, T. Katsuki, *Synlett*, **1996**, 1079.
96SL1143 V. Nair, S. Kumar, *Synlett*, **1996**, 1143.
96SL1173 C. E. Chase, J. A. Bender, F. G. West, *Synlett*, **1996**, 1173.
96SYN105 H-J. Lehmler, M. Nieger, E. Breitmaier, *Synthesis*, **1996**, 105.
96SYN297 I. E. Markó, D. J. Bayston, *Synthesis*, **1996**, 297.
96SYN307 C. J. Roxburgh, *Synthesis*, **1996**, 307.
96SYN327 H. Meier, A. Mayer, *Synthesis*, **1996**, 327.
96SYN519 J-Y. Mérour, L. Chichereau, E. Desarbre, P. Gadonneix, *Synthesis*, **1996**, 519.
96SYN647 H. Flörke, E. Schaumann, *Synthesis*, **1996**, 647.
96SYN652 N. D. Smith, P. J. Kocienski, S. D. A. Street, *Synthesis*, **1996**, 652.
96SYN863 T. Eicher, F. Servet, A. Speicher, *Synthesis*, **1996**, 863.
96SYN1095 D. Enders, O. F. Prokopenko, G. Raabe, J. Runsink, *Synthesis*, **1996**, 1095.
96SYN1280 G. Broggini, G. Zecchi, *Synthesis*, **1996**, 1280.
96T1205 E. Wada, W. Pei, H. Yasuoka, U. Chin, S. Kanemasa, *Tetrahedron*, **1996**, *52*, 1205.
96T2125 P. C. Bulman Page, R. D. Wilkes, E. S. Namwindwa, M. J. Witty, *Tetrahedron*, **1996**, *52*, 2125.
96T3117 L. A. White, R. C. Storr, *Tetrahedron*, **1996**, *52*, 3117.
96T9427 K. Nakatani, A. Okamoto, I. Saito, *Tetrahedron*, **1996**, *52*, 9427.

96T9713	A. Nishida, M. Kawahara, M. Nishida, O. Yonemitsu, *Tetrahedron*, **1996**, *52*, 9713.
96T12177	T. Honda, S. Yamane, F. Ishikawa, M. Katoh, *Tetrahedron*, **1996**, *52*, 12177.
96T12233	G. Capozzi, P. Fratini, S. Menichetti, C. Nativi, *Tetrahedron*, **1996**, *52*, 12233.
96T12247	G. Capozzi, P. Fratini, S. Menichetti, C. Nativi, *Tetrahedron*, **1996**, *52*, 12247.
96T12597	A. Bojilova, R. Nikolova, C. Ivanov, N. A. Rodios, A. Terzis, C. P. Raptopoulou, *Tetrahedron*, **1996**, *52*, 12597.
96T13623	H. Sharghi, F. Tamaddon, *Tetrahedron*, **1996**, *52*, 13623.
96T14081	W. E. Bauta, J. Booth, M. E. Bos, M. Delucca, L. Diorazio, T. J. Donohoe, C. Frost, N. Magnus, P. Magnus, J. Mendoza, P. Pye, J. G. Tarrant, S. Thom, F. Ujjainwalla, *Tetrahedron*, **1996**, *52*, 14081.
96T14247	S. Cossu, O. De Lucchi, *Tetrahedron*, **1996**, *52*, 14247.
96T14273	K. Singh, J. Singh, H. Singh, *Tetrahedron*, **1996**, *52*, 14273.
96TL93	V. J. Ram, A. Goel, *Tetrahedron Lett.*, **1996**, *37*, 93.
96TL123	A. S. Bell, C. W. G. Fishwick, J. E. Reed, *Tetrahedron Lett.*, **1996**, *37*, 123.
96TL141	N. A. Petasis, S-P. Lu, *Tetrahedron Lett.*, **1996**, *37*, 141.
96TL213	T. Nakata, S. Nomura, H. Matsukura, *Tetrahedron Lett.*, **1996**, *37*, 213.
96TL257	R. K. Haynes, S. C. Vonwiller, *Tetrahedron Lett.*, **1996**, *37*, 257.
96TL275	K. M. Foote, C. J. Hayes, G. Pattenden, *Tetrahedron Lett.*, **1996**, *37*, 275.
96TL303	P. Magnus, M. B. Roe, *Tetrahedron Lett.*, **1996**, *37*, 303.
96TL667	H. Shima, R. Kobayashi, T. Nabeshima, N. Furukawa, *Tetrahedron Lett.*, **1996**, *37*, 667.
96TL815	G. H. Posner, D. Wang, L. González, X. Tao, J. N. Cumming, D. Klinedinst, T. A. Shapiro, *Tetrahedron Lett.*, **1996**, *37*, 815.
96TL1073	K. Toshima, H. Yamaguchi, T. Jyojima, Y. Noguchi, M. Nakata, S. Matsumura, *Tetrahedron Lett.*, **1996**, *37*, 1073.
96TL1077	A. Abiko, S. Masamune, *Tetrahedron Lett.*, **1996**, *37*, 1077.
96TL1081	A. Abiko, S. Masamune, *Tetrahedron Lett.*, **1996**, *37*, 1081.
96TL1313	C. D. Gabbutt, B. M. Heron, J. D. Hepworth, M. M. Rahman, *Tetrahedron Lett.*, **1996**, *37*, 1313.
96TL1551	H-Y. Li, G. A. Boswell, *Tetrahedron Lett.*, **1996**, *37*, 1551.
96TL1647	E. A. Boyd, W. C. Chan, V. M. Loh, Jr., *Tetrahedron Lett.*, **1996**, *37*, 1647.
96TL1663	N. Haddad, I. Kuzmenkov, *Tetrahedron Lett.*, **1996**, *37*, 1663.
96TL1715	P. Perlmutter, E. Puniani, G. Westman, *Tetrahedron Lett.*, **1996**, *37*, 1715.
96TL1913	P. Charoenying, D. H. Davies, D. McKerrecher, R. J. K. Taylor, *Tetrahedron Lett.*, **1996**, *37*, 1913.
96TL2433	T. Takahashi, H. Tanaka, A. Matsuda, H. Yamada, T. Matsumoto, Y. Sugiura, *Tetrahedron Lett.*, **1996**, *37*, 2433.
96TL2463	T. Tsunoda, F. Ozaki, N. Shirakata, Y. Tamaoka, H. Yamamoto, S. Itô, *Tetrahedron Lett.*, **1996**, *37*, 2463.
96TL2667	A. Krief, L. Defrère, *Tetrahedron Lett.*, **1996**, *37*, 2667.
96TL2743	J. H. Byers, C. C. Whitehead, M. E. Duff, *Tetrahedron Lett.*, **1996**, *37*, 2743.
96TL2821	T. Ozturk, *Tetrahedron Lett.*, **1996**, *37*, 2821.
96TL2865	E. Alvarez, M. Delgado, M. T. Diaz, L. Hanxing, R. Pérez, J. D. Martin, *Tetrahedron Lett.*, **1996**, *37*, 2865.
96TL2997	Y. Ishibashi, S. Ohba, S. Nishiyama, S. Yamamura, *Tetrahedron Lett.*, **1996**, *37*, 2997.
96TL3015	G-Q. Lin, M. Zhong, *Tetrahedron Lett.*, **1996**, *37*, 3015.
96TL3059	I. Kadota, D. Hatakeyama, K. Seki, Y. Yamamoto, *Tetrahedron Lett.*, **1996**, *37*, 3059.
96TL3179	J. C. Carretero, N. Diaz, M. L. Molina, J. Rojo, *Tetrahedron Lett.*, **1996**, *37*, 3179.
96TL3687	P. Hayes, G. Dujardin, C. Maignan, *Tetrahedron Lett.*, **1996**, *37*, 3687.
96TL3741	K. Mori, T. Nakayama, H. Takikawa, *Tetrahedron Lett.*, **1996**, *37*, 3741.
96TL3755	P. Perlmutter, E. Puniani, *Tetrahedron Lett.*, **1996**, *37*, 3755.
96TL3841	P. Metz, D. Seng, B. Plietker, *Tetrahedron Lett.*, **1996**, *37*, 3841.
96TL3895	D. Bell, M. R. Davies, F. J. L. Finney, G. R. Green, P. M. Kincey, I. S. Mann, *Tetrahedron Lett.*, **1996**, *37*, 3895.
96TL4331	X-X. Shi, S. P. Khanapure, J. Rokach, *Tetrahedron Lett.*, **1996**, *37*, 4331.
96TL4447	C. Z. Ding, A. V. Miller, *Tetrahedron Lett.*, **1996**, *37*, 4447.
96TL4675	F. E. McDonald, J. L. Bowman, *Tetrahedron Lett.*, **1996**, *37*, 4675.
96TL4949	V-H. Nguyen, H. Nishino, K. Kurosawa, *Tetrahedron Lett.*, **1996**, *37*, 4949.
96TL5073	V. B. Birman, A. Chopra, C. A. Ogle, *Tetrahedron Lett.*, **1996**, *37*, 5073.
96TL5077	S. Gauthier, F. Labrie, *Tetrahedron Lett.*, **1996**, *37*, 5077.
96TL5243	H-J. Lim, G. A. Sulikowski, *Tetrahedron Lett.*, **1996**, *37*, 5243.
96TL5397	M. Wakayama, H. Nemoto, M. Shibuya, *Tetrahedron Lett.*, **1996**, *37*, 5397.
96TL5461	G. J. McGarvey, M. W. Stepanian, *Tetrahedron Lett.*, **1996**, *37*, 5461.

96TL5673	P. A. Krasutsky, I. V. Kolomitsin, R. M. Carlson, M. Jones, Jr., *Tetrahedron Lett.*, **1996**, *37*, 5673.
96TL6117	V. Alphand, N. Gaggero, S. Colonna, R. Furstoss, *Tetrahedron Lett.*, **1996**, *37*, 6117.
96TL6173	N. Hayashi, K. Fujiwara, A. Murai, *Tetrahedron Lett.*, **1996**, *37*, 6173.
96TL6181	K. P. Volcho, D. V. Korchagina, N. F. Salakhutdinov, V. A. Barkhash, *Tetrahedron Lett.*, **1996**, *37*, 6181.
96TL6351	S. Oi, K. Kashiwagi, E. Terada, K. Ohuchi, Y. Inoue, *Tetrahedron Lett.*, **1996**, *37*, 6351.
96TL6365	M. Morimoto, H. Matsukura, T. Nakata, *Tetrahedron Lett.*, **1996**, *37*, 6365.
96TL6499	F. J. Zawacki, M. T. Crimmins, *Tetrahedron Lett.*, **1996**, *37*, 6499.
96TL6511	A. G. Schultz, Y-J. Li, *Tetrahedron Lett.*, **1996**, *37*, 6511.
96TL6635	M. K. Schwaebe, R. D. Little, *Tetrahedron Lett.*, **1996**, *37*, 6635.
96TL7225	G. H. Posner, X. Tao, J. N. Cumming, D. Klinedinst, T. A. Shapiro, *Tetrahedron Lett.*, **1996**, *37*, 7225.
96TL7739	C. Castagnino, J. Vercauteren, *Tetrahedron Lett.*, **1996**, *37*, 7739.
96TL7955	G. Bertram, A. Scherer, W. Steglich, W. Weber, T. Anke, *Tetrahedron Lett.*, **1996**, *37*, 7955.
96TL8001	M. G. Dhara, S. K. De, A. K. Mallik, *Tetrahedron Lett.*, **1996**, *37*, 8001.
96TL8011	A. Krief, L. Defrère, *Tetrahedron Lett.*, **1996**, *37*, 8011.
96TL8015	A. Krief, L. Defrère, *Tetrahedron Lett.*, **1996**, *37*, 8015.
96TL8053	C. McNicholas, T. J. Simpson, N. J. Willett, *Tetrahedron Lett.*, **1996**, *37*, 8053.
96TL8321	A. Robinson, H-Y. Li, J. Feaster, *Tetrahedron Lett.*, **1996**, *37*, 8321.
96TL8679	Y. Hu, D. J. Skalitzky, S. D. Rychnovsky, *Tetrahedron Lett.*, **1996**, *37*, 8679.
96TL8883	K. Okuma, T. Yamamoto, T. Shirokawa, T. Kitamura, Y. Fujiwara, *Tetrahedron Lett.*, **1996**, *37*, 8883.
96TL8913	S. C. Gupta, A. Saini, S. Sharma, M. Kapoor, S. N. Dhawan, *Tetrahedron Lett.*, **1996**, *37*, 8913.
96TL8929	P. C. Bulman Page, M. Purdie, D. Lathbury, *Tetrahedron Lett.*, **1996**, *37*, 8929.

Chapter 7

Seven-Membered Rings

David J. Le Count
Formerly of Zeneca Pharmaceuticals, UK,
1, Vernon Avenue, Congleton, Cheshire, UK

7.1 INTRODUCTION

It may have been a long time in coming, but it arrived just in time to be included in this report of last year's heterocyclic chemistry. I refer, of course, to the second edition of Comprehensive Heterocyclic Chemistry, which is now available to all - at a price! 7-Membered rings feature in volume 9 <96CHECII(9)>. A further publication devoted to exclusively azepines has appeared in the form of a new addition to the Chemistry of Heterocyclic Compounds range of publications <96MI1>. One trap not to fall into in writing such a report as this is to concentrate on the synthesis of the systems under review at the cost of their reactions. To avoid this, at least in part, the reader's attention is drawn to a review on the ring contraction of heterocycles by sulfur extrusion <96AHC(65)40>. Considerable attention has been paid to the formation of 6-membered homocyclic and heterocyclic rings by sulfur extrusion from the corresponding 7-membered sulfur compound.

7.2 RING SYSTEMS CONTAINING ONE HETEROATOM

7.2.1 Azepines

In 1995, and regrettably missed in last year's review, Klötgen and Würthwein described the formation of the 4,5-dihydroazepine derivatives **2** by lithium induced cyclisation of the triene **1**, followed by acylation <95TL7065>. This work has now been extended to the preparation of a number of 1-acyl-2,3-dihydroazepines **4** from **3** <96T14801>. The formation of the intermediate anion and its subsequent cyclisation was followed by NMR spectroscopy and the stereochemistry of the final product elucidated by x-ray spectroscopy. The synthesis of optically active 2*H*-azepines **6** from amino acids has been described <96T10883>. The key step is the cyclisation of the amino acid derived alkene **5** with TFA. These azepines isomerise to the thermodynamically more stable 3*H*-azepines **7** in solution.

(1)

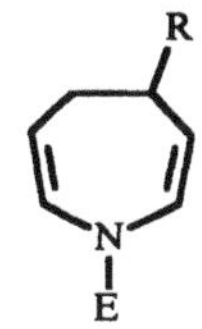

(2)

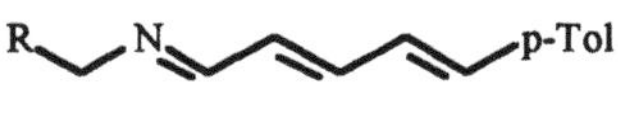

(3)

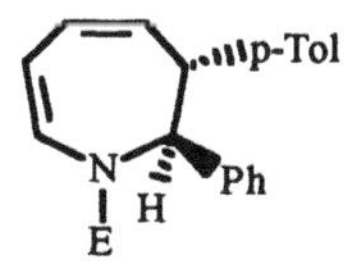

(4)

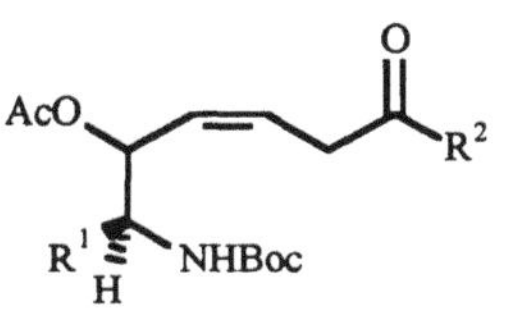

(5)

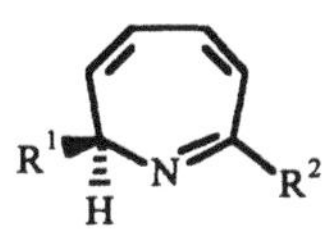

(6)

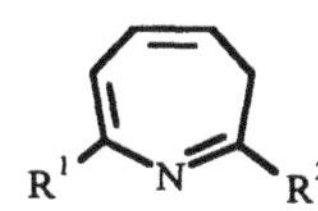

(7)

3*H*-azepines have also been prepared in moderate yield by the acid-catalysed thermolytic deamination of 1,6-disubstituted 1,6-diamino-1,3,5-hexatrienes. Structures were confirmed by x-ray analysis <96LA887>. Bis(trimethylsilyl)thioketene is reported to react with 1-methyl-pyrrolidin-2-one to afford the azepine-2-thione derivative **8** (Scheme 1) <96CC1621>.

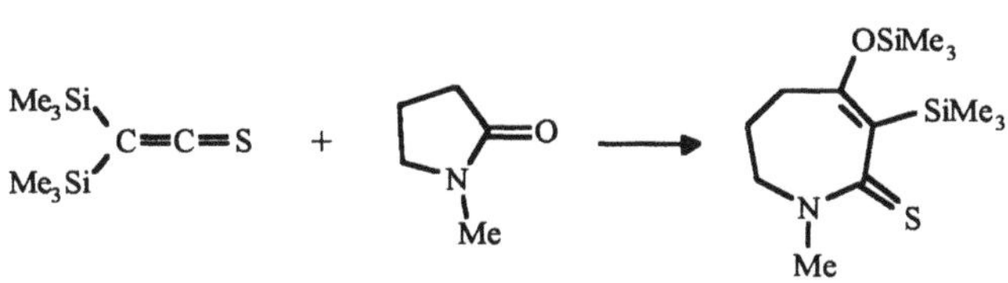

(8)

Scheme 1

A number of examples of ring-closing metathesis (RCM) as a route to fused azepine synthesis have been published in the year under review. In their studies on the synthesis of (-)-stemoamide Kinoshita and Mori have used the ruthenium catalysed cyclisation of the enyne **(9)** to form the pyrroloazepine derivative **10** <96JOC8356>, whilst in their efforts to prepare the same ring system **11** Martin and co-workers used molybdenum catalysis <96T7251>. Cyclisation to monocyclic azepine derivatives has also been undertaken on solid phase systems <96TL8249>. In the presence of tetrakis(triphenylphosphine)palladium the enamine **12** cyclises after 10 h to give a mixture of the isoindoline **13** and the dihydro-2-benzazepine **14** as illustrated in Scheme 2 <96CC2257>. A closer examination of the reaction revealed that **13** is the immediate product, which rearranges to **14** over a period of time. Yields of the azepine were in excess of 90%. 4-Ethoxy-2*H*,5*H*-1-benzazepin-2-one has been prepared by the reaction of ethyl *o*-aminophenyl acetate and ketenylidene(triphenyl)phosphorane, whereas ethyl *o*-hydroxyphenylacetate forms the corresponding benzoxepine <96JCS(P1)2799>.

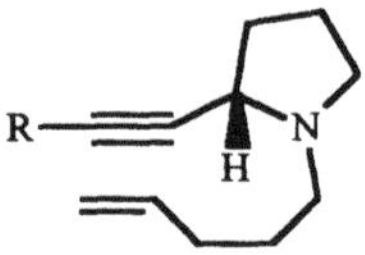

(9)

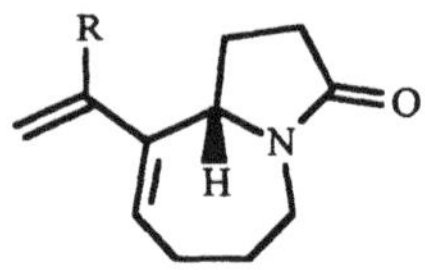

(10)

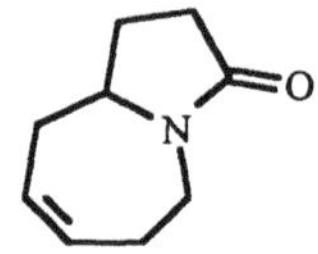

(11)

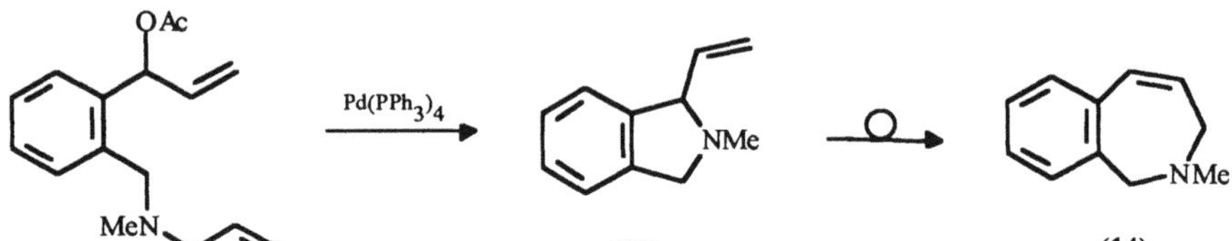

(12) **(13)** **(14)**

Scheme **2**

In a series of three papers, Noguchi and co-workers have reported their continuing studies on the formation of heterocycle-fused azepine systems <96T13081, 96T13097, 96T13111>. A typical example is the conversion of the aldehyde **15** into the azepines **16** and **17** (Scheme 3). The reaction also proceeds with imines when the dihydroazepine prior to bridging can be isolated. Mechanistic and stereochemical aspects of the reaction have been explored.

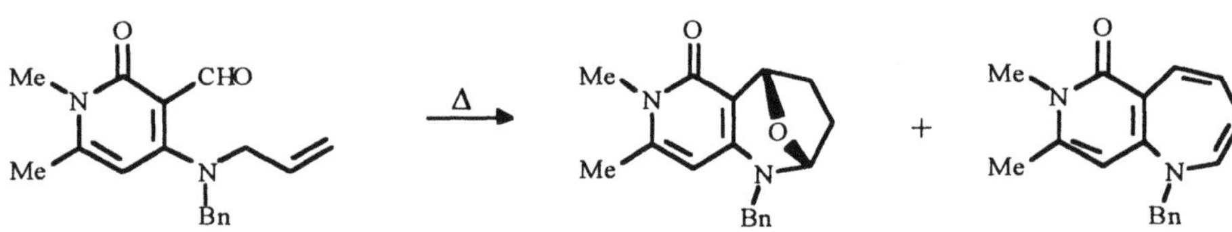

(15) **(16)** **(17)**

Scheme **3**

Indolones and isoindolones have been utilised in the synthesis of fused azepine derivatives. In the one reaction, rearrangement of the alkynes **18** to 2-benzazepine-1,5-diones **19** in the presence of Lewis acids has been reported <96TL393>. The yields vary from moderate to very good. Tricyclic azepines **20** are obtained by the reaction of the 4-[2'-(*p*-toluenesulfonyloxy)ethyl]-2-oxindole with imines <96JHC209>.

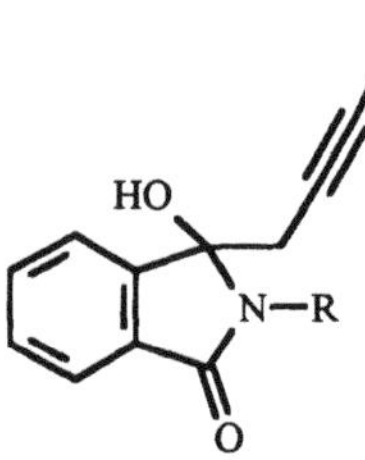

(18)

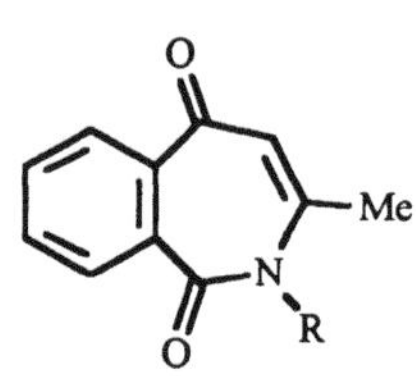

(19)

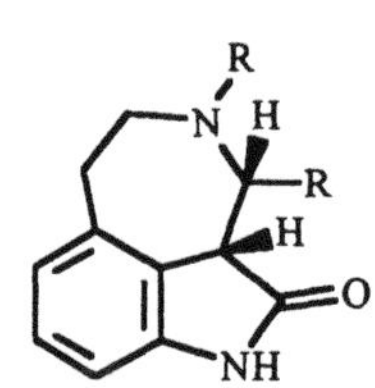

(20)

The work of Sharp and his many co-workers at Edinburgh cannot be underestimated. In a more recent communication they have extended the scope of his cyclisation of diene-conjugated nitrile ylides to triene homologues <96CC2739>. Thus cyclisation of the triene **21** afforded the cyclopropa[*c*]isoquinoline **22**, which on heating gave a mixture of **23** and **24**. The isomeric triene **25** also gave **24** as the sole product (Scheme 4). In this instance the intermediate cyclopropane could not be isolated.

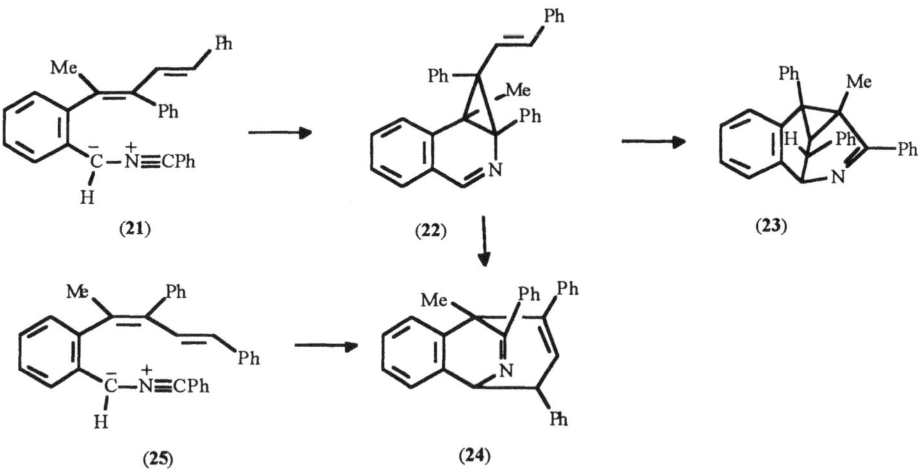

Scheme 4

The cycloaddition of alkynes and alkenes to nitrile oxides has been used in the synthesis of functionalised azepine systems <96JHC259>, <96T5739>. The concomitantly formed isoxazole (dihydroisoxazole) ring is cleaved by reduction in the usual way. Other routes to 1-benzazepines include intramolecular amidoalkylation <96SC2241> and intramolecular palladium-catalysed aryl amination and aryl amidation <96T7525>. Spiro-substituted 2-benzazepines have been prepared by phenolic oxidation (Scheme 5) <96JOC5857> and the same method has been applied to the synthesis of dibenzazepines <96CC1481>.

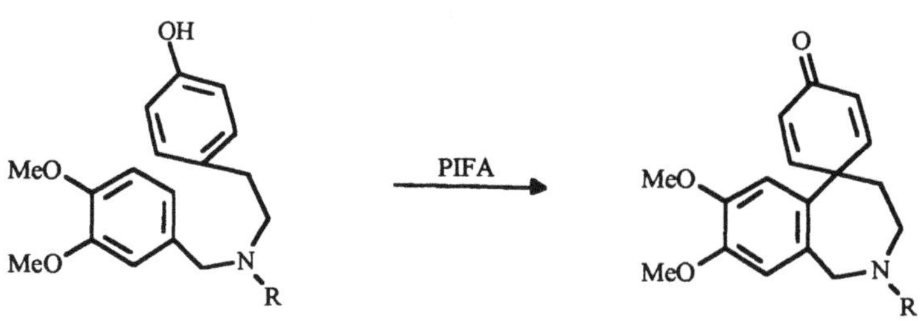

Scheme 5

The β-lactam-epoxide **26**, prepared from 1,5-cyclooctadiene by sequential [2 + 2] cycloaddition, protection and epoxidation, undergoes a ring opening/ring closure reaction to **27** with methyllithium (Scheme 6). This new bridged azepane is further converted to ± anatoxin-*a*, a toxic principle of blue-green algae (96T11637>.

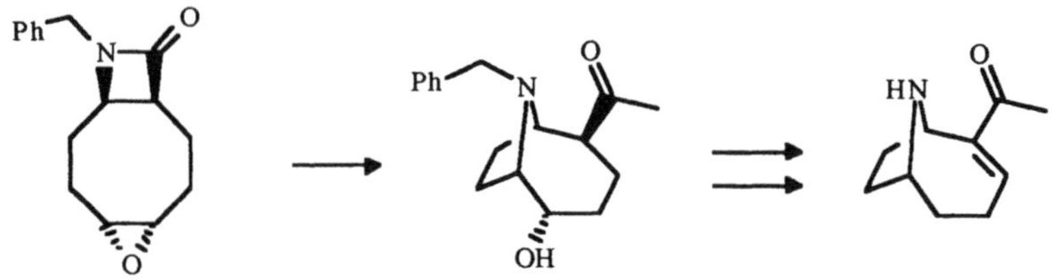

(26) (27)

Scheme 6

The insertion of nitrogen into a six-membered ring continues to be the source of azepines and azepanes. These take the form of the pyrolysis of aryl azides <96JHC1333>, inter- and intra-molecular Schmidt reactions <96JA7647, 96T3403, 96TL1531, 96JOC10>, 96SC1839> and the Beckmann rearrangement <96CC1071, 96SL29>.

Continuing his studies on the metallation of tetrahydro-2-benzazepine formamidines, Meyers has now shown that the previously unsuccessful deprotonation of 1-alkyl derivatives can be achieved with *sec*-butyllithium at -40 °C <96H(42)475>. In this way 1,1-dialkylated derivatives are now accessible. The preparation of 3*H*-benzazepines by chemical oxidation of 2,5- and 2,3-dihydro-1*H*-1-benzazepines has been reported <96T4423>. 3*H*-Diazepines are also formed by rearrangement of the 5*H*-tautomers which had been previously reported to be the products of electrochemical oxidation of 2,5-dihydro-1*H*-1-benzazepine <95T9611>. The synthesis and radical trapping activities of a number of benzazepine derived nitrones have been reported <96T6519, 96JBC3097>.

7.2.2 Oxepines

Studies aimed at syntheses of the cyclic polyether marine toxins ciguatoxin, hemibrevetoxin and the brevetoxins have again proved to be an impetus for the development of oxapane synthesis. Isobe has again utilised cyclisation of acetylene cobalt complexes to form the seven-membered ring, this time in the A/B/C rings of ciguatoxin <96SL351>. Martin and co-workers have developed two routes to *trans*-fused polyethers, namely cyclisation of the stannane derivative **28** to **29** with n-Bu_4NIO_4 and formation of the oxepane **31** by ring contraction of **30** in a Ramberg-Bäcklund olefination reaction <96TL2865, 96TL2869>. In their preparation of medium ring ethers Hirama and co-workers used a ring expansion strategy, illustrated by the formation of **33** from the furan derivative **32** <96SL1165>.

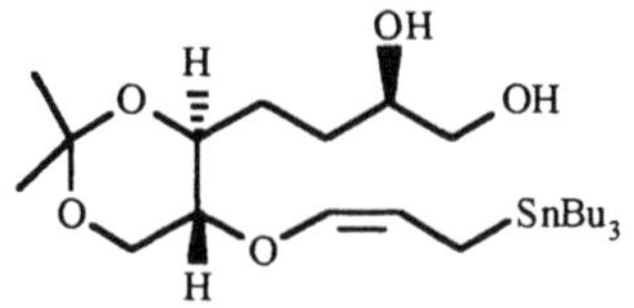

(28)

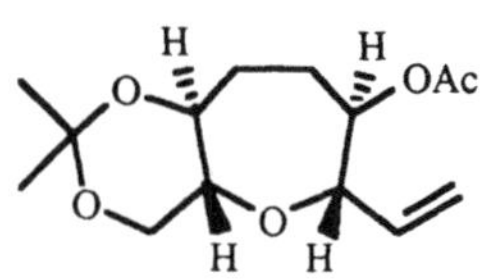

(29)

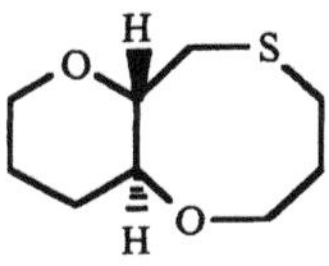
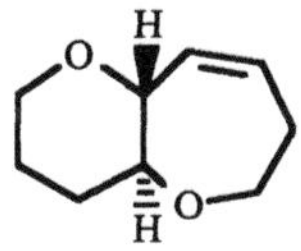
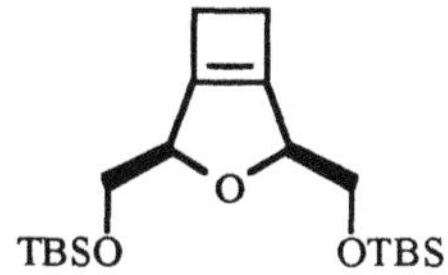

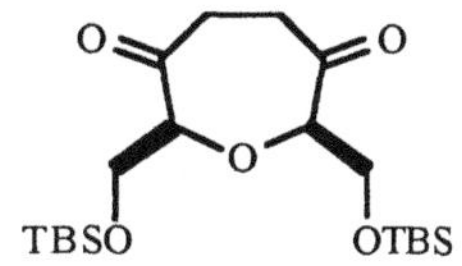

(30) **(31)** **(32)** **(33)**

In a series of papers culminating in the report of a total synthesis of hemibrevetoxin B, Nakata and co-workers utilised a zinc catalysed ring expansion of poly-substituted tetrahydropyrans, exemplified by the transformation of **34** to **35** (Scheme 7) <96TL213, 96TL217, 96TL6365>.

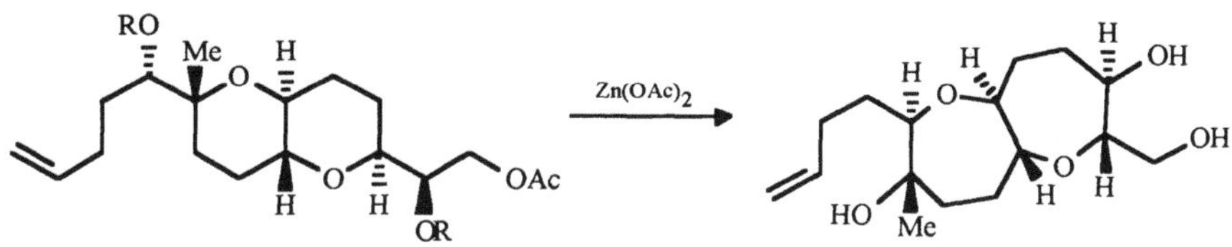

(34) **(35)**

Scheme 7

In other cyclisations to functionalised oxepanes, Rychnovsky and Dahanukar have shown that the epoxide **36** cyclises with BF_3 etherate and TMSCN to form the oxepane **37** as single product <96TL339>, and Evans and Roseman have prepared a series of cyclic ethers by radical cyclisation of the acylselenides **38** (Scheme 8) <96JOC2252>. The major product was always the *cis*-isomer and the best yields were obtained with $(TMS)_3SiH$.

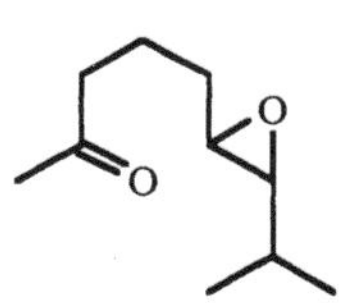
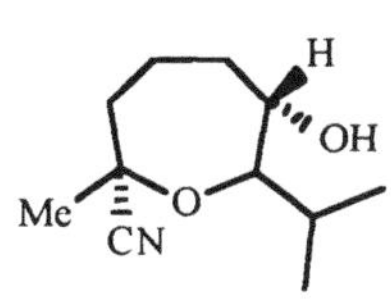

(36) **(37)**

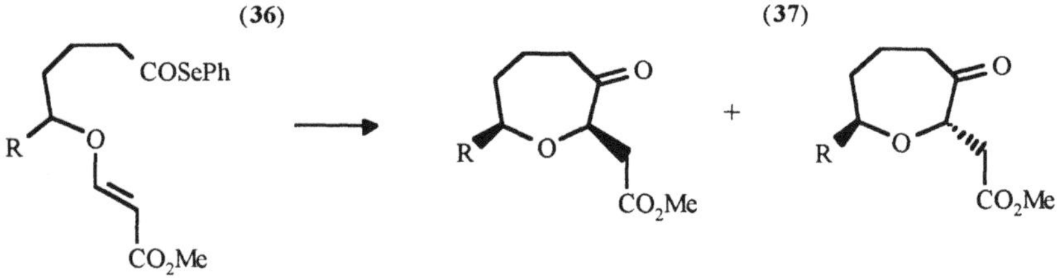

(38)

Scheme 8

Oxepanes have been obtained by the iodoetherification of unsaturated alcohols using bis(*sym*-collidine)iodine(I) hexafluorophosphate <96JOC5793>. Thus 6-hepten-1-ol affords 2-iodomethyloxepane in 95% yield. A similar reaction with, for example, (3Z,6Z)-3,6-octadien-1-ol affords 2-(iodomethyl)-2-methyl-4-oxepane.

2-Allylphenols react with carbon monoxide and hydrogen in the presence of catalytic quantities of palladium(II) to form a mixture of 5-, 6- and 7-membered lactones <96JA4264>. In most cases the 7-membered lactone is the major product. Grigg continues his cyclisation studies with a report of the formation of tetrahydro-2-benzoxepine derivatives by the cyclisation of (2-iodobenzyl)propargyl ethers with π-allylpalladium <96TL6565>. 2-Benzoxepin-5-ones are formed by insertion of a two-carbon fragment into phthalides by reaction with prop-2-ynylmagnesium bromide <96CC19>.

Eberbach has shown that pyrolysis of the system **39**, in which both A and B represent benzene rings, the product is the stable dihydrodibenzo[*c,e*]oxepine **40**, (A,B = benzo), in which the intermediate ylide undegoes a 1,7-electrocyclisation followed by a [1,5] H shift (Scheme 9) <85CB4035>. This work has now been extended by Sharp to include systems where B is a thiophene or pyridine, offering a route to the corresponding thieno- and pyrido-benzoxepines <96JCS(P1)515>.

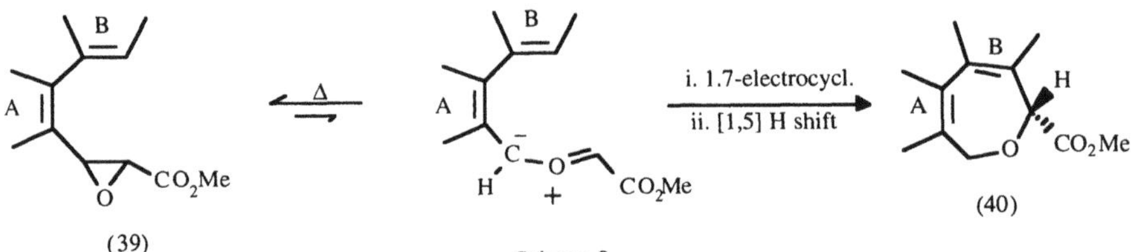

Scheme **9**

Pyridazine *N*-oxides undergo 1,3-dipolar addition with benzyne and a number of its analogues to form adducts which, with loss of nitrogen, form 1-benzoxepines and this work has now been extended to 1,2,4-triazine 1-oxides. In this case the product is 1,3-benzoxazepine <96H(43)2091>.

Magnesium monoperphthalate in acetonitrile is reported to be an effective reagent in the Baeyer-Villiger oxidation of cyclic ketones <96SC4591> and prochiral cyclohexanones are oxidised by "designer yeast" with high enantioselectivity <96JCS(P1)755, 96JOC7652>. A number of 2-substituted-2,5,6,7-tetrahydrooxepines have been shown to undergo kinetic resolution with ethylmagnesium bromide in the presence of zirconium catalysts <96JA4291>.

7.2.3 Atoms other than nitrogen or oxygen

Thiepin-1,1-dioxide undergoes a number of chromium(0) mediated [6π + 4π] cycloaddition reactions with a range of 1,3-dienes. The intermediate adduct undergoes a Ramberg-Bäcklund rearrangement to form new benzannulated products <96JOC7644>.

A versatile route to 3-benzoheteropines has been reported starting from *o*-phthalaldehyde, including the first preparations of 3-benzarsepines and the parent 3-benzothiepin and 3-benzoselenepins <96CC2183>. 1,7-Dihydro-1*H*-dibenzo[*c,e*]tellurepin has been prepared from 2,2'-bis(bromomethyl)biphenyl and potassium tellurocyanate and its complexes with palladium and ruthenium species have been studied; a number of mono- and binuclear complexes are formed <96RTC427>.

7.3 RING SYSTEMS CONTAINING TWO HETEROATOMS

7.3.1 Diazepines

1*H*-1,2-benzodiazepine undergoes [3π + 2π] cycloadditions with 3,5-dichloro-2,4,6-trimethylbenzonitrile to give a mixture of products with cycloaddition occuring at both the 2,3- and the 4,5-double bonds <96H(43)2179>.

Both 2-azido-6-trifluoromethylpyridine **41** and the isomeric tetrazolo[1,5-*a*]pyridine **42** ($R^1 = CF_3$, R^2 = H) yield 2-methoxy-4-trifluoromethyl-1*H*-1,3-diazepine **43** (R = OMe) by photolysis in methanol (Scheme 10) <96CC813>. If the photolysis is carried out in diisopropylamine-dioxane the product is the dialkylamino derivative **43** ($R = NPr^i_2$), which isomerises to the 3*H*-isomer on heating or during preparative chromatography. In contrast, photolysis of 6,8-dichlororotetrazolo[1,5-*a*]pyridine **42** ($R^1 = R^2$ = Cl) in secondary amines affords only 3*H*-derivative. Other examples of these rearrangements are given, together with full NMR data and x-ray data for the *N*-benzoyl derivative of **43**.

The ylide **44**, prepared from the corresponding diazine and tetracyanoethylene oxide, rearranges in methanol the give the 1,3-diazepine **45** <96TL1587>. The x-ray geometry for **45** is reported.

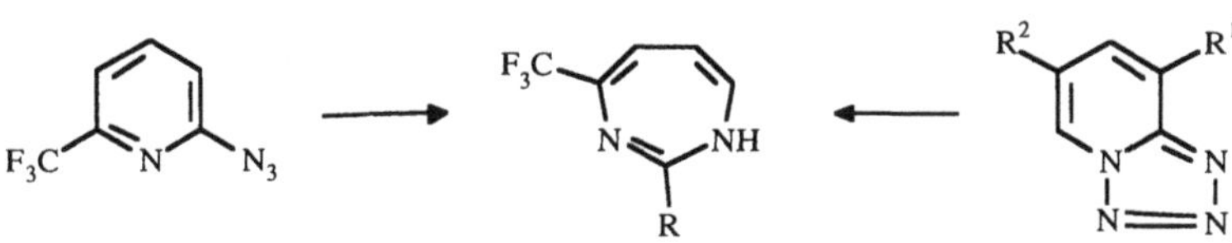

(41) (43) (42)

Scheme **10**

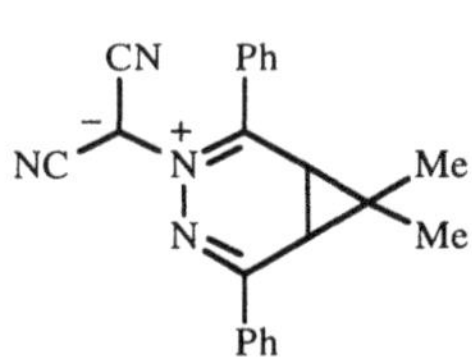

(44)

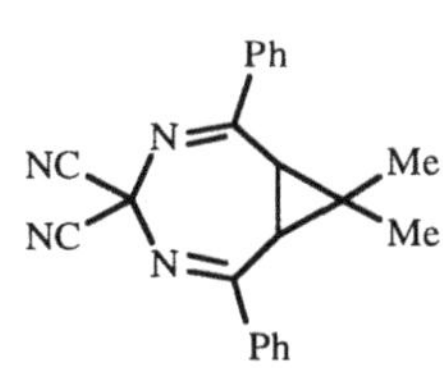

(45)

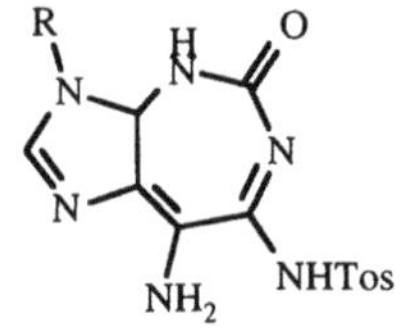

(46)

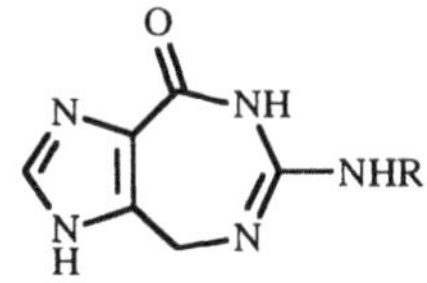

(47)

Most 1,3-diazepine syntheses employ the insertion of a single carbon fragment between the two nitrogen atoms. This is further exemplified by the use of tosyl isocyanate and methyl isothiocyanate in the preparation of the bicyclic imidazo[4,5-*e*][1,3]diazepines **46** and the imidazo[4,5-*d*][1,3]diazepines **47** respectively <96JHC855, 96JCS(P1)2257>.

1,2,3-Trihydro-5*H*-pyrrolo[1,2-*b*][2,4]benzodiazepin-5-one has been prepared by the catalytic reduction of 1-(2-ethoxycarbonylbenzyl)-2-nitropyrrole <96T4485>.

The monocyclic 1,4-diazepine, 1,4-diaza-5,7-cycloheptanedione, has in the past been reported to have been formed by the reaction of 1,2-diaminoethane with various malonic acid derivatives. It is now claimed that this derivative has never been formed, and may never be formed <96JPR121>. Oligomers and polymers are the sole products. However, reaction of 1,3-diaminopropane with the lactone **48** yields the monocyclic diazepine derivative **49** <96JHC703>.

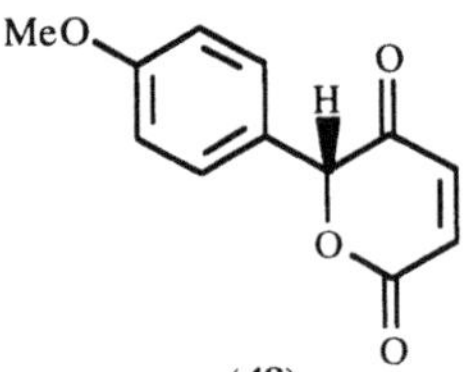

(**48**)

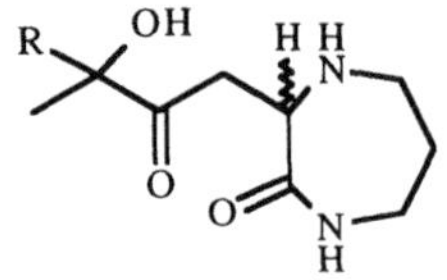

(**49**)

New routes to pyrrolo[2,1-*c*][1,4]benzodiazepines and their thiazolo[4,3-*c*] analogues have been reported <96CC385, 96TL6803, 96JHC275>. A number of new routes to pyrrolo[1,2-*a*]thieno[3,2-*f*][1,4]diazepines have also been published <96JHC75, 96JHC87>.

1,4-Benzodiazepine-2,5-diones have been prepared in a solid phase synthesis <96TL8081>. Polymer supported amino acids are reacted with *o*-nitrobenzoic acids or protected anthranilic acids which upon cyclisation and concomitant cleavage from the resin gives the products in high yields with excellent purity. Reaction of anthranilic acids with an aldehyde, amine and 1-isocyanocyclohexene, followed by acid mediated cyclisation of the intermediate Ugi four-component condensation product, provides a route to 1,4-benzodiazepine-2,5-diones, a reaction which should have potential in combinatorial syntheses <96JOC8935>. Amides formed from 3-aminopyrazine-2-carboxylic acid and α-amino acid methyl esters undergo an intramolecular aza-Wittig condensation to afford a number of pyrazino[2,3-*e*][1,4]diazepin-2,5-diones <96TL81>.

1,4-Benzodiazepin-2-ones are converted efficiently into the 3-amino derivatives by reaction with triisopropylbenzenesulfonyl (trisyl) azide followed by reduction <96TL6685>. Imines from these amines undergo thermal or lithium catalysed cycloaddition to dipolarophiles to yield 3-spiro-pyrrolidine derivatives <96T13455>. Thus, treatment of the imine **50** (R = naphthyl) with LiBr/DBU in the presence of methyl acrylate affords **51** in high yield.

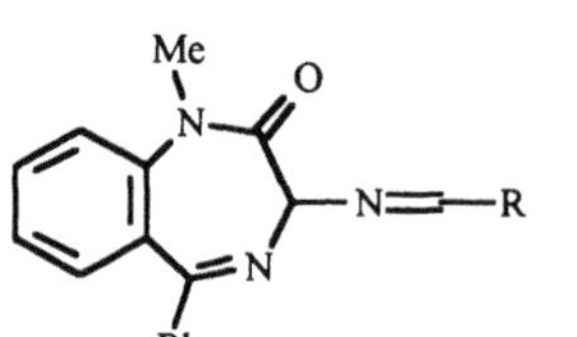

(**50**)

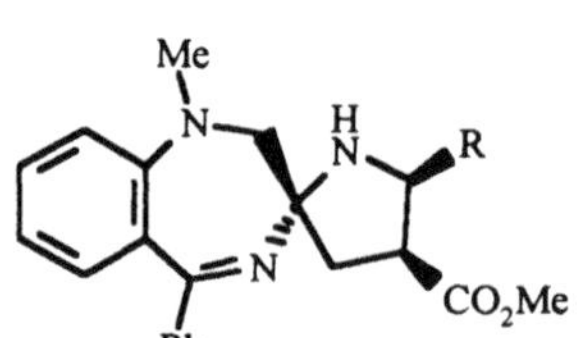

(**51**)

The [2π + 2π] and [3π + 2π] cyloadditions at the 1,2-double bond of 2-aryl-4-thiomethyl-3*H*-1,5-benzodiazepines with methoxyacetyl chloride and benzonitrile oxides respectively are reported <96JHC271, 96JHC1159>. NMR data are given for characterisation. CD measurements have been carried out on a number of chiral 2-methyl-1,3,4,5-tetrahydro-2*H*-1,5-benzodiazepines <96LA127>. It is suggested that the observed sign changes of Cotton effects indicate a conformational change of the 7-membered ring upon *N*-5 methylation.

3*H*-1,5-Benzodiazepines are frequently synthesised by the reaction of 1,2-diaminobenzenes and β-dicarbonyl compounds. It has now been demonstrated that the reaction with β-keto esters proceeds in a matter of minutes and in high yields by the use of microwave irradiation <96JCR(S)92>. 2-Phenyl-4-trichloromethyl-3*H*-1,5-benzodiazepines are readily prepared by heating with β-methoxy-β-aryl-trichloromethyl vinyl ketones in place of the usual dicarbonyl reagent <96TL9155>. Interestingly the preparation of the analogous trifluoromethyl derivatives with β-isobutoxy-β-trifluoromethyl vinyl ketones does not proceed under thermal conditions, but proceeds efficiently under microwave irradiation <96TL2845>. This method has also been used to prepare 2,3-dihydro-5-trifluoromethyl-1*H*-1,4-diazepine <96TL2845>.

7.3.2 Other systems with two heteroatoms

1,4-Dioxepan-5-ones have been prepared by the insertion of ketenes into *O,O*-acetals <96AG(E)1970>. The stereochemistry of the final product is defined by the equatorial arrangement of the substituents in the chair-shaped transition state (Scheme 11).

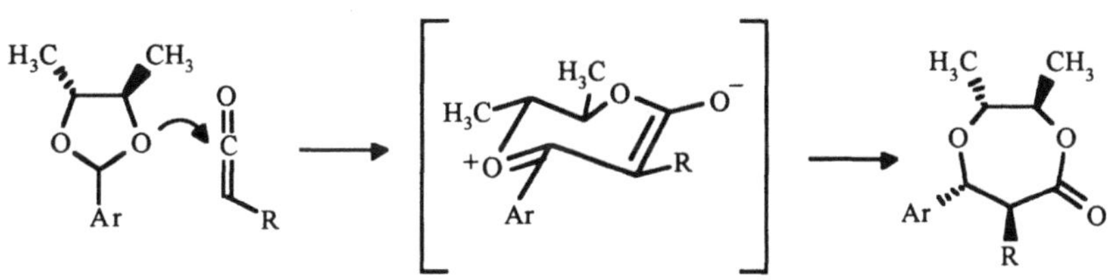

Scheme **11**

Treatment of the 1,2-oxazines **52** with carbon monoxide at 1000 psi in the presence of cobalt carbonyl brings about insertion of carbon monoxide to form the 1,3-oxazepines **53** <96TL2713>. A convenient route to β-lactams fused to oxepines is made available by alkene metathesis. Thus reaction of 4-acetoxyazetidin-2-one with allyl alcohol in the presence of zinc acetate, followed by *N*-allylation of the nitrogen affords the derivative **54** which cyclises by RCM to form the oxazepinone **55** <96CC2231>. The same communication describes a similar synthesis of 1,3-dioxepines.

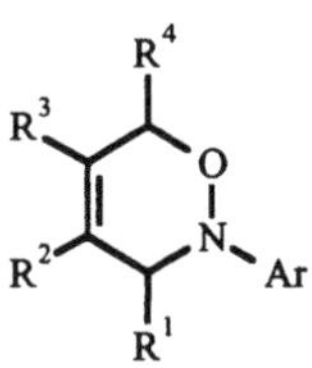

(52)

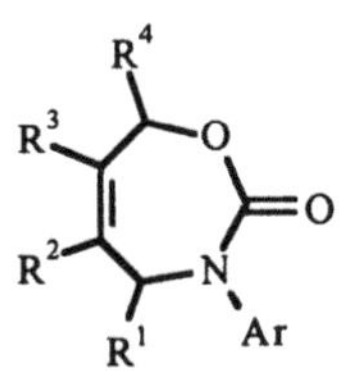

(53)

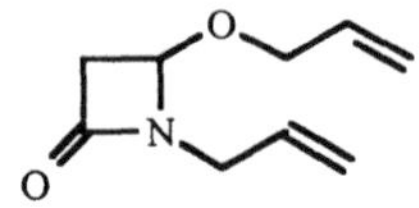

(54)

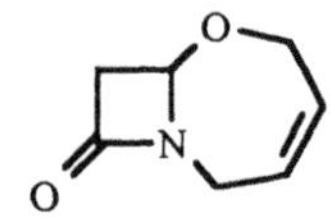

(55)

Sequential hydrostannation-palladium catalysed cyclisation of the allene **56** leads to formation of the vinyltetrahydobenzoxazepine **57** <96T13441>.

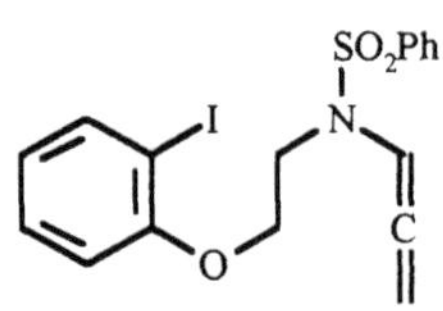

(56)

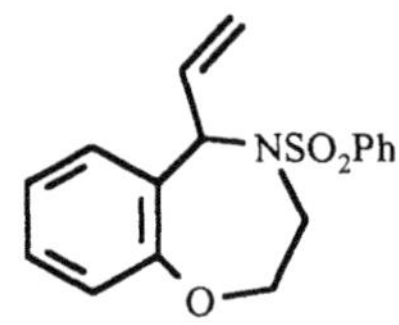

(57)

The Meisenheimer rearrangement of azetopyridoindoles **58** with *m*-CPBA results in the formation of the fused 1,2-oxazepines **59**, a reaction which has been used in the synthesis of a number 12-carbaeudistomin analogues <96CPB900> and oxidation of the *m*-cyclophane **60** with NBS in the presence of AIBN brings about transannular cyclisation to form the 1,4-oxazepines **61**, together with bromo derivatives and brominated starting material <96SL1061>. If the methoxycarbonyl group is absent, the main product is the corresponding indoline.

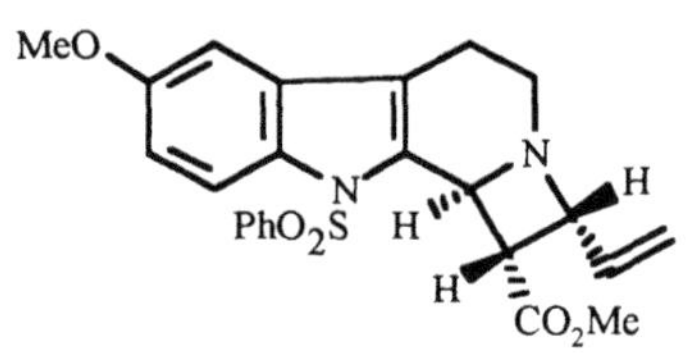

(58)

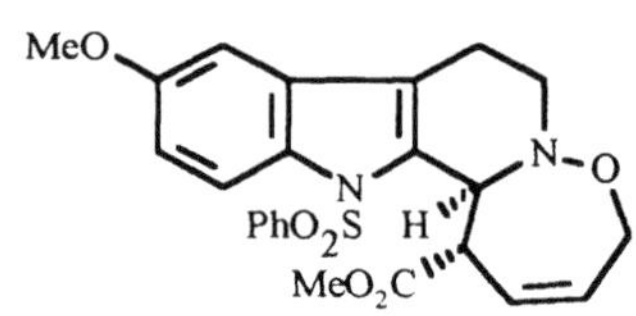

(59)

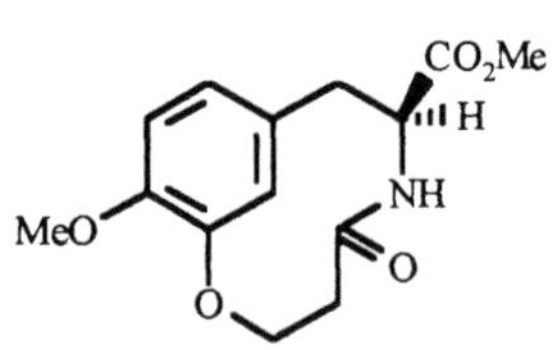

(60)

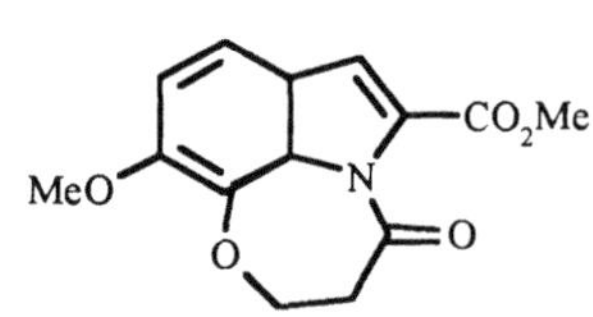

(61)

The pharmacological activity of the 1,5-benzothiazepine derivative diltiazem has given further impetus for synthetic routes to this ring system. A traditional preparation of the intermediate **62** by the reaction sequence shown in Scheme 12 has been subject to microwave studies, when the final product was obtained as a mixture of isomers <96TL6413>. Under optimised conditions irradiation in toluene at 390 watts for 20 minutes gave mainly the *cis*-isomer as the main product (*cis/trans* 9:1) in 75% yield. However, reaction at 490 watts in the presence of acetic acid resulted in a reversal of this ratio and a yield of 84%. The enantioselective synthesis of the ring system has also been studied using carbohydrates as chiral auxiliaries <96JCR(S)312>.

A study of the transformations of 1,2-benzoisothiazole 1,1-dioxides has revealed that the ethers **63** (R = Et, Me_3Si, X = SO_2) are transformed into the benzothiazepines **64** (R = Et, Me_3Si, X = SO_2) with 1-dimethylamino-1-propyne <96T3339>. These may be hydrolysed to the ketone **65** (X = SO_2) which in the solid state is in equilibrium with the enol **64** (R = H, X = SO_2) (Scheme 13). If the reaction is carried out with the isoindolone **63** (R = Et, X = CO), the product is the benzazepinone **64** (R = Et, X = CO).

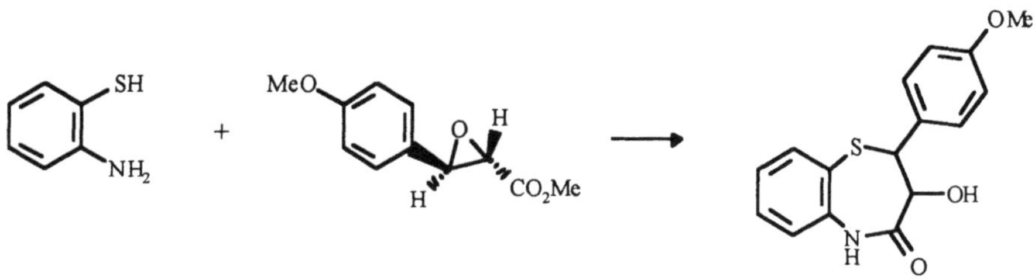

(62)

Scheme 12

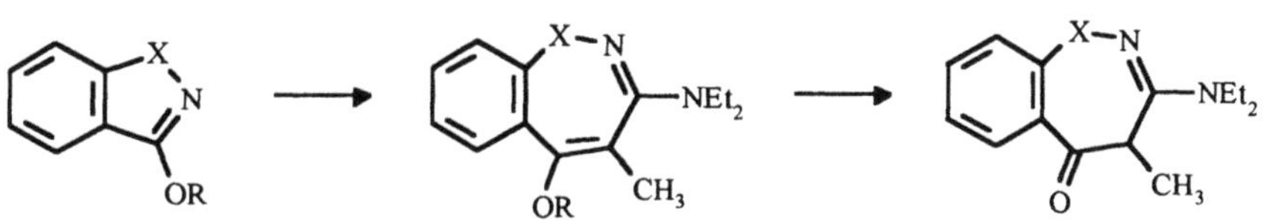

(63) (64) (65)

Scheme 13

The synthesis of a number of new pyrazolo[3,2-*c*][2,1]benzothiazepines and 2-phenylpyridazino[4,5-*b*][1,5]thiazepines is reported <96JHC151, 96JHC583>.

In a synthesis of 2,3-di(hetero)arylpyrido[3,2-*f*][1,4]thiazepines developed by Couture, 2-chloro-3-formylpyridine is reacted with arylmethylamines to form the imines. Deprotonation with LDA at –78 °C followed by treatment with non enolisable aryl thioesters gives the title compounds which may be further annulated by irradiation in benzene in the presence of iodine and propylene oxide <96S986> (Scheme 14).

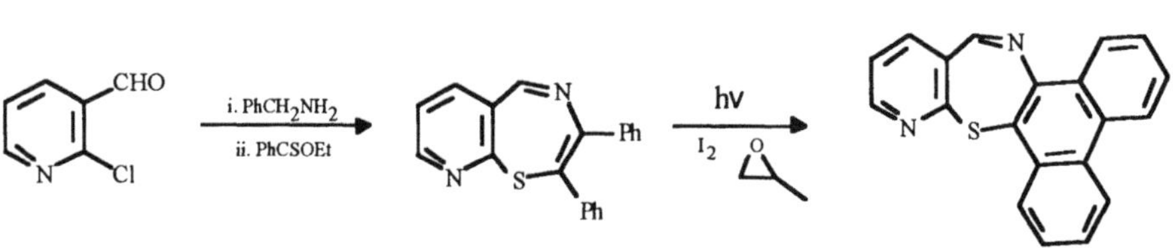

Scheme 14

The *N*-phosphino-1-azadiene **66** undergoes cycloaddition with DMAD in ether at –20 °C to form the bridged structure **67** <96AG(E)896>. The new compound is thermally unstable and isomerises to the alternative bridged structure **68** at 25 °C. X-ray analysis data have been interpreted to suggest that the isomer **69** co-exists in the solid state. The same ring system is also formed by protonation at the sp^2 carbon atom neighbouring phosphorus in **67** (Scheme 15).

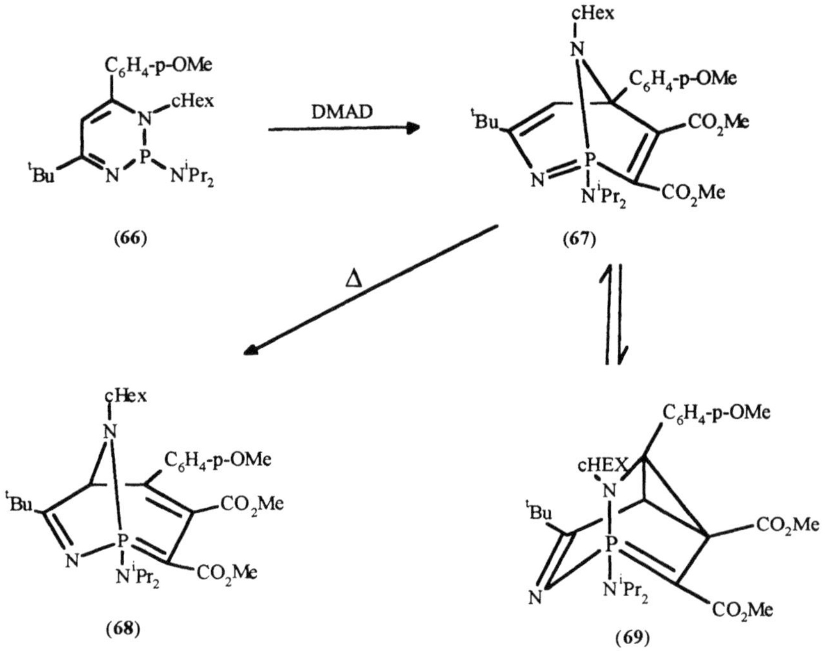

Scheme **15**

7.4 RING SYSTEMS CONTAINING THREE NITROGEN ATOMS

Reduction of the nitro group in **70** with zinc in hot sodium hydroxide results in formation of pyrrolo[2,1-*d*][1,2,5]benzotriazepine **71** by intramolecular coupling of the amino group with the newly formed nitroso group (Scheme 16). If the reduction is carried out under the less rigorous conditions of zinc in aqueous ammonium chloride the intermediate hydroxy compound is formed <96T10751>.

Treatment of the acylproline esters **72** with methylhydrazine results in the cyclisation with inversion to form the 1,3,5-triazepin-3,5-diones **73** <96CC85>. 4,5-Dihydro-1*H*-1,3,4-benzotriazepines have been prepared by the cyclisation of amidrazones with carbonyl compounds <95AP505, 96H(43)2549> and a route to 3,4-dihydro-1*H*-1,3,4-benzotriazepine-2,5-diones is reported <96JHC1131>.

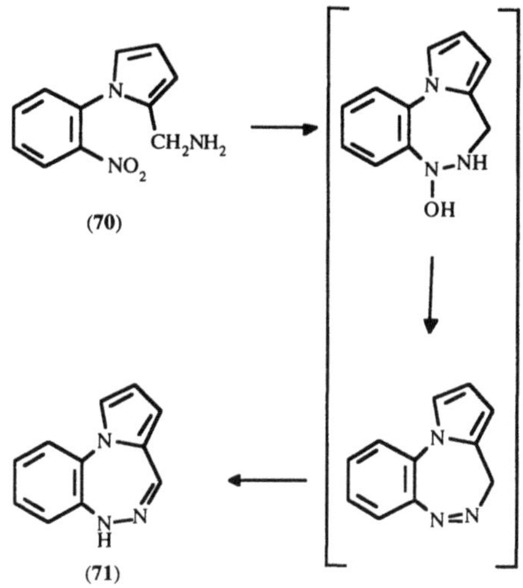

(**70**) (**71**)

Scheme **16**

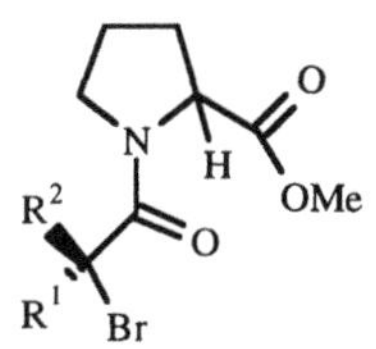

(**72**)

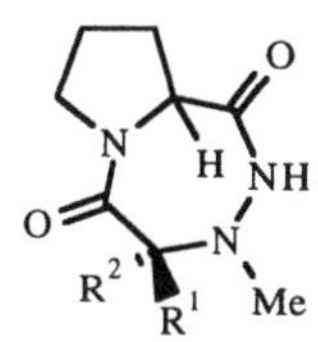

(**73**)

7.5 REFERENCES

85CB4035	W. Eberbach, U. Trostmann, *Chem. Ber.* **1985**, *118*, 4035.
95T9611	B. Kharraz, P. Uriac, L. Toupet, J. P. Hurvois, C. Moinet, A. Tallec, *Tetrahedron* **1995**, *51*, 9611.
95TL7065	S. Klötgen, E.-U. Würthwein, *Tetrahedron Lett.* **1995**, *36*, 7065.
96AG(E)896	J. Barluenga, M. Tomás, K. Bieger, S. García-Granda, R. Santiago-García, *Angew. Chem. Int. Ed. Engl.* **1996**, *35*, 896.
96AG(E)1970	J. Mulzer, D. Trauner, J. W. Bats, *Angew. Chem., Int. Ed. Engl.* **1996**, *35*, 1970.
96AHC(65)40	M. Bohle, J. Liebscher, *Adv. Heterocycl. Chem.* **1996**, *65*, 40.
96CC19	Y. Nagao, I.-Y. Jeong, W. S. Lee. S. Sano, 1996, *Chem. Commun.* **1996**, 19.
96CC85	M. M. Lenmann, S. L. Ingham, D. Gani, *Chem. Commun.* **1996**, 85.
96CC385	A. Kamal, N. V. Rao, *Chem. Commun.* **1996**, 385.
96CC813	A. Reisinger, C. Wentrup, *Chem. Commun.* **1996**, 813.
96CC1071	L.-X. Dai, R. Hayasaka, Y. Iwaki, K. A. Koyano, T. Tatsumi, *Chem. Commun.* **1996**, 1071.
96CC1481	Y. Kita, M. Gyoten, M. Ohtsubo, H. Tohma, T. Takada, *Chem. Commun.* **1996**, 1481.
96CC1621	T. Tsuchiya, A. Oishi, I. Shibuya, Y. Taguchi, K. Honda, *Chem. Commun.* **1996**, 1621.
96CC2183	S. Yakuiske, T. Kiharada, J. Kurita, T. Tsuchiya, 1996, *Chem. Commun.* **1996**, 2183.
96CC2231	A. G. M. Barrett, S. P. D. Baugh, V. C. Gibson, M. R. Giles, E. L. Marshall, P. A. Procopiou, *Chem. Commun.* **1996**, 2231.
96CC2257	M. Grellier, M. Pfeffer, *Chem. Commun.* **1996**, 2257.
96CC2739	J.-P. Strachan, J. T. Sharp, S. Parsons, *Chem. Commun.* **1996**, 2739.
96CHECII(9)	Comprehensive Heterocyclic Chemistry II, ed. A. R. Katritzky, C. W. Rees, E. F. V. Scriven, Pergamon, Oxford, 1996; Volume 9, ed. G. R. Newkome.
96CPB900	T. Kurihara, Y. Sakamoto, T. Kimura, H. Ohishi, S. Harusawa R. Yoneda, T. Suzutani, M. Azuma, *Chem. Pharm. Bull.* **1996**, *44*, 900.

96H(42)475	A. I. Meyers, R. Hutchings, *Heterocycles* **1996**, *42*, 475.
96H(43)2091	N. Kakusawa, K. Sakamoto, J. Kurita, T. Tsuchiya, *Heterocycles* **1996**, *43*, 2091.
96H(43)2179	P. Beltrame, E. Cadoni, M. M. Carnasciali, G. Gelli, A. Mugnoli, *Heterocycles* **1996**, *43*, 2179.
96H(43)2549	P. Frohberg, P. Nuhn, *Heterocycles* **1996**, *43*, 2549.
96JA4264	B. El Ali, K. Okuro, G. Vasapollo, H. Alper, *J. Amer. Chem. Soc.* 1996, *118*, 4264.
96JA4291	M. S. Visser, N. M. Heron, M. T. Didiuk, J. F. Saga, A. H. Hoveyda, *J. Amer. Chem. Soc.* **1996**, *118*, 4291.
96JA7647	F. Morís-Varas, X.-H. Qian, C.-H. Wong, *J. Amer. Chem. Soc.* **1996**, *118*, 7647.
96JBC3097	C. E. Thomas, D. F. Ohlweiler, A. A. Carr, T. R. Nieduzak, D. A. Hay, G. Adams, R. Vaz, R. C. Bernotas, *J. Biol. Chem.* **1996**, *271*, 3097.
96JCR(S)92	M. S. Khajavi, M. Hajihada, R. Naderi, *J. Chem. Res. (S)* **1996**, 92.
96JCR(S)312	A. Nangia, P. B. Rao, N. N. Madhavi, *J. Chem. Res. (S)* **1996**, 312.
96JCS(P1)515	D. F. O'Shea, J. T. Sharp, *J. Chem. Soc., Perkin 1* **1996**, 515.
96JCS(P1)755	J. D. Stewart, K. W. Reed, M. M. Kayser, *J. Chem. Soc., Perkin 1* **1996**, 755.
96JCS(P1)2257	P. K. Bridson, H. Huang, X. Lin, J. *Chem. Soc., Perkin 1* **1996**, 2257.
96JCS(P1)2799	J. Löffler, R. Schobert, *J. Chem. Soc., Perkin 1* **1996**, 2799.
96JHC75	M.-P. Foloppe, P. Sonnet, I. Bureau, S. Rault, M. Robba, *J. Heterocycl. Chem.* **1996**, *33*, 75.
96JHC87	M. Boulouard, S. Rault, P. Dallemagne, A. Alsaïdi, M. Robba. *J. Heterocycl. Chem.* **1996**, *33*, 87.
96JHC151	E. Arranz, J. A. Díaz, E. Morante, C. Pérez, S. Vega, *J. Heterocycl. Chem.* **1996**, *33*, 151.
96JHC209	J. F. Hayes, J. D. Hayler, T. C. Walsgrove, C. Wicks, *J. Heterocycl. Chem.* **1996**, *33*, 209.
96JHC259	M. L. Miller, P. S. Ray, *J. Heterocycl. Chem.* **1996**, *33*, 259.
96JHC271	R. Martínez, P. E. Hernández, E. Angeles, *J. Heterocycl. Chem.* **1996**, *33*, 271.
96JHC275	A.-C. Gillard, S. Rault, M. Boulourd, M. Robba, *J. Heterocycl. Chem.* **1996**, *33*, 275.
96JHC583	P. Mátyus, E. Zára-Kaczián, S. Boros, Z. Böcskei, *J. Heterocycl. Chem.* **1996**, *33*, 583.
96JHC703	C. D. Apostolopoulos, M. P. Georgiadis, E. A. Couladouros, *J. Heterocycl. Chem.* **1996**, *33*, 703.
96JHC855	A. M. Dias, M. F. J. R. P. Proença, *J. Heterocycl. Chem.* **1996**, *33*, 855.
96JHC1131	G. M. Karp, *J. Heterocycl. Chem.* **1996**, *33*, 1131.
96JHC1159	E. C. Cortés, A. M. M. Ambrosia, *J. Heterocycl. Chem.* **1966**, *33*, 1159.
96JHC1333	H. Y. Neela, S. Ramakumar, M. A. Viswamitra, *J. Heterocycl. Chem.* **1996**, *33*, 1333.
96JOC10	V. Gracias, G. L. Milligan, J. Aubé, *J. Org. Chem.* 1996, *61*, 10.
96JOC2252	P. A. Evans, J. D. Roseman, *J. Org. Chem.* **1996**, *61*, 2252.
96JOC5793	Y. Brunel, G. Rousseau, *J. Org. Chem.* **1996**, *61*, 5793.
96JOC5857	Y. Kita, T. Takada, M. Gyoten, H. Tohma, M. H. Zenk, J. Eichhorn, *J. Org. Chem.* **1996**, *61*, 5857.
96JOC7644	J. H. Rigby, N. C. Warshakoon, *J. Org. Chem.* **1996**, *61*, 7644.
96JOC7652	J. D. Stewart, K. W. Reed, J. Zhu, G. Chen, M. M. Kayser, *J. Org. Chem.* **1996**, *61* 7652.
96JOC8356	A. Kinoshita, M. Mori, *J. Org. Chem.* **1996**, *61*, 8356.
96JOC8935	T. A. Keating, R. W. Armstrong, *J. Org. Chem.* **1996**, *61*, 8935.
96JPR121	P. Imming, M. Resch, *J. Prakt. Chem.* **1996**, *338*, 121.
96LA127	F. Malik, M. Hasan, K. M. Khan, S. Perveen, G. Snatzke, H. Duddeck, W. Voelter, *Justus Liebigs Ann. Chem.* **1996**, 127.
96LA887	D. Kowalski, G. Erker, S. Kotila, *Justus Liebigs Ann. Chem.* **1996**, 887.
96MI1	G. R. Proctor, J. Redpath, "Monocyclic Azepines", John Wiley & Sons, New York, 1996.
96RTC427	A. Z. Al-Rubaie, A. Y. Al-Marzook, S. A. N. Al-Jadaan, *Recl. Trav. Chim. Pays-Bas* **1996**, *115*, 427.
96S986	A. Couture, E. Deniau, P. Grandclaudon, C. Simion, *Synthesis*, **1996**, 986.
96SC1839	R. Di Santo, R. Costi, S. Massa, M. Artico, *Synth. Commun.* **1996**, *26*, 1839.
96SC2241	F. J. Urban, R. Breitenbach, D. Gonyaw, S. E. Kelly, *Synth. Commun.* **1996**, *26*, 2241.
96SC4591	M. Hirano, S. Yakabe, A. Satoh, J. H. Clark, T. Morimoto, *Synth. Commun.* **1996**, *26*, 4591.
96SL29	E. Albertini, A. Barco, S. Bennetti, C. De Risi, G. P. Pollini, V. Zanirato, *Synlett.* **1996**, 29.
96SL351	S. Hosokawa, M. Isobe, *Synlett.* **1996**, 351.
96SL1061	R. Rexaie, J. B. Bremner, *Synth. Lett.* **1996**, 1061.
96SL1165	T. Oishi, M. Shoji, K. Maeda, N. Kumahara, M. Hirama, *Synlett.* **1996** 1165.
96T3339	R. A. Abramovitch, I. Shinkai, B. J. Mavunkel, K. M. More, S. O'Connor, G. H. Ooi, W. T. Pennington, P. C. Srinivasan, J. R. Stowers, *Tetrahedron* **1996**, *52*, 3339.

96T3403 C. J. Mossman, J. Aubé, *Tetrahedron* **1996**, *52*, 2403.
96T4423 B. Kharraz, P. Uriac, S. Sinbandhit, L. Toupet, *Tetrahedron* **1996**, *52*, 4423.
96T4485 J. Cobb, I. N. Demetropoulos, D. Korakas, S. Skoulika, G. Varvounis, *Tetrahedron* **1996**, *52*, 4485.
96T5739 M. L. Miller, P. S. Ray, *Tetrahedron*. **1996**, *52*, 5739.
96T6519 R. C. Bernotas, G. Adams, A. A. Carr, *Tetrahedron* **1996**, *52*, 6519.
96T7251 S. F. Martin, H.-J. Chen, A. K. Courtney, Y. Liao, M. Pätzel, M. N. Ramser, A. S. Wagman, *Tetrahedron* **1996**, *52*, 7251.
96T7525 J. P. Wolfe, R. A. Rennels, S. L. Buchwald, *Tetrahedron* **1996**, *52*, 7525.
96T10751 D. Korakas, A. Kimbaris, G. Varvounis, *Tetrahedron* **1996**, *52*, 10751.
96T10883 D. Hamprecht, J. Josten, W. Steglich, *Tetrahedron* **1996**, *52*, 10883.
96T11637 P. J. Parsons, N. P. Camp, J. P. Camp, J. M. Underwood, D. M. Harvey, *Tetrahedron*, **1996**, *52*, 11637.
96T13081 M. Noguchi, T. Mizukoshi, A. Kakehi, *Tetrahedron* **1996**, *52*, 13081.
96T13097 M. Noguchi, T. Mizukoshi, T. Uchida, Y. Kuroki, *Tetrahedron* **1996**, *52*, 13097.
96T13111 M. Noguchi, T. Mizukoshi, S. Nakagawa, A. Kakehi, *Tetrahedron* **1996**, *52*, 13111.
96T13441 R. Grigg, J. M. Sansano, *Tetrahedron*, **1996**, *52*, 13441.
96T13455 H. Ali Dondas, R. Grigg, M. Thornton-Pett, *Tetrahedron* **1996**, *52*, 13455.
96T14801 S. Klötgen, R. Fröhlich, E.-U. Würthwein, *Tetrahedron* **1996**, *52*, 14801.
96TL81 T. Okawa, S. Eguchi, *Tetrahedron Lett.* **1996**, *37*, 81.
96TL213 T. Nakata, S. Nomura, H. Matsukura, *Tetrahedron Lett.* **1996**, *37*, 213.
96TL217 T. Nakata, S. Nomura, H. Matsukura, M. Morimoto, *Tetrahedron Lett.* **1996**, *37*, 217.
96TL339 S. D. Rychnovsky, V. H. Dahanukar, *Tetrahedron Lett.* **1996**, *37*, 339.
96TL393 Y. Nagao, I-Y. Jeong, W. S. Lee, *Tetrahedron Lett.* **1966**, *37*, 393.
96TL1531 J. A. Wendt, J. Aubé, *Tetrahedron Lett.* **1996**, *37*, 1531.
96TL1587 P. Riebel, A. Weber, T. Troll, J. Sauer, J. Breu, H. Nöth, *Tetrahedron Lett.* **1996**, *37*, 1587.
96TL2713 K. Okuro, T. Dang, K. Khumtaveeporn, H. Alper, *Tetrahedron Lett.* **1996**, *37*, 2713.
96TL2845 A. C. S. Reddy, P. S. Rao, R. V. Venkataratnam, *Tetrahedron Lett.* **1996**, *37*, 2845.
96TL2865 E. Alvarez, M. Delgado, M. T. Diaz, L. Hanxing, R. Pérez, J. D. Martin, *Tetrahedron Lett.* **1996**, *37*, 2865.
96TL2869 J. L. Ravelo, A. Regueiro, E. Rodriguez, J. De Vera, J. D. Martin, *Tetrahedron Lett.* **1996**, *37*, 2869.
96TL6365 M. Morimoto, H. Matsukura, T. Nakata, *Tetrahedron Lett.* **1996**, *37*, 6365.
96TL6413 J. A. Vega, S. Cueto, A. Ramos, J. J. Vaquero, J. L. García-Navio, J. Alverez-Builla, *Tetrahedron Lett.* **1996**, *37*, 6413.
96TL6565 R. Grigg, V. Savic, *Tetrahedron Lett.* **1996**, *37*, 6565.
96TL6685 J. W. Butcher, N. J. Liverton, H. G. Selnick, J. M. Elliot, G. R. Smith, A. J. Tebben, D. A. Pribush, J. S. Wai, D. A. Claremon, *Tetrahedron Lett.* **1996**, *37*, 6685.
96TL6803 A. Kamal, B. S. P. Reddy, B. S. N. Reddy, *Tetrahedron Lett.* **1996**, *37*, 6803.
96TL8081 J. P. Mayer, J. Zhang, K. Bjergarde, D. M. Lenz, J. J. Gaudino, *Tetrahedron Lett.* **1996**, *37*, 8081.
96TL8249 J. H. Van Maarseveen, J. A. J. Den Hartog, V. Engelen, E. Finner, G. Visser, C. G. Kruse, *Tetrahedron Lett.* **1996**, *45*, 8249.
96TL9155 H. G. Bonacorso, A. T. Bittencourt, A. D. Wastowski, A. P. Wentz, N. Zanatta, M. A. P. Martins, *Tetrahedron Lett.* **1966**, *27*, 9155.

Chapter 8

Eight-Membered and Larger Rings

George R. Newkome
University of South Florida, Tampa, FL, USA

8.1 INTRODUCTION

In the first half of the nineties, there has been a continuing trend from synthetic studies of classical "crown ethers" towards the polyazamacromolecules and the introduction of multiple heteroatoms, including most recently the metal atom centers.

Numerous reviews and perspectives have appeared throughout the year that are of interest to the macroheterocyclic scientist: lanthanide complexes with cyclen <96JCS(D)3613>, dimeric tetrathiafulvenes <96AM203>, mono- and multilayers containing macrocyclic ionopheres <96AM615>, template-directed syntheses of rotaxanes <96CCCC1>, synthetic models for reaction center complexes assembled around transition metals <96CSR41>, pseudorotaxane superstructures <96CC1483>, pyridinocalix[4]arenes <96NJC465>, azaporphyrins <96CCR41>, synthesis of porphyrins with exocyclic rings from cycloalkenopyrroles <95MI1>, second sphere coordination <96CB981, 96MI8>, molecular recognition <96RTC109, 96PAC1279>, anion complexation <96RTC307>, the Mannich reaction in the preparation of azamacroheterocycles <96SL933>, metal ion interactions in mixed solvents <96BE383>, polymeric crown ethers <95MI3>, lariat ethers <96MI3>, crown ethers <96MI4>, complexation of organic cations <96MI5>, anion activation by crown ethers and cryptands <96MI6>, chromoionophores <96MI7>, crown ether metal complexes <96MI9>, molecular recognition of stable metal complexes <96PAC1225>, simultaneous binding of cations and anions <96MI10>, self-assembly in chemical systems <96PAC1225>, complexation mechanisms <96MI12>, complexation kinetics <96MI13>, troponoid thiacrown ethers <96CCR71>, armed crown ether complexes <96CCR1>, silicon-bridged macrocycles <96MI14>, alkylphosphoric acid armed crown ethers <96CCR97>, chiral and meso-crown ethers <96CCR199>, thallium(I) selective crown ethers <96CCR171>, lithium ion selective crown ethers <96CCR135>, ion recognition <96CCR151>, 1,4,8,11-tetraoxa-cyclotetradecane for lithium ion sensing and separation <93MI1>, photochromic crown ethers <96CCR41a>, crowned spiropyrans <96MI15>, chiral recognition <96MI16>, immobilized crown ethers <95ZFK2117>, polyrotaxanes <96MI17>, host-guest chemistry <95MI4>, crown and pseudocrown ethers as endo-receptors <96MI18>, immobilized crown ethers <95ZFK1735>, and even a special volume dedicated to macrocyclic metal complexes <96MI1> as well as to host design and recognition <95MI2>. The initial phases of the Herculean effort entitled: Comprehensive Supramolecular Chemistry has started to appear <96MI19>. The review entitled "Self-Assembly in Natural and Unnatural Systems" adds an important dimension to macroheterocycles as we enter the nanoscale regime <96AG(E)1155>.

Because of spatial limitations, only meso- and macrocycles possessing heteroatoms and/or subheterocyclic rings are reviewed; in general, lactones, lactams, and cyclic imides have been excluded. In view of the delayed availability of some articles appearing in previous years, several have been incorporated.

8.2 CARBON–OXYGEN RINGS

Novel macrocycles, derived from a dibenzo-18-crown-6, which incorporate either two cyclotriveratrylene <96JCS(P1)741> or *bis*-ferrocenyl <96P775> substructures have been reported. New derivatives of benzocrown ethers containing ω-aminoalkyl and 2,4,6-trisubstituted pyridinium units in the second coordination sphere have been generated <96LA959>. A new synthesis of (1,5)(2,4)-durenetetrayl-*bis*(18-crown-6) and *bis-m*-xylylene crowns has appeared <96SL430>. Preparations of pseudorotaxanes with secondary dialkylammonium salts and crown ethers have been reported <96CEJ709, 96CEJ729>. The process of unthreading and rethreading a [2]pseudorotaxane was induced by a chemical stimuli; this is a supramolecular prototype of a simple molecular machine <96AG(E)978>, also see: <96JA4931>.

Tethered *bis*-benzophenone or *bis*-benzaldehyde derivatives have been subjected to McMurry coupling conditions to create a series of crownophanes, which are crown/stilbene hybrids <96LA655>. The use of 1 mol% of the metathesis catalyst [$RuCl_2(PCy_3)(CHPh)$] with *bis*-allyl ethers of podands gave the corresponding macrocyclic crown ether in good yield at ambient temperatures <96SL1013>. The synthesis and initial evaluation of catalytic performance of cyclophane **1**, the first functional model of pyruvate oxidase that combines a well-defined binding site with flavin and thiazolium prosthetic groups, have been demonstrated <96AG(E)1341>.

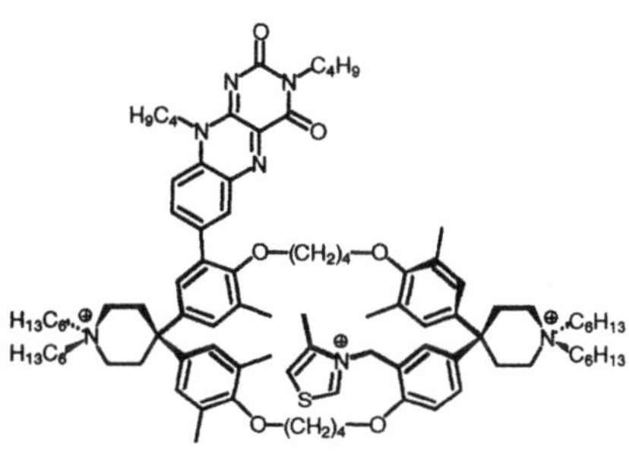

1

Chiral crown ethers possessing two chiral *cis*-1-phenylcyclohexane-1,2-diol moieties as well as a *p*-(2,4-dinitrophenylazo)phenol chromophore were prepared and with chiral alkylamines were observed to show enantiomeric selectivity <96JCS(P1)383>.

Calixcrowns have continued to be constructed due to their distinctive cavity characteristics: specifically, 1,3;2,4-*bis*crown-5-calix[4]arene <96AJC183>, bridged crowns of calix[5]arenes <96JCS(P2)1855, 96JOC8724>, chiral calix[4]arenes <96JCS(P1)1945>, *bis*(calix[4]arenes) threaded by two intramolecular bridges <96LA757>, calix[4]arene-*bis*-crowns-6 (binding properties) <96NJC453, 96TL6315>, *tris*(trimethylsilyl)oxacalix[3]arene <96NJC427>, 25,27-dialkoxycalix[4]arene-crowns-5 (conformers with better K^+/Na^+ selectivity) <96CEJ436>.

The structures and conformational properties of a simple hemicarcerand, created earlier by Cram, see: <96JA5590>, as well as the complexation and decomplexation with guest molecules have been computationally studied <96JA8056>.

The synthesis of several cyclic oligomers of furan and acetone containing four or more furan subunits has been re-examined; the [1.$_9$](2,5-furanophane) is the largest yet reported cyclic oligomer in this family <96TL4593>; chiral naphthafurophanes, prepared from furanophanes <96TL6201>, have been established <96TL6205>. The severely distorted 1,8-dioxa[8](2,7)pyrenophane has been prepared by the irreversible loss of hydrogen from the corresponding dihydropyrene <96AG(E)1320>.

8.3 CARBON–NITROGEN RINGS

Cyclophane-linked macrocyclic hosts, based on 2,11-diaza[3.3](2,6)pyridinophanes and pyridine connectors, have been shown to possess very electron-rich cavities due to the inwardly directed *N*-electrons <96JCS(P1)277>. The formation of [2]pseudorotaxanes from tetracationic cyclophanes has been reported <96AM37>. A new cyclophane (**2**) was synthesized by the cyclization of a tetrahalomethyl cyclophane with tosylamide monosodium salt; related belt-shaped and multibridged macrocycles were also reported <96JCS(P1)2061>. The longest, yet reported, molecular ribbon (**3**, n = 7) has been created by a repetitive procedure using this basic synthetic procedure <96CEJ832>. Treatment of 2,6-*bis*(halomethyl)pyridine with dimethyl sodiomalonate afforded the 24-membered (2,6)pyridinophane, which was hydrolyzed to the corresponding polycarboxylic acids <96LA297>. A conformational study of [3.3](2,6)pyridinophane has shown that its most stable isomer to be the syn (boat-boat) conformer <96LA1645>.

A facile procedure for the synthesis of *bis*polyazamacrocycles by the condensation of *N,N'*-di-BOC protected triazamacrocycles or *N,N',N''*-tri-BOC protected tetraaza-macrocycles with aromatic *bis*electrophiles was reported; the BOC protection is readily removed by the treatment with acid <96BSC(Fr)65>. Treatment of 1,5,9-triazacyclododecane with *tris*(3-chloropropyl)amine generated the 1,5,9,13-tetraaza-tricyclo[7.7.3.3^{5,13}]docosane, abbreviated [3^6]adamanzane, which is isolated in 27% yield as the inside monoprotonated bromide salt <96ACS294>. Facile syntheses of either symmetrical or unsymmetrical *bis*-tetraazamacrocycles have been shown starting from triprotected tetraazamacrocycles <96TL7711>. The synthesis and characterization of a series of *C*-substituted 1,4,7,10,13-pentaazacyclopentadecanes and their Mn(II) complexes have been investigated <96IC5213>.

A convergent 3+1 synthesis, from trinuclear protoheterophanes and a disubstituted 1,2,4-triazole, gave rise to the first [1$_4$]*meta*azolophanes and [1$_4$](*meta-ortho*)$_2$azolophanes possessing heterocyclic betaine building blocks <96T15171, 96T15189>. The first examples of [1$_6$]- and [1$_8$]heterophanes containing two 1,2,4-triazolate and two imidazolium heteroaromatic subunits have appeared <96SL285>.

When 1,2-*bis*(2-nitrophenyl)ethane is reduced with Pb-Na alloy in ethanol, the dibenzo-1,2-diazocine was not isolated but rather 11,12,23,24-tetrahydrotetrabenzo[*c,g,k,o*]-[1,2,9,10]tetraazacyclohexadecine (**4**) and the larger member in this family, 11,12,23,24,35,36-hexahydrohexabenzo[*c,g,k,o,s,w*][1,2,9,10,17,18]hexaazacyclotetracosine were isolated <96LA1213>.

8.4 CARBON–OXYGEN/CARBON–NITROGEN (CATENANES)

The template-directed preparation of cyclo*bis*(paraquat-4,4'-biphenylene (a "molecular square") has been achieved; the use of a macrocyclic hydroquinone-based polyether template incorporating an ester moiety in one polyether chain afforded a 1:1 mixture of two topologically stereoisomeric [3]catenanes <96CEJ877>.

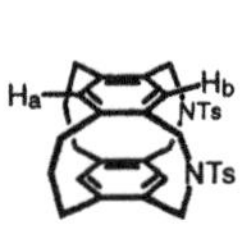

2

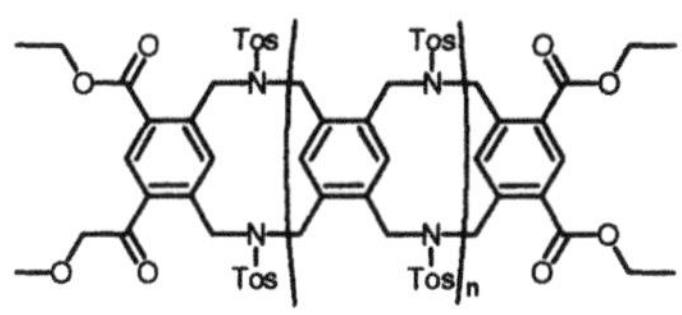

3 n = 1-7

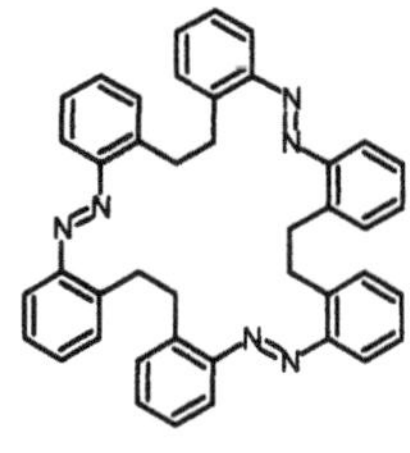

4

8.5 CARBON–SULFUR RINGS

In the synthesis of 1,3-dithiolan-2-ones from spirocyclic intermediates, via episulfides, substituted tetrathiacyclododecane and the related pentathiacyclopentadecane were isolated in good yields <96JCS(P1)289>. Preparation and molecular dynamics studies of 2,5,8,17,20,23-hexathia[9.9]-*p*-cyclophane have been reported <96P4203>. The syntheses and properties of thiocrowned 1,3-dithiole-2-thiones and their conversion to tetrathiafulvenes via treatment with triethylphosphine have been described <96LA551>.

When disodium (*Z*)-1,2-dicyanoethene-1,2-dithiolate and 1,3-dibromopropane are cyclized under high dilution conditions, the desired 14-membered macrocyclic tetrathioether was isolated <96LA1005>.

8.6 CARBON-SELENIUM RINGS

The synthesis of 1,5,9,13-tetraselenacyclohexadecan-3-ol from sodium propane-1,3-*bis*selenolate-2-ol and 4,8-diselenaundec-1,11-di-*p*-toluenesulfonate in 62% yield has been reported <96CJC533>. Macrocyclic polyselanides containing a 1,8-naphthaleno moiety have been synthesized <96JCS(P1)1783>.

8.7 CARBON–NITROGEN-OXYGEN RINGS

N,N'-Dialkyldiazacrown ethers and their precursors *bis*(alkylamino) derivatives of tri- and tetraethylene glycols were prepared <96CCCC622>. New hydroxy-bearing dibenzo-azocrown ethers have been conveniently prepared utilizing 1,3-*bis*(2-formylphenoxy)-2-propanol and a diamine, followed by reduction of the intermediate diimine <96P1197>. Fluorescent photoinduced electron transfer sensor **5** with monoaza-18-crown-6 and guanidinium receptor units demonstrated a fluorescence with γ-aminobutyric acid in a mixed aqueous solution <96CC2191, 96CC1967>. Diverse *N*-substituted (5-chloro-8-hydroxy-quinoline <96IC7229>, indole-terminated <96CC2147>, umbelliferone <96JOC7585>, 3-oxo (or aza)-3-substituted propylene <96JOC7578>, *p*-chlorophenol <96JHC933>, and 2-(4-substituted phenol) <96JOC5684> azacrown ethers have continued to proliferate due to their proposed utilitarian uses.

Cyclization of chiral diols possessing two methyl or benzyl moieties or both two methyl and phenyl substituents with 2,6-pyridinemethyl ditosylate in the presence of base afforded the symmetrical chiral diazapyridino-18-crown-6 or tetraazadipyridino-36-crown-12 macrocycles <96GCI159>. Chiral 2,16-dialkylpyridino-18-crown-6 species have been prepared in a traditional manner <96JOC7270, 96JOC8391>. A series of concave receptor molecules (**6**), functionalized with chiral aza-crown ether rings, has been constructed <96RTC357>; related cavities possessing a nearby catalytically active metal center have also been reported <96JOC4739>.

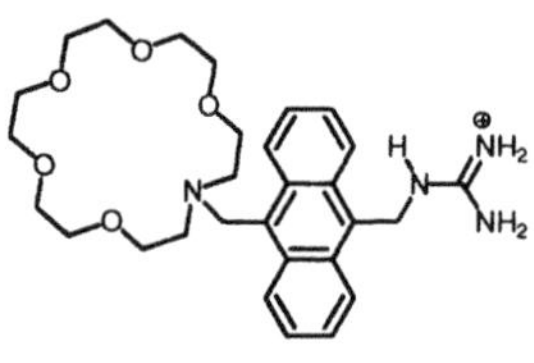

5

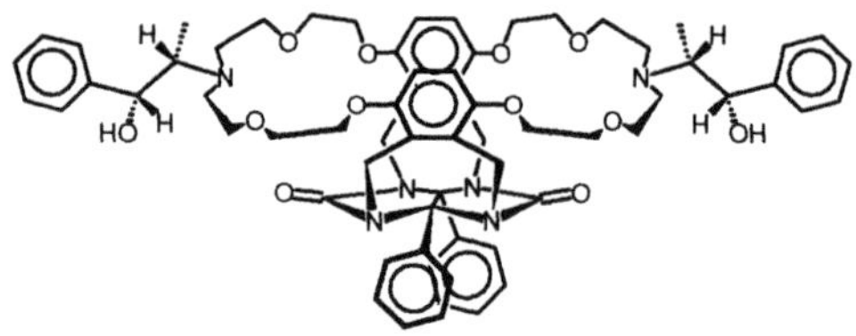

6

Partially fluorinated cryptands were synthesized from 1,3-*bis*(bromomethyl)-2-fluorobenzene and diaza-18-crown-6 or benzodiaza-18-crown-6 in good yields; these ligands formed very stable Ag(I) complexes <96CB1211>.

The synthesis of [3]- (figuratively shown as **7**) and a [5]rotaxane (**8**) with one central and two terminal porphyrins in the open configuration has been reported <96AG(E)906> also a rotaxane with two $Ru(terpy)_2$ stoppers has appeared <96CC1915>. A pseudorotaxane comprised of a macroring of 2,9-diphenyl-1,10-phenanthroline unit and a molecular string

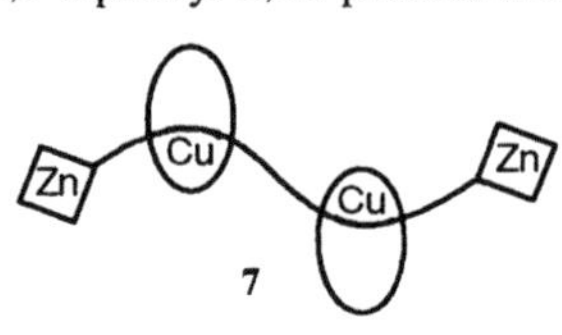

7

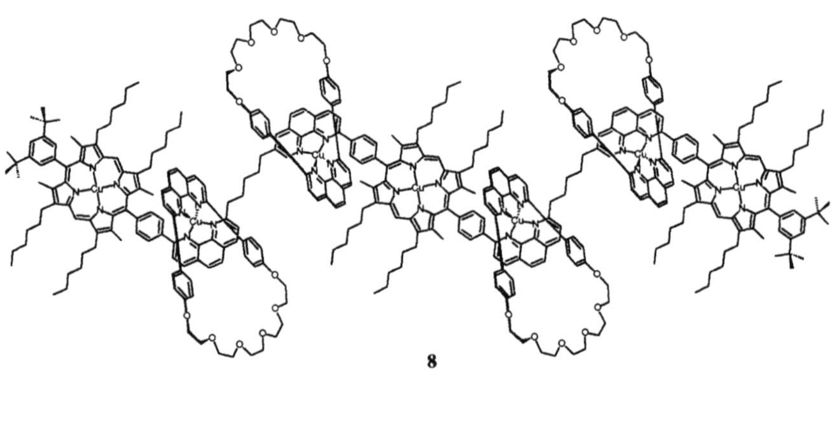

8

with two different coordination sites held together via a Cu(I) center displays electrochemically induced motion from center to center <96CC2005>. The first conducting polymetallorotaxanes have been synthesized by electrochemical polymerization of metallorotaxanes <96JA8713>.

A new type of calixarene-capped calixpyrrole (**9**) has been generated (32 %) in one-step from *p-tert*-butylcalix[4]arene tetramethylketone as the template <96TL7881>. A cylindrical calix[4]-*bis*-cryptand, in which the central calix[4]arene possesses two 1,3-alternating diaza-tetraoxa macrocycles on each face, has been synthesized <96TL8747>. A series of substituted 1,4-(2,6-pyridino)-bridged calix[6]arenes has prepared and studied <96LA1367>.

8.8 (CARBON-NITROGEN-OXYGEN)^n (n>1) RINGS (CATENANES)

The first synthesis of a crossed [2]-catenane and its topological isomer - a singly interlocked [2]-catenane - by means of a Cu(I) template procedure has been reported <96NJC685>. The dicopper(I) knotted compound was shown to be stable and that it can be demetalated in a step-wise manner; the singly metalated Cu(I) complex can be combined with a second metal and thus converted to a heterodinuclear specie <96AG(E)1119>. The crystallization of the diastereomeric salts of a dicopper(I) trefoil knot permitted separation into its enantiomers; demetalation and remetalation with Cu(I) afforded the starting dicopper(I) molecular knot with the same $[\alpha]_D$ value <96JA10932>. A *bis*(porphyrin) **10** has been created in which a catenane acts as a spacer between a donor and an acceptor porphyrins <96CC2441>. A series of "molecular composite knots" have been prepared from transition metal-assembled [Cu(I)] 1,10-phenanthroline precursors by means of a Glaser acetylene coupling reaction <96JA9110, 96T10921>.

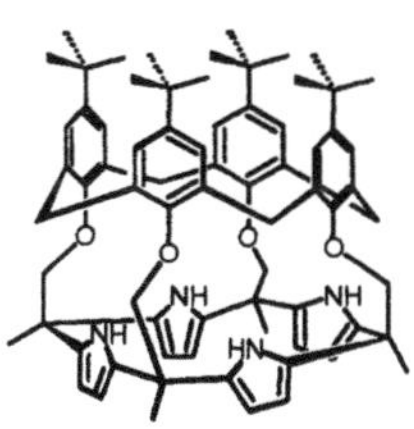

9

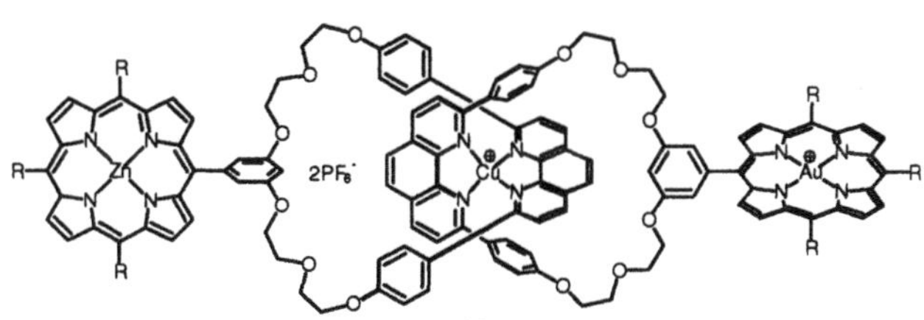

10

8.9 CARBON–SULFUR-OXYGEN RINGS

The reaction of 2,3-(4'-aminobenzo)-1,4-dioxa-7,10,13-trithiacyclopentadecane-2-ene, prepared from the corresponding nitro compound, was treated with *E,E*-dichloroglyoxime to generated a *E,E*-dioxime, which forms a stable Co(III) complex <96P3933>. The step-wise preparation of linear and mono/dimacrocyclic tetrathiafulvene, possessing aryl subunits has been reported <96CEJ624>. The synthesis and electrochemistry of unsymmetrical and functionalized TTF-crown ethers, as well as the spectroscopic studies of their metal complexes, have been delineated <96JCS(P1)1587>.

8.10 CARBON–NITROGEN–SULFUR RINGS

A simple and efficient preparation of three-directional *S,N*-macrocycles has been reported by the application of a protecting group strategy <96JCS(P1)1553>. Two thiacrown 1,10-phenanthrolinophanes have been synthesized from 2,9-di(chloromethyl)1,10-phenanthroline with the dithiol in the presence of Cs_2CO_3 <96JCS(D)3705>. A simple dithia pyrido-bridged 9,10-anthracenophane has been prepared and specifically designed as a luminescent chemosensor for “soft” cation recognition <96BSC(Fr)199>.

8.11 CARBON-SELENIUM-SULFUR

Condensation of dicesium 2-thioxo-1,3-dithiole-4,5-diselenolate with *bis*-alkylating polythioethers under high dilution conditions afforded the TTF-containing macrocycles possessing soft donor sites and 12-, 15-, and 18-membered rings <96JCS(P1)1995>.

8.12 CARBON–NITROGEN–PHOSPHORUS-OXYGEN RINGS

Semirigid phosphonamide ligands have been synthesized from the corand precursors by the reaction of 1,3-propanediol ditosylate or 1,2-dichloroethane <96JOC8904>.

8.13 CARBON-OXYGEN-METAL RINGS

A two-step synthetic procedure to isomeric redox-active metallocyclophanes $\{Mo(NO)Tp^*(4,4'\text{-}OC_6H_4\text{-}C_6H_4O)]_n$, where n = 3 or 4, has been reported; both triangular and square arrays have been observed <96P1553>.

8.14 CARBON–NITROGEN-METAL RINGS

Treatment of 4,4'-bipyridine with $[Pd(ethylenediamine)(NO_3)_2]$ was shown to form molecular squares <96MI2> and that they probably exist in equilibrium with molecular triangles <96CC1535>. Extension of this general procedure to incorporate octahedral rhenium corners and the use of pyrazine as well as 1,2-*bis*(4-pyridyl)ethylene has recently appeared <96IC4096>. The reaction of *R*-(+)-binap Pd(II) and Pt(II) *bis*(triflate) complexes with 2,6-diazaanthracene gave excellent yields of self-assembled, single, diastereomeric, stable molecular squares **11** <96AG(E)732>; for related nanoscale transition metal molecular squares, see: <96JA8221, 96JA8731>. Treatment of $[(AuOTf)_2(\mu\text{-dcypm})]$ with 1,4-di*iso*cyanidobenzene gave metallocycle **12**, possessing four gold(I) centers (dotted Au(I)..Au(I) distance is 3.133(3)Å <96CC1281>.

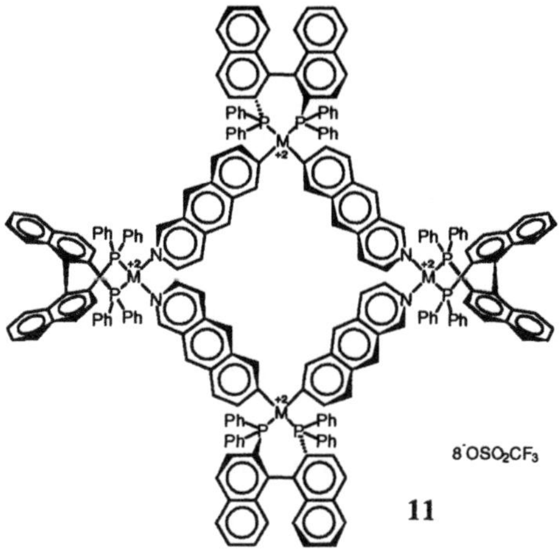

11

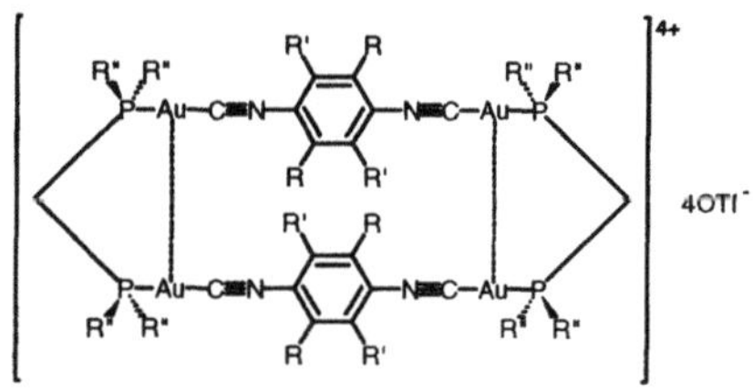

12

8.15 CARBON-NITROGEN-OXYGEN-METAL RINGS

Complexes of 1,1'':1',1'''-*bis*(1,4,10,13-tetraoxa-7,16-diazacyclooctadecane-7,16-dimethylene)*bis*ferrocene (**13**), prepared from ferrocene *bis*-acid chloride and diazamacrocycles, with diverse cations have been reported <96JOMC13>.

8.16 CARBON-PHOSPHORUS-METAL RINGS

Treatment $[Pt_2Cl_2(\mu\text{-dppm})_2]$ with 1 equivalent of $i\text{-}Pr_2PC_6H_4\text{-}C_6H_4Pi\text{-}Pr_2$ gave a product characterized as the dimeric cyclic complex $[Pt_2(\mu\text{-dppm})_2(\mu\text{-}i\text{-}Pr_2PC_6H_4\text{-}C_6H_4Pi\text{-}Pr_2]_2(BF_4)_2$ <96O5321>.

8.17 NITROGEN-OXYGEN-METAL RINGS

Lastly, an interesting approach to heteromacrocyclic chemistry is to create metallacrowns (**14**), which incorporate transition metal ions and nitrogen atoms into the methylene positions of the crown ethers, thus the metallacrown substitutes an $[M^{x+}NO]_n$ for $[CH_2CH_2O]_n$ repeat units <96IC6184>. "The [therein] reported 12-metallacrown-4 structural parameters compare favorably with those of 12-crown-4, an organic crown ether, as well as with those of the topologically similar alkali metal complexes of porphyrin and phthalocyanine dianions, solidifying the structural analogy between metallacrowns and crown ethers."

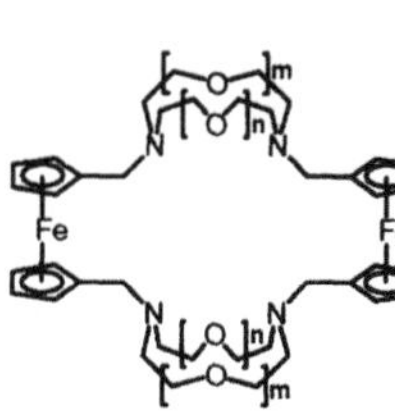

13

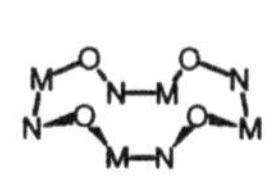

14

8.18 REFERENCES

93MI1	K. Kimura, *Trends Org. Chem.* **1993**, *4*, 203.
95CCR55	J. Vaugeois, M. Simard, J. D. Wuest, *Coord. Chem. Rev.* **1995**, *145*, 55.
95MI1	T. D. Lash, in "Advances in Nitrogen Heterocycles," C. J. Moody, ed., JAI Press, Greenwich, CT, Vol. 1, **1995**.
95MI2	*Topics in Current Chemistry*, E. Weber, ed., Springer-Varlag, New York, **1995**.
95MI3	E. E. Ergozhin, M.K. Kurmanaliev, *Izv. Nats. Akad. Nauk Resp. Kaz. Ser. Khim.* **1995**, 4.
95MI4	J. Dowden, J. D. Kilburn, P. Wright, *Contemp. Org. Synth.* **1995**, *2*, 289.
95ZFK2117	E. I. Grigor'ev, S.V. Nesterov, L I. Trakhtenberg, *Zh. Fiz. Khim.* **1995**, *69*, 2117.
95ZFK1735	E. I. Grigor'ev, S.V. Nesterov, L. I. Trakhtenberg, *Zh. Fiz. Khim.* **1995**, *69*, 1735.
96ACS294	J. Springborg, U. Pretzmann, C. E. Olsen, *Acta Chem. Scand.* **1996**, *50*, 294.
96AJC183	R. Abidi, Z. Asfari, J. M. Harrowfield, A.N. Sobolev, J. Vicens, *Aust. J. Chem.* **1996**, *49*, 183.
96AG(E)732	P. J. Stang, B. Olenyuk, *Angew. Chem., Int. Ed. Engl.* **1996**, *35*, 732.
96AG(E)906	N. Solladié, J.-C. Chambron, C. O. Dietrich-Buchecker, J.-P. Sauvage, *Angew. Chem., Int. Ed. Engl.* **1996**, *35*, 906.
96AG(E)978	R. Ballardini, V. Balzani, A. Credi, M. T. Gandolfi, S. J. Langford, S. Menzer, L. Prodi, J. F. Stoddart, M. Venturi, D. J. Williams, *Angew. Chem., Int. Ed. Engl.* **1996**, *35*, 978.
96AG(E)1119	C. O. Dietrich-Buchecker, J.-P. Sauvage, N. Armaroli, P. Ceroni, V. Balzani, *Angew. Chem., Int. Ed. Engl.* **1996**, *35*, 1119.
96AG(E)1155	D. Philp, J. F. Stoddart, *Angew. Chem., Int. Ed. Engl.* **1996**, *35*, 1155.
96AG(E)1320	G. J. Bodwell, J. N. Bridson, T. J. Houghton, J. W. J. Kennedy, M. R. Mannion, *Angew. Chem., Int. Ed. Engl.* **1996**, *35*, 1320.
96AG(E)1341	P. Mattei, F. Diederich, *Angew. Chem., Int. Ed. Engl.* **1996**, *35*, 1341.
96AM37	M. Asakawa, P. R. Ashton, G. R. Brown, W. Hayes, S. Menzer, J. F. Stoddart, A. J. P. White, D. J. Williams, *Adv. Mater.* **1996**, *8*, 37.
96AM203	T. Otsubo, Y. Aso, K. Takimiya, *Adv. Mater.* **1996**, *8*, 203.
96AM615	I. K. Ledner, M. C. Petty, *Adv. Mater.* **1996**, *8*, 615
96BSC(Fr)65	S. Brandès, C. Gros, F. Denat, P. Pullumbi, R. Guilard, *Bull. Soc. Chim. Fr.* **1996**, *133*, 65.
96BSC(Fr)199	I. Jacquet, J.-M. Lehn, P. Marsau, H. Andrianatoandro, Y. Barrans, J.-P. Desvergne, H. Bouas-Laurent, *Bull. Soc. Chim. Fr.* **1996**, *133*, 199.
96BE383	C. Kalidas, *Bull. Electrochem.* **1996**, *12*, 383.
96CB981	F. M. Raymo, J. F. Stoddart, *Chem. Ber.* **1996**, *129*, 981.
96CB1211	H. Plenio, R. Diodone, *Chem. Ber.* **1996**, *129*, 1211.
96CC1281	M. J. Irwin, L. M. Rendina, J. J. Vittal, R. J. Puddephatt, *J. Chem. Soc., Chem. Commun.* **1996**, 1281.
96CC1483	P. T. Glink, C. Schiavo, J. F. Stoddart, D. J. Williams, *J. Chem. Soc., Chem. Commun.* **1996**, 1483.
96CC1535	M. Fujita, O. Sasaki, T. Mitsuhashi, T. Fujita, J. Yazaki, K. Yamaguchi, K. Ogura, *J. Chem. Soc., Chem. Commun.* **1996**, 1535.
96CC1915	D. J. Cárdenas, P. Gaviña, J.-P. Sauvage, *J. Chem. Soc., Chem. Commun.* **1996**, 1915.
96CC1967	A. P. de Silva, H.Q.N. Gunaratne, T. Gunnlaugsson, M. Nieuwenhuizen, *J. Chem. Soc., Chem. Commun.* **1996**, 1967.
96CC2005	J.-P. Collin, P. Gavinã, J.-P. Sauvage, *J. Chem. Soc., Chem. Commun.* **1996**, 2005.
96CC2147	O. Murillo, E. Abel, G. E. M. Maguire, G. W. Gokel, *J. Chem. Soc., Chem. Commun.* **1996**, 2147.
96CC2191	A. P. de Silva, H. Q. N. Gunaratne, C. McVeigh, G. E. M. Maguire, P. R. S. Maxwell, E. O'Hanlon, *J. Chem. Soc., Chem. Commun.* **1996**, 2191.
96CC2441	D. B. Amabilino, J.-P. Sauvage, *J. Chem. Soc., Chem. Commun.* **1996**, 2441.
96CCCC1	M. Belohradsky, F. M. Raymo, J. F. Stoddart, *Collect. Czech. Chem. Commun.* **1996**, *61*, 1.
96CCCC622	H. Hosgören, M. Karakaplan, M. Togrul, *Collect. Czech. Chem. Commun.* **1996**, *61*, 622.
96CCR1	H. Tsukube, *Coord. Chem. Rev.* **1996**, *148*, 1.
96CCR41	P. A. Stuzhin, O. G. Khelevina, *Coord. Chem. Rev.* **1996**, *147*, 41.

96CCR41a K. Kimura, *Coord. Chem. Rev.* **1996**, *148*, 41.

96CCR71 A. Mori, K. Kubo, H. Takeshita, *Coord. Chem. Rev.* **1996**, *148*, 71.

96CCR97 Y. Habata, S. Akabori, *Coord. Chem. Rev.* **1996**, *148*, 97.

96CCR135 K. Kobiro, *Coord. Chem. Rev.* **1996**, *148*, 135.

96CCR151 T. Nabeshima, *Coord. Chem. Rev.* **1996**, *148*, 151.

96CCR171 M. Ouchi, T. Hakushi, *Coord. Chem. Rev.* **1996**, *148*, 171.

96CCR199 K. Naemura, Y. Tobe, T. Kaneda, *Coord. Chem. Rev.* **1996**, *148*, 199.

96CEJ436 A. Casnati, A. Pochini, R. Ungaro, C. Bocchi, F. Ugozzoli, R.J.M. Egberink, H. Struijk, R. Lugtenberg, F. de Jong, D.N. Reinhoudt, *Chem. Eur. J.* **1996**, *2*, 436.

96CEJ624 Z.-T. Li, P. C. Stein, J. Becher, D. Jensen, P. Mork, N. Svenstrup, *Chem. Eur. J.* **1996**, *2*, 624.

96CEJ709 P. R. Ashton, E. J. T. Chrystal, P. T. Glink, S. Menzer, C. Schiavo, N. Spencer, J. F. Stoddart, P. A. Tasker, A. J. P. White, D. J. Williams, *Chem. Eur. J.* **1996**, *2*, 709.

96CEJ729 P. R. Ashton, P. T. Glink, J. F. Stoddart, P. A. Tasker, A. J. P. White, D. J. Williams, *Chem. Eur. J.* **1996**, *2*, 729.

96CEJ832 S. Breidenbach, S. Ohren, F. Vögtle, *Chem. Eur. J.* **1996**, *2*, 832.

96CEJ877 M. Asakawa, P. R. Ashton, S. Menzer, F. M. Raymo, J. F. Stoddart, A. J. P. White, D. J. Williams, *Chem. Eur. J.* **1996**, *2*, 877.

96CJC533 I. Cordova-Reyes, H. Hu, J.-H. Gu, E. VandenHoven, A. Mohammed, S. Holdcroft, B.M. Pinto, *Can. J. Chem.* **1996**, *74*, 533.

96CSR41 A. Harriman, J.-P. Sauvage, *Chem. Soc. Rev.* **1996**, 41.

96GCI159 T. Wang, J. S. Bradshaw, R. M. Izatt, *Gazz. Chim. Ital.* **1996**, *126*, 159.

96IC4096 R. V. Sloane, J. T. Hupp, C. L. Stern, T. E. Albrecht-Schmitt, *Inorg. Chem.* **1996**, *35*, 4096.

96IC5213 D. P. Riley, S. L. Henke, P. J. Lennon, R. H. Weiss, W. L. Neumann, W. J. Rivers, Jr., K. W. Aston, K. R. Sample, H. Rahman, C.-S. Ling, J.-J. Shieh, D. H. Busch, W. Szulbinski, *Inorg. Chem.* **1996**, *35*, 5213.

96IC6184 B. R. Gibney, H. Wang, J. W. Kampf, V. L. Pecoraro, *Inorg. Chem.* **1996**, *35*, 6184.

96IC7229 A. V. Bordunov, J. S. Bradshaw, X. X. Zhang, N. K. Dalley, X. Kou, R. M. Izatt, *Inorg. Chem.* **1996**, *35*, 7229.

96JA4931 P. R. Ashton, R. Ballardini, V. Balzani, M. Belohradsky, M. T. Gandolfi, D. Philp, L. Prodi, F. M. Raymo, M. V. Reddington, N. Spencer, J. F. Stoddart, M. Venturi, D. J. Williams, *J. Am. Chem. Soc.* **1996**, *118*, 4931.

96JA5590 R. C. Helgeson, K. Paek, C. B. Knobler, E. F. Maverick, D. J. Cram, *J. Am. Chem. Soc.* **1996**, *118*, 5590.

96JA8056 C. Sheu, K. N. Houk, *J. Am. Chem. Soc.* **1996**, *118*, 8056.

96JA8221 B. Olenyuk, J. A. Whiteford, P. J. Stang, *J. Am. Chem. Soc.* **1996**, *118*, 8221.

96JA8713 S. S. Zhu, P. J. Carroll, T. M. Swager, *J. Am. Chem. Soc.* **1996**, *118*, 8713.

96JA8731 J. Manna, J. A. Whiteford, P. J. Stang, D. C. Muddiman, R. D. Smith, *J. Am. Chem. Soc.* **1996**, *118*, 8731.

96JA9110 R. F. Carina, C. Dietrich-Buchecker, J.-P. Sauvage, *J. Am. Chem. Soc.* **1996**, *118*, 9110.

96JA10932 G. Rapenne, C. Dietrich-Buchecker, J.-P. Sauvage, *J. Am. Chem. Soc.* **1996**, *118*, 10932.

96JCS(D)3613 D. Parker, J. A. G. Williams, *J. Chem. Soc., Dalton Trans.* **1996**, 3513.

96JCS(D)3705 A. J. Blake, F. Demartin, F. A. Devillanova, A. Garau, F. Isaia, V. Lippolis, M. Schröder, G. Verani, *J. Chem. Soc., Dalton Trans.* **1996**, 3705.

96JCS(P1)277 H. Takemura, S. Osada, T. Shinmyozu, T. Inazu, *J. Chem. Soc., Perkin Trans 1* **1996**, 277.

96JCS(P1)289 M. Barbero, I. Degani, S. Dughera, R. Fochi, L. Piscopo, *J. Chem. Soc., Perkin Trans 1* **1996**, 289

96JCS(P1)383 K. Naemura, K. Ueno, S. Takeuchi, K. Hirose, Y. Tobe, T. Kaneda, Y. Sakata, *J. Chem. Soc., Perkin Trans 1* **1996**, 383.

96JCS(P1)741 H. Iiara, S.-i. Watanabe, M. Yamada, O. Hoshino, *J. Chem. Soc., Perkin Trans 1* **1996**, 741.

96JCS(P1)1553 A. M. Groth, L. F. Lindoy, G. V. Meehan, *J. Chem. Soc., Perkin Trans 1* **1996**, 1553.

96JCS(P1)1587 R. Dieing, V. Morisson, A. J. Moore, L. M. Goldenberg, M. R. Bryce, J.-M. Raoul, M. C. Petty, J. Garín, M. Savirón, I. K. Lednev, R. E. Hester, J. N. Moore, *J. Chem. Soc., Perkin Trans 1*, 1587.

96JCS(P1)1783 H. Fujihara, M. Yabe, N. Furukawa, *J. Chem. Soc., Perkin Trans 1* **1996**, 1783.

96JCS(P1)1945 A. Ikeda, M. Yoshimura, P. Lhotak, S. Shinkai, *J. Chem. Soc., Perkin Trans 1* **1996**, 1945.

96JCS(P1)1995 M. Wagner, D. Madsen, J. Markussen, S. Larsen, K. Schaumburg, K.-H. Lubert, J. Becher, R.-M. Olk, *J. Chem. Soc., Perkin Trans 1* **1996**, 1995.

96JCS(P1)2061 S. Breidenbach, J. Harren, S. Neumann, M. Nieger, K. Rissanen, F. Vögtle, *J. Chem. Soc., Perkin Trans 1* **1996**, 2061.

96JCS(P2)1855 F. Arnaud-Neu, R. Arnecke, V. Böhmer, S. Fanni, J. L. M. Gordon, M.-J. Schwing-Weill, W. Vogt, *J. Chem. Soc., Perkin Trans 2* **1996**, 1855.

96JHC933 A. V. Bordunov, N. K. Dalley, X. Kou, J. S. Bradshaw, V. N. Pastushok, *J. Heterocycl. Chem.* **1996**, *33*, 933.

96JOC4739 H. K. A. C. Coolen, P. W. N. M. van Leeuwen, R. J. M. Nolte, *J. Org. Chem.* **1996**, *61*, 4739.

96JOC5684 K.-W. Chi, H.-C. Wei, T. Kottke, R. J. Lagow, *J. Org. Chem.* **1996**, *61*, 5684.

96JOC7270 P. C. Hellier, J. S. Bradshaw, J. J. Young, X. X. Zhang, R. M. Izatt, *J. Org. Chem.* **1996**, *61*, 7270.

96JOC7585 A. R. Katritzky, S. A. Belyakov, A. E. Sorochinsky, P. J. Steel, O. F. Schall, G. W. Gokel, *J. Org. Chem.* **1996**, *61*, 7585.

96JOC7578 A. R. Katritzky, O. V. Denisko, S. A. Belyakov, O. F. Schall, G. W. Gokel, *J. Org. Chem.* **1996**, *61*, 7578.

96JOC8391 Y. Habata, J. S. Bradshaw, J. J. Young, S. L. Castle, P. Huszthy, T. Pyo, M. L. Lee, R. M. Izatt, *J. Org. Chem.* **1996**, *61*, 8391.

96JOC8724 S. Pappalardo, M. F. Parisi, *J. Org. Chem.* **1996**, *61*, 8724.

96JOC8904 P. Delangle, J.-P. Dutasta, L. V. Oostenryck, B. Tinant, J.-P. Declercq, *J. Org. Chem.* **1996**, *61*, 8904.

96JOMC13 C. D. Hall, A. Leineweber, J. H. R. Tucker, D. J. Williams, *J. Organometal. Chem.* **1996**, *523*, 13.

96LA297 C. Meiners, M. Nieger, F. Vögtle, *Liebigs Ann.* **1996**, 297.

96LA551 M. Wagner, S. Zeltner, R.-M. Olk, *Liebigs Ann.* **1996**, 551.

96LA655 A. Fürstner, G. Seidel, C. Kopiske, C. Krüger, R. Mynott, *Liebigs Ann.* **1996**, 655.

96LA757 A. Siepen, A. Zett, F. Vögtle, *Liebigs Ann.* **1996**, 757.

96LA959 M. Barboiu, C. T. Supuran, C. Luca, G. Popescu, C. Barboiu, *Liebigs Ann.* **1996**, 959.

96LA1005 A. Spannenberg, H.-J. Holdt, K. Praefcke, J. Kopf, J. Teller, *Liebigs Ann.* **1996**, 1005.

96LA1213 E. Tauer, R. Machinek, *Liebigs Ann.* **1996**, 1213.

96LA1367 H. Ross, U. Lüning, *Liebigs Ann.* **1996**, 1367.

96LA1645 K. Sako, H. Tatemitsu, S. Onaka, H. Takemura, S. Osada, G. Wen, J. M. Rudzinski, T. Shinmyozu, *Liebigs Ann.* **1996**, 1645.

96MI1 *Inorg. Chim. Acta* **1996**, *246* (1/2)

96MI2 M. Fijita, *Comprehensive Supramolecular Chemistry*, ed. J.-M. Lehn, Elsevier, Oxford, **1996**, vol. 9, ch. 7.

96MI3 G. W. Gokel, O. F. Schall, *Comprehensive Supramolecular Chemistry*, ed. G. W. Gokel, Elsevier, Oxford, **1996**, vol. 1, pp 97-152.

96MI4 J. S. Bradshaw, R. M. Izatt, A. V. Bordunov, C. Y. Zhu, J. K. Hathaway, *Comprehensive Supramolecular Chemistry*, ed. G. W. Gokel, Elsevier, Oxford, **1996**, vol. 1, pp 35-95.

96MI5 G. W. Gokel, E. Ernesto, *Comprehensive Supramolecular Chemistry*, ed. G. W. Gokel, Elsevier, Oxford, **1996**, vol. 1, pp 511-535.

96MI6 B. Dietrich, *Comprehensive Supramolecular Chemistry*, ed. D. N. Reinhoudt, Elsevier, Oxford, **1996**, vol. 10, pp 361-387.

96MI7 T. Hayashita, M. Takagi, *Comprehensive Supramolecular Chemistry*, ed. G. W. Gokel, Elsevier, Oxford, **1996**, vol. 1, pp 635-669.

96MI8 S. J. Loeb, *Comprehensive Supramolecular Chemistry*, ed. G. W. Gokel, Elsevier, Oxford, **1996**, vol. 1, pp 733-753.

96MI9 R. D. Rogers, C. B. Bauer, *Comprehensive Supramolecular Chemistry*, ed. G. W. Gokel, Elsevier, Oxford, **1996**, vol. 1, pp 315-355.

96MI10 M. T. Reetz, *Comprehensive Supramolecular Chemistry*, ed. F. Vögtle, Elsevier, Oxford, **1996**, vol. 2, pp 553-562.

96MI11 J. Simon, T. Toupance, *Comprehensive Supramolecular Chemistry*, ed. D. N. Reinhoudt, Elsevier, Oxford, **1996**, vol. 10, pp 637-658.

96MI12	C. Detellier, *Comprehensive Supramolecular Chemistry*, ed. G. W. Gokel, Elsevier, Oxford, **1996**, vol. 1, pp 357-375.
96MI13	S. Petrucci, E. M. Eyring, G. Konya, *Comprehensive Supramolecular Chemistry*, eds. J. E. D. Davies, J. A. Ripmeester, Elsevier, Oxford, **1996**, vol. 8, pp 483-497.
96MI14	B. Koenig, GIT Fachz. Lab. **1996**, *40*, 528, 530.
96MI15	M. Inouye, *Senryo to Yakuhin*, **1996**, *41*, 69.
96MI16	M. Saeada, *Dojin News* **1996**, *78*, 10.
96MI17	H. W. Gibson, S. Liu, *Macromol. Sym.* **1996**, *102*, 55.
96MI18	V. Percec, G. Johansson, D. Schlueter, J. C. Ronda, G. Ungar, *Macromol. Symp.* **1996**, *101*, 43.
96MI19	*Comprehensive Supramolecular Chemistry*, eds. J. L. Atwood, D. D. MacNicol, J. E. Davies, F. Vögtle, Elsevier, Oxford, **1996**.
96NJC427	P. D. Hampton, C. E. Daitch, E. N. Duesler, *New J. Chem.* **1996**, *20*, 427.
96NJC453	F. Arnaud-Neu, Z. Asfari, B. Souley, J. Vicens, *New J. Chem.* **1996**, *20*, 453.
96NJC465	S. Pappalardo, *New J. Chem.* **1996**, *20*, 465.
96NJC685	J. F. Nierengarten, C. O. Dietrich-Buchecker, J.-P. Sauvage, *New J. Chem.* **1996**, *20*, 685.
96O5321	M. J. Irwin, G. Jia, J. J. Vittal, R. J. Puddephatt, *Organometallics* **1996**, *15*, 5321.
96P775	P. D. Beer, K. Y. Wild, *Polyhedron* **1996**, *15*, 775.
96P1197	B. Zhao, Y. J. Wu, J. C. Tao, H. Z. Yuan, X. A. Mao, *Polyhedron* **1996**, *15*, 1197.
96P1553	F. S. McQuillan, C. J. Jones, *Polyhedron* **1996**, *15*, 1553.
96P3933	Y. Gok, *Polyhedron* **1996**, *15*, 3933.
96P4203	L. Escriche, M.-P. Almajano, J. Casabó, F. Teixidor, J. C. Lockhart, G. A. Forsyth, R. Kivekäs, M. Sundberg, *Polyhedron* **1996**, *15*, 4203.
96PAC1225	A. L. Crumbliss, I. Batinic-Haberle, I. Spasojevic, *Pure Appld. Chem.* **1996**, *68*, 1225.
96PAC1255	S. J. Langford, J. F. Stoddart, *Pure Appld. Chem.* **1996**, *68*, 1255.
96PAC1279	M. T. Reetz, *Pure Appld. Chem.* **1996**, *68*, 1279.
96RJCC229	A. Yu. Tsivadze, V. I. Zhilov, S. V. Demin, *Russ. J. Coord. Chem.* **1996**, *22*, 229.
96RTC109	D. N. Reinhoudt, *Rec. Trav. Chim. Pays-Bas.* **1996**, *115*, 109.
96RTC307	J. Scheerder, J. F. J. Engbersen, D. N. Reinhoudt, *Rec. Trav. Chim. Pays-Bas.* **1996**, *115*, 307.
96RTC357	R. J. W. Schuurman, R. F. P. Grimbergen, H. W. Scheeren, R. J. M. Nolte, *Rec. Trav. Chim. Pays-Bas.* **1996**, *115*, 357.
96SL285	E. Alcalde, M. Gisbert, *Synlett* **1996**, 285.
96SL430	G. P. M. van Klink, T. Nomoto, E. van Wees, A. G. L. Ostheimer, O. S. Akkerman, F. Bickelhaupt, *Synlett* **1996**, 430.
96SL933	A. V. Bordunov, J. S. Bradshaw, V. N. Pastushok, R. M. Izatt, *Synlett* **1996**, 933.
96SL1013	B. König, C. Horn, *Synlett* **1996**, 1013.
96T15171	E. Alcalde, M. Alemany, M. Gisbert, *Tetrahedron* **1996**, *52*, 15171.
96T15189	E. Alcalde, M. Gisbert, C. Alvarez-Rúa, S. Garcia-Granda, *Tetrahedron* **1996**, *52*, 15189.
96T10921	J.-M. Kern, J.-P. Sauvage, J.-L. Weidmann, *Tetrahedron* **1996**, *52*, 10921.
96TL4593	F. H. Kohnke, G. L. La Torre, M. F. Parisi, S. Menzer, D. J. Williams, *Tetrahedron Lett.* **1996**, *37*, 4593.
96TL6201	P. Fonte, F. H. Kohnke, M. Parisi, D. J. Williams, *Tetrahedron Lett.* **1996**, *37*, 6201.
96TL6205	P. Fonte, F. H. Kohnke, M. F. Parisi, S. Menzer, D. J. Williams, *Tetrahedron Lett.* **1996**, *37*, 6205.
96TL6315	B. Pulpoka, Z. Asfari, J. Vicens, *Tetrahedron Lett.* **1996**, *37*, 6315.
96TL7711	I. Gardinier, A. Roignant, N. Oget, H. Bernard, J. J. Yaouanc, H. Handel, *Tetrahedron Lett.* **1996**, *37*, 7711.
96TL7881	P.A. Gale, J. L. Sessler, V. Lynch, P. I. Sansom, *Tetrahedron Lett.* **1996**, *37*, 7881.
96TL8747	B. Pulpoka, Z. Asfari, J. Vicens, *Tetrahedron Lett.* **1996**, *37*, 8747.

INDEX

www.ingramcontent.com/pod-product-compliance
Ingram Content Group UK Ltd.
Pitfield, Milton Keynes, MK11 3LW, UK
UKHW020233290926
15470UKWH00002B/519

9 780080 428017